A FIRST COURSE
IN LINEAR ALGEBRA

Second Edition

A FIRST COURSE IN LINEAR ALGEBRA

Second Edition

Hal G. Moore
Brigham Young University

Adil Yaqub
**University of California,
Santa Barbara**

HarperCollins*Publishers*

Sponsoring Editor: George Duda
Project Editor: Cathy Wacaser
Art Direction: Julie Anderson
Text Design: Paul Helenthal
Cover Design: Fern Logan
Cover Illustration/Photo: Fern Logan
Chapter Opener Art: Precision Graphics
Technical Artwork: Fineline Illustrations, Inc.
Production: Jeanie Berke, Linda Murray, and Helen Driller
Compositor: New England Typographic Service, Inc.
Printer and Binder: R. R. Donnelley & Sons Company
Cover Printer: Lehigh Press Lithographers

A First Course in Linear Algebra, Second Edition
Copyright © 1992 by HarperCollins Publishers Inc.

Library of Congress Cataloging-in-Publication Data

Moore, Hal G.
    A first course in linear algebra/Hal G. Moore & Adil Yaqub.—
2nd ed.
        p.   cm.
    Includes index.
    ISBN 0-673-38392-X
    1. Algebra, Linear.   Yaqub, Adil.   II. Title.
QA184.M64   1991
512′.5—dc20

91–37218
CIP

92 93 94 95 9 8 7 6 5 4 3 2 1

PREFACE

Why is elementary linear algebra a part of the mathematics core at most colleges and universities? The answers to this question are, no doubt, as many and as varied as there are courses or teachers. Condensed within these answers we find all the current discussions concerning the revitalization of the mathematics curriculum in the 1990s to provide a mathematically literate work force in the twenty-first century.* In them repose the many reasons for mathematics departments to teach mathematics to every student. Mathematical thought has a contribution to make to the education of all.[†] Linear algebra has as important a role to play as calculus does in the technological literacy of a citizen of the next century.

Objectives

This book is designed for an introductory course in linear algebra. It blends the requirements of problem solving, analytical thinking, computational technique, and applications needed by beginning linear algebra students. We are confident that instructors can use this material to design the sorts of courses they feel are needed for a more vital mathematical curriculum. It is only fair to point out that we feel that elementary linear algebra is precisely the place to start to teach university students *to think* mathematics —to use mathematical structure to think about and to solve mathematical problems. Too often, mathematics courses taken before linear algebra provided students with templates, canned processes, magic wands to wave, and bags of sacerdotal tricks. Many college students have come to expect to skip the exposition in their mathematics text while they match an assigned exercise with an example of precisely that kind. In this text we strive to give meaningful exposition of what is happening so that problem solving can take place on an informed basis.

---

* Board on Mathematical Sciences, *Renewing U.S. Mathematics, A Plan for the 1990s,* National Research Council, Washington D.C., 1990.

† Mathematical Sciences Education Board, *Everybody Counts,* National Research Council, Washington D.C., 1989.

Distinguishing Features

The Flavor of the Text   This is not merely a basic matrix theory text. It is also not just a finite dimensional vector spaces text. We hope that we have exploited the interplay between these two ideas so that students can move with ease between them. Thus, they can be equipped to learn more about either idea in advanced courses, if desired. Throughout the text we have emphasized the interplay between the algebraic and geometric concepts that are so much a part of linear algebra. We have given a substantial number of geometric examples whenever possible. At the end of each section there are many exercises—far more than anyone can do in one assignment. The instructor will want to carefully select those that reinforce the goals of the course, adding the particular emphasis that is desired: computation, abstraction, or application. The chapters are ordered so as to proceed from the familiar to the less familiar.

Mathematical Thought   Linear algebra is useful, its techniques are powerful, and it is widely applied. Because the applications are near at hand, elementary linear algebra is the ideal vehicle for introducing students to the incredible power of generalization and abstraction. Here they can learn to correlate facts and bring their power to bear on problems in both theory and application. We have presented many proofs for the student to study to see what the theoretical material embraces. We have, however, specifically *not included* some theorems and their proofs in the exposition until we have first asked the student to make conjectures and deduce them as exercises. We hope that this approach will transform students from observers of mathematics to participants in it. When a particular result is a fairly simple outcome we have asked the student to prove it, rather than merely to read our proof. This might frustrate a student who wishes to be programmed, but will make the material more lively for the student who wants to really be involved.

Computational Skill   Having said this, we hasten to add that there are ample computational exercises in this text as well. (Even the best of artists must get his or her hands "dirty" in the creation of art.) Mathematical exploration is an invaluable aid to arriving at mathematical ideas. Nevertheless, we have not, because of time and space limitations, groped too long for the numerical worms that can easily bog the subject down in the mud. We hope that the book includes an appropriate mix of computational exercises to fix ideas, simple demonstrations of theoretical results, and fairly elementary proofs for the students to do.

Technology   Modern technology provides an invaluable aid to solving computational linear algebra problems, but there is nothing so old as last year's software. Therefore, while we have specifically not tied this text to any particular computer or calculator program, the use of technological aid is encouraged to avoid much of the tedious computational labor for both student and

instructor. It is assumed that each person and institution will have their own current favorite (or whatever the budget will allow). Nevertheless, these aids bring with them their own interesting problems. This is not a text in those sorts of problems; we believe that linear algebra is very important in its own right, and a course based on this text should not be reduced to a computer literacy course. We recommend hand calculation at first until the *ideas* are mastered, then the use of technology to get through the computational hurdles on the way to the next idea.

Vocabulary   Included at the end of each section is a list of new terms. These terms are not just for this course, but will be encountered in all technical discourse. The endpapers also include a list of the important notation and principal theorems and definitions.

Review Exercises   Each chapter includes a section of Review Exercises. These focus on the basic ideas developed in that chapter and can be used to review for examinations.

Exploratory Projects   Also at the end of each chapter are the Supplementary Exercises and Projects. As most of today's mathematics are done in groups rather than by a single individual, these projects are designed as group projects. They can be assigned to lead the students to an active involvement in the subject. They are in no way exhaustive of all the projects that could be, and will be, set by the creative instructor. We hope that they are suggestive and helpful.

Appendices   In the Appendices at the end of the text we give a brief introduction to the complex numbers and to sets and functions. The third appendix also includes some of the computational algorithms developed in the text. These appendices provide an easy reference guide for the student to the more computational aspects of the material.

Solutions   Answers to half of the exercises—mostly the odd-numbered ones—are given at the end of the text. For about twenty percent of these exercises the answer includes a detailed sketch of the solution.

Course Flexibility

For the most part, we have assumed that students in this course have mastered the equivalent of two years of high school algebra. Most will, however, take this course after a semester or two of freshman calculus. For these students there are clearly labeled examples and exercises that will allow students to think beyond the ordinary Euclidean geometric world and apply calculus to a broader range of ideas. This additional calculus material may be omitted without disrupting the text's flow if the course precedes calculus.

Note that all of the recommendations presented in the Core Syllabus of the Linear Algebra Curriculum Study Group,* and more, are included in this text. *A First Course in Linear Algebra* contains more material than can be covered in any standard one-semester sophomore/junior level course. The instructor can, therefore, design the type of course (theoretical or applications oriented) that best fits the needs of the students. The heart of the material can be covered in a standard three-hour-per-week one-semester course, and one will probably want to include as many of each chapter's applications as time permits. With careful selection of material, the entire text could be covered in a semester course meeting four hours per week; a course that meets for a quarter will need to be more selective. The table gives some explicit suggestions for various course organizations.

Amount of Time Available	Cover Chapters
One Quarter—three meetings per week One Semester—two meetings per week	Chapters 1 (except Section 1.7), 2, 3, 4 (except Section 4.2 and Section 4.6) plus Sections 5.1, 5.2, 5.3, 6.1, 6.2
One Quarter—four meetings per week	Chapters 1 (except Section 1.7), 2, 3, 4 (except Section 4.2 and Section 4.6), 5, and Sections 6.1, 6.2
One Quarter—five meetings per week One Semester—three meetings per week	Chapters 1, 2, 3, 4 (except Section 4.6), 5, plus Sections 6.1, 6.2, 6.3, 6.5, plus selections from the rest of Chapter 6, and Section 4.6 as time permits.
One Semester—four meetings per week	Chapters 1, 2, 3, 4, 5, 6, and several projects.

The following diagram illustrates the interdependence of the chapters and suggests alternative paths through the material.

---

* See the Preliminary Report of the Linear Algebra Curriculum Study Group given at the Joint Mathematics Meetings in San Francisco, CA, January 17, 1991.

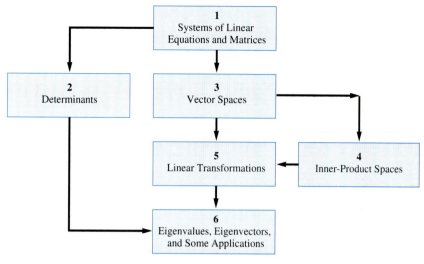

Interdependence of the Chapters

Supplements

The following supplements are available.

Convex Sets and Linear Programming This is an additional chapter, in pamphlet form, should one have the time or the inclination to make this topic part of the course.

Instructor's Solutions Manual The *Instructor's Solutions Manual* contains the following.

- Answers and solutions to all of the even-numbered exercises and projects.

- Some routines for handling linear algebra problems on HP 28S and 48SX calculators.

- A supplement containing some suggestions on how to do mathematical proofs.

Items 2 and 3 may be duplicated for distribution to the student and are also included in the *Student's Solutions Manual.*

Student's Solutions Manual The *Student's Solutions Manual* contains the following.

- The answers and, in many cases, detailed solutions to the odd-numbered exercises not given at the back of the book.

- Routines for handling linear algebra problems on HP 28S and 48SX calculators.

Computer Software Package

- MATRIX software is designed to run on all IBM PCs and compatibles running MS-DOS Version 3.1 or higher. This program is designed to help students learn what matrices are, what they can do, and how to work with them. It is available from the publisher or can be obtained by writing to Education Software, 195 Payne Avenue, Brentwood, California 94513.

- LINALG is a tool kit which has a "pull down" menu format and context-sensitive help. LINALG is a linear algebra program for use by instructors for classroom demonstrating as well as by students for exploring and solving problems. It is available in an IBM format and can be obtained by writing to the Department of Mathematics at the University of Arizona.

Acknowledgments

We appreciate the many people who have aided in the development and preparation of this text, particularly the staff of HarperCollins: George Duda, our developmental editor; Cathy Wacaser, project editor; Julie Anderson and Paul Helenthal, our designers. We would also like to express our sincere appreciation to Lynn E. Garner, Brigham Young University, for his material on using Hewlett-Packard calculators for linear algebra.

We are also indebted to the many suggestions for improvement and clarification that have come from our students and colleagues, particularly from the following reviewers.

Christopher Ennis, *Carleton College*
John Haverhalls, *Bradley University*
L. Kaufman, *University of Illinois-Chicago*
William Keane, *Boston College*
Kenneth Kramer, *CUNY, Queens College*
Robert Lax, *Louisiana State University*
Carl Leinbach, *Gettysburg College*
Gregory B. Passty, *Southwest Texas State University*
Jack Porter, *University of Kansas*
Steven Pruess, *Colorado School of Mines*
Donald Robinson, *Brigham Young University*
David Royster, *University of North Carolina, Charlotte*

Any defects that remain are our own responsibility. If you or your students have any comments or suggestions, please do not hesitate to write to either of us, or to the publisher. We welcome your response.

Hal G. Moore
Adil Yaqub

CONTENTS

This text is for you. It was written to you to teach you about the power and also about the beauty of linear algebra. Linear algebra is very useful mathematics, and it is *good* mathematics. It is the offspring of centuries of development in both algebra and geometry. We believe that mathematical thought has a contribution to make to your life and to your professional career. We also believe that elementary linear algebra is precisely the place for you to start to learn *to think* mathematics. As you use this book you will notice some important ancillary parts to help you. These include the following.

Vocabulary   At the end of each section in the text you will find a list of the new terms learned in that section. These terms are not just for this course, but are used in all technical discourse. We suggest that you incorporate them in your working vocabulary. It can also be a great help to you if you will carry along, as a bookmark, one or two 5″ by 8″ cards. Write down the new vocabulary and notational terms, together with their page references, as you encounter them. Use the endpapers of the text also as reference guides.

Review Exercises   Each chapter includes a section of Review Exercises. These focus on the major ideas developed in that chapter. Use them to review for examinations.

Appendices   The appendices at the end of the text are to help you as you study the material. Refer to them freely as you work your way through the material of the text.

▪ Appendices A.1 and A.2 review ideas that should have been part of your prerequisite background but, perhaps, were not; or, perhaps these ideas were covered in a prerequisite course that you took some time ago and have forgotten. Since we make free use of these notions in the text, we give brief summaries of the main points here. However, these are merely thumbnail sketches, as it were. If you need more information, you are referred to more elaborate treatments in standard textbooks. Most standard college algebra, trigonometry, and precalculus texts discuss the complex numbers. Many of these also discuss functions and sets.

- Appendix A.3 gathers together several of the linear algebra algorithms discussed in the text. They are presented in simplified form for convenient reference. You should consult the appropriate sections of the text for the full explanations of the procedures that are listed in this appendix. In many cases your college computer center, your personal computer, or your calculator will use algorithms different from those given in the text for actual numerical computations. You should always consult the appropriate manuals for these technological helps. These manuals should provide you with a discussion of the methods that are used and their specific limitations. Our algorithms are for hand calculation and do not necessarily yield the result in the shortest possible time.

- Appendix A.4 contains references. In it we have listed several books, papers, and software programs. You may wish to consult some of them.

- Appendix A.5 contains a listing of the Greek alphabet and frequently used symbols.

Answers   We have given the answers to the odd-numbered exercises, including fairly detailed solutions to some of these, at the end of the text. Use the answers and solutions wisely. You will not really *learn* the material if you *program* yourself in this way: (1) Do one exercise. (2) Look up the answer to it. (3) Correct it if it is wrong. (4) Do the next exercise. (5) Repeat the procedure. Rather than this, you should do as many of the exercises as you are sure of *before* you look at any answers. Then re-read the material for those about which you are unsure. Then, and only then, look up the answers to the entire exercise set.

Index   Use the index and the references in the endpapers of the book to locate definitions and concepts in the text.

Advice for Successful Study

Study regularly! Keep up your study day by day. New material builds on old. It is important that you don't fall behind. *It is better to keep up than to catch up.* If you are in the habit of studying for an impending crisis, create a daily crisis for yourself with this material. While you may be able to cram and pass an exam, all you will have in the end is a grade. Linear algebra is knowledge for now and for later. The exam is the beginning not the end.

Do the exercise! It has often been said that "Mathematics is not a spectator sport." By seeing how the calculations turn out, even though they at times seem tedious, you will discover the ideas that are embedded in the problem. You can't learn merely by observing them any more than you can become a tennis player by merely observing a tennis match. If you prove a theorem you'll know what it means and why it is true. Do all of the assigned exercises, not just the simple ones at the beginning of the set. Flex your mental muscles on the harder ones.

When you begin a study session, *supply yourself with some scratch paper.* Although we have tried to be clear, mathematics is written in telegraphic style. We want you to see the ideas and not get lost in technical details. For this reason, you may sometimes need to supply the in-between calculations for yourself. Don't be afraid to work a while to see why the next line follows. The phrase "it follows that" can literally mean that after about fifteen minutes of routine calculation, the result does indeed follow.

Read ahead! If you come to class already having read the material and having attempted the exercises you will know what questions to ask and will understand the answers. Listen to the questions of others. They may have seen something there that you missed. To paraphrase the poet, "If you can keep your head when all about you are losing theirs, perhaps you don't understand the problem." Try to answer yours and others' questions in your own words.

Adopt a teachable attitude! The value of abstraction is that it starts, as we do here, with the familiar and idealizes it. After looking at some examples, the definitions and axioms are given—from that moment you enter *the game.* While the intuitive ideas that got you there are a guide to what might be true, no longer is an intuitive feeling good enough. You must follow the rules of the game. If you are still back on the examples and view them as *absolute truth* beyond which there can be nothing, you have lost the game. There is a lot to learn from generalizations. It is not a bad approach to this course to adopt the attitude you have when reading science fiction. In doing mathematical fact, we will go beyond the ordinary, three-dimensional world of Newtonian mechanics. Don't be caught back in the seventeenth century in your approach to this material. Let your imagination run free.

Use the aids listed above and *go for it!*

Our students tell us that previous versions of this book have been clear and have helped them to learn *to think* mathematics. We would welcome your comments. Write to us with any suggestions or criticisms so that future editions can be improved.

A FIRST COURSE IN LINEAR ALGEBRA

Second Edition

1 SYSTEMS OF LINEAR EQUATIONS AND MATRICES

Linear algebra has many applications. Because of its notational economy and rules for analysis, it is often used to interpret raw data involved in problems of the physical, behavioral, management, and biological sciences. The nature of an actual model usually depends upon many factors and frequently involves an interplay between linear algebra, other branches of mathematics, and the science in which the problem is posed.

In this text we shall study those parts of linear algebra that will enable us to construct and manipulate some linear mathematical models. A convenient place to begin one's study of linear algebra is to consider the process of solving a system of linear equations. This leads us immediately to the idea of a matrix.

The algebra of matrices is studied and used to determine whether a given system of linear equations has a single solution, infinitely many solutions, or no solution at all. Matrices have far wider application, however, than simply to solve systems of linear equations. Therefore, we take a good look at matrix algebra. It develops that many, but by no means all, of the properties of high school algebra are also valid for matrix algebra: for example, the associative law for multiplication is still valid for matrix algebra. Nevertheless, the commutative law for multiplication badly fails for matrices. We also see that, unlike the case for real or complex number algebra, it is possible for the product of two *nonzero* matrices to be zero.

We give particular emphasis to determining whether a given matrix has a multiplicative inverse. One method for settling this question is to study elementary operations on matrices and, using these operations effectively, to find the matrix inverse when it exists. We consider how one can handle large matrices as matrices made of smaller matrices and introduce the much-used LU factorization of a matrix. We conclude with applications of matrices to networks and graphs. The techniques developed in this chapter are fundamental, and will be used throughout the course.

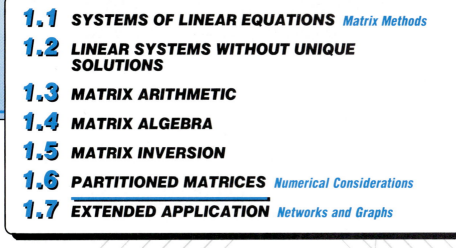

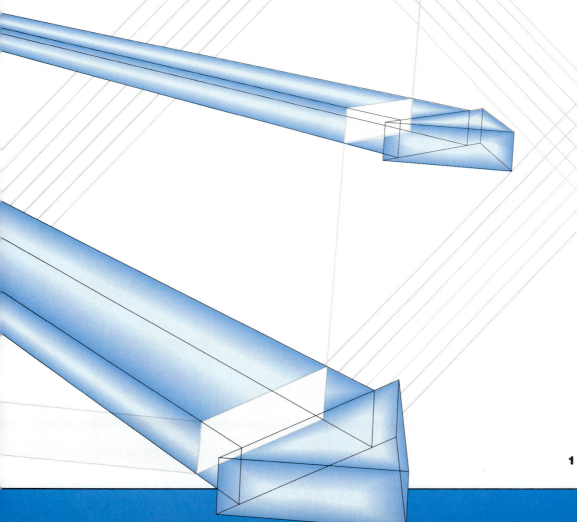

1.1 SYSTEMS OF LINEAR EQUATIONS Matrix Methods

To study the solution of a system of linear equations we must first introduce some general terminology.

A *linear equation* in n variables is an equation of the form

$$a_1 x_1 + a_2 x_2 + \cdots + a_n x_n = b$$

where b and the coefficients $a_1, a_2, \ldots a_n$ are constants (numbers) and $x_1, x_2, \ldots, x_n$ are the variables. A **system of linear equations** is any collection of such equations. We adopt a double subscript notation for the coefficients a_{ij}. The first subscript i denotes that this coefficient occurs in equation number i of the system, while the second subscript j indicates that this is the coefficient of the variable x_j in that equation. Thus,

$$
\begin{aligned}
a_{11} x_1 + a_{12} x_2 + \cdots + a_{1n} x_n &= b_1, \\
a_{21} x_1 + a_{22} x_2 + \cdots + a_{2n} x_n &= b_2, \\
\vdots \qquad \vdots \qquad\quad \vdots \qquad \vdots \\
a_{m1} x_1 + a_{m2} x_2 + \cdots + a_{mn} x_n &= b_m
\end{aligned}
\tag{1.1}
$$

is a system of m linear equations in n variables.

> **Definition**   A **solution** to a system of linear equations such as (1.1) is any ordered n-tuple
>
> $$(c_1, c_2, \ldots, c_n)$$
>
> of numbers such that *each* equation of the system is satisfied when the values $x_j = c_j, j = 1, 2, \ldots, n$, are substituted in the equation.

Let's consider a simple specific problem that might occur, and use it to introduce the techniques for solving a system of equations. As you learned in elementary algebra, the basic idea behind such techniques is to find a simpler system of equations that has the same solutions.

> **Definition**   Two systems of linear equations in n variables are called **equivalent** if they have precisely the *same set* of solutions.

Example 1   A milling company produces two kinds of feed. The first feed contains 5% protein and 15% carbohydrates. The second feed contains 15% protein and 25% carbohydrates. How many pounds of each feed should be combined to make 100 lb of a mixture that contains 10% protein? What is the carbohydrate content of this mixture?

This problem can be modeled by the following system of equations, where x_1 and x_2 represent the number of pounds of feeds *one* and *two,* respectively:

$$x_1 + \quad x_2 = 100 \qquad \text{Total pounds equation}$$
$$0.05x_1 + 0.15x_2 = 0.1(x_1 + x_2). \qquad \text{Protein equation} \qquad \textbf{(1.2)}$$

Here we have a system of two equations in the two unknowns x_1 and x_2. We can simplify the second equation in this system by multiplying it by 100 and collecting like terms. The system becomes

$$x_1 + \ x_2 = 100$$
$$-5x_1 + 5x_2 = 0.$$

To solve this system, *substitution* may be the easiest method. The first equation can be solved for x_1 in terms of x_2 and the constant 100; that is, $x_1 = 100 - x_2$. Substituting that solution for x_1 in the other equation gives us:

$$-5(100 - x_2) + 5x_2 = 0;$$

or, simplifying
$$-500 + 10x_2 = 0,$$

so
$$10x_2 = 500,$$

and
$$x_2 = 50.$$

Then, from the first equation, substituting for x_2,

$$x_1 = 100 - 50 = 50 \text{ lb.}$$

The pair of numbers (50, 50) is the solution to the system of equations (1.2). One uses 50 lb of each feed for the mixture. The carbohydrate content c of this mixture is computed from the equation $0.15x_1 + 0.25x_2 = c(x_1 + x_2)$. [Compare with the protein equation in (1.2).] For our solution, this becomes $0.15(50) + 0.25(50) = c(100)$. That is, $100c = 0.4(50) = 20$, or $c = 0.2$. The mixture is 20% carbohydrate.

The system of equations (1.2) in Example 1 is *equivalent* to the system

$$x_1 + \ x_2 = 100,$$
$$-5x_1 + 5x_2 = 0$$

which, in turn, is *equivalent* to the system

$$x_1 \quad\quad = 50,$$
$$x_2 = 50$$

because the solution set for each system is precisely the pair of numbers (50, 50). ∎

A second method for solving systems of linear equations involves *graphing.* It has its greatest value in allowing us to make a geometric picture of a situation; however, when the number of variables gets beyond three, most people severely strain their geometric intuition. Nevertheless, with a little imagination, we can still exploit these geometric ideas to give us additional insight. To illustrate the graphical method, consider the following system:

$$2x + 2y = 56,$$
$$3x + 4y = 96. \tag{1.3}$$

One may think of each of these equations as the equation of a line in 2-space. Thus, the system (1.3) represents two lines in the plane. Each line is the set of all those points (x, y) in the coordinate plane which satisfy the given equation, and the solution to the system is the point (16, 12) that lies on both lines; i.e., their intersection. Figure 1.1 illustrates this. In the case of three variables, a linear equation represents a plane, not a line (see Figure 1.6 in the next section).

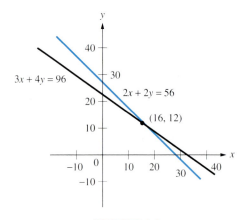

FIGURE 1.1

To illustrate the third solution method, *elimination,* the one we wish to generalize, consider the following three-variable example.

Example 2   Find all of the solutions to the system of equations

$$x_1 - 2x_2 + x_3 = 1,$$
$$x_1 + x_2 - x_3 = 2, \tag{1.4}$$
$$2x_1 - x_2 + x_3 = 1.$$

It is fairly easy to see that to use substitution or graphing to solve this system will require a lot of work. Using the method of elimination, we subtract the first equation from the second, and twice the first equation from the third to obtain the system

$$x_1 - 2x_2 + x_3 = 1,$$
$$3x_2 - 2x_3 = 1, \tag{a}$$
$$3x_2 - x_3 = -1.$$

Since adding (or subtracting) equals to (or from) equals doesn't change the equality, this system should be equivalent to (1.4). Now if we rearrange the equations in (a) by interchanging the second and third equations, the resulting system (b) is equivalent to the system (a). Thus,

$$\begin{aligned}
x_1 - 2x_2 + x_3 &= 1, \\
3x_2 - x_3 &= -1, \\
3x_2 - 2x_3 &= 1.
\end{aligned}$$
(b)

Subtracting the second equation of (b) from the third results in the equivalent system

$$\begin{aligned}
x_1 - 2x_2 + x_3 &= 1, \\
3x_2 - x_3 &= -1, \\
-x_3 &= 2,
\end{aligned}$$
(c)

which we say is in *triangular form.*

Definition A system of linear equations, using the notation of the general system (1.1), is in **triangular form** when $a_{ij} = 0$, for every $i > j$.

That is to say, the first equation in a system in triangular form has a nonzero coefficient for the first variable, while each equation after it in the list has a zero coefficient for x_1. The second equation has a nonzero coefficient for the next variable, and each equation after that in the list has a zero coefficient for that variable, and so on. The definition in the box above is the precise statement.

Although the system (c) above is in triangular form, the system in (1.4) and in the subsequent equivalent forms (a) and (b) are not in triangular form. Why?

Now, upon multiplying the second equation in (c) by 1/3 and multiplying the third equation by -1, we obtain another equivalent system (1.5) which is also in triangular form. This operation is valid, since "multiplying equals by equals gives equals."

$$\begin{aligned}
x_1 - 2x_2 + x_3 &= 1, \\
x_2 - \frac{1}{3}x_3 &= -\frac{1}{3}, \\
x_3 &= -2.
\end{aligned}$$
(1.5)

This system of equations (1.5) is said to be in *row echelon form.*

Definition A system of equations is in **row echelon form** if it is in *triangular form* and the first nonzero coefficient a_{ij} in each equation is "1."

The last equation in (1.5) tells us that $x_3 = -2$. By substituting this value in the second equation we can determine that $x_2 = -1$. Then, by substituting both values in the first equation, we find that $x_1 = 1$. This process is called **back substitution.** Therefore, the system of equations

$$\begin{aligned} x_1 &&&= 1, \\ &x_2 &&= -1, & \text{(1.6)} \\ &&x_3 &= -2, \end{aligned}$$

is equivalent to (1.5), and the solution to both is the triple $(1, -1, -2)$. A system of equations such as (1.6) is said to be in *reduced row echelon form.*   ■

> **Definition**   A system of equations is in **reduced row echelon form** if it is in *row echelon form,* and if every coefficient of the same variable with the initial "1" is a zero in all other equations.

As we have demonstrated with this example, to solve a system of linear equations one may successively replace it by equivalent systems with successively simpler equations, until finally a system appears that displays the solutions as in (1.6). The manipulations allowed in this process are the following three. You are asked in Exercises 53–55 to justify them.

1. Interchange the order of the equations; switch any two.
2. Multiply any equation by a nonzero constant.
3. Replace any equation by its sum with a nonzero constant multiple of any other equation in the system.

The procedure is to use these three **elementary operations** to replace a given system of equations with a system of equations that is equivalent, but easier to solve. We first use these elementary operations to find a system in triangular form, then one in row echelon form, and, finally, one such as (1.6), which is in *reduced row echelon form.* That each of these three elementary operations always yields a system of equations equivalent to the original one is fairly obvious. We do not state that fact formally as a theorem, although it is one. The proof of this contention for each of the individual elementary operations is left for you as an exercise. Notice in particular that each of these processes is reversible. For more on this topic, see the supplementary exercises to this chapter.

The process for finding the system of equations that is in echelon form and is equivalent to the given system is called **Gaussian elimination.** There is no one "best" tactic for doing this; one simply proceeds by judicious application of the elementary operations.

Gaussian elimination can be considerably simplified upon realizing that it is the coefficients in the successive equations that are important. The variables in the system act mostly as placeholders in this process. By detaching the coefficients and writing them in tabular form, we can decrease the amount of writing involved. Consider the following example.

Example 3   Use Gaussian elimination to solve the following system of linear equations:

$$\begin{aligned}
x_1 - 3x_2 + x_3 - x_4 &= 1, \\
2x_1 - 9x_2 - x_3 + x_4 &= -1, \\
x_1 - 2x_2 - 3x_3 + 4x_4 &= 5, \\
3x_1 - x_2 + 2x_3 - x_4 &= -2.
\end{aligned} \tag{1.7}$$

The left side of this system of equations can be abbreviated by the following array of its coefficients:

$$\begin{bmatrix}
1 & -3 & 1 & -1 \\
2 & -9 & -1 & 1 \\
1 & -2 & -3 & 4 \\
3 & -1 & 2 & -1
\end{bmatrix}. \tag{1.8}$$

Any such rectangular array is called a **matrix.** This particular matrix is called the **coefficient matrix** for the system of equations in (1.7). Each equation corresponds to a horizontal **row** in (1.8), while the coefficients of a given variable all lie in the same vertical **column** of (1.8). For example, the coefficients of x_3 in the system all lie in the third column, etc. This matrix in (1.8) is **square** in that it has the same number of rows as it has columns. It is a four by four (usually written 4×4) matrix. The numbers in a matrix are called either its **entries** or its **elements.**

The constant terms on the right side of the equations in (1.7) may also be written as a 4×1 matrix

$$\begin{bmatrix}
1 \\
-1 \\
5 \\
-2
\end{bmatrix}. \tag{1.9}$$

Combining the two matrices—the coefficient matrix (1.8) and the matrix of the constant terms of (1.9)—into a single matrix, results in a matrix that is called the **augmented matrix** of the given system of equations. Thus, the matrix

$$\left[\begin{array}{cccc|c}
1 & -3 & 1 & -1 & 1 \\
2 & -9 & -1 & 1 & -1 \\
1 & -2 & -3 & 4 & 5 \\
3 & -1 & 2 & -1 & -2
\end{array}\right] \tag{1.10}$$

is the augmented matrix of the system of equations (1.7).

By arranging the individual equations in a system of equations so that the coefficients of a given variable are all in the same column, the unique augmented matrix for that system of equations can be written easily.

Now the elementary operations on the *equations* of the system become operations on the *rows* of its augmented matrix. In our example, by adding to rows two, three, and four of the matrix (1.10) appropriate multiples of row one, we turn the first entries in all of these rows into zeros. That is, add -2

times row one to row two, -1 times row one to row three and -3 times row one to row four. The following matrix results:

$$
\begin{bmatrix}
1 & -3 & 1 & -1 & 1 \\
0 & -3 & -3 & 3 & -3 \\
0 & 1 & -4 & 5 & 4 \\
0 & 8 & -1 & 2 & -5
\end{bmatrix}. \tag{1.11}
$$

This matrix is the augmented matrix for an equivalent system of linear equations; write it out! But the equivalent system is not yet in triangular form; thus, we can continue. Multiply the second row by $-1/3$, and obtain the following matrix:

$$
\begin{bmatrix}
1 & -3 & 1 & -1 & 1 \\
0 & 1 & 1 & -1 & 1 \\
0 & 1 & -4 & 5 & 4 \\
0 & 8 & -1 & 2 & -5
\end{bmatrix}. \tag{1.12}
$$

Now proceed to change the entries in rows three and four and column two of the matrix in (1.12) to zeros (tantamount to eliminating the variable x_2 from the third and fourth equations). This is done by replacing row three of (1.12) by its sum with -1 times row two, and by replacing row four by its sum with -8 times row two. The following matrix results:

$$
\begin{bmatrix}
1 & -3 & 1 & -1 & 1 \\
0 & 1 & 1 & -1 & 1 \\
0 & 0 & -5 & 6 & 3 \\
0 & 0 & -9 & 10 & -13
\end{bmatrix}. \tag{1.13}
$$

This is the augmented matrix of yet a different system of equations equivalent to (1.7). Write it out! We've almost achieved a system in triangular form. It remains to replace the last row of (1.13) with a row in which the first three entries are zeros (modeling the elimination of x_3 from the fourth equation). To do this, add $-9/5$ times row three to row four. This gives the matrix

$$
\begin{bmatrix}
1 & -3 & 1 & -1 & 1 \\
0 & 1 & 1 & -1 & 1 \\
0 & 0 & -5 & 6 & 3 \\
0 & 0 & 0 & -\dfrac{4}{5} & -\dfrac{92}{5}
\end{bmatrix}. \tag{1.14}
$$

This matrix is the augmented matrix of a system of linear equations in triangular form. We can simplify it a little and put it into *row echelon form* by multiplying the fourth row by $-5/4$ and by multiplying the third row by $-1/5$, giving us the matrix

$$
\begin{bmatrix}
1 & -3 & 1 & -1 & 1 \\
0 & 1 & 1 & -1 & 1 \\
0 & 0 & 1 & -\dfrac{6}{5} & -\dfrac{3}{5} \\
0 & 0 & 0 & 1 & 23
\end{bmatrix}. \quad \blacksquare \tag{1.15}
$$

A matrix is in **row echelon form** when the first nonzero entry in any row is a one, when all entries in subsequent rows and in the same column with that leading one are zeros, and when all rows, if any, consisting entirely of zeros occur after those rows in which there are any nonzero entries. The matrix (1.15) is in row echelon form. Those in (1.8) through (1.14) are not. Why?

Now (1.15) represents the equivalent system of equations

$$x_1 - 3x_2 + x_3 - x_4 = 1,$$
$$x_2 + x_3 - x_4 = 1,$$
$$x_3 - \frac{6}{5}x_4 = -\frac{3}{5},$$
$$x_4 = 23.$$

We have the value 23 for x_4, and can use **back substitution** in the other equations to find the values of $x_1 = -12$, $x_2 = -3$, and $x_3 = 27$.

> **Definition**   We say that two matrices A and B are **row equivalent** if one can be obtained from the other by a finite sequence of elementary row operations.

Compare the following elementary row operations with the allowable operations on equations in a system of linear equations.

1. Interchange any two rows.
2. Multiply any row by a nonzero constant.
3. Replace any row with its sum with a nonzero constant multiple of any other row.

An algorithm for a step-by-step process that always yields the row echelon form for a given matrix, although not necessarily in the fewest number of steps, is given in Appendix A.3.

An alternative, and perhaps more systematic approach, to solving a system of equations would be to continue to perform elementary row operations on the augmented matrix of the system until we have eliminated all but the coefficient of x_1 from the first equation, all but the coefficient of x_2 from the second, all but the coefficient of x_3 from the third, and so on. This process is called a **Gauss-Jordan reduction** of the matrix.

To apply this to the solved system of Example 3, rather than use back substitution, we start with the matrix in (1.15). Add 6/5 times the last row of the matrix (1.15) to its third row, and 1 times the last row to both the second and the first rows, resulting in the matrix

$$\begin{bmatrix} 1 & -3 & 1 & 0 & 24 \\ 0 & 1 & 1 & 0 & 24 \\ 0 & 0 & 1 & 0 & 27 \\ 0 & 0 & 0 & 1 & 23 \end{bmatrix}. \qquad \textbf{(1.16)}$$

Now, to change the entry in row one and column three and the entry in row two and column three to zero (eliminating x_3 from equations one and two), we subtract row three from rows one and two, resulting in the matrix

$$\begin{bmatrix} 1 & -3 & 0 & 0 & -3 \\ 0 & 1 & 0 & 0 & -3 \\ 0 & 0 & 1 & 0 & 27 \\ 0 & 0 & 0 & 1 & 23 \end{bmatrix}. \tag{1.17}$$

Finally, we can change the entry in row one and column two to zero (eliminating x_2 from the first equation) by adding three times row two to row one. This results in the matrix

$$\begin{bmatrix} 1 & 0 & 0 & 0 & -12 \\ 0 & 1 & 0 & 0 & -3 \\ 0 & 0 & 1 & 0 & 27 \\ 0 & 0 & 0 & 1 & 23 \end{bmatrix}. \tag{1.18}$$

The matrix in (1.18) is in *reduced row echelon* form. It is the augmented matrix of a system of equations equivalent to (1.7), from which we can read the solution $(-12, -3, 27, 23)$ to that system; namely,

$$\begin{aligned} x_1 &= -12, \\ x_2 &= -3, \\ x_3 &= 27, \\ x_4 &= 23. \end{aligned}$$

A matrix is in **reduced row echelon form** if it is in row echelon form and all of the entries *above* the first "one" in each row are also all zeros. The matrix (1.18) is in reduced row echelon form. None of the preceding matrices are, however. Why?

The process of finding a matrix in reduced row echelon form that is row equivalent to a given matrix A is called the **Gauss-Jordan reduction of A.**

> A second algorithm given in Appendix A.3, at the end of the text, is a simplified flowchart that outlines the Gauss-Jordan reduction process for any matrix.

Example 4   Reduce the augmented matrix for the following system of equations to its reduced row echelon form. Find the solutions to the system (if any).

$$\begin{aligned} x + y + z &= 2, \\ 2x - y - z &= 1, \\ x - y - z &= 0, \\ 3x + 4y + 2z &= 3. \end{aligned}$$

The augmented matrix for this system is

$$\begin{bmatrix} 1 & 1 & 1 & 2 \\ 2 & -1 & -1 & 1 \\ 1 & -1 & -1 & 0 \\ 3 & 4 & 2 & 3 \end{bmatrix}.$$

Then, by using the Gauss-Jordan reduction techniques, we find that this matrix has reduced row echelon form:

$$\begin{bmatrix} 1 & 0 & 0 & 1 \\ 0 & 1 & 0 & -1 \\ 0 & 0 & 1 & 2 \\ 0 & 0 & 0 & 0 \end{bmatrix}.$$

Since the original system was *overdetermined,* the final row is, in this case, all zeros. The four equations might have put too many constraints on the three variables, and we could have had an inconsistent set of conditions with which to deal. We discuss this possibility further in the next section. Here, we have the augmented matrix for the system of equations

$$x \qquad\qquad = 1,$$
$$y \qquad\quad = -1,$$
$$z = 2,$$
$$0x + 0y + 0z = 0. \quad \blacksquare$$

The last equation is unnecessary, and we can safely ignore it. One of the equations in the original system was ultimately not essential to the solution $(1, -1, 2)$ for the system. *Overdetermined* and *underdetermined* systems can have other results that we will consider in Section 1.2.

Often nonlinear systems of equations may have linear form, and may be solved by the methods we have discussed by making an appropriate substitution that changes the variables and results in a system of linear equations.

Example 5   Solve the following system of equations using the matrix methods described and an appropriate substitution.

$$\frac{1}{x} + \frac{1}{y} - \frac{1}{z} = 0,$$

$$\frac{2}{x} - \frac{2}{y} + \frac{1}{z} = 3,$$

$$\frac{3}{x} - \frac{4}{y} + \frac{2}{z} = 4.$$

To solve this system of equations we first substitute $u = 1/x$, $v = 1/y$, and $w = 1/z$. Then the system results in the linear system

$$u + v - w = 0,$$
$$2u - 2v + w = 3,$$
$$3u - 4v + 2w = 4.$$

Its augmented matrix is

$$\begin{bmatrix} 1 & 1 & -1 & 0 \\ 2 & -2 & 1 & 3 \\ 3 & -4 & 2 & 4 \end{bmatrix}.$$

This is row equivalent to the following matrix in reduced row echelon form. (Verify this.)

$$\begin{bmatrix} 1 & 0 & 0 & 2 \\ 0 & 1 & 0 & 3 \\ 0 & 0 & 1 & 5 \end{bmatrix}.$$

Therefore, $u = 1/x = 2$, $v = 1/y = 3$, and $w = 1/z = 5$. So the solution to the original nonlinear system is

$$x = \frac{1}{2}, \qquad y = \frac{1}{3}, \qquad \text{and} \qquad z = \frac{1}{5}. \quad \blacksquare$$

Example 6   **Application—Kirchhoff's Laws.** The techniques we have developed to solve systems of linear equations can be applied to the analysis of direct-current electrical systems. When systems of resistors, batteries, and so on are connected into an *electrical network,* one typically uses one or both of the general principles formulated by the German physicist G. R. Kirchhoff (1824–1887) to analyze them. Figure 1.2 illustrates an electrical network containing two "meshes."

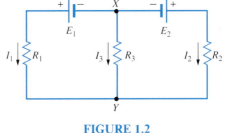

FIGURE 1.2

The formal statements of **Kirchhoff's laws** are the following:
1. The algebraic sum of the currents that meet at any junction of an electrical network is *zero.*
2. In going around any closed path of an electrical network, from any point of the network back to the same point, the algebraic sum of all the changes in potential (voltage) encountered on the way is *zero.*

In effect, the first law states that no electrical charge will accumulate at any point of a network. The second law is a statement that every point of the network has, at any given instant, a unique electrical potential (voltage).

The current in any part of the network is usually denoted by I, the potential (voltage) by E, and the resistance by R. We assume that **Ohm's law**

$$E = IR$$

holds. We also assume that the voltage drop caused by the current passing though several resistors connected in series is the sum of the individual voltage drops in each resistor. Then, to analyze the network given in Figure 1.2, we apply the first of Kirchhoff's laws to either junction X or Y and arrive at the equation

$$I_1 + I_2 + I_3 = 0.$$

Moving around the left-hand mesh of the network in Figure 1.2 counterclockwise, add all of the voltage drops algebraically. According to the second Kirchhoff law, we obtain the equation

$$E_1 - R_1 I_1 + R_3 I_3 = 0.$$

Now move around the right-hand mesh in a clockwise direction. This gives the equation

$$E_2 - R_2 I_2 + R_3 I_3 = 0.$$

In each case we have assumed that the current flows from $+$ to $-$. Suppose we know the resistance of the voltage. Then we have a system of three equations in the three unknowns I_1, I_2, and I_3:

$$\begin{aligned} I_1 + \quad I_2 + \quad I_3 &= 0 \\ R_1 I_1 \qquad\quad - R_3 I_3 &= E_1 \\ R_2 I_2 - R_3 I_3 &= E_2. \end{aligned}$$

The augmented matrix for this system is

$$\begin{bmatrix} 1 & 1 & 1 & 0 \\ R_1 & 0 & -R_3 & E_1 \\ 0 & R_2 & -R_3 & E_2 \end{bmatrix}.$$

Now suppose that $R_1 = 3\Omega$, $R_2 = 5\Omega$, and $R_3 = 2\Omega$ (ohms), while $E_1 = 3V$, and $E_2 = 6V$ (volts). To find the current at each resistor the augmented matrix becomes

$$\begin{bmatrix} 1 & 1 & 1 & 0 \\ 3 & 0 & -2 & 3 \\ 0 & 5 & -2 & 6 \end{bmatrix}.$$

Its reduced row echelon form is computed to be

$$\begin{bmatrix} 1 & 0 & 0 & \dfrac{9}{31} \\ 0 & 1 & 0 & \dfrac{24}{31} \\ 0 & 0 & 1 & \dfrac{-33}{31} \end{bmatrix}.$$

You should verify this. Thus, the currents are $I_1 = 9/31\ A$, $I_2 = 24/31\ A$, and $I_3 = -33/31\ A$ (amperes). The minus sign in I_3 indicates that it flows in the direction opposite to that indicated by the arrow in Figure 1.2. ∎

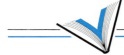

Vocabulary

We have considered a large number of concepts in this section. Some may be entirely new to you. The vocabulary developed here will be used throughout the text, so you should be sure you understand each of the concepts described by the terms in the following list. Review these ideas until you feel secure enough to build upon them.

linear equation	system of linear equations
solution to a system of equations	equivalent systems
triangular form	row echelon form
reduced row echelon form	back substitution
Gaussian elimination	elementary row operation
matrix (matrices)	coefficient matrix
augmented matrix	elements (entries) in a matrix
Gauss-Jordan reduction	row equivalent matrices

Exercises 1.1

In Exercises 1–4, interpret each of the given equations geometrically, sketch the graphs, and determine if a solution to the system exists. If so, approximate the solution from the graph.

1. $x - y = 3,$
   $2x + y = 6.$

2. $x - 3y = 7,$
   $2x + 6y = 8.$

3. $2x + y = 6,$
   $4x + 2y = 6.$

4. $x^2 + y^2 = 4,$
   $y = 2.$

In Exercises 5–12, write the augmented matrix for each of the given systems of equations.

5. $x - y = 4,$
   $x + y = 6.$

6. $x - 5y = 6,$
   $2x - 3y = 5.$

7. $x - y - z = -1,$
   $2x + 2y - z = 0,$
   $x + y + z = 3.$

8. $u - 2v - w = 0,$
   $2u + v + 2w = 7,$
   $3u + 4v - w = 4.$

9. Find the row echelon form for the augmented matrix of the system in Exercise 5 and solve the system by back substitution.

10. Find the row echelon form for the augmented matrix of the system in Exercise 6 and solve the system by back substitution.

11. Find the row-echelon form for the augmented matrix of the system in Exercise 7 and solve the system by back substitution.

12. Find the row echelon form for the augmented matrix of the system in Exercise 8 and solve the system by back substitution.

13. The following matrices are all row equivalent. Justify this statement by giving the reasons A is row equivalent to B, B is row equivalent to C, and so on.

$$A = \begin{bmatrix} 1 & -2 & -1 & -1 \\ 1 & 1 & -1 & 1 \\ 2 & 1 & 2 & -1 \end{bmatrix}, \quad B = \begin{bmatrix} 1 & 1 & -1 & 1 \\ 1 & -2 & -1 & -1 \\ 2 & 1 & 2 & -1 \end{bmatrix}, \quad C = \begin{bmatrix} 1 & 1 & -1 & 1 \\ 0 & -3 & 0 & -2 \\ 2 & 1 & 2 & -1 \end{bmatrix},$$

$$D = \begin{bmatrix} 1 & 1 & -1 & 1 \\ 0 & -3 & 0 & -2 \\ 0 & -1 & 4 & -3 \end{bmatrix}, \quad E = \begin{bmatrix} 1 & 1 & -1 & 1 \\ 0 & 1 & -4 & 3 \\ 0 & 0 & 12 & -7 \end{bmatrix}, \quad F = \begin{bmatrix} 1 & 1 & -1 & 1 \\ 0 & 1 & 0 & \frac{2}{3} \\ 0 & 0 & 1 & \frac{-7}{12} \end{bmatrix}.$$

In Exercises 14–30, write the augmented matrix for each of the given systems of equations, reduce the matrix to its reduced row echelon form, and solve the system.

14.
$$x_1 + x_2 - x_3 = 1,$$
$$x_1 - 2x_2 + x_3 = 2,$$
$$2x_1 + 3x_2 - 3x_3 = 1.$$

15.
$$x_1 - 3x_2 - 5x_3 = 1,$$
$$x_1 + x_2 + x_3 = 3,$$
$$2x_1 + x_2 + 2x_3 = 1.$$

16.
$$5x - 6y = 4,$$
$$7x + y = 13,$$
$$2x + 7y = 9.$$

17.
$$7x_1 + 3x_2 - 5x_3 = 4,$$
$$5x_1 + 5x_2 + 7x_3 = -2,$$
$$4x_1 - x_2 - 6x_3 = 5.$$

18.
$$2x_1 + 3x_2 + 4x_3 = 3,$$
$$3x_1 + 6x_2 - 2x_3 = 1,$$
$$5x_1 - 9x_2 + 6x_3 = 1.$$

19.
$$x + y - z = 3,$$
$$x - y + z = -1,$$
$$2x - 3y - 4z = 3.$$

20.
$$x + 2y + z = 2,$$
$$-x + 4y + z = 4,$$
$$3x + 5y - 3z = 0.$$

21.
$$x - y + z = 0,$$
$$x - 2y + z = 1,$$
$$3x + 2y - 4z = 0.$$

22.
$$5x - 6y - 2z = -5,$$
$$3x - 2y + z = 2,$$
$$4x + y + 2z = 1.$$

23.
$$x_1 + x_2 + x_3 - x_4 = -2,$$
$$2x_1 - x_2 + x_3 + x_4 = 0,$$
$$3x_1 + 2x_2 - x_3 - x_4 = 1,$$
$$-x_1 - 4x_2 - 3x_3 + 3x_4 = 5.$$

24.
$$2x_1 - x_2 + x_3 - x_4 = 3,$$
$$3x_1 - 2x_2 + 4x_3 - x_4 = 4,$$
$$5x_1 + 2x_2 - 3x_3 - 4x_4 = 0,$$
$$-x_1 - x_2 + 3x_3 + 4x_4 = 2.$$

25.
$$3a + 9b - 6c = 1,$$
$$2a + 3b + 12c = 3,$$
$$2a - 3b + 3c = 1.$$

26.
$$5u + 5v - w = -\frac{3}{2},$$
$$7u + v - 2w = 3,$$
$$3u + 2v + 3w = \frac{11}{2}$$
$$u + 6v + 4w = 1.$$

27.
$$x_1 - x_2 + x_3 - x_4 = 0,$$
$$x_1 + x_2 + x_3 + x_4 = 0,$$
$$x_1 - x_2 - x_3 - x_4 = 2,$$
$$x_1 + x_2 + x_3 - x_4 = 2.$$

28.
$$x_1 + 2x_2 + 2x_3 + x_4 = 1,$$
$$3x_1 - 6x_2 + 5x_3 + 2x_4 = 2,$$
$$2x_1 + 3x_2 - x_3 - x_4 = 1,$$
$$5x_1 - 5x_2 + 3x_3 - 2x_4 = -2.$$

29.
$$2u - v + w + 2s + 3t = 5,$$
$$3u + 2v + 2w + 2s + 4t = 9,$$
$$3u + 3v - 3w + 2s - 4t = 7,$$
$$5u - v - w + 3s - 7t = 1,$$
$$7u + v + 3w + 5s - 5t = 5.$$

30.
$$x_1 + x_2 + x_3 + x_4 + x_5 + x_6 = 0,$$
$$x_1 - x_2 - x_3 - x_4 - x_5 - x_6 = 2,$$
$$x_1 - x_2 + x_3 - x_4 + x_5 + x_6 = 4,$$
$$2x_1 + 2x_2 - 3x_3 + 4x_4 + x_5 - 2x_6 = -4,$$
$$3x_1 - 2x_2 + x_3 + x_4 - x_5 + 3x_6 = 1,$$
$$5x_1 + 5x_2 - 3x_3 - 3x_4 + 4x_5 + 3x_6 = 1.$$

31. The equation of a parabola in the plane has the form $y = ax^2 + bx + c$. Determine the equation of the parabola that passes through the three points $(0, 1)$, $(1, 2)$, and $(-1, 6)$. HINT: You will have three equations in the three unknown coefficients a, b, and c to solve.

32. Does the point $(2, 4)$ lie on the parabola of Exercise 31? Why, or why not?

33. Determine the third-degree polynomial function $y = ax^3 + bx^2 + cx + d$ that passes through the four points $(0, 1)$, $(1, 2)$, $(2, 3)$, and $(-1, 6)$.

In Exercises 34 and 35, find a matrix in reduced row echelon form that is equivalent to the given matrix. For what values of the variable x is this valid?

34.
$$\begin{bmatrix} 1 & x & -1 \\ x & 1 & -1 \\ -1 & 1 & x \end{bmatrix}.$$

35.
$$\begin{bmatrix} x-1 & 1 & 2 & 3 \\ 1 & x-1 & 1 & 0 \\ 3 & 2 & x-2 & 1 \\ 0 & 0 & 1 & x-2 \end{bmatrix}.$$

36. In a recent election the winning candidate received 52% of the vote, while his only opponent received 48%. If the winner received 1532 more votes than his opponent, how many votes did each receive? HINT: Set up the system of linear equations that models this, then use the augmented matrix to solve it.

37. In another election, the three candidates for office received 20%, 27%, and 53% of the vote, respectively. The winning candidate had a majority of 732 votes over the total of her opponents. How many votes did each candidate get?

38. A collection of 50 coins consists of nickels, dimes, and quarters. If there are three times as many nickels as dimes, and the total amount is $6.50, how many quarters are there?

39. A certain food contains 8 units per ounce of vitamin B_1 and 12 units per ounce of vitamin C. A second food contains 12 and 14 units per ounce of vitamins B_1 and C, respectively. How many ounces of each are required for a meal that will contain 100 units of vitamin B_1 and 120 units of vitamin C?

40. A person purchases a watch, a neck chain, and a ring for a total expenditure of $1000. The watch sold for $50 more than the neck chain. The ring sold for $25 more than the watch and neck chain together. What was the individual price of each item?

41. Use Kirchhoff's laws to find the electrical currents (in amperes) for the circuit depicted in Figure 1.3(a). (See Example 6.)

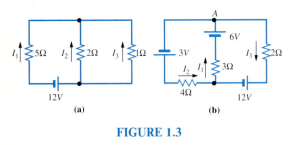

FIGURE 1.3

42. Use Kirchhoff's laws to find the electrical currents (in amperes) for the circuit depicted in Figure 1.3(b).

43. To balance a chemical equation, one must find whole numbers of molecules, a, b, c, etc., so the number of atoms of each of the elements on the left side of the equation equals the number of atoms on the right side. Balance the following reaction equation by finding a, b, and c: $a NH_3 + b O_2 \rightarrow 2NO + c H_2O$. HINT: There are $3a$ hydrogen atoms on the left side and $2c$ on the right, so $3a = 2c$. How many nitrogen and oxygen atoms are there?

44. See the previous exercise. Balance the following equation:

$$a MnSO_4 + b KMnO_4 + c H_2O$$
$$\rightarrow 5MnO_2 + d K_2SO_4 + e H_2SO_4.$$

45. Three alloys, x, y, and z, have the following percentages of lead, copper, and zinc.

Alloy	Lead	Copper	Zinc
x	50	20	30
y	40	30	30
z	30	0	70

How many grams of each alloy must be combined to obtain an alloy that is 44% lead, 18% copper, and 38% zinc?

In Exercises 46–51, make a substitution that will transform the given nonlinear system into a linear system. Then solve the system.

46.
$$\frac{1}{x} + \frac{1}{y} + \frac{3}{z} = 0,$$
$$\frac{2}{x} - \frac{2}{y} - \frac{2}{z} = -4,$$
$$\frac{3}{x} + \frac{3}{y} + \frac{3}{z} = 4.$$

47.
$$\sqrt{x} - \sqrt{y} + \sqrt{z} = 3,$$
$$2\sqrt{x} + \sqrt{y} - \sqrt{z} = 3,$$
$$\sqrt{x} - \sqrt{y} + 2\sqrt{z} = 7.$$

48.
$$x^3 - 3y^3 - z^3 = -4,$$
$$2x^3 + y^3 + z^3 = 9,$$
$$5x^3 - 3y^3 - z^3 = 0.$$

49.
$$3\sqrt{x^2 - 1} - 2y = \sqrt{3},$$
$$5\sqrt{x^2 - 1} - 4y = \sqrt{3}.$$

50.
$$\frac{2}{x - y} + \frac{1}{x + y} = 3,$$
$$\frac{5}{x - y} + \frac{6}{x + y} = 4.$$

51.
$$\frac{1}{3x - y} + \frac{2}{x + 3y} = 2,$$
$$\frac{10}{3x - y} - \frac{9}{x + 3y} = 7.$$

52. Determine a and b so the point $(1, 2)$ lies on the graphs of both the line $ax + by = 3$ and the circle $ax^2 + by^2 = 5$. Sketch both graphs. Are there any other points of intersection? If so, find them.

In Exercises 53–55, suppose we know that the n-tuple $(s_1, s_2, \ldots, s_n)$ is a solution to the following system of linear equations:

$$\begin{aligned}
a_{11}x_1 + a_{12}x_2 + \cdots + a_{1n}x_n &= b_1, \\
a_{21}x_1 + a_{22}x_2 + \cdots + a_{2n}x_n &= b_2, \\
&\ \ \vdots \\
a_{i1}x_1 + a_{i2}x_2 + \cdots + a_{in}x_n &= b_i \\
&\ \ \vdots \\
a_{j1}x_1 + a_{j2}x_2 + \cdots + a_{jn}x_n &= b_j \\
&\ \ \vdots \\
a_{m1}x_1 + a_{m2}x_2 + \cdots + a_{mn}x_n &= b_m.
\end{aligned}$$
(†)

53. Show that $(s_1, s_2, \ldots, s_n)$ is also a solution to the system:

$$\begin{aligned}
a_{11}x_1 + a_{12}x_2 + \cdots + a_{1n}x_n &= b_1, \\
a_{21}x_1 + a_{22}x_2 + \cdots + a_{2n}x_n &= b_2, \\
&\ \ \vdots \\
a_{j1}x_1 + a_{j2}x_2 + \cdots + a_{jn}x_n &= b_j, \\
&\ \ \vdots \\
a_{i1}x_1 + a_{i2}x_2 + \cdots + a_{in}x_n &= b_i, \\
&\ \ \vdots \\
a_{m1}x_1 + a_{m2}x_2 + \cdots + a_{mn}x_n &= b_m,
\end{aligned}$$

which is derived from the system (†) by interchanging equations i and j. HINT: What is a solution to the system of equations? [Substitute for the variables and see (†).]

54. If k is any nonzero constant, show that $(s_1, s_2, \ldots, s_n)$ is a solution to the following system:

$$\begin{aligned}
a_{11}x_1 + a_{12}x_2 + \cdots + a_{1n}x_n &= b_1, \\
a_{21}x_1 + a_{22}x_2 + \cdots + a_{2n}x_n &= b_2, \\
&\ \ \vdots \\
ka_{i1}x_1 + ka_{i2}x_2 + \cdots + ka_{in}x_n &= kb_i, \\
&\ \ \vdots \\
a_{j1}x_1 + a_{j2}x_2 + \cdots + a_{jn}x_n &= b_j, \\
&\ \ \vdots \\
a_{m1}x_1 + a_{m2}x_2 + \cdots + a_{mn}x_n &= b_m;
\end{aligned}$$

which is derived from (†) by multiplying equation i, any $i = 1, 2, \ldots, m$, by the constant k.

55. If k is any constant, show that $(s_1, s_2, \ldots, s_n)$ is a solution to the following system:

$$\begin{aligned}
a_{11}x_1 &+ a_{12}x_2 &+ \cdots + &a_{1n}x_n &= b_1, \\
a_{21}x_1 &+ a_{22}x_2 &+ \cdots + &a_{2n}x_n &= b_2, \\
&&&& \vdots \\
a_{i1}x_1 &+ a_{i2}x_2 &+ \cdots + &a_{in}x_n &= b_i, \\
&&&& \vdots \\
(a_{j1} + ka_{i1})x_1 + (a_{j2} + ka_{i2})x_2 + \cdots + (a_{jn} + ka_{in})x_n &= b_j + kb_i, \\
&&&& \vdots \\
a_{m1}x_1 &+ a_{m2}x_2 &+ \cdots + &a_{mn}x_n &= b_m;
\end{aligned}$$

which is derived from (†) by adding k times equation i to equation j, for any $i, j = 1, 2, \ldots, m, i \neq j$.

56. If you have a computer available and any software package such as *Mathematica*™, *Matrix Works*™, *Matlab*™, *Lintek*™, *Linear Kit*™, *Theorist*™, *Matrix*™, etc., use the incorporated Gauss-Jordan reduction methods available there to solve Exercises 27–30.

1.2 LINEAR SYSTEMS WITHOUT UNIQUE SOLUTIONS

Not every system of n linear equations in n variables has a unique solution. That is, there may not always be exactly one n-tuple of numbers that will simultaneously satisfy all of the equations of the system. And, of course, if the number of equations is different from the number of variables, the solution is even more involved. A given system of linear equations may have no solution at all, or may have more than one solution. In this section we informally discuss three general types of systems where a unique solution might not exist. Also, we will see that the matrix methods developed in Section 1.1 will help us here.

Dependent Systems

It is entirely possible that a given system of equations does not have one unique solution. This is true even if there are the same number of equations as there are variables. Consider, for example, the simple 2×2 system of equations:

$$x - y = 2,$$
$$3x - 3y = 6.$$

Every point on the line $y = x - 2$ is a solution to this system, as you can see by sketching the graph of these two equations (see Figure 1.4). They both have the same line as their graphical representation.

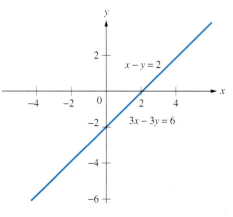

FIGURE 1.4

It is fairly clear that the second equation in this system is merely a multiple of the first. It gives no new information. Such a system is called **dependent,** but we shall defer a formal definition of dependence until a later chapter; here, we'll simply consider some examples of dependent systems. In systems with more than two variables, the dependence may be less apparent than in the example we just discussed. Consider the following.

Example 1

Solve the system of equations:

$$\begin{aligned} x_1 + 2x_2 - x_3 &= 4, \\ 2x_1 - x_2 + 2x_3 &= -1, \\ 3x_1 + x_2 + x_3 &= 3. \end{aligned} \tag{1.19}$$

It would require careful examination of this system of equations to reveal that the third equation in the system is the sum of the first two equations, and therefore gives no new constraint. This fact becomes obvious to us when we row reduce the augmented matrix for the system. Indeed, the augmented matrix for this system

$$\begin{bmatrix} 1 & 2 & -1 & 4 \\ 2 & -1 & 2 & -1 \\ 3 & 1 & 1 & 3 \end{bmatrix}$$

can be row reduced to the triangular matrix

$$\begin{bmatrix} 1 & 2 & -1 & 4 \\ 0 & -5 & 4 & -9 \\ 0 & 0 & 0 & 0 \end{bmatrix}$$

whose final row of zeros replaced the third row. This matrix is then the augmented matrix for a system of linear equations consisting of just two nonzero equations. The matrix can be reduced further to row echelon form

$$\begin{bmatrix} 1 & 0 & \dfrac{3}{5} & \dfrac{2}{5} \\ 0 & 1 & -\dfrac{4}{5} & \dfrac{9}{5} \\ 0 & 0 & 0 & 0 \end{bmatrix},$$

which is the augmented matrix for the system of equations

$$\begin{aligned} x_1 + \frac{3}{5}x_3 &= \frac{2}{5}, \\ x_2 - \frac{4}{5}x_3 &= \frac{9}{5}. \end{aligned}$$

This system has infinitely many solutions, since x_3 can be any real number. Thus, if $x_3 = t$, the solutions to this system have the parametric representation

$$\begin{aligned} x_1 &= \frac{2 - 3t}{5}, \\ x_2 &= \frac{9 + 4t}{5}, \\ x_3 &= t. \end{aligned} \tag{1.20}$$

The final row of zeros that first appeared in the triangular form of the augmented matrix of our system indicated that the system was dependent, and that there would be an infinite number of solutions. Since a linear equation in three variables can be graphically represented by a plane in 3-space, each equation in the system (1.19) is the equation of a different plane. These three planes do not intersect in a single point, as they would if a unique solution for the system existed. Rather, they intersect in a line; there are infinitely many points of 3-space common to all three planes. The line of this intersection has the parametric representation given in (1.20). See Figure 1.6 on page 23 for a graphical illustration.   ∎

Of course, infinitely many solutions to a system are not restricted to dependent systems of equations. The system consisting of the single equation

$$x - y = 1$$

has infinitely many solutions: $x = 1 + t$, $y = t$. Yet it is not a dependent system. Often, if there are fewer equations than unknowns, either initially, or because of dependence, the system of equations will have infinitely many solutions. Systems of linear equations with fewer equations than unknowns are sometimes called *underdetermined systems.*

On the other hand, if there are more equations in a given system of equations than there are variables, the system is called an *overdetermined system.* A variety of things might occur in the case of an overdetermined system including a unique solution, no solution, or infinitely many solutions. Before we turn to consider an example of an overdetermined system with a unique solution, we make the following observation.

Computer/Calculator Use and Caution:

From this point on you are free to use a hand-held calculator or computer to find the reduced row echelon form for any of the matrices that appear in the remainder of this text. In fact, such use is encouraged as a way to simplify what can be tedious computational problems. A fairly large number of software programs are available to choose from, and improved products appear almost monthly.

One problem that can be encountered when using a calculator or computer to reduce a matrix involves determining when a computed entry should be zero. The computing device might give an entry as a very small number; e.g., 4.52E-20, because of round-off errors. If the computer should use this entry as a pivot in a future step, the result can be chaotic. For this reason more advanced programs replace all sufficiently small computed entries with zero. Determining what is sufficiently small for any given problem depends on the magnitude of the nonzero entries in the matrix. When you use a computing device, be sure to observe the procedures given in that program for dealing with round-off error.

Example 2   A certain diet food concentrate contains 8% protein and 25% carbohydrate, by weight. A second diet food contains 13% protein and 30% carbohydrate. How much of each diet food should be used to form 100 lb of a product containing 10% protein and 27% carbohydrate?

This problem can be modeled by the following system of three equations in the two variables f and s, denoting the number of pounds of the first and second diet food, respectively, to be used.

$$\begin{aligned} f + \quad s &= 100 & \text{Weight} \\ 0.08f + 0.13s &= (0.10)100 & \text{Protein content} \\ 0.25f + 0.30s &= (0.27)100 & \text{Carbohydrate content} \end{aligned} \qquad (1.21)$$

Since there are more equations than there are variables in this overdetermined system, it is quite possible that no solution exists. Each of the equations in (1.21) corresponds to a line in the f, s plane. A unique solution to (1.21) would be the common point of intersection of the three lines. In this case intersection of two of them would have been enough. Three arbitrary lines in the plane could also intersect in three different points, or not at all. (You ought to sketch these possible situations.)

The augmented matrix for this system (1.21) will be a square matrix (3×3). Take care to remember that the last column is the column of constants; this matrix is not the coefficient matrix for a 3×3 system. The matrix is

$$\begin{bmatrix} 1 & 1 & | & 100 \\ 0.08 & 0.13 & | & 10 \\ 0.25 & 0.30 & | & 27 \end{bmatrix}. \qquad (1.22)$$

It can be determined that the reduced echelon form of this matrix is as follows:

$$\begin{bmatrix} 1 & 0 & | & 60 \\ 0 & 1 & | & 40 \\ 0 & 0 & | & 0 \end{bmatrix}.$$

You should verify this.

The final row of zeros indicates that the original system (1.21) is indeed dependent, since one of the equations gave no new information. In this case we have a unique solution as indicated by the matrix: 60 lb of the first diet food concentrate should be mixed with 40 lb of the second concentrate to form the desired product. ■

Systems with No Solutions

In the above argument we mentioned the possibility that there might be no solution to an overdetermined system.

> **Definition**   A system of linear equations is called **inconsistent** if the equations of the system express contradictory conditions; that is, if the system cannot possibly have any solutions.

For example:
$$x + y = 1,$$
$$x + y = 2$$

is an obviously inconsistent system. If we sketch the graphs of these two equations, as indicated in Figure 1.5, we see that the two lines are parallel. Hence, there can be no point common to both of them and, therefore, no solution to the given system exists.

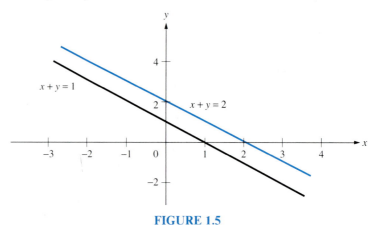

FIGURE 1.5

The use of matrices to solve systems of linear equations helps to illuminate otherwise obscure inconsistencies in the system. We illustrate this with the next example.

Example 3 Show that the following system of linear equations is inconsistent:
$$x - 2y + z = 3,$$
$$2x + 3y - 2z = 5,$$
$$3x + y - z = 6.$$

The augmented matrix for this system is the 3×4 matrix
$$\begin{bmatrix} 1 & -2 & 1 & 3 \\ 2 & 3 & -2 & 5 \\ 3 & 1 & -1 & 6 \end{bmatrix},$$

which has the following row echelon form (verify this, and note the cautionary statement preceding Example 2):
$$\begin{bmatrix} 1 & -2 & 1 & 3 \\ 0 & 7 & -4 & -1 \\ 0 & 0 & 0 & -2 \end{bmatrix}.$$

The last row of this matrix corresponds to the equation
$$0 = 0x + 0y + 0z = -2,$$

which is clearly inconsistent. Therefore, the original system, being equivalent to the one whose augmented matrix we have, is also inconsistent. ∎

Some of the geometric interpretations of conceivable situations for various systems of three linear equations in three variables are illustrated by Figure 1.6. Other configurations than those given in Figure 1.6—for example, exactly two of the planes coinciding—are also feasible. Of course, it is impossible to draw pictures such as these when the number of variables is larger than three. However, the algebraic concepts and the resulting solutions are analogous.

We make the following summary statements. They are, in fact, theorems, but they will be much easier to prove when we have more theoretical machinery available to us. Meanwhile, you may want to write out statements that justify each of them:

(a) An $m \times n$ system of linear equations is inconsistent if, and only if, the row echelon form of its $m \times (n + 1)$ augmented matrix has at least one row all of whose entries, *except for the entry in the last column,* are zero. (See Example 3.)

(b) A consistent $m \times n$ system of linear equations has a unique solution if, and only if, $m \geq n$ and the row echelon form of its augmented matrix has n "nonzero rows"; that is, n rows in which a nonzero entry appears in the first n columns. (See Example 2.)

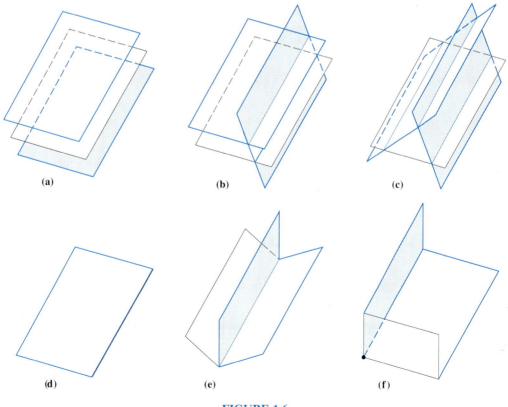

(a)   (b)   (c)

(d)   (e)   (f)

FIGURE 1.6

(c) A consistent $m \times n$ system of linear equations has infinitely many solutions that depend on $n - r$ independent parameters, if and only if, the echelon form of its augmented matrix has exactly $r < n$ "nonzero rows" (again, r rows in which a nonzero entry appears in the first n columns). (See Example 1.)

Example 4   Find all of the solutions, if any, to the system of equations

$$x + y + z + w = 0,$$
$$x - y - z - w = 1,$$
$$x + y - z + w = -1,$$
$$4x + 2z = 3,$$
$$3x - y + 3z - w = 4.$$

This is an overdetermined system where $m = 5$ and $n = 4$. Its augmented matrix is

$$\begin{bmatrix} 1 & 1 & 1 & 1 & 0 \\ 1 & -1 & -1 & -1 & 1 \\ 1 & 1 & -1 & 1 & -1 \\ 4 & 0 & 2 & 0 & 3 \\ 3 & -1 & 3 & -1 & 4 \end{bmatrix}.$$

Once again this matrix happens to be square—5×5—but don't let that mislead you. Remember that it is an *augmented* matrix, not a *coefficient* matrix. A row echelon form for this matrix, as you should work out, is

$$\begin{bmatrix} 1 & 1 & 1 & 1 & 0 \\ 0 & 1 & 0 & 1 & -1 \\ 0 & 0 & 1 & 0 & \frac{1}{2} \\ 0 & 0 & 0 & 0 & 0 \\ 0 & 0 & 0 & 0 & 0 \end{bmatrix},$$

which has $r = 3$ "nonzero rows." This is the augmented matrix of the following system of equations, which is equivalent to the original system:

$$x + y + z + w = 0,$$
$$y + w = -1,$$
$$z = \frac{1}{2},$$

plus two consistent equations of the form

$$0x + 0y + 0z + 0w = 0.$$

We of course ignore the consistent zero equations, since they give no information about the variables x, y, z, and w. We see from the remaining three equations that w can be any real number at all, set $w = t$. Thus, the system has infinitely many solutions, which involve $4 - 3 = 1$ independent parameter, denoted by t. All of the solutions have the form $\{(1/2, -1 - t, 1/2, t): t$ any real number$\}$. ■

It is probably fairly clear to you that in an overdetermined system where the number of equations exceeds the number of variables, the danger of too many constraints is increased. But even an underdetermined system, where the number of equations is less than the number of variables, might still fail to have a solution. For example, consider the system

$$2x + 3y - 4z = 8,$$
$$2x + 3y - 4z = 9.$$

These two equations can be thought of as representing two parallel planes in 3-space.

In summary then, all of the necessary information about a given system can be deduced from the row echelon form of its augmented matrix. Our matrix methods give us the required solutions if they exist, and demonstrate it when they don't.

Homogeneous Systems

We turn now to a type of system that will always have a solution. It is always consistent, no matter how many equations and how many variables are involved.

Definition   A system of equations is called **homogeneous** if the right side of each equation (the constant term) is zero.

Such a system as

$$
\begin{aligned}
a_{11}x_1 + a_{12}x_2 + \cdots + a_{1n}x_n &= 0, \\
a_{21}x_1 + a_{22}x_2 + \cdots + a_{2n}x_n &= 0, \\
&\ \vdots \\
a_{m1}x_1 + a_{m2}x_2 + \cdots + a_{mn}x_n &= 0,
\end{aligned}
\tag{1.23}
$$

always has the **trivial solution,** $x_1 = x_2 = x_3 = \ldots = x_n = 0$, as is easily seen. It is also possible that a homogeneous system may have additional **nontrivial solutions.** This is the case in the following example.

Example 5   Find *all* of the solutions to the following homogeneous system of equations:

$$
\begin{aligned}
x_1 + x_2 + x_3 + x_4 &= 0, \\
x_1 - x_2 + x_3 - x_4 &= 0, \\
x_1 + x_2 - x_3 - x_4 &= 0, \\
3x_1 + x_2 + x_3 - x_4 &= 0.
\end{aligned}
\tag{1.24}
$$

The augmented matrix for this system of equations (1.24) is

$$\left[\begin{array}{cccc|c} 1 & 1 & 1 & 1 & 0 \\ 1 & -1 & 1 & -1 & 0 \\ 1 & 1 & -1 & -1 & 0 \\ 3 & 1 & 1 & -1 & 0 \end{array}\right].$$

Since the final column of zeros will not change when we perform any elementary operation on this matrix, we can just as well omit it and work merely with the *coefficient matrix*

$$\left[\begin{array}{cccc} 1 & 1 & 1 & 1 \\ 1 & -1 & 1 & -1 \\ 1 & 1 & -1 & -1 \\ 3 & 1 & 1 & -1 \end{array}\right].$$

We row reduce this matrix to obtain the following reduced row echelon form:

$$\left[\begin{array}{cccc} 1 & 0 & 0 & -1 \\ 0 & 1 & 0 & 1 \\ 0 & 0 & 1 & 1 \\ 0 & 0 & 0 & 0 \end{array}\right].$$

Remember now, this is the *coefficient* matrix, not the *augmented* matrix, for a homogeneous system. So we have the following system of equations which is equivalent to system (1.24):

$$\begin{aligned} x_1 \phantom{{}+x_2} \phantom{{}+x_3} - x_4 &= 0, \\ x_2 \phantom{{}+x_3} + x_4 &= 0, \\ x_3 + x_4 &= 0. \end{aligned}$$

Thus, set $x_4 = t$, and the solutions to the system (1.24) are any of the 4-tuples $(t, -t, -t, t)$ for any real number t.   ∎

Example 6   Find *all* of the solutions to the following homogeneous system of equations:

$$\begin{aligned} x_1 + 2x_2 + x_3 + x_4 &= 0, \\ x_1 - x_2 + 2x_3 - x_4 &= 0, \\ x_1 + x_2 - x_3 - 2x_4 &= 0, \\ 3x_1 + x_2 + 2x_3 - x_4 &= 0. \end{aligned}$$

As we remarked in Example 5, it is necessary to use only the *coefficient* matrix for a homogeneous system of equations, because elementary row operations will not change the final zero column of the *augmented* matrix. The coefficient matrix for this system is

$$\left[\begin{array}{cccc} 1 & 2 & 1 & 1 \\ 1 & -1 & 2 & -1 \\ 1 & 1 & -1 & -2 \\ 3 & 1 & 2 & -1 \end{array}\right].$$

This matrix has the special matrix, denoted by I_4, as its reduced row echelon form, where

$$I_4 = \begin{bmatrix} 1 & 0 & 0 & 0 \\ 0 & 1 & 0 & 0 \\ 0 & 0 & 1 & 0 \\ 0 & 0 & 0 & 1 \end{bmatrix}.$$

You should work this out. Hence, the system equivalent to the original one is simply

$$\begin{aligned} x_1 &= 0, \\ x_2 &= 0, \\ x_3 &= 0, \\ x_4 &= 0. \end{aligned}$$

Thus, the trivial solution $(0, 0, 0, 0)$ is the only possible solution to the given system of equations. ■

The special $n \times n$ matrix

$$I_n = \begin{bmatrix} 1 & 0 & 0 & \cdots & 0 \\ 0 & 1 & 0 & \cdots & 0 \\ 0 & 0 & 1 & \cdots & 0 \\ \cdots & \cdots & \cdots & \cdots & \cdots \\ 0 & 0 & 0 & \cdots & 1 \end{bmatrix},$$

which has "ones" down its diagonal and zeros everywhere else, tells us something about a homogeneous system of equations having only the trivial solution (all variables must equal zero), namely:

> Whenever the coefficient matrix of an $n \times n$ homogeneous system of equations is row equivalent to I_n, the system always has the unique trivial solution $(0, 0, 0, \ldots, 0)$.

In a homogeneous system of equations (1.23), if there are more equations than variables, $m > n$, such an overdetermined system might be dependent, yet still have only the trivial solution. In this case its coefficient matrix will have reduced row echelon form

$$\begin{bmatrix} I_n \\ O \end{bmatrix},$$

with the special diagonal matrix I_n in the top block and $m - n$ rows of zero at the bottom. What do you suppose the coefficient matrix for an underdetermined *(m < n)* homogeneous system looks like? Can an underdetermined homogeneous system of linear equations have only the trivial solution?

Vocabulary

In this section we have encountered additional new concepts and terms. Since the vocabulary terms introduced in this section and the previous ones are used throughout the text, you should be certain you understand each of the concepts described by the words in the following list. Review these ideas until you feel secure in your ability to apply them throughout this course.

dependent system of linear
  equations
consistent system
underdetermined system
unique solution
nontrivial solution
parametric representation of a
  solution

inconsistent system
overdetermined system
homogeneous system
trivial solution
parameter(s)

Exercises 1.2

In Exercises 1–6, the augmented matrix (see 1.10 in Section 1.1) for a system of linear equations has been changed into its row echelon form. Decide in each case whether the corresponding system is consistent or inconsistent and, if it has a solution, find all of the possible solutions.

1. $\begin{bmatrix} 1 & 3 & 5 \\ 0 & 1 & 2 \\ 0 & 0 & 1 \end{bmatrix}$.

2. $\begin{bmatrix} 1 & 2 & -1 \\ 0 & 1 & -1 \\ 0 & 0 & 0 \end{bmatrix}$.

3. $\begin{bmatrix} 1 & -2 & -2 & 1 \\ 0 & 0 & 1 & 4 \\ 0 & 0 & 0 & 0 \end{bmatrix}$.

4. $\begin{bmatrix} 1 & -3 & 2 & -4 \\ 0 & 1 & -2 & 3 \\ 0 & 0 & 1 & -2 \end{bmatrix}$.

5. $\begin{bmatrix} 1 & -3 & 2 & -4 \\ 0 & 0 & 1 & 2 \\ 0 & 0 & 0 & 1 \\ 0 & 0 & 0 & 0 \end{bmatrix}$.

6. $\begin{bmatrix} 1 & -1 & 3 & 0 & 5 \\ 0 & 1 & -2 & 0 & -7 \\ 0 & 0 & 0 & 1 & 9 \\ 0 & 0 & 0 & 0 & 0 \end{bmatrix}$.

In Exercises 7–12, the augmented matrix of a system of linear equations has been transformed to its reduced row echelon form. In each case find all of the solutions to the corresponding system, if there are any.

7. $\begin{bmatrix} 1 & 0 & 0 & 2 \\ 0 & 1 & 0 & 3 \\ 0 & 0 & 1 & -3 \end{bmatrix}$.

8. $\begin{bmatrix} 1 & 4 & 0 & -2 \\ 0 & 0 & 1 & 3 \\ 0 & 0 & 0 & 1 \end{bmatrix}$.

9. $\begin{bmatrix} 1 & -2 & 0 & 3 \\ 0 & 0 & 1 & -1 \\ 0 & 0 & 0 & 0 \end{bmatrix}$.

10. $\begin{bmatrix} 1 & 2 & 0 & 1 & -3 \\ 0 & 0 & 1 & 2 & 4 \end{bmatrix}$.

11. $\begin{bmatrix} 1 & 1 & -2 & 0 & 3 \\ 0 & 0 & 0 & 1 & -2 \\ 0 & 0 & 0 & 0 & 0 \\ 0 & 0 & 0 & 0 & 0 \end{bmatrix}$.

12. $\begin{bmatrix} 0 & 1 & 0 & -1 \\ 0 & 0 & 1 & -2 \\ 0 & 0 & 0 & 0 \end{bmatrix}$.

In Exercises 13–27, use a Gauss-Jordan reduction of the appropriate matrix to find all of the solutions, if any exist, to the given system of linear equations.

13. $\begin{aligned} x + y + z &= 0, \\ x - y + z &= 0, \\ x - 3y + z &= 0. \end{aligned}$

14. $\begin{aligned} x + y + z &= 0, \\ x - y + z &= 0, \\ x + y - z &= 0. \end{aligned}$

15. $\begin{aligned} x_1 - x_2 + x_3 &= 0, \\ 2x_1 + 3x_2 - 5x_3 &= 0, \\ 3x_1 + 2x_2 - 4x_3 &= 0. \end{aligned}$

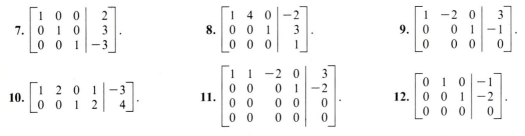

16. $x_1 + x_2 + x_3 = 0,$
$x_1 + 2x_2 + 3x_3 = 0,$
$2x_1 + 3x_2 + 2x_3 = 0.$

17. $x + y = w,$
$w + x = y,$
$w + z = 2x,$
$x - y = z.$

18. $s - t - u - v = 0,$
$3s - t - u - v = 0,$
$4s + t + u + v = 0,$
$s + t + u + v = 0,$
$5s - 3t + 2u - 4v = 0.$

19. $x - 2y - 2z = -1,$
$2x + 3y - z = 1,$
$x + y - z = 0.$

20. $x - 7y = 17,$
$2x + 3y = 0,$
$3x - 4y = 17.$

21. $x_1 - x_2 - x_3 = 3,$
$2x_1 - 2x_2 + 3x_3 = 5.$

22. $x_1 + x_2 - x_3 = 1,$
$2x_1 - x_2 + x_3 = 4,$
$x_1 + 4x_2 - 4x_3 = 3.$

23. $x_1 + 2x_2 - 2x_3 = 0,$
$3x_1 - x_2 + 3x_3 = 1,$
$4x_1 + x_2 + x_3 = 2.$

24. $x_1 - x_2 - x_3 = 1,$
$x_1 + x_2 + x_3 = -1,$
$2x_1 + 2x_2 - x_3 = 6.$

25. $u + v - w - x = 0,$
$u - v + w - x = 0,$
$u + v - w + x = 0.$

26. $x_1 + x_2 - 4x_3 + 2x_4 = 0,$
$-x_1 + x_2 - x_3 + 2x_4 + x_5 = 0,$
$x_1 + 2x_2 - x_3 + 4x_4 - x_5 = 0,$
$2x_1 - x_2 + 3x_3 - x_4 + x_5 = 0.$

27. $x_1 - 3x_2 + x_3 = 1,$
$2x_1 + x_2 - x_3 = 2,$
$x_1 + 4x_2 - 2x_3 = 1,$
$5x_1 - 8x_2 + 2x_3 = 5.$

28. Suppose the pair of numbers (c_1, c_2) is the solution to the 2×2 homogeneous system of equations

$$a_{11}x_1 + a_{12}x_2 = 0,$$
$$a_{21}x_1 + a_{22}x_2 = 0.$$

Show that for every real number k, the pair of numbers (kc_1, kc_2) is also a solution. (HINT: Substitute these values for the variables.)

29. Given the situation in Exercise 28, suppose the pair of real numbers (d_1, d_2) is also a solution to the given 2×2 homogeneous system different from (c_1, c_2). Show that $(d_1 + c_1, d_2 + c_2)$ is also a solution.

30. Given the $m \times n$ system of homogeneous equations in (1.23), suppose that $(s_1, s_2, \ldots, s_n)$ and $(t_1, t_2, \ldots, t_n)$ are two distinct solutions. Show that for every real number k, $(s_1 + kt_1, s_2 + kt_2, \ldots, s_n + kt_n)$ is also a solution.

31. Use the results of Exercise 30 to prove that if a given homogeneous system of equations has *two* distinct solutions, it has *infinitely many* solutions.

32. The following homogeneous system of linear equations does *not* have a unique solution. What can you say about the number a? Explain.

$$(a - 1)x_1 - 3x_2 + 5x_3 = 0,$$
$$ax_2 - 3x_3 = 0,$$
$$(a + 1)x_3 = 0.$$

33. Find all of the values for a in Exercise 32 for which the given homogeneous system of equations has a unique solution. Explain.

34. Are there any values for the number a in Exercise 32 for which the given system has no solution? Explain.

35. Explain why an overdetermined system of homogeneous linear equations such as (1.23) with $m > n$ must be a dependent system. (HINT: Consider row reduction of the augmented matrix.) Does that mean that only the trivial solution $(0, 0, 0, \ldots, 0)$ is possible?

36. Discuss what can occur in an underdetermined system of homogeneous linear equations, $m < n$.

37. If you have access to a computer, use the Gauss-Jordan reduction software methods available to solve Exercises 22–27.

1.3 *MATRIX ARITHMETIC*

In the preceding sections we saw how matrices are useful in solving systems of linear equations. This is not their only use, however. In fact, matrices have so many applications that they form an object of interest to mathematicians and scientists alike. In this section and the one following, we look at the basic algebra of matrices. This algebra has many similarities with and also some surprising differences from the familiar algebra of the real and complex numbers.

To begin, remember some notation already introduced. A *matrix* is any rectangular array of elements. The entries are called *scalars* and, for our purposes, will be either the real or the complex numbers. Usually uppercase letters, *A, B, C,* etc., denote matrices, while the scalar (number) entry in row i and column j of the matrix A, for example, is written a_{ij}. Thus the $m \times n$ (m rows, n columns) matrix A would appear as the array

$$A = \begin{bmatrix} a_{11} & a_{12} & a_{13} & \cdots & a_{1n} \\ a_{21} & a_{22} & a_{23} & \cdots & a_{2n} \\ a_{31} & a_{32} & a_{33} & \cdots & a_{3n} \\ \vdots & \vdots & \vdots & \cdots & \vdots \\ a_{m1} & a_{m2} & a_{m3} & \cdots & a_{mn} \end{bmatrix}. \qquad \textbf{(1.25)}$$

Often this notation is shortened to $A = (a_{ij})$. The matrices

$$A = \begin{bmatrix} 2 & -1 \\ 0 & 1 \\ 1 & 3 \end{bmatrix} \quad \text{and} \quad B = \begin{bmatrix} 2 & -i \\ 1+i & 0 \\ 1 & 3-i \end{bmatrix}$$

are both 3×2 matrices.

Matrix Equality

Two matrices A and B will be called *equal* if, and only if, they are exactly the same shape and have exactly the same entries. That is, $A = (a_{ij})$ is equal to $B = (b_{ij})$ if, and only if, both A and B are $m \times n$ matrices and $a_{ij} = b_{ij}$; $i = 1, 2, \ldots, m; j = 1, 2, \ldots n,$ for every scalar entry in the two matrices.

$$A = B \text{ if, and only if, } a_{ij} = b_{ij}.$$

Therefore, the two matrices

$$A = \begin{bmatrix} 2 & -1 \\ 0 & 1 \\ 1 & 3 \end{bmatrix} \quad \text{and} \quad B = \begin{bmatrix} 1 & 3 \\ 0 & 1 \\ 2 & -1 \end{bmatrix}$$

are *not* equal matrices, even though they are row equivalent. Given that $Z = A$, where Z is an "unknown" matrix $Z = (z_{ij})$, we can immediately conclude that $z_{11} = 2$, $z_{12} = -1$, $z_{21} = 0$, $z_{22} = 1$, $z_{31} = 1$, and $z_{32} = 3$.

Example 1 Solve the matrix equation

$$\begin{bmatrix} x & 2 & x-y \\ y & z & -1 \\ -1 & 3 & x+z \end{bmatrix} = \begin{bmatrix} 1 & y & -1 \\ 2 & 3 & y-z \\ -x & z & 4 \end{bmatrix}.$$

For these two matrices to be equal, the corresponding scalar entries must be identical; so, reading across the rows of each, it must follow that

$$\begin{array}{lll} x = 1, & 2 = y, & -1 = x - y, \\ y = 2, & 3 = z, & y - z = -1, \\ -1 = -x, & 3 = z, & x + z = 4. \end{array}$$

Since this system of (scalar) equations is consistent, $x = 1$, $y = 2$, $z = 3$ solves the matrix equation, and both of the above matrices equal the matrix

$$\begin{bmatrix} 1 & 2 & -1 \\ 2 & 3 & -1 \\ -1 & 3 & 4 \end{bmatrix}. \quad \blacksquare$$

Matrix Addition

Two matrices of the same shape, say $m \times n$, may be added together in a natural way to form a new matrix called their sum. Thus, if $A = (a_{ij})$ and $B = (b_{ij})$ are two $m \times n$ matrices, the **sum** $A + B$ is the matrix whose entry in row i and column j is the sum of the entries in those positions in A and B:

$$A + B = (a_{ij} + b_{ij}).$$

Example 2 Consider the three matrices

$$A = \begin{bmatrix} 1 & 1 & 0 \\ 2 & -1 & 3 \\ -1 & 0 & 4 \end{bmatrix}, \quad B = \begin{bmatrix} 2 & -1 & 1 \\ 0 & 5 & -7 \\ 4 & 1 & -2 \end{bmatrix}, \quad \text{and} \quad C = \begin{bmatrix} 1 & -1 & 2 \\ 0 & 1 & -5 \end{bmatrix}.$$

Since A and B are both 3×3 matrices, their sum exists and is computed as

$$A + B = \begin{bmatrix} 1+2 & 1-1 & 0+1 \\ 2+0 & -1+5 & 3-7 \\ -1+4 & 0+1 & 4-2 \end{bmatrix} = \begin{bmatrix} 3 & 0 & 1 \\ 2 & 4 & -4 \\ 3 & 1 & 2 \end{bmatrix}.$$

On the other hand, since C is 2×3, the sums $A + C$ and $B + C$ are not defined.   ■

> Matrix addition is defined only when both terms in the sum are matrices with the same number of rows and columns.

Since matrix addition is defined in terms of addition of the scalar entries in each matrix, it is not very surprising that matrix addition satisfies the same basic properties as does the addition of real or complex numbers. We have the following theorem summarizing the basic addition rules of matrix algebra.

> **Theorem 1.1**   Let A, B, C be $m \times n$ matrices with real (or complex) number entries. Then:
> (1) $A + B = B + A$ (the commutative law);
> (2) $A + (B + C) = (A + B) + C$ (the associative law);
> (3) there exists an $m \times n$ matrix O such that for each A:
> $A + 0 = 0 + A = A$ (the zero matrix)
> (4) for each matrix A there is a unique matrix $-A$ called the negative of A, such that $A + (-A) = 0$.

The proof of this theorem is a straightforward application of the definition of matrix addition and the corresponding properties of real and complex number algebra. The **zero matrix,** defined in part (3) of the theorem, is the matrix with m rows and n columns, all of whose entries are zero. There is a zero matrix of each size. Thus, the 3×2 *zero matrix* is

$$O = \begin{bmatrix} 0 & 0 \\ 0 & 0 \\ 0 & 0 \end{bmatrix}.$$

To prove part (4), it is easy to see that the negative of a matrix A is the matrix $-A = (-a_{ij})$. The proofs of parts (1) and (2) are left as exercises for you.
Using part (4), it is easy to define **subtraction** for matrices by:

$$A - B = A + (-B) = (a_{ij} - b_{ij}).$$

Example 3   Let the matrices A and B be as follows:

$$A = \begin{bmatrix} 1 & -1 & 3 \\ 4 & -2 & 0 \\ 5 & 0 & -1 \\ 2 & 4 & 1 \end{bmatrix}, \qquad B = \begin{bmatrix} 0 & 2 & -1 \\ -3 & 4 & 2 \\ 1 & 0 & 1 \\ 0 & 0 & 6 \end{bmatrix}.$$

Then $-B = \begin{bmatrix} 0 & -2 & 1 \\ 3 & -4 & -2 \\ -1 & 0 & -1 \\ 0 & 0 & -6 \end{bmatrix},$ and $A - B = \begin{bmatrix} 1 & -3 & 4 \\ 7 & -6 & -2 \\ 4 & 0 & -2 \\ 2 & 4 & -5 \end{bmatrix}.$ ■

Multiplication by a Scalar

For a given matrix $A = (a_{ij})$, it would be convenient to write $A + A = 2A$, and $A + A + A = 3A$, etc. It is easy to see from the definitions of matrix addition and the negative of a matrix that to do this for any integer n, $nA = (na_{ij})$; that is, multiply each entry by the integer n. Now suppose that r is any scalar, not just an integer. Then, to be consistent with this practice, we define the product of the scalar r and the $m \times n$ matrix A by

$$rA = (ra_{ij})$$

for $i = 1, 2, \ldots, m; j = 1, 2, \ldots, n.$

Example 4   You should have no trouble verifying that for A and B as in Example 3,

$$\frac{2}{5}A = \begin{bmatrix} \dfrac{2}{5} & -\dfrac{2}{5} & \dfrac{6}{5} \\ \dfrac{8}{5} & -\dfrac{4}{5} & 0 \\ 2 & 0 & -\dfrac{2}{5} \\ \dfrac{4}{5} & \dfrac{8}{5} & \dfrac{2}{5} \end{bmatrix} \qquad \text{and} \qquad \pi B = \begin{bmatrix} 0 & 2\pi & -\pi \\ -3\pi & 4\pi & 2\pi \\ \pi & 0 & \pi \\ 0 & 0 & 6\pi \end{bmatrix}. \quad ■$$

Note that the product of a matrix and a scalar is *not* the same thing as a type two elementary operation discussed in Section 1.1. In that case a single row of the matrix was multiplied by the scalar r. However, the matrix rA is obtained by multiplying *every* entry of A by r.

For convenience, we define Ar, multiplication of the matrix on the *right* by a scalar, as follows.

Multiply each entry a_{ij} on the right by r: $Ar = (a_{ij}r)$.

The following theorem is not difficult to verify, and it lists the basic algebraic properties of this product and their relationship to matrix addition.

Theorem 1.2   Let A and B be $m \times n$ real (or complex) matrices, and let r and s be real (or complex) scalars. Then:

(1) $r(A + B) = rA + rB$.   (2) $(r + s)A = rA + sA$.
(3) $(rs)A = r(sA)$.   (4) $rA = Ar$.
(5) $1A = A$.

Proof   The proof of this theorem is again a straightforward application of the definitions. For example, in part (1) $r(A + B)$ is the matrix whose entry in row i and column j is $r(a_{ij} + b_{ij}) = ra_{ij} + rb_{ij}$. But this latter sum is precisely the entry in row i and column j of the matrix sum $rA + rB$. Part (4) is justified by the fact that the scalars are real or complex numbers. So for each scalar entry a_{ij} in A, $ra_{ij} = a_{ij}r$. All of these properties depend upon the properties of real and complex number algebra and are left to you to finish.   ❏

Example 5   Solve the matrix equation $2A + 3X = B$, assuming that A and B are given $n \times n$ matrices.

Using the properties of matrix algebra of Theorems 1.1 and 1.2, we have

$$3X = B - 2A,$$

so

$$X = \frac{1}{3}(B - 2A) = \frac{1}{3}B - \frac{2}{3}A. \quad ■$$

Example 6   Write the complex matrix

$$A = \begin{bmatrix} 2 & -i \\ 1 + i & 0 \\ 1 & 3 - i \end{bmatrix}$$

in the form $M + iK$ where M and K are both real matrices; that is, all of their entries are real numbers. Describe how this can be done in general.

Let

$$M = \begin{bmatrix} 2 & 0 \\ 1 & 0 \\ 1 & 3 \end{bmatrix} \quad \text{and} \quad K = \begin{bmatrix} 0 & -1 \\ 1 & 0 \\ 0 & -1 \end{bmatrix}.$$

It is clear that

$$A = \begin{bmatrix} 2 & -i \\ 1 + i & 0 \\ 1 & 3 - i \end{bmatrix} = \begin{bmatrix} 2 & 0 \\ 1 & 0 \\ 1 & 3 \end{bmatrix} + i \begin{bmatrix} 0 & -1 \\ 1 & 0 \\ 0 & -1 \end{bmatrix}.$$

Let $A = ((x + iy)_{rs})$ be any $m \times n$ complex matrix, $r = 1, 2, \ldots, m$; $s = 1, 2, \ldots, n$. Then the two matrices $B = (x_{rs})$ and $C = (y_{rs})$ are such that $A = B + iC$. ∎

Matrix Multiplication

The third important matrix operation is that of matrix multiplication. One is naturally tempted to say that to multiply two matrices A and B, simply multiply their corresponding entries. This definition of product (the Hadamard product) is, however, not nearly so useful as the more complicated standard matrix product which is motivated by function composition. Before we actually give the definition, let us look at one situation where a matrix product is useful. Consider the following system of linear equations:

$$\begin{aligned}
x_1 + 2x_2 - x_3 + x_4 &= 4, \\
2x_1 - 3x_2 + x_3 - x_4 &= -1, \\
5x_1 + x_2 + 2x_3 + 2x_4 &= 7, \\
x_1 - x_2 - x_3 - x_4 &= 0.
\end{aligned}$$

(1.26)

The coefficient matrix for this system of equations is the matrix

$$A = \begin{bmatrix} 1 & 2 & -1 & 1 \\ 2 & -3 & 1 & -1 \\ 5 & 1 & 2 & 2 \\ 1 & -1 & -1 & -1 \end{bmatrix}.$$

Let us also consider two 4×1 (column) matrices X and B, where

$$X = \begin{bmatrix} x_1 \\ x_2 \\ x_3 \\ x_4 \end{bmatrix} \quad \text{and} \quad B = \begin{bmatrix} 4 \\ -1 \\ 7 \\ 0 \end{bmatrix}.$$

It would be most convenient to represent the system of equations (1.26) by the simple matrix equation

$$AX = B.$$

(1.27)

Then, a nice matrix algebra would suggest that knowing the coefficient matrix A and the matrix B, we could solve for the matrix X by multiplying both sides of equation (1.27) by the multiplicative inverse of A as follows:

$$A^{-1}(AX) = A^{-1}B,$$
$$X = A^{-1}B.$$

This happy state of affairs is, unfortunately, not always possible in matrix algebra, as you may have already guessed, since not all systems of equations have unique solutions. Nevertheless, for this to be possible at all, we need a definition of matrix multiplication that allows the product of the matrix A

and the matrix X to be a 4×1 matrix equal to the 4×1 matrix B. The entries in the product AX should be equal to the entries of B. The given equations in the system (1.26) state such equalities. That is,

$$AX = \begin{bmatrix} x_1 + 2x_2 - x_3 + x_4 \\ 2x_1 - 3x_2 + x_3 - x_4 \\ 5x_1 + x_2 + 2x_3 + 2x_4 \\ x_1 - x_2 - x_3 - x_4 \end{bmatrix} = \begin{bmatrix} 4 \\ -1 \\ 7 \\ 0 \end{bmatrix} = B.$$

The first row of AX is obtained from A and X by multiplying the corresponding entries of the first *row* of A by the first *column* of X and adding them together. Thus,

$$AX = \begin{bmatrix} 1 & 2 & -1 & 1 \\ 2 & -3 & 1 & -1 \\ 5 & 1 & 2 & 2 \\ 1 & -1 & -1 & -1 \end{bmatrix} \cdot \begin{bmatrix} x_1 \\ x_2 \\ x_3 \\ x_4 \end{bmatrix} = \begin{bmatrix} x_1 + 2x_2 - x_3 + x_4 \\ \cdots \\ \cdots \\ \cdots \end{bmatrix}.$$

The second entry in AX will be the sum of the products of corresponding entries in the second row of A and those of the first column of X, and so on.

Matrix multiplication is useful in contexts other than this one. We will see several applications in this book. In particular, in Chapter 5 we see how matrix multiplication is related to the composition of certain mappings. A definition of matrix multiplication that covers all such applications is the following.

Let A and B be two matrices as follows:

$$A = \begin{bmatrix} a_{11} & a_{12} & a_{13} & \cdots & a_{1n} \\ a_{21} & a_{22} & a_{23} & \cdots & a_{2n} \\ a_{31} & a_{32} & a_{33} & \cdots & a_{3n} \\ \vdots & \vdots & \vdots & \cdots & \vdots \\ a_{m1} & a_{m2} & a_{m3} & \cdots & a_{mn} \end{bmatrix}, \qquad B = \begin{bmatrix} b_{11} & b_{12} & b_{13} & \cdots & b_{1k} \\ b_{21} & b_{22} & b_{23} & \cdots & b_{2k} \\ b_{31} & b_{32} & b_{33} & \cdots & b_{3k} \\ \vdots & \vdots & \vdots & \cdots & \vdots \\ b_{n1} & b_{n2} & b_{n3} & \cdots & b_{nk} \end{bmatrix};$$

where A is $m \times n$ and B is $n \times k$. Then the **product matrix** AB is the $m \times k$ matrix

$$AB = \begin{bmatrix} c_{11} & c_{12} & c_{13} & \cdots & c_{1k} \\ c_{21} & c_{22} & c_{23} & \cdots & c_{2k} \\ c_{31} & c_{32} & c_{33} & \cdots & c_{3k} \\ \vdots & \vdots & \vdots & \cdots & \vdots \\ c_{m1} & c_{m2} & c_{m3} & \cdots & c_{mk} \end{bmatrix}.$$

The entry in row i and column j of this product is the *sum* of the products of corresponding entries from row i of the matrix A and clumn j of the matrix B. Thus,

$$c_{ij} = a_{i1}b_{1j} + a_{i2}b_{2j} + a_{i3}b_{3j} + \ldots + a_{in}b_{nj}. \tag{1.28}$$

By making use of the summation notation Σ we can write equation (1.28) more succinctly as

$$c_{ij} = \sum_{r=1}^{n} a_{ir} b_{rj}$$

for each $i = 1, 2, \ldots, m$ and $j = 1, 2, \ldots, k$.

If you are not familiar with the *sigma* notation, Σ, you should consult a college algebra or calculus text. Some of the basic properties are reviewed in Exercises 35–39 at the end of this section. Later we shall return to the system of equations given in (1.26) and the question of whether there is a multiplicative inverse for a given matrix. Let us first concentrate on the operation of matrix multiplication whose definition it motivated.

Example 7 Let A and B be the matrices

$$A = \begin{bmatrix} 1 & 0 & 1 & -1 \\ -1 & 2 & 4 & 2 \\ 3 & 2 & -1 & 0 \end{bmatrix}, \qquad B = \begin{bmatrix} 1 & 3 \\ 0 & 1 \\ 2 & -1 \\ -1 & 5 \end{bmatrix}.$$

Then, since A is 3×4 and B is 4×2, the product AB is a 3×2 matrix. To determine its entries we use (1.28). For example, to find the entry c_{21} in the second row and the first column of the product matrix $C = AB$, we focus our attention on *row 2* of A and *column 1* of B. Multiply corresponding entries together and add them up:

$$c_{21} = (-1)(1) + (2)(0) + (4)(2) + (2)(-1) = 5.$$

Thus,

$$AB = \begin{bmatrix} 1 & 0 & 1 & -1 \\ \mathbf{-1} & \mathbf{2} & \mathbf{4} & \mathbf{2} \\ 3 & 2 & -1 & 0 \end{bmatrix} \begin{bmatrix} \mathbf{1} & 3 \\ \mathbf{0} & 1 \\ \mathbf{2} & -1 \\ \mathbf{-1} & 5 \end{bmatrix} = \begin{bmatrix} - & - \\ \mathbf{5} & - \\ - & - \end{bmatrix}.$$

The remaining entries c_{ij} of the product matrix $AB = (c_{ij})$ are computed similarly. For example,

$$c_{11} = (1)(1) + (0)(0) + (1)(2) + (-1)(-1) = 4,$$
$$c_{32} = (3)(3) + (2)(1) + (-1)(-1) + (0)(5) = 12,$$

and so on. In fact

$$AB = \begin{bmatrix} 4 & -3 \\ 5 & 5 \\ 1 & 12 \end{bmatrix}.$$

Note particularly that if we wish to multiply two matrices A and B together, it is necessary that the number of entries in a given *row* of the first matrix-factor A (that is, the number of columns of A) must be the same as the number of entries in a given *column* of the second matrix-factor B (that is, the number of rows of B). Thus,

> If A is $m \times n$ and B is $n \times k$, then AB is $m \times k$.

The product matrix BA for the two matrices above in Example 7 is therefore *not defined.* ∎

The basic properties of matrix multiplication are discussed in the next section. We conclude this section with an example that shows a simple application of matrix multiplication to something other than solving a system of linear equations. In this example, matrix multiplication is used to model the possible spread of an infection. You will meet many other examples where matrix multiplication is an appropriate model, both in this text and in your professional career.

Example 8

Suppose two people have contracted a contagious disease. A second group of five people have been in possible contact with the two infected persons. After careful questioning of the second group of people, the 2×5 matrix A below is constructed. The entry $a_{ij} = 1$ if the jth person in the second group has been in contact with the ith infected person. We set $a_{ij} = 0$ if there has been no contact. Suppose the results are

$$A = \begin{bmatrix} 1 & 1 & 0 & 0 & 1 \\ 0 & 1 & 1 & 1 & 0 \end{bmatrix}.$$

Suppose further that a third group of four people has been in possible contact with the second group of five people. The contact matrix in this case is the matrix B below, with $b_{jk} = 1$ or 0, depending on whether person k of the third group has or has not, respectively, been in contact with person j of the second group. Suppose the results are as follows:

$$B = \begin{bmatrix} 1 & 1 & 0 & 1 \\ 0 & 1 & 0 & 1 \\ 0 & 0 & 1 & 0 \\ 0 & 0 & 1 & 1 \\ 0 & 1 & 1 & 0 \end{bmatrix}.$$

Now suppose that we are interested in the indirect or "second order" contacts of the third group of individuals with the two infected persons of the first group. This can be seen to be the matrix product $C = AB$. For example, person number *four* of the third group has had contact with persons *one, two,* and *four* of the second group, so his total indirect contact with infected persons is given by the expression $(1)(1) + (1)(1) + (0)(0) + (0)(1) + (1)(0) = 2$.

The first factor in each term indicates his contacts with a second group person, the second factor in each term indicates whether that person had contact with an infected person. This corresponds to the definition of matrix multiplication. Therefore, the matrix

$$C = AB = \begin{bmatrix} 1 & 3 & 1 & 2 \\ 0 & 1 & 2 & 2 \end{bmatrix}$$

contains all the results, and is a second-order contact matrix.   ■

Vocabulary

The additional new concepts and terms developed in this section and in the previous sections are used throughout the rest of the text. Thus, you should be sure you understand each of the concepts described by the terms in the following list. Review these ideas until you feel you have a thorough understanding of each one.

matrix (matrices)
elements, entries in a matrix
equal matrices
matrix addition
associative law
negative of a matrix
matrix multiplication

scalar
matrix row or column
row equivalent matrices
commutative law
the zero matrix
multiplying a matrix by a scalar
summation notation Σ (see Exercises 35–39)

Exercises 1.3

1. Solve the matrix equation $\begin{bmatrix} x & y \\ z & w \end{bmatrix} = \begin{bmatrix} 1 & 2x \\ 3y & 1 \end{bmatrix}$.

Each of Exercises 2–24 refers to the following particular matrices:

$$A = \begin{bmatrix} 1 & 1 \\ -1 & 2 \\ 0 & 3 \end{bmatrix}, B = \begin{bmatrix} 2 & -1 \\ 4 & 1 \\ 6 & 0 \end{bmatrix}, C = \begin{bmatrix} 2 & 1 & -1 \\ 3 & 1 & 4 \end{bmatrix}, D = \begin{bmatrix} 0 & 4 \\ -2 & 0 \end{bmatrix}, E = \begin{bmatrix} 1 & 0 & -1 \\ -2 & 1 & 3 \\ 1 & 1 & 0 \end{bmatrix}, F = \begin{bmatrix} 3 & 1 & -2 \\ 4 & -1 & 7 \\ 0 & 1 & 2 \end{bmatrix}.$$

2. Solve the matrix equation $2A - X = B$, X is a matrix.

3. Solve the matrix equation $3B + 2X = A$, X is a matrix.

4. Solve the matrix equation $2E + xF = O$, x is a scalar.

In Exercises 5–20 refer to the matrices given above and compute, if possible, each of the following:

5. $A + 2B$.

6. $E - F$.

7. $\frac{1}{3}E + \frac{2}{5}F$.

8. $A - C$.

9. AB.

10. AC.

11. AF.

12. AD.

13. BC.

14. $2AD$.

15. πAD.

16. $4eCF$.

17. $D^2 = DD$.

18. F^2.

19. E^3.

20. $5D^3$.

21. Show that the matrix D satisfies the equation $D^4 + 8D^2 = 0$.

22. Show that the matrix

$$N = \begin{bmatrix} 0 & 1 & -1 \\ 0 & 0 & 1 \\ 0 & 0 & 0 \end{bmatrix}$$

satisfies $N \neq O$, $N^2 \neq O$, but $N^3 = 0$. N is called *nilpotent (of index 3).*

23. Using the matrices A, C, and E, verify that $A(CE) = (AC)E$.

24. Illustrate that, in general, matrix multiplication is not commutative by showing that EF is not the same matrix as FE.

In the remaining exercises we no longer refer to the six matrices A–F given above. Each matrix referred to is either a general matrix, or is defined in the exercise.

25. Show that if $A + Z = Z + A = A$, for every $m \times n$ matrix A, then $Z = O$ (the $m \times n$ zero matrix). (HINT: Use Theorems 1.1 and 1.2 on pages 32 and 34 and the definition of $-A$.)

26. Show that the matrix $-A$ is indeed unique for a given $m \times n$ matrix A by demonstrating that if $A + B = B + A = O$, then $B = -A$.

27. Show that if r is any real or complex scalar and A and B are any 3×3 matrices, then $A(rB) = r(AB)$. See Exercises 43–45 below.

28. Prove part (1) of Theorem 1.1.   **29.** Prove part (2) of Theorem 1.1.

30. Prove part (2) of Theorem 1.2.   **31.** Prove part (3) of Theorem 1.2.

32. Show that if S is an $n \times n$ matrix in which all the entries in row i are zero, and if T is any $n \times m$ matrix, then ST also has only zeros as entries in row i.

33. Show that for any two matrices S and T whose product ST is defined, the ith row of the matrix product ST is the product of the ith row of S and the matrix T.

34. For the matrices S and T in Exercise 33, show that the jth column of ST is the product of S and the jth column of the matrix T.

Exercises 35–39 review the properties of the summation notation Σ. Remember that $\sum\limits_{i=1}^{n} a_i$ means add up the numbers a_i, where i is successively equal to 1, 2, . . . , n.

Thus, for example, $\sum\limits_{i=1}^{5} a_i = a_1 + a_2 + a_3 + a_4 + a_5$. In particular, $\sum\limits_{i=1}^{7} 2i =$

$2(1) + 2(2) + 2(3) + 2(4) + 2(5) + 2(6) + 2(7) = 2[1 + 2 + 3 + 4 + 5 + 6 + 7]$

$= 2\sum\limits_{i=1}^{7} i = 56.$ *Verify each of the following properties of the sigma notation.*

35. $\sum\limits_{i=1}^{n} r a_i = r \sum\limits_{i=1}^{n} a_i$, for any scalar r. HINT: Write it out as in the example above.

36. $\sum\limits_{i=1}^{n} a_i = \sum\limits_{k=1}^{n} a_k = \sum\limits_{j=1}^{n} a_j = \sum\limits_{\mu=1}^{n} a_\mu$, etc.

37. $\sum\limits_{i=1}^{n} (a_i + b_i) = \sum\limits_{i=1}^{n} a_i + \sum\limits_{i=1}^{n} b_i$.

38. $\sum\limits_{i=1}^{n} \left[\sum\limits_{j=1}^{m} a_{ij} \right] = \sum\limits_{j=1}^{m} \left[\sum\limits_{i=1}^{n} a_{ij} \right]$.

39. $\sum\limits_{k=1}^{n} a_{jk} \left[\sum\limits_{s=1}^{m} b_{ks} c_{sj} \right] = \sum\limits_{s=1}^{m} \left[\sum\limits_{k=1}^{n} a_{jk} b_{ks} \right] c_{sj}$.

In Exercises 40–45 consider the real matrices C, E, and F of Exercises 2–24 above. Write out the (complex) matrices.

40. $E + iF.$ **41.** $E - iF.$ **42.** $3E + 2iF.$ **43.** $iCF.$ **44.** $CiE.$ **45.** $FiE.$

46. Let

$$A = \begin{bmatrix} 0 & 0 & 0 & 1 \\ 0 & 0 & 1 & 0 \\ 0 & 1 & 0 & 0 \\ 1 & 0 & 0 & 0 \end{bmatrix}.$$

Compute the matrices A^2, A^3, and A^4. What can you say about the matrix A^n, for any given positive integer n?

1.4 *MATRIX ALGEBRA*

We saw in Theorems 1.1 and 1.2 (pages 32 and 34) that many of the rules of ordinary real- or complex-number algebra carry over to matrix addition and multiplication by a scalar. This followed largely because these matrix operations involve entry-by-entry operations with the scalar entries. There are, however, some significant differences between matrix algebra and the ordinary real-number algebra with which you are familiar. These differences involve matrix multiplication, one of the most striking of them being the *failure* of the commutative law, **ab = ba**, for multiplication of matrices. You may have noticed already from your work with the exercises of the previous section that the matrix products *AB* and *BA*, even when both are defined, are likely to be different matrices.

Example 1   Let

$$A = \begin{bmatrix} 1 & -1 \\ 2 & 1 \end{bmatrix}, \quad \text{and} \quad B = \begin{bmatrix} 2 & 0 \\ -1 & 1 \end{bmatrix}.$$

Then

$$AB = \begin{bmatrix} 3 & -1 \\ 3 & 1 \end{bmatrix}, \quad \text{while} \quad BA = \begin{bmatrix} 2 & -2 \\ 1 & 2 \end{bmatrix}. \quad \blacksquare$$

Thus, even in so simple a case as this

AB and *BA* are not equal.

Although the commutative law of multiplication *does not hold for matrices,* many of the properties of ordinary real-number multiplication *are* true for multiplication of matrices. The following theorem lists some basic properties for matrix algebra.

Theorem 1.3   Let A, B, and, C be matrices of appropriate sizes so that the indicated operations are defined. Then:
(1) $(AB)C = A(BC)$ (associative law);
(2) $A(B + C) = AB + AC$ (left distributive law);
(3) $(B + C)A = BA + CA$ (right distributive law);
(4) For any scalar r, $r(AB) = (rA)B = A(rB)$.

This theorem follows directly from the definitions; however, the proof of each part is a bit more tedious than were the proofs of the parts of Theorems 1.1 and 1.2. To illustrate, we prove part (1) and leave the rest as exercises for you. We use summation signs to condense the calculations; see Exercises 35–38 of the previous section.

Proof of part (1)   Let $A = (a_{ij})$ be an $m \times n$ matrix, $B = (b_{jk})$ an $n \times t$ matrix, and $C = (c_{kh})$ a $t \times s$ matrix. Then the element in *row i, column h* of the matrix product $(AB)C = (d_{ih})$ is given by definition as

$$d_{ih} = \sum_{\lambda=1}^{t} \left[\sum_{j=1}^{n} (a_{ij}b_{j\lambda})c_{\lambda h} \right] = \sum_{\lambda=1}^{t} \sum_{j=1}^{n} a_{ij}(b_{j\lambda}c_{\lambda h}),$$

by the associative law for real numbers. Now use the rule of Exercise 38 to interchange the summation signs, and that of Exercise 35 to factor a_{ij} out of the inside sum, so that

$$d_{ih} = \sum_{j=1}^{n} a_{ij} \sum_{\lambda=1}^{t} b_{j\lambda}c_{\lambda h}.$$

The interchange of the summation signs indicates merely a rearrangement of the terms in the sum and a different factoring. This last expression for d_{ih}, however, is the element in *row i, column h* of the matrix product $A(BC)$. Therefore, the two matrices $A(BC)$ and $(AB)C$ are element by element equal, as desired.   ◘

Example 2   To illustrate the properties of matrix multiplication, consider the following matrices:

$$A = \begin{bmatrix} 1 & -1 \\ 2 & 1 \\ -3 & 0 \end{bmatrix}, \quad B = \begin{bmatrix} 2 & 0 \\ -1 & 1 \end{bmatrix} \quad \text{and} \quad C = \begin{bmatrix} 3 & -2 \\ 1 & 0 \end{bmatrix}.$$

Then

$$AB = \begin{bmatrix} 3 & -1 \\ 3 & 1 \\ -6 & 0 \end{bmatrix}, \quad AC = \begin{bmatrix} 2 & -2 \\ 7 & -4 \\ -9 & 6 \end{bmatrix}, \quad \text{and} \quad B + C = \begin{bmatrix} 5 & -2 \\ 0 & 1 \end{bmatrix}.$$

Now, as indicated in the theorem,

$$AB + AC = \begin{bmatrix} 5 & -3 \\ 10 & -3 \\ -15 & 6 \end{bmatrix} = A(B + C).$$

Notice here that neither BA nor CA is defined, and that BC and CB are different matrices:

$$BC = \begin{bmatrix} 6 & -4 \\ -2 & 2 \end{bmatrix} \neq \begin{bmatrix} 8 & -2 \\ 2 & 0 \end{bmatrix} = CB. \quad \blacksquare$$

Additional properties of matrix algebra are given in the following theorem.

> **Theorem 1.4** Let A, B, and C be matrices of appropriate sizes to perform the indicated operations, r and s are scalars. Then:
> (1) $(-r)A = -(rA)$;   (5) $(A - B)C = AC - BC$;
> (2) $-(-A) = A$;   (6) $(r - s)A = rA - sA$;
> (3) $A - B = -(B - A)$;   (7) $r(A - B) = rA - rB$;
> (4) $A(B - C) = AB - AC$;   (8) $OA = O$, and $AO = O$.

The proof of Theorem 1.4 is a straightforward application of the previous three theorems and the definitions. Most of it we leave to you as exercises; however, we will prove part (8).

Proof of part (8) Since the zero matrix O has only the scalar 0 for its entries, for any appropriate-sized matrix A we have

$$OA = (c_{ij}), \quad \text{where each} \quad c_{ij} = \sum_{k=1}^{n} 0_{ik} a_{kj} = 0.$$

Similarly, AO has only zero entries; hence, $OA = AO = O$. $\quad \blacksquare$

Example 3 Refer to the matrices A, B, and C of Example 2 above.
The 2×2 zero matrix is the matrix

$$O = \begin{bmatrix} 0 & 0 \\ 0 & 0 \end{bmatrix},$$

and

$$OB = \begin{bmatrix} 0 & 0 \\ 0 & 0 \end{bmatrix} = BO.$$

Using appropriate-sized zero matrices, you can also readily verify that $CO = OC = O$ and $O_{4\times3} A = O_{4\times2}$, while $AO_{2\times2} = O_{3\times2}$. Note also that

$$B - C = \begin{bmatrix} -1 & 2 \\ -2 & 1 \end{bmatrix} \text{ and } A(B - C) = \begin{bmatrix} 1 & 1 \\ -4 & 5 \\ 3 & -6 \end{bmatrix} = AB - AC. \quad \blacksquare$$

Zero Divisors

We turn now to another deviation from the ordinary real-number algebra, where it is true that if $ab = 0$, then $a = 0$ or $b = 0$. This is false in matrix algebra, as you can see from the following example.

> **Definition** If A and B are nonzero matrices such as $AB = O$ we call each of them a (proper) **zero divisor.**

Technically, the matrix $A \neq O$ is a *left zero divisor,* if there is a nonzero matrix B such that $AB = O$. A *right zero divisor* has a similar definition.

Example 4 Let

$$A = \begin{bmatrix} -6 & 3 \\ 4 & -2 \end{bmatrix} \quad \text{and} \quad B = \begin{bmatrix} 1 & -2 \\ 2 & -4 \end{bmatrix}.$$

Then it is easily verified that

$$AB = \begin{bmatrix} 0 & 0 \\ 0 & 0 \end{bmatrix},$$

while neither A nor B is a zero matrix. ∎

Multiplicative Cancellation

Since it is possible for non-zero matrices to be zero divisors, it becomes *impossible* for the multiplicative cancellation law always to hold for matrix algebra. That is, just because $AB = AC$ and A is a nonzero matrix *does not guarantee* that $B = C$. Consider the following example.

Example 5 Given the following 2×2 matrices, A, B, and C, notice that while $AB = AC$, B is not equal to C:

$$A = \begin{bmatrix} 1 & -1 \\ -2 & 2 \end{bmatrix}, \quad B = \begin{bmatrix} 3 & -5 \\ 4 & -5 \end{bmatrix}, \quad \text{and} \quad C = \begin{bmatrix} -1 & 1 \\ 0 & 1 \end{bmatrix}.$$

Thus,

$$AB = \begin{bmatrix} -1 & 0 \\ 2 & 0 \end{bmatrix} = AC.$$

$A \neq O$, and $B \neq C$. Therefore,

> Multiplicative cancellation fails for matrices. ∎

Let $A = (a_{ij})$ be an $n \times n$ matrix. The entries a_{ij}, whose row and column indices are the same, are said to lie on the **main diagonal** of A. Thus, the main diagonal of A is $a_{11}, a_{22}, a_{33}, \ldots, a_{nn}$, and is shaded in the following square matrix.

$$A = \begin{bmatrix} a_{11} & a_{12} & a_{13} & \cdots & a_{1n} \\ a_{21} & a_{22} & a_{23} & \cdots & a_{2n} \\ a_{31} & a_{32} & a_{33} & \cdots & a_{3n} \\ \vdots & \vdots & \vdots & \cdots & \vdots \\ a_{n1} & a_{n2} & a_{n3} & \cdots & a_{nn} \end{bmatrix}. \tag{1.29}$$

The Identity Matrix

For each positive integer n, the $n \times n$ matrix I_n (which we met in Section 1.2), whose main diagonal entries are each *one* and the rest of whose entries are *zero*, is called the $(n \times n)$ **identity matrix**. The reason for this title is because for any $n \times k$ matrix A, we always have $I_n A = A$, and for any $m \times n$ matrix M, $MI_n = M$. The verification of these facts is left to you as exercises.

$$I_n = \begin{bmatrix} 1 & 0 & 0 & \cdots & 0 \\ 0 & 1 & 0 & \cdots & 0 \\ 0 & 0 & 1 & \cdots & 0 \\ \cdots & \cdots & \cdots & \cdots & \cdots \\ 0 & 0 & 0 & \cdots & 1 \end{bmatrix}.$$

When the size of the identity matrix is clear from the context, it is usually written simply as I. The three by three identity matrix is the following:

$$I = \begin{bmatrix} 1 & 0 & 0 \\ 0 & 1 & 0 \\ 0 & 0 & 1 \end{bmatrix}.$$

> **Definition** A square matrix D (such as the identity matrix I), whose only nonzero entries (if any) lie on the main diagonal, is called a **diagonal matrix**.

Diagonal matrices are very frequently written in one-line form as $D = \text{diag}\,[a_{11}, a_{22}, \ldots, a_{nn}]$. Thus, $I_5 = \text{diag}\,[1, 1, 1, 1, 1]$; $\pi I_4 = \text{diag}\,[\pi, \pi, \pi, \pi]$; etc., and

$$M = \text{diag}\,[2, 3, -1, 4, 0] = \begin{bmatrix} 2 & 0 & 0 & 0 & 0 \\ 0 & 3 & 0 & 0 & 0 \\ 0 & 0 & -1 & 0 & 0 \\ 0 & 0 & 0 & 4 & 0 \\ 0 & 0 & 0 & 0 & 0 \end{bmatrix}.$$

Transpose

If A is any $m \times n$ matrix, $A = (a_{ij})$. Interchanging the rows and columns of A is called **transposing** the matrix.

> **Definition**   The $n \times m$ matrix A^T, whose entry in row s and column t is the same as the entry a_{ts}, in row t and column s of the matrix A, is called the **transpose** of A.

For example, the transpose of the matrix

$$M = \begin{bmatrix} 1 & -1 & 2 \\ 0 & 3 & -2 \\ 4 & 1 & -1 \end{bmatrix} \quad \text{is} \quad M^T = \begin{bmatrix} 1 & 0 & 4 \\ -1 & 3 & 1 \\ 2 & -2 & -1 \end{bmatrix}.$$

Observe that the rows of M^T are precisely the columns of M.

We have the following theorem listing the properties of the transpose.

> **Theorem 1.5**   Let A be any $m \times n$ matrix and B be any $n \times k$ matrix. Then
> (1) $(A^T)^T = A$;   (2) $(A + B)^T = A^T + B^T$;
> (3) $(AB)^T = B^T A^T$;   (4) for any scalar r, $(rA)^T = rA^T$;
> (5) If A is a diagonal matrix, then $A = A^T$.

We leave the proofs of most of the parts of this theorem as exercises for you. However we do prove part (3).

Proof of part (3)   Notice first that if $k \neq m$, $A^T B^T$ is not even defined. Thus, the fact that the transpose of AB must reverse the order of multiplication is not surprising. Let $A = (a_{ij})$ and $B = (b_{jt})$. Then the product matrix $AB = (c_{it})$ is $m \times k$, where

$$c_{it} = \sum_{j=1}^{n} a_{ij} b_{jt}.$$

Now the element γ_{ti} in *row* t and *column* i of the transpose $(AB)^T$ matrix is this element c_{it} from row i and column t of AB. At the same time, note that the element α_{ji} in row j and column i of A^T is the same as the element a_{ij} of the matrix A, and similarly, the element β_{tj} of B^T is the same as b_{jt} of B. Therefore,

$$c_{it} = \sum_{j=1}^{n} a_{ij} b_{jt} = \sum_{j=1}^{n} \beta_{tj} \alpha_{ji} = \gamma_{ti}.$$

Thus, the elements of $(AB)^T$ are those of the product matrix $B^T A^T$.   ∎

Part (5) is a simple observation that should not be difficult for you to verify. Its converse is false (see Exercise 19). The proofs of the rest of the theorem are similar to the proof we just gave for part (3), and are left to you.

Matrix Inverses

As we noted above, the identity matrix I acts like the number 1 does for multiplication of real numbers. In real-number algebra, if $a \neq 0$, there is always a real number b, called the **inverse** of a, so that $ab = 1$. This is not always the case for matrix algebra, as we see in the following example.

Example 6   Let A be the matrix A for Example 5. We show that there is *no* matrix K so that $AK = I$.

Let us write a possible such matrix K as

$$\begin{bmatrix} k_{11} & k_{12} \\ k_{21} & k_{22} \end{bmatrix}.$$

Since

$$A = \begin{bmatrix} 1 & -1 \\ -2 & 2 \end{bmatrix},$$

we have

$$AK = \begin{bmatrix} k_{11} - k_{21} & k_{12} - k_{22} \\ -2k_{11} + 2k_{21} & -2k_{12} + 2k_{22} \end{bmatrix} = I = \begin{bmatrix} 1 & 0 \\ 0 & 1 \end{bmatrix}.$$

This means we must find solutions to the system of linear equations

$$
\begin{aligned}
k_{11} && - \; k_{21} && &= 1, \\
&& k_{12} && - \; k_{22} &= 0, \\
-2k_{11} && + \; 2k_{21} && &= 0, \\
&& -2k_{12} && + \; 2k_{22} &= 1.
\end{aligned}
$$

The augmented matrix for this system is the matrix

$$M = \begin{bmatrix} 1 & 0 & -1 & 0 & 1 \\ 0 & 1 & 0 & -1 & 0 \\ -2 & 0 & 2 & 0 & 0 \\ 0 & -2 & 0 & 2 & 1 \end{bmatrix},$$

which row reduces to

$$\begin{bmatrix} 1 & 0 & -1 & 0 & 1 \\ 0 & 1 & 0 & -1 & 0 \\ 0 & 0 & 0 & 0 & 2 \\ 0 & 0 & 0 & 0 & 1 \end{bmatrix}.$$

The last two rows of this reduced matrix show that our system is inconsistent and has no solution. Therefore, *there can be no matrix K so that* $AK = I$. ■

We deal further with the idea of inverses of matrices in the next section.

Row and Column Matrices

Let us consider an arbitrary $m \times n$ matrix $A = (a_{ij})$.

$$A = \begin{bmatrix} a_{11} & a_{12} & \cdot & \cdot & \cdot & a_{1n} \\ a_{21} & a_{22} & \cdot & \cdot & \cdot & a_{2n} \\ \cdot & \cdot & \cdot & \cdot & \cdot & \cdot \\ \cdot & \cdot & \cdot & \cdot & \cdot & \cdot \\ \cdot & \cdot & \cdot & \cdot & \cdot & \cdot \\ a_{m1} & a_{m2} & \cdot & \cdot & \cdot & a_{mn} \end{bmatrix}.$$

Definition Each of the rows of A is a $1 \times n$ matrix. We denote them by a'_i.

Thus, for example, $a'_3 = (a_{31}, a_{32}, \ldots, a_{3n})$. Similarly, we have:

Definition Each of the columns of the matrix A is an $m \times 1$ matrix which we denote by a_j.

Thus, for example,

$$a_4 = \begin{bmatrix} a_{14} \\ a_{24} \\ \cdot \\ \cdot \\ \cdot \\ a_{m4} \end{bmatrix}.$$

Now let $B = (b_{jt})$ be any $n \times k$ matrix. We use similar notation for the columns and rows of B. One can readily calculate that the jth column of the product matrix AB is computed as the matrix product Ab_j. It is also easy to calculate that the entry c_{it}, in row i column t of the matrix $AB = (c_{it})$ is given by the formula $c_{it} = a'_i b_t$.

It is often convenient to write the matrix A as $[a_1 : a_2 : \ldots : a_n]$ in terms of its columns.

Example 7 Let A and B be the following matrices.

$$A = \begin{bmatrix} 1 & -2 & 0 & 2 \\ 1 & -1 & 4 & 0 \\ -3 & 1 & 0 & 1 \end{bmatrix} \qquad B = \begin{bmatrix} 1 & 0 & 1 \\ -1 & 3 & 4 \\ 2 & -1 & 0 \\ 0 & 3 & -2 \end{bmatrix}$$

Then, the entry c_{32} in row 3 and column 2 of the product matrix AB is

$$a'_3 b_2 = (-3, 1, 0, 1) \begin{bmatrix} 0 \\ 3 \\ -1 \\ 3 \end{bmatrix} = 6.$$

And the second column of the matrix product AB is $A\mathbf{b_2}$.

$$\begin{bmatrix} 1 & -2 & 0 & 2 \\ 1 & -1 & 4 & 0 \\ -3 & 1 & 0 & 1 \end{bmatrix} \begin{bmatrix} 0 \\ 3 \\ -1 \\ 3 \end{bmatrix} = \begin{bmatrix} 0 \\ -7 \\ 6 \end{bmatrix}. \quad \blacksquare$$

We conclude this section with a lemma that illustrates the power of matrix algebra. The lemma is about solutions to systems of equations (see Section 1.2); however, you will see no actual linear equations in the proof, only the matrix representation.

> **Lemma 1.6** Every system of linear equations has either no solution, or exactly one solution, or infinitely many solutions.

Proof Denote the given system of linear equations by the matrix equation $AX = B$. If this matrix equation has no solution, the lemma is satisfied, so suppose that $AX = B$ has a solution $X = X'$. If X' is the only solution, the lemma is again satisfied, so suppose that there are two different solutions X_1 and X_2. Then, $X_1 \neq X_2$ but $AX_1 = AX_2 = B$. Thus,

$$O = AX_1 - AX_2 = A(X_1 - X_2) = O.$$

Set $X_0 = X_1 - X_2$, and let r be any scalar. Observe that $AX_0 = O$. We claim that

$$X'' = X_1 + rX_0$$

is a solution to the given system, for any real or complex number r; hence, infinitely many solutions exist to $AX = B$. This claim is easily verified by computing

$$AX'' = A(X_1 + rX_0) = AX_1 + rAX_0 = B + rO = B. \quad \square$$

Vocabulary

Once again, the vocabulary terms introduced in this section are basic to the study of linear algebra and are used throughout the text. Be sure you understand and could apply each of the concepts described by the terms in the following list.

associative laws

commutative laws

zero divisors

main diagonal

column matrix

matrix representation of a system
   of linear equations

distributive laws

noncommutative multiplication

the identity matrix I_n

transpose

row matrix

Exercises 1.4

In Exercises 1–10 calculate the given matrix product.

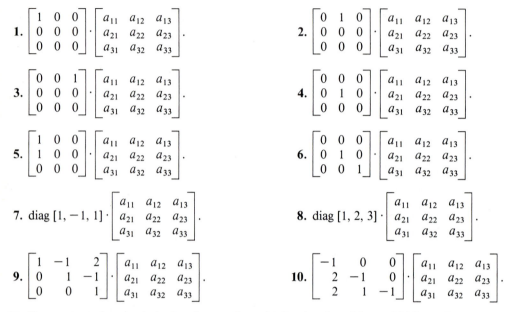

1. $\begin{bmatrix} 1 & 0 & 0 \\ 0 & 0 & 0 \\ 0 & 0 & 0 \end{bmatrix} \cdot \begin{bmatrix} a_{11} & a_{12} & a_{13} \\ a_{21} & a_{22} & a_{23} \\ a_{31} & a_{32} & a_{33} \end{bmatrix}$.

2. $\begin{bmatrix} 0 & 1 & 0 \\ 0 & 0 & 0 \\ 0 & 0 & 0 \end{bmatrix} \cdot \begin{bmatrix} a_{11} & a_{12} & a_{13} \\ a_{21} & a_{22} & a_{23} \\ a_{31} & a_{32} & a_{33} \end{bmatrix}$.

3. $\begin{bmatrix} 0 & 0 & 1 \\ 0 & 0 & 0 \\ 0 & 0 & 0 \end{bmatrix} \cdot \begin{bmatrix} a_{11} & a_{12} & a_{13} \\ a_{21} & a_{22} & a_{23} \\ a_{31} & a_{32} & a_{33} \end{bmatrix}$.

4. $\begin{bmatrix} 0 & 0 & 0 \\ 0 & 1 & 0 \\ 0 & 0 & 0 \end{bmatrix} \cdot \begin{bmatrix} a_{11} & a_{12} & a_{13} \\ a_{21} & a_{22} & a_{23} \\ a_{31} & a_{32} & a_{33} \end{bmatrix}$.

5. $\begin{bmatrix} 1 & 0 & 0 \\ 1 & 0 & 0 \\ 0 & 0 & 0 \end{bmatrix} \cdot \begin{bmatrix} a_{11} & a_{12} & a_{13} \\ a_{21} & a_{22} & a_{23} \\ a_{31} & a_{32} & a_{33} \end{bmatrix}$.

6. $\begin{bmatrix} 0 & 0 & 0 \\ 0 & 1 & 0 \\ 0 & 0 & 1 \end{bmatrix} \cdot \begin{bmatrix} a_{11} & a_{12} & a_{13} \\ a_{21} & a_{22} & a_{23} \\ a_{31} & a_{32} & a_{33} \end{bmatrix}$.

7. $\text{diag}\,[1, -1, 1] \cdot \begin{bmatrix} a_{11} & a_{12} & a_{13} \\ a_{21} & a_{22} & a_{23} \\ a_{31} & a_{32} & a_{33} \end{bmatrix}$.

8. $\text{diag}\,[1, 2, 3] \cdot \begin{bmatrix} a_{11} & a_{12} & a_{13} \\ a_{21} & a_{22} & a_{23} \\ a_{31} & a_{32} & a_{33} \end{bmatrix}$.

9. $\begin{bmatrix} 1 & -1 & 2 \\ 0 & 1 & -1 \\ 0 & 0 & 1 \end{bmatrix} \cdot \begin{bmatrix} a_{11} & a_{12} & a_{13} \\ a_{21} & a_{22} & a_{23} \\ a_{31} & a_{32} & a_{33} \end{bmatrix}$.

10. $\begin{bmatrix} -1 & 0 & 0 \\ 2 & -1 & 0 \\ 2 & 1 & -1 \end{bmatrix} \cdot \begin{bmatrix} a_{11} & a_{12} & a_{13} \\ a_{21} & a_{22} & a_{23} \\ a_{31} & a_{32} & a_{33} \end{bmatrix}$.

11. Demonstrate the associative law for matrix multiplication for arbitrary 2×2 matrices A, B, and C.

12. Show that for an arbitrary 3×3 matrix A, the three by three identity matrix I, and any scalar r, $rIA = IrA$.

13. Show that for the following matrices A and B, $AB = I_2$, but $BA \neq I_3$:

$$A = \begin{bmatrix} 1 & 2 & -1 \\ 2 & 0 & 1 \end{bmatrix}; \qquad B = \begin{bmatrix} 1 & -3 \\ -1 & 5 \\ -2 & 7 \end{bmatrix}.$$

14. For the matrices A and B of Exercise 13 show that the second column of the product matrix BA is equal to Ba_2.

15. For the matrices A and B of Exercise 13, find A^T and B^T.

16. For the matrices A and B of Exercise 13, find $(AB)^T$ and $(BA)^T$.

17. For the matrices A and B of Exercise 13, find $A + B^T$ and $A^T + B$.

18. For the matrices A and B of Exercise 13, find $(A + B^T)^T$.

19. An $n \times n$ matrix A is called symmetric if $A^T = A$. Give an example of a 3×3 symmetric matrix that is not a diagonal matrix.

20. Show that for any $n \times n$ matrix A, the matrix $A + A^T$ is symmetric.

21. Show that for any $n \times n$ matrix A, the matrix, AA^T is symmetric.

22. Show that for any square $(n \times n)$ matrix A, $IA = AI = A$.

23. Let D be a 3×3 diagonal matrix and A an arbitrary 3×3 matrix. Prove or disprove that $AD = DA$.

In Exercises 24–37 complete the proofs of Theorems 1.3, 1.4, and 1.5 (pages 42, 43, and 46) by supplying the proof of the part indicated.

24. Theorem 1.4 (1).   25. Theorem 1.4 (2).   26. Theorem 1.4 (3).

27. Theorem 1.4 (4).   28. Theorem 1.4 (5).   29. Theorem 1.4 (6).

30. Theorem 1.4 (7).   31. Theorem 1.5 (1).   32. Theorem 1.5 (2).

33. Theorem 1.5 (4).   34. Theorem 1.5 (5).   35. Theorem 1.3 (2).

36. Theorem 1.3 (3).   37. Theorem 1.3 (4).

---

38. Suppose that three men, Tom, Dick, and Harry, do piecework; that is, they are paid for each unit turned out. Suppose they manufacture three different products and the amount paid for each product is given by the following table (matrix):

Product	Wage per Unit $
1	5.50
2	7.75
3	6.15

The output of each worker for a given day is contained in the following table (matrix):

Product	1	2	3
Tom	4	5	5
Dick	3	6	4
Harry	5	3	7

Use matrix multiplication to determine how much each worker earned on that day.

A matrix $A = (a_{ij})$ is called **upper triangular** if all of its entries below the main diagonal are zeros; that is, if $a_{ij} = 0$ when $i > j$. It is called **lower triangular** if all the entries above the main diagonal are zero; that is, if $a_{ij} = 0$ when $i < j$.

39. Give an example of a 4×4 upper triangular matrix whose diagonal elements are all 2.

40. Give an example of a 4×4 lower triangular matrix whose diagonal elements are all -1.

41. Give an example of a nonzero matrix that is both *upper* triangular and *lower* triangular.

42. Prove that if A is an $n \times n$ upper triangular matrix, then A^T is lower triangular.

43. Prove that the sum of two upper triangular matrices is an upper triangular matrix.

44. Prove that the sum of two lower triangular matrices is a lower triangular matrix.

45. Prove that the product of two upper triangular matrices is an upper triangular matrix.

46. Prove that the product of two lower triangular matrices is a lower triangular matrix.

In Exercises 47 and 48, let A be the following upper triangular matrix:

$$\begin{bmatrix} 0 & 1 & 2 \\ 0 & 0 & -1 \\ 0 & 0 & 0 \end{bmatrix}.$$

47. Show that $A^2 \neq 0$, but $A^3 = 0$.

48. Show that $(A^T)^3 = 0$.

A square matrix N is called **nilpotent** of index k if, for some positive integer k, $N^{k-1} \neq 0$, but $N^k = 0$. (The zero matrix is nilpotent of index 1.)

49. Prove that any $n \times n$ upper (or lower) triangular matrix, all of whose main diagonal elements are zero, is nilpotent.

50. Prove that if N is any $n \times n$ nonzero nilpotent matrix, then there is no matrix M such that $NM = I_n$. (HINT: look at $N^k M = N^{k-1}(NM)$.)

51. Suppose that A is an $n \times n$ nilpotent matrix. Prove that A^T is also nilpotent. Is the converse of this statement true? Give reasons.

52. Is the product of two nilpotent matrices always nilpotent? If so, prove it. If not, give a counterexample; that is, give two matrices A and B so that $A^k = 0$, $B^m = 0$, but AB is not nilpotent.

53. Repeat Exercise 52 for the sum $A + B$ of two nilpotent matrices A and B.

---

54. This exercise shows how matrix multiplication might be applied to some sociometric experiments to interpret the data. In a given experiment, members of a group therapy session are asked which other members of the group they had positive feelings toward. The responses were collected into a response diagram as given in Figure 1.7. An arrow going from person i to person j means that person i has a positive feeling about person j. Convert the diagram in Figure 1.7 into a matrix $C = (c_{ij})$, where $c_{ij} = 1$ if person i has a positive feeling for person j, and $c_{ij} = 0$, otherwise. (Make the diagonal elements $c_{ii} = 0$, even though a person may have positive feelings for himself/herself). Let the column matrix U be all ones. Note that $U^T C$ gives a *popularity* score for each person in the group. Interpret CU.

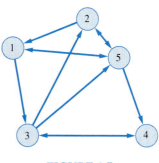

FIGURE 1.7

1.5 *MATRIX INVERSION*

It is a fact of real- and complex-number algebra that whenever a is a *nonzero* real or complex number, then the equation $ax = 1$ has a unique solution $x = a^{-1} = 1/a$. The number a^{-1} is called the **multiplicative inverse** of the nonzero number a. By contrast, it is *not always true* that a given nonzero matrix A has a multiplicative inverse.

> **Definition**   When A is a square ($n \times n$) matrix for which there *does exist* an $n \times n$ matrix B such that $AB = BA = I_n$, we call A an **invertible** or **nonsingular** matrix and say that the matrix B is its **inverse**. In this event we write $B = A^{-1}$.

When such a matrix B *fails to exist* we say that A is **singular** or **not invertible**.

Example 1   Given the two matrices

$$A = \begin{bmatrix} 1 & -3 \\ 2 & -5 \end{bmatrix} \quad \text{and} \quad B = \begin{bmatrix} -5 & 3 \\ -2 & 1 \end{bmatrix},$$

the matrix B is the *inverse* of the matrix A because

$$AB = \begin{bmatrix} 1 & -3 \\ 2 & -5 \end{bmatrix} \cdot \begin{bmatrix} -5 & 3 \\ -2 & 1 \end{bmatrix} = \begin{bmatrix} 1 & 0 \\ 0 & 1 \end{bmatrix} = I_2,$$

and

$$BA = \begin{bmatrix} -5 & 3 \\ -2 & 1 \end{bmatrix} \cdot \begin{bmatrix} 1 & -3 \\ 2 & -5 \end{bmatrix} = \begin{bmatrix} 1 & 0 \\ 0 & 1 \end{bmatrix} = I_2. \quad \blacksquare$$

Remember that for A^{-1} to exist, it must work on *both sides* of A; that is $AA^{-1} = A^{-1}A = I$. That this is not a trivial comment is seen in Exercise 13 of the previous section; there $AB = I$, but $BA \neq I$. Try it. Here is a second example of this situation.

Example 2   Let

$$A = \begin{bmatrix} 1 & -1 & 1 \\ -2 & 0 & 1 \end{bmatrix} \quad \text{and} \quad B = \begin{bmatrix} 1 & -1 \\ 2 & -2 \\ 2 & -1 \end{bmatrix}.$$

Then note that

$$AB = \begin{bmatrix} 1 & -1 & 1 \\ -2 & 0 & 1 \end{bmatrix} \begin{bmatrix} 1 & -1 \\ 2 & -2 \\ 2 & -1 \end{bmatrix} = \begin{bmatrix} 1 & 0 \\ 0 & 1 \end{bmatrix} = I_2.$$

But

$$BA = \begin{bmatrix} 1 & -1 \\ 2 & -2 \\ 2 & -1 \end{bmatrix} \begin{bmatrix} 1 & -1 & 1 \\ -2 & 0 & 1 \end{bmatrix} = \begin{bmatrix} 3 & -1 & 0 \\ 6 & -2 & 0 \\ 4 & -2 & 1 \end{bmatrix} \neq I_3.$$

Therefore, it is *not true* that B is the inverse of A, $B \neq A^{-1}$. In fact, in this case A^{-1} does not exist. A is a singular (not invertible) matrix. $\quad \blacksquare$

You may have noticed that this could perhaps have been expected, since we have defined inverses only for square matrices. However, *not even all square matrices have inverses*. In Example 4 of the preceding section, the matrix

$$B = \begin{bmatrix} 1 & -2 \\ 2 & -4 \end{bmatrix}$$

was shown to satisfy the matrix equation $AB = O$, with

$$A = \begin{bmatrix} -6 & 3 \\ 4 & -2 \end{bmatrix}.$$

Had B^{-1} existed, we would obtain $A = AB(B^{-1}) = OB^{-1} = O$, contrary to the fact that $A \neq O$. So *B has no inverse.*

It is reasonable to ask if every *invertible* (= nonsingular) matrix has just *one* inverse. In other words, is our use of the symbol A^{-1} justified by the fact that the inverse for a matrix, if it has one, is unique? We give the affirmative answer in the following theorem.

Theorem 1.7   If B and C are both $n \times n$ matrix inverses for the $n \times n$ matrix A, then $B = C$.

Proof   If $AB = BA = I_n$ and $CA = AC = I_n$, then $C = CI_n = C(AB) = (CA)B = I_n B = B$. This proof yields a little more information. We have the following corollary.   ❏

Corollary   If $AB = I$ and $CA = I$, for the $n \times n$ matrices A, B, and C, then $B = C = A^{-1}$.

For square matrices even more is true. The situation of Example 2 where the matrix A has a *right* inverse but not a *left* inverse cannot occur. Although the proof is a bit involved, and we omit it (see, however, Supplementary Exercise 18, at the end of the chapter), the following theorem holds. (This fact makes checking considerably simpler.)

Theorem 1.8   If A and B are $n \times n$ matrices and $AB = I$ or $BA = I$, then A and B are both nonsingular and $B = A^{-1}$, and $A = B^{-1}$.

Therefore, to determine whether an $n \times n$ matrix A has an inverse, it suffices to find an $n \times n$ matrix B for which *either* $AB = I$ or $BA = I$.

The fact that a nonsingular (invertible) matrix has an inverse can now be used to solve a system of linear equations.

Example 3   Use matrix algebra and Example 1 to solve the system of equations

$$x - 3y = 4,$$
$$2x - 5y = 9.$$

We write this system in matrix form as the matrix equation $AX = C$, where

$$A = \begin{bmatrix} 1 & -3 \\ 2 & -5 \end{bmatrix}, \qquad X = \begin{bmatrix} x \\ y \end{bmatrix}, \qquad \text{and} \qquad C = \begin{bmatrix} 4 \\ 9 \end{bmatrix}.$$

Since A is an invertible matrix, we may solve the matrix equation by multiplying both sides by A^{-1} as follows: $A^{-1}(AX) = A^{-1}C$, so $(A^{-1}A)X = IX = X = A^{-1}C$. We found A^{-1} in Example 1, so

$$X = \begin{bmatrix} x \\ y \end{bmatrix} = A^{-1}C = \begin{bmatrix} -5 & 3 \\ -2 & 1 \end{bmatrix} \cdot \begin{bmatrix} 4 \\ 9 \end{bmatrix} = \begin{bmatrix} 7 \\ 1 \end{bmatrix},$$

or $x = 7$, $y = 1$. This you can easily see is the solution. Examples involving larger systems are solved in the same way. So long as A is invertible (nonsingular) $X = A^{-1}C$ is the solution. ∎

We collect many of the algebraic facts about matrix inverses in the following theorem.

Theorem 1.9   If A and B are $n \times n$ nonsingular matrices, then
   (i) $(A^{-1})^{-1} = A$;
   (ii) the matrix AB is also nonsingular and $(AB)^{-1} = B^{-1}A^{-1}$;
   (iii) for any nonzero scalar r, $(rA)^{-1} = \dfrac{1}{r} A^{-1}$.

Proof   Since $AA^{-1} = A^{-1}A = I$, part (i) follows directly from the definition and Theorem 1.7. For part (ii) note that

$$(AB)(B^{-1}A^{-1}) = A(BB^{-1})A^{-1} = AIA^{-1} = AA^{-1} = I.$$

So part (ii) follows from Theorems 1.7 and 1.8: $B^{-1}A^{-1}$ is the unique inverse of AB. We leave part (iii) as an exercise. ∎

The results in part (ii) can be extended to the product of any finite number of invertible matrices; for example, $(ABC)^{-1} = C^{-1}B^{-1}A^{-1}$, provided all three inverses exist.

Let us now combine the properties of matrix multiplication (as given in Theorems 1.3 and 1.9) with the Gauss-Jordan reduction techniques for solving systems of linear equations to obtain an algorithm for finding the inverse A^{-1} of a given invertible (nonsingular) matrix A. You recall that in Section 1.1 we found that a system of n linear equations in n unknowns has a unique solution if the augmented matrix of the system has a reduced row echelon form

$$\begin{bmatrix} 1 & 0 & 0 & \cdots & 0 & c_1 \\ 0 & 1 & 0 & \cdots & 0 & c_2 \\ 0 & 0 & 1 & \cdots & 0 & c_3 \\ \cdot & \cdot & \cdot & \cdots & \cdot & \cdot \\ 0 & 0 & 0 & \cdots & 1 & c_n \end{bmatrix}.$$

That is, the coefficient matrix A of the linear system $AX = C$ will be reducible to the $n \times n$ identity matrix I_n. We now note that each of the elementary row operations used to accomplish this reduction can be performed by multiplying A on the left by a certain nonsingular matrix called an *elementary matrix*. If $E_1, E_2, \ldots, E_k$ is a sequence of these elementary matrices such that

$$E_k E_{k-1} \ldots E_2 E_1 A = I,$$

then the product $(E_k E_{k-1} \ldots E_2 E_1)$ must equal A^{-1}. This leads us to the algorithm for computing the inverse of the matrix; namely, **compute the product of these elementary matrices.** Let's begin with a definition.

> **Definition** An $n \times n$ matrix E is called an **elementary matrix** if it is obtained by a single elementary row operation on the identity matrix I_n.

Since there are three basic elementary row operations, there are three basic types of elementary matrices. Each of these is nonsingular.

Type I Interchanging Rows. The elementary matrix $E_I(i, j)$ is obtained by interchanging rows i and j of the identity matrix I_n. For example if $n = 4$, $i = 1$, and $j = 3$, we have

$$E_I(1, 3) = \begin{bmatrix} 0 & 0 & 1 & 0 \\ 0 & 1 & 0 & 0 \\ 1 & 0 & 0 & 0 \\ 0 & 0 & 0 & 1 \end{bmatrix}.$$

Note that to again obtain I_n, we need only interchange rows i and j back. You can verify that

$$(E_I(i, j))^{-1} = E_I(i, j).$$

> Each Type I elementary matrix is its own inverse.

For the case $n = 4$ of our example, you can easily compute that $E_I(1, 3) E_I(1, 3) = I_4$. As an exercise (see Supplementary Exercises) we ask you to show that performing a Type I elementary row operation on a matrix A is the same as *premultiplying* (multiplying on the left) the matrix A by the elementary matrix $E_I(i, j)$. Thus, $E_I(i, j)A$ interchanges rows i and j of A.

Example 4 Let

$$A = \begin{bmatrix} 2 & -1 & 4 & 6 \\ 0 & 1 & 3 & 5 \\ 1 & 0 & -1 & 2 \\ 0 & 1 & 0 & 5 \end{bmatrix}.$$

Then

$$E_I(1, 3)A = \begin{bmatrix} 0 & 0 & 1 & 0 \\ 0 & 1 & 0 & 0 \\ 1 & 0 & 0 & 0 \\ 0 & 0 & 0 & 1 \end{bmatrix} \cdot \begin{bmatrix} 2 & -1 & 4 & 6 \\ 0 & 1 & 3 & 5 \\ 1 & 0 & -1 & 2 \\ 0 & 1 & 0 & 5 \end{bmatrix} = \begin{bmatrix} 1 & 0 & -1 & 2 \\ 0 & 1 & 3 & 5 \\ 2 & -1 & 4 & 6 \\ 0 & 1 & 0 & 5 \end{bmatrix} = B.$$

Notice that the row equivalent matrix B is obtained from the matrix A by interchanging rows *one* and *three* of A. This Type I operation was done by multiplying A *on the left* by the appropriate elementary matrix. ∎

Type II Multiplying a Row by a Nonzero Constant c.   The elementary matrix $E_{II}(c, i)$ is obtained by multiplying row i of the $n \times n$ identity matrix I_n by the nonzero constant c. Thus, if $n = 4$, $c = -3$, and $i = 2$, we have

$$E_{II}(-3, 2) = \begin{bmatrix} 1 & 0 & 0 & 0 \\ 0 & -3 & 0 & 0 \\ 0 & 0 & 1 & 0 \\ 0 & 0 & 0 & 1 \end{bmatrix}.$$

Again, as an exercise you can verify that to perform this operation on a row of an arbitrary matrix A, one multiplies A *on the left* (premultiply) by the elementary matrix $E_{II}(c, i)$.

Example 5   Suppose that

$$K = \begin{bmatrix} 2 & 1 & 0 \\ 4 & 2 & 6 \\ 1 & 3 & -1 \end{bmatrix}.$$

Then to multiply the second row of K by $1/2$, we multiply K on the left by the Type II elementary matrix $E_{II}(1/2, 2)$. Thus

$$E_{II}\left(\frac{1}{2}, 2\right) K = \begin{bmatrix} 1 & 0 & 0 \\ 0 & \frac{1}{2} & 0 \\ 0 & 0 & 1 \end{bmatrix} \cdot \begin{bmatrix} 2 & 1 & 0 \\ 4 & 2 & 6 \\ 1 & 3 & -1 \end{bmatrix} = \begin{bmatrix} 2 & 1 & 0 \\ 2 & 1 & 3 \\ 1 & 3 & -1 \end{bmatrix}. \quad ∎$$

To find the *inverse* for a Type II elementary matrix $E_{II}(c, i)$, one notes that it is merely necessary to multiply row i by $1/c$ in order to obtain the identity matrix I_n again, so

$$(E_{II}(c, i))^{-1} = E_{II}\left(\frac{1}{c}, i\right).$$

You should also verify this.

Type III Substituting for a Given Row Its Sum with a Nonzero Constant Multiple of Another Row.   The elementary matrix $E_{III}(i, k, j)$ is obtained from the identity matrix I_n by adding k times row j to row i and substituting this sum in place of the original row i. Consider the 4×4 example, with $i = 2$, $k = -3$, and $j = 1$ (add -3 times row *one* to row *two*).

$$E_{III}(2, -3, 1) = \begin{bmatrix} 1 & 0 & 0 & 0 \\ -3 & 1 & 0 & 0 \\ 0 & 0 & 1 & 0 \\ 0 & 0 & 0 & 1 \end{bmatrix}.$$

Again one can demonstrate that premultiplying a given matrix A by the elementary matrix $E_{III}(i, k, j)$ has the effect of performing this row operation on A. The inverse of $E_{III}(i, k, j)$ is seen to be $E_{III}(i, -k, j)$ as you can demonstrate. That is,

$$(E_{III}(i, k, j))^{-1} = E_{III}(i, -k, j).$$

Example 6   If A is the matrix A in Example 4, then to subtract twice row *three* from row *one* we multiply A on the left by the Type III elementary matrix $E_{III}(1, -2, 3)$. Therefore,

$$E_{III}(1, -2, 3) \cdot A = \begin{bmatrix} 1 & 0 & -2 & 0 \\ 0 & 1 & 0 & 0 \\ 0 & 0 & 1 & 0 \\ 0 & 0 & 0 & 1 \end{bmatrix} \cdot \begin{bmatrix} 2 & -1 & 4 & 6 \\ 0 & 1 & 3 & 5 \\ 1 & 0 & -1 & 2 \\ 0 & 1 & 0 & 5 \end{bmatrix}$$

$$= \begin{bmatrix} 0 & -1 & 6 & 2 \\ 0 & 1 & 3 & 5 \\ 1 & 0 & -1 & 2 \\ 0 & 1 & 0 & 5 \end{bmatrix}. \quad \blacksquare$$

The entire process of reducing a matrix A to its row echelon form, say N (or to any equivalent form, for that matter), can be accomplished by multiplying A *on the left* by finitely many elementary matrices.

At this point we abandon our intuitive definition, given in Section 1.1 for *row equivalent matrices:* that is, "A is row equivalent to B if B obtained from A by a finite number of elementary row operations." We replace it with a new matrix algebra definition.

Definition   The $m \times n$ matrix B is **row equivalent** to the $m \times n$ matrix A if there exist finitely many elementary matrices $E_1, E_2, \ldots, E_k$ (of any of the three types) so that $B = (E_1 \cdot E_2 \cdots E_k)A$.

We ask you, as an exercise, to write out a proof that this definition is logically equivalent to the former one.

Example 7   Let us show that the matrix K of Example 5 has reduced row echelon form I_3, and is therefore nonsingular. In the process we'll find K^{-1} as the product of the elementary matrices involved.

1. From Example 5 we have

$$E_1 K = E_{\mathrm{II}}\left(\frac{1}{2}, 2\right) K = \begin{bmatrix} 1 & 0 & 0 \\ 0 & \frac{1}{2} & 0 \\ 0 & 0 & 1 \end{bmatrix} \cdot \begin{bmatrix} 2 & 1 & 0 \\ 4 & 2 & 6 \\ 1 & 3 & -1 \end{bmatrix} = \begin{bmatrix} 2 & 1 & 0 \\ 2 & 1 & 3 \\ 1 & 3 & -1 \end{bmatrix}.$$

2. Then, as one tactic, let's make the *third* row of $E_1 K$ the *first* row of our next equivalent matrix. Thus,

$$E_2(E_1 K) = E_{\mathrm{I}}(1, 3)(E_1 K)$$

$$= \begin{bmatrix} 0 & 0 & 1 \\ 0 & 1 & 0 \\ 1 & 0 & 0 \end{bmatrix} \cdot \begin{bmatrix} 2 & 1 & 0 \\ 2 & 1 & 3 \\ 1 & 3 & -1 \end{bmatrix} = \begin{bmatrix} 1 & 3 & -1 \\ 2 & 1 & 3 \\ 2 & 1 & 0 \end{bmatrix}.$$

3. Next, let us subtract row *three* from row *two*. This means that we multiply the matrix $E_2(E_1 K)$ of Step 2 by the elementary matrix $E_3 = E_{\mathrm{III}}(2, -1, 3)$

$$E_3(E_2 E_1 K) = \begin{bmatrix} 1 & 0 & 0 \\ 0 & 1 & -1 \\ 0 & 0 & 1 \end{bmatrix} \cdot \begin{bmatrix} 1 & 3 & -1 \\ 2 & 1 & 3 \\ 2 & 1 & 0 \end{bmatrix} = \begin{bmatrix} 1 & 3 & -1 \\ 0 & 0 & 3 \\ 2 & 1 & 0 \end{bmatrix}.$$

4. Now let's interchange rows *two* and *three*. To do this we multiply the matrix $E_3(E_2 E_1 K)$ of Step 3 by the elementary matrix $E_4 = E_{\mathrm{I}}(2, 3)$ as follows:

$$E_4(E_3 E_2 E_1 K) = \begin{bmatrix} 1 & 0 & 0 \\ 0 & 0 & 1 \\ 0 & 1 & 0 \end{bmatrix} \cdot \begin{bmatrix} 1 & 3 & -1 \\ 0 & 0 & 3 \\ 2 & 1 & 0 \end{bmatrix} = \begin{bmatrix} 1 & 3 & -1 \\ 2 & 1 & 0 \\ 0 & 0 & 3 \end{bmatrix}.$$

5. Now multiply row *three* of this new matrix by $1/3$. Then follow that by subtracting 2 times row *one* from row *two*. This will require two elementary matrices, one of Type II and the other Type III. Thus,

$$E_5(E_4 E_3 E_2 E_1 K) = \begin{bmatrix} 1 & 0 & 0 \\ 0 & 1 & 0 \\ 0 & 0 & \frac{1}{3} \end{bmatrix} \cdot \begin{bmatrix} 1 & 3 & -1 \\ 2 & 1 & 0 \\ 0 & 0 & 3 \end{bmatrix} = \begin{bmatrix} 1 & 3 & -1 \\ 2 & 1 & 0 \\ 0 & 0 & 1 \end{bmatrix}; \text{ and}$$

$$E_6(E_5 E_4 E_3 E_2 E_1 K) = \begin{bmatrix} 1 & 0 & 0 \\ -2 & 1 & 0 \\ 0 & 0 & 1 \end{bmatrix} \cdot \begin{bmatrix} 1 & 3 & -1 \\ 2 & 1 & 0 \\ 0 & 0 & 0 \end{bmatrix} = \begin{bmatrix} 1 & 3 & -1 \\ 0 & -5 & 2 \\ 0 & 0 & 1 \end{bmatrix}.$$

This is a matrix in triangular form equivalent to K. To finish reducing K to reduced row echelon form we could adopt the following tactics.

6. Add the *third* row to the *first* row, and then follow up by adding -2 times the *third* row to the *second* row. Two more elementary matrices, both of Type III, accomplish this.

$$E_8(E_7(E_6E_5E_4E_3E_2E_1K)) = \begin{bmatrix} 1 & 0 & 0 \\ 0 & 1 & -2 \\ 0 & 0 & 1 \end{bmatrix} \cdot \begin{bmatrix} 1 & 0 & 1 \\ 0 & 1 & 0 \\ 0 & 0 & 1 \end{bmatrix}$$

$$\cdot \begin{bmatrix} 1 & 3 & -1 \\ 0 & -5 & 2 \\ 0 & 0 & 1 \end{bmatrix} = \begin{bmatrix} 1 & 3 & 0 \\ 0 & -5 & 0 \\ 0 & 0 & 1 \end{bmatrix}.$$

7. The final two steps needed to obtain I_3 involve multiplying the *second* row of the matrix above by $-1/5$, and then following that by adding -3 times this new *second* row to the *first* row. These are done with first a Type II elementary matrix E_9, followed by a Type III elementary matrix E_{10}.

$$E_9(E_8E_7E_6E_5E_4E_3E_2E_1K) = \begin{bmatrix} 1 & 0 & 0 \\ 0 & -\dfrac{1}{5} & 0 \\ 0 & 0 & 1 \end{bmatrix} \cdot \begin{bmatrix} 1 & 3 & 0 \\ 0 & -5 & 0 \\ 0 & 0 & 1 \end{bmatrix}$$

$$= \begin{bmatrix} 1 & 3 & 0 \\ 0 & 1 & 0 \\ 0 & 0 & 1 \end{bmatrix}.$$

And

$$E_{10}(E_9E_8E_7E_6E_5E_4E_3E_2E_1K) = \begin{bmatrix} 1 & -3 & 0 \\ 0 & 1 & 0 \\ 0 & 0 & 1 \end{bmatrix} \cdot \begin{bmatrix} 1 & 3 & 0 \\ 0 & 1 & 0 \\ 0 & 0 & 1 \end{bmatrix}$$

$$= \begin{bmatrix} 1 & 0 & 0 \\ 0 & 1 & 0 \\ 0 & 0 & 1 \end{bmatrix} = I_3.$$

We make no claim that this is the only sequence of elementary matrices that will accomplish this reduction. We have employed one set of tactics, perhaps not even the best, to do the job. Nevertheless, we have shown that

$$E_{10}E_9E_8E_7E_6E_5E_4E_3E_2E_1(K) = I_3.$$

Therefore, (by Theorem 1.8) K is indeed a nonsingular matrix. Its inverse must be the product of the ten elementary matrices used in this process. Therefore,

$$K^{-1} = E_{10}E_9E_8E_7E_6E_5E_4E_3E_2E_1.$$

You can multiply these ten matrices together and see that

$$K^{-1} = \begin{bmatrix} \dfrac{2}{3} & -\dfrac{1}{30} & -\dfrac{1}{5} \\[2ex] -\dfrac{1}{3} & \dfrac{1}{15} & \dfrac{2}{5} \\[2ex] -\dfrac{1}{3} & \dfrac{1}{6} & 0 \end{bmatrix}.$$

This rather long, complicated procedure gives an indication of the proof of the following theorem. ∎

Theorem 1.10 An $n \times n$ matrix A is nonsingular (invertible) if, and only if, A is row equivalent to the identity matrix I_n.

Proof A is row equivalent to the identity matrix I_n if, and only if, there exist finitely many elementary matrices $E_1, E_2, \ldots, E_k$ so that $I_n = E_1 \cdot E_2 \cdots E_k(A)$. Therefore, A is invertible and $A^{-1} = E_1 E_2 \cdots E_k$. Conversely, if A is invertible, then any system of equations $AX = B$ has the unique solution $X = A^{-1}B$, so the coefficient matrix A has I_n as its reduced row echelon form. Thus A is row equivalent to I_n. ◻

Rather than actually computing each elementary matrix in turn, collecting them, and multiplying them together at the end to find A^{-1}, it is easier to perform the row operations directly both on A and on I_n at the same time. This gives us the following algorithm for computing the inverse of a square matrix.

The Matrix Inversion Algorithm

Step 1 Given the $n \times n$ matrix A, adjoin the identity matrix I_n at the right of A. That is, augment A with these n additional columns, forming the $n \times 2n$ "block" matrix

$$[A : I_n]. \tag{1.30}$$

Step 2 Apply row operations to this two-part matrix $[A : I_n]$ so as to change the matrix A in the first block into its reduced row echelon form U. This results in the block matrix

$$[U : B].$$

If A is nonsingular, then U will be I_n, and B will be A^{-1}. The final result, in this case, is

$$[I_n : A^{-1}]. \tag{1.31}$$

This algorithm works because, at each stage of the process of row reducing the "A block" of the matrix in (1.30), the product of the elementary matrices involved in the process is being accumulated in the "I_n block" (the right half of the big augmented matrix). That is, after s row operations we have transformed the matrix of (1.30) into the matrix

$$[E_s E_{s-1} \cdots E_2 E_1 A : E_s E_{s-1} \cdots E_2 E_1].$$

At the final stage of this process the reduced row echelon form of A is in the left block and the accumulated product of elementary matrices is in the right block. When A is invertible (nonsingular) the left block is I_n and the right block is A^{-1}, as desired. A simplified flowchart of this algorithm is depicted in Chart 2 of the Appendix. An alternative justification for this algorithm is suggested by Exercise 29. (Also see Exercise 30.)

Example 8   Find A^{-1}, if it exists, for the matrix

$$A = \begin{bmatrix} 1 & -1 & 2 \\ 0 & 2 & -1 \\ 3 & -1 & 4 \end{bmatrix}.$$

Step 1   Form the block matrix $[A : I_n]$.

$$[A : I_n] = \begin{bmatrix} 1 & -1 & 2 & 1 & 0 & 0 \\ 0 & 2 & -1 & 0 & 1 & 0 \\ 3 & -1 & 4 & 0 & 0 & 1 \end{bmatrix}.$$

Step 2   Row reduce $[A : I_n]$ until it has the form $[U : B]$, where U is the reduced row echelon form for A. If A is nonsingular U will equal I_n and B will be the desired inverse A^{-1} for A. We'll list a possible sequence of row reductions here.

(a) Add (-3) times row *one* to row *three:*

$$\begin{bmatrix} 1 & -1 & 2 & 1 & 0 & 0 \\ 0 & 2 & -1 & 0 & 1 & 0 \\ 0 & 2 & -2 & -3 & 0 & 1 \end{bmatrix}.$$

(b) Add (-1) times row *two* to row *three:*

$$\begin{bmatrix} 1 & -1 & 2 & 1 & 0 & 0 \\ 0 & 2 & -1 & 0 & 1 & 0 \\ 0 & 0 & -1 & -3 & -1 & 1 \end{bmatrix}.$$

(c) Multiply row *three* by (-1):

$$\begin{bmatrix} 1 & -1 & 2 & 1 & 0 & 0 \\ 0 & 2 & -1 & 0 & 1 & 0 \\ 0 & 0 & 1 & 3 & 1 & -1 \end{bmatrix}.$$

The left block is a triangular form for A. Continue to get *reduced row echelon form* for A.

(d) Add row *three* to row *two,* and follow that by adding (-2) times row *three* to row *one:*

$$\begin{bmatrix} 1 & -1 & 0 & -5 & -2 & 2 \\ 0 & 2 & 0 & 3 & 2 & -1 \\ 0 & 0 & 1 & 3 & 1 & -1 \end{bmatrix}.$$

(e) Multiply row *two* by $\frac{1}{2}$:

$$\begin{bmatrix} 1 & -1 & 0 & -5 & -2 & 2 \\ 0 & 1 & 0 & \frac{3}{2} & 1 & -\frac{1}{2} \\ 0 & 0 & 1 & 3 & 1 & -1 \end{bmatrix}.$$

(f) Add row *two* to row *one:*

$$\begin{bmatrix} 1 & 0 & 0 & -\frac{7}{2} & -1 & \frac{3}{2} \\ 0 & 1 & 0 & \frac{3}{2} & 1 & -\frac{1}{2} \\ 0 & 0 & 1 & 3 & 1 & -1 \end{bmatrix}.$$

We now have the identity matrix in the first block as the reduced row echelon form for $A,$ so the right three columns contain $A^{-1}.$ Thus,

$$A^{-1} = \begin{bmatrix} -\frac{7}{2} & -1 & \frac{3}{2} \\ \frac{3}{2} & 1 & -\frac{1}{2} \\ 33 & 1 & -1 \end{bmatrix}.$$

Now you can check that $AA^{-1} = I_3.$ ■

Even if we do not know in advance whether a given $n \times n$ matrix A is invertible (nonsingular) or not, we can use the inversion algorithm anyway. If A is singular (not invertible) we will be *unable* to reduce it to the identity. Thus, at some point in the reduction process, we'll obtain at least one row of zeros in the left-hand block of $[U' : B'],$ our large working matrix obtained from $[A : I_n].$ There is no point in proceeding all the way to the reduced row echelon form for A then, since we know that A will not be invertible. U' will always have at least one row of zeros when A is singular. Can you prove that?

Example 9   Is the matrix K below singular or nonsingular? If it is nonsingular find $K^{-1}.$

$$K = \begin{bmatrix} 2 & -1 & 1 \\ 1 & 2 & -1 \\ 4 & 3 & -1 \end{bmatrix}.$$

We apply the matrix inversion algorithm and form the matrix

$$[K : I_3] = \begin{bmatrix} 2 & -1 & 1 & 1 & 0 & 0 \\ 1 & 2 & -1 & 0 & 1 & 0 \\ 4 & 3 & -1 & 0 & 0 & 1 \end{bmatrix}.$$

Interchange rows *one* and *two* to obtain

$$[E_1 K : E_1] = \begin{bmatrix} 1 & 2 & -1 & 0 & 1 & 0 \\ 2 & -1 & 1 & 1 & 0 & 0 \\ 4 & 3 & -1 & 0 & 0 & 1 \end{bmatrix}.$$

Now add (-2) times row *one* to row *two* and (-4) times row *one* to row *three* to obtain

$$[U'' : B''] = \begin{bmatrix} 1 & 2 & -1 & 0 & 1 & 0 \\ 0 & -5 & 3 & 1 & -2 & 0 \\ 0 & -5 & 3 & 0 & -4 & 1 \end{bmatrix}.$$

Since the second and third rows in the U'' block are both the same, it's clear that by subtracting row *two* from row *three* we shall have a block matrix $[U' : B']$ in which the last row of the U' block is all *zeros*. Therefore, the reduced row echelon form for K *cannot* be I_3. Thus, K is a singular matrix. The block matrix in question is

$$[U' : B'] = \begin{bmatrix} 1 & 2 & -1 & 0 & 1 & 0 \\ 0 & -5 & 3 & 1 & -2 & 0 \\ 0 & 0 & 0 & -1 & -2 & 1 \end{bmatrix}.$$

The matrix B' in the right-hand block remains the product of all the elementary matrices used in this reduction. But, since K is singular, it doesn't tell us anything else (see, however, Section 1.6). The matrix K^{-1} *does not exist.* ■

Example 10 Suppose the United States government decides to levy a corporate tax based on a fixed percentage of a corporation's profits. The tax rate varies, depending upon the country in which the profit was earned. The Giget Corporation does business in both the United States and Japan. The tax rate for the United States is r_{ss} for profits earned in the United States and r_{sj} for profits earned in Japan. Suppose also that the Japanese government has a similar sort of tax. Its rates for profits earned in Japan are r_{jj} and for profits earned in the United States are r_{js}. We can arrange these tax rates in the following tax matrix:

$$R = \begin{bmatrix} r_{ss} & r_{sj} \\ r_{js} & r_{jj} \end{bmatrix}.$$

The amounts of taxes paid and profits earned by the Giget Corporation can also be shown, respectively, by the two column matrices:

$$T = \begin{bmatrix} t_s \\ t_j \end{bmatrix} \quad \text{and} \quad P = \begin{bmatrix} p_s \\ p_j \end{bmatrix}.$$

The nature of the notation is fairly obvious. This situation can be modeled by the matrix equation

$$T = RP,$$

since $\qquad t_s = \text{U.S. taxes} = r_{ss}p_s + r_{sj}p_j,$

and $\qquad t_j = \text{Japanese taxes} = r_{js}p_s + r_{jj}p_j.$

Suppose that for 1990 Giget paid 50 million yen in U.S. taxes and 35 million yen in Japanese taxes. Find the 1990 profits of the Giget Corporation if the tax rate matrix R is as follows:

$$R = \begin{bmatrix} 0.03 & 0.05 \\ 0.07 & 0.02 \end{bmatrix}.$$

If the matrix R is nonsingular, then the answer to our question is given by using simple matrix algebra on the equation $T = RP$. Thus, the following matrix equation gives the solution:

$$P = R^{-1}T.$$

We use the matrix reduction algorithm to find R^{-1}. Set up the block matrix

$$[R : I_2] = \begin{bmatrix} 0.03 & 0.05 & 1 & 0 \\ 0.07 & 0.02 & 0 & 1 \end{bmatrix}.$$

This matrix is row equivalent to each of the following. This is one possible row reduction process motivated specifically for hand calculation. You might do it differently, particularly if you use a calculator or a computer. Nevertheless, try to supply the reasons for each stage.

$$\begin{bmatrix} 3 & 5 & 100 & 0 \\ 7 & 2 & 0 & 100 \end{bmatrix} \rightarrow \begin{bmatrix} 3 & 5 & 100 & 0 \\ 1 & -8 & -200 & 100 \end{bmatrix}$$

$$\rightarrow \begin{bmatrix} 1 & -8 & -200 & 100 \\ 3 & 5 & 100 & 0 \end{bmatrix} \rightarrow \begin{bmatrix} 1 & -8 & -200 & 100 \\ 0 & 29 & 700 & -300 \end{bmatrix}$$

$$\rightarrow \begin{bmatrix} 1 & -8 & -200 & 100 \\ 0 & 1 & \dfrac{700}{29} & \dfrac{-300}{29} \end{bmatrix} \rightarrow \begin{bmatrix} 1 & 0 & \dfrac{-200}{29} & \dfrac{500}{29} \\ 0 & 1 & \dfrac{700}{29} & \dfrac{-300}{29} \end{bmatrix}.$$

Since R^{-1} exists, we can use it now, as indicated, to compute the profits.

$$P = R^{-1}T = \begin{bmatrix} \dfrac{-200}{29} & \dfrac{500}{29} \\ \dfrac{700}{29} & \dfrac{-300}{29} \end{bmatrix} \begin{bmatrix} 50 \\ 35 \end{bmatrix} = \frac{1}{29} \begin{bmatrix} 7500 \\ 24500 \end{bmatrix}.$$

So the profits in the United States were valued at about 258.6 million yen, and the profits in Japan were valued at 844.8 million yen. ∎

Vocabulary

Review these ideas until you feel secure in your ability to apply them throughout this course.

multiplicative inverse	singular matrix
nonsingular matrix	invertible matrix
matrix inverse	Type I elementary matrix
Type II elementary matrix	Type III elementary matrix
row equivalent matrices	matrix inversion algorithm

Exercises 1.5

In Exercises 1–8 find the inverse of each of the given elementary matrices.

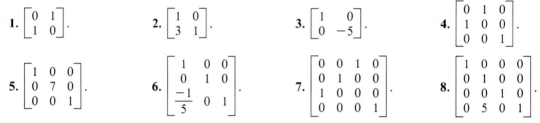

1. $\begin{bmatrix} 0 & 1 \\ 1 & 0 \end{bmatrix}$.
2. $\begin{bmatrix} 1 & 0 \\ 3 & 1 \end{bmatrix}$.
3. $\begin{bmatrix} 1 & 0 \\ 0 & -5 \end{bmatrix}$.
4. $\begin{bmatrix} 0 & 1 & 0 \\ 1 & 0 & 0 \\ 0 & 0 & 1 \end{bmatrix}$.

5. $\begin{bmatrix} 1 & 0 & 0 \\ 0 & 7 & 0 \\ 0 & 0 & 1 \end{bmatrix}$.
6. $\begin{bmatrix} 1 & 0 & 0 \\ 0 & 1 & 0 \\ -\frac{1}{5} & 0 & 1 \end{bmatrix}$.
7. $\begin{bmatrix} 0 & 0 & 1 & 0 \\ 0 & 1 & 0 & 0 \\ 1 & 0 & 0 & 0 \\ 0 & 0 & 0 & 1 \end{bmatrix}$.
8. $\begin{bmatrix} 1 & 0 & 0 & 0 \\ 0 & 1 & 0 & 0 \\ 0 & 0 & 1 & 0 \\ 0 & 5 & 0 & 1 \end{bmatrix}$.

In Exercises 9–14 determine if the given matrix is singular or nonsingular. Find the inverse for each matrix that is nonsingular.

9. $\begin{bmatrix} 7 & 0 & 0 \\ 0 & -3 & 0 \\ 0 & 0 & 4 \end{bmatrix}$.
10. $\begin{bmatrix} 2 & -1 & 3 \\ 0 & 1 & -4 \\ 0 & 0 & -5 \end{bmatrix}$.
11. $\begin{bmatrix} 4 & 0 & 0 \\ -2 & 3 & 0 \\ 6 & -1 & 5 \end{bmatrix}$.

12. $\begin{bmatrix} 1 & 0 & -1 & 2 \\ 0 & 3 & -5 & 9 \\ 0 & 0 & 7 & 1 \\ 0 & 0 & 0 & -2 \end{bmatrix}$.
13. $\begin{bmatrix} 1 & 0 & 0 & 1 \\ 0 & 1 & 0 & 1 \\ 1 & 0 & 1 & 0 \\ 1 & 1 & 0 & 1 \end{bmatrix}$.
14. $\begin{bmatrix} 1 & 0 & 1 & 0 \\ 0 & 1 & 0 & 1 \\ -2 & 3 & -2 & 3 \\ 1 & 1 & 1 & 1 \end{bmatrix}$.

In Exercises 15–18 write the given system of equations as a matrix equation, $AX = B$. Solve the system by solving the matrix equation $X = A^{-1}B$; that is, find A^{-1} and then multiply on the left of B.

15. $2x_1 - x_2 + 3x_3 = 2,$
$x_2 - 4x_3 = 5,$
$2x_1 - x_2 - 2x_3 = 7.$

16. $x_1 - x_2 + x_3 = 5,$
$x_1 + x_2 - x_3 = -1,$
$4x_1 - 3x_2 + 2x_3 = -3.$

17. $x_1 + 2x_2 - x_3 = 4,$
$2x_1 - 3x_2 + x_3 = -1,$
$5x_1 + 7x_2 + 2x_3 = -1.$

18. $x_1 + 2x_2 + x_3 - 2x_4 = 4,$
$x_1 - x_2 - x_3 - 3x_4 = 4,$
$2x_1 + 2x_2 + 3x_3 - 3x_4 = 4,$
$3x_1 + 5x_2 + 5x_3 - x_4 = 4.$

In Exercises 19–22 find the inverse of each of the given matrices, if it exists.

19. $\begin{bmatrix} 1 & 0 & 1 & 0 \\ 1 & -1 & 1 & 3 \\ 5 & 2 & 1 & 1 \\ 2 & 0 & -3 & 9 \end{bmatrix}$.
20. $\begin{bmatrix} 1 & -1 & -1 & 1 \\ 3 & 1 & 1 & 1 \\ 2 & -1 & -1 & 1 \\ 1 & 2 & 3 & 1 \end{bmatrix}$.
21. $\begin{bmatrix} 1 & 1 & 1 & 1 \\ 1 & 2 & -1 & 2 \\ 1 & -1 & 2 & 1 \\ 1 & 3 & 3 & 2 \end{bmatrix}$.
22. $\begin{bmatrix} 1 & 1 & 2 & 1 \\ 0 & -2 & 0 & 0 \\ 1 & 2 & -1 & 2 \\ 0 & 3 & 2 & 1 \end{bmatrix}$.

23. Write out the formal proof of the Corollary to Theorem 1.7 (page 54).

24. Prove part (iii) of Theorem 1.9 (page 55).

25. Prove that A is an invertible (nonsingular) matrix if, and only if, the homogeneous system of linear equations, $AX = O$, has only the trivial solution $X = O$.

26. Use mathematical induction to prove that if A is any $n \times n$ invertible (nonsingular) matrix, then for each positive integer k, the matrix A^k is invertible and $(A^k)^{-1} = (A^{-1})^k$.

27. Prove that an $n \times n$ nilpotent matrix is always singular. (See the definition before Exercise 49 in Section 1.4.)

28. Prove that if A is nonsingular, then so is its transpose A^T, and $(A^T)^{-1} = (A^{-1})^T$. (See Theorem 1.5, page 46.)

29. Show that if A is a 3×3 nonsingular matrix, then the columns of A^{-1} are the three solutions to the three systems of equations

$$A \begin{bmatrix} x_{11} \\ x_{21} \\ x_{31} \end{bmatrix} = \begin{bmatrix} 1 \\ 0 \\ 0 \end{bmatrix}; \quad A \begin{bmatrix} x_{12} \\ x_{22} \\ x_{32} \end{bmatrix} = \begin{bmatrix} 0 \\ 1 \\ 0 \end{bmatrix}; \quad \text{and} \quad A \begin{bmatrix} x_{13} \\ x_{23} \\ x_{33} \end{bmatrix} = \begin{bmatrix} 0 \\ 0 \\ 1 \end{bmatrix}.$$

30. Use the results of Exercise 29 to suggest a justification for the matrix-inversion algorithm different from that given in the text.

31. Prove or disprove the following statement: If A and B are two $n \times n$ nonsingular matrices, then the matrix $A + B$ is also nonsingular.

32. Show that the reduced row echelon form of the matrix A of Example 4 is indeed I_4 and find all of the elementary matrices used to perform row operations to reduce A to I_4.

33. Use the matrix reduction algorithm to show that any 2×2 matrix

$$A = \begin{bmatrix} a_{11} & a_{12} \\ a_{21} & a_{22} \end{bmatrix}$$

is nonsingular if, and only if, the number $\Delta = a_{11}a_{22} - a_{12}a_{21} \neq 0$, in which case

$$A^{-1} = \frac{1}{\Delta} \begin{bmatrix} a_{22} & -a_{12} \\ -a_{21} & a_{11} \end{bmatrix}.$$

In Exercises 34 and 35, suppose that the multinational Macht Corporation does business in the United States, Japan, and the European Economic Community. Each entity taxes the profits of the corporation according to the following tax matrix (table), where a_{ij} represents the tax rate charged by entity i on the profits of a corporation earned in country j.

$$\begin{array}{c} \text{U.S.} \\ \text{Japan} \\ \text{EEC} \end{array} \begin{bmatrix} 0.09 & 0.06 & 0.05 \\ 0.11 & 0.01 & 0.12 \\ 0.05 & 0.08 & 0.04 \end{bmatrix}.$$

34. Find the amount of tax owed if the profits for 1992 are given in thousands of dollars by the matrix $[850, 710, 787]^T$.

35. Find the amount of the profits of 1990 if the taxes paid in thousands of dollars are given by the matrix $[9, 11, 8]^T$.

In Exercises 36–38 prove that for $n \times n$ elementary matrices the statements are true.

36. $(E_{\mathrm{I}}(i, j))^{-1} = E_{\mathrm{I}}(i, j)$.

37. If $c \neq 0$, then $(E_{\mathrm{II}}(c, i))^{-1} = E_{\mathrm{II}}\left(\dfrac{1}{c}, i\right)$.

38. $(E_{\mathrm{III}}(i, k, j))^{-1} = E_{\mathrm{III}}(i, -k, j)$.

39. Show that, for $n \times n$ matrices A and B, if A is a singular matrix (not invertible) then the matrix product BA is also a singular matrix. (HINT: Suppose otherwise, and use elementary matrices to reduce BA to I. Then look at the product of these elementary matrices and B.)

40. Given the conditions of Exercise 39, show that the product matrix AB is also singular.

In Exercises 41–44 use the definition of row equivalence for $m \times n$ matrices that is given in this section.

41. Prove that every $m \times n$ matrix A is row equivalent to itself.

42. Prove that if A is row equivalent to B, then B is row equivalent to A.

43. Prove that if A is row equivalent to B and B is row equivalent to C, then A is row equivalent to C.

44. Prove that each $m \times n$ matrix A is row equivalent to one, and only one, $m \times n$ matrix U in *reduced* row echelon form.

45. What can you say about the statement in Exercise 44 if the word *reduced* is omitted?

*An upper triangular matrix is called **strictly upper triangular** if all of the entries on its main diagonal are zero, and similarly for **strictly lower triangular** matrices. (For the definition of upper and lower triangular matrices, see the paragraph preceding Exercise 39 in Section 1.4.)*

46. Prove that a *strictly* upper triangular matrix is always singular.

47. Prove that a *strictly* lower triangular matrix is always singular.

48. Let A be an $n \times n$ upper (or lower) triangular matrix, and let k be any *nonzero* real or complex number. Discuss when $kI_n - A$ is nonsingular.

49. Let

$$A = \begin{bmatrix} 1 & 2 & -3 \\ 0 & 1 & 4 \\ 0 & 0 & 1 \end{bmatrix}.$$

Does A^{-1} exist? If so find it.

50. Let

$$A = \begin{bmatrix} 1 & 1 & 1 \\ 1 & 1 & 1 \\ 1 & 1 & 1 \end{bmatrix}.$$

Is A singular or nonsingular? Give reasons.

51. An $n \times n$ matrix E is called **idempotent** if $E^2 = E$. Show that the matrix

$$E = \begin{bmatrix} 1 & 0 & 0 \\ 0 & 0 & 0 \\ 0 & 0 & 0 \end{bmatrix}$$

is idempotent.

52. Find all of the $n \times n$ nonsingular idempotent matrices E, if any exist. Give reasons.

53. Find all $n \times n$ matrices that are *both* idempotent and nilpotent, if any exist. Give reasons. (HINT: Use the matrix equation definitions.)

54. Let

$$A = \begin{bmatrix} 0 & 0 & 1 \\ 0 & 1 & 0 \\ 1 & 0 & 0 \end{bmatrix}.$$

Does A^{-1} exist? If so, find it. Generalize this problem to any such $n \times n$ matrix.

55. If you are using a computer or calculator to do matrix operations, find the inverse of each of the matrices A_ε, where the number ε takes on the various values ± 0.1, ± 0.3, ± 0.5, ± 0.7, ± 0.9, ± 1. Make a table of the answers. Does your computer/calculator give you an inverse for a singular matrix? How can you tell?

$$A_\varepsilon = \begin{bmatrix} 1.1 & 2.1 & 0.4 & 0.1 \\ 3.1 & 5.2 & 1+\varepsilon & 2.1 \\ 2.0 & 3.1 & 1.6 & 1+\varepsilon \\ 1-\varepsilon & \varepsilon & 1.6 & 2.0 \end{bmatrix}.$$

1.6 PARTITIONED MATRICES Numerical Considerations

It is often useful to think of a matrix as being made up of smaller matrices. We have already encountered this idea in several places in the previous sections of this chapter. For example, the augmented matrix of the system of equations $Ax = b$ can be written in the form $[A : b]$. The Matrix Inversion Algorithm begins with the matrix $[A : I]$ and ends with the matrix $[U : B]$, which is $[I : A^{-1}]$ when A is nonsingular. We have also discussed writing a given $n \times n$ matrix A in terms of its columns as $A = [a_1 : a_2 : \ldots : a_n]$. Partitioning a matrix can be a useful aid to computation, particularly for very large matrices. Often, a large matrix overloads the capabilities of a computer or calculator program. By splitting the matrix up into smaller blocks, each operation can then be performed. We consider the method for doing that here. The ideas don't require extremely large matrices to illustrate.

Example 1 Consider the matrix

$$A = \begin{bmatrix} 3 & 0 & 0 & 0 & 0 \\ 0 & 3 & 0 & 0 & 0 \\ 2 & 0 & -1 & 2 & 4 \\ 0 & 2 & 3 & -1 & -2 \end{bmatrix}.$$

It seems rather natural to subdivide A into submatrix blocks, as indicated, and to consider it to be made up of four submatrices as follows:

$$A = \begin{bmatrix} A_{11} & A_{12} \\ A_{21} & A_{22} \end{bmatrix} = \begin{bmatrix} 3I_2 & O_{2\times 3} \\ 2I_2 & A_{22} \end{bmatrix}.$$

This sort of partitioning is particularly helpful when multiplying two large matrices. In Section 1.4 we discussed partitioning a matrix into its rows or columns and indicated how this could be used in matrix multiplication. That is, if the $m \times k$ matrix $B = (b_{jt})$ is written in column form as

$$B = [b_1 : b_2 : \ldots : b_k]$$

and M is any $n \times m$ matrix, then the $n \times k$ matrix product MB is seen to be the matrix

$$MB = [Mb_1 : Mb_2 : \ldots : Mb_k]. \quad \blacksquare$$

This idea can be extended to cover other sorts of partitioning into blocks, as illustrated in the next example.

Example 2 Let A be the matrix of Example 1, and suppose that b is the following 5×5 matrix:

$$B = \begin{bmatrix} 1 & -1 & 3 & -1 & 0 \\ 4 & -2 & 1 & 0 & -1 \\ 0 & 1 & -1 & 0 & 1 \\ 4 & -1 & 0 & 0 & 1 \\ 1 & 1 & -2 & 0 & 1 \end{bmatrix} = \begin{bmatrix} B_{11} & B_{12} \\ B_{21} & B_{22} \end{bmatrix} = \begin{bmatrix} B_{11} & -I_2 \\ B_{21} & B_{22} \end{bmatrix},$$

where we have written B as partitioned into four submatrices. Since our aim is to multiply A and B by multiplying submatrix blocks, we partitioned the rows of B consistent with size requirements of the way A is partitioned. The top two blocks B_{11} and B_{12} of B are the first two rows of B. This corresponds to the fact that the two left-hand blocks of the partitioned matrix A of Example 1 each have two columns. The third and fourth blocks, B_{21} and B_{22}, of B have 3 rows each, corresponding to the fact that the blocks involving the remaining columns of the partitioned matrix A have three columns each. Since the partitioned matrices have submatrix blocks of appropriate sizes to multiply, we can multiply them as if they were individual elements. This results in the matrix product

$$AB = \begin{bmatrix} A_{11} & A_{12} \\ A_{21} & A_{22} \end{bmatrix} \cdot \begin{bmatrix} B_{11} & B_{12} \\ B_{21} & B_{22} \end{bmatrix} = \begin{bmatrix} A_{11}B_{11} + A_{12}B_{21} & A_{11}B_{12} + A_{12}B_{22} \\ A_{21}B_{11} + A_{22}B_{21} & A_{21}B_{12} + A_{22}B_{22} \end{bmatrix}$$

$$= \begin{bmatrix} 3I_2 & O_{2 \times 3} \\ 2I_2 & A_{22} \end{bmatrix} \cdot \begin{bmatrix} B_{11} & -I_2 \\ B_{21} & B_{22} \end{bmatrix} = \begin{bmatrix} 3I_2 B_{11} + OB_{21} & 3I_2(-I_2) + OB_{22} \\ 2I_2 B_{11} + A_{22}B_{21} & 2I_2(-I_2) + A_{22}B_{22} \end{bmatrix}$$

$$= \begin{bmatrix} 3B_{11} & -3I_2 \\ 2B_{11} + A_{22}B_{21} & -2I_2 + A_{22}B_{22} \end{bmatrix}.$$

The only nontrivial products involved in this multiplication are the products of the blocks $A_{22}B_{21}$ and $A_{22}B_{22}$, which can be computed in the usual manner. You can verify that

$$A_{22}B_{21} = \begin{bmatrix} 12 & 1 & -7 \\ -6 & 2 & 1 \end{bmatrix},$$

so that

$$2B_{11} + A_{22}B_{21} = \begin{bmatrix} 14 & -1 & -1 \\ 2 & -2 & 3 \end{bmatrix}.$$

Similarly,

$$A_{22}B_{22} = \begin{bmatrix} 0 & 5 \\ 0 & 0 \end{bmatrix},$$

so that

$$-2I_2 + A_{22}B_{22} = \begin{bmatrix} -2 & 5 \\ 0 & -2 \end{bmatrix}.$$

Therefore,

$$AB = \begin{bmatrix} 3B_{11} & -3I_2 \\ 2B_{11} + A_{22}B_{21} & -2I_2 + A_{22}B_{22} \end{bmatrix} = \begin{bmatrix} 3 & -3 & 9 & -3 & 0 \\ 12 & -6 & 3 & 0 & -3 \\ 14 & -1 & -1 & -2 & 5 \\ 2 & -2 & 3 & 0 & -2 \end{bmatrix}. \quad \blacksquare$$

> In the general case, let A be any $m \times n$ matrix and suppose that $n = n_1 + n_2 + \ldots + n_k$, and $m = m_1 + m_2 + \ldots + m_s$. Then the matrix A can be partitioned into ks submatrices $A_{\lambda\mu}$, $\mu = 1, 2, \ldots, k$, and $\lambda = 1, 2, \ldots, s$; where each submatrix is $m_\lambda \times n_\mu$.

Block Multiplication

Let us then make use of this example to formally define how one multiplies two matrices, once they have been compatibly partitioned.

Let A and B be two matrices such that A is $m \times n$ and B is $n \times p$. If $n = n_1 + n_2 + \ldots + n_k$, let the columns of A and the rows of B be partitioned in the *same way* so that the blocks A_{si} are $m_s \times n_i$ and the blocks B_{it} are $n_i \times p_t$. The rows of A and the columns of B are partitioned in any manner whatsoever. Then the two matrices may be written in block form as follows:

$$A = \begin{bmatrix} A_{11} & A_{12} & \cdots & A_{1k} \\ A_{21} & A_{22} & \cdots & A_{2k} \\ \vdots & \vdots & \cdots & \vdots \\ A_{s1} & A_{s2} & \cdots & A_{sk} \end{bmatrix}, \quad B = \begin{bmatrix} B_{11} & B_{12} & \cdots & B_{1t} \\ B_{21} & B_{22} & \cdots & B_{2t} \\ \vdots & \vdots & \cdots & \vdots \\ B_{k1} & B_{k2} & \cdots & B_{kt} \end{bmatrix}.$$

Then the matrix product AB can be computed as if the blocks were elements. Thus, AB has the form:

$$AB = \begin{bmatrix} C_{11} & C_{12} & \cdots & C_{1t} \\ C_{21} & C_{22} & \cdots & C_{2t} \\ \vdots & \vdots & \cdots & \vdots \\ C_{s1} & C_{s2} & \cdots & C_{st} \end{bmatrix},$$

where

$$C_{ij} = \sum_{r=1}^{k} A_{ir} B_{rj}.$$

This is the same form as that taken by the *element-by-element multiplication* of A and B, and is justified by actual element-by-element multiplication. In many cases, however, this block form is much easier to calculate, as Example 2 and the following example illustrate.

Example 3 Compute M^3 for the following matrix.

$$M = \begin{bmatrix} 1 & 0 & 0 & 0 & 0 & 0 \\ 0 & 1 & 0 & 0 & 0 & 0 \\ 0 & 0 & 1 & 0 & 0 & 0 \\ 0 & 0 & 0 & \dfrac{1}{2} & 0 & \dfrac{1}{2} \\ 0 & 0 & 0 & 0 & 3 & 0 \\ 0 & 0 & 0 & 0 & 0 & 3 \end{bmatrix}.$$

Solution By partitioning M into the indicated blocks, we can write M as the matrix

$$M = \begin{bmatrix} I_3 & O \\ O & M_{22} \end{bmatrix},$$

so that

$$M^3 = MMM = \begin{bmatrix} I_3 & O \\ O & M_{22} \end{bmatrix} \cdot \begin{bmatrix} I_3 & O \\ O & M_{22} \end{bmatrix} \cdot \begin{bmatrix} I_3 & O \\ O & M_{22} \end{bmatrix}$$

$$= \begin{bmatrix} I_3 & O \\ O & M_{22}^3 \end{bmatrix}.$$

Therefore, we need only compute M_{22}^3, where

$$M_{22} = \begin{bmatrix} \dfrac{1}{2} & 0 & \dfrac{1}{2} \\ 0 & 3 & 0 \\ 0 & 0 & 3 \end{bmatrix}.$$

Let us partition M_{22} as indicated:

$$M_{22} = \begin{bmatrix} \dfrac{1}{2} & 0 & \dfrac{1}{2} \\ 0 & 3 & 0 \\ 0 & 0 & 3 \end{bmatrix}.$$

So

$$M_{22} = \begin{bmatrix} \frac{1}{2}I & M' \\ O & 3I \end{bmatrix}$$

and

$$M_{22}^2 = \begin{bmatrix} \left(\frac{1}{2}I\right)^2 & \frac{1}{2}IM' + M'3I \\ O & 9I^2 \end{bmatrix} = \begin{bmatrix} \frac{1}{4}I & M'' \\ O & 9I \end{bmatrix},$$

where

$$M'' = \frac{1}{2}IM' + M'3I = \begin{bmatrix} 0 & \frac{7}{4} \end{bmatrix}.$$

Therefore,

$$M_{22}^3 = \begin{bmatrix} \frac{1}{8}I & \frac{1}{4}IM' + 3IM'' \\ O & 27I^3 \end{bmatrix} = \begin{bmatrix} \frac{1}{8} & 0 & \frac{43}{8} \\ 0 & 27 & 0 \\ 0 & 0 & 27 \end{bmatrix}.$$

Thus,

$$M^3 = \left[\begin{array}{ccc|ccc} 1 & 0 & 0 & 0 & 0 & 0 \\ 0 & 1 & 0 & 0 & 0 & 0 \\ 0 & 0 & 1 & 0 & 0 & 0 \\ \hline 0 & 0 & 0 & \frac{1}{8} & 0 & \frac{43}{8} \\ 0 & 0 & 0 & 0 & 27 & 0 \\ 0 & 0 & 0 & 0 & 0 & 27 \end{array} \right].$$

which you can verify by direct calculation. ∎

Often, when solving a system of linear equations, $Ax = b$, if the system is particularly large or if there are several systems each having the same coefficient matrix A, but different columns b_j, it is convenient to factor a matrix in a special way.

In finding the inverse for an $n \times n$ matrix A, an alternative view of the matrix reduction algorithm suggests viewing the problem as solving n systems of linear equations $Ax = e_i$, where e_i is the ith column of the identity matrix I. That is, e_i is an $n \times 1$ matrix with a "1" in the ith row and zeros in all other rows ($i = 1, 2, \ldots, n$). Then the augmented matrix

$$[A : I] = [A : e_1 : e_2 : \cdots : e_n].$$

This same method can be applied to a series of linear equations $Ax = b_j$, all with the same coefficient matrix A. Simply write the augmented matrix $[A : b_1 : b_2 : \cdots : b_k]$ and row reduce A to the identity matrix I; the solutions then appear in the remaining columns.

Example 4 Solve the following four systems of linear equations:

$$
\begin{array}{ll}
\text{(1)} & \text{(2)} \\
\begin{aligned}
x_1 - 3x_2 + 4x_3 &= 1, \\
2x_1 - x_2 - 3x_3 &= -2, \\
3x_1 + 2x_2 - x_3 &= 11.
\end{aligned}
&
\begin{aligned}
x - 3y + 4z &= 1, \\
2x - y - 3z &= 4, \\
3x + 2y - z &= 0.
\end{aligned}
\end{array}
$$

$$
\begin{array}{ll}
\text{(3)} & \text{(4)} \\
\begin{aligned}
u - 3v + 4w &= 2, \\
2u - v - 3w &= -1, \\
3u + 2v - w &= 5.
\end{aligned}
&
\begin{aligned}
r - 3s + 4t &= 0, \\
2r - s - 3t &= -3, \\
3r + 2s - t &= 2.
\end{aligned}
\end{array}
$$

The coefficient matrix A is the same for each of these systems, namely,

$$
A = \begin{bmatrix} 1 & -3 & 4 \\ 2 & -1 & -3 \\ 3 & 2 & -1 \end{bmatrix}.
$$

Set

$$
b_1 = \begin{bmatrix} 1 \\ -2 \\ 11 \end{bmatrix}; \quad
b_2 = \begin{bmatrix} 1 \\ 4 \\ 0 \end{bmatrix}; \quad
b_3 = \begin{bmatrix} 2 \\ -1 \\ 5 \end{bmatrix}; \quad \text{and} \quad
b_4 = \begin{bmatrix} 0 \\ -3 \\ 2 \end{bmatrix},
$$

and form the combined augmented matrix

$$
[A : b_1 : b_2 : b_3 : b_4] = \left[\begin{array}{rrr|r|r|r|r} 1 & -3 & 4 & 1 & 1 & 2 & 0 \\ 2 & -1 & -3 & -2 & 4 & -1 & -3 \\ 3 & 2 & -1 & 11 & 0 & 5 & 2 \end{array}\right].
$$

By row reducing A to the identity matrix I_3, we obtain the partitioned matrix

$$
\left[\begin{array}{rrr|r|r|r|r} 1 & 0 & 0 & 2.50 & 0.48 & 1.32 & 0.20 \\ 0 & 1 & 0 & 2.50 & -1.05 & 0.96 & 1.09 \\ 0 & 0 & 1 & 1.50 & -0.66 & 0.89 & 0.77 \end{array}\right].
$$

The last four columns of this matrix display calculator approximations of the solutions to the four given equations. You can verify that they are indeed solutions. ∎

The LU Decomposition: Another Numerical Approach

In actual practice, the above suggestion for solving a collection of systems of linear equations $Ax = b_i$ may be impractical. It may, in fact, be impossible for a single computer run. There could be several reasons for this; for example, the overall size of the system may be too large, or the matrices b_i may be generated over time. Given that, let us look at a variation. This idea was first worked out by Alan Turing in 1948 in developing the stored program computer. It is practical and is currently part of many prepackaged computer or calculator matrix operations.

Consider the $n \times n$ matrix A, and let us make the assumption that A can be row reduced to an upper-triangular matrix U in the usual way, but *without any interchanges of its rows.* For large matrices this reduction requires a lot of work, and we should not wish to have to repeat it later for any new data. To solve any subsequent systems $Ax = b$, we must keep a record of the entire sequence $E_k \cdots E_3 E_2 E_1$ of elementary matrices that were applied to A to produce the upper triangular matrix U. Since we are not interchanging rows, and since there is no need to change signs on any row, we may assume that each of the elementary matrices involved is Type III. Therefore, each of the elementary matrices is lower triangular.

Since $E_k \cdots E_3 E_2 E_1 A = U$, and since each of the elementary matrices is invertible, we must have

$$A = E_1^{-1} E_2^{-1} E_3^{-1} \cdots E_k^{-1} U.$$

Each of the inverses E_i^{-1} of a Type III elementary E_i matrix is also lower triangular. For example, if

$$E_i = \begin{bmatrix} 1 & 0 & 0 & \cdots & 0 \\ 0 & 1 & 0 & \cdots & 0 \\ 0 & 0 & 1 & \cdots & 0 \\ . & . & . & \cdots & . \\ -a_{11} & . & . & .1. & 0 \\ . & . & . & \cdots & . \\ 0 & 0 & 0 & \cdots & 1 \end{bmatrix}, \quad \text{Row } i$$

then

$$E_i^{-1} = \begin{bmatrix} 1 & 0 & 0 & \cdots & 0 \\ 0 & 1 & 0 & \cdots & 0 \\ 0 & 0 & 1 & \cdots & 0 \\ . & . & . & \cdots & . \\ a_{11} & . & . & .1. & 0 \\ . & . & . & \cdots & . \\ 0 & 0 & 0 & \cdots & 1 \end{bmatrix}. \quad \text{Row } i$$

The product $E_1^{-1} E_2^{-1} E_3^{-1} \cdots E_k^{-1}$ is therefore a lower triangular matrix L with *ones* on the main diagonal. Below the main diagonal of L are the *negatives* of the multipliers of those rows of A that were added to a subsequent row in reaching the upper triangular matrix U. The matrix A is the product of this lower triangular matrix L and the upper triangular matrix U.

$$A = LU.$$

This product is called the *LU decomposition (factorization)* of A.

We illustrate this in the following example.

Example 5   Let

$$A = \begin{bmatrix} 1 & -1 & 3 \\ 2 & 3 & -1 \\ 4 & 1 & 2 \end{bmatrix}.$$

To form the *LU* decomposition, we proceed to row reduce A to triangular form by adding -2 times row *one* to row *two*, resulting in the matrix

$$A' = \begin{bmatrix} 1 & -1 & 3 \\ 0 & 5 & -7 \\ 4 & 1 & 2 \end{bmatrix}.$$

The matrix $L_1 = E_1^{-1}$, which undoes this elementary operation, is the identity matrix with the addition of $-(-2) = 2$ to its 2, 1 position. Thus,

$$L_1 = \begin{bmatrix} 1 & 0 & 0 \\ 2 & 1 & 0 \\ 0 & 0 & 1 \end{bmatrix}.$$

You can easily verify that $A = L_1 A'$.

Continuing, in the matrix A', add -4 times row *one* to row *three*, obtaining the matrix

$$A'' = \begin{bmatrix} 1 & -1 & 3 \\ 0 & 5 & -7 \\ 0 & 5 & -10 \end{bmatrix}.$$

This places a $-(-4) = 4$ in the 3, 1 position of the matrix $L_2 = E_1^{-1}E_2^{-1}$, so

$$L_2 = \begin{bmatrix} 1 & 0 & 0 \\ 2 & 1 & 0 \\ 4 & 0 & 1 \end{bmatrix}.$$

We see that $A = L_2 A''$.

Finally, add -1 times row *two* of A'' to its row *three*, yielding the upper triangular matrix

$$U = \begin{bmatrix} 1 & -1 & 3 \\ 0 & 5 & -7 \\ 0 & 0 & -3 \end{bmatrix}.$$

This puts a $-(-1) = 1$ in the 3, 2 position of $L_3 = E_1^{-1}E_2^{-1}E_3^{-1}$, so the resulting matrix is

$$L = L_3 = \begin{bmatrix} 1 & 0 & 0 \\ 2 & 1 & 0 \\ 4 & 1 & 1 \end{bmatrix}.$$

Thus, $A = LU$, as desired. ∎

Let us apply this by solving several systems of linear equations, $Ax = b_i$, all with the same coefficient matrix A. While reducing A to the upper triangular matrix U, we store the information needed for the row reduction in the lower triangular matrix L. Then, upon substituting LU for A, we have

$$LU\boldsymbol{x} = \boldsymbol{b}_i.$$

The lower triangular systems

$$L\boldsymbol{y}_i = \boldsymbol{b}_i$$

may be solved by forward substitution; see below to learn how this can be done quickly. Then, since U is upper triangular, one can use back substitution to solve

$$U\boldsymbol{x} = \boldsymbol{y}_i.$$

The matrix L is a *record* of row reductions that may be applied to the individual column matrices $\boldsymbol{b}_i$, thereby obtaining the solutions $\boldsymbol{y}_i$ without actually doing a forward substitution. We demonstrate this in the next example. If additional equations occur later, we have stored all of the information we need to solve those additional systems in the two triangular matrices L and U.

Example 6 Consider the four systems of linear equations $A\boldsymbol{x} = \boldsymbol{b}_i$, where A is the matrix of Example 5

$$A = \begin{bmatrix} 1 & -1 & 3 \\ 2 & 3 & -1 \\ 4 & 1 & 2 \end{bmatrix},$$

and the right members of the equations are the following:

$$\boldsymbol{b}_1 = \begin{bmatrix} -1 \\ 2 \\ 1 \end{bmatrix}; \quad \boldsymbol{b}_2 = \begin{bmatrix} 1 \\ -4 \\ 3 \end{bmatrix}; \quad \boldsymbol{b}_3 = \begin{bmatrix} -2 \\ 0 \\ -5 \end{bmatrix}; \quad \text{and} \quad \boldsymbol{b}_4 = \begin{bmatrix} 0 \\ 3 \\ -2 \end{bmatrix}.$$

Using the matrix $L = (l_{ij})$ of Example 5 as a record of the row reductions, we simply apply these to the matrices $\boldsymbol{b}_i$. Since

$$l_{21} = 2$$

multiply row *one* of each $\boldsymbol{b}_i$ by -2 and add to row *two*, resulting in

$$\boldsymbol{b}'_1 = \begin{bmatrix} -1 \\ 4 \\ 1 \end{bmatrix}; \quad \boldsymbol{b}'_2 = \begin{bmatrix} 1 \\ -6 \\ 3 \end{bmatrix}; \quad \boldsymbol{b}'_3 = \begin{bmatrix} -2 \\ 4 \\ -5 \end{bmatrix}; \quad \text{and} \quad \boldsymbol{b}'_4 = \begin{bmatrix} 0 \\ 3 \\ -2 \end{bmatrix},$$

respectively. Then, since

$$l_{31} = 4$$

multiply row *one* of each $\boldsymbol{b}'_i$ by -4 and add to row *three*. Hence,

$$\boldsymbol{b}''_1 = \begin{bmatrix} -1 \\ 4 \\ 5 \end{bmatrix}; \quad \boldsymbol{b}''_2 = \begin{bmatrix} 1 \\ -6 \\ -1 \end{bmatrix}; \quad \boldsymbol{b}''_3 = \begin{bmatrix} -2 \\ 4 \\ 3 \end{bmatrix}; \quad \text{and} \quad \boldsymbol{b}''_4 = \begin{bmatrix} 0 \\ 3 \\ -2 \end{bmatrix}.$$

Finally, since

$$l_{32} = 1$$

multiply row *two* of each b''_i by -1 and add to row *three*. This gives the columns

$$y_1 = \begin{bmatrix} -1 \\ 4 \\ 1 \end{bmatrix}; \quad y_2 = \begin{bmatrix} 1 \\ -6 \\ 5 \end{bmatrix}; \quad y_3 = \begin{bmatrix} -2 \\ 4 \\ -1 \end{bmatrix}; \quad \text{and} \quad y_4 = \begin{bmatrix} 0 \\ 3 \\ -5 \end{bmatrix}.$$

Now, putting these results together with the matrix U of Example 5, we have the augmented (partitioned) matrices $[U : y_1]$, $[U : y_2]$, $[U : y_3]$, $[U : y_4]$. Each of the systems $Uy_i = b_i$ can be solved by back substitution, or by further reduction of the augmented matrix. Thus,

$$[U : y_1] = \begin{bmatrix} 1 & -1 & 3 & -1 \\ 0 & 5 & -7 & 4 \\ 0 & 0 & -3 & 1 \end{bmatrix}.$$

So,

$$x_3 = \frac{-1}{3}; \quad x_2 = \frac{7x_3 + 4}{5} = \frac{1}{3}; \quad \text{and} \quad x_1 = x_2 - 3x_3 - 1 = \frac{1}{3}$$

is the solution to the first system. For the second system, the matrix

$$[U : y_2] = \begin{bmatrix} 1 & -1 & 3 & 1 \\ 0 & 5 & -7 & -6 \\ 0 & 0 & -3 & 5 \end{bmatrix}$$

tells us that $x_3 = -5/3$; $x_2 = -53/15$; and $x_1 = 37/15$. We leave the other two systems as exercises. ∎

The *LU* decomposition of a matrix A is also a matrix algebra example of the computer science fact that computer *programs* can be viewed as a special form of *data*. Presently, programs are stored as a string of 0's and 1's just like other data. The matrix L contains the elimination multipliers for the "Gaussian reduction program." Matrix multiplication makes it possible for matrices to be both data and programs, as is the case with computers. It is important to remark that:

> The *LU* factorization of a given matrix A is not unique.

You can see from Example 5 that the order in which row operations are performed will change the nature of either L or U or both. Nevertheless, any LU factorization can be used as we did in Example 6. It is also an unfortunate fact that:

> Not every $n \times n$ matrix A has an LU decomposition.

See Exercises 22 and 23.

In storing the LU factorization for a given matrix A in a computer, the upper triangular matrix U and the lower triangular matrix L are often combined into a single matrix. Remember that A is L times U, not $L + U$. Nevertheless, if the "ones" are omitted from the main diagonal of L, the rest of L can fill in the lower triangle of U, replacing the zeros that were there.

Thus, an **L/U display** for the matrix

$$A = \begin{bmatrix} 1 & -1 & 3 \\ 2 & 3 & -1 \\ 4 & 1 & 2 \end{bmatrix},$$

of Example 5 is the matrix

$$\begin{bmatrix} 1 & -1 & 3 \\ 2 & 5 & -7 \\ 4 & 1 & -3 \end{bmatrix}.$$

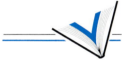 **Vocabulary**

Each of the concepts described by the terms in the following list is fundamental. Review these ideas until you feel secure enough to use them.

partitioned matrix	matrix blocks
block multiplication	LU decomposition
LU factorization	L/U display
submatrix	

Exercises 1.6

1. Use block multiplication to compute the matrix A^2, given the matrix

$$A = \begin{bmatrix} 3 & 0 & 1 & 2 & -1 \\ 0 & 3 & 0 & 1 & 2 \\ 0 & 0 & 1 & 0 & 0 \\ 0 & 0 & 0 & 1 & 0 \\ 0 & 0 & 0 & 0 & 1 \end{bmatrix}.$$

2. Given the matrix A of Exercise 1, compute A^4.

3. Using the matrix A of Exercise 1 and the following matrix B, find AB by block multiplication.

$$B = \begin{bmatrix} 1 & 2 & 4 & 1 & 5 & 1 \\ 1 & -1 & 3 & 0 & -2 & 0 \\ 4 & 1 & -1 & 0 & 1 & 0 \\ 0 & 0 & 1 & 1 & 0 & 0 \\ 1 & 0 & 1 & 0 & 1 & 0 \end{bmatrix}.$$

4. Let C and M be the matrices below: Compute CM, using block multiplication.

$$C = \begin{bmatrix} 2 & 3 & 1 & 0 & 0 \\ 3 & -1 & 0 & 0 & 0 \\ 1 & 0 & -2 & 0 & 1 \\ -1 & 0 & 0 & 1 & 0 \\ 0 & -1 & 1 & 0 & 1 \end{bmatrix}, \quad M = \begin{bmatrix} 1 & 0 & 0 \\ -2 & 0 & 0 \\ 0 & 3 & -1 \\ -1 & 0 & 0 \\ 1 & 0 & 0 \end{bmatrix}.$$

5. For the matrix C of Exercise 4, compute C^2.   6. For the matrix C of Exercise 4, compute C^3.

7. For the matrix M of Example 3, compute M^4.   8. For the matrix M of Example 3, compute M^5.

9. Let K be the block matrix diag $[J_1, J_2, J_3, I_2, -I_3]$, where I_2 is the 2×2 identity matrix and $-I_3$ is the negative of the 3×3 identity matrix and

$$J_1 = \begin{bmatrix} 2 & 1 & 0 \\ 0 & 2 & 1 \\ 0 & 0 & 2 \end{bmatrix}, \quad J_2 = \begin{bmatrix} \frac{1}{3} & 1 \\ 0 & \frac{1}{3} \end{bmatrix}, \quad \text{and} \quad J_2 = \begin{bmatrix} \frac{1}{2} & 1 & 0 \\ 0 & \frac{1}{2} & 1 \\ 0 & 0 & \frac{1}{2} \end{bmatrix}.$$

Write the matrix K in expanded form as a 13×13 matrix.

10. Find K^2 for the matrix K of Exercise 9.

11. Given the submatrices J_1, J_2, and J_3 from Exercise 9 above, write the matrix $G = $ diag $[1, -1, J_1, 2, J_3, J_2, 0, 0]$ in expanded form.

12. Finish the solution in Example 6 to the linear system $Ax = b_3$, using the given LU decomposition of A.

13. Finish the solution in Example 6 to the linear system $Ax = b_4$, using the given LU decomposition of A.

14. Find an LU decomposition for the matrix A of Example 4.

15. Use the results of Exercise 14 to confirm the solution to the system of equations

$$\begin{aligned} r - 3s + 4t &= 0 \\ 2r - s - 3t &= -3 \\ 3r + 2s - t &= 2 \end{aligned}$$

given in Example 4.

16. Use the results of Exercise 14 to solve the following system of equations:

$$\begin{aligned} x - 3y + 4z &= 12, \\ 2x - y - 3z &= -3, \\ 3x + 2y - z &= 7. \end{aligned}$$

17. Use the results of Example 5 to solve the system of equations

$$Ax = \begin{bmatrix} 1 \\ -3 \\ 0 \end{bmatrix}. \text{ (A is the matrix of Example 5.)}$$

18. Use the results of Example 5 to solve the system of equations

$$Ax = \begin{bmatrix} -2 \\ 4 \\ 1 \end{bmatrix}. \text{ (Here too, A is the matrix of Example 5.)}$$

19. Find an LU factorization $A = LU$ for the matrix

$$A = \begin{bmatrix} 2 & 3 & 1 & 0 \\ 0 & 3 & -1 & 0 \\ 0 & 1 & 1 & -1 \\ 3 & 1 & 3 & -2 \end{bmatrix}.$$

20. Use the results of Exercise 19 to solve the homogeneous system of equations $Ax = 0$.

21. Write the LU factorization of A in Exercise 19 as a single matrix in an L/U display.

22. Show that if the matrix $A = \begin{bmatrix} 0 & 1 \\ 1 & 0 \end{bmatrix}$ could be written as $A = LU$, with $L = \begin{bmatrix} l_{11} & O \\ l_{21} & l_{22} \end{bmatrix}$ and $U = \begin{bmatrix} u_{11} & u_{12} \\ O & u_{22} \end{bmatrix}$, then either L or U must be singular, whereas A is nonsingular. (Why is this a contradiction?)

23. Show that there exists a Type I elementary matrix E so that EA, for the matrix A in Exercise 22, does have an LU decomposition.

*In Exercises 24–28, let A be an $n \times k$ and let B be an $m \times s$ matrix. Define the **direct sum**, $A \oplus B$, of A and B as the $(n + m) \times (k + s)$ block matrix*

$$A \oplus B = \begin{bmatrix} A & O \\ O & B \end{bmatrix},$$

where O is the zero matrix of appropriate size.

24. Write $A \oplus B$ for $A = \begin{bmatrix} -1 & -3 \\ 5 & 8 \end{bmatrix}$ and $B = \begin{bmatrix} -1 & -1 \\ 1 & 1 \end{bmatrix}$.

25. Repeat Exercise 24 for $A = I_3$ and $B = \begin{bmatrix} 1 & -2 & 4 & 0 \\ 3 & 5 & 1 & 2 \\ 3 & -1 & 0 & 1 \end{bmatrix}$.

26. Let A, B, and C be $n \times n$ matrices. Does $C(A \oplus B) = CA \oplus CB$?

27. Let A, B, C, and D be $n \times n$ matrices. Does $(C \oplus D)(A \oplus B) = CA \oplus DB$?

28. Suppose that $K = A \oplus B$. Prove that K is nonsingular if, and only if, *both A and B are* nonsingular. Find K^{-1}.

1.7 EXTENDED APPLICATION *Networks and Graphs*

In this section we illustrate how some simple linear algebra techniques are used in a branch of mathematics called *graph theory.* Although this theory is several hundreds of years old, it has recently been found to be very helpful in interpreting situations arising in the behavioral and management sciences.

The term *graph* in the context of *graph theory* is not used in quite the same sense as it is in other branches of mathematics. In graph theory we have the following formal definition.

> **Definition**   A **graph** is a nonempty collection of a finite number of points, denoted P_1, P_2, ... P_n called **vertices,** together with a finite number of **edges,** $P_i P_j$, connecting some of the vertices in pairs.

Example 1   Figure 1.8 contains several examples of *graphs* in this sense.

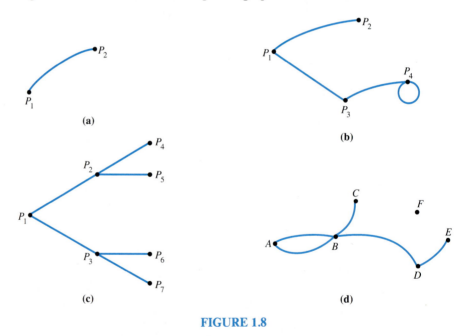

(a)

(b)

(c)

(d)

FIGURE 1.8

Note that, in Figure 1.8, graph (b) has an edge from vertex P_4 back to P_4; such an edge is called a **loop.** Note also that in graph (d), vertex F has no edge connecting it to any other vertex, while vertices A and B are connected by two distinct edges. This is possible in a graph. In these examples, the edge $P_i P_j$ for each graph is considered to be the same as the edge $P_j P_i$; *the order in which they are mentioned is immaterial.* This exactly mirrors many applications. Several cities connected by highways or air routes are examples of graphs in which the order of the end points is unimportant (see Figure 1.9).

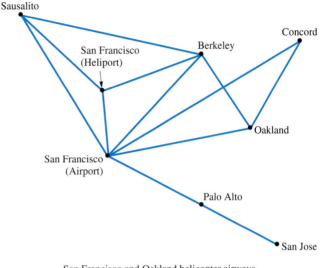

San Francisco and Oakland helicopter airways

FIGURE 1.9

However, in other applications the order of the vertices may indeed be important, as in hierarchical structures and in many communication models. This leads us to a new concept. An **oriented graph** is a nonempty collection of a finite number of vertices P_i together with a finite number of directed edges P_iP_j joining some or all of these. ■

Definition   An oriented graph is called a **directed graph** or a **digraph** if it contains no loops and has at most one edge from P_i to P_j for each i and j. (A second edge connecting P_1 and P_j must go from P_j to P_i.)

Example 2   The graphs in Figure 1.10 (on the next page) illustrate oriented graphs. Note that graph (e) fails to be a digraph because it contains a loop. Note also that graph (d) could be the model of a one-way street grid in a downtown area.

It is often convenient to represent an oriented graph by a matrix. A **matrix representation** $A = (a_{ij})$ for a given oriented graph has the same number of rows and columns as there are vertices in the graph. This matrix is called an *incidence matrix* by some authors and the *adjacency matrix* of the graph by others. The entries a_{ij} in the matrix representation of a given oriented graph are the number of directed edges from the vertex P_i to the vertex P_j. If

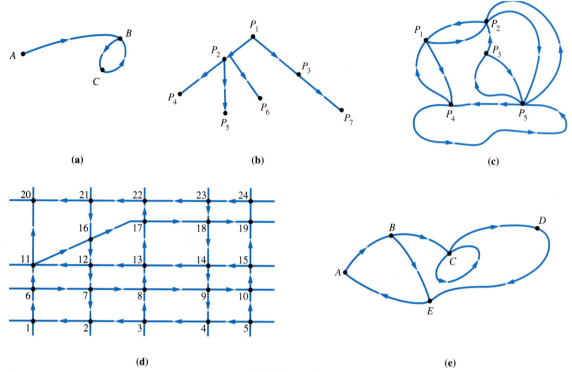

FIGURE 1.10

the graph is a digraph, then all of the entries in this matrix representation are either zero or one. The matrix representations for four of the graphs in Figure 1.10 are as follows:

$$
\begin{array}{c c}
& \begin{array}{c c c} A & B & C \end{array} \\
\begin{array}{c} A \\ B \\ C \end{array} & \left[\begin{array}{c c c} 0 & 1 & 0 \\ 0 & 0 & 1 \\ 0 & 1 & 0 \end{array}\right]
\end{array}, \qquad
\begin{array}{c c}
& \begin{array}{c c c c c c c} 1 & 2 & 3 & 4 & 5 & 6 & 7 \end{array} \\
\begin{array}{c} 1 \\ 2 \\ 3 \\ 4 \\ 5 \\ 6 \\ 7 \end{array} & \left[\begin{array}{c c c c c c c} 0 & 1 & 1 & 0 & 0 & 0 & 0 \\ 0 & 0 & 0 & 1 & 1 & 1 & 0 \\ 0 & 0 & 0 & 0 & 0 & 0 & 1 \\ 0 & 0 & 0 & 0 & 0 & 0 & 0 \\ 0 & 0 & 0 & 0 & 0 & 0 & 0 \\ 0 & 0 & 0 & 0 & 0 & 0 & 0 \\ 0 & 0 & 0 & 0 & 0 & 0 & 0 \end{array}\right]
\end{array},
$$

(a)   (b)

$$
\begin{array}{c c}
& \begin{array}{c c c c c} 1 & 2 & 3 & 4 & 5 \end{array} \\
\begin{array}{c} 1 \\ 2 \\ 3 \\ 4 \\ 5 \end{array} & \left[\begin{array}{c c c c c} 0 & 1 & 0 & 1 & 0 \\ 1 & 0 & 0 & 0 & 1 \\ 0 & 1 & 0 & 0 & 1 \\ 1 & 0 & 0 & 0 & 1 \\ 0 & 1 & 1 & 1 & 0 \end{array}\right]
\end{array}, \qquad
\begin{array}{c c}
& \begin{array}{c c c c c} A & B & C & D & E \end{array} \\
\begin{array}{c} A \\ B \\ C \\ D \\ E \end{array} & \left[\begin{array}{c c c c c} 0 & 1 & 0 & 0 & 0 \\ 0 & 0 & 1 & 0 & 1 \\ 0 & 0 & 1 & 1 & 0 \\ 0 & 0 & 0 & 0 & 1 \\ 1 & 0 & 0 & 0 & 0 \end{array}\right]
\end{array}.
$$

(c)   (e)

You may wish to write out the 24×24 matrix representation for the street grid in Figure 1.10(d). Any matrix representation of a digraph will have zeros on its main diagonal. (Why?)

In applying graph theory to the study of organizations, an important concept is *dominance*. This arises when, with any two people or groups or teams, one dominates or influences the other. One may model this situation with a digraph in which the vertices represent the people or groups in the organization, while the directed edges flow from the "person" with influence to the "person" being influenced. ■

Example 3 A juvenile gang consisting of five members—*H, J, K, L,* and *M*—is interviewed by a social worker. All possible pairs of members of the gang are interviewed, and it is noted which member of each pair is the more influential. The following digraph is formed (see Figure 1.11) summarizing these notes. The matrix representation for the graph is the given matrix *A*.

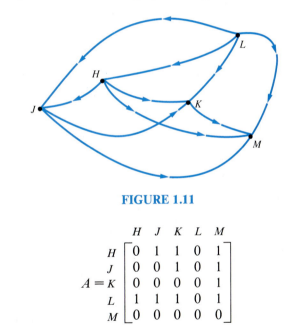

FIGURE 1.11

$$
A = \begin{array}{c} \\ H \\ J \\ K \\ L \\ M \end{array}
\begin{array}{c} \begin{array}{ccccc} H & J & K & L & M \end{array} \\
\begin{bmatrix} 0 & 1 & 1 & 0 & 1 \\ 0 & 0 & 1 & 0 & 1 \\ 0 & 0 & 0 & 0 & 1 \\ 1 & 1 & 1 & 0 & 1 \\ 0 & 0 & 0 & 0 & 0 \end{bmatrix} \end{array}
$$

By adding the row entries of the matrix *A*, we find that in one stage *H* dominates three people, *J* dominates two people, *K* dominates one person, *L* dominates all four of the other members of the gang, and *M* dominates none of them. Since *L* dominates everyone else, *L* is called the *consensus leader* of the gang.

To determine the leader of an organization when there is no consensus leader, the idea of *n-stage dominance* is used. Let us illustrate this with the following example involving athletic teams as the "persons." ■

Example 4 The following lists the results of football games played in a six-team conference.

Who should be the conference representative in a postseason bowl game?

Team	Defeated	Lost to
a	d, e, f	b, c
b	a, c, d	e, f
c	a, d, f	b, e
d	e, f	a, b, c
e	b, c	a, d, f
f	b, e	a, c, d

Since three teams are tied with identical 3:2 records, holding a three-way play-off or drawing straws could provide the answer. An alternative answer is given by the concept of n-stage dominance. Let us set up the matrix representation for the dominance digraph directly and order the rows and columns as in the above list:

$$
M = \begin{array}{c} \\ a \\ b \\ c \\ d \\ e \\ f \end{array}
\begin{array}{c} \begin{array}{cccccc} a & b & c & d & e & f \end{array} \\
\left[\begin{array}{cccccc}
0 & 0 & 0 & 1 & 1 & 1 \\
1 & 0 & 1 & 1 & 0 & 0 \\
1 & 0 & 0 & 1 & 0 & 1 \\
0 & 0 & 0 & 0 & 1 & 1 \\
0 & 1 & 1 & 0 & 0 & 0 \\
0 & 1 & 0 & 0 & 1 & 0
\end{array} \right] \end{array}.
$$

The number of games won by each team is the sum of the entries in its row. What is the number of games lost? From this example it is fairly clear that there is no transitivity of dominance; team a beat team e; team e beat b; but team a did not beat team b. However, because a beat e and e beat b, we say that a has two-stage dominance over b. We could use the concept of two-stage dominance to select the conference representative in the bowl game. The square of the matrix representation of the dominance digraph is the matrix that represents the two-stage dominance. Here, in this example, the two-stage matrix is

$$
M' = M^2 = \begin{array}{c} \\ a \\ b \\ c \\ d \\ e \\ f \end{array}
\begin{array}{c} \begin{array}{cccccc} a & b & c & d & e & f \end{array} \\
\left[\begin{array}{cccccc}
0 & 2 & 1 & 0 & 2 & 1 \\
1 & 0 & 0 & 2 & 2 & 3 \\
0 & 1 & 0 & 1 & 3 & 2 \\
0 & 2 & 1 & 0 & 1 & 0 \\
2 & 0 & 1 & 2 & 0 & 1 \\
1 & 1 & 2 & 1 & 0 & 0
\end{array} \right] \end{array}
\begin{array}{c} 6 \\ 8 \\ 7 \\ 4 \\ 6 \\ 5 \end{array}.
$$

The fact that $m'_{12} = 2$ means that team a has two-stage dominance over team b via two different paths:

$$a \to e \to b \quad \text{and} \quad a \to f \to b.$$

The value of $m'_{26} = 3$ indicates that team b has two-stage dominance over team f via three different paths; namely,

$$b \to a \to f, \quad b \to c \to f, \quad \text{and} \quad b \to d \to f. \quad \blacksquare$$

As an exercise, sketch the digraph of this situation and notice that the entries in $M' = M^2$ give the number of two-edge paths from P_i to P_j; that is, those that pass through an intermediate vertex. In general:

> **Definition** Let M be the matrix representation for a dominance graph, and $M^k = (a_{ij})$ be the kth power of M. Then the entries a_{ij} indicate the number of k-stage dominance of "person" i over "person" j.

If you wish, apply this analysis to Example 4 to send team b to the bowl game as the conference representative, because its number of two-stage dominance (8) is larger than that of any of the other teams in the conference. It is the "leader" at stage two.

Probably a more accurate determination could be made by taking the *sum* of the first- and second-stage dominance for each individual (team). This number is often called the **power** of the individual in the dominance digraph, and is the matrix sum $M + M^2$. For the situation in Example 4 this becomes

$$M + M^2 = \begin{array}{c} \\ a \\ b \\ c \\ d \\ e \\ f \end{array}
\begin{array}{c}
\begin{array}{cccccc} a & b & c & d & e & f \end{array} \\
\left[\begin{array}{cccccc}
0 & 2 & 1 & 1 & 3 & 2 \\
2 & 0 & 1 & 3 & 2 & 3 \\
1 & 1 & 0 & 2 & 3 & 3 \\
0 & 2 & 1 & 0 & 2 & 1 \\
2 & 1 & 2 & 2 & 0 & 1 \\
1 & 2 & 2 & 1 & 1 & 0
\end{array}\right]
\end{array}
\begin{array}{c}
\text{Power} \\
9 \\
11 \\
10 \\
6 \\
8 \\
7
\end{array} .$$

This idea can be generalized to the sum $M + M^2 + M^3 + \cdots + M^k$, for dominance through k-stages, a sort of **k-power**.

Communication Networks

As a second example from the behavioral sciences of how graphs and their matrices may be used to model ideas, let us consider communication networks.

> **Definition** A **communication network** consists of a finite set A_1, A_2, ..., A_n of individuals (people, companies, cities, etc.) such that between some pair of individuals there is a one- or two-way communication link.

Examples of two-way communication links are two-way radio, telephone, highways, airlines, and so on. A one-way link could be television, a messenger, a signal light, one-way radio, and the like. Such a communication network can be represented by an oriented graph. A good example is the airways link in Figure 1.9.

Example 5 The four towns of Delta, Jacobtown, Cooperton, and Spring City are connected by direct telephone lines in the manner indicated in the oriented graph of Figure 1.12. The double arrows indicate two-way communication and are the equivalent of two edges each, one each way.

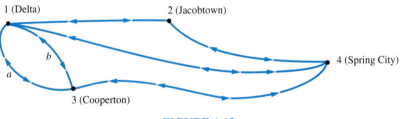

1 (Delta) 2 (Jacobtown) 4 (Spring City) 3 (Cooperton) b a

FIGURE 1.12

The matrix representation of a communication graph is often called a *communication matrix*. In a **communication matrix** $M = (m_{ij})$, the value of each entry m_{ij} is the number of direct-communication links (one-way) from individual A_i to individual A_j, in that direction. Otherwise, the value of m_{ij} is zero, including, for convenience, the values of m_{ii} (even though one can surely communicate with oneself). The communication matrix for Example 5 is

$$
M = \begin{array}{c} \\ \\ \\ \\ \end{array}
\overset{\begin{array}{cccc} D & J & C & S \end{array}}{\begin{bmatrix} 0 & 1 & 2 & 1 \\ 1 & 0 & 0 & 1 \\ 2 & 0 & 0 & 1 \\ 1 & 1 & 1 & 0 \end{bmatrix}}
\begin{array}{l} \text{Delta} \\ \text{Jacobtown} \\ \text{Cooperton} \\ \text{Spring City} \end{array}
$$

As in the case of dominance described above, the square of the matrix M gives two-stage communication links; that is, M^2 tells us the number of communication links between two individuals (towns) that pass through exactly one other individual (town). Thus, for the link of Example 5 the matrix

$$
M^2 = \begin{bmatrix} 6 & 1 & 1 & 3 \\ 1 & 2 & 3 & 1 \\ 1 & 3 & 5 & 2 \\ 3 & 1 & 2 & 3 \end{bmatrix} = (m'_{ij})
$$

indicates that Cooperton has three communication links with Jacobtown because $m'_{32} = 3$. These links are:

$$\text{Cooperton} \rightarrow \text{Delta (a)} \rightarrow \text{Jacobtown,}$$
$$\text{Cooperton} \rightarrow \text{Delta (b)} \rightarrow \text{Jacobtown,}$$
$$\text{Cooperton} \rightarrow \text{Spring City} \rightarrow \text{Jacobtown.}$$

How should the fact that $m'_{11} = 6$ be interpreted?

The cube M^3 of the communication matrix would indicate the number of communication links between individuals (towns) which pass through exactly two intermediary individuals (towns), and so on. ∎

A communication network in which every link is two-way is called a **perfect communication network.** Example 5 is an illustration of a perfect communication network. The matrix representation for a perfect communication network is always symmetric (why?). See the matrix M of Example 5.

We need an additional graph theory concept.

> **Definition** An oriented graph is called **connected** if between any two distinct vertices there is a path consisting of a positive integral number of edges. If such a path does not exist between every pair of vertices, the graph is **disconnected.**

To be effective, any communication network should have a connected graph. Any individual in a communication graph is called a **liaison** if the removal of that person from the graph (together with any connecting edges) disconnects the graph.

Example 6 Consider the perfect communication network in a company's executive council depicted in Figure 1.13. Note that individual C is a liaison, since removal of C from the network disconnects it. How about individual B?

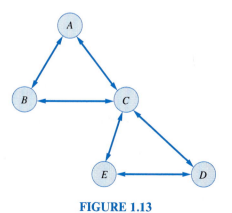

FIGURE 1.13

The matrix representation for the graph in Figure 1.13 is

$$M = \begin{array}{c} \\ A \\ B \\ C \\ D \\ E \end{array} \begin{array}{c} \begin{array}{ccccc} A & B & C & D & E \end{array} \\ \begin{bmatrix} 0 & 1 & 1 & 0 & 0 \\ 1 & 0 & 1 & 0 & 0 \\ 1 & 1 & 0 & 1 & 1 \\ 0 & 0 & 1 & 0 & 1 \\ 0 & 0 & 1 & 1 & 0 \end{bmatrix} \end{array}. \quad \blacksquare$$

It is obvious that in general (not necessarily perfect) communication networks, different individuals could play roles of varying importance. Some individuals may only send messages, while others only receive messages, and still others may both send and receive messages. To analyze these situations we shall first examine subsets of communication networks. We shall be interested in dividing a communication network into subsystems, so that within the given subsystem *every individual has a two-way communication link,* not necessarily direct, *with every other individual in the subsystem.* We also want each of the subsystems that has this property to be as large as possible. We shall discover that such subsystems form *equivalence classes* within the large network in the sense that every individual in the network belongs to exactly one class, and that between classes there is at most a one-way communication link. This is a particular manifestation of an important mathematical idea, that of an **equivalence relation** defined on a particular set.

> **Definition**   Let $\mathscr{C} = \{A_1, A_2, \ldots, A_n\}$ be a communication network with individuals A_i, $i = 1, 2, \ldots, n$. We define a relation $\approx$ on $\mathscr{C}$ as follows: $A_i \approx A_j$ if, and only if, there is a two-way communication link (not necessarily in one step) between A_i and A_j, or else $i = j$.

We leave it as Exercise 34 for you to show that the relation $\approx$ has the following properties:

Reflexive: $A_i \approx A_i$ for every A_i is the set $\mathscr{C}$.

Symmetric: If $A_i \approx A_j$, then $A_j \approx A_i$.

Transitive: If $A_i \approx A_j$ and $A_j \approx A_k$, then $A_i \approx A_k$.

> **Definition**   Any relation defined on a given set that has these three properties; that is, it is reflexive, symmetric, and transitive, is called an **equivalence relation** on that set.

Here $\approx$ is an equivalence relation on the communication network $\mathscr{C}$.

It is a property common to all equivalence relations that they partition the set on which they are defined into disjoint subsets called **equivalence classes** (modulo the given relation). Such is the case with our communication network $\mathscr{C}$. We demonstrate this explicitly for this case. The proof in the general case is entirely analogous. (See Supplementary Exercises and Projects: Project 2.)

Let $[A_i]$ denote the collection of all those individuals A_j in $\mathscr{C}$ that have a two-way communication link with A_i, that is, $[A_i] = \{A_j | A_j \rightleftharpoons A_i\}$. Now since $A_i \rightleftharpoons A_i$, the class $[A_i]$ is not empty. Each member of $\mathscr{C}$ belongs to some such class, because $\rightleftharpoons$ is reflexive. Suppose first that $A_k \rightleftharpoons A_i$. If $A_s \in [A_k]$, then $A_s \rightleftharpoons A_k$; so by the transitivity of the relation $\rightleftharpoons$, we also know that $A_s \rightleftharpoons A_i$. From this we conclude that $A_s \in [A_i]$, so that $[A_k] \subseteq [A_i]$. But, on the other hand, if $A_r \in [A_i]$, then $A_r \rightleftharpoons A_i$, and $A_i \rightleftharpoons A_k$ (because $\rightleftharpoons$ is symmetric). Thus we conclude that $A_r \rightleftharpoons A_k$, so $A_r \in [A_k]$, and $[A_i] \subseteq [A_k]$. Therefore, $[A_k] = [A_i]$.

Suppose now that $[A_i]$ and $[A_j]$ are two of these communication subclasses, and that individual A has a two-way communication link with both A_i and A_j; that is, suppose that $A \in [A_i] \cap [A_j] \neq \varnothing$. Then $A \rightleftharpoons A_i$ and $A \rightleftharpoons A_j$. But then we have just demonstrated that $[A] = [A_i]$ and $[A] = [A_j]$, so if $[A_i]$ and $[A_j]$ are not disjoint, they are equal. We have proved, then, that

> The equivalence relation $\rightleftharpoons$ partitions the communication network $\mathscr{C}$ into disjoint classes in which each individual has a two-way communication link with every other individual in its class.

To complete our analysis of the communication network $\mathscr{C}$, we must show that there is at most a one-way link between distinct equivalence classes $[A]$ and $[B]$. To see this, suppose there are two classes $[A_i]$ and $[A_j]$ such that $[A_i] \neq [A_j]$. Then, clearly, $[A_i] \cap [A_j] = \varnothing$. Suppose that there is some individual X in $[A_i]$ who can send a message to some individual Y in $[A_j]$, and also that there is some individual U in $[A_j]$ who can send a message to some individual V in $[A_i]$; thus, there is a two-way link between $[A_i]$ and $[A_j]$. Since X and V are in $[A_i]$, they have a two-way link between them; $X \rightleftharpoons V$. Similarly for Y and U in $[A_j]$; $Y \rightleftharpoons U$. Therefore, the following communication link would exist:

$$X \rightarrow Y \rightarrow U \rightarrow V \rightarrow X,$$

which is a *two-way* communication link between Y and X. Hence, they are in the same class, contrary to the fact that $[A_i] \cap [A_j] = \varnothing$. Thus, no such link can exist, and at most, one-way communication between distinct classes is possible.

The value of splitting the communication network in a large organization with equivalence classes is fairly obvious, since once these classes are identified, the flow of information in the organization can be optimized. (What disadvantages do you see in doing this?) Let us illustrate the advantages with a small-scale example.

Example 7 The network of intelligence agents in the (fictitious country of) Platonia has six key members who need to be given certain information. The agency's director does not know where they all are. Instead, she has information of the following form: Hanks knows where Jacobson is; Kimball knows where Scarlati is; and so on. The director sets up the following communication matrix M, which summarizes the information available.

$$
M = \begin{array}{c} \\ \text{Hanks} \\ \text{Jacobson} \\ \text{Kimball} \\ \text{Scarlati} \\ \text{Russell} \\ \text{Le Bouf} \end{array}
\begin{array}{c} \begin{array}{cccccc} H & J & K & S & R & L \end{array} \\
\left[\begin{array}{cccccc}
0 & 1 & 0 & 0 & 0 & 0 \\
0 & 0 & 0 & 0 & 1 & 0 \\
0 & 0 & 0 & 1 & 0 & 0 \\
1 & 0 & 0 & 0 & 0 & 1 \\
0 & 0 & 0 & 0 & 0 & 0 \\
0 & 0 & 1 & 0 & 0 & 0
\end{array} \right] \end{array}
$$

Here, $m_{ij} = 1$ means that $i \neq j$ and person i knows where person j is and can send him or her a message.

Next, we try to identify all those individuals to whom each of the individuals can send messages through channels. This information is obtained from the matrices M, M^2, M^3, ..., that detail the one-link, two-link, three-link, etc. communication connections. This list is constructed by adding to it in successive steps until no new individual can be added to the list, or until at least one individual can, in successive stages, send a message to *all* the other agents. We also presume that an individual is able to communicate with himself. The lists and the stages at which contact can be made are given in the following table.

Agent	Stage 0 or 1	Stages 0–2	Stages 0–3	Final
Hanks	*H, J*	*H, J, R*	*H, J, R*	*H, J, R*
Jacobson	*J, R,*	*J, R,*	*J, R*	*J, R*
Kimball	*K, S*	*H, K, S, L*	*H, J, K, S, L*	*H, J, K, S, R, L*
Scarlati	*H, S, L*	*H, J, K, S, L*	*H, J, K, S, R, L*	*H, J, K, S, R, L*
Russell	*R*	*R*	*R*	*R*
Le Bouf	*K, L*	*K, S, L*	*H, K, S, L*	*H, J, K, S, R, L*

The information in the table is obtained from the following matrices.

$$
M^2 = \begin{array}{c} \\ H \\ J \\ K \\ S \\ R \\ L \end{array} \begin{array}{cccccc} H & J & K & S & R & L \\ \left[\begin{array}{cccccc} 0 & 0 & 0 & 0 & 1 & 0 \\ 0 & 0 & 0 & 0 & 0 & 0 \\ 1 & 0 & 0 & 0 & 0 & 1 \\ 0 & 1 & 1 & 0 & 0 & 0 \\ 0 & 0 & 0 & 0 & 0 & 0 \\ 0 & 0 & 0 & 1 & 0 & 0 \end{array}\right] \end{array}, \quad
M^3 = \begin{array}{c} \\ H \\ J \\ K \\ S \\ R \\ L \end{array} \begin{array}{cccccc} H & J & K & S & R & L \\ \left[\begin{array}{cccccc} 0 & 0 & 0 & 0 & 0 & 0 \\ 0 & 0 & 0 & 0 & 0 & 0 \\ 0 & 1 & 1 & 0 & 0 & 0 \\ 0 & 0 & 0 & 1 & 1 & 0 \\ 0 & 0 & 0 & 0 & 0 & 0 \\ 1 & 0 & 0 & 0 & 0 & 1 \end{array}\right] \end{array}.
$$

After three stages we see that a message sent to Scarlati will be relayed to everyone, so it is only necessary to contact that agent to communicate with all of them. After four stages a message sent to Kimball will reach everyone, and after five stages a message sent to Le Bouf will have been relayed to everyone. (Verify this.)

Let us continue our analysis further by assuming that we need a list of those from whom the agents can *receive* messages and, finally, the equivalence classes of this network. The *receive-from lists* can be computed by again constructing a communication matrix, or by merely reading the *send list* backward. That is, if person A_j is on A_i's send list, put A_i on A_j's receive-from list. The equivalence classes are then the intersection of the two lists (receive and send). This is given in the following table.

Agent	Send to	Receive from	Equivalence class
Hanks	*H, J, R*	*H, K, S, L*	$\{H\}$
Jacobson	*J, R*	*H, J, K, S, L*	$\{J\}$
Kimball	*H, J, K, S, R, L*	*K, S, L*	$\{K, S, L\}$
Scarlati	*H, J, K, S, R, L*	*K, S, L*	$\{K, S, L\}$
Russell	*R*	*H, J, K, S, R, L*	$\{R\}$
Le Bouf	*H, J, K, S, R, L*	*K, S, L*	$\{K, S, L\}$

Let us use all of these data to sketch the oriented graph of the network of the equivalence classes as Figure 1.14. This graph, which is a digraph, gives a fairly accurate picture of the intercommunication within this intelligence network. ∎

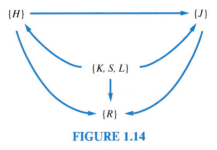

FIGURE 1.14

Vocabulary

The vocabulary in this section is rather extensive, and we have merely indicated some of the ideas. Our main motivation was to demonstrate a readily accessible application of the linear algebra ideas developed to this point. Each of the concepts described by the terms in the following list was discussed. You should review these ideas.

graph
vertex (vertices)
oriented graph
matrix representation
n-stage dominance
communication network
liaison
reflexive relation
equivalence relation
equivalence classes
edge(s)

loop
digraph
consensus leader
power
communication matrix
connected graph
disconnected graph
symmetric relation
transitive relation
partition of a set
perfect communication network

Exercises 1.7

In Exercises 1–6, decide if the given matrix could be a matrix representation of the digraph of a situation in which there is dominance.

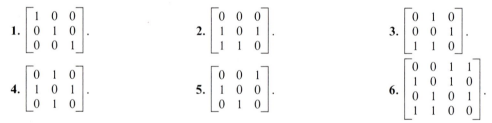

1. $\begin{bmatrix} 1 & 0 & 0 \\ 0 & 1 & 0 \\ 0 & 0 & 1 \end{bmatrix}$.

2. $\begin{bmatrix} 0 & 0 & 0 \\ 1 & 0 & 1 \\ 1 & 1 & 0 \end{bmatrix}$.

3. $\begin{bmatrix} 0 & 1 & 0 \\ 0 & 0 & 1 \\ 1 & 1 & 0 \end{bmatrix}$.

4. $\begin{bmatrix} 0 & 1 & 0 \\ 1 & 0 & 1 \\ 0 & 1 & 0 \end{bmatrix}$.

5. $\begin{bmatrix} 0 & 0 & 1 \\ 1 & 0 & 0 \\ 0 & 1 & 0 \end{bmatrix}$.

6. $\begin{bmatrix} 0 & 0 & 1 & 1 \\ 1 & 0 & 1 & 0 \\ 0 & 1 & 0 & 1 \\ 1 & 1 & 0 & 0 \end{bmatrix}$.

7. Given the matrix of Exercise 1, find all of the two-stage dominances and the total dominance in one and two stages (the power) for each individual. Is there a leader?

8. Given the matrix of Exercise 2, find all of the two-stage dominances and the total dominance in one and two stages (the power) for each individual. Is there a leader?

9. Given the matrix in Exercise 3, find all of the two-stage dominances and the total dominance in one and two stages (the power) for each individual. Is there a leader?

10. Given the matrix of Exercise 4, find all of the two-stage dominances and the total dominance in one and two stages (the power) for each individual. Is there a leader?

11. Given the matrix of Exercise 5, find all of the two-stage dominances and the total dominance in one and two stages (the power) for each individual. Is there a leader?

12. Given the matrix of Exercise 6, find all of the two-stage dominances and the total dominance in one and two stages (the power) for each individual. Is there a leader?

In Exercises 13–16, consider the dominance digraph in Figure 1.15.

13. Write the matrix representation of this digraph.

14. Is there a consensus leader? Who?

15. Find the sum of one- and two-stage dominances (power).

16. Is there a leader? Who?

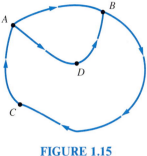

FIGURE 1.15

---

17. A tennis match played among four people ends with the following result: Kelly has beaten both O'Brien and Thompson; Thompson has beaten O'Brien; Carter has beaten Kelly, O'Brien, and Thompson. Find the power of each player and rank them accordingly.

18. Given the following statements, draw a reasonable digraph with vertices representing each statement. Then compute the power of each vertex.
   a. Tensions increase in the Middle East.
   b. Oil prices increase.
   c. Money spent for geothermal exploration increases.
   d. Coal-generated electricity increases.
   e. Air pollution increases.
   f. The hole in the ozone layer enlarges.

19. Sketch the digraph of the football games of Example 4 and compare the paths through two vertices with the entries μ_{ij} of the matrix M^2, given the matrix M of that example.

20. Sketch a digraph corresponding to the following communication matrix:

$$M = \begin{bmatrix} 0 & 1 & 0 & 1 \\ 1 & 0 & 1 & 0 \\ 0 & 1 & 0 & 1 \\ 1 & 0 & 1 & 0 \end{bmatrix}.$$

21. Is the network of Exercise 20 a perfect communication network? Why?

22. Construct the communication matrix for the business communication network described by

the graph in Figure 1.16. Is this a perfect communication network?

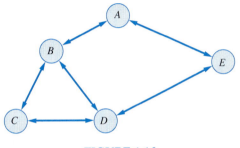

FIGURE 1.16

23. Identify any liaison individuals in the network of Exercise 22.

24. Are there any liaison individuals in the network of Exercise 20?

25. Consider five neighbors who gossip with each other. The incidence matrix representing this situation is as given below. Sketch the communication network. If the fourth person moves out of the neighborhood, is it still possible for a rumor to spread among the remaining neighbors?

$$G = \begin{bmatrix} 0 & 1 & 1 & 0 & 1 \\ 1 & 0 & 0 & 1 & 0 \\ 0 & 0 & 1 & 0 & 1 \\ 1 & 1 & 1 & 0 & 1 \\ 1 & 0 & 1 & 1 & 0 \end{bmatrix}.$$

26. Sketch the one-way communication digraph for Example 7. Is this a connected graph? Are there any liaison individuals?

Exercises 27–33 refer to the oriented graph in Figure 1.17 below. This is a diagram of possible commuter helicopter routes in the San Francisco Bay Area. See Figure 1.9 in this section for the names of the stops.

27. Write the matrix representation for this graph.

28. Write the matrix representation for the *fly-to* graph.

29. Write the matrix representation for the *fly-from* graph.

30. How many stages are necessary to fly *from* (4) *to* (6)?

31. How many stages are necessary to fly *from* (8) *to* (5)?

32. Is the graph of Figure 1.17 connected?

33. Identify any liaison cities.

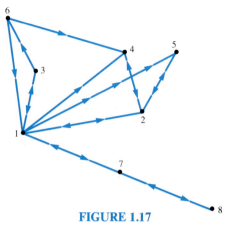

FIGURE 1.17

---

34. Demonstrate that the relation ⇌ defined in this section for the general communication network 𝒞, discussed in this section after Example 6, is indeed reflexive, symmetric, and transitive.

35. Show that if 𝒞 is a communication network of n individuals, and if individual A_i can communicate at all with individual A_j, then this can be done in not more than $n - 1$ stages.

36. In an interest survey given to student Tom Jones, the concept of *paired comparison* was used. It was discovered that he preferred playing golf to fishing, watching sports on TV, playing chess, or boating. He prefers playing basketball to boating, TV sports, chess, golf, or fishing. He prefers boating to TV sports, and he prefers TV sports to playing chess. He prefers to play chess rather than to go boating. What is his favorite pastime? What is the power of each pastime?

37. Use a digraph to model the authority structure of your family. Compare the power of each member.

38. Use a digraph to model the authority structure of your hometown.

CHAPTER 1 SUMMARY

In this chapter, we discussed some matrix methods of solving systems of linear equations. We also saw that some of these systems may have a unique solution, while others may have more than one solution or no solution at all. Matrix algebra was introduced, involving the two fairly natural operations of matrix addition and the multiplication by a scalar. The important, though not obvious, concept of the product of an $m \times n$ matrix A and an $n \times p$ matrix B to form the $m \times p$ matrix product AB was motivated and discussed. We also noted that the product BA *need not* be possible, even if AB is defined. Further, we saw that even when both products do exist, it is unlikely that $AB = BA$. The basic properties of this matrix algebra were studied. Many, *but by no means all* of these basic laws are the same as for the real and complex number algebra with which you are familiar. A notable exception, in addition to the noncommutativity of multiplication, $AB \neq BA$, is that even though $A \neq 0$ and $B \neq 0$, AB may well be zero. The transpose A^T of a matrix A was also defined, as well as the inverse A^{-1}, if it exists. We saw that often a nonzero matrix A fails to have an inverse. But when A^{-1} does exist, we learned a method for finding it.

This method used the concept of elementary matrices by means of which we (sometimes) were able to reduce an $n \times n$ matrix A to the identity matrix I_n. That is, we were able to find elementary matrices $E_1, E_2, \ldots, E_m$ such that $(E_m E_{m-1} \cdots E_2 E_1)A = I_n$. When this happened to be the case, of course, A^{-1} exists and, in fact, $A^{-1} = (E_m E_{m-1} \cdots E_2 E_1)$. This chapter concluded with techniques for partitioning a large matrix into blocks and working with the smaller blocks, as well as the idea of the *LU* decomposition of a matrix. Finally, we applied the concepts of the chapter to networks and graphs.

CHAPTER 1 REVIEW EXERCISES

1. Write the augmented matrix for the system of equations

$$\begin{aligned}
x_1 - x_2 + x_3 &= 1, \\
x_1 + 2x_2 - x_3 &= 2, \\
2x_1 - 3x_2 + 3x_3 &= 1.
\end{aligned}$$

Reduce the matrix to its *reduced row echelon form* and solve the system, if it has a solution.

2. Use a Gauss-Jordan reduction of the approximate matrix to find all of the solutions to the following system of equations, if any exist.

$$\begin{aligned}
-x_1 - 3x_2 + x_3 &= 1, \\
-2x_1 + x_2 - x_3 &= 2, \\
-x_1 + 4x_2 - 2x_3 &= 1, \\
-5x_1 - 8x_2 + 2x_3 &= 5.
\end{aligned}$$

3. The homogeneous system of equations

$$\begin{aligned}
(1-a)x_1 + x_2 - a)x_3 &= 0, \\
ax_2 + 2x_3 &= 0, \\
(1+a)x_3 &= 0,
\end{aligned}$$

does not have a unique solution. Find all possible values for a.

4. For each of the values of a you found in Exercise 3, find *all* of the solutions of the homogeneous system of equations.

5. For what values of a does the homogeneous system of equations in Exercise 3 have a *unique* solution. What is this unique solution?

6. Let

$$A = \begin{bmatrix} -5 & 2 \\ 3 & 4 \end{bmatrix} \quad \text{and} \quad B = \begin{bmatrix} 7 & 0 \\ 5 & -1 \end{bmatrix}.$$

Find $A + B$, $A - B$, AB, BA, and $3A$. Are AB and BA equal?

7. Let

$$A = \begin{bmatrix} 1 & 0 & 0 \\ 2 & 1 & 0 \\ -1 & 3 & 1 \end{bmatrix}.$$

Does A^{-1} exist? If so, find it.

8. Let

$$A = \begin{bmatrix} 0 & 0 & 0 \\ 2 & 0 & 0 \\ -1 & 3 & 0 \end{bmatrix}.$$

Find A^2 and A^3.
Does A^{-1} exist? Give reasons.

9. Given the matrix

$$A = \begin{bmatrix} 1 & 2 & 3 \\ 4 & 5 & 6 \\ 7 & 8 & 9 \end{bmatrix},$$

use the matrix inversion algorithm of Section 1.5 to determine whether A^{-1} exists, and if it does, find it.

10. Let A be an $n \times n$ matrix with real number entries. Suppose that $A^2 = I_n$. Does A^{-1} exist? If so, find A^{-1} (in terms of A).

11. Suppose that A is any $n \times n$ nonsingular matrix with real number entries. Show that its transpose A^T is also nonsingular.

12. Suppose that A is any $n \times n$ nilpotent matrix (i.e., $A^k = 0$, for some positive integer $k \geq 1$). Prove that A^T is also nilpotent.

CHAPTER 1 SUPPLEMENTARY EXERCISES AND PROJECTS

In Exercises 1 and 2, suppose we have the two $m \times n$ systems of linear equations

(a)

$$
\begin{aligned}
a_{11}x_1 + a_{12}x_2 + \cdots + a_{1n}x_n &= b_1, \\
a_{21}x_1 + a_{22}x_2 + \cdots + a_{2n}x_n &= b_2, \\
&\ \vdots \\
a_{i1}x_1 + a_{i2}x_2 + \cdots + a_{in}x_n &= b_i, \\
&\ \vdots \\
a_{j1}x_1 + a_{j2}x_2 + \cdots + a_{jn}x_n &= b_j, \\
&\ \vdots \\
a_{m1}x_1 + a_{m2}x_2 + \cdots + a_{mn}x_n &= b_m,
\end{aligned}
$$

and

(b)

$$
\begin{aligned}
a_{11}x_1 + a_{12}x_2 + \cdots + a_{1n}x_n &= c_1, \\
a_{21}x_1 + a_{22}x_2 + \cdots + a_{2n}x_n &= c_2, \\
&\ \vdots \\
a_{i1}x_1 + a_{i2}x_2 + \cdots + a_{in}x_n &= c_i, \\
&\ \vdots \\
a_{j1}x_1 + a_{j2}x_2 + \cdots + a_{jn}x_n &= c_j, \\
&\ \vdots \\
a_{m1}x_1 + a_{m2}x_2 + \cdots + a_{mn}x_n &= c_m.
\end{aligned}
$$

Suppose also that the first system (a) has the solution $(r_1, r_2, \ldots r_n)$ and the second system (b) has the solution $(s_1, s_2, \ldots s_n)$.

1. Show that for any real number k the n-tuple $(kr_1, kr_2, \ldots kr_n)$ is a solution to the system

$$
\begin{aligned}
a_{11}x_1 + a_{12}x_2 + \cdots + a_{1n}x_n &= kb_1, \\
a_{21}x_1 + a_{22}x_2 + \cdots + a_{2n}x_n &= kb_2, \\
&\ \vdots \\
a_{i1}x_1 + a_{i2}x_2 + \cdots + a_{in}x_n &= kb_i, \\
&\ \vdots \\
a_{j1}x_1 + a_{j2}x_2 + \cdots + a_{jn}x_n &= kb_j, \\
&\ \vdots \\
a_{m1}x_1 + a_{m2}x_2 + \cdots + a_{mn}x_n &= kb_m.
\end{aligned}
$$

2. Show that the following system has a solution, and find it.

$$
\begin{aligned}
a_{11}x_1 + a_{12}x_2 + \cdots + a_{1n}x_n &= b_1 + c_1, \\
a_{21}x_1 + a_{22}x_2 + \cdots + a_{2n}x_n &= b_2 + c_2, \\
&\ \vdots \\
a_{i1}x_1 + a_{i2}x_2 + \cdots + a_{in}x_n &= b_i + c_i, \\
&\ \vdots \\
a_{j1}x_1 + a_{j2}x_2 + \cdots + a_{jn}x_n &= b_j + c_j, \\
&\ \vdots \\
a_{m1}x_1 + a_{m2}x_2 + \cdots + a_{mn}x_n &= b_m + c_m.
\end{aligned}
$$

Exercises 3–7 refer to an $m \times n$ system of linear equations [such as (a) above] that is called a **reduced linear system** when it has the following characteristics:

(i) The equations are arranged so that each of the first r ($0 \leq r \leq m$) equations has at least one nonzero coefficient, but the last $m - r$ equations are trivial; i.e.,
$0x_1 + 0x_2 + \cdots + 0x_n = 0$.

(ii) For $i = 1, 2, \ldots, r$, the leading coefficient in the ith equation is a one. The variable x_t having that $a_{it} = 1$ in that equation is called the **ith basic variable** of the system.

(iii) If $i < j$ and x_t and x_s are the ith and jth basic variables, respectively, then $t < s$.

(iv) The ith basic x_t variable has a nonzero coefficient only in the ith equation; that is, $a_{jt} = 0$ when $j \neq i$.

3. Write an example of a 7×5 reduced linear system with $r = 4$.

4. Prove that every $m \times n$ reduced linear system is consistent; i.e., it has at least one solution.

5. Prove than an $m \times n$ reduced linear system with r nontrivial equations has a unique solution if $r = n$.

6. Prove that an $m \times n$ reduced linear system with r nontrivial equations has infinitely many solutions if $r < n$.

7. Prove that an $m \times n$ reduced linear system must have infinitely many solutions if $m < n$.

8. Show that the effect of each on the elementary row operations on a matrix can be reversed by an operation of the same type.

9. Prove that a Type I elementary row operation of an $m \times n$ matrix A is accomplished by premultiplying A by a Type I elementary matrix E_I; $(E_I A)$.

10. Prove that a Type II elementary row operation on an $m \times n$ matrix A is accomplished by premultiplying A by a Type II elementary matrix E_{II}; $(E_{II} A)$.

11. Prove that a Type III elementary row operation on an $m \times n$ matrix A is accomplished by premultiplying A by a Type III elementary matrix E_{III}; $(E_{III} A)$,

12. Prove that the formal definition of row equivalence of matrices given in Section 1.5 is logically equivalent to the definition given in Section 1.1. That is, prove that if A and B are row equivalent using one definition, they are row equivalent using the other and conversely. (Use Exercises 9–11.)

13. Prove that two consistent $m \times n$ systems of linear equations are equivalent (have the same set of solutions) if, and only if, their augmented matrices have the same reduced row echelon form.

Exercises 14–17 refer to the upper triangular matrix $A = \begin{bmatrix} a & b & c \\ 0 & a & d \\ 0 & 0 & a \end{bmatrix}$.

14. Find all real numbers a, if any, for which the matrix A is invertible. Find A^{-1} if it exists.

15. For the given matrix A, find all values of a, if any, for which A is a nilpotent matrix. In each case find the index of nilpotency k $(A^k = 0$, but $A^{k-1} \neq 0)$.

16. Find all matrices A in the given form that are both idempotent and nilpotent.

17. Find all matrices A in the given form that are both idempotent and invertible. Explain.

18. Prove Theorem 1.8 (page 54) by showing that for any $n \times n$ matrix A, $AB = I$ if, and only if, $BA = I$. HINTS: Note *first* that it is enough to show that if $AB = I$, then $BA = I$. The converse follows by interchanging the roles of A and B. *Second,* note that if $AB = I$, then the system of equations $A\mathbf{x} = \mathbf{c}$ has the solution $\mathbf{x} = B\mathbf{c}$ for every $n \times 1$ matrix $\mathbf{c}$. *Third,* let $U = E_1 E_2 \ldots E_k A$ be the reduced row echelon form of A, with the E_i being elementary matrices. If U is not the identity matrix, show that there is a contradiction. Conclude the proof by using the corollary to Theorem 1.7.

Project 1. (Theoretical) Skew-Symmetric Matrices

> **Definition**   A matrix A is called **skew-symmetric** if $A^T = -A$.

a. Prove that a skew-symmetric matrix must be square ($n \times n$).

b. If $A = (a_{ij})$ is a skew-symmetric matrix, prove that each diagonal element $a_{ii} = 0$.

c. Show that if A is both symmetric and skew-symmetric, then $A = O$, the zero matrix.

d. If A is any $n \times n$ matrix, show that the matrix $A - A^T$ is skew-symmetric.

e. If A and B are symmetric (see Exercises 19–21 in Section 1.4), show that $A + B$ is symmetric. What if A and B are skew-symmetric?

f. If A and B are symmetric (see Exercises 19–21 in Section 1.4), show that AB is symmetric if, and only if, A and B commute ($AB = BA$). What if A and B are skew-symmetric?

g. Show that every $n \times n$ matrix A is the sum of a symmetric and a skew symmetric matrix. HINT: Look at the two matrices $\frac{1}{2}(A + A^T)$ and $\frac{1}{2}(A - A^T)$.

Project 2. (Theoretical) Equivalence Classes Partition a Set

Let S be an arbitrary set and let $\sim$ be an equivalence relation on S; that is, $\sim$ is a reflexive, symmetric, and transitive relation on the set S (see Section 1.7). For each $a \in S$, define the set

$$[a] = \{b \in S \mid b \sim a\}.$$

That is, let [a] be the subset of S containing all those elements that are equivalent to a.

a. Prove that $[a]$ is always nonempty. It is the equivalence class (modulo $\sim$) to which the element a belongs.

b. Prove that if $b \sim a$, then $[b] \subseteq [a]$; therefore, $[b] = [a]$.

c. Prove that if $c \in [a] \cap [b]$, then $[c] = [a] = [b]$.

d. Prove that either $[a] \cap [b] = \varnothing$ or $[a] = [b]$.

e. Note that the collection of all distinct equivalence classes $\{[a_i]\}$ forms a **partition** of the set S in the sense that S is the union of these distinct disjoint classes.

f. Let S be the set $\mathbb{Z}$ of all integers $\{0, \pm 1, \pm 2, \ldots\}$. Define $n \sim m$ if, and only if, the number $m - n$ is a multiple of 5. Show that this is an equivalence relation. What are the integers in each equivalence class? How many classes are there?

Project 3. (Numerical) Inverting a Large Block Matrix

Let A be the 16×16 block matrix

$$\begin{bmatrix} A_{11} & (1.1)I_4 & O & O \\ O & A_{22} & (2.1)I_4 & O \\ O & O & A_{33} & (-3.2)I_4 \\ O & O & O & A_{44} \end{bmatrix},$$

where each block is a 4 × 4 matrix with

$$A_{11} = \begin{bmatrix} 2+\varepsilon & 1.3 & -1.7 & 2.4 \\ 1.1 & -2.4 & 0.5 & -1.8 \\ 1.1 & 3.7 & -2.2 & 4+\varepsilon \\ 3.0 & 5.0 & -3.9 & 6.3 \end{bmatrix}, \quad A_{22} = \begin{bmatrix} 3.1 & 1.4 & -2.6 & 1.5 \\ -1.4 & 1-\varepsilon & 3.5 & -2.2 \\ 2.7 & 1.7 & 0.9 & -0.7 \\ \varepsilon & 1.4 & 4.4 & -2.9 \end{bmatrix},$$

$$A_{33} = \begin{bmatrix} 2\varepsilon & 0 & 0 & 0 \\ 0 & -2\varepsilon & 0 & 0 \\ 0 & 0 & 3\varepsilon & 0 \\ 0 & 0 & 0 & -3\varepsilon \end{bmatrix}, \quad and \quad A_{44} = \begin{bmatrix} 0 & 0 & 0 & 1 \\ 0 & 0 & 1 & 0 \\ 0 & 1 & 0 & 0 \\ 1 & 0 & 0 & 0 \end{bmatrix} = J_4.$$

Find A^{-1}, if it exists, for the following values of ε:

a. $0.1, -0.1, 0.2, -0.2, 0.5$, and -0.5. **b.** $0.7, -0.7, 0.9, -0.9, 1.0$, and -1.1.

c. $1.3, -1.3, 1.5, -1.5, 2.0, -2.1$, and -3.2.

Project 4. (Numerical) Quasi-magic Squares

An $n \times n$ matrix is a quasi-magic square if it has only nonnegative integer entries arranged so that each row and column add up to the same number s. For example, the following is a 4 × 4 quasi-magic square:

$$\begin{bmatrix} 16 & 2 & 3 & 13 \\ 5 & 11 & 10 & 8 \\ 9 & 7 & 6 & 12 \\ 4 & 14 & 15 & 1 \end{bmatrix}.$$

a. Verify that in the above example $s = 34$.

b. Prove that every $n \times n$ Type I elementary matrix is a quasi-magic square. What is s?

c. Write down a 3 × 3 quasi-magic square with $s = t$ for any given positive integer t.
HINT: Let the first row be $(1, 0, t - 1)$.

d. Suppose that A and B are $n \times n$ quasi-magic squares with sums s_A and s_B, respectively. Prove that $A + B$ is an $n \times n$ magic square with sum $s_A + s_B$.

e. Suppose that A and B are $n \times n$ quasi-magic squares with sums s_A and s_B, respectively. Prove that AB is an $n \times n$ magic square with sums $s_A s_B$.

f. An $n \times n$ quasi-magic square with $s = 1$ is sometimes called a **probability matrix.** Find the inverse of the following probability matrix, if it exists.

$$\begin{bmatrix} \dfrac{1}{2} & \dfrac{1}{2} & 0 \\ \dfrac{1}{4} & 0 & \dfrac{3}{4} \\ \dfrac{1}{4} & \dfrac{1}{2} & \dfrac{1}{4} \end{bmatrix}$$

g. A **magic square** is a quasi-magic square in which the diagonal sums are also equal to s. Write down a 3 × 3 and a 4 × 4 magic square.

2 DETERMINANTS

In this chapter we consider the notion of a *determinant,* a *number* that is associated with a square matrix. Possibly as early as 1100 B.C., the Chinese used a method to solve a 2×2 system of linear equations that is very much like the determinant method we describe here. Legend has it that the seventeenth century Japanese mathematician Seki Kowa also used this method in 1683. Ten years later, Leibniz gave a rule for solving linear systems that was much like that of the Chinese. His rule was amplified in 1750 by G. Cramer and is now known as *Cramer's rule.* While this method is very efficient for solving a small and consistent system of linear equations, its usefulness diminishes rapidly as the number of variables or the complexity of the system increases. Our purpose in studying the determinant is not for its use in solving equations; the methods of Chapter 1 are, indeed, better. Rather it is for the determinant's theoretical application to the study of matrices. In many cases in this chapter, we merely outline the proofs of the relevant facts of determinant theory rather than make the long digressions necessary to give more formal proofs. However, we supply the outline of the key features in the general arguments for most of what we discuss.

Even though determinants of square matrices are formally defined using the concept of a permutation, and this definition is often the most useful for proving theorems, nevertheless determinants are usually computed by applying the so-called *Laplace expansion* in combination with certain interesting properties that they possess. This sometimes greatly simplifies the actual computations involved and is often more efficient than to merely use the definition alone. We shall see how these properties and the Laplace expansion are related to our work in the previous chapter involving the row reduction of matrices. It should also be noted that the determinant can be used to determine whether or not a given square matrix has an inverse and is related to that inverse, when it exists. Indeed we develop a formula for the inverse of a nonsingular matrix that involves the determinant and is quite nice when applied to very small matrices.

In subsequent chapters we use determinants from time to time, although most of the work can be done without them. In Chapter 6 we see that the characteristic polynomial of a matrix is given as the determinant of a related matrix and that the determinant is always equal to the product of the eigenvalues of the matrix.

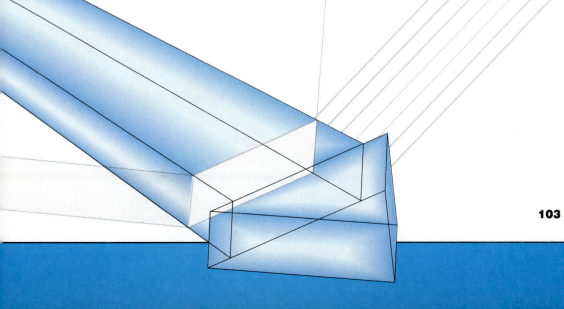

2.1 THE DETERMINANT FUNCTION AND CRAMER'S RULE

The general form of a system of two linear equations in the two variables x_1 and x_2 is

$$a_{11}x_1 + a_{12}x_2 = k_1,$$
$$a_{21}x_1 + a_{22}x_2 = k_2. \tag{2.1}$$

When this system is solved by the methods of Chapter 1, the augmented matrix has the following reduced row echelon form, provided, of course, that $a_{11}a_{22} - a_{12}a_{21} \neq 0$. (Please verify this for yourself.)

$$\begin{bmatrix} 1 & 0 & \dfrac{k_1 a_{22} - k_2 a_{12}}{a_{11}a_{22} - a_{12}a_{21}} \\ 0 & 1 & \dfrac{a_{11}k_2 - a_{21}k_1}{a_{11}a_{22} - a_{12}a_{21}} \end{bmatrix} \tag{2.2}$$

Thus, if
$$x_1 = \frac{k_1 a_{22} - k_2 a_{12}}{a_{11}a_{22} - a_{12}a_{21}}, \tag{2.3}$$

and
$$x_2 = \frac{a_{11}k_2 - a_{21}k_1}{a_{11}a_{22} - a_{12}a_{21}}, \tag{2.4}$$

the pair of numbers (x_1, x_2) is the solution to the system of equations (2.1).

You should confirm that these two numbers do indeed satisfy the system (2.1) **provided that the number** $\Delta = a_{11}a_{22} - a_{12}a_{21}$ **is not 0.** This number Δ involves the sum of products of the entries in the coefficient matrix for the system and is called the **determinant** of that matrix, and sometimes of the system. The determinant of a matrix is a real (complex) number written $\Delta = \det A$ or $\Delta = |A|$, associated with the matrix in a special way. We give a precise definition for the determinant below, but let us look at some examples first.

- If A is the 1×1 matrix $[a]$, then the determinant of A is the *number* a; $(\Delta = a)$.

Therefore, the determinant of the matrix [5] is the number 5.

- If A is the 2×2 matrix $A = \begin{bmatrix} a_{11} & a_{12} \\ a_{21} & a_{22} \end{bmatrix}$, then, as we noted above, $\det A = \Delta = a_{11}a_{22} - a_{12}a_{21}$.

Note that this number, det A, can be computed by the following diagonal scheme:

$$\det \begin{bmatrix} a_{11} & a_{12} \\ a_{21} & a_{22} \end{bmatrix} = a_{11}a_{22} - a_{12}a_{21}. \tag{2.5}$$

Example 1   The determinant $\Delta = $ of the 2×2 matrix $\begin{bmatrix} 2 & 3 \\ -1 & 2 \end{bmatrix}$ is the *number*

$$\Delta = (2)(2) - (-1)(3) = 7.$$

Consider again the formulas (2.3) and (2.4) for the solution to the system (2.1). Both denominators of the fractions are equal to the determinant of the coefficient matrix of the system; i.e., to (2.5). We observe that the numerators of the fractions in (2.3) and (2.4) can also be considered to be determinants of certain 2×2 matrices. That is,

$$\det \begin{bmatrix} k_1 & a_{12} \\ k_2 & a_{22} \end{bmatrix} = k_1 a_{22} - a_{12} k_2, \tag{2.6}$$

and

$$\det \begin{bmatrix} a_{11} & k_1 \\ a_{21} & k_2 \end{bmatrix} = a_{11} k_2 - k_1 a_{21} \tag{2.7}$$

are the respective numerators. ■

Thus the solution (x_1, x_2) can be cast in the following useful form known as *Cramer's rule*.

2 × 2 Cramer's Rule

The solution of the system

$$a_{11} x_1 + a_{12} x_2 = k_1,$$
$$a_{21} x_1 + a_{22} x_2 = k_2,$$

where $a_{11}, a_{12}, a_{21}, a_{22}, k_1$, and k_2 are all known real numbers, is given by

$$x_1 = \frac{\det \begin{bmatrix} k_1 & a_{12} \\ k_2 & a_{22} \end{bmatrix}}{\det \begin{bmatrix} a_{11} & a_{12} \\ a_{21} & a_{22} \end{bmatrix}}, \qquad x_2 = \frac{\det \begin{bmatrix} a_{11} & k_1 \\ a_{21} & k_2 \end{bmatrix}}{\det \begin{bmatrix} a_{11} & a_{12} \\ a_{21} & a_{22} \end{bmatrix}}, \tag{2.8}$$

provided that

$$\Delta = \det \begin{bmatrix} a_{11} & a_{12} \\ a_{21} & a_{22} \end{bmatrix} \neq 0.$$

Example 2   Use Cramer's rule to find the solution of the system

$$2x_1 - 3x_2 = 1,$$
$$5x_1 + \ x_2 = 11.$$

Here

$$\Delta = \det \begin{bmatrix} a_{11} & a_{12} \\ a_{21} & a_{22} \end{bmatrix} = \det \begin{bmatrix} 2 & -3 \\ 5 & 1 \end{bmatrix} = (2)(1) - (-3)(5) = 17 \neq 0.$$

Thus we may apply Cramer's rule to obtain

$$x_1 = \frac{\det \begin{bmatrix} 1 & -3 \\ 11 & 1 \end{bmatrix}}{\Delta} = \frac{(1)(1) - (-3)(11)}{17} = 2,$$

and

$$x_2 = \frac{\det \begin{bmatrix} 2 & 1 \\ 5 & 11 \end{bmatrix}}{\Delta} = \frac{(2)(11) - (1)(5)}{17} = 1,$$

so the solution to the given system is the pair (2, 1).

If in (2.8) $\Delta = \det \begin{bmatrix} a_{11} & a_{12} \\ a_{21} & a_{22} \end{bmatrix} = 0$, then the given system of equations will have either an *infinite* number of solutions, as in the following example:

$$\begin{aligned} x_1 + x_2 &= 1, \\ 2x_1 + 2x_2 &= 2; \end{aligned}$$

or the given system will have no solution at all, as in

$$\begin{aligned} x_1 + x_2 &= 1, \\ 2x_1 + 2x_2 &= 3. \end{aligned}$$

In the former case, the system is *consistent and dependent,* while in the latter case, the system is *inconsistent.* See Section 1.2. Of course, the formulas given in (2.8) become meaningless and cannot be used whenever $\Delta = 0$. ∎

Cramer's rule also works for larger systems. A Gauss-Jordan reduction of the augmented matrix for a general system of three linear equations in three variables will yield results similar to those in (2.2) but will contain more terms. These formulas, which you can work out if you wish, will involve the determinant of the 3×3 matrix

$$A = \begin{bmatrix} a_{11} & a_{12} & a_{13} \\ a_{21} & a_{22} & a_{23} \\ a_{31} & a_{32} & a_{33} \end{bmatrix}. \tag{2.9}$$

The **determinant** of this matrix A is also written $\det A = \Delta$, (or $|A| = \Delta$), where Δ is the number

$$\begin{aligned} \Delta = a_{11}a_{22}a_{33}a_{12}a_{23}a_{31} + a_{13}a_{21}a_{32} \\ - a_{13}a_{22}a_{31} - a_{12}a_{21}a_{33} - a_{11}a_{23}a_{32}. \end{aligned} \tag{2.10}$$

The values given for the determinants of both 2×2 and 3×3 matrices involve the sum of signed products of elements from the matrix. Such products are called *elementary products* when they are made up, as these are, of exactly one element from each row and each column of the matrix. Formally, we have the following definition.

> **Definition** If A is any $n \times n$ matrix, an **elementary product from** A is any product of n elements from A no two of which come from the same row or same column.

Some of the elementary products involved in the determinants we've seen have positive signs and some have negative signs preceding them. They are *signed elementary products.* To describe exactly which sign is to be used for a given elementary product requires a short digression into the theory of permutations.

Permutations—The Definition of Determinant

Let S be a finite set. Since S is finite, we may just as well think of it as the set of integers *one* through n:

$$S = \{1, 2, 3, 4, \ldots, n\}.$$

A **permutation of** S is any one-to-one mapping (function) β of S onto itself. If we denote the value $\beta(i)$, of the function β at i, by k_i for each i in S, then the *result* of the function β is often written in the form you saw in college algebra, that is, as $\{k_1, k_2, \ldots, k_n\}$.

In elementary texts, this list of functional values is often identified directly with the permutation β. We do this and write $\beta = \{k_1, k_2, \ldots, k_n\}$.

There are $n! = n(n-1) \cdots 2 \cdot 1$ distinct permutations of an n-element set. Why? (See Exercise 35.)

It is customary to denote the set consisting of *all* of the permutations of an n-element set by S_n. This set has $n!$ elements.

A permutation β in S_n is said to have an *inversion* whenever $\beta(i) > \beta(j)$ for $i < j$. Thus the permutation $\sigma = \{1, 3, 2, 4\}$ has one inversion because 3 precedes 2.

> **Definition** A permutation is called **even** or **odd** according to whether the number of inversions in its result $\{k_1, k_2, \ldots, k_n\}$ is an even or an odd integer, respectively. (Of course, zero is an even number, so no inversions makes a permutation even.)

Example 3 The permutation $\rho = \{2, 1, 4, 3\}$ is even because there are two inversions: $\rho(1) > \rho(2)$ and $\rho(3) > \rho(4)$. The permutation $\sigma = \{1, 3, 2, 4\}$ is odd because $\sigma(2) > \sigma(3)$.

What about $\tau = \{3, 4, 2, 1\}$? It is odd because there are five inversions. Work them out. ∎

Let's look at all the possible elementary products of the 3×3 matrix A of (2.9) and consider which column number permutations are even and which are odd. We list all $6 = 3!$ of the products in Table 2.1.

TABLE 2.1

Elementary product	Permutation	Odd or even
$a_{11}a_{22}a_{33}$	$\{1, 2, 3\}$	Even
$a_{12}a_{23}a_{31}$	$\{2, 3, 1\}$	Even
$a_{13}a_{21}a_{32}$	$\{3, 1, 2\}$	Even
$a_{13}a_{22}a_{31}$	$\{3, 2, 1\}$	Odd
$a_{12}a_{21}a_{33}$	$\{2, 1, 3\}$	Odd
$a_{11}a_{23}a_{32}$	$\{1, 3, 2\}$	Odd

We are now in a position to define the determinant of an $n \times n$ matrix. For a given permutation β of the set $S = \{1, 2, 3, 4 \ldots, n\}$ let

$$\text{sgn}\,(\beta) = \begin{cases} 1 & \text{when } \beta \text{ is even} \\ -1 & \text{when } \beta \text{ is odd.} \end{cases}$$

Notice that in the formulas (2.5) and (2.10) given for the determinants of 2×2 and 3×3 matrices, respectively, those elementary products whose column subscripts are *even* permutations have *plus* signs while those which are *odd* permutations have *minus* signs.

Given the $n \times n$ matrix

$$A = \begin{bmatrix} a_{11} & a_{12} & \cdots & a_{1n} \\ a_{21} & a_{22} & \cdots & a_{2n} \\ \cdot & \cdot & \cdots & \cdot \\ a_{n1} & a_{n2} & \cdots & a_{nn} \end{bmatrix}.$$

We define the determinant of A to be the *number* $\Delta = \det A = |A|$, given by

Definition $\det A = |A| = \displaystyle\sum_{\beta \varepsilon S_n} \text{sgn}(\beta) a_{1\beta(1)} a_{2\beta(2)} \cdots a_{n\beta(n)}.$

The elementary product $a_{1\beta(1)} a_{2\beta(2)} \cdots a_{n\beta(n)}$ with the plus or minus sign determined by $\text{sgn}\,(\beta)$ is called a **signed elementary product.**

Thus, the determinant of an $n \times n$ matrix A is the sum of all possible signed elementary products from A.

Example 4   As we have already discussed,

$$\det \begin{bmatrix} a_{11} & a_{12} \\ a_{21} & a_{22} \end{bmatrix} = a_{11}a_{22} - a_{12}a_{21}, \qquad (\{2, 1\} \text{ is odd}), \qquad (2! = 2).$$

And

$$\det \begin{bmatrix} a_{11} & a_{12} & a_{13} \\ a_{21} & a_{22} & a_{23} \\ a_{31} & a_{32} & a_{33} \end{bmatrix} = \begin{aligned} & a_{11}a_{22}a_{33} + a_{12}a_{23}a_{31} + a_{13}a_{21}a_{32} \\ & - a_{13}a_{22}a_{31} - a_{12}a_{21}a_{33} - a_{11}a_{23}a_{32}. \end{aligned}$$

Refer to Table 2.1 to see how the signed elementary products were calculated in the 3×3 case. ∎

The 4×4 matrix A below will have $4! = 24$ products in the expansion of its determinant. Why? In Exercise 16 we ask you to compute a table such as Table 2.1 for all of the elementary products from A. There will be 24 entries in that table. In Exercise 17 you are asked to write out the number det A, where

$$A = \begin{bmatrix} a_{11} & a_{12} & a_{13} & a_{14} \\ a_{21} & a_{22} & a_{23} & a_{24} \\ a_{31} & a_{32} & a_{33} & a_{34} \\ a_{41} & a_{42} & a_{43} & a_{44} \end{bmatrix}.$$

The theory of permutations is a basic idea in many branches of mathematics. In the study of finite algebraic groups permutations play a major role. For our purposes, however, we shall not spend any more time on this interesting subject. We merely need to make some observations about permutations as we go along. If you are interested, you are referred to the supplementary projects at the end of this chapter for additional properties.

In the case of a 3×3 matrix a diagonal scheme, such as that given in (2.5) for the 2×2 case, helps to compute the determinant.

$$\det A = \det \begin{bmatrix} a_{11} & a_{12} & a_{13} \\ a_{21} & a_{22} & a_{23} \\ a_{31} & a_{32} & a_{33} \end{bmatrix} = \begin{array}{ccccc} a_{11} & a_{12} & a_{13} & a_{11} & a_{12} \\ a_{21} & a_{22} & a_{23} & a_{21} & a_{22} \\ a_{31} & a_{32} & a_{33} & a_{31} & a_{32} \end{array} \quad \textbf{(2.11)}$$

Notice that we have recopied the first two columns of the matrix A at the right of this scheme. The products down the main diagonals are positive, whereas those up the "sinistral" diagonals are negative.

Example 5   Evaluate $\det \begin{bmatrix} 2 & -4 & -2 \\ -1 & 5 & -4 \\ 7 & 2 & -4 \end{bmatrix}$.

Using the diagonal scheme of (2.11), we recopy the first two columns of the matrix immediately to the right and obtain

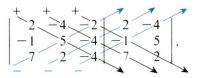

Thus the determinant is

$$(2)(5)(-4) + (-4)(-4)(7) + (-2)(-1)(2)$$
$$- (7)(5)(-2) - (2)(-4)(2) - (-4)(-1)(-4)$$
$$= -40 + 112 + 4 - (-70) - (-16) - (-16)$$
$$= 178. \quad \blacksquare$$

Example 6   Verify that the determinant of the 3×3 matrix A given in (2.9) can be computed in terms of three 2×2 determinants according to the following scheme:

$$\det A = a_{11} \det \begin{bmatrix} a_{22} & a_{23} \\ a_{32} & a_{33} \end{bmatrix} - a_{12} \det \begin{bmatrix} a_{21} & a_{23} \\ a_{31} & a_{33} \end{bmatrix}$$
$$+ a_{13} \det \begin{bmatrix} a_{21} & a_{22} \\ a_{31} & a_{32} \end{bmatrix}. \tag{2.12}$$

A direct calculation shows that the right side of equation (2.12) is

$$a_{11}(a_{22}a_{33} - a_{23}a_{32}) - a_{12}(a_{21}a_{33} - a_{23}a_{31}) + a_{13}(a_{21}a_{32} - a_{22}a_{31})$$

which, when multiplied out and rearranged, is precisely the expression in (2.10) for the determinant of A. $\blacksquare$

Caution   Unfortunately, no simple diagonal scheme, analogous to (2.5) and (2.11), exists for computing the determinant of a 4×4 *or larger* matrix.

However, in the next section we look at an expansion scheme for $n \times n$ matrices that is patterned after Example 6. This method is much easier to apply than to use the definition directly to compute the determinant of a matrix.

We conclude this section by extending Cramer's rule to the 3×3 case, since we know how to compute easily the determinant of a 3×3 matrix.

3×3 *Cramer's Rule*

For a given system of three linear equations

$$a_{11}x_1 + a_{12}x_2 + a_{13}x_3 = b_1,$$
$$a_{21}x_1 + a_{22}x_2 + a_{23}x_3 = b_2, \tag{2.13}$$
$$a_{13}x_1 + a_{32}x_2 + a_{33}x_3 = b_3,$$

the matrix of coefficients is

$$A = \begin{bmatrix} a_{11} & a_{12} & a_{13} \\ a_{21} & a_{22} & a_{23} \\ a_{31} & a_{32} & a_{33} \end{bmatrix}.$$

Let $\Delta = \det A$ as given in equation (2.10), and then consider also the three matrices

$$B_1 = \begin{bmatrix} b_1 & a_{12} & a_{13} \\ b_2 & a_{22} & a_{23} \\ b_3 & a_{32} & a_{33} \end{bmatrix}, \quad B_2 = \begin{bmatrix} a_{11} & b_1 & a_{13} \\ a_{21} & b_2 & a_{23} \\ a_{31} & b_3 & a_{33} \end{bmatrix}, \quad \text{and} \quad B_3 = \begin{bmatrix} a_{11} & a_{12} & b_1 \\ a_{21} & a_{22} & b_2 \\ a_{31} & a_{32} & b_3 \end{bmatrix},$$

where in each $B_i (i = 1, 2, 3)$ the column of constant terms has replaced the column of coefficients of x_i in the matrix A. Set

$$\Delta_1 = \det B_1, \qquad \Delta_2 = \det B_2, \qquad \Delta_3 = \det B_3.$$

Then, similar to (2.8), the solution to the general 3×3 system of linear equations (2.13) is given by

$$x_1 = \frac{\Delta_1}{\Delta} = \frac{\det B_1}{\det A}, \qquad x_2 = \frac{\Delta_2}{\Delta} = \frac{\det B_2}{\det A}, \qquad x_3 = \frac{\Delta_3}{\Delta} = \frac{\det B_3}{\det A}, \qquad \textbf{(2.14)}$$

provided $\Delta \neq 0$.

These are the values for the variables x_1, x_2, and x_3 that we would actually have computed if we had augmented the matrix A with the column of constants and then used the Gauss-Jordan reduction method to find an equivalent matrix in reduced row echelon form.

Let us remind you again that, when we are computing the determinants Δ_i in the numerators of the fractions in (2.14), the column of constants from the system appears in the same position in the matrix B_i as do the coefficients of the corresponding variable x_i in the matrix A. This is similar to the case of the 2×2 system. [See equation (2.8).]

The following example illustrates the use of Cramer's rule to solve a 3×3 system of linear equations.

Example 7 Find all the solutions, if any, to the system of equations

$$4x_2 + 3x_3 = -2,$$
$$-2x_1 + 5x_2 + -2x_3 = 1, \qquad \textbf{(2.15)}$$
$$3x_1 + 4x_2 + 5x_3 = 6.$$

Before we can apply Cramer's rule we must first evaluate the determinant Δ of the coefficient matrix A of this system. Thus

$$\Delta = \det A = \det \begin{bmatrix} 0 & 4 & 3 \\ -2 & 5 & -2 \\ 3 & 4 & 5 \end{bmatrix} = 0 - 24 - 24 - 45 + 40 - 0 = -53.$$

Since $\Delta \neq 0$, Cramer's rule applies. Therefore, the solution is found as follows:

$$x_1 = \frac{\det\begin{bmatrix} -2 & 4 & 3 \\ 1 & 5 & -2 \\ 6 & 4 & 5 \end{bmatrix}}{\Delta} = \frac{-50 - 48 + 12 - 90 - 20 - 16}{-53} = \frac{-212}{-53} = 4,$$

and

$$x_2 = \frac{\det\begin{bmatrix} 0 & -2 & 3 \\ -2 & 1 & -2 \\ 3 & 6 & 5 \end{bmatrix}}{\Delta}$$

$$= \frac{0 + 12 - 36 - 9 - 20 - 0}{-53} = \frac{12 - 65}{-53} = \frac{-53}{-53} = 1.$$

To find the appropriate value of x_3 one can either use the formula of Cramer's rule again or else substitute the values of x_1 and x_2 already found into one of the equations of the system and solve for x_3. If we decide to use Cramer's rule again, we have

$$x_3 = \frac{\det\begin{bmatrix} 0 & 4 & -2 \\ -2 & 5 & 1 \\ 3 & 4 & 6 \end{bmatrix}}{\Delta} = \frac{0 + 12 + 16 - (-30) - 0 - (-48)}{-53} = \frac{106}{-53} = -2.$$

Therefore, $(4, 1, -2)$ is the solution to the system (2.15). You should check that these values do satisfy each of the equations in the system when substituted for x_1, x_2, and x_3, respectively.   ■

Cramer's rule for solving a system of linear equations can be generalized to any $n \times n$ system of equations $A\mathbf{x} = \mathbf{b}$ in the obvious way. But the determinant of an $n \times n$ matrix A is a sum of $n!$ products, and when $n > 3$, this becomes extremely tedious to compute. The methods of Chapter 1 are much more efficient. For any system of equations except a 2×2 system, where it is very efficient, Cramer's rule is largely a historical artifact.

Vocabulary

In this section we have again encountered several new concepts and terms. Since this vocabulary is basic, be sure that you understand each of the concepts described by the terms in the following list. Review these ideas until you feel secure enough to use them.

determinant of a 2×2 matrix	determinant of a 3×3 matrix
elementary product	inversion
signed elementary product	permutation
even permutation	odd permutation
determinant of an $n \times n$ matrix	Cramer's rule

Exercises 2.1

In Exercises 1–9 use the diagonal schemes to evaluate the determinant of the given matrix.

1. $\det \begin{bmatrix} 1 & 2 \\ -3 & 4 \end{bmatrix}$.

2. $\det \begin{bmatrix} 3 & 1 \\ 0 & -5 \end{bmatrix}$.

3. $\det \begin{bmatrix} -1 & 0 \\ 1 & -6 \end{bmatrix}$.

4. $\det \begin{bmatrix} 1 & \tan x \\ \tan x & -1 \end{bmatrix}$.

5. $\det \begin{bmatrix} a & b \\ -b & a \end{bmatrix}$.

6. $\det \begin{bmatrix} 2 & 3 & 4 \\ 1 & 5 & 3 \\ -3 & 6 & 9 \end{bmatrix}$.

7. $\det \begin{bmatrix} 1 & -1 & 3 \\ 2 & 4 & 0 \\ -1 & 1 & 9 \end{bmatrix}$.

8. $\det \begin{bmatrix} 1 & 4 & -1 \\ 0 & 3 & 1 \\ 0 & 0 & -2 \end{bmatrix}$.

9. $\det \begin{bmatrix} 1 & 0 & 0 \\ -2 & -3 & 0 \\ 4 & 1 & 2 \end{bmatrix}$.

In Exercises 10–15 state whether the given permutation β from S_n is even or odd. Find sgn (β) in each case.

10. $\beta = \{1, 3, 2, 4\}$.

11. $\beta = \{5, 1, 2, 3, 4\}$.

12. $\beta = \{3, 1, 2, 5, 4\}$.

13. $\beta = \{5, 2, 3, 4, 1\}$.

14. $\beta = \{5, 4, 3, 2, 1\}$.

15. $\beta = \{1, 4, 2, 5, 3, 6\}$.

16. Make a table in the form of Table 2.1 that lists all the elementary products from the general 4×4 matrix. List the permutations and whether they are even or odd.

17. Write out the expansion of the determinant of the general 4×4 matrix A.

18. Using the results of Exercise 16, compute the determinant of the matrix

$$M = \begin{bmatrix} 1 & 2 & 0 & 3 \\ 0 & -1 & 4 & -2 \\ 2 & 0 & -1 & -3 \\ -2 & 1 & 0 & 1 \end{bmatrix}.$$

19. Use the fact that real (and complex) number multiplication is commutative to notice that the elementary product $a_{12}a_{23}a_{31} = a_{31}a_{12}a_{23}$, the right-hand side being written in column number order. Rewrite each of the elementary products in Table 2.1 in column number order and check the permutations of the row indices for even or odd. Would the signed elementary product be the same in this form as it was when the row numbers were in order?

20. Repeat Exercise 19 for the table of elementary products you made in Exercise 16.

In Exercises 21–23 verify by direct computation.

21. $\det \begin{bmatrix} a_{11} & a_{12} \\ a_{21} & a_{22} \end{bmatrix} = \det \begin{bmatrix} a_{11} & a_{21} \\ a_{12} & a_{22} \end{bmatrix}$.

22. $\det \begin{bmatrix} a_{11} & a_{12} & a_{13} \\ a_{21} & a_{22} & a_{23} \\ a_{31} & a_{32} & a_{33} \end{bmatrix} = \det \begin{bmatrix} a_{11} & a_{21} & a_{31} \\ a_{12} & a_{22} & a_{32} \\ a_{13} & a_{23} & a_{33} \end{bmatrix}$.

23. $\det \begin{bmatrix} a_1 + a'_1 & a_2 + a'_2 \\ b_1 & b_2 \end{bmatrix} = \det \begin{bmatrix} a_1 & a_2 \\ b_1 & b_2 \end{bmatrix} + \det \begin{bmatrix} a'_1 & a'_2 \\ b_1 & b_2 \end{bmatrix}$.

In Exercises 24–29 use Cramer's rule to solve each of the given systems of equations.

24. $2x_1 - 3x_2 = 1,$
$\quad 5x_1 + 4x_2 = -1.$

25. $3x_1 + 4x_2 = 7,$
$\quad 4x_1 + 5x_2 = 9.$

26. $-4x_1 + 3x_2 \qquad = 5,$
$\qquad\quad x_2 - 3x_3 = -4,$
$\quad 2x_1 - 4x_2 + 5x_3 = 0.$

27. $x_1 - x_2 + x_3 = -1,$
$\quad 2x_1 + x_2 - 2x_3 = 1,$
$\quad 3x_1 - 2x_2 + 3x_3 = -2.$

28. $\dfrac{2}{a} - \dfrac{3}{b} + \dfrac{5}{c} = 3,$
$\qquad \dfrac{2}{b} - \dfrac{1}{c} = 2,$
$\quad -\dfrac{4}{a} + \dfrac{7}{b} + \dfrac{2}{c} = 0.$

29. $x \cos\theta + y \sin\theta = 3,$
$\quad -x \sin\theta + y \cos\theta = 5,$
where θ is a given positive real number.

In Exercises 30–34 compute the determinant of each of the given complex matrices.

30. $\begin{bmatrix} -i & -3 \\ 5i & 8-i \end{bmatrix}.$

31. $\begin{bmatrix} 1+i & -3i \\ -4i & 1-2i \end{bmatrix}.$

32. $\begin{bmatrix} -3i & 3-3i \\ 1+2i & 8-3i \end{bmatrix}.$

33. $\begin{bmatrix} 1-i & -3 & 5i \\ 8-i & 2+i & i \\ -1+2i & -1 & 0 \end{bmatrix}.$

34. $\begin{bmatrix} 1 & -3+i & 2i \\ 1-i & 2+i & i \\ 0 & -1+i & -1-i \end{bmatrix}.$

---

35. Prove that an $n \times n$ matrix A has $n!$ elementary products. (HINT: How many choices are possible for an element from the first row? Then how many from the second? etc.)

36. A boy has $3.45 consisting of nickels, dimes, and quarters. He has exactly 27 coins, and the number of nickels is twice the number of dimes. How many of each type of coin does he have?

37. When painters A and B paint a house together, they need 60/11 days; when painters B and C paint the house together, they need 20/3 days; and when painters A and C paint the house together, they need 6 days. Other things being equal, how many days would each painter require to paint the house if he must work alone?

38. A man deposits $7,500 for a year. Of this he deposits $x at bank A, which pays an 8% annual rate of interest. He deposits $y at bank B, which pays a 7% annual rate of interest, and $z at bank C, which pays a 9% annual rate of interest. At the end of one year his combined interest from banks A and B is exactly $70 more than the interest received from bank C. His total interest that year from all three banks is $610. How much does he have deposited in each bank?

39. The system of equations

$$ax_1 - 2x_2 - x_3 = 0,$$
$$(a+1)x_2 + 4x_3 = 0,$$
$$(a-1)x_3 = 0,$$

has more than one solution. Find the value of a. HINT: Cramer's rule does not apply here. Why?

40. Describe the set of all solutions of the resulting system of equations for each value of a obtained in Exercise 39.

41. Show that for all *real* numbers a the only solution of the system

$$(a+1)x_1 - x_2 = 0,$$
$$2x_1 + (a-1)x_2 = 0,$$

is the trivial one: $x_1 = x_2 = 0.$

42. The Plastic Products Company makes plastic plates and plastic cups. Both require time on two machines: cups require 1 hr on Machine I and 2 hr on Machine II; plates require 3 hr on Machine I and 1 hr on Machine II. Each machine can operate 15 hr a day. How many of each product can be manufactured in such a 15-hr day under these conditions?

Exercises 43–46 refer to the following: The business analyst of Commercial Products Manufacturing Company wants to find an equation that can be used to project the sales for a relatively new cleaning product. For the years 1988–1990 the amounts of sales in millions of dollars are given by the following table.

1988	1989	1990
115	164	223

43. Use the year 1988 as reference, call it 0 on the horizontal axis, and divide the vertical axis into tens of millions of dollars. Sketch a sales graph as a point graph. The 1990 data will be the point (2, 22.3).

44. Find the equation of a straight line $y = mx + b$ passing through the points representing the 1988 and 1989 data. Is the 1990 point near this line?

Remember $m = \dfrac{y_2 - y_1}{x_2 - x_1}$.

45. For the parabola $y = ax^2 + bx + c$, find the equation that passes through all three points.

46. Use the results of Exercises 44 and 45 to project the sales for 1991 as points on the respective graphs. In which projection do you have the most confidence? Why?

47. Show that the area of a triangle ABC, where $A: (x_1, y_1)$, $B: (x_2, y_2)$, and $C: (x_3, y_3)$ are three distinct points in the plane, is equal to the absolute value of

$$\frac{1}{2} \det \begin{bmatrix} x_1 & y_1 & 1 \\ x_2 & y_2 & 1 \\ x_3 & y_3 & 1 \end{bmatrix}.$$

HINT: From the points A, B, and C drop perpendicular lines to the x axis. Then use the formula for the area of a trapezoid.

48. Show that the equation of a line L passing through the two distinct points $A: (x_1, y_1)$ and $B: (x_2, y_2)$ can be written as the determinantal equation

$$\det \begin{bmatrix} x & y & 1 \\ x_1 & y_1 & 1 \\ x_2 & y_2 & 1 \end{bmatrix} = 0.$$

HINT: See Exercise 47.

2.2 LAPLACE EXPANSION OF DETERMINANTS

The formal definition of a determinant, involving the sum of signed elementary products, can become cumbersome as a method for evaluating a given determinant. It does, however, lead us to a more efficient method known as the *Laplace expansion,* or the *cofactor expansion,* which we discuss here. We start by describing this step-by-step process, which is similar to that illustrated in Example 6 of the previous section. Let

$$A = \begin{bmatrix} a_{11} & a_{12} & a_{13} & \cdots & a_{1n} \\ a_{21} & a_{22} & a_{23} & \cdots & a_{2n} \\ a_{31} & a_{32} & a_{33} & \cdots & a_{3n} \\ \vdots & \vdots & \vdots & \cdots & \vdots \\ a_{n1} & a_{n2} & a_{n3} & \cdots & a_{nn} \end{bmatrix}. \tag{2.16}$$

We shall see that the determinant of A can be written as a linear expression involving the determinants of $(n-1) \times (n-1)$ submatrices of A. We first introduce two new terms: the *minor* and the *cofactor* of any entry in a matrix A.

Definition   The **minor** of the entry a_{ij} in row i and column j of the matrix A of (2.16) is the determinant of the submatrix $A(i|j)$ obtained from A by deleting row i and column j.

Definition   The **cofactor** A_{ij} of the entry a_{ij} of the matrix A is $(-1)^{i+j}$ times the minor of a_{ij}; that is, $A_{ij} = (-1)^{i+j} \det A(i|j)$.

Example 1   Find the minors of each of the entries of the matrix

$$A = \begin{bmatrix} 1 & -2 & 3 \\ 2 & -1 & 0 \\ 4 & 5 & -1 \end{bmatrix}.$$

Since the entry $a_{11} = 1$ occurs in row 1 and column 1 of A, we delete this row and column from A to obtain the submatrix

$$A(1|1) = \begin{bmatrix} -1 & 0 \\ 5 & -1 \end{bmatrix}.$$

Therefore, by definition, the minor of a_{11} is the determinant of this submatrix, namely, the number $(-1)(-1) - (5)(0) = 1$.

Similarly, we can compute the minors of each of the entries of the matrix A. For example, the minor of the entry a_{23}, ($a_{23} = 0$) is the determinant of the submatrix

$$A(2|3) = \begin{bmatrix} 1 & -2 \\ 4 & 5 \end{bmatrix}, \qquad \text{which is} \qquad (1)(5) - (4)(-2) = 13.$$

You should verify that the following table is correct.

Element/Value	Minor
$a_{11} = 1$	1
$a_{12} = -2$	-2
$a_{13} = 3$	14
$a_{21} = 2$	-13
$a_{22} = -1$	-13
$a_{23} = 0$	13
$a_{31} = 4$	3
$a_{32} = 5$	-6
$a_{33} = -1$	3

■   (2.17)

Remember that the cofactor A_{ij} of the entry a_{ij} of A is given by

$$A_{ij} = (-1)^{i+j} (\text{minor of } a_{ij}) = (-1)^{i+j} \det A(i|j).$$

Example 2 Find the cofactors of the matrix

$$A = \begin{bmatrix} 1 & -2 & 3 \\ 2 & -1 & 0 \\ 4 & 5 & -1 \end{bmatrix}.$$

Note that this is the same matrix as that in Example 1, so we already have the minors of each of its entries given by the table (2.17). Therefore, the co-factors are

$$A_{11} = (-1)^{1+1} \text{ (minor of } a_{11}) = (+1)(1) = 1;$$
$$A_{12} = (-1)^{1+2} \text{ (minor of } a_{12}) = (-1)(-2) = 2.$$

Similarly, you can verify that

$$A_{13} = 14; \quad A_{21} = 13; \quad A_{22} = -13; \quad A_{23} = -13;$$
$$A_{31} = 3; \quad A_{32} = 6; \quad A_{33} = 3. \quad \blacksquare$$

We can now state the *Laplace expansion theorem,* the process for finding the determinant of an $n \times n$ matrix in terms of the cofactors of its entries. We only sketch its rather tedious proof.

Theorem 2.1 (Laplace Expansion Theorem) Let A be any $n \times n$ matrix ($n \geq 2$). Then for any $i = 1, 2, \ldots, n$,

$$\Delta = \det A = a_{i1}A_{i1} + a_{i2}A_{i2} + \cdots + a_{in}A_{in} = \sum_{j=1}^{n} a_{ij}A_{ij}. \quad \textbf{(2.18)}$$

In other words, the determinant of A is the sum of n terms obtained by multiplying each entry a_{ij} in (a given) row i by its cofactor A_{ij}.

Outline of Proof It is enough to verify that each signed elementary product from A that involves the element a_{ik} appears as one of the terms in the expression $a_{ik}A_{ik}$, and conversely that the terms (products) in the expansion of $a_{ik}A_{ik}$ include all of those signed elementary products from A which involve the element a_{ik}. This verification, although not particularly difficult conceptually, is quite tedious and involves several steps that are not in the mainstream of our work here. We omit these details. ❑

Example 3 Illustrate Theorem 2.1 for the matrix of Examples 1 and 2.

Here
$$A = \begin{bmatrix} 1 & -2 & 3 \\ 2 & -1 & 0 \\ 4 & 5 & -1 \end{bmatrix}.$$

Combining the results of Example 2 and equation (2.18), we have, using the *first* row of A,

$$\Delta = \det A = a_{11}A_{11} + a_{12}A_{12} + a_{13}A_{13} = (1)(1) + (-2)(2) + (3)(14) = 39.$$

If we were to use the *third* row of A, we would have

$$\Delta = \det A = a_{31}A_{31} + a_{32}A_{32} + a_{33}A_{33} = (4)(3) + (5)(6) + (-1)(3) = 39.$$

You should do the expansion along the *second* row and verify that the same value results. ∎

Not only can one find the determinant of A by using the Laplace expansion along one of its rows as in (2.18) but one may also expand *down a column*. To arrive at this fact, we begin to look at some properties of the determinant function. We have the following important theorem about the determinant of the transpose A^T of A.

Theorem 2.2 For any $n \times n$ matrix A,

$$\det A^T = \det A. \tag{2.19}$$

Outline of Proof Again we shall not write out the tedious details, but one can verify that, by using the definition, exactly the same signed elementary products from A appear in the sum of signed elementary products that give $\det A^T$, the determinant of the transpose. Since the rows of A^T correspond to the columns of A, for $\det A^T$ the elementary products will appear as if they were in column number (from A) order with their row numbers (from A) permuted. Yet these products will have the same sgn value, 1 or -1, as they did in the expansion of $\det A$. This fact is a permutation exercise, which we omit. Hence the two sums are equal. See Exercises 19 and 20 of the previous section for the 3×3 and 4×4 cases.

For example, $\det \begin{bmatrix} 2 & 3 \\ 7 & -1 \end{bmatrix} = \det \begin{bmatrix} 2 & 7 \\ 3 & -1 \end{bmatrix}$. In fact, you can quickly verify that $\det \begin{bmatrix} a & b \\ c & d \end{bmatrix} = \det \begin{bmatrix} a & c \\ b & d \end{bmatrix}$. ◛

Theorem 2.2 leads us to the following corollary to Theorem 2.1, which allows us to use the *columns* of a matrix to expand its determinant if we wish.

Corollary Let A be any $n \times n$ matrix. Then, for any $j = 1, 2, \ldots, n$,

$$\Delta = \det A = a_{1j}A_{1j} + \cdots + a_{nj}A_{nj} = \sum_{i=1}^{n} a_{ij}A_{ij}. \tag{2.20}$$

This formula is called the *Laplace expansion of the determinant down a column of A*. That it is true follows immediately from the preceding two theorems, since the columns of A are precisely the rows of A^T.

Example 4   Illustrate (2.20) for the matrix

$$A = \begin{bmatrix} 2 & 3 & 1 \\ -1 & 0 & 4 \\ 5 & -2 & -3 \end{bmatrix}.$$

To expand down column *one* means, of course, to take $j = 1$ in (2.20). We get

$$\Delta = \det A = a_{11}A_{11} + a_{21}A_{21} + a_{31}A_{31}$$
$$= 2(8) + (-1)(7) + 5(12) = 69.$$

A more judicious choice would be to expand down column *two,* thereby actually eliminating one term (since $a_{22} = 0$, we don't need to bother with $A_{22} = -11$). In this case we get

$$\Delta = \det A = a_{12}A_{12} + a_{22}A_{22} + a_{32}A_{32}$$
$$= 3(17) + (0)A_{22} + (-2)(-9) = 69.$$

You can verify that the same number also results when the *third* column is used:

$$\det A = a_{13}A_{13} + a_{23}A_{23} + a_{33}A_{33}. \quad \blacksquare$$

Example 5   Let $A = \begin{bmatrix} 1 & 0 & -2 & -1 \\ 2 & 4 & 1 & 3 \\ 5 & -2 & 3 & -1 \\ 1 & -4 & 3 & -5 \end{bmatrix}$. Then its transpose is

$$A^T = \begin{bmatrix} 1 & 2 & 5 & 1 \\ 0 & 4 & -2 & -4 \\ -2 & 1 & 3 & 3 \\ -1 & 3 & -1 & -5 \end{bmatrix}.$$

Using the Laplace expansion (2.18) "across the first row" of the matrix A, we get

$$\det A = (1) \det \begin{bmatrix} 4 & 1 & 3 \\ -2 & 3 & -1 \\ -4 & 3 & -5 \end{bmatrix} - (0) \det \begin{bmatrix} 2 & 1 & 3 \\ 5 & 3 & -1 \\ 1 & 3 & -5 \end{bmatrix}$$

$$+ (-2) \det \begin{bmatrix} 2 & 4 & 3 \\ 5 & -2 & -1 \\ 1 & -4 & -5 \end{bmatrix} - (-1) \det \begin{bmatrix} 2 & 4 & 1 \\ 5 & -2 & 3 \\ 1 & -4 & 3 \end{bmatrix}$$

$$= 1(-60 + 4 - 18 + 36 - 10 + 12) - 0(\ )$$
$$- 2(20 - 4 - 60 + 6 + 100 - 8) + 1(-12 + 12 - 20 + 2 - 60 + 24)$$
$$= (-36) - 2(54) + (-54) = -198.$$

If we use (2.18) again to compute det A^T "across the first row" of A^T, we find that

$$\det A^T = (1) \det \begin{bmatrix} 4 & -2 & -4 \\ 1 & 3 & 3 \\ 3 & -1 & -5 \end{bmatrix} - (2) \det \begin{bmatrix} 0 & -2 & -4 \\ -2 & 3 & 3 \\ -1 & -1 & -5 \end{bmatrix}$$

$$+ (5) \det \begin{bmatrix} 0 & 4 & -4 \\ -2 & 1 & 3 \\ -1 & 3 & -5 \end{bmatrix} - (1) \det \begin{bmatrix} 0 & 4 & -2 \\ -2 & 1 & 3 \\ -1 & 3 & -1 \end{bmatrix}$$

$$= -198 \text{ also.} \quad \blacksquare$$

You can readily compute this. Observe that det $A = $ det A^T, in agreement with Theorem 2.2 (page 118).

As you can infer from this example, the Laplace expansion of a determinant can still be a very tedious computation. However, use of a calculator or a computer can speed up the work.

We close this section with a theorem that, in combination with the Laplace expansion and with our work with row reduction in the previous chapter, makes the computation of a given determinant much easier. This also has theoretical significance. How it can help with computing a determinant will be discussed in the next section.

Recall the definitions of triangular matrices from Chapter 1. The matrix $A = (a_{ij})$ is **upper-triangular** if, and only if, $a_{ij} = 0$ whenever $j < i$. The matrix A is **lower-triangular** when $a_{ij} = 0$ for all $j > i$. The matrix is called **triangular** when it is either upper-triangular or lower-triangular. (If it is both, it's a diagonal matrix.) For example, the matrices

$$A = \begin{bmatrix} 2 & -1 & 3 \\ 0 & 4 & 7 \\ 0 & 0 & -2 \end{bmatrix} \quad \text{and} \quad B = \begin{bmatrix} 2 & 0 & 0 \\ 1 & 3 & 0 \\ 5 & 1 & -5 \end{bmatrix}$$

are, respectively, upper- and lower-triangular. Our theorem concerning the determinant of triangular matrices is as follows:

Theorem 2.3   If $A = (a_{ij})$ is any $n \times n$ (upper- or lower-) triangular matrix, then

$$\det A = a_{11} a_{22} a_{33} \cdots a_{nn}. \tag{2.21}$$

That is, the determinant of a triangular matrix is equal to the product of its main diagonal elements.

Proof   This theorem follows readily from the definition of the determinant. Any elementary product, other than $a_{11}a_{22}a_{33} \cdots a_{nn}$, from a triangular matrix must contain a zero factor. Why? That product is therefore zero. The only possible nonzero term in the sum of the signed elementary products is the product of the diagonal elements. ◻

One may alternatively write a proof by induction on the size of the matrix, using Theorem 2.1 (page 117) and the corollary to Theorem 2.2. See Exercise 36.

Example 6   Find the determinant of the 5×5 matrix

$$K = \begin{bmatrix} 3 & 2 & 0 & -1 & 5 \\ 0 & 1 & 4 & 1 & 2 \\ 0 & 0 & -1 & -2 & 1 \\ 0 & 0 & 0 & 2 & -1 \\ 0 & 0 & 0 & 0 & 5 \end{bmatrix}.$$

A direct application of Theorem 2.3 to this upper-triangular matrix gives us

$$\Delta = \det K = (3)(1)(-1)(2)(5) = -30. \quad \blacksquare$$

As other direct applications of Theorem 2.3, we have the following corollaries.

Corollary 1   If $A = \text{diag}\,[d_1, d_2, \ldots, d_n]$, then $\det A = d_1 \cdot d_2 \cdots d_n$.

The special case of the identity matrix is stated for emphasis as Corollary 2.

Corollary 2   For the $n \times n$ identity matrix I_n, $\det I_n = 1$.

Vocabulary

In this Section we have again encountered several new concepts and terms. Since this vocabulary is basic, be sure that you understand each of the concepts described by the terms in the following list. Review these ideas until you feel secure enough to use them.

minor of a_{ij}
submatrix
expansion across a row
determinant of transpose
determinant of a triangular matrix

cofactor of a_{ij}
Laplace expansion
expansion down a column
triangular matrix

Exercises 2.2

In Exercises 1–10 use the Laplace expansion across any row or down any column of the given matrix to compute its determinant.

1. $\begin{bmatrix} 1 & 0 & 1 \\ 2 & 1 & 0 \\ 0 & 1 & -1 \end{bmatrix}$.

2. $\begin{bmatrix} -1 & 1 & 1 \\ 2 & 2 & 1 \\ 4 & 0 & -1 \end{bmatrix}$.

3. $\begin{bmatrix} 1 & 0 & 1 & -1 \\ 0 & 1 & -1 & 2 \\ 2 & 1 & -2 & 1 \\ 0 & 1 & 0 & 3 \end{bmatrix}$.

4. $\begin{bmatrix} 1 & 0 & -1 & 1 \\ 2 & 1 & -1 & 4 \\ 1 & 0 & 1 & 3 \\ 0 & 1 & -1 & 0 \end{bmatrix}$.

5. $\begin{bmatrix} \sin x & \cos x \\ -\cos x & \sin x \end{bmatrix}$.

6. $\begin{bmatrix} e^x & -\sin x \\ \sin x & e^{-x} \end{bmatrix}$.

7. $\begin{bmatrix} 4 & 2 & 4 & 6 \\ 2 & 1 & 1 & 1 \\ 2 & 4 & -1 & -1 \\ 2 & 1 & 1 & 1 \end{bmatrix}$.

8. $\begin{bmatrix} 1 & 3 & 5 & -1 \\ 2 & 1 & -3 & 2 \\ 4 & 1 & -3 & 1 \\ 0 & 1 & 2 & 3 \end{bmatrix}$.

9. $\begin{bmatrix} 1 & 0 & 1 & -1 & 1 \\ 2 & 0 & 1 & -1 & 2 \\ 2 & 0 & 1 & -2 & 1 \\ 4 & 0 & 1 & 0 & 3 \\ 1 & 2 & 3 & 4 & 5 \end{bmatrix}$.

10. $\begin{bmatrix} 1 & 2 & 0 & -1 & 2 \\ -2 & 0 & 1 & 3 & 0 \\ 0 & 1 & 3 & -1 & 0 \\ 4 & 1 & -4 & 0 & 1 \\ 0 & 1 & 1 & 0 & 0 \end{bmatrix}$.

11. Demonstrate by using the 3×3 diagonalization scheme of the previous section that $\det A = \det A^T$ for the following matrix:

$$A = \begin{bmatrix} 1 & 0 & 2 \\ -3 & 5 & 1 \\ -1 & 3 & -4 \end{bmatrix}.$$

In Exercises 12–16 use Theorem 2.3 (page 120), and its corollaries, to compute the determinant of the given triangular matrix.

12. I_4.

13. $3I_5$.

14. diag $[2, 4, 6, 8]$.

15. $\begin{bmatrix} 2 & 4 & 6 & 8 & 10 \\ 0 & 2 & 4 & 6 & 8 \\ 0 & 0 & 2 & 4 & 6 \\ 0 & 0 & 0 & 2 & 4 \\ 0 & 0 & 0 & 0 & 2 \end{bmatrix}$.

16. $\begin{bmatrix} 1 & 0 & 0 & 0 & 0 & 0 \\ 3 & 1 & 0 & 0 & 0 & 0 \\ 5 & 3 & 1 & 0 & 0 & 0 \\ 7 & 5 & 3 & 1 & 0 & 0 \\ 9 & 7 & 5 & 3 & 1 & 0 \\ 11 & 9 & 7 & 5 & 3 & 1 \end{bmatrix}$.

In Exercises 17–25 compute the determinant of the given elementary matrix (see Section 1.5).

17. $E_1 = \begin{bmatrix} 0 & 1 & 0 \\ 1 & 0 & 0 \\ 0 & 0 & 1 \end{bmatrix}$.

18. $E_2 = \begin{bmatrix} 1 & 0 & 0 \\ 0 & 0 & 1 \\ 0 & 1 & 0 \end{bmatrix}$.

19. $E_3 = \begin{bmatrix} 0 & 0 & 1 \\ 0 & 1 & 0 \\ 1 & 0 & 0 \end{bmatrix}$.

20. $E_4 = \begin{bmatrix} 1 & 0 & 0 \\ 0 & -3 & 0 \\ 0 & 0 & 1 \end{bmatrix}$.

21. $E_5 = \begin{bmatrix} \frac{1}{2} & 0 & 0 \\ 0 & 1 & 0 \\ 0 & 0 & 1 \end{bmatrix}$.

22. $E_6 = \begin{bmatrix} 1 & 0 & 0 \\ 0 & 1 & 0 \\ 0 & 0 & -\frac{1}{6} \end{bmatrix}$.

23. $E_7 = \begin{bmatrix} 1 & 0 & 0 \\ 0 & 1 & 0 \\ -5 & 0 & 1 \end{bmatrix}$.

24. $E_8 = \begin{bmatrix} 1 & 0 & 0 \\ \frac{2}{3} & 1 & 0 \\ 0 & 0 & 1 \end{bmatrix}$.

25. $E_9 = \begin{bmatrix} 1 & -3 & 0 \\ 0 & 1 & 0 \\ 0 & 0 & 1 \end{bmatrix}$.

Exercises 26–29 refer to the following matrix:

$$A = \begin{bmatrix} 1 & 3 & 3 \\ 2 & -1 & 4 \\ 3 & -2 & 0 \end{bmatrix}.$$

26. Compute det A.

27. Row reduce A to an upper triangular matrix U. Compute det U.

28. Find the row echelon form N for A and compute det N.

29. Find the reduced row echelon form J for A and compute det J.

30. Repeat Exercise 26 for the matrix

$$B = \begin{bmatrix} 2 & 3 & 0 \\ 1 & -2 & 1 \\ 4 & 0 & -2 \end{bmatrix}.$$

31. Repeat Exercise 27 for the matrix B of Exercise 30.

32. Repeat Exercise 28 for the matrix B of Exercise 30.

33. Repeat Exercise 29 for the matrix B of Exercise 30.

34. Let A and B be the two matrices referred to in Exercises 26–33. Show by multiplying the two matrices and computing the determinants involved that

$$\det(AB) = (\det A)(\det B).$$

35. Show that for these same matrices $\det(A + B) \neq \det A + \det B$.

36. Prove Theorem 2.3 by mathematical induction on the size of the matrix. HINT: The theorem is trivially true when $n = 1$. (Although not necessary for the proof, the cases when $n = 2$ and $n = 3$ are easy and worth doing to see how they go.) For the inductive step assume that the theorem is true for every $(n - 1) \times (n - 1)$ upper-triangular matrix. Then expand the determinant of the $n \times n$ upper-triangular matrix U down the first column using the corollary to Theorem 2.2 (page 118). (For a lower-triangular matrix L one uses the first row.)

37. Prove by induction on the size of the matrix that any $n \times n$ matrix A, $n \geq 2$, with two identical rows has zero determinant.

38. Let A be the 3×3 matrix

$$A = \begin{bmatrix} 2 & 3 & 0 \\ 1 & -2 & 1 \\ 1 & 3 & -2 \end{bmatrix}.$$

Compute $\Delta' = a_{11}A_{31} + a_{12}A_{32} + a_{13}A_{33}$; that is, multiply the elements of the first row of A by the cofactors of the corresponding elements in the third row, and then add up these products.

39. Show that the value of Δ' in Exercise 38 is the value of the determinant of the following matrix

A'. (Show $\Delta' = \det A'$.) Use the properties of determinants to justify your answer.

$$A' = \begin{bmatrix} 2 & 3 & 0 \\ 1 & -2 & 1 \\ 2 & 3 & 0 \end{bmatrix}.$$

40. If A is any 4×4 matrix, $A = (a_{ij})$, what do you conclude is the numerical value of

$$\Delta' = \sum_{i=1}^{4} a_{1i}A_{ji},$$

for $j \neq 1$? Why?

41. Compute the determinant of the following complex matrix:

$$\begin{bmatrix} i & 0 & 1 & -i \\ 1 - 2i & i & 1 - i & 1 + i \\ i & -1 + 2i & -1 & 1 + i \\ 0 & 0 & 0 & 1 \end{bmatrix}.$$

42. Compute the determinant of the following complex matrix:

$$\begin{bmatrix} 1 + i & 0 & 0 & -i \\ 1 - 2i & i & 1 - i & 1 + i \\ i & -1 + 2i & -1 & 1 + i \\ 1 + 2i & 2 - i & 3i & 2 + 2i \end{bmatrix}.$$

43. Let l be a line in the plane drawn through the distinct points $P(a, b)$ and $Q(c, d)$. Show that the equation of l can be written in the following determinantal form:

$$\det \begin{bmatrix} x & y & 1 \\ a & b & 1 \\ c & d & 1 \end{bmatrix} = 0.$$

HINT: Use the "two-point form of the line" you learned in previous courses.

44. Solve the following equation for x:

$$\det A = \det \begin{bmatrix} 1 + x & 1 & 1 \\ 1 & 1 + x & 1 \\ 1 & 1 & 1 + x \end{bmatrix} = 0.$$

45. Generalize the results of Exercises 39 and 40 to $n \times n$ matrices.

2.3 PROPERTIES OF DETERMINANTS

In this section we develop some interesting and useful properties of determinants. In the previous section (Theorem 2.3) we found that the determinant of a triangular matrix is the product of its main diagonal elements. In Chapter 1 the Gauss-Jordan process of row reducing a given matrix was developed. It was shown there that reducing a matrix to triangular form can be accomplished by premultiplying it by appropriate elementary matrices. Here we see how these ideas can be combined to reduce the tedious work of computing the determinant of any square matrix M. The row echelon form U of a given matrix M is always upper-triangular. We relate det M to det U. We must begin with a word of caution, however, since **det M will not generally equal det U**. Consider the following example (see also Exercises 26–33 of the preceding section).

Example 1   Let $M = \begin{bmatrix} 2 & -1 \\ 5 & 6 \end{bmatrix}$. Reduce M to its row echelon form U by elementary row operations. Then compare the values of det M and det U.

By appropriate elementary row operations we obtain the following matrices equivalent to M:

(a) $E_1 M = \begin{bmatrix} 2 & -1 \\ 1 & 8 \end{bmatrix}$,

(b) $E_2 E_1 M = \begin{bmatrix} 1 & 8 \\ 2 & -1 \end{bmatrix}$,

(c) $E_3 E_2 E_1 M = \begin{bmatrix} 1 & 8 \\ 0 & -17 \end{bmatrix}$,

(d) $U = E_4 E_3 E_2 E_1 M = \begin{bmatrix} 1 & 8 \\ 0 & 1 \end{bmatrix}$. ∎

Now it is easy to see that det $M = 17$ while det $U = 1$. How does one compare these two numbers? We show that the determinants of the elementary matrices in steps (a) through (d) actually relate these two numbers. Let us find the determinants of each type of elementary matrix.

Type I The Interchange of Two Rows.   If we interchange row i with row j of the $n \times n$ identity matrix I, we obtain a Type I elementary matrix E. By repeated application of the Laplace expansion (Theorem 2.1) of the determinant of E we must eventually arrive at

$$\det E = (1)(1)(1) \cdots (1) \det \begin{bmatrix} 0 & 1 \\ 1 & 0 \end{bmatrix} = -1.$$

We have sketched the first part of the proof of the following lemma:

Lemma 2.4   The determinant of Type I elementary matrix E is -1. Furthermore, if the $n \times n$ matrix B is obtained from a given $n \times n$ matrix A by an interchange of two rows, then $B = EA$ and

$$\det (EA) = \det B = -\det A = (\det E)(\det A). \tag{2.22}$$

To complete the proof requires technical details involving permutations. We sketch how it goes but leave the details as part of Supplementary Project 2 at the end of the chapter. Roughly, then, suppose that the $n \times n$ matrix B is obtained from a given $n \times n$ matrix A by an interchange of two rows, say row i with row j, $i < j$. From the definition of det B we have the following:

$$\det B = \sum_{\sigma} \operatorname{sgn}(\sigma)\, a_{1\sigma(1)} a_{2\sigma(2)} \cdots a_{j\sigma(j)} \cdots a_{i\sigma(i)} \cdots a_{n\sigma(n)}\,.$$

But the row numbers in each of these signed elementary products, while in the proper order for the matrix B, are not in ascending order as elements of the first matrix A. In the expression for the determinant for A the element $a_{j\sigma(j)}$ would *follow* instead of *precede* the element $a_{i\sigma(i)}$ in the product. If we rearrange this product so as to correspond to the order of elements in A, the permutation σ, which gives the column numbers in each elementary product, changes to a different permutation σ'. The new permutation σ' involves one additional transposition. (See Supplementary Project 2.) So the permutation involved changes from an even to an odd or from an odd to an even one. Thus $\operatorname{sgn}(\sigma) = -\operatorname{sgn}(\sigma')$, a change in sign, and we have the result det $B = -\det A$, as desired.

Example 2

$$\det \begin{bmatrix} 2 & -1 \\ 5 & 6 \end{bmatrix} = -\det \begin{bmatrix} 5 & 6 \\ 2 & -1 \end{bmatrix}.$$

This could be easily computed directly by expanding both determinants, since

$$(2)(6) - (5)(-1) = -[(5)(-1) - (2)(6)].$$

A less computational approach is to apply Lemma 2.4. The right-hand matrix is the result of interchanging rows *one* and *two* of the left-hand matrix.

In a similar way we see that

$$\det \begin{bmatrix} 1 & 2 & 4 \\ 0 & 3 & 5 \\ 1 & -3 & 3 \end{bmatrix} = -\det \begin{bmatrix} 1 & -3 & 3 \\ 0 & 3 & 5 \\ 1 & 2 & 4 \end{bmatrix},$$

because of the interchange of rows *one* and *three*. Why does

$$\det \begin{bmatrix} -1 & 2 & 1 \\ 0 & 3 & 1 \\ 1 & -2 & 3 \end{bmatrix} = -\det \begin{bmatrix} 2 & -1 & 1 \\ 3 & 0 & 1 \\ -2 & 1 & 3 \end{bmatrix}? \quad \blacksquare$$

Type II Multiplying Row i by a Nonzero Constant.   If row i of the identity matrix I_n is multiplied by the nonzero constant k we have the elementary matrix $E = \operatorname{diag}[1, \ldots, 1, k, 1, \ldots, 1]$, where k is in row i. By Theorem 2.3 (page 120) we immediately conclude that det $E = k$. Now suppose that A is any $n \times n$ matrix and that $B = EA$ is the same matrix as A except that row i of B is k times row i of A. Then, by using a Laplace expansion along row i of B, we have that

$$\det B = \det (EA) = ka_{i1}A_{i1} + ka_{i2}A_{i2} + \cdots + ka_{in}A_{in} = k(\det A).$$

This proves the following lemma.

Lemma 2.5   If E is an $n \times n$ Type II elementary matrix, then $\det E = k$. And if A is any $n \times n$ matrix, then

$$\det (EA) = (\det E)(\det A) = k(\det A). \tag{2.23}$$

Example 3

$$\det \begin{bmatrix} 1 & 2 & 4 & 0 \\ 3 & 5 & 1 & 2 \\ 5 & 10 & 15 & -10 \\ 0 & -1 & 2 & 4 \end{bmatrix} = 5 \det \begin{bmatrix} 1 & 2 & 4 & 0 \\ 3 & 5 & 1 & 2 \\ 1 & 2 & 3 & -2 \\ 0 & -1 & 2 & 4 \end{bmatrix}.$$

$$\underset{K}{} \qquad\qquad\qquad \underset{M}{}$$

This is because row three of the left-hand matrix, K, is 5 times row three of the right-hand matrix, M.

If we look at this lemma in a slightly different way, we can say that we have *factored* 5 *out of row three.* Do not, however, confuse this property of determinants with multiplying the matrix by 5. If M is the right-hand matrix, then $5M$ is not the left-hand matrix K. Write out $5M$ and notice that $\det (5M) = 625(\det M)$; that is, $\det (5M) = 5^4 \cdot \det M$. ■

We now turn to the final type of elementary operation: replacing one row of a matrix with its sum with a nonzero constant multiple of another row.

Type III Adding a Multiple of One Row to Another.   Suppose that row j of the identity matrix is replaced with its sum with k times row i. Then the resulting elementary matrix E is either upper-triangular (if $j < i$) or lower-triangular (if $j > i$). In any case the diagonal elements of E remain all ones; so $\det E = 1$, by Theorem 2.3.

Now suppose that the $n \times n$ matrix B is obtained from the $n \times n$ matrix A by this elementary operation: $B = EA$. Then we have

$$\det B = \sum_{\sigma} \operatorname{sgn}(\sigma)\, a_{1\sigma(1)}a_{2\sigma(2)} \cdots (a_{j\sigma(j)} + ka_{i\sigma(i)}) \cdots a_{n\sigma(n)}$$

$$= \sum_{\sigma} \operatorname{sgn}(\sigma)\, a_{1\sigma(1)}a_{2\sigma(2)} \cdots a_{j\sigma(j)} \cdots a_{n\sigma(n)}$$

$$+ k\sum_{\sigma} \operatorname{sgn}(\sigma)\, a_{1\sigma(1)}a_{2\sigma(2)} \cdots a_{i\sigma(j)} \cdots a_{i\sigma(i)} \cdots a_{n\sigma(n)}$$

$$= \det A + k \det A^{\#},$$

where the matrix $A^{\#}$ is the following matrix whose ith and jth rows are the same.

$$A^{\#} = \begin{bmatrix} a_{11} & a_{12} & \cdots & a_{1n} \\ \cdot & \cdot & & \cdot \\ a_{i1} & a_{i2} & \cdots & a_{in} \\ \cdot & & & \\ a_{i1} & a_{i2} & \cdots & a_{in} \\ \cdot & & & \\ a_{n1} & a_{n2} & \cdots & a_{nn} \end{bmatrix} \tag{2.24}$$

It follows that det $A^{\#} = 0$ (see Exercise 37 of Section 2.2). Therefore, det $B = \det A$ and we have sketched the proof of the following lemma.

Lemma 2.6   If A is an $n \times n$ matrix and E is a Type III elementary matrix, then det $E = 1$ and

$$\det EA = \det A = (\det E)(\det A). \tag{2.25}$$

Example 4

$$\det \begin{bmatrix} 2 & 6 & 7 \\ -1 & -3 & 1 \\ 5 & 8 & 10 \end{bmatrix} = \det \begin{bmatrix} 2+2(-1) & 6+2(-3) & 7+2(1) \\ -1 & -3 & 1 \\ 5 & 8 & 10 \end{bmatrix}$$

$$= \det \begin{bmatrix} 0 & 0 & 9 \\ -1 & -3 & 1 \\ 5 & 8 & 10 \end{bmatrix} = 0 + 0 + 9 \det \begin{bmatrix} -1 & -3 \\ 5 & 8 \end{bmatrix}$$

$$= 9(-8+15) = 9(7) = 63. \quad \blacksquare$$

The results of Lemmas 2.4, 2.5, and 2.6 are often stated as **properties of determinants.** If the $n \times n$ matrix B is obtained from the matrix A by an elementary row operation, then it is of the form EA, and we have

---

Property 1. Interchanging rows (columns) changes the sign of the determinant.
Property 2. Multiplying a row (column) by a nonzero constant multiplies the determinant by that constant.
Property 3. Adding to a row (column) a nonzero constant multiple of another row (column) does not change the determinant.

In stating these properties, we have made use of Theorem 2.2 (page 118) to note that an elementary *row* operation on A^T is the same as doing the same operation to a *column* of the matrix A.

Now let us return to Example 1. We see that the elementary matrices in that example have the following determinants: $\det E_1 = 1$, $\det E_2 = -1$, $\det E_3 = 1$, and $\det E_4 = -1/17$. So [see(d)] $\det U = (-1/17)(1)(-1)(1)(\det M)$, by repeated use of the three lemmas. Since $\det U = 1$, we see that $\det M = 17$.

These three lemmas also enable us to prove the following very important theorem.

Theorem 2.7   If A and B are $n \times n$ matrices, then

$$\det (AB) = (\det A)(\det B).$$

Proof   Suppose first that A is singular. Then there exist elementary matrices $E_1, E_2, \cdots, E_m$ so that the row echelon form $U = E_1 E_2 \cdots E_m A$ of A has at least one row of zeros; see Theorem 1.10 (page 61). Then $\det U = 0$, from Theorem 2.3 (page 120), because U is upper-triangular and has a zero on its main diagonal. From our three lemmas, we have

$$0 = \det U = (\det E_1)(\det E_2) \cdots (\det E_m)(\det A).$$

Since no elementary matrix has a zero determinant, it must follow that $\det A = 0$. Furthermore, if A is singular, so is the matrix AB singular (see Section 1.5, Exercise 40). Hence $\det (AB) = 0$ by the above argument with A now replaced by the matrix AB. Combining this with $\det A = 0$, we get

$$0 = \det (AB) = (\det A)(\det B).$$

Now suppose that A is nonsingular. Then, by Theorem 1.10, A is row equivalent to the identity matrix I; that is, there exist elementary matrices so that

$$A = E_1 E_2 \cdots E_k I.$$

Thus,

$$AB = E_1 E_2 \cdots E_k IB = E_1 E_2 \cdots E_k B.$$

Therefore, by the above three lemmas,

$$\det (AB) = (\det E_1)(\det E_2) \ldots (\det E_k)(\det B) = (\det A)(\det B).$$

So, in either case, *the determinant of the product is the product of the determinants,* as asserted. ◘

As a corollary we have the following. One part follows immediately from the above proof.

Corollary   The $n \times n$ matrix A is singular if, and only if, $\det A = 0$.

To finish the proof, note that we have shown above that the determinant of a singular matrix is zero. Conversely, if det $A = 0$ and A were nonsingular, we would have, from Theorem 2.7, $1 = \det I = \det (AA^{-1}) = (\det A)(\det A^{-1}) = 0$, a contradiction.

Example 5 Note that the coefficient matrix for the homogeneous system of equations

$$\begin{aligned} x_1 + x_2 - 2x_3 &= 0 \\ 2x_1 - 3x_2 + x_3 &= 0 \\ 3x_1 - 2x_2 - x_3 &= 0 \end{aligned}$$

is singular and $\det \begin{bmatrix} 1 & 1 & -2 \\ 2 & -3 & 1 \\ 3 & -2 & -1 \end{bmatrix} = 0$, as you can readily verify. ∎

The following two additional properties of determinants are useful and are special cases of the properties already given. You can readily write out their proofs as Exercises 21 and 22.

> **Property 4.** If any two rows (columns) of a matrix A are identical, then det $A = 0$. **(2.26)**
> **Property 5.** If a row (column) of a matrix A is entirely zero, then det $A = 0$.

Example 6 Notice that

$$\det \begin{bmatrix} 1 & 2 & 4 & 0 \\ 3 & 5 & 1 & 2 \\ 0 & 0 & 0 & 0 \\ 3 & 8 & 7 & 1 \end{bmatrix} = 0, \quad \text{and} \quad \det \begin{bmatrix} 1 & 2 & 4 & 2 \\ 3 & 5 & 1 & 5 \\ 0 & 2 & 0 & 2 \\ 3 & 8 & 7 & 8 \end{bmatrix} = 0.$$

For the first matrix we use Property 5. Notice that columns *two* and *four* of the second matrix are identical, so we use Property 4. ∎

Example 7 Compute the determinant of the matrix

$$A = \begin{bmatrix} 4 & 2 & 4 & 6 \\ 2 & 1 & 1 & 1 \\ 2 & 4 & -1 & -1 \\ -2 & 1 & 1 & 1 \end{bmatrix}.$$

Observe that

$$\det A = 4 \det \begin{bmatrix} 1 & 1 & 2 & 3 \\ 1 & 1 & 1 & 1 \\ 1 & 4 & -1 & -1 \\ -1 & 1 & 1 & 1 \end{bmatrix}$$ by using **Property 2** twice, first on row *one* and then on column *one*

$$= 4 \det \begin{bmatrix} 1 & 1 & 2 & 3 \\ 0 & 0 & -1 & -2 \\ 0 & 3 & -3 & -4 \\ 0 & 2 & 3 & 4 \end{bmatrix}$$ by using **Property 3** three times

$$= 4(1) \det \begin{bmatrix} 0 & -1 & -2 \\ 3 & -3 & -4 \\ 2 & 3 & 4 \end{bmatrix}$$ by expanding down column *one*—see (2.20)

$$= 4(1)(-2) \det \begin{bmatrix} 0 & -1 & 1 \\ 3 & -3 & 2 \\ 2 & 3 & -2 \end{bmatrix}$$ by using **Property 2** on the *third* column

$$= (-8) \det \begin{bmatrix} 0 & 0 & 1 \\ 3 & -1 & 2 \\ 2 & 1 & -2 \end{bmatrix}$$ by using **Property 3**—add column *three* to column *two*

$$= (-8)(1) \det \begin{bmatrix} 3 & -1 \\ 2 & 1 \end{bmatrix}$$ expanding along row *one*—see (2.18)

$$= (-8)[3 - (-2)] = -40.$$ see (2.5)   ∎

We state an additional property of determinants that has important theoretical consequences. We leave the proof to you as a part of Supplementary Project 1 [see part (b)].

> **Property 6.** Let A be the $n \times n$ matrix
> $$A = [\mathbf{a}_1, \mathbf{a}_2, \ldots, \mathbf{a}_{k-1}, \mathbf{b}_k + \mathbf{c}_k, \mathbf{a}_{k+1}, \ldots, \mathbf{a}_n]$$
> where $\mathbf{a}_i$ denotes the ith column of A. Then
> $$\det A = \det [\mathbf{a}_1, \mathbf{a}_2, \ldots, \mathbf{a}_{k-1}, \mathbf{b}_k, \mathbf{a}_{k+1}, \ldots, \mathbf{a}_n]$$
> $$+ \det [\mathbf{a}_1, \mathbf{a}_2, \ldots, \mathbf{a}_{k-1}, \mathbf{c}_k, \mathbf{a}_{k+1}, \ldots, \mathbf{a}_n].$$

Example 8   Show that

$$\det \begin{bmatrix} a_{11} & a_{12} + x_1 & a_{13} \\ a_{21} & a_{22} + x_2 & a_{23} \\ a_{31} & a_{32} + x_3 & a_{33} \end{bmatrix} = \det \begin{bmatrix} a_{11} & a_{12} & a_{13} \\ a_{21} & a_{22} & a_{23} \\ a_{31} & a_{32} & a_{33} \end{bmatrix} + \det \begin{bmatrix} a_{11} & x_1 & a_{13} \\ a_{21} & x_2 & a_{23} \\ a_{31} & x_3 & a_{33} \end{bmatrix}.$$

One may do this fairly easily by direct calculation. It is clear that

$$\det A = a_{11}(a_{22} + x_2)a_{33} + (a_{12} + x_1)a_{23}a_{31} + a_{13}a_{21}(a_{32} + x_3)$$
$$- a_{13}(a_{22} + x_2)a_{31} - (a_{12} + x_1)a_{21}a_{33} - a_{11}a_{23}(a_{32} + x_3)$$
$$= (a_{11}a_{22}a_{33} + a_{12}a_{23}a_{31} + a_{13}a_{21}a_{32} - a_{13}a_{22}a_{31} - a_{12}a_{21}a_{33} - a_{11}a_{23}a_{32})$$
$$+ (a_{11}x_2a_{33} + x_1a_{23}a_{31} + a_{13}a_{21}x_3 - a_{13}x_2a_{31} - x_1a_{21}a_{33} - a_{11}a_{23}x_3).$$

Note that this is exactly the sum of the two determinants on the right side of the equation.   ∎

Let us conclude this section with an example that relates our work with determinants to solving homogeneous systems of equations.

Example 9   Show, using determinants, that the only solution to the following homogeneous system of equations

$$2x_1 - 3x_2 + x_3 = 0$$
$$2x_2 + 5x_3 = 0$$
$$4x_1 - 5x_2 \quad\quad = 0$$

is the trivial solution: $x_1 = x_2 = x_3 = 0$.

We evaluate the determinant Δ of the coefficient matrix A to obtain

$$\Delta = \det A = \det \begin{bmatrix} 2 & -3 & 1 \\ 0 & 2 & 5 \\ 4 & -5 & 0 \end{bmatrix} = 2 \det \begin{bmatrix} 1 & -3 & 1 \\ 0 & 2 & 5 \\ 2 & -5 & 0 \end{bmatrix}$$

$$= 2 \det \begin{bmatrix} 1 & -3 & 1 \\ 0 & 2 & 5 \\ 0 & 1 & -2 \end{bmatrix} = -18 \neq 0.$$

Thus, from the corollary to Theorem 2.7 (page 128) we conclude that the coefficient matrix A is nonsingular. Hence the matrix equation $AX = 0$ has the unique solution $X = A^{-1}0 = 0$. So we have just the values $x_1 = x_2 = x_3 = 0$ as solutions. One could also get this same result using Cramer's rule. How?   ∎

The argument in this example can be generalized to cover any homogeneous system of n linear equations in n variables. If we combine this argument with the results in Chapter 1, we obtain the following interesting theorem.

Theorem 2.8   Suppose that the determinant of the coefficient matrix A of an $n \times n$ homogeneous system of linear equations $AX = 0$ is not equal to zero. Then the system has only the trivial solution $X = 0$; i.e., $x_1 = x_2 = \cdots = x_n = 0$. If, on the other hand, $\det A = 0$, the system has additional nontrivial solutions. In fact, in the latter case there are infinitely many solutions.

Vocabulary

No new terms were introduced in this section. As a convenient reference, however, we have collected below the main *properties of determinants* that were developed in this and in the previous section.

- The determinant of a triangular matrix is the product of its main diagonal elements.
- $\det A^T = \det A$.
- Interchanging rows (columns) changes the sign of the determinant.
- Multiplying a row (column) by a nonzero constant multiplies the determinant by that constant.
- Adding to a row (column) a nonzero constant multiple of another row (column) does not change the determinant.
- $\det (AB) = (\det A)(\det B)$.
- If any two rows (columns) of a matrix A are identical, then $\det A = 0$.
- If a row (column) of a matrix A is entirely zero, then $\det A = 0$.
- A is singular if, and only if, $\det A = 0$. Consequently,
- A is nonsingular if, and only if, $\det A \neq 0$.
- $\det [\mathbf{a}_1, \mathbf{a}_2, \ldots, \mathbf{a}_{k-1}, \mathbf{b}_k + \mathbf{c}_k, \mathbf{a}_{k+1}, \ldots, \mathbf{a}_n]$
  $= \det [\mathbf{a}_1, \mathbf{a}_2, \ldots, \mathbf{a}_{k-1}, \mathbf{b}_k, \mathbf{a}_{k+1}, \ldots, \mathbf{a}_n]$
  $+ \det [\mathbf{a}_1, \mathbf{a}_2, \ldots, \mathbf{a}_{k-1}, \mathbf{c}_k, \mathbf{a}_{k+1}, \ldots, \mathbf{a}_n]$,
  where $\mathbf{a}_i$ denotes the ith column of the matrix A.

Exercises 2.3

In Exercises 1–7, use the properties of determinants to show that the given equalities are true. Do these exercises without computing the determinants.

1. $\det \begin{bmatrix} 1 & 2 & 4 \\ 0 & 3 & 5 \\ 1 & 2 & 3 \end{bmatrix} = -\det \begin{bmatrix} 1 & 4 & 2 \\ 0 & 5 & 3 \\ 1 & 3 & 2 \end{bmatrix}$.

2. $\det \begin{bmatrix} 1 & 5 & 4 \\ 3 & -3 & 6 \\ 2 & 4 & 7 \end{bmatrix} = 3 \det \begin{bmatrix} 1 & 5 & 4 \\ 1 & -1 & 2 \\ 2 & 4 & 7 \end{bmatrix}$.

3. $\det \begin{bmatrix} 1 & 3 & 4 \\ 2 & 5 & 7 \\ 3 & 2 & 8 \end{bmatrix} = \det \begin{bmatrix} 1 & 3 & 4 \\ 0 & -1 & -1 \\ 0 & -7 & -4 \end{bmatrix} = \det \begin{bmatrix} 1 & 3 & 4 \\ 2-2[1] & 5-2[3] & 7-2[4] \\ 3-3[1] & 2-3[3] & 8-3[4] \end{bmatrix}$.

4. $\det \begin{bmatrix} 1 & 2 & 1 \\ 2 & 3 & 2 \\ 3 & 2 & 3 \end{bmatrix} = 0$.

5. $\det \begin{bmatrix} 1 & 0 & 4 \\ -5 & 0 & 3 \\ 1 & 0 & 3 \end{bmatrix} = 0$.

6. $\det \begin{bmatrix} 3 & 1 & 7 \\ 0 & 4 & 9 \\ 0 & -5 & 8 \end{bmatrix} = 3 \det \begin{bmatrix} 4 & 9 \\ -5 & 8 \end{bmatrix}$.

7. $\det \begin{bmatrix} a & b & c & d \\ 0 & e & f & g \\ 0 & 0 & h & i \\ 0 & 0 & 0 & k \end{bmatrix} = aehk$.

8. Show that $\det \begin{bmatrix} 1 & 1 & 1 \\ a & b & c \\ a^2 & b^2 & c^2 \end{bmatrix} = (a-b)(b-c)(c-a)$.

9. Show that $\det \begin{bmatrix} 1 & x & x^2 & x^3 \\ 1 & y & y^2 & y^3 \\ 1 & z & z^2 & z^3 \\ 1 & w & w^2 & w^3 \end{bmatrix} = (x-y)(x-z)(x-w)(y-z)(y-w)(z-w).$

This determinant is called a (4×4) **Vandermonde determinant.** The pattern can be extended to $n \times n$ matrices.

In Exercises 10–19, evaluate the given determinant and simplify the results where appropriate.

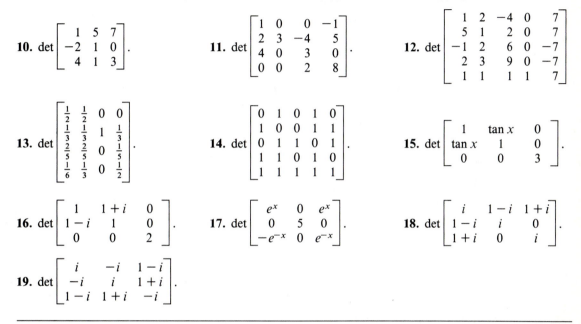

10. $\det \begin{bmatrix} 1 & 5 & 7 \\ -2 & 1 & 0 \\ 4 & 1 & 3 \end{bmatrix}$.

11. $\det \begin{bmatrix} 1 & 0 & 0 & -1 \\ 2 & 3 & -4 & 5 \\ 4 & 0 & 3 & 0 \\ 0 & 0 & 2 & 8 \end{bmatrix}$.

12. $\det \begin{bmatrix} 1 & 2 & -4 & 0 & 7 \\ 5 & 1 & 2 & 0 & 7 \\ -1 & 2 & 6 & 0 & -7 \\ 2 & 3 & 9 & 0 & -7 \\ 1 & 1 & 1 & 1 & 7 \end{bmatrix}$.

13. $\det \begin{bmatrix} \frac{1}{2} & \frac{1}{2} & 0 & 0 \\ \frac{1}{3} & \frac{1}{3} & 1 & \frac{1}{3} \\ \frac{2}{5} & \frac{2}{5} & 0 & \frac{1}{5} \\ \frac{1}{6} & \frac{1}{3} & 0 & \frac{1}{2} \end{bmatrix}$.

14. $\det \begin{bmatrix} 0 & 1 & 0 & 1 & 0 \\ 1 & 0 & 0 & 1 & 1 \\ 0 & 1 & 1 & 0 & 1 \\ 1 & 1 & 0 & 1 & 0 \\ 1 & 1 & 1 & 1 & 1 \end{bmatrix}$.

15. $\det \begin{bmatrix} 1 & \tan x & 0 \\ \tan x & 1 & 0 \\ 0 & 0 & 3 \end{bmatrix}$.

16. $\det \begin{bmatrix} 1 & 1+i & 0 \\ 1-i & 1 & 0 \\ 0 & 0 & 2 \end{bmatrix}$.

17. $\det \begin{bmatrix} e^x & 0 & e^x \\ 0 & 5 & 0 \\ -e^{-x} & 0 & e^{-x} \end{bmatrix}$.

18. $\det \begin{bmatrix} i & 1-i & 1+i \\ 1-i & i & 0 \\ 1+i & 0 & i \end{bmatrix}$.

19. $\det \begin{bmatrix} i & -i & 1-i \\ -i & i & 1+i \\ 1-i & 1+i & -i \end{bmatrix}$.

20. Prove that if A is any $n \times n$ nilpotent matrix (remember $A^k = 0$ for some positive integer k), then det $A = 0$.

21. Use Lemma 2.4 to prove that the determinant of an $n \times n$ matrix with two identical rows is zero. Compare with Exercise 37 of Section 2.2. HINT: Interchange the two identical rows.

22. Prove Property 5, that the determinant of a matrix with a row (or column) of zeros is zero.

23. Under precisely what conditions does an $n \times n$ upper-triangular matrix U have zero determinant?

24. Under precisely what conditions does an $n \times n$ lower-triangular matrix L have zero determinant?

25. Let A and B be two arbitrary 2×2 matrices. Confirm Theorem 2.7 (page 128) by directly

computing the matrix product AB and all three of the determinants, to show that det $(AB) =$ (det A)(det B).

26. Let A be any $n \times n$ matrix and let k be any scalar (real or complex number). Use mathematical induction and the properties of determinants to show that det $(kA) = k^n$ det A.

27. Use Theorem 2.7 to prove that if A is an $n \times n$ nonsingular matrix, then det $A \neq 0$. (HINT: $AA^{-1} = I$.)

28. Use Theorem 2.7 and Exercise 27 to prove that if A is an $n \times n$ nonsingular matrix, then det $(A^{-1}) = 1/\det A$.

29. Let A be an $n \times n$ matrix with the property that $AA^T = I$. Prove that det $A = 1$ or -1.

30. Let A be an $n \times n$ idempotent matrix ($A^2 = A$). What is det A? HINT: Use Theorem 2.7.

31. Prove that if $A = (a_{ij})$ is an $n \times n$ matrix, then for any two different rows, *row i* and *row j*, $i \neq j$, $i, j = 1, 2, \ldots, n$, of A it is true that $a_{i1}A_{j1} + a_{i2}A_{j2} + \cdots + a_{in}A_{jn} = 0$. Note that this resembles the Laplace expansion of det A, but it is *along row i* using the *cofactors* of row j. HINT: See the matrix $A^{\#}$ in (2.24) stated preceding Lemma 2.6 (page 127).

32. Let $T_5 = \begin{bmatrix} a & b & b & b & b \\ b & a & b & b & b \\ b & b & a & b & b \\ b & b & b & a & b \\ b & b & b & b & a \end{bmatrix}$.

Show that det $T_5 = (a - b)^4(a + 4b)$.

33. Generalize the result of Exercise 32 to any such matrix $T = (t_{ij})$ where

$$t_{ij} = \begin{cases} a & \text{if } j = i \\ b & \text{otherwise.} \end{cases}$$

That is, show that

$$\det T_n = (a - b)^{n-1}[a + (n - 1)b].$$

34. The $n \times n$ matrix $K_n = (k_{ij})$ where $k_{ij} = 0$ for $|j - i| > 1$ is called a **tridiagonal matrix**. Let K_n be

a tridiagonal matrix with the property that $k_{ii} = a$, $k_{i,i+1} = b$, and $k_{i,i-1} = c$. Show that det $K_n = a$ det $K_{n-1} - bc$ det K_{n-2} for all $n \geq 3$. Note that

$$K_4 = \begin{bmatrix} a & b & 0 & 0 \\ c & a & b & 0 \\ 0 & c & a & b \\ 0 & 0 & c & a \end{bmatrix}.$$

35. Show that for all real numbers a and b the trivial solution is the only solution to the homogeneous system of linear equations

$$\begin{aligned} x_1 + \qquad ax_2 + \qquad bx_3 &= 0, \\ (a + 1)x_2 + \qquad 2x_3 &= 0, \\ (b^2 + 1)x_2 + (1 - a)x_3 &= 0. \end{aligned}$$

36. Solve the determinantal equation

$$\det \begin{bmatrix} 3 & x + 1 \\ 2 & x - 1 \end{bmatrix} = 0 \text{ for } x.$$

37. Solve the determinantal equation

$$\det \begin{bmatrix} x + 1 & 1 & 1 \\ 1 & x + 1 & 1 \\ 1 & 1 & x + 1 \end{bmatrix} = 0 \text{ for } x.$$

---

*Exercises 38–45 are for students who have studied calculus. Let f be a differentiable function; we denote its derivative at any point x_0 of its domain by $f'(x_0)$. Let x_0 be a fixed number in the common domains of the functions f, g, h, ... We define the **Wronskian** $w(f, g, h, \ldots, x_0)$, at x_0, to be the value of the determinant of the indicated square matrix:*

$$w(f, g, h, \ldots, x_0) = \det \begin{bmatrix} f(x_0) & g(x_0) & h(x_0) & \cdots \\ f'(x_0) & g'(x_0) & h'(x_0) & \cdots \\ f''(x_0) & g''(x_0) & h''(x_0) & \cdots \\ \cdots & \cdots & \cdots & \cdots \end{bmatrix}.$$

38. Evaluate $w(e^x, e^{2x}, 0)$.

39. Evaluate $w(\sin x, \sin 2x, \pi/4)$.

40. Evaluate $w(1, \cos 2x, \sin 2x, 0)$.

41. Evaluate $w(e^x, e^{x^2}, e^{x^3}, 0)$.

42. Evaluate $w(e^x, e^{x^2}, e^{x^3}, \frac{1}{2})$.

43. Evaluate $w(\sin x, \sin^2 x, \sin^3 x, \sin^4 x, \pi/4)$.

---

44. Let $F(x)$ be the 2×2 matrix of differentiable functions $\begin{bmatrix} f_{11}(x) & f_{12}(x) \\ f_{21}(x) & f_{22}(x) \end{bmatrix}$. Let $D(x) = \det F(x)$.

$$\frac{dD}{dx} = \det \begin{bmatrix} f'_{11}(x) & f_{12}(x) \\ f'_{21}(x) & f_{22}(x) \end{bmatrix} + \det \begin{bmatrix} f_{11}(x) & f'_{12}(x) \\ f_{21}(x) & f'_{22}(x) \end{bmatrix}.$$

Show that $(f'_{ij}(x)$ denotes the derivative of $f_{ij})$.

45. State a generalization of Exercise 44 to any $n \times n$ such functional matrix $F(x)$. (Can you prove it?)

2.4 DETERMINANTS AND MATRIX INVERSION

In the last section we saw that the existence of a nonzero determinant determined that a given square matrix was nonsingular (invertible). In this section we describe a second method for computing the inverse of a given matrix, when it exists, which involves its determinant and the determinant of certain of its submatrices. Recall, from Section 2.2, that if $A = (a_{ij})$ is an $n \times n$ matrix, then the **cofactor** A_{ij} of *each entry* a_{ij} in A is defined by:

> **Definition** Cofactor of $a_{ij} = A_{ij} = (-1)^{i+j} \det A\,(i\,|\,j)$

where $A\,(i\,|\,j)$ denotes the $(n-1) \times (n-1)$ submatrix obtained from A by deleting row i and column j, and $\det A\,(i\,|\,j)$ is the minor of the element a_{ij}.

The **cofactor matrix** of $A = (a_{ij})$, denoted by Cof A, is the $n \times n$ matrix whose entries are the cofactors of the corresponding entries of A:

> **Definition** Cof $A = (A_{ij})$.

Example 1 Write the cofactor matrix for the 3×3 matrix

$$A = \begin{bmatrix} -1 & 2 & 4 \\ 0 & 3 & 5 \\ 2 & -2 & 3 \end{bmatrix}.$$

It is an easy, though long, exercise to calculate the cofactors for each of the elements of A. For example,

$$A_{31} = (-1)^{3+1} \det A(3|1) = (-1)^{3+1} \det \begin{bmatrix} 2 & 4 \\ 3 & 5 \end{bmatrix} = 10 - 12 = -2.$$

Continuing in this manner, you can verify that

$$\text{Cof } A = \begin{bmatrix} 19 & 10 & -6 \\ -14 & -11 & 2 \\ -2 & 5 & -3 \end{bmatrix}. \quad \blacksquare$$

For our purposes, however, it is the *transpose* of the cofactor matrix that we want. This matrix is called the **adjugate** (or **adjoint**) of A, and is denoted by adj A. Formally, then,

> **Definition** adj $A = (\text{Cof } A)^T$.

Example 2   Find adj A for the matrix A of Example 1.

We transpose Cof A of Example 1 to obtain

$$\text{adj } A = \begin{bmatrix} 19 & -14 & -2 \\ 10 & -11 & 5 \\ -6 & 2 & -3 \end{bmatrix}. \quad \blacksquare$$

Now let us turn to a description of the relationship between the matrix A, its inverse, if it has one, and its determinant. We have the following theorem.

Theorem 2.9   Let A be any $n \times n$ matrix. Then A^{-1} exists if, and only if, det $A \neq 0$, in which case,

$$A^{-1} = \frac{1}{\det A} \cdot \text{adj } A. \tag{2.27}$$

The first part of this theorem was proved in the previous section. It is a restatement of the corollary to Theorem 2.7 (page 128). To prove the rest of the theorem, namely, (2.27), let A be any $n \times n$ matrix and recall the Laplace expansion of det A across row i ($i = 1, 2, \ldots, n$):

$$\Delta = \det A = a_{i1}A_{i1} + a_{i2}A_{i2} + \cdots + a_{in}A_{in} = \sum_{j=1}^{n} a_{ij}A_{ij}. \tag{2.28}$$

If $i \neq j$, we have the following (see Exercise 31 of Section 2.3) from the matrix $A^{\#}$ of (2.24):

$$\det A^{\#} = a_{i1}A_{j1} + a_{i2}A_{j2} + \cdots + a_{in}A_{jn} = \sum_{k=1}^{n} a_{ik}A_{jk} = 0. \tag{2.29}$$

Let us now look at the product of the two matrices A and adj A. We have

$$A\,(\text{adj } A) = \begin{bmatrix} a_{11} & a_{12} & a_{13} & \cdots & a_{1n} \\ a_{21} & a_{22} & a_{23} & \cdots & a_{2n} \\ a_{31} & a_{32} & a_{33} & \cdots & a_{3n} \\ \vdots & \vdots & \vdots & \cdots & \vdots \\ a_{n1} & a_{n2} & a_{n3} & \cdots & a_{nn} \end{bmatrix} \begin{bmatrix} A_{11} & A_{21} & A_{31} & \cdots & A_{n1} \\ A_{12} & A_{22} & A_{32} & \cdots & A_{n2} \\ A_{13} & A_{23} & A_{33} & \cdots & A_{n3} \\ \vdots & \vdots & \vdots & \cdots & \vdots \\ A_{1n} & A_{2n} & A_{3n} & \cdots & A_{nn} \end{bmatrix}.$$

The entry m_{ij} in row i and column j of the product matrix $A(\text{adj } A)$ is

$$m_{ij} = a_{i1}A_{j1} + a_{i2}A_{j2} + a_{i3}A_{j3} + \cdots + a_{in}A_{jn}$$

which is exactly the middle part of equation (2.29) above. So when $i \neq j$, $m_{ij} = 0$. When $i = j$, then m_{ij} is precisely [see (2.28)] the determinant of A.

Therefore, the product matrix is a diagonal matrix with the number $(\det A)$ as each of its main diagonal entries. Thus,

$$A(\text{adj }A) = \begin{bmatrix} \det A & 0 & 0 & \dots & 0 \\ 0 & \det A & 0 & \dots & 0 \\ 0 & 0 & \det A & \dots & 0 \\ \vdots & \vdots & \vdots & \vdots & \vdots \\ 0 & 0 & 0 & \dots & \det A \end{bmatrix},$$

or

$$A(\text{adj }A) = (\det A)I_n. \tag{2.30}$$

Note that if $\det A = 0$, then $A(\text{adj }A) = 0$ by (2.30). On the other hand, if $\det A \neq 0$, then multiply both sides of (2.30) by the number $1/\det A$. We get

$$\left[\frac{1}{\det A}\right] A(\text{adj }A) = A\left[\frac{1}{\det A}\right](\text{adj }A) = I_n.$$

Therefore, we have the desired formula (2.27); namely,

$$A^{-1} = \left[\frac{1}{\det A}\right](\text{adj }A).$$

Example 3   Verify equation (2.30) and find the inverse, if it exists, of the matrix
$$M = \begin{bmatrix} -1 & -3 \\ 5 & 8 \end{bmatrix}.$$

It is precisely with 2×2 matrices, such as this one, that equation (2.27) for the matrix inverse is particularly useful. You can readily calculate that

$$\text{Cof }M = \begin{bmatrix} 8 & -5 \\ 3 & -1 \end{bmatrix}.$$

Therefore, $\text{adj }M = \begin{bmatrix} 8 & 3 \\ -5 & -1 \end{bmatrix}$, and $M(\text{adj }M) = \begin{bmatrix} 7 & 0 \\ 0 & 7 \end{bmatrix}$, verifying (2.30),

since $\det M = 7$. Thus $M^{-1} = \dfrac{1}{7}\begin{bmatrix} 8 & 3 \\ -5 & -1 \end{bmatrix} = \begin{bmatrix} \dfrac{8}{7} & \dfrac{3}{7} \\ \dfrac{-5}{7} & \dfrac{-1}{7} \end{bmatrix}$. Check this by

showing $MM^{-1} = I$. ∎

Example 4 Use Theorem 2.9 (page 136) to find the inverse of the matrix of Example 1, if it exists.

We calculate that det $A = -23$, so A is nonsingular from Theorem 2.9. We found adj A in Example 2; so from (2.27) we have

$$A^{-1} = \frac{1}{-23} \text{ adj } A = \frac{1}{-23} \begin{bmatrix} 19 & -14 & -2 \\ 10 & -11 & 5 \\ -6 & 2 & -3 \end{bmatrix} = \begin{bmatrix} \dfrac{-19}{23} & \dfrac{14}{23} & \dfrac{2}{23} \\ \dfrac{-10}{23} & \dfrac{11}{23} & \dfrac{-5}{23} \\ \dfrac{6}{23} & \dfrac{-2}{23} & \dfrac{3}{23} \end{bmatrix}.$$

Now you can compute AA^{-1} to verify that this is indeed the inverse of the matrix A of Example 1. Notice that $A(\text{adj } A) = \text{diag}\,[-23, -23, -23]$, as (2.30) asserts. ∎

Make particular note of the considerable number of calculations involved in the process of this example. You might wish to calculate A^{-1} using the matrix reduction algorithm of Chapter 1, and compare the effort required for each of the two methods. What do you think the comparison would be for a 4×4 matrix?

Example 5 Let $K = \begin{bmatrix} 2 & 3 & 0 \\ -1 & 4 & 7 \\ 0 & 11 & 14 \end{bmatrix}$. Does K^{-1} exist? If so, find it.

It is a simple calculation to show that det $K = 0$. Therefore, K is a singular matrix and K^{-1} *does not exist.* (See Theorem 2.9.) ∎

Example 6 Let $A = \begin{bmatrix} a & b \\ c & d \end{bmatrix}$. Here a, b, c, and d are real or complex numbers such that $ad - bc \neq 0$. Show that A^{-1} exists and, in fact,

$$A^{-1} = \frac{1}{ad - bc} \begin{bmatrix} d & -b \\ -c & a \end{bmatrix}.$$

This is a nice little formula that many people like to use for the rapid calculation of the inverse of a 2×2 matrix. To see that the formula holds, we first note that, since det $A = ad - bc \neq 0$, A^{-1} does exist. Then one can either perform the matrix multiplication AA^{-1} and obtain I_2, or one can derive the formula from (2.27); that is, calculate adj A as follows:

$$\text{adj} \begin{bmatrix} a & b \\ c & d \end{bmatrix} = \begin{bmatrix} +d & -c \\ -b & +a \end{bmatrix}^T = \begin{bmatrix} d & -b \\ -c & a \end{bmatrix}.$$

Theorem 2.9 then gives the desired result. Such a formula for each size matrix could be concocted in this manner, but the work required for anything beyond the 2×2 case makes this method inferior to the matrix inversion algorithm we learned in Chapter 1. We suggest that you use the inversion algorithm for any hand calculation of the inverse of any matrix larger than 2×2. Computer software programs are, of course, the most rapid means to arrive at the inverse of a given $n \times n$ nonsingular matrix. These may, however, involve some numerical difficulties and problems that are not covered in this course. When using a computer, it may sometimes help to also compute the value of the determinant of the given matrix A. The determinant can also, however, be contaminated by computer round-off errors. In some cases the computer could produce "an answer" for A^{-1} or a "large" value for the determinant even though the given matrix A is singular rather than nonsingular. Nevertheless the determinant may be of some help. ∎

Vocabulary

In this section we have encountered just two new terms. As before, be sure that you understand both of these concepts as well as the theorems of this section. We've also listed Equation (2.30) below. It will reappear in our later work.

cofactor matrix   adjugate (adjoint) adj A

$A(\text{adj } A) = (\det A)I_n$

Exercises 2.4

1. Find det K, where $K = \begin{bmatrix} 1 & 3 \\ -4 & 2 \end{bmatrix}$. Is K singular or nonsingular? Calculate $K(\text{adj } K)$.

Find K^{-1}, if it exists.

In Exercises 2–5 use equation (2.30) and Theorem 2.9 to find $A(\text{adj } A)$ and the inverse A^{-1} for each of the following matrices, if the inverse exists:

2. $A = \begin{bmatrix} -1 & -3 \\ 5 & 8 \end{bmatrix}$.   **3.** $A = \begin{bmatrix} -2 & 1 \\ 5 & 3 \end{bmatrix}$.   **4.** $A = \begin{bmatrix} 0 & 1 \\ 5 & -2 \end{bmatrix}$.   **5.** $A = \begin{bmatrix} 3 & 1 \\ -6 & -2 \end{bmatrix}$.

In Exercises 6–8 compute, using Theorem 2.9 (or the formula of Example 6), the inverse (if it exists) for each of the given complex matrices. (See Exercises 30–32 of Section 2.1.)

6. $\begin{bmatrix} -i & -3 \\ 5i & 8-i \end{bmatrix}$.   **7.** $\begin{bmatrix} 1+i & -3i \\ -4i & 1-2i \end{bmatrix}$.   **8.** $\begin{bmatrix} -3i & 3-3i \\ 1+2i & 8-3i \end{bmatrix}$.

9. Find det A, adj A, $A(\text{adj } A)$, and A^{-1} (if it exists) for the matrix

$$A = \begin{bmatrix} 1 & 3 & -1 \\ 4 & 0 & 1 \\ 2 & 1 & 3 \end{bmatrix}.$$

10. Given the following matrix A, does A^{-1} exist? If so, find it. If not, give a reason why.

$$A = \begin{bmatrix} 2 & 4 & 1 \\ -1 & 3 & 5 \\ 1 & 17 & 17 \end{bmatrix}.$$

11. Given the following matrix A, does A^{-1} exist? If so, find it. Also find adj A if you have not already done so.

$$A = \begin{bmatrix} 1 & 0 & 0 & 0 \\ 0 & 0 & 1 & 0 \\ 0 & 0 & 0 & 1 \\ 0 & 1 & 0 & 0 \end{bmatrix}.$$

12. Given $A = \begin{bmatrix} a & b & c & d \\ 0 & e & f & g \\ 0 & 0 & h & i \\ 0 & 0 & 0 & k \end{bmatrix}$, with $a \neq 0$, $e \neq 0$,

$h \neq 0$, and $k \neq 0$. Show that A^{-1} exists.

13. Let $A = \begin{bmatrix} a & b & c \\ 0 & a & d \\ 0 & 0 & a \end{bmatrix}$. Suppose that $a \neq 0$. Show

that A^{-1} exists. Also find A^{-1}.

14. Suppose that in the matrix in Exercise 13, $a = 0$. Show that A is not invertible and that $A^3 = 0$.

15. Suppose that A is any $n \times n$ nilpotent matrix; that is, $A^k = 0$ for some $k \geq 1$. Prove that A has no inverse.

16. Find the inverse, if it exists, of the following complex matrix by whichever of the two methods described in this text you wish to use.

$$M = \begin{bmatrix} i & 0 & 1 & -i \\ 1-2i & i & 1-i & 1+i \\ i & -1+2i & -1 & 1+i \\ 0 & 0 & 0 & 1 \end{bmatrix}.$$

17. Prove that if an $n \times n$ matrix A is nonsingular, then so is its transpose A^T.

18. Suppose that A and B are nonzero $n \times n$ matrices such that $AB = 0$. Prove that both A and B are singular.

19. Suppose that A and B are nonzero $n \times n$ matrices such that AB is singular. Is it necessarily true that both A and B are singular? Give reasons.

20. Can the product of two nonsingular matrices be singular? Give reasons.

21. Can the product of two singular matrices be nonsingular? Given reasons.

22. Suppose that the $n \times n$ matrix A is nonsingular. Prove that $\det(\text{adj } A) = (\det A)^{n-1}$.

23. Repeat Exercise 22 when A is singular. Conclude that for every $n \times n$ matrix A, $\det(\text{adj } A) = (\det A)^{n-1}$.

24. Suppose that A, B, and C are nonzero $n \times n$ matrices such that $AB = AC$. Can we conclude that $B = C$? What if A is nonsingular?

25. Prove that an $n \times n$ matrix A is nonsingular if, and only if, adj A is nonsingular.

26. Suppose that A is an $n \times n$ nonsingular matrix, $n > 2$, with real entries, and that adj $A = A$. Prove that $\det A = \pm 1$.

27. In Exercise 26, for what values of n is $\det A$ guaranteed to be equal to 1? Why?

28. What can one say about the determinants of the $n \times n$ matrices A and B if it is known that $B = P^{-1}AP$, for some nonsingular matrix P?

29. Let T_5 be the matrix of Exercise 32 in Section 2.3. What conditions on the numbers a and b makes T_5 nonsingular? Specifically, can $a = 0$?

30. Let K_4 be the tridiagonal matrix in Exercise 34 in Section 2.3. What conditions on the numbers a and b and c make K_4 nonsingular? Specifically, can $a = 0$?

31. If you have learned to use the software available in your college (department) computer center, use the computer routine to compute $\det M$ for the following 6×6 matrix.

$$M = \begin{bmatrix} 1 & -2 & 4 & 0 & 3 & -5 \\ 1 & 2 & -3 & 0 & 4 & 2 \\ 1 & 0 & 1 & 2 & 5 & -8 \\ 0 & 3 & 1 & 0 & 2 & 1 \\ 0 & -1 & 6 & 4 & 1 & 0 \\ -2 & 0 & 6 & 1 & 0 & -4 \end{bmatrix}.$$

32. If you are using a computer, find M^{-1} for the matrix M in Exercise 31, if it exists.

33. If you are using a computer, solve each of the systems of equations $M\mathbf{x} = \mathbf{b}_i$ for the following column matrices $\mathbf{b}_i$, $i = 1$, 2, 3, 4. Discuss the method you are using.

$$\mathbf{b}_1 = \begin{bmatrix} 1 \\ 0 \\ 0 \\ 2 \\ 0 \\ 1 \end{bmatrix}; \mathbf{b}_2 = \begin{bmatrix} 1 \\ 2 \\ 1 \\ 2 \\ 0 \\ -1 \end{bmatrix}; \mathbf{b}_3 = \begin{bmatrix} -1 \\ 3 \\ 6 \\ -2 \\ 3 \\ 1 \end{bmatrix}; \mathbf{b}_4 = \begin{bmatrix} 3 \\ 1 \\ 0 \\ -3 \\ 0 \\ -1 \end{bmatrix}.$$

CHAPTER 2 SUMMARY

In this chapter, we defined the concept of a determinant of an $n \times n$ matrix. We saw that in small systems of equations (two equations in two variables or maybe three equations in three variables) determinants can be used effectively to find the solutions, if they exist. This involved Cramer's rule. However, for large n, Cramer's rule is not efficient. We discussed the Laplace expansion of a determinant along with some basic properties of determinants. In particular we learned that det $(AB) = (\text{det } A)(\text{det } B)$. We introduced the concept of cofactors of an entry in a matrix, the cofactor matrix and the adjugate (classical adjoint) adj A of an $n \times n$ matrix A. We derived the important relationship $A(\text{adj } A) = (\text{det } A)I_n$ and noted that A^{-1} exists if, and only if, det $A \neq 0$. From this we derived the numerical formula $A^{-1} = \dfrac{1}{\text{det } A} \text{adj } A$.

CHAPTER 2 REVIEW EXERCISES

1. Find det $\begin{bmatrix} 2 & 3 & 0 \\ -1 & 1 & 0 \\ 4 & 7 & 3 \end{bmatrix}$ and

det $\begin{bmatrix} 3 & 7 & 5 & 1 \\ 0 & 1 & 2 & 3 \\ 0 & 0 & 4 & 7 \\ 0 & 0 & 0 & 8 \end{bmatrix}$.

2. Use the formula $A^{-1} = \dfrac{1}{\text{det } A} \text{adj } A$ to find inverses, for the two matrices in Exercise 1, if they exist. Check your answers either with a computer or by using the matrix reduction algorithm of Chapter 1.

3. Use Cramer's rule to determine the solutions, if any exist, to the following system of equations

$$2x - 3y + 4z = 1,$$
$$x + 2y - z = 4,$$
$$3x + y + z = 1.$$

4. Use Cramer's rule to determine the solutions, if any exist, to the following system of equations

$$5x - 3y + 2z = 4,$$
$$2y - 5z = 1,$$
$$3x + y = 2.$$

5. Suppose that A is an $n \times n$ matrix such that $A^2 = I_n$. Find all possible values of det A.

6. Suppose that A is an $n \times n$ matrix such that $A^2 = -I_n$. Find all possible values of det A.

7. Let $A = J_3 = \begin{bmatrix} 0 & 0 & 1 \\ 0 & 1 & 0 \\ 1 & 0 & 0 \end{bmatrix}$.

Does A^{-1} exist? If so, find it.

8. Find det $\begin{bmatrix} 1 & 3 & 3^2 & 3^3 \\ 1 & 5 & 5^2 & 5^3 \\ 1 & 2 & 2^2 & 2^3 \\ 1 & 7 & 7^2 & 7^3 \end{bmatrix}$.

9. Suppose that A is an $n \times n$ matrix such that adj $A = -A$. Find all possible values of det A.

10. Let A be an $n \times n$ matrix such that $A^T = -A$. Find all possible values of det A if n is an *odd* positive integer.

CHAPTER 2 SUPPLEMENTARY EXERCISES AND PROJECTS

1. Show that, for each positive integer n, det $A = 0$ when A is the $n \times n$ matrix

$$
A = \begin{bmatrix}
1-n & 1 & 1 & \cdots & 1 \\
1 & 1-n & 1 & \cdots & 1 \\
\cdots & \cdots & \cdots & \cdots & \cdots \\
1 & 1 & 1 & \cdots & 1-n
\end{bmatrix}.
$$

2. Let A be an $m \times n$ matrix. The dimension of the largest nonsingular square submatrix of A is called the **determinantal rank** of the matrix A. Verify that the matrix

$$
A = \begin{bmatrix}
1 & 2 & 4 \\
0 & 3 & 5 \\
1 & 2 & 3 \\
0 & 1 & 1
\end{bmatrix}
$$

   has determinantal rank 3.

3. Verify that the matrix

$$
\begin{bmatrix}
1 & 2 & 4 & 0 & -1 \\
1 & -1 & 1 & -1 & 1 \\
2 & 1 & 5 & -1 & 0 \\
3 & 0 & 6 & -2 & 1
\end{bmatrix}
$$

   has determinantal rank 2 by showing that all of its 4×4 and its 3×3 submatrices are singular, but it has at least one nonsingular 2×2 submatrix.

4. Prove that the determinantal rank of any $n \times n$ elementary matrix is n.

5. Prove that if A and B are row equivalent matrices, they have the same determinantal rank.

6. Let A be an $n \times n$ matrix with determinantal rank r. Show that if $r = n$, then adj A has determinantal rank n.

7. Let A be an $n \times n$ matrix with determinantal rank r. Show that if $r \leq n - 2$, then adj $A = 0$.

8. Let A be an $n \times n$ matrix with determinantal rank r. Show that if $r = n - 1$, then adj A has determinantal rank 1. HINT: Show that the columns of adj A each belong to the set of solutions to the homogeneous system $AX = 0$. Then show that each such solution is a scalar multiple, kX_0, of a particular solution X_0.

9. The determinants of the square submatrices of a given $m \times n$ matrix A are called its **subdeterminants.** If A has at least one nonsingular $k \times k$ submatrix, then it has a nonzero subdeterminant. Verify that the subdeterminants for the matrix of Exercise 2 are as follows: 1×1: $\{0, 1, 2, 3, 4, 5\}$; 2×2: $\{-5, -3, -2, -1, 0, 1, 3, 5\}$; 3×3: $\{-3, -2, 2\}$.

10. Find all of the subdeterminants for the matrix of Exercise 3.

11. Let A be an $m \times n$ matrix. For each k satisfying $1 \le k \le \min\{m, n\}$, the greatest common divisor f_k of all the $k \times k$ subdeterminants of A is called a **determinantal divisor** of A. Verify that the determinantal divisors of the matrix A of Exercise 2 are the following: $f_1 = 1$; $f_2 = 1$; $f_3 = 1$.

12. Refer to Exercise 11. Verify that the determinantal divisors of the matrix

$$A = \begin{bmatrix} 0 & 2 & 3 \\ 1 & 0 & 2 \\ 4 & 1 & 5 \end{bmatrix} \text{ are } f_1 = 1, f_2 = 1, f_3 = 9.$$

13. Prove that row equivalent matrices have the same determinantal divisors.

14. Prove that if A is an $n \times n$, $n \ge 2$, nonsingular matrix, then adj (adj A) $= (\det A)^{n-2}A$.

15. Prove that if A and B are $n \times n$ nonsingular matrices, then adj $(AB) = (\text{adj } B)(\text{adj } A)$.

16. Show that the only solution to the homogeneous system of linear equations

$$\begin{aligned} x_1 + 2x_2 + 2^2x_3 + 2^3x_4 &= 0, \\ x_1 + 3x_2 + 3^2x_3 + 3^3x_4 &= 0, \\ x_1 + 4x_2 + 4^2x_3 + 4^3x_4 &= 0, \\ x_1 + 5x_2 + 5^2x_3 + 5^3x_4 &= 0. \end{aligned}$$

is the trivial one: $x_1 = x_2 = x_3 = x_4 = 0$. HINT: See Problem 9 of Section 2.3.

17. Suppose that p is a *prime* integer. Prove that the determinant of the $(p-1) \times (p-1)$ matrix

$$\begin{bmatrix} 1 & 1 & 1 & \cdots & 1 \\ 1 & 2 & 2^2 & \cdots & 2^{p-2} \\ 1 & 3 & 3^2 & \cdots & 3^{p-2} \\ \vdots & \vdots & \vdots & \vdots & \vdots \\ 1 & p-1 & (p-1)^2 & \cdots & (p-1)^{p-2} \end{bmatrix}$$

is relatively prime to p. (That is, its greatest common divisor with p is 1.)

18. Let $D = \text{diag}\ [d_1, d_2, \ldots, d_n]$. Find adj D.

19. For which values of $d_1, d_2, \ldots, d_n$ is the matrix D in Exercise 18 nonsingular? Why?

20. Suppose that the matrix D of Exercise 18 is nonsingular. Find D^{-1}.

21. Find the values of $d_1, d_2, \ldots, d_n$ in the matrix D in Exercise 18 such that adj D is nonsingular.

22. Suppose that the matrix adj D is nonsingular, for D the diagonal matrix in Exercise 18. Prove that then D itself is nonsingular.

23. Let A be an $n \times n$ matrix with integer entries. Suppose that for some integer $m \ge 1$, $A^m = I_n$. Prove that $\det A = 1$ or -1. For which exponents m must $\det A = 1$? Explain.

Project 1. (Theoretical) The Determinant as a Multilinear Function

Let D be a function whose domain is the set of all $n \times n$ matrices A, with entries from $\mathbb{R}$ (or $\mathbb{C}$), and whose codomain is $\mathbb{R}$ (or $\mathbb{C}$, respectively). Furthermore, let D satisfy the following three properties, here $i = 1, 2, \ldots, n$:

1. If $\mathbf{a}_i$ is any column of A and $\mathbf{a}_i = k\mathbf{b}_i + r\mathbf{c}_i$, k and r in $\mathbb{R}$ (or $\mathbb{C}$), then $D(A) = D(\mathbf{a}_1, \mathbf{a}_2, \ldots, \mathbf{a}_i, \ldots, \mathbf{a}_n) = D(\mathbf{a}_1, \mathbf{a}_2, \ldots, k\mathbf{b}_i + r\mathbf{c}_i, \ldots, \mathbf{a}_n) = kD(\mathbf{a}_1, \mathbf{a}_2, \ldots, \mathbf{b}_i, \ldots, \mathbf{a}_n) + rD(\mathbf{a}_1, \mathbf{a}_2, \ldots, \mathbf{c}_i, \ldots, \mathbf{a}_n)$.
2. If $\mathbf{a}_i = \mathbf{a}_{i+1}$; i.e., if two adjacent columns of A are equal, then $D(A) = 0$.
3. $D(I) = 1$.

a. Prove that, for each c in $\mathbb{R}$ (or $\mathbb{C}$), $D(\mathbf{a}_1, \mathbf{a}_2, \ldots, c\mathbf{a}_i, \cdots \mathbf{a}_n) = cD(\mathbf{a}_1, \mathbf{a}_2, \ldots, \mathbf{a}_i, \ldots, \mathbf{a}_n)$.

b. Prove that $D(\mathbf{a}_1, \mathbf{a}_2, \ldots, \mathbf{b}_i + \mathbf{c}_i, \ldots, \mathbf{a}_n) = D(\mathbf{a}_1, \mathbf{a}_2, \ldots, \mathbf{b}_i, \ldots, \mathbf{a}_n) + D(\mathbf{a}_1, \mathbf{a}_2, \ldots, \mathbf{c}_i, \ldots, \mathbf{a}_n)$.

c. Prove that $D(O) = 0$; O is the $n \times n$ zero matrix.

d. Prove that $D(\mathbf{a}_1, \mathbf{a}_2, \ldots, \mathbf{a}_i, \cdots \mathbf{a}_n) = D(\mathbf{a}_1, \mathbf{a}_2, \ldots, \mathbf{a}_i + r\mathbf{a}_{i+1}, \ldots, \mathbf{a}_n)$.

e. Prove that $D(\mathbf{a}_1, \mathbf{a}_2, \ldots, \mathbf{a}_i, \mathbf{a}_{i+1}, \ldots, \mathbf{a}_n) = -D(\mathbf{a}_1, \mathbf{a}_2, \ldots, \mathbf{a}_{i+1}, \mathbf{a}_i, \cdots \mathbf{a}_n)$.

f. Prove that if $\mathbf{a}_i = \mathbf{a}_k$ for $i \neq k$, $i, k = 1, 2, \ldots, n$, then $D(A) = 0$.

g. Prove that $D(\mathbf{a}_1, \mathbf{a}_2, \ldots, \mathbf{a}_i + r\mathbf{a}_k, \ldots, \mathbf{a}_n) = D(A)$, for any $i \neq k$.

h. Prove that $D(\mathbf{a}_1, \mathbf{a}_2, \ldots, \mathbf{a}_i, \ldots, \mathbf{a}_k, \ldots, \mathbf{a}_n) = -D(\mathbf{a}_1, \mathbf{a}_2, \ldots, \mathbf{a}_k, \ldots, \mathbf{a}_i, \ldots, \mathbf{a}_n)$.

i. Show that if the function $D(A)$ exists, $D(A) = \det A$.

Project 2. (Theoretical) Permutations

We are concerned with permutations π of the finite set $S = \{1, 2, \ldots, n\}$, where $n > 1$. See Section 2.1. The set of all such permutations is denoted by S_n. The permutation ε which has the property that $\varepsilon(i) = i$ for all i in S is called the identity permutation.

a. Prove that sgn $\varepsilon = 1$.

b. A permutation τ in S_n is called a **transposition** if, and only if, τ interchanges two elements of S and leaves all of the other elements of S fixed. For example, if $n = 5$, $\{1, 2, 4, 3, 5\}$ is a transposition. Prove that sgn $\tau = -1$.

c. If α and β are two permutations in S_n, denote their composition, as functions, by $\alpha\beta$; that is, for each $i = 1, 2, \ldots, n$, $(\alpha\beta)(i) = \alpha(\beta(i))$. Show that $\alpha\beta$ is in S_n.

d. Make a "multiplication table" for the permutations in S_2.

e. Make a "multiplication table" for the permutations in S_3.

f. Show that for every permutation $\sigma \in S_n$, $\varepsilon\sigma = \sigma$.

g. Prove that each permutation in S_n can be obtained from the identity permutation ε by a sequence of transpositions. That is, for each σ in S_n, there exist some transpositions $\alpha_1, \alpha_2, \ldots, \alpha_k$ so that each $\sigma = \alpha_1\alpha_2 \cdots \alpha_k\varepsilon$.

h. Refer to part (g). Show that sgn $\sigma = (-1)^k$.

i. Use the results so far to justify the details of the proof of Lemma 2.4 (page 124).

j. Show that for each permutation π in S_3 there is a unique permutation π^{-1} in S_3 such that $\pi\pi^{-1} = \varepsilon$, the identity permutation.

k. Generalize part (j) to S_n, for any positive integer n; that is, show that for each permutation π in S_n there is a unique permutation π^{-1} in S_n such that $\pi\pi^{-1} = \varepsilon$, the identity permutation.

l. Prove that ε is always a product of an even number of transpositions.

m. Prove that if τ is a transposition, then τ^{-1} is also a transposition.

n. Suppose that $\sigma \in S_n$ is any permutation and $\sigma = \alpha_1\alpha_2 \cdots \alpha_k = \beta_1\beta_2 \cdots \beta_m$; where each α_i and β_j, $i = 1, 2, \ldots, k$, and $j = 1, 2, \ldots, m$, is a transposition. Prove that k and m are either both even or both odd integers.

o. Prove that for each $\sigma \in S_n$, sgn $\sigma^{-1} =$ sgn σ (σ^{-1} and σ are both even permutations or they are both odd permutations). HINT: See parts (g), (k), and (l).

p. Prove that exactly half of the permutations in S_n are even (and the other half are odd).

q. If π is a permutation in S_n, then applying π to the row numbers in the $n \times n$ identity matrix I_n results in the **permutation** matrix P_π. For example, if $\pi = \{2, 1, 4, 3\}$, then

$$P_\pi = \begin{bmatrix} 0 & 1 & 0 & 0 \\ 1 & 0 & 0 & 0 \\ 0 & 0 & 0 & 1 \\ 0 & 0 & 1 & 0 \end{bmatrix}.$$

Prove that det $P_\pi =$ sgn π.

r. If π is a permutation in S_n and Q_π is the matrix obtained by applying π to the column numbers of I_n, show that $P_\pi^T = Q_\pi$ and that $P_\pi Q_\pi = I_n$. See part (q).

s. Use the results of parts (q) and (r) to prove that det $Q_\pi =$ sgn π.

Project 3. *Special Matrices*

Let J and A be the following 4 × 4 matrices. J has ones on the sinistral diagonal and zeros elsewhere. A has every entry equal to 1.

$$J = \begin{bmatrix} 0 & 0 & 0 & 1 \\ 0 & 0 & 1 & 0 \\ 0 & 1 & 0 & 0 \\ 1 & 0 & 0 & 0 \end{bmatrix}, \quad A = \begin{bmatrix} 1 & 1 & 1 & 1 \\ 1 & 1 & 1 & 1 \\ 1 & 1 & 1 & 1 \\ 1 & 1 & 1 & 1 \end{bmatrix}.$$

a. Is J singular or nonsingular? If nonsingular, find J^{-1}. Why is A obviously singular?

b. Find det J and adj A.

c. What is the determinantal rank (see Supplementary Exercise 2) of the matrix A? Explain.

d. What is the determinantal rank of the matrix adj A?

e. Find the determinantal divisors of the matrix J.

f. Find the determinantal divisors of the matrix A.

g. What can you say about a 5×5 matrix J that has *ones* only on its sinistral diagonal? What if J is $n \times n$?

h. What can you say about a 5×5 matrix A that has every entry equal to one? What if A is $n \times n$?

Project 4.

*Suppose that P_1: (x_1, y_1), P_2: (x_2, y_2), . . . , P_n: (x_n, y_n), are points in the plane for which all of the x_i are distinct real numbers. Use the ideas in Exercise 16 above and the **Vandermonde determinant** (Exercise 9 of Section 2.3) to construct a polynomial $f(x)$ of degree $n - 1$ such that the points P_1, P_2, . . . , P_n all lie on the graph of $y = f(x)$.*

a. Prove that this polynomial $f(x)$ is unique.

b. Find the unique polynomial of degree 1 passing through the points $(1, 2)$ and $(-3, 4)$.

c. Find the unique polynomial of degree 2 passing through the points $(0, 3)$, $(2, 0)$, and $(1, 1)$.

d. Find the unique polynomial of degree 3 passing through the points $(2, 1)$, $(1, 0)$, $(0, 0)$ and $(-1, 2)$.

Project 5. Application—A Message Code

A very common elementary code is made by assigning each letter of the alphabet a positive integer value and then sending a message as a string of integers. For example, suppose that we use the following encoding table:

A	B	C	D	E	F	G	H	I	J	K	L	M	N	O	P	Q	R	S	T	U	V	W	X	Y	Z
6	12	16	24	4	19	8	5	20	13	17	7	1	21	2	25	10	3	23	14	11	22	15	26	9	18

Then the message COME IN OUT OF THE COLD could be sent to the intelligence agents of Example 7 in Section 1.7 as the string of integers

$$16, 2, 1, 4, 20, 21, 2, 11, 14, 2, 19, 14, 5, 4, 16, 2, 7, 24.$$

This simple code, however, is not very difficult to break. It can be disguised further by using matrix multiplication. If A is a matrix whose entries are all integers and whose determinant is ± 1, then $A^{-1} = \pm$ adj A, so the entries of A^{-1} will also be integers.

One can put the simply coded message into a matrix. There are 18 symbols in the message given above, not counting spaces between words. Thus, we can use a 3×6 matrix B, such as the following one, to represent the message.

$$B = \begin{bmatrix} 16 & 2 & 1 & 4 & 20 & 21 \\ 2 & 11 & 14 & 2 & 19 & 14 \\ 5 & 4 & 16 & 2 & 7 & 24 \end{bmatrix}. \quad \text{Let } A = \begin{bmatrix} 1 & 2 & -2 \\ 5 & -2 & 3 \\ -1 & -1 & 1 \end{bmatrix}.$$

The given matrix A has determinant det A = −1, so the product matrix AB gives a disguised coded message.

$$AB = \begin{bmatrix} 1 & 2 & -2 \\ 5 & -2 & 3 \\ -1 & -1 & 1 \end{bmatrix} \cdot \begin{bmatrix} 16 & 2 & 1 & 4 & 20 & 21 \\ 2 & 11 & 14 & 2 & 19 & 14 \\ 5 & 4 & 16 & 2 & 7 & 24 \end{bmatrix}$$

$$= \begin{bmatrix} 10 & 16 & 3 & 4 & 32 & 1 \\ 91 & 0 & 25 & 22 & 83 & 14 \\ -13 & -9 & 1 & 4 & -32 & -11 \end{bmatrix}.$$

The message is then sent as the string of integers

10, 16, −3, 4, 44, 1, 91, 0, 25, 22, 83, 149, −13, −9, 1, −4, −32, −11.

a. Compute adj A, and A^{-1}, and decode the encoded string just given. Is it the original message?

b. Write a message of your own, and use the given matrix A and the encoding table given to encode it. Then decode the result using A^{-1} from part (a). Use 0 for enough blank spaces to make your message have $3k$ characters.

c. Begin with the identity matrix I and perform Type III elementary operations, using integer multiples only on it, to construct a different transforming matrix C, which will have integer entries and determinant ± 1. Use this new matrix to encode the message you sent in part (b). Check by finding C^{-1} and decoding your coded message.

d. Make a new coding table by adding (−11) to each of the entries in the given coding table. Use the given matrix A to transform the message. Check by decoding it.

e. Must the transforming matrix be 3×3? Use the following matrix

$$K = \begin{bmatrix} 3 & 0 & -1 & 10 \\ 0 & -1 & 0 & 3 \\ -2 & -3 & 1 & 3 \\ 2 & 3 & -2 & -6 \end{bmatrix}$$

as the transforming matrix and the coding table originally given. Find K^{-1} and use it to decode a message received as

25, 15, 278, 19, 111, 48, −13, −14, 39, 8, 6, −14, −55, −43, −53,
21, −42, −74, 52, 14, −14, −54, −9, 74.

f. Use the matrix K of part (e) to encode your own message.

g. Write and send a message using the matrix K of part (e) and any encoding table you wish.

h. Start with 0 for a blank space, $A = 1$, $B = 2$, . . . , $Z = 26$. Can one make an original encoding table using a 3×3 matrix? How?

3 VECTOR SPACES

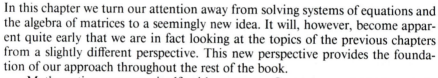

In this chapter we turn our attention away from solving systems of equations and the algebra of matrices to a seemingly new idea. It will, however, become apparent quite early that we are in fact looking at the topics of the previous chapters from a slightly different perspective. This new perspective provides the foundation of our approach throughout the rest of the book.

Mathematics concerns itself with patterns. Geometry and algebra study seemingly different patterns, but the interplay between algebra and geometry has led to the mathematical advances of the past thousand years. Geometry and algebra are the parents of linear algebra. From each the subject inherits much of its terminology, sometimes different names for the same thing, and much of its applicability. The patterns that we consider here are fundamental not only in linear algebra but also to much advanced mathematics and its many applications. We begin with the geometry of ordinary 3-space and the Euclidean plane but we go on to show how the vector ideas encountered there are special manifestations of highly important and far-reaching patterns. Just as we learned in Chapter 1 that matrices have an algebra that is similar in many ways to ordinary real and complex number algebra, but with some striking differences, so in this chapter we learn that there is a vector-space algebra that has very interesting properties.

Vector spaces are of fundamental importance in mathematics and occur in great profusion. Examples can be found not only in geometry but in the study of solutions of certain systems of linear equations, matrices, function spaces, polynomials, transformations, and so on. Central to the study of vector spaces are important concepts such as linear combinations, linear dependence, linear independence, spanning, bases, and dimension. These basic concepts are the theme of this chapter and play an important role throughout the text.

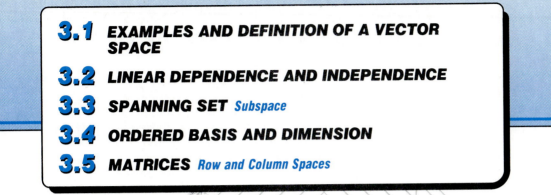

3.1 *EXAMPLES AND DEFINITION OF A VECTOR SPACE*

Many physical quantities such as length, area, density, and mass are completely described by a single real number. On the other hand, consider the plight of your college-admission officer, who must compare student applicants described by several different numbers: age, high-school GPA, test scores in mathematics, language, general ability, perhaps even numbers that assign priorities to such things as geography, sex, national origin, and financial need. Even many physical concepts need both a magnitude and a direction to describe them (force and velocity are just two examples); such concepts are frequently modeled by what are called **vectors.** Let us begin with a geometric example and then look at the more general pattern that emerges.

Example 1   **Geometric vectors in the plane**   Consider those directed line segments in the plane that are called *vectors* in elementary mechanics. They require at least two numbers to describe them, their direction and magnitude, as we shall see. Geometrically the direction is indicated by an arrowhead on the segment and the magnitude by the length of the segment as in Figure 3.1.

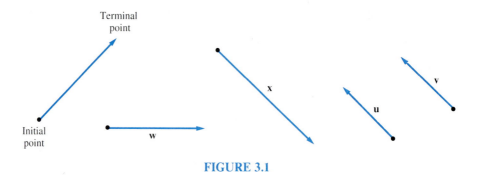

Terminal point

Initial point

x

u

v

w

FIGURE 3.1

It is generally agreed in mechanics that two of these directed line segments are *equivalent vectors* if they have the same direction and the same magnitude, even if they have different initial and terminal points. Therefore, the vectors **u** and **v** in Figure 3.1 are equivalent, but neither is equivalent to the vector **w** there.   ∎

Two operations are possible with these geometric vectors. One may

1. *Multiply* a vector by a real number α (which is called a **scalar** to distinguish it from the vector) and
2. *Add* two vectors together.

To multiply a vector by a scalar, one simply multiplies its length by that scalar. Thus $2\mathbf{v}$ is a vector twice as long as $\mathbf{v}$ while $\frac{1}{2}\mathbf{v}$ is a vector half as long. See Figure 3.2(a). The vectors $\mathbf{v}$, $2\mathbf{v}$, and $\frac{1}{2}\mathbf{v}$ all have the same direction as does $\mathbf{v}$, while $-2\mathbf{v}$ is a vector twice as long as $\mathbf{v}$ but in the opposite direction. Addition of vectors is defined by the familiar *parallelogram rule*.

> The sum of two vectors $\mathbf{u}$ and $\mathbf{v}$ that act at the same point P is the diagonal, beginning at P, of the parallelogram having $\mathbf{u}$ and $\mathbf{v}$ as adjacent sides.

See Figure 3.2(b).

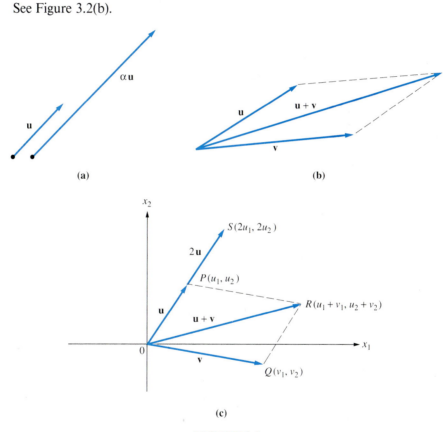

FIGURE 3.2

Using the agreement about equivalent vectors described previously, one may describe these geometric vectors algebraically by putting the initial point of all vectors at the origin of a Cartesian coordinate system. Then the terminal point of any vector $\mathbf{u}$ is a point of the coordinate plane described by the pair (u_1, u_2). Conversely, any pair of real numbers in this way describes a geometric vector. See Figure 3.2(c).

Thus it turns out that the *coordinates of the sum of two vectors are the sums of the coordinates of the individual vectors.* And also *the coordinates of the product of a vector and a scalar α are α times the original coordinates.* You can verify this geometrically: Therefore, if, as in Figure 3.2(c), we have $\mathbf{u} = (u_1, u_2)$ and $\mathbf{v} = (v_1, v_2)$, then we have the general rules

$$\mathbf{u} + \mathbf{v} = (u_1 + v_1, u_2 + v_2);$$
$$2\mathbf{u} = (2u_1, 2u_2);$$

and

$$\alpha\mathbf{u} = (\alpha u_1, \alpha u_2).$$

Similar rules apply to vectors in 3-space. We defer the problem of trying to sketch 3-space vectors on the flat page until Example 3; we then see that the same rules for addition of vectors and multiplication by a scalar apply also.

We return to this most familiar manifestation of the vector idea a little later. It is not, however, the only way in which vectors arise naturally. We look at another example from the previous chapters.

Example 2   **Row vectors of a matrix**   Although it may not be immediately apparent, similar operations occur when we consider the process of row-reducing a given matrix. Let A be the following matrix:

$$A = \begin{bmatrix} 1 & 2 & -1 \\ 3 & 1 & 4 \\ 5 & 6 & -7 \end{bmatrix} \begin{matrix} \mathbf{r}, \\ \mathbf{s}, \\ \mathbf{t}. \end{matrix}$$

Note that we have denoted the three rows of the matrix A by $\mathbf{r}$, $\mathbf{s}$, and $\mathbf{t}$, respectively. Now we proceed to reduce A to an upper-triangular matrix U by elementary row operations. To do so, we perform a vector arithmetic with the rows of A. In this case each row is a triple of real numbers. We can think of them as *row vectors* of the matrix A. Here

$$\mathbf{r} = (1, 2, -1), \qquad \mathbf{s} = (3, 1, 4), \qquad \text{and} \qquad \mathbf{t} = (5, 6, -7).$$

As a first approximation to U, let us change A to the matrix M_1 by performing the Type III elementary operations indicated to the right of M_1 with the rows of A.

$$M_1 = \begin{bmatrix} 1 & 2 & -1 \\ 0 & -5 & 7 \\ 0 & -4 & -2 \end{bmatrix} \begin{matrix} \mathbf{r}, \\ \mathbf{s} - 3\mathbf{r}, \\ \mathbf{t} - 5\mathbf{r}. \end{matrix}$$

For the second stage we perform additional row operations on M_1 to obtain the matrix M_2. The necessary row operations (row vector arithmetic) are indicated at the right of M_2:

$$M_2 = \begin{bmatrix} 1 & 2 & -1 \\ 0 & -1 & 9 \\ 0 & 2 & 1 \end{bmatrix} \quad \begin{array}{l} \mathbf{r}, \\ (\mathbf{s} - 3\mathbf{r}) - (\mathbf{t} - 5\mathbf{r}) = \mathbf{s} + 2\mathbf{r} - \mathbf{t}, \\ -\frac{1}{2}(\mathbf{t} - 5\mathbf{r}) = \dfrac{5\mathbf{r} - \mathbf{t}}{2}. \end{array}$$

Finally M_2 is row equivalent to the following upper-triangular matrix:

$$U = \begin{bmatrix} 1 & 2 & -1 \\ 0 & 1 & -9 \\ 0 & 0 & 19 \end{bmatrix} \quad \begin{array}{l} \mathbf{r}, \\ -(\mathbf{s} + 2\mathbf{r} - \mathbf{t}) = \mathbf{t} - 2\mathbf{r} - \mathbf{s}, \\ \dfrac{5\mathbf{r} - \mathbf{t}}{2} - 2(\mathbf{t} - 2\mathbf{r} - \mathbf{s}) = \dfrac{13\mathbf{r} + 4\mathbf{s} - 5\mathbf{t}}{2}. \end{array}$$

The equations at the right of U indicate how its rows (row vectors) are related to the rows (row vectors) of the original matrix A. Thus we see that the third row of U is the vector $\frac{1}{2}(13\mathbf{r} + 4\mathbf{s} - 5\mathbf{t})$, a combination of multiples of the three rows of A. We also return again to this example. ∎

Both of these examples, out of many possible, suggest that these aggregates called **vectors** have an algebra. We have added them and multiplied them by scalars. As we learned earlier that matrices have an algebra that was similar in many ways to ordinary real and complex number algebra, but with some striking differences, so also will an algebra of vectors have interesting properties. We formalize a vector algebra. The basic rules for vector algebra are distilled from our experience with geometric and analytic vectors and from operations with the rows of matrices. However, we make a significant abstraction at this point and state the fundamental rules as axioms, that is, statements we accept as true and upon which we base our discourse. From now on a **vector** will be any object that obeys the following fundamental rules, an element of a **vector space.** All of our preceding examples will be particular manifestations of sets that are vector spaces.

Definition of a Vector Space (Axioms)

Let V be any set of objects (line segments in the plane, rows of a matrix, functions, matrices, etc., might be examples) or *elements*. We call these objects **vectors** provided the set V of them satisfies the following conditions. Further let $\mathbb{R}$ denote the set of real numbers and $\mathbb{C}$ the set of complex numbers. We call elements of either $\mathbb{R}$ or $\mathbb{C}$ **scalars.** The set V is called a *real,* respectively, *complex* **vector space** provided that *all* of the following axioms hold for the elements of the set V.

Vector Addition
For **u**, **v**, and **w** any vectors in V,

1. **u** + **v** is a unique vector in V. (*Closure law*)
2. **u** + **v** = **v** + **u**. (*Commutative law*)
3. (**u** + **v**) + **w** = **u** + (**v** + **w**). (*Associative law*)
4. There is a vector **0** in V, called the **zero vector,** with the property that **u** + **0** = **u**, for every vector **u** in V. (*Additive identity*)
5. For each vector **u** in V there is a vector −**u** with the property that **u** + (−**u**) = **0**. (*Additive inverse*) −**u** is called the **negative of u.**

Multiplication by a Scalar
If α and β are any scalars (real or complex numbers), **u** and **v** any vectors in V then

6. α**u** is a unique vector in V. (*Closure*)
7. α(**u** + **v**) = α**u** + α**v**.
8. (α + β)**u** = α**u** + β**u**.
9. ($\alpha\beta$)**u** = α(β**u**).
10. 1**u** = **u**, where 1 is the real (complex) number "one."

There are many different sets of objects that are vector spaces. We consider quite a few examples here. Throughout your career you will no doubt discover many other examples to which vector-space theory with all of its power can apply. In each case a *vector* will be an object that admits the operations outlined by these axioms.

Definition   A *vector* is any element of (object in) a *vector space*.

Example 1   **Revisited. Analytic vectors—2-space:** $\mathbb{R}^2$ Let us return to the set of all points in the coordinate plane. That is, for our set V let the objects (elements) be all ordered pairs (x, y) where x and y are any real numbers. We define vector addition and multiplication by a scalar (= real number) for these pairs as follows.

$$(x_1, y_1) + (x_2, y_2) = (x_1 + x_2, y_1 + y_2), \tag{3.1}$$
$$\alpha(x_1, y_1) = (\alpha x_1, \alpha y_1). \tag{3.2}$$

Now one checks the above axioms one by one and verifies that our set V does indeed satisfy them all. V is a vector space. In particular note that the zero vector (Axiom 4) is the pair $(0, 0)$ and that $-(x_1, y_1) = (-x_1, -y_1)$ (Axiom 5). Let us verify Axiom 3 in detail for this vector space as an illustration.

Suppose that we have three ordered pairs, as follows:

$$\mathbf{v}_1 = (x_1, y_1), \qquad \mathbf{v}_2 = (x_2, y_2), \qquad \text{and} \qquad \mathbf{v}_3 = (x_3, y_3).$$

Then, by our definition (3.1) of addition, $\mathbf{v}_1 + \mathbf{v}_2 = (x_1 + x_2, y_1 + y_2)$, so that, again by (3.1),

$$
\begin{aligned}
(\mathbf{v}_1 + \mathbf{v}_2) + \mathbf{v}_3 &= (x_1 + x_2, y_1 + y_2) + (x_3, y_3) \\
&= ((x_1 + x_2) + x_3, (y_1 + y_2) + y_3) \\
&= (x_1 + (x_2 + x_3), y_1 + (y_2 + y_3)) = \mathbf{v}_1 + (\mathbf{v}_2 + \mathbf{v}_3).
\end{aligned}
$$

Here we used the *associative law* for real number algebra $(a + b) + c = a + (b + c)$ in each component of the ordered pair, because they were real numbers. Thus these pairs satisfy the associative law for vector addition (Axiom 3). ∎

In a similar way each axiom must be verified. You should have no trouble doing this making use of the corresponding rules of ordinary algebra. Thus, our set V is a vector space *because it satisfies all ten of the axioms.*

We mentioned earlier that this example can also be interpreted geometrically. This geometric interpretation gives rise to the **geometric vectors** we first discussed. We can think of the ordered pair $\mathbf{u} = (x_1, x_2)$ as the directed line segment (arrow) $\overrightarrow{OP}$ from the origin $O = (0, 0)$ to the point P whose coordinates are (x_1, x_2). Or we can think of it as an equivalent arrow between two points $Q(u_1, u_2)$ and $R(v_1, v_2)$ with the same direction and magnitude. You will want to verify that if $x_1 = v_1 - u_1$, and $x_2 = v_2 - u_2$, then the two vectors $\overrightarrow{OP}$ and $\overrightarrow{QR}$ are indeed equal. (See Figures 3.1 and 3.2 also.) Then, as we remarked at the beginning of Example 1, our vector addition (3.1) corresponds to the parallelogram rule for these line segments (arrows).

Similarly, multiplication by a scalar corresponds to the geometric interpretation previously given. For example, if $\overrightarrow{OP}_1 = \mathbf{u} = (x_1, x_2)$, then $2\,\overrightarrow{OP}_1 = 2\,\mathbf{u} = (2x_1, 2x_2)$, by (3.2). See Figure 3.3.

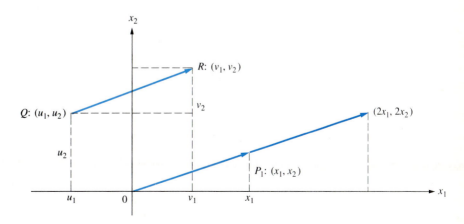

FIGURE 3.3

The vectors $\overrightarrow{OP}_2 = \frac{1}{2}\overrightarrow{OP}_1$ and $\overrightarrow{OP}_3 = (-1)\overrightarrow{OP}_1$ are illustrated in Figure 3.4(a). Notice that to multiply $\overrightarrow{OP}_1$ by the *negative* scalar (-1) is interpreted as extending it in the *opposite direction* to its original length. To multiply by $(-\alpha)$ means change to the *opposite direction* and multiply its length by $|\alpha|$. Sketch other pictures illustrating other vector sums and scalar products.

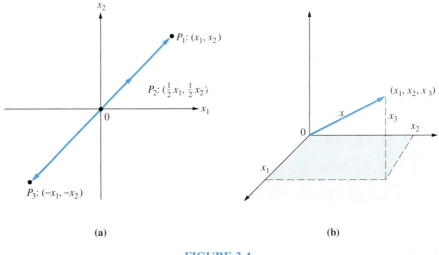

(a)   (b)

FIGURE 3.4

Example 3   **3-space**   This example generalizes Example 1. Let V be the set of all points in geometric 3-space. We most often describe V analytically by using a Cartesian coordinate system consisting of three mutually perpendicular lines intersecting at the origin. See Figure 3.4(b). Then we may think of V as consisting of all possible ordered triples (x_1, x_2, x_3) of real numbers. Define addition and multiplication by a scalar (real number) α for these ordered triples as follows:

$$(x_1, x_2, x_3) + (y_1, y_2, y_3) = (x_1 + y_1, x_2 + y_2, x_3 + y_3),$$
$$\alpha(x_1, x_2, x_3) = (\alpha x_1, \alpha x_2, \alpha x_3).$$

It should not be difficult for you to check to see that V satisfies each of the 10 axioms for a vector space. Indeed the zero vector (Axiom 4) is the triple $(0, 0, 0)$ and the negative (Axiom 5) of the vector $\mathbf{x} = (x_1, x_2, x_3)$ is the vector $-\mathbf{x} = (-x_1, -x_2, -x_3)$. Go ahead and check each of the rest of the axioms to verify that $V = \{(x_1, x_2, x_3): x_1, x_2, \text{ and } x_3 \in \mathbb{R}\}$ is indeed a vector space.   ∎

Example 4   **n-space $\mathbb{R}^n$**   This example generalizes both Example 1 and Example 3. Here one can only imagine the geometric picture for the positive integer $n > 3$. However, the analytic idea is easily generalized. Let n be any positive integer. Consider the set

$$V_n = \{(x_1, x_2, \ldots, x_n): \text{each } x_i \in \mathbb{R}\}, \qquad i = 1, 2, \ldots, n.$$

That is, V_n is the set of all *n-tuples* $(x_1, x_2, \ldots, x_n)$ of real numbers. We define addition of objects in V_n and multiplication by a scalar similar to the way we did in Example 3; namely,

$$(x_1, x_2, \ldots, x_n) + (y_1, y_2, \ldots, y_n) = (x_1 + y_1, x_2 + y_2, \ldots, x_n + y_n),$$
$$\alpha(x_1, x_2, \ldots, x_n) = (\alpha x_1, \alpha x_2, \ldots, \alpha x_n).$$

Of course, the vector space in Example 3 is the case when $n = 3$. As we did there, we suggest that you verify that each of the 10 vector-space axioms is satisfied by V_n for any positive integer n. What will the zero vector look like? Clearly $-\mathbf{x} = (-x_1, -x_2, \ldots, -x_n)$, as you can verify.

This vector space is called **Cartesian *n*-space**. We represent this space hereafter by the symbol $\mathbb{R}^n$. Note then that the vector spaces in Examples 1 and 3 are Cartesian 2-space $\mathbb{R}^2$ and Cartesian 3-space $\mathbb{R}^3$, respectively.

$\mathbb{R}^n = \{(x_1, x_2, \ldots, x_n): \text{each } x_i \in \mathbb{R}\}, i = 1, 2, \ldots, n, \text{where}$

$(x_1, x_2, \ldots, x_n) + (y_1, y_2, \ldots, y_n) = (x_1 + y_1, x_2 + y_2, \ldots, x_n + y_n)$

and $\qquad \alpha(x_1, x_2, \ldots, x_n) = (\alpha x_1, \alpha x_2, \ldots, \alpha x_n), \qquad$ all α in $\mathbb{R}$. ∎

Example 5 **Polynomial Space P_n** Consider the set V to be the set of all polynomials in an indeterminate x of degree n or less with real coefficients, together with the zero polynomial. Let vector (i.e., polynomial) addition and multiplication by a scalar (real number) be defined by the usual rules of high school algebra. We claim that V is a vector space and that *polynomials are vectors.* To verify this, one checks each of the 10 vector-space axioms. First recall that a polynomial of degree n or less can always be expressed in the form

$$p(x) = a_0 + a_1 x + a_2 x^2 + \cdots + a_n x^n. \tag{3.3}$$

Suppose also that

$$q(x) = b_0 + b_1 x + b_2 x^2 + \cdots + b_n x^n$$

is another polynomial in V. Then, as we said

$$p(x) + q(x) = (a_0 + b_0) + (a_1 + b_1)x + (a_2 + b_2)x^2 + \cdots + (a_n + b_n)x^n.$$

Moreover, $\alpha p(x) = \alpha a_0 + \alpha a_1 x + \alpha a_2 x^2 + \cdots + \alpha a_n x^n$, for each scalar α. As you verify each axiom you will note that the zero vector is the zero polynomial

$$0(x) = 0 + 0x + 0x^2 + \cdots + 0x^n$$

all of whose coefficients are zero (Axiom 4) and (Axiom 5)

$$-p(x) = -a_0 - a_1 x - a_2 x^2 - \cdots - a_n x^n.$$

Convince yourself that all of the rest of the axioms are satisfied.

Caution When we say a polynomial is zero, $p(x) = 0$, we mean that $p(x)$ is the zero polynomial, $0 + 0x + 0x^2 + \cdots + 0x^n$. We do not mean that, for some real number(s) x, a *polynomial function $p(x)$*, in the form of the polynomial $p(x)$, takes on the value of the real number zero.

The vector space P_n consists of polynomials, not of polynomial functions. Functions do form vector spaces, and we'll discuss them below. Here we should not confuse the two. ■

Throughout the rest of this book the vector space of all polynomials with real coefficients and degree at most n will be denoted by P_n. When we want the set of polynomials with complex coefficients, we'll use the symbol $\mathbb{P}_n$. You should read Example 5 again and see that all that we have said about the vectors (polynomials) in P_n carries over for those in $\mathbb{P}_n$ as well.

Example 6 **Matrices as Vectors: $\mathbf{M}_{m \times n}(\mathbb{R})$ and $\mathbf{M}_{m \times n}(\mathbb{C})$** Consider the set $M_{m \times n}(\mathbb{R})$ of all $m \times n$ matrices with *real* number entries. According to the results of Theorems 1.1 and 1.2 in Section 1.4 this set is a real vector space. We ask you to verify each of the 10 axioms for this set as an exercise. A similar exercise asks you to verify that the set $M_{m \times n}(\mathbb{C})$ of all $m \times n$ matrices with complex number entries is a complex vector space. What is the zero vector in each of these spaces? When $m = n$, we'll use the shorter notation $M_n(\mathbb{R})$ and $M_n(\mathbb{C})$ for the vector spaces of *square* real and complex matrices.

Notice that these vector spaces are different from a vector space of row vectors of a given matrix discussed in Example 2. Here the *vectors are* the *matrices* themselves, *not* their individual rows. The rows are vectors in a different vector space, $\mathbb{R}^n$ (or $\mathbb{C}^n$). ■

Example 7 **The Solution Space of a Homogeneous System of Equations** Consider any homogeneous system of equations. This can be written in matrix notation as

$$AX = \mathbf{0}, \tag{3.4}$$

where A is an $n \times m$ matrix, X is an $m \times 1$ matrix, and $\mathbf{0}$ is the $n \times 1$ matrix of zeros;

$$X = \begin{bmatrix} x_1 \\ x_2 \\ \vdots \\ x_m \end{bmatrix} \quad \text{and} \quad \mathbf{0} = \begin{bmatrix} 0 \\ 0 \\ \vdots \\ 0 \end{bmatrix}.$$

In the important special case where A is an $n \times n$ matrix, note that if A is nonsingular, the only solution to the system (3.4) is the trivial solution, $\mathbf{0}$. However, if $\mathbf{A}$ is singular, then in addition to the trivial solution there are infinitely many nontrivial solutions. In previous exercises you should have shown that if X_1 and X_2 are two such solutions then so is $X_1 + X_2$ and αX_1, for every scalar α. These facts follow easily from (3.4) since $A(X_1 + X_2) = AX_1 + AX_2$ and $A(\alpha X_1) = \alpha AX_1$. Now you should verify that the set of *all* solutions to a homogeneous system of equations (3.4) does indeed satisfy each of the 10 vector-space axioms.

This vector space is called the **solution space** $S(A)$ of the given matrix A. For example, in addition to the trivial solution, the system

$$
\begin{aligned}
x_1 - x_2 - 3x_3 + 2x_4 &= 0, \\
2x_1 + x_2 - x_3 + x_4 &= 0, \\
x_1 + 2x_2 + 2x_3 - x_4 &= 0, \\
3x_1 + 3x_2 + x_3 \qquad &= 0,
\end{aligned}
\tag{3.5}
$$

has infinitely many nontrivial solutions. Among them are the two solutions

$$
X_1 = \begin{bmatrix} 4 \\ -5 \\ 3 \\ 0 \end{bmatrix}
\quad \text{and} \quad
X_2 = \begin{bmatrix} -1 \\ 1 \\ 0 \\ 1 \end{bmatrix},
$$

as you can readily verify. Verify also that

$$
X_1 + X_2 = \begin{bmatrix} 3 \\ -4 \\ 3 \\ 1 \end{bmatrix}
\quad \text{and} \quad
\alpha X_1 = \begin{bmatrix} 4\alpha \\ -5\alpha \\ 3\alpha \\ 0 \end{bmatrix}
$$

are solutions for any α. The vector

$$
\alpha X_1 + \beta X_2 = \begin{bmatrix} 4\alpha - \beta \\ -5\alpha + \beta \\ 3\alpha \\ \beta \end{bmatrix}
$$

is also a solution, for every α and β. You should check this for the specific case of (3.5). Returning to the general case where A is an $n \times m$ matrix, we call the solution $m \times 1$ matrices X to a given homogeneous system of equations $AX = \mathbf{0}$, as in (3.4), *column vectors,* or *solutions vectors* to the system. ∎

Example 8   **Complex n-space $\mathbb{C}^n$**   Let $\mathbb{C}^n$ be the set of ordered n-tuples $(z_1, z_2, \ldots, z_n)$ of complex numbers. It is not difficult to see that each of the axioms for a vector space is satisfied by this set. The verification is based on complex-number algebra and is almost identical to the vertification for the vector space $\mathbb{R}^n$ of Example 4. You should work this verification out. ∎

Example 9

Space of Continuous Functions $C[a, b]$ Let $C[a, b]$ denote the set of all real-valued continuous functions defined on the closed interval $[a, b]$ of the real line. Let f, g, h, etc., denote functions in $C[a, b]$; define addition and scalar multiplication of functions as follows:

- $(f + g)(x) = f(x) + g(x)$ for all x in the interval $[a, b]$; that is, the *sum function* $f + g$ is that function whose value at x is the sum of the individual values of f and g at x.
- $(\alpha f)(x) = \alpha(f(x))$ for all real numbers α; that is, the function αf is that function whose value at any x is α times the value of f at x.

Now, from the properties of functions one can readily see that each of the axioms for a vector space is satisfied by $C[a, b]$. Notice, in particular, that the zero vector is the function $O(x) \equiv 0$. You should check each axiom and see how they are fulfilled by this set. This is a very important vector space in advanced mathematical analysis. See Figure 3.5. ■

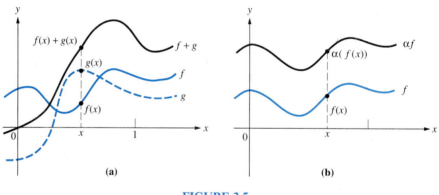

FIGURE 3.5

We conclude this section with a final example, a definition, and a theorem.

Example 10

The Zero Vector Space Let V consist of the single element $\mathbf{0}$. One can view this as the zero function from $C[a, b]$, the zero matrix from $M_n(\mathbb{R})$ or $M_n(\mathbb{C})$, the zero real or complex number, the zero polynomial, or an n-tuple consisting of only zeros, or whatever zero vector one is interested in. $V = \{\mathbf{0}\}$ is a vector space in its own right. It trivially satisfies all of the axioms. We call it the zero vector space. If V is a vector space and $V \neq \{\mathbf{0}\}$, we say "V is a nonzero vector space." ■

Now, in view of Example 6, we might expect that the usual algebraic properties of matrix addition and scalar multiplication that were contained in Theorem 1.4 (page 43) have their analogs in *any* vector space V. Indeed many of them do, as we shall see. We first of all define **vector subtraction** by the statement

---

Definition $\mathbf{u} - \mathbf{v} = \mathbf{u} + (-\mathbf{v})$.

Now we conclude with the following theorem, whose proof is similar to the proof of Theorem 1.4. We leave the proof of each part to you as some rather routine exercises. One comment is in order. In the theorem for clarity we have written 0_S for the zero (real or complex number) scalar, and $\mathbf{0}_V$ for the zero vector from V. In the future we omit the subscripts. Quite often we merely write 0 for either of them and assume that the context makes it clear which zero we are referring to. Boldface $\mathbf{0}$ will also often be used for the zero vector, and merely 0 for the zero scalar.

Theorem 3.1   If V is any real (or complex) vector space, then for all vectors $\mathbf{u}$ and $\mathbf{v}$ in V and for all scalars α, β, etc.

  (i) $(-1)\mathbf{u} = -\mathbf{u}$.
 (ii) $-(-\mathbf{u}) = \mathbf{u}$.
(iii) $\mathbf{u} - \mathbf{v} = -(\mathbf{v} - \mathbf{u})$.
 (iv) $0_S\mathbf{v} = \mathbf{0}_v$, where 0_S is the zero scalar and $\mathbf{0}_v$ is the zero vector.
  (v) $\alpha\mathbf{0}_v = \mathbf{0}_v$ for any scalar α.
 (vi) If $\alpha\mathbf{v} = \mathbf{0}_v$, then $\alpha = 0_S$ or $\mathbf{v} = \mathbf{0}_v$.
(vii) $(\alpha - \beta)\mathbf{u} = \alpha\mathbf{u} - \beta\mathbf{u}$.
(viii) $\alpha(\mathbf{u} - \mathbf{v}) = \alpha\mathbf{u} - \alpha\mathbf{v}$.
 (ix) If $\mathbf{v}_1, \mathbf{v}_2, \ldots, \mathbf{v}_n$ is any finite collection of vectors from V, then the vector $\mathbf{u} = c_1\mathbf{v}_1 + c_2\mathbf{v}_2 + \cdots + c_n\mathbf{v}_n$ is a vector in V, for any scalars $c_1, c_2, \ldots, c_n$.

The proof of part (ix) follows from Axioms 6 and 1 by induction on the positive integer n.

Vocabulary

In this section we have again encountered new ideas and terms. As in all previous sections, be sure that you understand the concepts as well as the theorems. Memorize the definition of a vector space—all 10 axioms, and the special vector spaces given. Listed below are the new terms and symbols to add to your linear algebra vocabulary.

geometric vector
parallelogram rule
vector space
zero vector
$M_{m\times n}(\mathbb{R})$ and $M_{m\times n}(\mathbb{C})$
$M_n(\mathbb{R})$ and $M_n(\mathbb{C})$
$C[a, b]$

row vector
vector
scalar
Cartesian n-space $\mathbb{R}^n$
P_n and $\mathbb{P}_n$
$\mathbb{C}^n$
solution space $S(A)$

Exercises 3.1

1. Let the points P_1:(1, 2) and P_2:(−1, 1) of the co-ordinate plane be given. Sketch the two vectors $\overrightarrow{OP_1}$ and $\overrightarrow{OP_2}$ from the origin to each point.

2. Sketch the sum of the two vectors $\overrightarrow{OP_1}$ and $\overrightarrow{OP_2}$ in Exercise 1.

3. Sketch the vector from the point P_2 to the point P_1 of Exercise 1. Show that it is the difference of the two vectors $\overrightarrow{OP_1}$ and $\overrightarrow{OP_2}$.

4. Represent each of the vectors in Exercises 1, 2, and 3 by vectors (ordered pairs) in $\mathbb{R}^2$.

In Exercises 5–18 find $\mathbf{u} + \mathbf{v}$, $3\mathbf{u}$, $\frac{2}{3}\mathbf{v}$, $\mathbf{u} - \mathbf{v}$, *and* $\frac{1}{2}\mathbf{u} - \frac{3}{2}\mathbf{v}$ *for the given vectors* $\mathbf{u}$ *and* $\mathbf{v}$.

5. $\mathbf{u} = (2, 1)$, $\mathbf{v} = (-1, -2)$ from $\mathbb{R}^2$.

6. $\mathbf{u} = (1, 1, 1)$, $\mathbf{v} = (2, -1, -3)$ from $\mathbb{R}^3$.

7. $\mathbf{u} = (1, -1, -1)$, $\mathbf{v} = (5, 2, -2)$ from $\mathbb{R}^3$.

8. $\mathbf{u} = (1, -1, -1, 1)$, $\mathbf{v} = (-3, 5, 2, -2)$ from $\mathbb{R}^4$.

9. $\mathbf{u} = (1, -1, 0, -1, 1)$, $\mathbf{v} = (-3, 5, 2, -2, 0)$ from $\mathbb{R}^5$.

10. $\mathbf{u} = (-i, -1 + i)$, $\mathbf{v} = (2 - i, 1 - 2i)$ from $\mathbb{C}^2$.

11. $\mathbf{u} = (1, -i, -1 + i)$, $\mathbf{v} = (2i, 2 - i, 1 - 2i)$ from $\mathbb{C}^3$.

12. $\mathbf{u} = 2x^2 - 3x + 4$, $\mathbf{v} = x^2 - 7x + 3$ from P_2.

13. $\mathbf{u} = 4x^3 - 2x^2 - 3x + 4$, $\mathbf{v} = x^3 - 2x^2 - 7x + 3$ from P_3.

14. $\mathbf{u} = 4x^3 - 2ix^2 - (3 + i)x + (4 - i)$, $\mathbf{v} = (1 + i)x^3 + (1 - 2i)x^2 - 7ix + 3$ from $\mathbb{P}_3$.

15. $\mathbf{u} = \begin{bmatrix} 1 & -1 \\ 2 & 1 \end{bmatrix}$, $\mathbf{v} = \begin{bmatrix} 2 & 0 \\ -1 & 1 \end{bmatrix}$ from $M_2(\mathbb{R})$.

16. $\mathbf{u} = \begin{bmatrix} -6 & 3 \\ 4 & -2 \end{bmatrix}$, $\mathbf{v} = \begin{bmatrix} 1 & -2 \\ 2 & -4 \end{bmatrix}$ from $M_2(\mathbb{R})$.

17. $\mathbf{u} = e^x$, $\mathbf{v} = e^{-x}$ from $C[-1, 1]$.

18. $\mathbf{u} = e^x \cos x$, $\mathbf{v} = e^{-x} \sin x$ from $C[-\pi, \pi]$.

---

19. Sketch the two vectors in Exercise 6.

20. Complete the verification that the axioms for a vector space are satisfied for $\mathbb{R}^3$.

21. Complete the verification that the axioms for a vector space are satisfied for $\mathbb{C}^3$.

22. Complete the verification that the axioms for a vector space are satisfied for $\mathbb{P}_3$.

23. Complete the verification that the axioms for a vector space are satisfied for $M_{m \times n}(\mathbb{R})$.

24. Complete the verification that the axioms for a vector space are satisfied for $M_{m \times n}(\mathbb{C})$.

25. Complete the verification that the axioms for a vector space are satisfied for $C[a, b]$.

In Exercises 26–44 consider each of the given sets and the operations of vector addition and multiplication by a scalar that are proposed. Determine if the given set is a vector space under these conditions. That is, verify that all of the axioms are satisfied or else state which axiom(s) fail to hold.

26. W is the set of all ordered pairs (x, y) of real numbers that satisfy the equation $2x + 3y = 0$.

27. W' is the set of all ordered pairs (x, y) of real numbers that satisfy the equation $2x + 3y = 1$.

28. U is the set of all ordered triples (x, y, z) of real numbers that satisfy the equation $2x + 3y - z = 0$.

29. U' is the set of all ordered triples (x, y, z) of real numbers that satisfy the equation $2x + 3y - z = c$. What must the value of c be for U' to be a vector space?

30. H is the set of all ordered pairs (x, y) of real numbers. Define *addition* as $(x, y) + (x', y') = (x + x', 0)$; and *multiplication* by a scalar as $\alpha(x, y) = (\alpha x, \alpha y)$.

31. H is the set of all ordered pairs (x, y) of real numbers. Define *addition* as $(x, y) + (x', y') = (x', y)$; and *multiplication* by a scalar as $\alpha(x, y) = (\alpha x, 0)$.

32. H is the set of all ordered triples (x, y, z) of real numbers. Define *addition* as $(x, y, z) + (x', y', z') = (x + x', 0, z + z')$; and *multiplication* by a scalar as $\alpha(x, y, z) = (\alpha x, \alpha y, 0)$.

33. Let $\mathbb{Q}$ be the set of all the rational numbers. Let $\mathbb{Q}$ serve as both the set of vectors with ordinary addition of rational numbers and the set of scalars with ordinary multiplication of rational numbers.

34. Let $\mathbb{R}$ be the set of all the real numbers. Let $\mathbb{R}$ serve as both the set of vectors with ordinary addition of real numbers and the set of scalars with ordinary multiplication of real numbers.

35. Let $\mathbb{Z}$ be the set of all the ordinary integers. Let $\mathbb{Z}$ serve as both the set of vectors with ordinary addition of integers and the set of scalars with ordinary multiplication of integers.

36. Let $\mathbb{C}$ be the set of all the complex numbers. Let $\mathbb{C}$ serve as both the set of vectors with ordinary addition of complex numbers and the set of scalars with ordinary multiplication of complex numbers.

37. Let V be the collection of all 2×2 matrices of the form $\begin{bmatrix} r & 0 \\ 0 & s \end{bmatrix}$ with addition and scalar multiplication as in $M_2(\mathbb{R})$.

38. Let V be the collection of all 2×2 matrices of the form $\begin{bmatrix} r & 1 \\ 1 & s \end{bmatrix}$ with addition and scalar multiplication as in $M_2(\mathbb{R})$.

39. Let K be the collection of all those functions f in $C[0, 1]$ for which $f(1) = 0$.

40. Let H be the collection of all those functions f in $C[0, 1]$ for which $f(0) = 1$.

41. Let $D[a, b]$ be the collection of all real-valued *differentiable* functions defined on the closed interval $[a, b]$ with the same operations as in $C[a, b]$.

42. Let $D^2[a, b]$ be the collection of all real-valued *twice differentiable* functions defined on the closed interval $[a, b]$ with the same operations as in $C[a, b]$.

43. Let $\mathbb{I}[a, b]$ be the collection of all real-valued *integrable* functions defined on the closed interval $[a, b]$ with the same operations as in $C[a, b]$.

44. Let $\mathbb{F}[a, b]$ be the collection of *all* real-valued functions defined on the closed interval $[a, b]$ with the same operations as in $C[a, b]$.

---

45. Prove that in any vector space V the zero vector $\mathbf{0}$ is unique. HINT: Suppose that there are two zero vectors $\mathbf{0}_1$ and $\mathbf{0}_2$. Then what is $\mathbf{0}_1 + \mathbf{0}_2$?

46. Verify that the solution space $S(A)$ in Example 7, for the homogeneous system $AX = 0$, where A is a given $n \times n$ matrix, is indeed always a vector space.

---

47. Prove Theorem 3.1 (i), page 161.

48. Prove Theorem 3.1 (ii).

49. Prove Theorem 3.1 (iii).

50. Prove Theorem 3.1 (iv).

51. Prove Theorem 3.1 (v).

52. Prove Theorem 3.1 (vi).

53. Prove Theorem 3.1 (vii).

54. Prove Theorem 3.1 (viii).

3.2 LINEAR DEPENDENCE AND INDEPENDENCE

Let us look back at two examples in Section 3.1. First of all, in Example 2 we reduced the matrix

$$A = \begin{bmatrix} 1 & 2 & -1 \\ 3 & 1 & 4 \\ 5 & 6 & -7 \end{bmatrix} \begin{array}{l} \mathbf{r}, \\ \mathbf{s}, \\ \mathbf{t}, \end{array}$$

with row vectors $\mathbf{r}$, $\mathbf{s}$, and $\mathbf{t}$ as indicated, to the upper-triangular matrix

$$U = \begin{bmatrix} 1 & 2 & -1 \\ 0 & 1 & -9 \\ 0 & 0 & 19 \end{bmatrix} \begin{array}{l} \mathbf{u} = \mathbf{r}, \\ \mathbf{v} = \mathbf{t} - 2\mathbf{r} - \mathbf{s}, \\ \mathbf{w} = \dfrac{13\mathbf{r} + 4\mathbf{s} - 5\mathbf{t}}{2}. \end{array}$$

The row vectors $\mathbf{u}$, $\mathbf{v}$, and $\mathbf{w}$ of the matrix U were derived from the row vectors of the matrix A by the equations

$$
\begin{align}
\mathbf{u} &= \mathbf{r}, \\
\mathbf{v} &= \mathbf{t} - 2\mathbf{r} - \mathbf{s}, \tag{3.6} \\
\mathbf{w} &= \frac{13}{2}\mathbf{r} + 2\mathbf{s} - \frac{5}{2}\mathbf{t}.
\end{align}
$$

For the second example consider Example 7 in Section 3.1. There we considered the solution space $S(A)$ for a homogeneous system of equations. In the numerical example given there we found that the system

$$
\begin{align}
x_1 - x_2 - 3x_3 + 2x_4 &= 0, \\
2x_1 + x_2 - x_3 + x_4 &= 0, \\
x_1 + 2x_2 + 2x_3 - x_4 &= 0, \tag{3.7} \\
3x_1 + 3x_2 + x_3 &= 0,
\end{align}
$$

had the two particular solutions

$$X_1 = \begin{bmatrix} 4 \\ -5 \\ 3 \\ 0 \end{bmatrix} \quad \text{and} \quad X_2 = \begin{bmatrix} -1 \\ 1 \\ 0 \\ 1 \end{bmatrix}.$$

You were asked to show that for any two real numbers α and β, the column vector (4×1 matrix)

$$\alpha X_1 + \beta X_2 = \begin{bmatrix} 4\alpha - \beta \\ -5\alpha + \beta \\ 3\alpha \\ \beta \end{bmatrix}$$

is also a solution. The coefficient matrix for this system of equations is the matrix

$$K = \begin{bmatrix} 1 & -1 & -3 & 2 \\ 2 & 1 & -1 & 1 \\ 1 & 2 & 2 & -1 \\ 3 & 3 & 1 & 0 \end{bmatrix} \begin{matrix} \mathbf{r}_1, \\ \mathbf{r}_2, \\ \mathbf{r}_3, \\ \mathbf{r}_4. \end{matrix}$$

You can compute that

$$K(\alpha X_1 + \beta X_2) = \begin{bmatrix} 1 & -1 & -3 & 2 \\ 2 & 1 & -1 & 1 \\ 1 & 2 & 2 & -1 \\ 3 & 3 & 1 & 0 \end{bmatrix} \begin{bmatrix} 4\alpha - \beta \\ -5\alpha + \beta \\ 3\alpha \\ \beta \end{bmatrix} = \begin{bmatrix} 0 \\ 0 \\ 0 \\ 0 \end{bmatrix}.$$

More generally, suppose that we are given any homogeneous system of equations $KX = 0$, with any two particular solutions X_1 and X_2 to that system. Then for any two real numbers s and t the column vector

$$X = sX_1 + tX_2 \tag{3.8}$$

is also a solution. This follows from the rules of matrix algebra (see Theorem 1.4, page 43) since

$$K(sX_1 + tX_2) = KsX_1 + KtX_2 = sKX_1 + tKX_2 = s0 + t0 = 0,$$

because both X_1 and X_2 are solutions. So $X = sX_1 + tX_2$ is always a solution.

Now note from these two examples that the row vectors of the matrix U were written as combinations of the row vectors of the matrix A, as indicated in equation (3.6). Similarly, the solution to the system (3.5) of equations is a combination of the two solutions X_1 and X_2 as given by equation (3.8). Combinations of vectors such as these are called **linear combinations.** In general, we have the following definition.

Definition Let V be any vector space; let $\mathbf{v}_1, \mathbf{v}_2, \ldots, \mathbf{v}_n$ be vectors from V, and let $c_1, c_2, \ldots, c_n$ be any scalars. The vector

$$\mathbf{u} = \sum_{i=1}^{n} c_i \mathbf{v}_i = c_1 \mathbf{v}_1 + c_2 \mathbf{v}_2 + \cdots + c_n \mathbf{v}_n \tag{3.9}$$

is called a **linear combination** of the n vectors $\mathbf{v}_1, \mathbf{v}_2, \ldots, \mathbf{v}_n$.

The axioms for a vector space tell us that this vector $\mathbf{u}$ is a vector in the vector space V. See Theorem 3.1 (ix). Thus, the rows of the matrix U above are *linear combinations* of the row vectors $\mathbf{r}$, $\mathbf{s}$, and $\mathbf{t}$ of the matrix A. The solution X of the system of equations $KX = 0$ of (3.7) is a *linear combination* of the particular solution vectors X_1 and X_2 from its solution space $S(K)$.

Now let us take a look at the row vectors of the 4×4 matrix K above; call them $\mathbf{r}_1, \mathbf{r}_2, \mathbf{r}_3,$ and $\mathbf{r}_4$. Although you may not have noticed it before, it should be clear that

$$\mathbf{r}_1 = \mathbf{r}_2 - \mathbf{r}_3 \quad \text{and} \quad \mathbf{r}_4 = \mathbf{r}_2 + \mathbf{r}_3.$$

Of course, you did notice that the homogeneous system (3.7) was *dependent*. Otherwise it would have had only the trivial solution. That is, the solution space $S(K)$ would have been merely the *zero vector space*. (See Example 10 in Section 3.1.)

The individual rows of the matrix K are dependent on each other; in particular, the row vectors $\mathbf{r}_1$ and $\mathbf{r}_4$ depend on the row vectors $\mathbf{r}_2$ and $\mathbf{r}_3$ as indicated. Note that

$$\mathbf{r}_1 - \mathbf{r}_2 + \mathbf{r}_3 = \mathbf{0} \quad \text{and} \quad \mathbf{r}_4 - \mathbf{r}_2 - \mathbf{r}_3 = \mathbf{0}.$$

In generalizing these ideas we are led to the following definition.

Definition   Let V be a vector space and let $S = \{\mathbf{v}_1, \mathbf{v}_2, \ldots, \mathbf{v}_n\}$ be a set of vectors from V. We say that the vectors $\mathbf{v}_1, \mathbf{v}_2, \ldots, \mathbf{v}_n$ are **linearly dependent,** or that the set S is a **linearly dependent set,** if there exist scalars $c_1, c_2, \ldots, c_n$, *not all of which are zero,* such that

$$c_1\mathbf{v}_1 + c_2\mathbf{v}_2 + \cdots + c_n\mathbf{v}_n = \mathbf{0}; \tag{3.10}$$

otherwise, vectors $\mathbf{v}_1, \mathbf{v}_2, \ldots, \mathbf{v}_n$ (or the set S) are **linearly independent.**

Thus the row vectors $\mathbf{r}_1, \mathbf{r}_2, \mathbf{r}_3,$ and $\mathbf{r}_4$ of K are *linearly dependent.* But the row vectors $\mathbf{r}, \mathbf{s},$ and $\mathbf{t}$ of the matrix A of Example 2 in Section 3.1 are *linearly independent,* as you can verify (Exercise 13).

A very useful formulation of the concept of *linear independence* is given in the following lemma.

Lemma 3.2   A set $S = \{\mathbf{v}_1, \mathbf{v}_2, \ldots, \mathbf{v}_n\}$ of vectors from the vector space V is a **linearly independent set** if, and only if, whenever any linear combination $c_1\mathbf{v}_1 + c_2\mathbf{v}_2 + \cdots + c_n\mathbf{v}_n$ of the vectors in S is zero, then **all** of the scalar coefficients $c_1, c_2, \ldots, c_n$ are zero. In other words, the vectors $\mathbf{v}_1, \mathbf{v}_2, \ldots, \mathbf{v}_n$ are linearly independent when the equation

$$c_1\mathbf{v}_1 + c_2\mathbf{v}_2 + \cdots + c_n\mathbf{v}_n = \mathbf{0}$$

forces each and every one of the c_i to be zero: $c_1 = c_2 = \cdots = c_n = 0$.

The lemma follows immediately from the definition since, if there were but one of the c_i different from zero and yet $c_1\mathbf{v}_1 + c_2\mathbf{v}_2 + \cdots + c_n\mathbf{v}_n = \mathbf{0}$, then the vectors would be linearly *dependent* rather than *independent*. In what follows we frequently use the formulation of linear independence given in this lemma.

Example 1   Are the three vectors $x_1 = (0, 1, 1)$, $x_2 = (1, 0, 1)$, and $x_3 = (1, 1, 0)$ in Cartesian 3-space $\mathbb{R}^3$ linearly dependent or are they linearly independent?

To decide which, let us use Lemma 3.2. Suppose that we have the linear combination

$$\begin{aligned} 0 &= c_1x_1 + c_2x_2 + c_3x_3 \\ &= c_1(0, 1, 1) + c_2(1, 0, 1) + c_3(1, 1, 0) \\ &= (0, 0, 0). \end{aligned} \qquad \textbf{(3.11)}$$

Recalling the definitions of vector addition and scalar multiplication in $\mathbb{R}^3$ discussed in Example 3 of Section 3.1, we see that equation (3.11) becomes

$$\begin{aligned} (0, c_1, c_1) + (c_2, 0, c_2) + (c_3, c_3, 0) &= (0 + c_2 + c_3, c_1 + 0 + c_3, c_1 + c_2 + 0) \\ &= (0, 0, 0). \end{aligned}$$

This last vector equation will be satisfied if, and only of, the unknown scalars c_1, c_2, and c_3 satisfy the following system of equations:

$$\begin{aligned} c_2 + c_3 &= 0, \\ c_1 + \quad c_3 &= 0, \\ c_1 + c_2 \quad &= 0. \end{aligned}$$

We will have linear dependence if this system has any nontrivial solution. If, however, only the trivial solution is possible, then Lemma 3.2 would assure us that the given vectors x_1, x_2, and x_3 are linearly independent. Note that the coefficient matrix for this particular homogeneous system of linear equations is

$$M = \begin{bmatrix} 0 & 1 & 1 \\ 1 & 0 & 1 \\ 1 & 1 & 0 \end{bmatrix}$$

and M is nonsingular. (Note that det $M = 2 \neq 0$.) Therefore, the only solution to this system is $X = M^{-1}0 = 0$. In other words, each of the scalar coefficients c_1, c_2, and c_3 in equation (3.11) *must* be zero. Equation (3.11) forces all the scalars to be zero, so the three vectors $x_1 = (0, 1, 1)$, $x_2 = (1, 0, 1)$, and $x_3 = (1, 1, 0)$ are *linearly independent.*   ∎

Example 2   Are the three vectors $x_1 = (1, 2, -1)$, $x_2 = (3, 0, 4)$, and $x_3 = (1, -4, 6)$ in Cartesian 3-space $\mathbb{R}^3$ linearly dependent, or are they linearly independent?

Again, consider the linear combination

$$\begin{aligned} 0 &= c_1x_1 + c_2x_2 + c_3x_3 \\ &= c_1(1, 2, -1) + c_2(3, 0, 4) + c_3(1, -4, 6) \\ &= (0, 0, 0). \end{aligned} \qquad \textbf{(3.12)}$$

Our objective is to decide whether or not equation (3.12) forces the scalar coefficients to be zero or if nonzero values are possible. Again using the definitions of vector addition and scalar multiplication in $\mathbb{R}^3$ we see that equation (3.12) becomes

$$(c_1, 2c_1, -c_1) + (3c_2, 0, 4c_2) + (c_3, -4c_3, 6c_3) = (0, 0, 0)$$

or

$$(c_1 + 3c_2 + c_3, 2c_1 - 4c_3, -c_1 + 4c_2 + 6c_3) = (0, 0, 0).$$

The last vector equation will be satisfied if, and only if, the unknown scalars c_1, c_2, and c_3 satisfy the following homogeneous system of equations:

$$
\begin{aligned}
c_1 + 3c_2 + c_3 &= 0, \\
2c_1 \quad\quad - 4c_3 &= 0, \\
-c_1 + 4c_2 + 6c_3 &= 0.
\end{aligned}
$$

We will have linear dependence if this system has any nontrivial solution. If only the trivial solution exists, then Lemma 3.2 would assure us that the given vectors x_1, x_2, and x_3 are independent. However, in this case nontrivial solutions do exist. (How can one tell? See Chapters 1 and 2.) For example, $c_3 = 1$, $c_1 = 2$, and $c_2 = -1$ form one such nontrivial solution. Thus,

$$2(1, 2, -1) - 1(3, 0, 4) + 1(1, -4, 6) = (0, 0, 0).$$

Therefore, by definition, the vectors x_1, x_2, and x_3 are *linearly dependent*.   ■

The term **linear dependence** suggests that in a linearly dependent collection of vectors some vectors "depend" on the others. This was the case with the row vectors of the coefficient matrix K for the system of equations (3.7). There we had the relationships

$$\mathbf{r}_1 = \mathbf{r}_2 - \mathbf{r}_3 \quad\text{and}\quad \mathbf{r}_4 = \mathbf{r}_2 + \mathbf{r}_3.$$

In general, suppose that the set

$$S = \{\mathbf{v}_1, \mathbf{v}_2, \ldots, \mathbf{v}_n\}$$

is a linearly *dependent* set. Then there will exist scalars $c_1, c_2, \ldots, c_n$, not all of which are zero, such that

$$c_1\mathbf{v}_1 + c_2\mathbf{v}_2 + \cdots + c_n\mathbf{v}_n = 0. \tag{3.13}$$

Since at least one of the scalars is not zero, let us assume (renumbering everything if necessary) that it is $c_1 \neq 0$. Multiply both sides of equation (3.13) by the scalar $1/c_1$, and solve it for the vector $\mathbf{v}_1$. We have, then,

$$\mathbf{v}_1 = \frac{-c_2}{c_1}\mathbf{v}_2 + \frac{-c_3}{c_1}\mathbf{v}_3 + \cdots + \frac{-c_n}{c_1}\mathbf{v}_n, \tag{3.14}$$

so the vector $\mathbf{v}_1$ is a linear combination (that is, does "depend" on) of the rest of the vectors

$$\{\mathbf{v}_2, \mathbf{v}_3, \ldots, \mathbf{v}_n\}.$$

In the exercises at the end of this section we ask you to develop some other interesting consequences of the definition of linear independence or dependence of a set of vectors from V.

Example 3   Consider the three geometric vectors in the plane $\mathbb{R}^2$ shown in Figure 3.6.

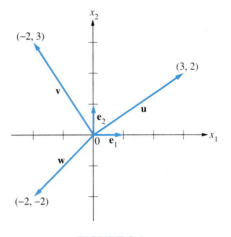

FIGURE 3.6

We first see that these three vectors are linearly dependent and later that two of them are linearly independent. To find some appropriate scalars that will show that the three vectors together form a linearly dependent set, write

$$c_1\mathbf{u} + c_2\mathbf{v} + c_3\mathbf{w} = \mathbf{0}.$$

Then, using the analytic form for each vector, this becomes

$$\mathbf{0} = c_1\mathbf{u} + c_2\mathbf{v} + c_3\mathbf{w} = c_1(3, 2) + c_2(-2, 3) + c_3(-2, -2) = (0, 0),$$

or

$$(3c_1, 2c_1) + (-2c_2, 3c_2) + (-2c_3, -2c_3) = (0, 0).$$

This results in the vector equation

$$(3c_1 - 2c_2 - 2c_3, 2c_1 + 3c_2 - 2c_3) = (0, 0),$$

or the two scalar equations

$$3c_1 - 2c_2 - 2c_3 = 0,$$
$$2c_1 + 3c_2 - 2c_3 = 0.$$

This is an underdetermined system whose general solution is $c_1 = 10t/13$, $c_2 = 2t/13$, $c_3 = t$, for any real number t. Since it is not required that $t = 0$, there are surely nonzero constants c_1, c_2, and c_3 so that $\mathbf{0} = c_1\mathbf{u} + c_2\mathbf{v} + c_3\mathbf{w}$. Hence the three vectors $\mathbf{u}, \mathbf{v}$, and $\mathbf{w}$ are indeed linearly dependent. In particular, letting $t = 13$, we have $10\mathbf{u} + 2\mathbf{v} + 13\mathbf{w} = \mathbf{0}$. We can also solve for $\mathbf{w}$ and have $\mathbf{w} = -(10/13)\mathbf{u} - (2/13)\mathbf{v}$.

Thus the vectors $\mathbf{u}, \mathbf{v}$, and $\mathbf{w}$ form a *linearly dependent* set of vectors. However, the two-vector set $\{\mathbf{u}, \mathbf{v}\}$ is *linearly independent*. To see this, suppose that there were scalars c_1 and c_2 such that

$$c_1\mathbf{u} + c_2\mathbf{v} = \mathbf{0}.$$

If so, we would get the system of equations

$$3c_1 - 2c_2 = 0,$$
$$2c_1 + 3c_2 = 0.$$

But this system's only solution is the trivial one: $c_1 = c_2 = 0$ (verify this).

Finally, if $\mathbf{x} = (x_1, x_2)$ is any vector in the plane the vector $\mathbf{x}$ "depends" on $\mathbf{u}$ and $\mathbf{v}$. To see that this is true consider the equation

$$c_1 \mathbf{u} + c_2 \mathbf{v} = \mathbf{x}.$$

Here we get

$$c_1(3, 2) + c_2(-2, 3) = (x_1, x_2)$$

resulting in the system of equations

$$3c_1 - 2c_2 = x_1,$$
$$2c_1 + 3c_2 = x_2.$$

The solution to this system is

$$c_1 = \frac{3x_1 + 2x_2}{13} \quad \text{and} \quad c_2 = -\frac{2x_1 - 3x_2}{13}.$$

Hence the set of vectors $\{\mathbf{u}, \mathbf{v}, \mathbf{x}\}$ is linearly dependent and

$$\mathbf{x} = \frac{3x_1 + 2x_2}{13}\mathbf{u} - \frac{2x_1 - 3x_2}{13}\mathbf{v}. \quad \blacksquare$$

Note that this computation shows that any vector $\mathbf{x}$ in the plane $\mathbb{R}^2$ is a *linear combination* of the vectors $\mathbf{u}$ and $\mathbf{v}$. We expand this idea to general vector spaces in the next section.

Example 4 Consider the set consisting of the three polynomials $p(x) = 1 + x$, $q(x) = 1 - x$, and $h(x) = 1 - x^2$, from P_2. Is this set linearly independent or linearly dependent?

Even though polynomials can be multiplied together to form new polynomials, and in this case $h(x) = p(x)q(x)$, this fact is irrelevant when we consider *linear independence or dependence.* The question here is: Do there exist scalars c_1, c_2, and c_3, not all zero, such that

$$c_1 p(x) + c_2 q(x) + c_3 h(x) = 0$$

(the zero polynomial)? Let us write out the polynomial on the left side and equate its coefficients with the zero polynomial's coefficients (all zeros). Thus,

$$
\begin{aligned}
c_1 p(x) + c_2 q(x) + c_3 h(x) &= c_1(1 + x) + c_2(1 - x) + c_3(1 - x^2) \\
&= (c_1 + c_2 + c_3) + (c_1 - c_2)x - c_3(x^2) \\
&= 0 + 0x + 0x^2.
\end{aligned}
$$

From this we conclude that

$$c_1 + c_2 + c_3 = 0$$
$$c_1 - c_2 \qquad = 0$$
$$\qquad - c_3 = 0.$$

This system of homogeneous equations (solved by the methods of Chapter 1 or Chapter 2 or however you wish) has only the trivial solution $c_1 = c_2 = c_3 = 0$. Hence the given set of three polynomials is *linearly independent*. ▪

Example 5 Let $\{(1, -1, -i), (i, 1, 1 - i), (1 + 2i, -3i, 1 - i)\}$ be a set of three vectors from $\mathbb{C}^3$. Is this set linearly dependent or linearly independent?

The procedure should be familiar. Do there exist scalars, not all zero, such that

$$c_1(1, -1, -i) + c_2(i, 1, 1 - i) + c_3(1 + 2i, -3i, 1 - i) = (0, 0, 0)?$$

However, here the scalars can be complex numbers, rather than real numbers. Expanding, we have

$$c_1(1, -1, -i) + c_2(i, 1, 1 - i) + c_3(1 + 2i, -3i, 1 - i)$$
$$= (c_1 + c_2 i + c_3 + 2c_3 i, -c_1 + c_2 - 3c_3 i, -c_1 i + c_2 - c_2 i + c_3 - c_3 i).$$

So we must have

$$c_1 + c_2 i \qquad + c_3 + 2c_3 i = 0,$$
$$-c_1 + c_2 \qquad - 3c_3 i = 0,$$
$$-c_1 i + c_2 - c_2 i + c_3 - c_3 i = 0.$$

The coefficient matrix for this homogeneous system of equations is the following complex matrix:

$$\begin{bmatrix} 1 & i & 1 + 2i \\ -1 & 1 & -3i \\ -i & 1 - i & 1 - i \end{bmatrix},$$

which has reduced row echelon form

$$\begin{bmatrix} 1 & 0 & 2i \\ 0 & 1 & -i \\ 0 & 0 & 0 \end{bmatrix}.$$

Thus, there do exist nonzero coefficients $c_1 = (-2i)t$, $c_2 = it$, and $c_3 = t$, for any nonzero complex number t, so that the resulting linear combination is the zero vector. In particular, $c_1 = -2i$, $c_2 = i$, $c_3 = 1$; or $c_1 = -2$, $c_2 = 1$, $c_3 = -i$ are possible nonzero values for these scalars. These three vectors therefore form a *linearly dependent* set. ■

Vocabulary

In this section three very fundamental ideas were introduced. Be sure that you understand these concepts as well as Lemma 3.2 (page 166). It is very important to understand and memorize the definitions of linear dependence and linear independence. They are basic to the rest of our exposition. The three new terms to add to your linear algebra vocabulary are the following:

linear combination   linearly independent
linearly dependent

Exercises 3.2

Show that each of the following sets of vectors from the given vector space is linearly independent.

1. $\{(1, 0), (0, 1)\} \subseteq \mathbb{R}^2$.

2. $\{(1, 1), (1, -1)\} \subseteq \mathbb{R}^2$.

3. $\{(1, 0, 0), (0, 1, 0), (0, 0, 1)\} \subseteq \mathbb{R}^3$.

4. $\{(1, 1, 1), (1, -1, 1), (1, 1, -1)\} \subseteq \mathbb{R}^3$.

5. $\{(1, i), (i, 1)\} \subseteq \mathbb{C}^2$.

6. $\{(1, 0, 0, 0), (0, 1, 0, 0), (0, 0, 1, 0)\} \subseteq \mathbb{R}^4$.

7. $\{(1, 0, 0, 0, 0), (0, 0, 0, 0, 1), (1, 1, 1, 1, 1)\} \subseteq \mathbb{R}^5$.

8. $\{(1, i, -i), (0, 1, i)\} \subseteq \mathbb{C}^3$.

9. $\{1, x, x^2, x^3\} \subseteq P_3$.

10. $\{1 + x, 1 + x^2, 1 + x^3\} \subseteq P_3$.

11. $\{\cos x, \sin x\} \subseteq C[-\pi, \pi]$.

12. $\{e^x, e^{2x}\} \subseteq C[0, 1]$.

13. The *rows* of the matrix $A = \begin{bmatrix} 1 & 2 & -1 \\ 3 & 1 & 4 \\ 5 & 6 & -7 \end{bmatrix}$ as vectors in $\mathbb{R}^3$.

14. The *rows* of the matrix $B = \begin{bmatrix} 1 & 0 & -1 \\ 0 & 3 & 1 \\ -1 & 2 & 0 \end{bmatrix}$ as vectors in $\mathbb{R}^3$.

15. The set of matrices $\left\{ \begin{bmatrix} 1 & 0 \\ 0 & 0 \end{bmatrix}, \begin{bmatrix} 0 & 1 \\ 0 & 0 \end{bmatrix}, \begin{bmatrix} 0 & 0 \\ 1 & 0 \end{bmatrix}, \begin{bmatrix} 0 & 0 \\ 0 & 1 \end{bmatrix} \right\} \subseteq M_2(\mathbb{R})$.

16. The set of matrices $\left\{ \begin{bmatrix} 1 & 1 \\ 0 & 0 \end{bmatrix}, \begin{bmatrix} 0 & 1 \\ 1 & 0 \end{bmatrix}, \begin{bmatrix} 0 & 0 \\ 1 & 1 \end{bmatrix} \right\} \subseteq M_2(\mathbb{R})$.

In Exercises 17–33 decide whether the given set of vectors is linearly independent or linearly dependent in the indicated vector space.

17. $\{(1, 1), (2, 2)\} \subseteq \mathbb{R}^2$.

18. $\{(2, 1), (2, -1)\} \subseteq \mathbb{R}^2$.

19. $\{(1, 1, 1), (-1, 1, 0), (0, 2, 1)\} \subseteq \mathbb{R}^3$.

20. $\{(1, 1, 1), (1, -1, 1), (1, 3, 1)\} \subseteq \mathbb{R}^3$.

21. $\{(1, i), (i, -1)\} \subseteq \mathbb{C}^2$.

22. $\{(1, 1, 1, 0), (1, -1, 0, 0), (1, 3, 1, 0)\} \subseteq \mathbb{R}^4$.

23. $\{(1, 0, 1, 0, 0), (0, 1, 0, 1, 1), (1, 1, 1, 1, 1)\} \subseteq \mathbb{R}^5$.

24. $\{(1 + i, i, -i), (-i, 1 - i, i)\} \subseteq \mathbb{C}^3$.

25. $\{1 + x, x + x^2, 1 + x^2\} \subseteq P_2$.

26. $\{1 + 2x^2, 1 - x^2, 1 - 2x^2\} \subseteq P_2$.

27. $\{1 - x, 1 + 2x, 1 - 5x\} \subseteq P_1$.

28. $\{1, \cos^2 x, \sin^2 x\} \subseteq C[-\pi, \pi]$.

29. The rows of the matrix $A = \begin{bmatrix} 1 & 2 & 4 \\ 0 & 3 & 5 \\ -1 & 1 & 1 \end{bmatrix}$ as vectors in $\mathbb{R}^3$.

30. The rows of the matrix $B = \begin{bmatrix} 1 & 1 & -1 \\ 2 & 3 & 1 \\ -1 & 2 & 1 \end{bmatrix}$ as vectors in $\mathbb{R}^3$.

31. The set of matrices $\left\{ \begin{bmatrix} 1 & i \\ 1 & i \end{bmatrix}, \begin{bmatrix} 1 & i \\ 0 & 0 \end{bmatrix}, \begin{bmatrix} 0 & 1 \\ i & 0 \end{bmatrix}, \begin{bmatrix} 1 & 0 \\ 0 & i \end{bmatrix} \right\} \subseteq M_2(\mathbb{C})$.

32. The set of matrices $\left\{ \begin{bmatrix} 1 & 2 \\ 2 & 1 \end{bmatrix}, \begin{bmatrix} -1 & 1 \\ 1 & -1 \end{bmatrix}, \begin{bmatrix} 2 & 2 \\ 1 & 1 \end{bmatrix} \right\} \subseteq M_2(\mathbb{R})$.

33. The set of matrices $\left\{ \begin{bmatrix} 1 & 2 \\ 2 & -1 \end{bmatrix}, \begin{bmatrix} 1 & 1 \\ 1 & 1 \end{bmatrix}, \begin{bmatrix} 0 & 0 \\ 0 & 0 \end{bmatrix}, \begin{bmatrix} 2 & -2 \\ 1 & 1 \end{bmatrix} \right\} \subseteq M_2(\mathbb{R})$.

---

34. Given the vectors $\sin(x + \theta)$ and $\cos x$ in $C[-\pi, \pi]$, determine the value(s) of θ for which they will form a linearly *dependent* set. Can you interpret this geometrically?

35. Given the two functions $f_1(x) = 2x$ and $f_2(x) = |x|$, show that the set $\{f_1, f_2\}$ is linearly *independent* in the vector space $C[-1, 1]$ but linearly *dependent* in $C[0, 1]$.

36. Show that a set consisting of any single nonzero vector is linearly independent.

37. Show that any set consisting of two nonzero vectors is linearly dependent if, and only if, one is a scalar multiple of the other.

38. Show that any set of three vectors from $\mathbb{R}^2$ is linearly dependent.

39. Show that any finite set of vectors that contains the zero vector is linearly dependent.

40. Prove that if the set $\{v_1, v_2, v_3\}$ of vectors from the vector space V is linearly independent, then the subset $\{v_1, v_2\}$ is also a linearly independent set of vectors.

41. Prove that if the set $\{v_1, v_2, \ldots, v_n\}$ of vectors from the vector space V is linearly independent, then any subset containing one or more of these vectors is also linearly independent.

42. Prove that if the set $\{v_1, v_2, \ldots, v_n\}$ of vectors from the vector space V is linearly *independent*, then any vector u that can be expressed as a linear combination of the vectors in the set can be so expressed in *exactly one way.* That is, such an expression has unique scalars. Contrast this with the case when the set is *dependent.* HINT: Suppose that

$$u = \sum_{i=1}^{n} \alpha_i v_i = \sum_{i=1}^{n} \beta_i v_i.$$

Subtract $u - u$ and conclude that $\alpha_i = \beta_i$ for each $i = 1, 2, \ldots, n$.

43. Prove or disprove the following assertion: If the set of vectors $\{u, v\}$ is linearly independent and the set of vectors $\{v, w\}$ is linearly independent, then all three vectors u, v, and w form a linearly independent set.

Exercises 44–47 are for those who have studied calculus.

44. Let V be the vector space $D_2[0, 1]$ of all twice-differentiable real-valued functions on the unit interval $[0, 1]$. Given the set of three functions $\{f_1(x), f_2(x), f_3(x)\}$ from V, their *Wronskian* is defined as the **determinant** of the matrix

$$\begin{bmatrix} f_1(x) & f_2(x) & f_3(x) \\ f'_1(x) & f'_2(x) & f'_3(x) \\ f''_1(x) & f''_2(x) & f''_3(x) \end{bmatrix}.$$

Show that the set is linearly independent if their Wronskian is not the zero vector in $D_2[0, 1)]$; that is, if there exists some x_0 in $[0, 1]$ so that

$$\det \begin{bmatrix} f_1(x_0) & f_2(x_0) & f_3(x_0) \\ f'_1(x_0) & f'_2(x_0) & f'_3(x_0) \\ f''_1(x_0) & f''_2(x_0) & f''_3(x_0) \end{bmatrix} \neq 0.$$

45. Use the results of Exercise 44 to prove the linear independence of the set $\{e^x, e^{-x}, e^{2x}\}$ over $\mathbb{R}$. (The set of scalars is $\mathbb{R}$.)

46. Use the results of Exercise 44 to prove the linear independence of the set $\{e^x, xe^x, x^2e^x\}$ over $\mathbb{R}$. (The set of scalars is $\mathbb{R}$.)

47. Use the results of Exercise 44 to prove the linear independence of the set $\{e^x, \sin x, \cos x\}$ over $\mathbb{R}$. (The set of scalars is $\mathbb{R}$.)

3.3 SPANNING SET Subspace

As we noted above, it is a direct consequence of the axioms for a vector space that any linear combination of vectors from a given vector space is again a vector in that vector space. Thus, if $\mathbf{v}_1, \mathbf{v}_2, \ldots, \mathbf{v}_n$ are vectors in the vector space V, the vector

$$\mathbf{w} = c_1\mathbf{v}_1 + c_2\mathbf{v}_2 + \cdots + c_n\mathbf{v}_n$$

is also in V.

Example 1   In Example 3 of Section 3.2 we saw that any vector in the plane (Cartesian 2-space, $\mathbb{R}^2$) is a linear combination of the two given vectors $\mathbf{u} = (3, 2)$ and $\mathbf{v} = (-2, 3)$. (See Figure 3.6.)   ■

Example 2   Given the three vectors $\mathbf{v}_1 = (1, 1, -1)$, $\mathbf{v}_2 = (1, 0, 1)$, and $\mathbf{v}_3 = (1, 3, -5)$ in Cartesian 3-space, $\mathbb{R}^3$, then the vector

$$\mathbf{u} = (3, -4, 11) = 2(1, 1, -1) + 3(1, 0, 1) - 2(1, 3, -5) = 2\mathbf{v}_1 + 3\mathbf{v}_2 - 2\mathbf{v}_3$$

is a linear combination of the vectors, $\mathbf{v}_1$, $\mathbf{v}_2$, and $\mathbf{v}_3$, and is indeed a vector in $\mathbb{R}^3$. Notice that any vector $\mathbf{x}$ in $\mathbb{R}^3$ that is of the form $(r + s + t,\ r + 3t,\ -r + s - 5t)$ *is a linear combination of* the vectors $\mathbf{v}_1$, $\mathbf{v}_2$, and $\mathbf{v}_3$, because $\mathbf{x}$ is $r\mathbf{v}_1 + s\mathbf{v}_2 + t\mathbf{v}_3$.   ■

Example 3   Consider the solution space $S(K)$ for the homogeneous system of equations $KX = 0$ discussed in the previous section. [See (3.7).] Recall that

$$K = \begin{bmatrix} 1 & -1 & -3 & 2 \\ 2 & 1 & -1 & 1 \\ 1 & 2 & 2 & -1 \\ 3 & 3 & 1 & 0 \end{bmatrix}. \tag{3.15}$$

We have already demonstrated that any linear combination $sX_1 + tX_2$ of the two solution vectors

$$X_1 = \begin{bmatrix} 4 \\ -5 \\ 3 \\ 0 \end{bmatrix} \quad \text{and} \quad X_2 = \begin{bmatrix} -1 \\ 1 \\ 0 \\ 1 \end{bmatrix}$$

is also a solution. Let us now demonstrate that *every* solution to the system has the form

$$X = sX_1 + tX_2. \tag{3.16}$$

That is, every solution is a linear combination of these two particular solutions. Suppose that

$$H = \begin{bmatrix} h_1 \\ h_2 \\ h_3 \\ h_4 \end{bmatrix}$$

is *any* solution of the system $KX = 0$. The coefficient matrix K for the system can be reduced by the Gauss-Jordan process to obtain the following equivalent upper-triangular matrix U:

$$U = \begin{bmatrix} 1 & 0 & -\dfrac{4}{3} & 1 \\ 0 & 1 & \dfrac{5}{3} & -1 \\ 0 & 0 & 0 & 0 \\ 0 & 0 & 0 & 0 \end{bmatrix}.$$

Thus the general solution to the system must satisfy the equations

$$h_1 = \frac{4}{3}h_3 - h_4,$$

$$h_2 = -\frac{5}{3}h_3 + h_4.$$

Therefore,

$$H = \begin{bmatrix} \dfrac{4}{3}h_3 - h_4 \\ -\dfrac{5}{3}h_3 + h_4 \\ h_3 \\ h_4 \end{bmatrix} = \frac{h_3}{3}\begin{bmatrix} 4 \\ -5 \\ 3 \\ 0 \end{bmatrix} + h_4\begin{bmatrix} -1 \\ 1 \\ 0 \\ 1 \end{bmatrix}. \quad \blacksquare$$

That is, any solution to $KX = 0$ is a linear combination $X = (h_3/3)X_1 + h_4X_2$ of the solutions X_1 and X_2. Hence *every* vector of the solution space $S(K)$ is a linear combination of the two vectors X_1 and X_2. We say that these two vectors **span** (or *generate*) the solution space $S(K)$. These examples lead us to the following definition:

Definition A given set $\{v_1, v_2, \ldots, v_n\}$ of vectors in a vector space V is said to **span** (or **generate**) the vector set $W \subseteq V$ if every vector x in W can be written as a linear combination of the vectors $v_1, v_2, \ldots, v_n$, that is,

$$x = c_1v_1 + c_2v_2 + \cdots + c_nv_n$$

where $c_1, c_2, \ldots, c_n$, are suitably chosen scalars.

Example 4   The collection of the three vectors $v_1 = (1, 1)$, $v_2 = (1, -1)$, and $v_3 = (-1, 0)$ *spans* all of $\mathbb{R}^2$. You can readily verify that any vector $x = (a, b)$ in $\mathbb{R}^2$ can be written, using the rules for addition and multiplication by a scalar in $\mathbb{R}^2$, in the form

$$x = \frac{a+b+1}{2}(1, 1) + \frac{a-b+1}{2}(1, -1) + (-1, 0). \quad \blacksquare$$

Example 5   Does the collection consisting of three vectors $e_1 = (1, 0, 0)$, $e_2 = (0, 1, 0)$, and $e_3 = (0, 0, 1)$ from $\mathbb{R}^3$ span $\mathbb{R}^3$? (See Figure 3.7.)

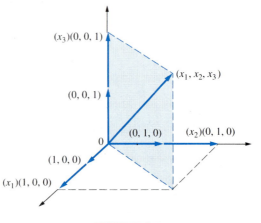

FIGURE 3.7

We recall that every vector in $\mathbb{R}^3$ has the form (x_1, x_2, x_3) where the entries x_1, x_2, and x_3 are real numbers. Using the definitions of vector addition and scalar multiplication given in Example 3 of Section 3.1, we see that

$$x_1(1, 0, 0) + x_2(0, 1, 0) + x_3(0, 0, 1) = (x_1, 0, 0) + (0, x_2, 0) + (0, 0, x_3)$$
$$= (x_1, x_2, x_3).$$

And, therefore, the vectors $e_1 = (1, 0, 0)$, $e_2 = (0, 1, 0)$, and $e_3 = (0, 0, 1)$ span $\mathbb{R}^3$. $\quad \blacksquare$

Example 6   Do the vectors $(1, 0, -1, 0)$, $(1, 1, 2, 0)$, $(2, -1, 0, 0)$, and $(3, 1, -1, 0)$ in Cartesian 4-space $\mathbb{R}^4$ span $\mathbb{R}^4$?

Recall that a typical vector in Cartesian 4-space $\mathbb{R}^4$ (see Section 3.1, Example 4) has the form (x_1, x_2, x_3, x_4), where x_1, x_2, x_3, x_4 are arbitrary real numbers. Now a glance at the four given vectors shows that the last coordinate (or entry) of each of them is zero. Moreover, in view of the way vector addition and scalar multiplication are defined in $\mathbb{R}^4$, we see that any linear combination of these vectors will be of the form

$$c_1(1, 0, -1, 0) + c_2(1, 1, 2, 0) + c_3(2, -1, 0, 0) + c_4(3, 1, -1, 0) \quad \textbf{(3.17)}$$

and must therefore have a zero as its fourth coordinate. A vector in $\mathbb{R}^4$ such as $(0, 0, 0, 1)$ cannot possibly be written in the form (3.17) since its fourth coordinate is *one,* not *zero.* Thus the given vectors *do not* span all of $\mathbb{R}^4$.

What do they span? Consider the collection W of all those vectors in $\mathbb{R}^4$ that have the form

$$\mathbf{w} = (w_1, w_2, w_3, 0).$$

That is, W is the subset of $\mathbb{R}^4$ consisting of all those vectors from $\mathbb{R}^4$ whose fourth component is zero. By checking each of the vector-space axioms, you can verify that W is a real-vector space in its own right with the same operations as those in $\mathbb{R}^4$. Any vector $\mathbf{w}$ in W can indeed be expressed as a linear combination of the four given vectors. To see this, write

$$\mathbf{w} = c_1(1, 0, -1, 0) + c_2(1, 1, 2, 0) + c_3(2, -1, 0, 0) + c_4(3, 1, -1, 0).$$

Therefore,

$$(w_1, w_2, w_3, 0) = (c_1 + c_2 + 2c_3 + 3c_4, c_2 - c_3 + c_4, -c_1 + 2c_2 - c_4, 0).$$

We find the scalars c_1, c_2, c_3, c_4 by solving the system of equations that results from equating the components; namely,

$$
\begin{aligned}
c_1 + c_2 + 2c_3 + 3c_4 &= w_1, \\
c_2 - c_3 + c_4 &= w_2, \\
-c_1 + 2c_2 \qquad - c_4 &= w_3.
\end{aligned}
$$

We skip the details of the solution process, which is thoroughly discussed in Chapter 1, and just give the answer. There are an infinite number of solutions to this system. For each real number t we see that

$$c_1 = \frac{1}{5}(2w_1 + 4w_2 - 3w_3 - 13t), \qquad c_2 = \frac{1}{5}(w_1 + 2w_2 + w_3 - 4t),$$

$$c_3 = \frac{1}{5}(w_1 - 3w_2 + w_3 + t), \qquad c_4 = t.$$

Therefore, the vectors $(1, 0, -1, 0)$, $(1, 1, 2, 0)$, $(2, -1, 0, 0)$ and $(3, 1, -1, 0)$ *span* the vector space W. ■

Since W is a subset of the vector space $\mathbb{R}^4$, and is a vector space in its own right under the same operations as in $\mathbb{R}^4$, we call W a **subspace** of $\mathbb{R}^4$. The general definition for subspace is as follows.

Definition   Let V be a vector space. A nonempty subset W of V is called a **subspace** of V if W is itself a vector space with respect to the same operations of V.

Example 7   Find the subspace of P_3 spanned by the two vectors (polynomials) $\mathbf{u}_1 = 1 - x$ and $\mathbf{u}_2 = 1 + x^3$.

The above statement implies that the set spanned by these two vectors is, in fact, a subspace. We'll prove that fact in a more general setting after Theorem 3.3 below. However, to describe the set spanned by these two, we merely want all linear combinations of the two vectors, that is, the collection of all polynomials of the form

$$a(1 - x) + b(1 + x^3) = a - ax + b + bx^3 = (a + b) - ax + bx^3,$$

for any real numbers a and b. It is not difficult to see that the polynomial $2 - x + x^3$ is one of these while $1 + x^2$ is not. Why?   ∎

Example 8   Consider the following subsets of the plane $\mathbb{R}^2$:

$$V = \mathbb{R}^2 = \text{set of all points } (x_1, x_2); \qquad x_1, x_2 \text{ real};$$
$$W_1 = \text{set of all points } (x_1, 0);$$
$$W_2 = \text{set of all points } (0, x_2);$$
$$W_3 = \text{set of all points } (x, x).$$

In geometric terms V is the plane, W_1 is the x_1 axis, and W_2 is the x_2 axis (see Figure 3.8). It can be verified that W_1 is a vector space in its own right, and so are W_2 and W_3. Convince yourself that *all of the axioms* in the definition of a vector space given in Section 3.1 are satisfied for W_1, W_2, and W_3. In verifying this assertion, observe that the closure axioms for both vector addition and scalar multiplication hold in each subset. For example, in W_1

$$(x_1, 0) + (x'_1, 0) = (x_1 + x'_1, 0) = \text{an element in } W_1;$$
$$k(x_1, 0) = (kx_1, 0) = \text{an element in } W_1.$$

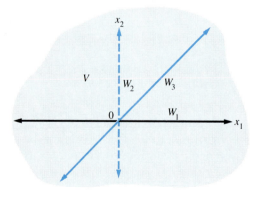

FIGURE 3.8

Also $(0, 0)$ is in W_1, and the negative of any element in W_1 is again in W_1. All of the equational axioms (i.e., commutative, associative, distributive, etc.) are true for all elements of V, so they are just as true if we select these elements only from W_1, since every element of W_1 is also an element of V. Similar arguments are used to verify that W_2 and W_3 are subspaces of V.

It is easy to see that the subspace W_1 is spanned by the vector $\mathbf{e}_1 = (1, 0)$ since any vector $\mathbf{w} = (x_1, 0)$ in W_1 is a linear combination of the form

$$\mathbf{w} = x_1 \mathbf{e}_1 = x_1(1, 0).$$

Similarly the vector space W_2 is spanned by the vector $\mathbf{e}_2 = (0, 1)$, while the vector space W_3 is spanned by the vector $\mathbf{v} = (1, 1)$; that is, $(x, x) = x\mathbf{v}$. ∎

With an eye on Example 8, we ask the following question: When is a nonempty subset W of a vector space V actually a **subspace** of V? Our discussion makes the following theorem quite plausible.

Theorem 3.3 Let V be a vector space and let W be a nonempty subset of V. Then W is a **subspace** of V if, and only if, the following two conditions hold in W:

(i) If $\mathbf{w}_1$ and $\mathbf{w}_2$ are both in W, then so is $\mathbf{w}_1 + \mathbf{w}_2$ in W;
(ii) If $\mathbf{w}$ is in W and k is any scalar, then $k\mathbf{w}$ is also in W.

Obviously the two conditions of this theorem are necessary for W to be a subspace, since they are two of the axioms (1 and 6) for a vector space. That these two conditions are also sufficient follows from the fact that W is a subset of V and so inherits many of its properties: associativity and commutativity of vector addition, etc. The zero vector of V will always be in W because $0\mathbf{w}$ is in W by (ii). Similarly $-\mathbf{w} = (-1)\mathbf{w}$ is in W. We ask you to supply all of the details of the proof of this theorem as Exercise 56.

Example 9 Let V be Cartesian 3-space $\mathbb{R}^3$ and let $W =$ the set of all these vectors (x_1, x_2, x_3) in V for which

$$x_1 + x_2 + x_3 = 0. \tag{3.18}$$

Is W a subspace of V?

We first of all observe that the zero vector $(0, 0, 0)$ is in W, so W is not empty. Because of Theorem 3.3, it suffices to show that both conditions (i) and (ii) of that theorem are satisfied by W. Thus, suppose that

$$(x_1, x_2, x_3) \quad \text{and} \quad (x'_1, x'_2, x'_3) \quad \text{are both in } W. \tag{3.19}$$

Then by the definition of W, we have

$$x_1 + x_2 + x_3 = 0 \quad \text{and} \quad x'_1 + x'_2 + x'_3 = 0.$$

Upon adding these two equation we get

$$(x_1 + x'_1) + (x_2 + x'_2) + (x_3 + x'_3) = 0.$$

The definition of W then, in view of this last equation, tells us that

$$(x_1 + x'_1, x_2 + x'_2, x_3 + x'_3) \text{ is in } W,$$

and, therefore (why?)

$$(x_1, x_2, x_3) + (x'_1, x'_2, x'_3) \qquad \text{is in } W. \tag{3.20}$$

Statements (3.19) and (3.20) satisfy condition (i) in Theorem 3.3.

Now suppose that (x_1, x_2, x_3) is in W and that k is any scalar. Then, again by the definition of W [see (3.18)], $x_1 + x_2 + x_3 = 0$; consequently,

$$k(x_1 + x_2 + x_3) = kx_1 + kx_2 + kx_3 = 0$$

so that the vector $k(x_1, x_2, x_3) = (kx_1, kx_2, kx_3)$ is in W. So our subset W also satisfies condition (ii) of Theorem 3.3. Hence W is a subspace of V.

Now it is not difficult to verify that the subspace W is actually spanned by the two vectors

$$\mathbf{w}_1 = (1, -1, 0) \qquad \text{and} \qquad \mathbf{w}_2 = (1, 0, -1).$$

Clearly every $\mathbf{x} = (x_1, x_2, x_3)$ in W can be written as the linear combination

$$\mathbf{x} = (-x_2)\mathbf{w}_1 + (-x_3)\mathbf{w}_2 = (-x_2 - x_3, x_2, x_3) = (x_1, x_2, x_3),$$

because of (3.18). ■

Later we find a method for discovering spanning sets for a given vector space or subspace. These two vectors were merely an illustration. We do, however, have the following corollary to Theorem 3.3.

> **Corollary** If S is any collection of vectors from the vector space V, then the set $\mathbf{L}(S)$ consisting of all finite linear combinations of vectors from S is a subspace of V.

Proof The proof of this is straightforward, since each element of $\mathbf{L}(S)$ is a linear combination of finitely many elements of S. Let $\mathbf{u}$ and $\mathbf{v}$ be vectors in $\mathbf{L}(S)$. Then there is a finite collection of vectors $\{\mathbf{v}_1, \mathbf{v}_2, \ldots, \mathbf{v}_k\}$ from S so that both $\mathbf{u}$ and $\mathbf{v}$ are linear combinations of these vectors; that is,

$$\mathbf{u} = a_1\mathbf{v}_1 + a_2\mathbf{v}_2 + \cdots + a_k\mathbf{v}_k \qquad \text{and} \qquad \mathbf{v} = b_1\mathbf{v}_1 + b_2\mathbf{v}_2 + \cdots + b_k\mathbf{v}_k.$$

Therefore, the vector $\mathbf{u} + \mathbf{v} = (a_1 + b_1)\mathbf{v}_1 + (a_2 + b_2)\mathbf{v}_2 + \cdots + (a_k + b_k)\mathbf{v}_k$ is in $\mathbf{L}(S)$, satisfying part (i) of Theorem 3.3. The verification of part (ii) is similar. For any scalar c, the vector $c\mathbf{u} = c(a_1\mathbf{v}_1 + a_2\mathbf{v}_2 + \cdots + a_k\mathbf{v}_k) = ca_1\mathbf{v}_1 + ca_2\mathbf{v}_2 + \cdots + ca_k\mathbf{v}_k$ is in $\mathbf{L}(S)$. ◻

> **Definition** The set $\mathbf{L}(S)$ is called the **linear span of S,** and the set S is a **spanning set** for the subspace $\mathbf{L}(S)$.

A spanning set for a particular vector space is by no means a unique thing. A given nonzero vector space can have any number of different spanning sets.

Example 10   Cartesian 2-space $\mathbb{R}^2$ is spanned (or generated) by each of the following sets of vectors in addition to the set in Example 1 above.

$$S_1 = \{(1, 1), (1, -1), (-1, 1), (0, 1)\},$$
$$S_2 = \{(1, 1), (1, 0), (0, 1)\},$$
$$S_3 = \{(1, 1), (1, -1)\}.$$

To see that S_1 is a spanning set we remember that any vector in $\mathbb{R}^2$ has the form (x_1, x_2), where x_1 and x_2 are real numbers. So we seek certain scalars, so that

$$(x_1, x_2) = c_1(1, 1) + c_2(1, -1) + c_3(-1, 1) + c_4(0, 1)$$
$$= (c_1 + c_2 - c_3, c_1 - c_2 + c_3 + c_4).$$

Many such scalars are available to us. Any solution to the system of equations

$$c_1 + c_2 - c_3 \qquad = x_1,$$
$$c_1 - c_2 + c_3 + c_4 = x_2$$

will do. For example, $c_1 = \frac{1}{2}(x_1 + x_2 - 2)$, $c_2 = \frac{1}{2}(x_1 - x_2 + 2)$, $c_3 = 0$, $c_4 = 2$ works. So S_1 is indeed a spanning set for $\mathbb{R}^2$; thus $\mathbf{L}(S_1) = \mathbb{R}^2$.

In a similar way, if $k_1 = 1$, $k_2 = x_1 - 1$, $k_3 = x_2 - 1$, we have an arbitrary vector (x_1, x_2) from $\mathbb{R}^2$ represented as a linear combination of the vectors in S_2 as (verify this)

$$(x_1, x_2) = k_1(1, 1) + k_2(1, 0) + k_3(0, 1).$$

So $\mathbb{R}^2$ is spanned by the set S_2; thus $\mathbf{L}(S_2) = \mathbb{R}^2$.

Finally, in a similar way, $\mathbf{L}(S_3) = \mathbb{R}^2$ since

$$(x_1, x_2) = \frac{x_1 + x_2}{2}(1, 1) + \frac{x_1 - x_2}{2}(1, -1).$$

Thus, each of the three distinct sets S_1, S_2, and S_3 is a spanning set for $\mathbb{R}^2$. ∎

This example raises some very interesting questions that will be answered in the next section. Does a *minimal* spanning set exist for a vector space? That is, does there exist, among all of the sets of vectors that span a given vector space V, one or more sets with the least number of vectors in it? Is there any significance to the fact that in Example 10 only the set S_3 is a linearly *independent* set? (You should verify that the sets S_1 and S_2 are linearly *dependent*, while S_3 is linearly *independent*.)

Example 11   Let $V = \mathbb{R}^3$, and consider the two vectors $\mathbf{u} = (1, 1, 1)$ and $\mathbf{v} = (1, 1, -1)$. The geometric depiction of these two 3-space vectors is Figure 3.9.

The subspace $L(\mathbf{u}, \mathbf{v})$ spanned (or generated) by these two vectors is the collection of all vectors in $\mathbb{R}^3$ that are linear combinations of $\mathbf{u}$ and $\mathbf{v}$, hence that have the form

$$\mathbf{x} = (x_1, x_2, x_3) = \alpha\mathbf{u} + \beta\mathbf{v} = (\alpha + \beta, \alpha + \beta, \alpha - \beta) = (a, a, b).$$

(The equation of this plane is $x_1 = x_2$. Note that there is no restriction on x_3.) ∎

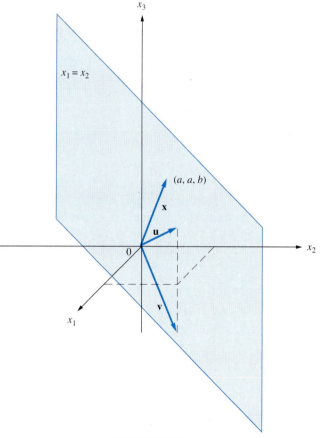

FIGURE 3.9

Example 12   Let $V = P_2$, the vector space of polynomials of degree two or less. Let

$$u(x) = 1 - x \quad \text{and} \quad v(x) = 1 + x^2.$$

Then the subspace $L(u, v)$ is the collection of all polynomials that are linear combinations of these two polynomials, namely, of the form

$$p(x) = cu(x) + dv(x) = c - cx + d + dx^2$$
$$= (c + d) - cx + dx^2.$$

Note that the polynomials $f(x) = -1 + x$ and $g(x) = 3 - 2x + x^2$ belong to $\mathbf{L}(u, v)$, whereas the polynomial $2 + x^2$ does not. You should do the computations involved in checking the truth of these assertions. ■

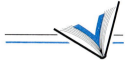

Vocabulary

In this section we have introduced four new basic ideas. Be sure that you understand the concepts as well as the theorems of this section. It is necessary to understand the definition of a spanning set. Listed below are the new terms to add to your linear algebra vocabulary.

spans (generates)   spanning set
subspace   linear span $\mathbf{L}(S)$

Exercises 3.3

In Exercises 1–29 use Theorem 3.3 (page 179) to determine which of the given sets of vectors are subspaces of the indicated vector space.

1. All vectors in $\mathbb{R}^2$ of the form $(0, t)$. (See Example 8.)

2. All vectors in $\mathbb{R}^2$ of the form $(-t, t)$.

3. All vectors in $\mathbb{R}^3$ of the form $(0, 0, t)$.

4. All vectors in $\mathbb{R}^3$ of the form (t, s, t).

5. All vectors in $\mathbb{R}^3$ of the form (r, s, t), where $r + 1 = s + t$.

6. All vectors in $\mathbb{R}^3$ of the form (r, s, t), where $r - s = s - t$.

7. All polynomials (vectors) in P_2 of the form $a + bx^2$.

8. All polynomials in P_2 of the form $a + ax + ax^2$.

9. All polynomials in P_3 of the form $a_0 + a_1x + a_2x^2 + a_3x^3$, where $a_0 + a_1 + a_2 + a_3 = 0$.

10. All polynomials in P_3 of the form $a_0 + a_1x + a_2x^2 + a_3x^3$, whose coefficients a_0, a_1, a_2, a_3 are all integers.

11. All 2×2 upper triangular matrices in $M_2(\mathbb{R})$. [Recall that $M_n(\mathbb{R})$ is the vector space of *all $n \times n$ matrices with real number entries*—see Section 3.1, Example 6.]

12. All 2×2 symmetric matrices A in $M_2(\mathbb{R})$. (Recall, A is symmetric if $A^T = A$.)

13. All 2×2 matrices A in $M_2(\mathbb{R})$ for which $\det A = 0$.

14. All 2×2 matrices A in $M_2(\mathbb{R})$ for which $\det A = 1$.

15. All 2×2 matrices A in $M_2(\mathbb{R})$ for which $\det A \neq 0$.

16. The subset W_3 of Example 8.

17. All vectors (z_1, z_2) in $\mathbb{C}^2$ (see Section 3.1, Example 8) for which z_1 is a real number.

18. All vectors (z_1, z_2) in $\mathbb{C}^2$ for which z_2 is a real number.

19. All vectors (z_1, z_2) in $\mathbb{C}^2$ for which z_2 is a pure imaginary number bi.

20. All vectors (z_1, z_2) in $\mathbb{C}^2$ for which $z_2 = \overline{z_1}$, the complex conjugate of z_1.

21. All those functions $f(x)$ in the vector space $C[0, 1]$ of continuous functions on $[0, 1]$ (see Section 3.1, Example 9) for which $f(x) = 0$ for some x in $[0, 1]$.

22. All those functions $f(x)$ in $C[0, 1]$ for which $f(x) \geq 0$ for all x in $[0, 1]$.

23. All those functions $f(x)$ in $C[0, 1]$ for which $f(0) = 0$.

24. All those functions $f(x)$ in $C[0, 1]$ for which $f(0) = 1$.

25. All those functions $f(x)$ in $C[0, 1]$ for which $f'(x)$ exists for every x in $[0, 1]$. (This set is $C_d[0, 1] =$ differentiable functions on $[0, 1]$.)

26. All those functions $f(x)$ in $C[0, 1]$ for which $f'(x) \geq 0$ (positive first derivative) for every x in $[0, 1]$.

27. All those functions $f(x)$ in $C[0, 1]$ for which $f''(x)$ exists for every x in $[0, 1]$.

28. All those functions $f(x)$ in $C[0, 1]$ for which $f'(x) + f(x) = 0$, for every x in $[0, 1]$.

29. All those functions $f(x)$ in $C[0, 1]$ for which $f''(x) + f(x) = 0$, for every x in $[0, 1]$.

In Exercises 30–48 determine whether or not the given vector **x** *is in the subspace* **L**(S) *spanned by the given set S of vectors by writing* **x** *as a linear combination of the vectors in S, if that is possible.*

30. $\mathbf{x} = (1, -2)$ $S = \{(1, 1), (1, -1)\}$.

31. $\mathbf{x} = (1, -2)$ $S = \{(-1, 1), (1, -1)\}$.

32. $\mathbf{x} = (i, -2i)$ $S = \{(1, i), (i, 1)\}$.

33. $\mathbf{x} = (i, -2i)$ $S = \{(i, i), (1, 1)\}$.

34. $\mathbf{x} = (1, -2, -2)$ $S = \{(1, 0, 1), (0, 1, -1)\}$.

35. $\mathbf{x} = (2, -1, -1)$ $S = \{(1, 0, 1), (0, 1, 1)\}$.

36. $\mathbf{x} = (-1, 2, 3)$ $S = \{(1, 1, 0), (0, 1, 1)\}$.

37. $\mathbf{x} = (1, 5, -3)$ $S = \{(1, -1, 3), (2, 4, 0)\}$.

38. $\mathbf{x} = (0, 0, 0)$ $S = \{(2, 1, -1), (0, 2, 3)\}$.

39. $\mathbf{x} = (1, -2i, 1 - 2i)$ $S = \{(i, 0, 1 + i), (0, 1 - i, 1 + i)\}$.

40. $\mathbf{x} = (1 - 2i, 1 + i, -2i)$ $S = \{(i, i, i), (1, 1, 1)\}$.

41. $\mathbf{x} = (-2 + i, -i, 1 - 2i)$ $S = \{(i, i, i), (1, 1, 1)\}$.

42. $\mathbf{x} = 2 - x^2$ $S = \{1, x\}$.

43. $\mathbf{x} = 2 - x^2$ $S = \{1, x^2\}$.

44. $\mathbf{x} = 2 - 3x^2$ $S = \{x, x^2\}$.

45. $\mathbf{x} = 2 - 3x + x^2$ $S = \{1 + x, 1 + x^2\}$.

46. $\mathbf{x} = 1 + 2x - x^2$ $S = \{1 + x, 1 - x^2\}$.

47. $\mathbf{x} = \begin{bmatrix} 1 & 2 \\ 0 & -1 \end{bmatrix}$ $S = \left\{ \begin{bmatrix} 2 & 4 \\ 0 & 1 \end{bmatrix}, \begin{bmatrix} 1 & 2 \\ 0 & 2 \end{bmatrix} \right\}$.

48. $\mathbf{x} = \begin{bmatrix} 2 & -1 \\ -1 & -2 \end{bmatrix}$ $S = \left\{ \begin{bmatrix} 1 & 1 \\ 0 & 1 \end{bmatrix}, \begin{bmatrix} 1 & -1 \\ 0 & -1 \end{bmatrix} \right\}$.

In Exercises 49–55, determine whether or not the given sets S of vectors span all of $\mathbb{R}^3$.

49. $S = \{(1, 0, 0), (0, 1, 1), (1, 0, 1)\}$.

50. $S = \{(1, 0, 1), (0, 1, 1), (2, -1, 1)\}$.

51. $S = \{(1, 0, 1), (2, 0, 1), (4, 0, -1)\}$.

52. $S = \{(2, 1, -1), (0, 2, 3), (1, -1, 0), (0, 1, 1)\}$.

53. $S = \{(1,2,3), (3, 2, 1), (2, 0, -1), (2, 2, 2)\}$.

54. The three rows of the matrix $M = \begin{bmatrix} 1 & 2 & 4 \\ 0 & 3 & 5 \\ 1 & 2 & 3 \end{bmatrix}$.

55. The three columns of the matrix M in Exercise 54.

---

56. Supply all of the computational details of the proof of Theorem 3.3 (page 179).

57. Show that the set of all the solutions to the system of equations $KX = B$, when B is *not* zero, do not form a vector space.

58. Find an equation for the plane in $\mathbb{R}^3$ spanned by the vectors $\mathbf{u} = (1, 0, 1)$ and $\mathbf{v} = (1, 1, 0)$.

59. Find an equation for the plane in $\mathbb{R}^3$ spanned by the vectors $\mathbf{u} = (1, 0, 1)$ and $\mathbf{v} = (1, -1, 0)$.

60. Find an equation for the plane in $\mathbb{R}^3$ spanned by the vectors $\mathbf{u} = (1, 2, 3)$ and $\mathbf{v} = (1, -1, 1)$.

61. Prove that $W \subseteq V$ is a subspace of V if, and only if, for every $\mathbf{u} \in W$ and $\mathbf{v} \in W$ and every pair of scalars α and β, $\alpha\mathbf{u} + \beta\mathbf{v} \in W$.

62. Show that if W_1 and W_2 are subspaces of a given vector space V, then their union $W_1 \cup W_2$ is a subspace of V if, and only if, $W_1 \subseteq W_2$ or $W_2 \subseteq W_1$.

3.4 ORDERED BASIS AND DIMENSION

In the previous section we posed the question: Does a vector space have a minimal spanning set S? By *minimal* we mean a spanning set containing as few vectors as possible. In Example 10 of that section we discovered that the set

$$S_3 = \{(1, 1), (1, -1)\}$$

does span $\mathbb{R}^2$; that is, $\mathbf{L}(S_3) = \mathbb{R}^2$. Let us now show that S_3 is *minimal* in our sense for $\mathbb{R}^2$. We do this by demonstrating, in Example 1 below, that no *one* vector in $\mathbb{R}^2$ can span all of $\mathbb{R}^2$. Since S_3 does indeed span $\mathbb{R}^2$, it is a spanning set with the least possible number of vectors (two).

Example 1 No *one* vector from $\mathbb{R}^2$ can span all of $\mathbb{R}^2$.

Let $\mathbf{u} = (a, b)$ be any nonzero vector from $\mathbb{R}^2$. Every vector in the subspace $\mathbf{L}(\mathbf{u})$ of $\mathbb{R}^2$ spanned by $\mathbf{u}$ is of the form $k\mathbf{u} = k(a, b) = (ka, kb)$, for some real number k. This is because $\mathbf{L}(\mathbf{u})$ consists of all linear combinations of the single vector $\mathbf{u}$. Now if $a \neq 0$, the vector $\mathbf{v} = (0, 1)$ fails to belong to $\mathbf{L}(\mathbf{u})$, while if $b \neq 0$, the vector $\mathbf{w} = (1, 0)$ fails to be in $\mathbf{L}(\mathbf{u})$. So there are vectors in $\mathbb{R}^2$ that are not in $\mathbf{L}(\mathbf{u})$, for any single nonzero vector $\mathbf{u}$ in $\mathbb{R}^2$. ∎

Example 2   Show that the set $S = \{1, x\}$ is a spanning set for the vector space P_1 of all polynomials of degree one or less. (See Example 5 in Section 3.1.) Show also that no spanning set for P_1 can consist of fewer than two vectors; i.e., there can be no spanning set containing but a single polynomial from P_1.

A typical element of P_1 is the polynomial

$$a_0 + a_1 x = a_0(1) + a_1(x).$$

which clearly lies in $\mathbf{L}(S)$, for the given set S. Now let us consider any particular polynomial $u(x) = a + bx$ in P_1. Then if $a \neq 0$, the polynomial x is not in $\mathbf{L}(u(x))$. If $b \neq 0$, the polynomial 1 is not in $\mathbf{L}(u(x))$. Therefore, no set consisting of a single polynomial can span all of P_1. (Verify the claims made in this argument.) ∎

It is natural to say that a spanning set S is *minimal* for a given vector space if any proper subset of S *fails* to span the space. On the other hand, as we have noted in Section 3.3., if there are *more* vectors in a particular spanning set H than the smallest number needed to span the given vector space, that spanning set H is linearly dependent. The set we seek, therefore, is a *linearly independent* set that *spans* the vector space. To make this precise, we state the formal definition of a basis for a vector space.

Let V be any vector space different from the zero vector space ($V \neq \{\mathbf{0}\}$).

Definition   A set S of vectors, $S = \{\mathbf{v}_1, \mathbf{v}_2, \ldots, \mathbf{v}_n\}$ is called a **basis** for (of) V, if

(i) S spans V; i.e., $\mathbf{L}(S) = V$; and
(ii) S is linearly independent.

So a basis for a vector space is a *linearly independent spanning set* for V. For example, the set $S_3 = \{(1, 1), (1, -1)\}$, discussed above, is a *basis* for $\mathbb{R}^2$. We have already shown that it is a *spanning set*: $\mathbf{L}(S) = \mathbb{R}^2$. To see that S is *linearly independent,* we suppose that there are scalars α_1 and α_2 in $\mathbb{R}$ such that

$$\alpha_1(1, 1) + \alpha_2(1, -1) = \mathbf{0} = (0, 0).$$

Then we must have

$$\alpha_1 + \alpha_2 = 0$$

and

$$\alpha_1 - \alpha_2 = 0.$$

This 2×2 homogeneous system of linear equations has *only the trivial solution* $\alpha_1 = 0$, $\alpha_2 = 0$. From this we conclude that S is a linearly independent set of vectors, hence a basis for $\mathbb{R}^2$.

It turns out that every real or complex vector space V has a basis, finite or infinite. For completeness we say that the zero vector space $\{0\}$ has the empty set for a basis. These two statements will be accepted without proof.

Example 3

Do the vectors $e_1 = (1, 0, 0)$, $e_2 = (0, 1, 0)$, and $e_3 = (0, 0, 1)$ in Cartesian 3-space $\mathbb{R}^3$ form a basis E for $\mathbb{R}^3$?

In Example 5 of Section 3.3 we proved that the given vectors *span* $\mathbb{R}^3$. Now we test for linear independence. Thus, suppose that for scalars c_1, c_2, and c_3

$$c_1 e_1 + c_2 e_2 + c_3 e_3 = 0.$$

Then

$$c_1(1, 0, 0) + c_2(0, 1, 0) + c_3(0, 0, 1) = 0 = (0, 0, 0). \qquad \textbf{(3.21)}$$

Computing the left side of this equation, we obtain

$$(c_1, 0, 0) + (0, c_2, 0) + (0, 0, c_3) = 0 = (0, 0, 0),$$
$$(c_1, c_2, c_3) = (0, 0, 0);$$

and hence, $c_1 = 0$, $c_2 = 0$, $c_3 = 0$. We have thus shown that equation (3.21) forces all of the scalars $c_1 = c_2 = c_3 = 0$; hence the vectors $e_1 = (1, 0, 0)$, $e_2 = (0, 1, 0)$, and $e_3 = (0, 0, 1)$ form a linear independent set. Therefore, the set $E = \{e_1, e_2, e_3\}$ is a basis for $\mathbb{R}^3$. ∎

A basis for a vector space is the kind of *minimal* spanning set for that space that we were looking for. However,

---

A basis for a vector space is by no means unique.

The two sets $S = \{(1, 1), (1, -1)\}$ and $E = \{(1, 0), (0, 1)\}$ are both bases for $\mathbb{R}^2$. The facts that they each contain *two* linearly independent vectors, and that no set with just *one* vector (as seen in Example 1) can be a basis, since it does not span $\mathbb{R}^2$, are crucial here. We make this more precise, and thereby more easily applied. First of all we have the following lemma.

Lemma 3.4 Let V be a vector space and let $S = \{v_1, v_2, \ldots, v_n\}$ be a basis for V. Then every set containing more than n vectors from V is linearly dependent and hence cannot be a basis for V.

The proof of this lemma is an application of the machinery that we have developed to this point. Therefore, even though it is a little bit tedious, we write it out.

Proof   Suppose that $T = \{\mathbf{w}_1, \mathbf{w}_2, \ldots, \mathbf{w}_m\}$ with $m > n$ is any collection of (more than n) vectors from V. We shall show that T must be linearly dependent. Because S is a basis for V it must span V. Every vector in V, in particular each vector in T, is a linear combination of the basis vectors in S. That is, for each $j = 1, 2, \ldots, m$ there must be scalars a_{ij}, $i = 1, 2, \ldots, n$, so that

$$\mathbf{w}_j = a_{1j}\mathbf{v}_1 + a_{2j}\mathbf{v}_2 + \cdots + a_{nj}\mathbf{v_n}. \tag{3.22}$$

To show that the vectors $\mathbf{w}_j$ are linearly dependent we must find scalars c_j (not all zero) such that

$$c_1\mathbf{w}_1 + c_2\mathbf{w}_2 + \cdots + c_m\mathbf{w}_m = \mathbf{0}. \tag{3.23}$$

By substituting the values of $\mathbf{w}_j$ from (3.22) into (3.23) we have

$$\sum_{j=1}^{m} c_j (a_{1j}\mathbf{v}_1 + a_{2j}\mathbf{v}_2 + \cdots + a_{nj}\mathbf{v}_n) = \mathbf{0}. \tag{3.24}$$

By multiplying (3.24) out and collecting the coefficients for the vectors $\mathbf{v}_1, \mathbf{v}_2, \ldots, \mathbf{v}_n$, we have

$$(c_1 a_{11} + c_2 a_{12} + \cdots + c_m a_{1m})\mathbf{v}_1 + (c_1 a_{21} + c_2 a_{22} + \cdots + c_m a_{2m})\mathbf{v}_2$$
$$+ \cdots + (c_1 a_{n1} + c_2 a_{n2} + \cdots + c_m a_{nm})\mathbf{v}_n = \mathbf{0}.$$

But S is a basis, so the vectors $\mathbf{v}_1, \mathbf{v}_2, \ldots, \mathbf{v}_n$ are linearly independent and, hence, each coefficient in the above equation is zero. This fact gives rise to the following homogeneous system of equations:

$$\begin{aligned}
c_1 a_{11} + c_2 a_{12} + \cdots + c_m a_{1m} &= 0, \\
c_1 a_{21} + c_2 a_{22} + \cdots + c_m a_{2m} &= 0, \\
\vdots \qquad \vdots \qquad \vdots \\
c_1 a_{n1} + c_2 a_{n2} + \cdots + c_m a_{nm} &= 0.
\end{aligned} \tag{3.25}$$

Therefore, to show that T is indeed linearly *dependent* we need only show that this system (3.25) of equations has a *nontrivial* solution $(c_1, c_2, \ldots, c_m)$. That is, not all of the c_i are zero. But *in this system there are fewer equations than there are unknowns. It is an underdetermined system.* So (see Section 1.2, statement C) *there must be a nontrivial solution.* This completes the proof of the lemma.   ◨

This lemma contains the main components of the proof of the following important theorem that *every finite basis for a given vector space contains the same number of vectors.*

> **Theorem 3.5**   Let $S = \{\mathbf{v}_1, \mathbf{v}_2, \ldots, \mathbf{v}_n\}$ and $T = \{\mathbf{w}_1, \mathbf{w}_2, \ldots, \mathbf{w}_m\}$ be two bases for a vector space V. Then $m = n$.

Proof From the lemma we have $m \leq n$, or T would not be linearly independent. But it is a basis. Likewise, $n \leq m$ or S would not be linearly independent; but it is a basis. Hence $m = n$. ❏

Therefore, the number of vectors in a basis for a vector space with a *finite* basis is an invariant of the vector space. This number is called the *dimension* of the vector space. A vector space with a finite number of vectors in any of its bases is called **finite-dimensional.**

> **Definition** The **dimension** of a finite-dimensional vector space is the number of vectors in a basis for that space. This number is the same no matter which basis is used.

The zero vector space $\{\mathbf{0}\}$ has dimension zero, and we have observed already that the vector space $\mathbb{R}^2$ has dimension *two* and $\mathbb{R}^3$ has dimension *three*. These are special cases of the following theorem.

> **Theorem 3.6** The dimension of Cartesian n-space $\mathbb{R}^n$ is n.

Proof Let $\mathbf{x} = (\xi_1, \xi_2, \ldots, \xi_n)$ be any vector in $\mathbb{R}^n$. Let us single out the following set E of special vectors from $\mathbb{R}^n$.

> $$E = \{\mathbf{e}_1 = (1, 0, 0, \ldots, 0), \mathbf{e}_2 = (0, 1, 0, \ldots, 0), \ldots,$$
> $$\mathbf{e}_n = (0, 0, 0, \ldots, 1)\}.$$

In other words, each vector $\mathbf{e}_i$ in E is an n-tuple from $\mathbb{R}^n$ with a *one* as its ith coordinate and zeros elsewhere. Now notice that

$$\mathbf{x} = (\xi_1, \xi_2, \ldots, \xi_n)$$
$$= \xi_1(1, 0, 0, \ldots, 0) + \xi_2(0, 1, 0, \ldots, 0) + \cdots + \zeta_n(0, 0, 0, \ldots, 1).$$

This equation shows that the set E is a *spanning set* for $\mathbb{R}^n$; $\mathbf{L}(E) = \mathbb{R}^n$.

Next consider linear independence. The argument used in Example 3 shows that the equation $c_1\mathbf{e}_1 + c_2\mathbf{e}_2 + \cdots + c_n\mathbf{e}_n = (0, 0, \ldots, 0) = \mathbf{0}$ forces each of the coefficients $c_1, c_2, \ldots, c_n$ to be zero. So we can conclude that the set E is linearly independent. Therefore, E is a linearly independent spanning set, hence is a basis for $\mathbb{R}^n$. ❏

> **Definition** The basis E, as given in the above proof and ordered in that way, is called the **standard basis** for $\mathbb{R}^n$.

Corollary   The plane (i.e., Cartesian 2-space) is of dimension *two,* and has standard basis

$$E = \{\mathbf{e}_1 = (1, 0), \mathbf{e}_2 = (0, 1)\}.$$

Additionally, Cartesian 3-space is of dimension *three,* and has standard basis

$$E = \{\mathbf{e}_1 = (1, 0, 0), \mathbf{e}_2 = (0, 1, 0), \mathbf{e}_3 = (0, 0, 1)\}.$$

A given vector space has many bases, not just one. Note that the set S_3 discussed before Example 1 is a basis for $\mathbb{R}^2$ different from E. What all bases for the same finite-dimensional vector space have in common is that *they contain the same number of vectors.* Bases are different even when they contain exactly the same vectors, but the vectors are listed differently. The basis $S_3 = \{(1, 1), (1, -1)\}$ is different as a basis for $\mathbb{R}^2$ from the basis $S'_3 = \{(1, -1), (1, 1)\}$.

Most of the time this is a distinction that is not critical. Nevertheless, in some of our work such as the matrix representation of a vector or a transformation, it makes considerable difference. When the order of the vectors in a basis is important, we sometimes use the term **ordered basis.**

Example 4   Show that the set of vectors $G = \{(0, 1, 1), (1, 0, 1), (1, 1, 0)\}$ in Cartesian 3-space $\mathbb{R}^3$ is a basis for $\mathbb{R}^3$.

In Exercise 35 we ask you to show that G is a *linearly independent* set of vectors. Given that fact, we have to show only that they span $\mathbb{R}^3$. To this end suppose that $\mathbf{x} = (\alpha_1, \alpha_2, \alpha_3)$ is any vector in $\mathbb{R}^3$. We wish to find scalars x_1, x_2, and x_3 such that

$$x_1(0, 1, 1) + x_2(1, 0, 1) + x_3(1, 1, 0) = (\alpha_1, \alpha_2, \alpha_3) = \mathbf{x}. \qquad \textbf{(3.26)}$$

Now, if we simplify the left side of (3.26) we see that it becomes

$$(0, x_1, x_1) + (x_2, 0, x_2) + (x_3, x_3, 0) = (\alpha_1, \alpha_2, \alpha_3),$$

or

$$(x_2 + x_3, x_1 + x_3, x_1 + x_2) = (\alpha_1, \alpha_2, \alpha_3).$$

This last equation gives rise to the following system of linear equations:

$$x_2 + x_3 = \alpha_1,$$
$$x_1 \qquad + x_3 = \alpha_2,$$
$$x_1 + x_2 \qquad = \alpha_3.$$

When we solve this system for the unknown scalars x_1, x_2, and x_3 we get (verify this)

$$x_1 = \frac{-\alpha_1 + \alpha_2 + \alpha_3}{2}; \qquad x_2 = \frac{\alpha_1 - \alpha_2 + \alpha_3}{2}; \qquad \text{and} \qquad x_3 = \frac{\alpha_1 + \alpha_2 - \alpha_3}{2}.$$

Hence scalars x_1, x_2, and x_3 that satisfy (3.26) exist and, therefore, the set of vectors $G = \{(0, 1, 1), (1, 0, 1), (1, 1, 0)\}$ spans $\mathbb{R}^3$. Thus G is a basis for $\mathbb{R}^3$.   ∎

Example 5   The "standard basis" for the vector space $M_n = M_n(\mathbb{R})$ of all $n \times n$ matrices over the real numbers consists of the n^2 matrices E_{ij} whose entry in row i and column j is a 1 (where $i, j, = 1, 2, \ldots, n$) with all of the other entries zero. For example, in the space M_2 this *standard basis* consists of the following four matrices:

$$E_{11} = \begin{bmatrix} 1 & 0 \\ 0 & 0 \end{bmatrix}, \qquad E_{12} = \begin{bmatrix} 0 & 1 \\ 0 & 0 \end{bmatrix}, \qquad E_{21} = \begin{bmatrix} 0 & 0 \\ 1 & 0 \end{bmatrix}, \qquad E_{22} = \begin{bmatrix} 0 & 0 \\ 0 & 1 \end{bmatrix}.$$

That the matrices E_{ij} do indeed form a basis for M_n may be verified as follows. An arbitrary matrix $A = (a_{ij})$ from M_n may be written in the form

$$A = (a_{ij}) = \sum_{i,\,j=1}^{n} a_{ij} E_{ij} .$$

So the set E of these matrices E_{ij} spans M_n. Also, if

$$\sum_{i,\,j=1}^{n} c_{ij} E_{ij} = 0 \qquad \text{(the zero matrix)},$$

then it must be true that each $c_{ij} = 0$. (See Exercise 33.) And the set E is *linearly independent.* Thus E is a basis as we asserted. ∎

Example 6   The "standard basis" for the vector space P_n of all real polynomials of degree $\leq n$ is the set of vectors (polynomials)

$$E = \{1, x, x^2, \ldots, x^n\}.$$

That this set is *linearly independent* and spans P_n is left for you to prove in Exercise 19. ∎

Let us now consider three theorems. The proofs of each of these are left to you as exercises, since doing the proofs will reinforce your understanding of what a basis for a vector space is. Moreover, each theorem contains useful information regarding a basis. The first two tell us that if we know that a vector space V has dimension n and we have a set S of *exactly n* vectors from V, we need only check to see if S *spans V* or if S is *linearly independent,* not both.

> **Theorem 3.7**   If V is an n-dimensional vector space and $S = \{v_1, v_2, \ldots, v_n\}$ is a linearly independent set of n vectors from V, then S is a basis for V.

To prove this theorem, see Exercise 29.

> **Theorem 3.8** If V is an n-dimensional vector space and $S = \{\mathbf{v}_1, \mathbf{v}_2, \ldots, \mathbf{v}_n\}$ is a set of n vectors that spans V, then S is a basis for V.

To prove this theorem, see Exercise 30.

The final theorem tells us that every linearly independent subset of an n-dimensional vector space can be enlarged by adjoining additional linearly independent vectors to it to form a basis for V. We ask you to do this proof as Exercise 31.

> **Theorem 3.9** If $S = \{\mathbf{v}_1, \mathbf{v}_2, \ldots, \mathbf{v}_k\}$ is a linearly independent set of k vectors from the n-dimensional vector space V, and $k < n$, then there exist vectors $\mathbf{v}_{k+1}, \mathbf{v}_{k+2}, \ldots, \mathbf{v}_n$ in V such that the set $B = \{\mathbf{v}_1, \mathbf{v}_2, \ldots, \mathbf{v}_k, \mathbf{v}_{k+1}, \mathbf{v}_{k+2}, \ldots, \mathbf{v}_n\}$ is a basis for V.

As you prove this theorem notice that $\mathbf{L}(S)$ is a proper subspace of V when $k < n$. Select a vector $\mathbf{v}_{k+1}$ which is in V but not in $\mathbf{L}(S)$. Show that the set $S \cup \{\mathbf{v}_{k+1}\}$ is still linearly independent.

It is instructive to give a geometric interpretation of the results of Example 9 in Section 3.3. We saw there that the set of all points (x_1, x_2, x_3) in Cartesian 3-space $\mathbb{R}^3$ that satisfy the equation $x_1 + x_2 + x_3 = 0$ form a subspace W of $\mathbb{R}^3$. It is well known that the locus of all points (x_1, x_2, x_3) in Cartesian 3-space that satisfy this particular equation, $x_1 + x_2 + x_3 = 0$, is a plane passing through the origin. Thus it is not surprising that the dimension of this plane is *two*. If we were to consider the collection of all points that satisfy not only this equation but also *another* one, say, $x_1 - x_2 - 2x_3 = 0$, we would expect the resulting subspace U consisting of these points (x_1, x_2, x_3) in $\mathbb{R}^3$ which satisfy *both* of these equations to be of dimension *one*. (Why?) This is consistent with the fact that the two planes

$$x_1 + x_2 + x_3 = 0 \quad \text{and} \quad x_1 - x_2 - 2x_3 = 0,$$

intersect in a straight line whose geometric dimension is, of course, *one*.

We conclude this section by classifying *all* of the subspaces of the plane (i.e., of Cartesian 2-space $\mathbb{R}^2$). Thus, recall that $\mathbb{R}^2 = \{(x_1, x_2), \text{ where } x_1 \text{ and } x_2 \text{ are real numbers}\}$. In view of the result of Exercise 28, and because the dimension of $\mathbb{R}^2$ is *two* (why?), it follows that every subspace W of $\mathbb{R}^2$ must have dimension 0, 1, or 2. It can be seen that the only subspace of dimension 0 is the zero vector space $\{\mathbf{0}\}$. Moreover, the only subspace $\mathbb{R}^2$ of dimension 2 is $\mathbb{R}^2$ itself. [See Exercise 28(b).] So we are left with the task of identifying *all* of the subspaces W or $\mathbb{R}^2$ of dimension *one*. To do that, let w be any one-dimensional subspace of $\mathbb{R}^2$, and suppose that the vector (a_1, a_2) forms a basis for W. Note that not both of a_1 and a_2 are zero (why?).

Let (x_1, x_2) be *any* vector in W. Then $(x_1, x_2) = c(a_1, a_2)$, for some scalar c. This equation implies that (why?): $x_1 = ca_1$, and $x_2 = ca_2$, and hence

$$a_2 x_1 = ca_1 a_2 = a_1 x_2.$$

That is, for all vectors (x_1, x_2) in W,

$$a_2 x_1 - a_1 x_2 = 0.$$

We have thus shown that W is the set of all vectors (x_1, x_2) in $\mathbb{R}^2$ that are such that

$$a_2 x_1 - a_1 x_2 = 0.$$

Interpreted geometrically, this says that every subspace W of $\mathbb{R}^2$ of dimension one is a straight line passing through the origin. These results on subspaces of $\mathbb{R}^2$ are summarized in the following chart.

Dimension	Description
0	Origin
1	A line through the origin
2	The whole plane

Vocabulary

In this section we have again encountered important new ideas. Be sure that you understand the concepts as well as the theorems of this section. It is especially necessary to understand the definition of a basis (ordered basis) and to be certain of what *dimension* means. Listed below are the new terms to add to your linear algebra vocabulary.

basis   ordered basis
standard basis   dimension
n-dimensional   dimension zero

Exercises 3.4

1. Show that the set $S = \{x_1 = (1, -2, 1), x_2 = (1, 0, 1), x_3 = (-1, 0, 1)\}$ is a basis for Cartesian 3-space, $\mathbb{R}^3$. Do this by showing that $\mathbf{L}(S) = \mathbb{R}^3$ and that S is linearly independent.

2. Is the set of vectors $K = \{(1, 0, 0), (0, 0, 1), (1, 1, 1), (1, -1, 1)\}$ a spanning set for $\mathbb{R}^3$?

3. Is the set of vectors K in Exercise 2 a basis set for $\mathbb{R}^3$? Justify your answer.

4. Show that the set $H = \{v_1 = (1, 2), v_2 = (1, -2), v_3 = (2, 3)\}$ spans $\mathbb{R}^2$ but is not a basis for $\mathbb{R}^2$.

*In Exercises 5–12 show that the given set of vectors is a **basis** for the indicated vector space by showing that it both spans the space and is linearly independent.*

5. In $\mathbb{R}^2$, the set $\{(1, 1), (-10, 11)\}$.

6. In $\mathbb{R}^2$, the set $\{(\pi, 0), (1, e)\}$.

7. In $\mathbb{R}^3$, the set $\{(1, 1, -1), (1, -1, 1), (2, 1, 2)\}$.

8. In $\mathbb{R}^3$, the set $\{(0, 1, -1), (1, -1, 0), (2, 0, 2)\}$.

9. In P_2, the set $\{1 - x, 1 + x, 1 - x^2\}$.

10. In P_2, the set $\{1 + 2x, -x, 3x^2\}$.

11. In $M_2(\mathbb{R})$, the set of matrices $\left\{ \begin{bmatrix} 1 & 0 \\ 0 & 0 \end{bmatrix}, \begin{bmatrix} 0 & 1 \\ 0 & 0 \end{bmatrix}, \begin{bmatrix} 0 & 0 \\ 1 & 0 \end{bmatrix}, \begin{bmatrix} 0 & 0 \\ 0 & 1 \end{bmatrix} \right\}$.

12. In $M_2(\mathbb{R})$, the set of matrices $\left\{ \begin{bmatrix} 1 & 1 \\ 1 & 0 \end{bmatrix}, \begin{bmatrix} 0 & 1 \\ 1 & 1 \end{bmatrix}, \begin{bmatrix} 1 & 0 \\ 1 & 1 \end{bmatrix}, \begin{bmatrix} 1 & 1 \\ 0 & 1 \end{bmatrix} \right\}$.

---

13. Show that the set of vectors $S = \{(1, 1, -1), (0, 2, 1)\}$ is linearly independent in $\mathbb{R}^3$ but is not a basis. Extend it to a basis for $\mathbb{R}^3$.

14. Let W be the subset of $\mathbb{R}^3$ that consists of all vectors (x, y, z) for which $x - y + z = 0$. Verify that W is a subspace of $\mathbb{R}^3$ and find a basis for it.

15. Let W be the subset of $\mathbb{R}^3$ that consists of all vectors (x, y, z) for which $x + 2y - 3z = 0$. Verify that W is a subspace of $\mathbb{R}^3$ and find a basis for it.

16. Show that the set S of Exercise 13 is a basis for a subspace of $\mathbb{R}^3$. Describe this subspace geometrically.

17. Let Ω be the set of all vectors (x, y, z) in $\mathbb{R}^3$ that satisfy both of the equations $x - y - z = 0$ and $x + y + z = 0$. Show that Ω is a subspace of $\mathbb{R}^3$. What is its dimension? Find a basis for Ω. Interpret the results geometrically.

18. Let Γ be the set of all vectors (x, y, z) in $\mathbb{R}^3$ that satisfy both of the equations $x - 2y + z = 0$ and $x + 2y - z = 0$. Show that Γ is a subspace of $\mathbb{R}^3$. What is its dimension? Interpret the results geometrically. Find a basis for Γ.

19. Carry out the verification of Example 6 that the set $\{1, x, x^2, \dots, x^n\}$ is a basis for P_n. What is the dimension of P_n?

20. Describe the subspace of P_2 for which the set $G = \{1 - x^2, 1\}$ is a basis.

21. Describe the subspace of P_2 for which the set $K = \{x + x^2, 2x\}$ is a basis.

22. Extend the set G of Exercise 20 to a basis for P_2.

23. Extend the set K of Exercise 21 to a basis for P_2.

24. Show that the set P_0 of all *constant* polynomials is a subspace of P_n, for any positive integer

$n = 1, 2, \dots$. What is its dimension? Give a basis for this space.

25. Let A be the following 3×3 real matrix:

$$\begin{bmatrix} 1 & 2 & 4 \\ 0 & 3 & 5 \\ 2 & 1 & 3 \end{bmatrix}.$$

What is the dimension of the space $S(A)$ of all solutions to the equation $AX = 0$?

26. Show that every nonzero vector space V has at least two subspaces, namely, the zero subspace $\{0\}$ and the space V itself.

27. Suppose that a vector space V is known to have *exactly* two subspaces. Show that V has dimension 1. HINT: What must be true about any two nonzero vectors in V?

28. Prove the following theorem: If V is an n-dimensional real vector space and W is a subspace of V, then
(a) The dimension of $W \leq n$.
(b) If dim $W = n$, then $W = V$.

29. Prove Theorem 3.7 (page 191). HINT: Let $\mathbf{x}$ be an arbitrary vector in V. Use the linear independence of the set S, and the dimension of V, to prove that $\mathbf{x}$ must be a linear combination of the vectors of S. Therefore, S is also a spanning set for V.

30. Prove Theorem 3.8 (page 192). HINT: If S were not a linearly independent set, what does this do to the dimension of V, since S is a spanning set?

31. Prove Theorem 3.9 (page 192).

32. Write a "standard basis" for $M_3(\mathbb{R})$.

33. For each $i, j = 1, 2, \ldots, n$, let E_{ij} be the $n \times n$ matrix with a *one* in row i, column j and *zeros* elsewhere. Show that the set $E = \{E_{ij} : i, j = 1, 2, \ldots, n\}$ is a basis for $M_n(\mathbb{R})$. What is the dimension of $M_n(\mathbb{R})$?

34. What is the dimension of $\mathbb{C}^3$? Give a basis to justify your answer.

35. Show that the set $G = \{(0, 1, 1), (1, 0, 1), (1, 1, 0)\}$ of vectors from $\mathbb{R}^3$ is linearly independent in $\mathbb{R}^3$.

36. Is the converse of Exercise 27 true? That is, if a vector space V has dimension 1, does this force V to have *exactly two* subspaces? Give reasons.

37. Let P be the set of all polynomials (including the zero polynomial) of **arbitrary** degree and with real coefficients. Show that P is a vector space with respect to the usual operations of polynomial addition and scalar multiplication.

38. Show that the vector space P of Exercise 37 is *not* finite-dimensional. HINT: Show that if $B = \{f_1, f_2, \ldots, f_k\}$ is any finite set of polynomials in P, where k is any positive integer, then there exists a polynomial g in P which is not in the linear span of B; that is, g is not a linear combination of the vectors (polynomial) $f_1, f_2, \ldots, f_k$.

39. Show that the space $C[0, 1]$ cannot have a *finite* basis. HINT: Is the space P of Exercise 37 a subspace of $C[0, 1]$?

40. Prove that any vector $\mathbf{v}$ in a finite-dimensional vector space V can be written as a *unique* linear combination of the vectors in a basis for V. HINT: Write a basis and suppose that $\mathbf{v}$ can be written in two ways as a linear combination of the basis vectors.

3.5 MATRICES *Row and Column Spaces*

In several examples of the preceding sections we considered the rows of a given $m \times n$ matrix A as vectors in $\mathbb{R}^n$ or in $\mathbb{C}^n$. We could equally well have considered the *columns* of A. They are vectors in the space of $m \times 1$ matrices, which we often denote by $\mathbb{R}^m$ or by $\mathbb{C}^m$ depending on whether the scalars are real or complex numbers. In this section we focus our attention on two important subspaces derived from the rows and the columns of a given matrix. These are called the *row space* and the *column space*, respectively, of the matrix A.

Let A be an $m \times n$ matrix with real or complex number entries, say,

$$A = \begin{bmatrix} a_{11} & a_{12} & a_{13} & \cdots & a_{1n} \\ a_{21} & a_{22} & a_{23} & \cdots & a_{2n} \\ a_{31} & a_{32} & a_{33} & \cdots & a_{3n} \\ \vdots & \vdots & \vdots & \cdots & \vdots \\ a_{m1} & a_{m2} & a_{m3} & \cdots & a_{mn} \end{bmatrix}. \tag{3.27}$$

Consider the rows of A:

$$\mathbf{r}_1 = (a_{11}, a_{12}, a_{13}, \ldots, a_{1n})$$
$$\mathbf{r}_2 = (a_{21}, a_{22}, a_{23}, \ldots, a_{2n})$$
$$\vdots$$
$$\mathbf{r}_m = (a_{m1}, a_{m2}, a_{m3}, \ldots, a_{mn}).$$

Then the n-tuples $\mathbf{r}_1, \mathbf{r}_2, \ldots, \mathbf{r}_m$ are m elements (vectors) of $V = \mathbb{R}^n$ or $\mathbb{C}^n$. They will span some subspace of V.

> **Definition**   The subspace of n-space $\mathbb{R}^n$ or $\mathbb{C}^n$ spanned by the m rows of the $m \times n$ matrix A is called the **row space** of the matrix A. The dimension of this subspace is called the **row rank** of the matrix A.

Example 1   Find the row space and the row rank of the 3×3 matrix

$$A = \begin{bmatrix} 1 & -1 & 0 \\ 2 & 3 & 1 \\ 3 & -2 & 4 \end{bmatrix}.$$

By definition, the row space of A is the subspace of $\mathbb{R}^3$ spanned by the rows of A. Thus the row space of A is

$$\{k_1(1, -1, 0) + k_2(2, 3, 1) + k_3(3, -2, 4): k_1, k_2, k_3, \text{ any real numbers}\}.$$

Moreover, since the row rank of A is the dimension of its row space, the row rank is equal to the *maximum* number of linearly independent vectors in the spanning set

$$S = \{(1, -1, 0), (2, 3, 1), (3, -2, 4)\}.$$

To calculate this number, the row rank of A, we could proceed as follows: Clearly the set $\{(1, -1, 0)\}$, consisting of the single nonzero vector, is linearly independent. What about the set $\{(1, -1, 0), (2, 3, 1)\}$? Since neither of these two vectors is a scalar multiple of the other, they form a linearly independent set as well (why?). Finally we ask: Is the set S itself linearly independent? We leave it to you to verify that this is indeed the case—you can row reduce A (why?) or find det A (why?) or use the linear dependence/independence algorithm as in Example 1 of Section 3.2. Since S is independent and spans the row space, it is a *basis* for the row space. The dimension is *three;* hence the row rank of A is 3. ■

The concepts of the **column space** and the **column rank** of a matrix A are defined in a similar way. To do so, we first must note that the collection $M_{m \times 1}$ of all $m \times 1$ matrices is a vector space. (See Section 3.1.) We'll call these matrices **column vectors** and henceforth use the following notations:

$$\overline{\mathbb{R}^m} = M_{m \times 1}(\mathbb{R})$$

$$= \text{the set of all column vectors} \begin{bmatrix} x_1 \\ x_2 \\ \vdots \\ x_m \end{bmatrix}, \text{ all } x_i \text{ } real \text{ numbers.}$$

$$\overline{\mathbb{C}^m} = M_{m \times 1}(\mathbb{C})$$

$$= \text{the set of all column vectors} \begin{bmatrix} x_1 \\ x_2 \\ \cdot \\ \cdot \\ \cdot \\ x_m \end{bmatrix}, \text{all } x_i \text{ } complex \text{ numbers.}$$

> **Definition**   The subspace of *m*-space $\overline{\mathbb{R}^m}$ or $\overline{\mathbb{C}^m}$ spanned by the *n* columns of the $m \times n$ matrix *A* is called the **column space** of the matrix *A*. The dimension of this subspace is called the **column rank** of the matrix *A*.

We illustrate this definition with the following example.

Example 2   Find the column space and the column rank of the following matrix.

$$A = \begin{bmatrix} 1 & -1 & 0 \\ 2 & 3 & 1 \\ 3 & -2 & 4 \end{bmatrix}.$$

By definition the column space of *A* is the subspace of $\overline{\mathbb{R}^3}$, the space of all 3×1 column matrices (column vectors), spanned by the columns of *A*. Thus the column space consists of all linear combinations of these columns, expressions of the form

$$\alpha \begin{bmatrix} 1 \\ 2 \\ 3 \end{bmatrix} + \beta \begin{bmatrix} -1 \\ 3 \\ -2 \end{bmatrix} + \gamma \begin{bmatrix} 0 \\ 1 \\ 4 \end{bmatrix} \qquad \text{where } \alpha, \beta, \text{ and } \gamma \text{ are any real numbers.}$$

Moreover, the column rank of *A* is the dimension of this subspace. Hence the column rank is the maximum number of linearly independent vectors in the set

$$\begin{bmatrix} 1 \\ 2 \\ 3 \end{bmatrix}, \quad \begin{bmatrix} -1 \\ 3 \\ -2 \end{bmatrix}, \quad \begin{bmatrix} 0 \\ 1 \\ 4 \end{bmatrix}.$$

Using the same arguments as in Example 1, we can verify that these three column vectors are linearly independent vectors. Therefore, the *column rank* of *A* is three.   ∎

In the case of the matrix *A* of Examples 1 and 2 the column rank and the row rank were the same, namely, three. Is this just a peculiarity of this matrix *A,* or does that always happen? The answer is that for any matrix the column rank and the row rank are equal, as we shall see in Theorem 3.12 (page 201).

However, it is *not* the case that the rank is always equal to the size of the matrix. As you can easily deduce from the definition, if A is an $m \times n$ matrix, the row rank could be any integral value $k \leq m$ and the column rank any integer (also k, by what we just said) where $k \leq n$. Before we prove Theorem 3.12, and some important preliminary results, let us see how we can more easily determine the row rank of a given matrix. The easiest way is to make use of the Gauss-Jordan reduction techniques discussed in Chapter 1. By reducing a matrix A to its row echelon form, we can determine a basis for the row space. In fact, by writing any finite collection of vectors as the rows of a matrix A, we can determine a basis for the space they span. This process is justified by the following lemma and theorem.

Lemma 3.10   Let A be any $m \times n$ matrix and E be an $m \times m$ elementary matrix. The row space of A is the same as the row space of EA. In other words, elementary row operations do no change the row space of a matrix.

Proof   Designate the row vectors of A by $\mathbf{r}_1, \mathbf{r}_2, \ldots, \mathbf{r}_m$. Elementary matrices are of three types. If EA is a row interchange, we have merely reordered the m vectors $\mathbf{r}_1, \mathbf{r}_2, \ldots, \mathbf{r}_m$, which does not change the space that they span. If E multiplies a row, say, row i, by a nonzero constant k, it is quite clear that the set of new row vectors $\mathbf{r}_1, \mathbf{r}_2, \ldots, k\mathbf{r}_i, \ldots, \mathbf{r}_m$ will span the same subspace of V as did the original set. Finally, if E is a Type III elementary matrix, the row vectors of EA are the set of vectors

$$\{\mathbf{r}_1, \mathbf{r}_2, \ldots, \mathbf{r}_i, \ldots, \mathbf{r}_j + k\mathbf{r}_i, \ldots, \mathbf{r}_m\}.$$

Any linear combination $\mathbf{x} = \alpha_1\mathbf{r}_1 + \alpha_2\mathbf{r}_2 + \cdots + \alpha_i\mathbf{r}_i + \cdots + \alpha_j(\mathbf{r}_j + k\mathbf{r}_i) + \cdots + \alpha_m\mathbf{r}_m$ of these vectors is a linear combination

$$\mathbf{x} = \alpha_1\mathbf{r}_1 + \alpha_2\mathbf{r}_2 + \cdots + (\alpha_i + k\alpha_j)\mathbf{r}_i + \cdots + \alpha_j\mathbf{r}_j + \cdots + \alpha_m\mathbf{r}_m$$

of the row vectors of A. Hence the row space of $EA \subseteq$ row space of A. Conversely, any vector $\mathbf{x} = \alpha_1\mathbf{r}_1 + \alpha_2\mathbf{r}_2 + \cdots + \alpha_i\mathbf{r}_i + \cdots + \alpha_j\mathbf{r}_j + \cdots + \alpha_m\mathbf{r}_m$, in the row space of A, is in the row space of EA. This is true because we may write

$$\mathbf{x} = \alpha_1\mathbf{r}_1 + \alpha_2\mathbf{r}_2 + \cdots + \beta_i\mathbf{r}_i + \cdots + \beta_j(\mathbf{r}_j + k\mathbf{r}_i) + \cdots + \alpha_m\mathbf{r}_m,$$

where $\beta_j = \alpha_j$ and $\beta_i = (\alpha_i - k\alpha_j)$. Hence row space of $A \subseteq$ row space of EA, and thus these two row spaces are equal.   ❏

Now, if B is any $m \times m$ nonsingular matrix, we can write B as a finite product of elementary matrices $B = E_1 E_2 \cdots E_s$. Thus, by induction on the positive integer s, the row rank of BA is the same as that of A. We have the following theorem.

> **Theorem 3.11**   Let A be an $m \times n$ matrix. Then
>
> (i)   The nonzero row vectors in a row echelon form U of A form a basis for the row space of A.
> (ii)   If B is any $m \times m$ nonsingular matrix, the row rank of BA is the same as the row rank of A.
> (iii)   If C is any $n \times n$ nonsingular matrix, the column rank of AC is the same as the column rank of A.
> (iv)   The nonzero row vectors in a row echelon form L of A^T form a basis for the column space of A.

We ask you to furnish the details of the proof of this theorem in the exercises. In doing so use Lemma 3.10 and take note of the facts that $(AC)^T = C^T A^T$ and C^T can be written as a product of elementary matrices. Let's look at some examples.

Example 3   Find the row rank of the matrix

$$A = \begin{bmatrix} 1 & 0 & 1 & 2 \\ 2 & 1 & 0 & 3 \\ 1 & -1 & 3 & 3 \end{bmatrix}.$$

If we row reduce A to its row echelon form, we obtain the matrix (verify this)

$$U = \begin{bmatrix} 1 & 0 & 1 & 2 \\ 0 & 1 & -2 & -1 \\ 0 & 0 & 0 & 0 \end{bmatrix}.$$

Thus, the two nonzero row vectors $\mathbf{r}_1 = (1, 0, 1, 2)$ and $\mathbf{r}_2 = (0, 1, -2, -1)$ of U are a basis for the row space of A, and the row rank of A is *two*. ∎

Now we illustrate how to use this method to determine a basis for a vector space (subspace) spanned by any given finite collection of vectors from $\mathbb{R}^n$ in the next example.

Example 4   Find a basis for the subspace of $\mathbb{R}^3$ spanned by the set of vectors

$$S = \{(1, 3, 0), (0, 2, 4), (1, 5, 4), (1, 1, -4)\}.$$

To do this, we manufacture a matrix A in which these are the row vectors, thereby changing the question to one of finding a basis for the row space of the matrix A. So

$$A = \begin{bmatrix} 1 & 3 & 0 \\ 0 & 2 & 4 \\ 1 & 5 & 4 \\ 1 & 1 & -4 \end{bmatrix}$$

and has, after a Gauss-Jordan reduction, the row echelon form

$$U = \begin{bmatrix} 1 & 3 & 0 \\ 0 & 1 & 2 \\ 0 & 0 & 0 \\ 0 & 0 & 0 \end{bmatrix}.$$

And we conclude that the linear span of the set S is of dimension 2 and has a basis consisting of the two vectors $(1, 3, 0)$ and $(0, 1, 2)$. ∎

Example 5 Find a basis for the column space of the matrix

$$M = \begin{bmatrix} 1 & 2 & 4 & 0 & -3 \\ 0 & 1 & -5 & -1 & 2 \\ 0 & 1 & 3 & -1 & 0 \\ 2 & 0 & -1 & -1 & 0 \end{bmatrix}.$$

If we transpose M, its column vectors become the row vectors of M^T. Thus the *column* space of M is the same as the *row* space of M^T. Now you can verify that

$$M^T = \begin{bmatrix} 1 & 0 & 0 & 2 \\ 2 & 1 & 1 & 0 \\ 4 & -5 & 3 & -1 \\ 0 & -1 & -1 & -1 \\ -3 & 2 & 0 & 0 \end{bmatrix}$$

has row echelon form

$$U = \begin{bmatrix} 1 & 0 & 0 & 2 \\ 0 & 1 & 0 & 3 \\ 0 & 0 & 1 & -2 \\ 0 & 0 & 0 & 1 \\ 0 & 0 & 0 & 0 \end{bmatrix}.$$

So the row vectors $\{(1, 0, 0, 2), (0, 1, 0, 3), (0, 0, 1, -2), (0, 0, 0, 1)\}$ form a basis for the *row* space of M^T. Therefore, the column vectors

$$\begin{bmatrix} 1 \\ 0 \\ 0 \\ 2 \end{bmatrix}, \quad \begin{bmatrix} 0 \\ 1 \\ 0 \\ 3 \end{bmatrix}, \quad \begin{bmatrix} 0 \\ 0 \\ 1 \\ -2 \end{bmatrix}, \quad \text{and} \quad \begin{bmatrix} 0 \\ 0 \\ 0 \\ 1 \end{bmatrix}$$

form a basis for the *column* space of M. ∎

This example illustrates the following fact:

> The **column space** of an $m \times n$ matrix A is the same as the **row space** of its $n \times m$ transpose A^T.

The processes for determining the row and column ranks of a given matrix A are essentially the same, except for changing from row vectors **r** to column vectors $\mathbf{r}^T$. The matrix M in Example 5 has column rank 4. Its row rank is also 4, since a row echelon form for M is

$$U' = \begin{bmatrix} 1 & 2 & 4 & 0 & -3 \\ 0 & 1 & -5 & -1 & 2 \\ 0 & 0 & 1 & 0 & -\dfrac{1}{4} \\ 0 & 0 & 0 & 1 & -\dfrac{27}{20} \end{bmatrix}.$$

As we stated in Examples 1 and 2, the fact that M has equal row and column ranks is not just a fluke. Indeed, we have the following important theorem.

> **Theorem 3.12**   Let A be any $m \times n$ real (or complex) matrix. Then the row rank of A is equal to the column rank of A.

Before we finally write the proof of this theorem, let us consider one more example.

Example 6   Find the row and column ranks of the matrix

$$A = \begin{bmatrix} 1 & 2 & 0 & -1 \\ 3 & 1 & 2 & 5 \\ 1 & -3 & 2 & 7 \end{bmatrix}.$$

We proceed to reduce A to its row echelon form and obtain the matrix

$$U = \begin{bmatrix} 1 & 2 & 0 & -1 \\ 0 & 1 & \dfrac{-2}{5} & \dfrac{-8}{5} \\ 0 & 0 & 0 & 0 \end{bmatrix}.$$

So the row rank of A is *two*. Let us now consider the column rank. The conceptually easiest thing to do is to transpose A and compute the row rank of A^T. An alternate method is suggested in Supplementary Project 1. Note that

$$A^T = \begin{bmatrix} 1 & 3 & 1 \\ 2 & 1 & -3 \\ 0 & 2 & 2 \\ -1 & 5 & 7 \end{bmatrix}$$

can be reduced to the following row echelon form:

$$W = \begin{bmatrix} 1 & 3 & 1 \\ 0 & 1 & 1 \\ 0 & 0 & 0 \\ 0 & 0 & 0 \end{bmatrix}.$$

Thus A^T has row rank *two;* therefore, A has column rank *two*. ∎

Let us now prove the theorem.

Proof of Theorem 3.12   This proof is an important application of the material of this chapter. Let $A = (a_{ij})$ be any $m \times n$ matrix where $i = 1, 2, \ldots, m$, and $j = 1, 2, \ldots, n$ [see (3.27)]. Denote the row vectors of A by $\mathbf{r}_1, \mathbf{r}_2, \ldots, \mathbf{r}_m$, respectively, and assume that the row space $\mathbf{L}_r = L(\{\mathbf{r}_1, \mathbf{r}_2, \ldots, \mathbf{r}_m\})$ of A has dimension k. Then $\mathbf{L}_r$ has a basis consisting of k vectors, say, $S = \{\mathbf{b}_1, \mathbf{b}_2, \ldots, \mathbf{b}_k\}$, so $L(S) = \mathbf{L}_r$. Suppose that for each $\mu = 1, 2, \ldots, k$

$$\mathbf{b}_\mu = (b_{\mu 1}, b_{\mu 2}, \ldots, b_{\mu n}).$$

Since S is a basis for $\mathbf{L}_r$, each row vector of A is expressible as a linear combination of the vectors in S. That is, for each $i = 1, 2, \ldots, m$ and appropriate scalars $c_{i1}, \ldots, c_{ik}$,

$$\mathbf{r}_i = c_{i1}\mathbf{b}_1 + c_{i2}\mathbf{b}_2 + \cdots + c_{ik}\mathbf{b}_k. \tag{3.28}$$

Now two vectors in $\mathbb{R}^n$ or $\mathbb{C}^n$ are equal if and only if their corresponding components (entries in the n-tuple) are equal. Therefore, the jth components on each side of (3.28) are equal for each $i = 1, 2, \ldots, m$. So for all i and $j = 1, 2, \ldots, n$ we have

$$a_{ij} = c_{i1}b_{1j} + c_{i2}b_{2j} + \cdots + c_{ik}b_{kj}. \tag{3.29}$$

Writing out all the columns in the system of equations (3.29), we get the vector (matrix) equation

$$\begin{bmatrix} a_{1j} \\ a_{2j} \\ \vdots \\ a_{mj} \end{bmatrix} = b_{1j}\begin{bmatrix} c_{11} \\ c_{21} \\ \vdots \\ c_{m1} \end{bmatrix} + b_{2j}\begin{bmatrix} c_{12} \\ c_{22} \\ \vdots \\ c_{m2} \end{bmatrix} + \cdots + b_{kj}\begin{bmatrix} c_{1k} \\ c_{2k} \\ \vdots \\ c_{mk} \end{bmatrix}. \tag{3.30}$$

However, the column vector on the left side of (3.30) is the jth column vector $\mathbf{a}_j$ of the matrix A. Since (3.30) holds for each $j = 1, 2, \ldots, n$, we see that *all* column vectors of A lie in the subspace of $\mathbb{R}^m$ or $\mathbb{C}^m$ spanned by the k-vectors

$$\begin{bmatrix} c_{11} \\ c_{21} \\ \vdots \\ c_{m1} \end{bmatrix}, \begin{bmatrix} c_{12} \\ c_{22} \\ \vdots \\ c_{m2} \end{bmatrix}, \ldots, \begin{bmatrix} c_{1k} \\ c_{2k} \\ \vdots \\ c_{mk} \end{bmatrix}.$$

Hence the column rank of A is at most k, which is the *row* rank of A. Thus,

$$\text{column rank of } A \leq \text{row rank of } A. \tag{3.31}$$

Now if we do the same thing for the transpose A^T of A, we find that

$$\text{column rank of } A^T \leq \text{row rank of } A^T. \tag{3.32}$$

But

$$\text{column rank of } A^T = \text{row rank of } A,$$

and

$$\text{row rank of } A^T = \text{column rank of } A.$$

Therefore, (3.32) becomes

$$\text{row rank of } A \leq \text{column rank of } A. \tag{3.33}$$

Combining (3.31) and (3.33), we obtain the desired equality: row rank = column rank. ∎

Definition The common value of the row rank and column rank of the matrix A is called simply the **rank** of A.

We leave it as Exercise 35 for you to prove the following theorem for square matrices:

Theorem 3.13 If A is an $n \times n$ matrix, then A is nonsingular if, and only if, A has rank n.

Our work in Chapter 1 with systems of linear equations was somewhat informal in view of the fact that we did not at that time have the language and machinery of vector spaces available to us. Now we have developed the necessary mathematical structure to place some of that work on a more formal footing. Suppose that we have a system of linear equations whose matrix form is

$$A\mathbf{x} = \mathbf{b},$$

where $A = (a_{ij}) \in M_{m \times n}$, $\mathbf{x} = (x_1, x_2, \ldots, x_n)^T$, and $\mathbf{b} = (b_1, b_2, \ldots, b_m)^T$. Then we may write out this system of equations much as we did in the proof of Theorem 3.12. $A\mathbf{x} = \mathbf{b}$ becomes the following linear combination of column vectors:

$$x_1 \begin{bmatrix} a_{11} \\ a_{21} \\ \vdots \\ a_{m1} \end{bmatrix} + x_2 \begin{bmatrix} a_{12} \\ a_{22} \\ \vdots \\ a_{m2} \end{bmatrix} + \cdots + x_n \begin{bmatrix} a_{1n} \\ a_{2n} \\ \vdots \\ a_{mn} \end{bmatrix} = \begin{bmatrix} b_1 \\ b_2 \\ \vdots \\ b_m \end{bmatrix}. \tag{3.34}$$

To say that the system of linear equations $A\mathbf{x} = \mathbf{b}$ has a solution $\mathbf{x} = (x_1, x_2, \ldots, x_n)^T$ is to say that there exist scalars $x_1, x_2, \ldots, x_n$ such that the vector **b** is a linear combination of the column vectors $\mathbf{a}_i$ of the matrix A. [See (3.34).] But this says that **b** belongs to the column space of A. Therefore,

> A system of linear equations $A\mathbf{x} = \mathbf{b}$ has a solution if, and only if, **b** lies in the column space of the coefficient matrix A.

If we were to add the column vector **b** to the collection of column vectors of A, we would therefore not change the dimension of the resulting subspace of $\mathbb{R}^m$ or $\mathbb{C}^m$. Were it otherwise, no such scalars x_i could exist. Therefore, the matrix A and the augmented matrix $[A:\mathbf{b}]$ must have the same column rank, and hence by Theorem 3.12 must have the same rank. This proves the following classical theorem.

> **Theorem 3.14**   A system of linear equations $A\mathbf{x} = \mathbf{b}$ has a solution if, and only if, the rank of the augmented matrix $[A:\mathbf{b}]$ is the same as the rank of the coefficient matrix A.

Vocabulary

In this section we have again encountered new ideas and terms. Be sure that you understand the concepts as well as the theorems of this section. Listed below are the new terms to add to your linear algebra vocabulary.

row space of A row rank of A
column space of A column rank of A
rank of A

Exercises 3.5

In Exercises 1–9 find the row rank of A by finding a row echelon form U for A. Then find the column rank of A by finding a row echelon form for A^T.

1. $\begin{bmatrix} 1 & 2 \\ 4 & 8 \end{bmatrix}.$

2. $\begin{bmatrix} 1 & 2 & 4 \\ 2 & 4 & 7 \\ 1 & 2 & 3 \end{bmatrix}.$

3. $\begin{bmatrix} 1 & 2 & 0 \\ 2 & 4 & 0 \\ 3 & 6 & 0 \end{bmatrix}.$

4. $\begin{bmatrix} 1 & 2 & -1 \\ -1 & 4 & -3 \\ 3 & 0 & 1 \end{bmatrix}.$

5. $\begin{bmatrix} 1 & 2 & 4 & 0 \\ 3 & 5 & 1 & 2 \\ 3 & 0 & 1 & 0 \end{bmatrix}.$

6. $\begin{bmatrix} 1 & 0 & -1 & 0 \\ 3 & -1 & 1 & 2 \\ 3 & 0 & 1 & 0 \end{bmatrix}.$

7. $\begin{bmatrix} 1 & -2 & -1 \\ 0 & 1 & 1 \\ 1 & 2 & 3 \\ 0 & 1 & -1 \end{bmatrix}.$

8. $\begin{bmatrix} 1 & -1 & 0 \\ -1 & 0 & 5 \\ 0 & 1 & -2 \\ -1 & 1 & 0 \end{bmatrix}.$

9. $\begin{bmatrix} -1 & 1 & 0 & 5 \\ 0 & 1 & -2 & -1 \\ 1 & 0 & -3 & 0 \\ -1 & 0 & -2 & 0 \end{bmatrix}.$

In Exercises 10–18 find a basis for the row space for the matrix in the given exercise.

10. Exercise 1.   **11.** Exercise 2.   **12.** Exercise 3.   **13.** Exercise 4.   **14.** Exercise 5.

15. Exercise 6.   **16.** Exercise 7.   **17.** Exercise 8.   **18.** Exercise 9.

In Exercises 19–27 find a basis for the column space for the matrix in the given exercise.

19. Exercise 1.   **20.** Exercise 2.   **21.** Exercise 3.   **22.** Exercise 4.   **23.** Exercise 5.

24. Exercise 6.   **25.** Exercise 7.   **26.** Exercise 8.   **27.** Exercise 9.

28. Write out the details of the proof of Lemma 3.10 (page 198).

29. Prove that the rank of an $m \times n$ matrix is at most the smaller of m and n.

30. Can a singular $n \times n$ matrix have rank n? Give reason.

31. Prove part (i) of Theorem 3.11 (page 199).

32. Prove part (ii) of Theorem 3.11.

33. Prove part (iii) of Theorem 3.11.

34. Prove part (iv) of Theorem 3.11.

35. Prove Theorem 3.13 (page 203).

36. Prove that the rank of AB is at most the smaller of rank A and rank B.

37. Suppose that N is an $n \times n$ *nilpotent* matrix. Prove that the rank of A is less than n. If $N^k = 0$ but $N^{k-1} \neq 0$, is the rank of $N = k - 1$? Give reasons.

38. Prove that the $n \times n$ matrix A has nonzero determinant if, and only if, rank $A = n$.

39. Prove that the $n \times n$ matrix A has rank n if, and only if, A is row equivalent to the identity matrix I_n.

40. Prove that if A and B are row equivalent matrices, they have the same rank.

41. Prove that the $n \times n$ homogeneous system of linear equations $A\mathbf{x} = \mathbf{0}$ has a nontrivial solution if, and only if, rank $A < n$. HINT: See the argument preceding Theorem 3.14 (page 204).

42. In the vector space P_2 of all polynomials of degree two or less, represent each polynomial $ax^2 + bx + c$ by the vector (a, b, c) in $\mathbb{R}^3$. By means of this identification and the results of this section find a basis for the subspace of P_2 spanned by the four polynomials $1 + x^2$; $1 - x^2$; $1 + x$; $1 - x$.

43. Show that if A is an $n \times n$ matrix and the solution space $S(A)$ for the homogeneous system of equations $AX = 0$ has dimension k, then $n = r + k$, where $r =$ rank of A.

44. An $m \times n$ matrix A is said to have a *left inverse* if, and only if, there exists an $n \times m$ matrix M so that $MA = I_n$. Prove that if A has a left inverse, then the row vectors of A span $\mathbb{R}^n$. HINT: Write $MA = I_n$ as a linear combination of row vectors of A or, perhaps easier to see, the column vectors of $A^T M^T = I_n$, in terms of the column vectors of A^T as in the proof of Theorem 3.14.

45. Do the converse of Exercise 44. Prove that *if* the row vectors of an $m \times n$ matrix A span $\mathbb{R}^n$, *then* A has a left inverse.

In Exercises 46–49 find the rank of the given matrix A and the rank of adj A.

46. $A = \begin{bmatrix} 0 & 1 & 0 & 0 \\ 0 & 0 & 1 & 0 \\ 0 & 0 & 0 & 1 \\ 0 & 0 & 0 & 0 \end{bmatrix}$ **47.** $A = \begin{bmatrix} 0 & 0 & 0 & 1 \\ 0 & 0 & 1 & 0 \\ 0 & 1 & 0 & 0 \\ 1 & 0 & 0 & 0 \end{bmatrix}$ **48.** $A = \begin{bmatrix} 1 & 0 & 0 & 0 \\ 0 & 1 & 0 & 0 \\ 0 & 0 & 1 & 0 \\ 0 & 0 & 0 & 0 \end{bmatrix}$.   **49.** $A = \begin{bmatrix} 1 & 1 & 1 & 1 \\ 1 & 1 & 1 & 1 \\ 1 & 1 & 1 & 1 \\ 1 & 1 & 1 & 1 \end{bmatrix}$.

50. Let $\mathbf{v}_1, \mathbf{v}_2, \ldots, \mathbf{v}_n$ be vectors in $\mathbb{R}^n$. Prove that the set $\{\mathbf{v}_1, \mathbf{v}_2, \ldots, \mathbf{v}_n\}$ is a linearly independent set if, and only if, the determinant of the $n \times n$ matrix whose *rows* are $\mathbf{v}_1, \mathbf{v}_2, \ldots, \mathbf{v}_n$ is nonzero.

Chapter 3 Summary

In this chapter, we studied the important concept of a vector space and found that this rather ubiquitous algebraic structure included not only physical or geometric vectors but other important mathematical objects, such as functions, polynomials, and matrices, as well. General concepts such as linear dependence and linear independence of vectors, the span of a set of vectors, and the notion of a subspace of a vector space were also studied. The important concept of a basis for a vector space was then considered. In particular we learned that every vector space has a basis in terms of which every vector in the space can be expressed as a finite linear combination. We learned that a vector space can have many different bases, but if a given vector space V has one basis consisting of exactly n vectors, then every basis for V must consist of exactly n vectors. This invariant number n is called the dimension of the vector space. We noted that $\mathbb{R}^n$ and $\mathbb{C}^n$ have dimension n, as expected, whereas the dimension of other spaces was less transparent. Of particular note are the concepts of row and column space of an $m \times n$ matrix. The row rank of a matrix is defined to be the dimension of its row space, with the column rank being defined similarly. We also learned that no matter what the relative sizes of m and n, the row and column rank for a given $m \times n$ matrix were the same number, called simply the rank of A.

Chapter 3 Review Exercises

1. Are the three vectors $(1, 2, -1)$, $(3, 0, 2)$, and $(-1, 1, 0)$ linearly independent in $\mathbb{R}^3$? Give reasons.

2. Are the vectors $(1, 2, -1)$, $(3, 0, 2)$, $(-1, 1, 0)$, and $(1, 1, 1)$ linearly independent in $\mathbb{R}^3$? Give reasons.

3. Are the vectors $1 + x$, $1 - x$, $1 - x^2$, and $1 + x + x^2$ linearly independent in P_2? Give reasons.

4. Are the vectors $\begin{bmatrix} 1 & 0 \\ -1 & 0 \end{bmatrix}$, and $\begin{bmatrix} 2 & 0 \\ 3 & 0 \end{bmatrix}$ linearly independent in $M_2(\mathbb{R})$? Give reasons.

5. Show that any set of vectors containing the zero vector is linearly dependent.

6. Show that the set of all vectors (x_1, x_2, x_3) in $\mathbb{R}^3$ that are such that $2x_1 - x_2 + 3x_3 = 0$ forms a subspace of $\mathbb{R}^3$. What is the dimension of this subspace? Find a basis for this subspace.

7. Construct a basis for $\mathbb{R}^3$ that contains the vectors $(1, 2, -1)$ and $(-3, 0, 5)$.

8. Can a set of five vectors in $\mathbb{R}^4$ ever be linearly independent? Give reasons.

9. The vectors $(1, 2, a)$, $(3, -1, 1 - a)$, and $(4, 0, 5 + 2a)$ are linearly dependent in $\mathbb{R}^3$. Find a.

10. Find the row rank of the matrix
$$A = \begin{bmatrix} 1 & 2 & -3 & 0 \\ 4 & 1 & 0 & 1 \\ -1 & 2 & 1 & 2 \end{bmatrix}.$$
Also find the column rank and verify that the two ranks are equal.

11. What is the rank of the matrix
$$J = \begin{bmatrix} 0 & 0 & 1 \\ 0 & 1 & 0 \\ 1 & 0 & 0 \end{bmatrix}?$$

12. What is the rank of the matrix
$$J = \begin{bmatrix} 1 & 1 & 1 & 1 \\ 1 & 1 & 1 & 1 \\ 1 & 1 & 1 & 1 \\ 1 & 1 & 1 & 1 \end{bmatrix}?$$

13. Suppose that the $n \times n$ matrix A satisfies $A^2 = -I_n$. What is the rank of A?

14. Suppose that for the $n \times n$ matrix A det $A \neq 0$. What is the rank of A?

CHAPTER 3 SUPPLEMENTARY EXERCISES AND PROJECTS

1. Prove the following: A homogeneous system of m linear equations in n unknowns, $A\mathbf{x} = \mathbf{0}$, has a unique solution (the trivial one $\mathbf{x} = \mathbf{0}$) if the rank of A is n. If the rank r of A is less than n, the solution space $S(A)$ has dimension $n - r$. HINT: Consider the reduced row echelon form of A.

2. Prove that a consistent $m \times n$ system of linear equations $A\mathbf{x} = \mathbf{b}$ has a unique solution if the rank of A is n. HINT: Show that the number of solutions to $A\mathbf{x} = \mathbf{b}$ is the same as that for the related homogeneous system $A\mathbf{x} = \mathbf{0}$.

3. Let A be an $n \times n$ real matrix and let $\{\mathbf{u}_1, \mathbf{u}_2, \ldots, \mathbf{u}_k\}$ be a linearly independent subset of $\mathbb{R}^n$. Show that A is nonsingular if, and only if, the set $\{A\mathbf{u}_1, A\mathbf{u}_2, \ldots, A\mathbf{u}_k\}$ is linearly independent.

4. Prove that the column vectors of an $n \times n$ upper-triangular matrix $A = (a_{ij})$ are linearly independent if, and only if, $a_{ii} \neq 0$ for each $i = 1, 2, \ldots, n$.

5. Prove that for any $A \in M_{m \times n}(\mathbb{C})$, rank $A =$ rank $A^T =$ rank $A^* =$ rank $\overline{A}$.

6. Prove the Frobenius inequality: If $A \in M_{m \times k}(\mathbb{C})$, $B \in M_{k \times s}(\mathbb{C})$, and $C \in M_{s \times n}(\mathbb{C})$, then rank $AB +$ rank $BC \leq$ rank $B +$ rank ABC.

7. Suppose that A is a 2×2 matrix in $M_2(\mathbb{Z})$; that is, A has integer entries. Prove that A has an inverse in $M_2(\mathbb{Z})$ if, and only if, det $A = 1$ or -1.

8. Generalize the result in Exercise 7 to $n \times n$ matrices in $M_n(\mathbb{Z})$. Prove your assertion.

9. What is the dimension of the set $\mathbb{C}$ of all complex numbers considered as a *real* vector space? What is its dimension as a *complex* vector space? Is the set of real numbers $\mathbb{R}$ a vector space over the rational numbers $\mathbb{Q}$? Is it finite-dimensional over $\mathbb{Q}$?

10. The set $\mathbb{Z}_p$ of all residue classes of integers modulo a prime p is a *finite* field. Let V be an n-dimensional vector space whose scalars come only from $\mathbb{Z}_p$. How many vectors are there in V? Since both are n-dimensional, contrast this with the number of vectors in $\mathbb{R}^n$.

Project 1. (Numerical) An Alternate Method for Determining a Basis for the Column Space of a Matrix

Use a computer or a calculator to do the computations called for in this project.

a. Show that if one *row* reduces the matrix A (not A^T) to a row echelon form U, and notes which *columns* of U containing leading 1's (or the pivots) in the Gaussian reduction correspond to linearly independent columns of A, then these columns of A (not of U, see part (b)) form a basis for the *column* space of A. HINT: Show that the rest of the columns of A must be linear combinations of these columns.

b. Consult Example 6 in Section 3.5. Show that the linear span of $S = \{(1, 0, 0)^T, (2, 1, 0)^T\}$ *does not equal* the linear span of $B = \{(1, 3, 1)^T, (0, 1, 1)^T\} =$ the column space of A (which is the linear span of the four columns of A). Both spaces have dimension two, of course. Note that S consists of the two columns of U with leading 1's, while B consists of the transposes of the two linearly independent rows of W, the row echelon form of A^T.

c. Use the matrices of Example 6 in Section 3.5. Show that the column space of A is spanned by the columns *one* and *two,* since this is where the leading ones are in the row echelon form U of A.

d. Use the method of this project to find a basis for the column space of the matrix

$$A = \begin{bmatrix} 1 & -2 & 1 & 1 & 2 \\ -1 & 3 & 0 & 2 & -2 \\ 0 & 1 & 1 & 3 & 4 \\ 1 & 2 & 5 & 13 & 5 \end{bmatrix}.$$

whose row echelon form is $U = \begin{bmatrix} 1 & -2 & 1 & 1 & 2 \\ 0 & 1 & 1 & 3 & 0 \\ 0 & 0 & 0 & 0 & 1 \\ 0 & 0 & 0 & 0 & 0 \end{bmatrix}.$

That is, show that columns *one, two,* and *five* of A form a basis for the column space of A. Show also that columns *one, two,* and *five* of U **do not** form a basis for the column space of A.

e. Use the matrix A of part (d). Find A^T. Row reduce A^T and find a different basis for the column space of A. Show that both bases span the same subspace of $\mathbb{R}^4$.

f. Use the method suggested by this project to find a basis for the subspace W of $\mathbb{R}^5$ spanned by the following vectors; that is,

  1. Write each of the vectors below as the columns $\mathbf{k}_i$ of a matrix K.
  2. Row reduce K and note the columns with leading 1's in the row echelon form U of K.
  3. The corresponding columns of K are the desired basis.

  $\mathbf{k}_1 = (2, 3, 0, -1, 1); \mathbf{k}_2 = (1, -1, 0, 1, 1); \mathbf{k}_3 = (-2, 4, 11, 1, 0); \mathbf{k}_4 = (3, -5, 11, 0, 1);$
  $\mathbf{k}_5 = (3, 2, 0, 0, 2); \mathbf{k}_6 = (5, 0, 1, -2, -1); \mathbf{k}_7 = (1, -1, 1, -1, 1).$

g. Use the method of part (f) to determine the two-dimensional subspace of $\mathbb{R}^3$ spanned by the four vectors $\mathbf{v}_1 = (1, 2, 3)$, $\mathbf{v}_2 = (1, -1, 0)$, $\mathbf{v}_3 = (-2, 1, 3)$, and $\mathbf{v}_4 = (3, -2, -3)$. Sketch this plane in 3-space. Also sketch the plane spanned by the columns with leading 1's in the row echelon form U of the constructed matrix K. Contrast the two.

Project 2. New Subspaces from Old

a. Let V be a vector space and let W_1 and W_2 be any two subspaces of V. Show that $W_1 \cap W_2 = \{\mathbf{x} \in V: \mathbf{x} \in W_1 \text{ and } \mathbf{x} \in W_2\}$ is a subspace of V.

b. Extend part (a) to any collection of subspaces of V; i.e., prove that $\cap_\mu W_\mu$ is a subspace of V, if each W_μ is a subspace.

c. Show by an example from $\mathbb{R}^3$ that in general $W_1 \cup W_2 = \{\mathbf{x} \in V: \mathbf{x} \in W_1 \text{ or } \mathbf{x} \in W_2 \text{ (or both)}\}$ is not a subspace of V.

d. If W and U are both subspaces of finite-dimensional vector space V let $W + U = \{\mathbf{w} + \mathbf{u}: \mathbf{w} \in W \text{ and } \mathbf{u} \in U\}$. Prove or disprove that $W + U$ is a subspace of V. HINT: If $\mathbf{x}, \mathbf{y} \in W + U$, then $\mathbf{x} = \mathbf{w}_1 + \mathbf{u}_1$ and $\mathbf{y} = \mathbf{w}_2 + \mathbf{u}_2$. Use Theorem 3.3 (page 179).

e. Show that, for subspaces W_1, W_2, and W_3 of the finite-dimensional vector space V, it is not necessarily true that $W_1 \cap (W_2 + W_3) = (W_1 \cap W_2) + (W_1 \cap W_3)$.

In parts (f)–(k), W_1 and W_2 are nonzero subspaces of the finite-dimensional vector space V.

f. If $W_1 \cap W_2 \neq \{\mathbf{0}\}$ explain how a basis B for $W_1 \cap W_2$ can be enlarged to separate bases S for W_1 and T for W_2.

g. If $\mathbf{x} \in W_1 + W_2$, show how to write $\mathbf{x}$ as a linear combination of the vectors in $S \cup T$.

h. Is $S \cup T$ (part (f)) always a basis for $W_1 + W_2$? Explain.

i. Combine the results of parts (f), (g), and (h) to prove that $\dim(W_1) + \dim(W_2) = \dim(W_1 + W_2) + \dim(W_1 \cap W_2)$.

j. Show that if W and U are both three-dimensional subspaces of the five-dimensional vector space V then $W \cap U \neq \{\mathbf{0}\}$.

k. The vector space V is said to be the **direct sum** of W_1 and W_2, written $W_1 \oplus W_2$, when $V = W_1 + W_2$ and $W_1 \cap W_2 = \{\mathbf{0}\}$. Prove that if $V = W_1 \oplus W_2$ and S is a basis for W_1 and T is a basis for W_2, then $S \cup T$ is a basis for V.

Project 3. *Matrix Polynomials*

Let A be a fixed 3×3 matrix and consider the set $G = \{I, A, A^2, \ldots, A^8, A^9\}$.

a. Prove that G is a linearly dependent subset of $M_3(\mathbb{R})$.

b. Use the result of part (a) to show that there exists an integer m, $1 \leq m \leq 9$, and scalars $k_0, k_1, k_2, \ldots, k_{m-1}$ so that $A^m = k_0 I + k_1 A + k_2 A^2 + \cdots + k_{m-1} A^{m-1}$, and the set $H = \{I, A, A^2, \ldots, A^{m-1}\}$ is linearly independent.

c. Generalize arguments above to show that each $n \times n$ real matrix A must satisfy a polynomial equation of the form $p(A) = k_0 I + k_1 A + k_2 A^2 + \cdots + k_{m-1} A^{m-1} + A^m = 0$, for some $m \leq n^2$.

d. If the $m \times n$ matrix A has rank r, does the matrix A^2 necessarily have rank r? Give reasons.

e. Let A be the 4×4 matrix

$$\begin{bmatrix} 1 & -2 & 1 & 1 \\ 2 & -1 & 3 & 0 \\ 2 & -2 & 0 & 1 \\ 1 & 3 & 4 & 1 \end{bmatrix}.$$

Find the set H of part (b) and the polynomial $p(A)$ of part (c) for this matrix A.

f. Let t be a variable. Compute the polynomial $p(t) = \det(tI - A)$ for the matrix A of part (e).

g. Show that $p(A) = 0$ (see part (c)), where $p(t)$ is the polynomial of part (f); that is, substitute the matrix A in place of the variable t in the polynomial of part (f), taking care to identify the constant term k of $p(t)$ with the scalar matrix kI. Compare this polynomial with the polynomial of part (e). Use a computer or a calculator for the calculations involved.

Project 4. (Theoretical) Matrix Equivalence

Let A be an $m \times n$ real matrix whose rank is r.

a. Show that if M is the reduced row echelon form of A, there exists a nonsingular matrix Q such that $M = QA$. HINT: Consider the elementary matrices. What is the size of Q?

b. Prove that the reduced row echelon form K of the matrix M^T, the transpose of the matrix M of part (a), has the block form diag $[I_r : O]$.

c. Let P be a nonsingular matrix such that $P^T M^T = K$ (of part (b)). Why does P exist? Show that $QAP = K$. What is the size of P?

> An $m \times n$ matrix B is said to be **equivalent** (not row-equivalent) to an $m \times n$ matrix A if there exist two nonsingular matrices Q and P so that $B = QAP$. See parts (a)–(c).

d. Prove that every $m \times n$ matrix A is equivalent to itself, and that if B is equivalent to A, then A is equivalent to B. That is, equivalence is a reflexive and symmetric relation on the vector spaces $M_{m\times n}(\mathbb{R})$ and $M_{m\times n}(\mathbb{C})$.

e. Show that equivalence is a transitive relation on $M_{m\times n}$ by proving that if A is equivalent to B and B is equivalent to C then A is equivalent to C. Thus equivalence is an "equivalence relation" on $M_{m\times n}$.

f. Prove that two matrices are equivalent if and only if they have the same rank. HINT: Expand the argument in parts (a)–(c).

g. If A and B are equivalent, are A^2 and B^2 necessarily equivalent? HINT: See Project 3, part (d).

h. Under what conditions is AB equivalent to BA? Give reasons.

Project 5. (Theoretical)*

Let A be an $m \times n$ matrix of rank n with $m > n$.

a. Show that the matrix $A^T A$ has rank n. Conclude that $(A^T A)$ is nonsingular.

---

*(See G. Williams, Overdetermined Systems of Linear Equations, *American Mathematical Monthly*, **97** (6), June–July 1990, pp. 511–513.)

b. Modify the overdetermined system of linear equations $AX = B$ by multiplying both sides of the equation by A^T. Is the solution $X' = (A^TA)^{-1}A^TB$ of the modified system of equations a solution to the original system? Do all of this for the following system of equations.

$$x_1 + 2x_2 = -1,$$
$$2x_1 - x_2 = 3,$$
$$2x_1 - 3x_2 = 2.$$

c. The matrix $(A^TA)^{-1}A^T$ is called the **pseudoinverse** or the **Moore-Penrose** inverse of A. Compute this pseudoinverse for the matrix

$$A = \begin{bmatrix} 1 & -2 & 1 & 1 \\ 2 & -1 & 3 & 0 \\ 2 & -2 & 0 & 1 \\ 1 & 3 & 4 & 1 \\ 1 & -1 & 2 & 4 \end{bmatrix}.$$

d. Does the system $AX = (1, 3, -1, 4, 2)^T$ have a solution for the matrix A of part (c)? Compute the solution X' to the modified system $A^TAX = A^T(1, 3, -1, 4, 2)^T$.

e. Show that for the general overdetermined system $AX = B$, where A is an $m \times n$ matrix of rank n with $m > n$, $A^T(AX' - B) = 0$.

f. In the next chapter we call two column vectors **u** and **v** *orthogonal* if $u^Tv = 0$. Conclude from part (e) that the vector $AX' - B$ is orthogonal to the column space of the matrix A.

g. Note that in general the overdetermined system $AX = B$ has no solution and the solution X' to the modified system is in the sense of part (f) the best we can get. The vector $AX' - B$ is called the **error vector** and the number $[(AX' - B)^T(AX' - B)]^{1/2}$ is called the **error**. Why is $(AX' - B)^T(AX' - B)$ a positive real number?

h. Compute the error for the situation in part (d).

4 INNER-PRODUCT SPACES

In this chapter we continue our discussion of the relationship between conventional geometric concepts and the idea of a vector space. As we mentioned at the beginning of Chapter 3, many geometric ideas are most usefully expressed in terms of algebraic vectors with real numbers as scalars. Ordinary Euclidean geometry, as it has developed over the centuries, has exploited the important concepts of length, distance between two points, lines, and the angle between two rays or line segments. These concepts can appropriately be modeled by real two- and three-dimensional vector spaces in which unit vectors along the coordinate axes serve as the basis for the vector space $\mathbb{R}^2$ or $\mathbb{R}^3$. Equally important, however, is the fact that these ideas can be generalized to vector spaces of more than three dimensions and can involve complex number scalars as well. Hence we also examine the generalized concepts of distance and angle in $\mathbb{R}^n$ and in $\mathbb{C}^n$ and eventually in other more general vector spaces. Furthermore the conventional coordinate axes and their placement are arbitrary and can be generalized, as they must be in many applications, to other bases for 2- and 3-space and ultimately for n-space.

Our investigation is not limited to how such ideas as ordinary Euclidean length, distance, and angle can be extended to ideas involving norms in more abstract vector spaces. Rather we also consider how those elementary ideas of analytic geometry which involve a Cartesian coordinate system in the plane or 3-space can be extended to the concept of an orthonormal basis for a given real or complex vector space. We shall see explicitly how to construct an orthonormal basis from a given basis for $\mathbb{R}^2$ or $\mathbb{R}^3$. We then extend this idea to any vector space endowed with an inner product. We further consider some of the important inequalities involving inner products and norms, including the very important Cauchy-Schwarz inequality. In the final section we conclude with an application of these concepts to the problem of finding a polynomial curve that in a certain sense "best fits" given experimental data points.

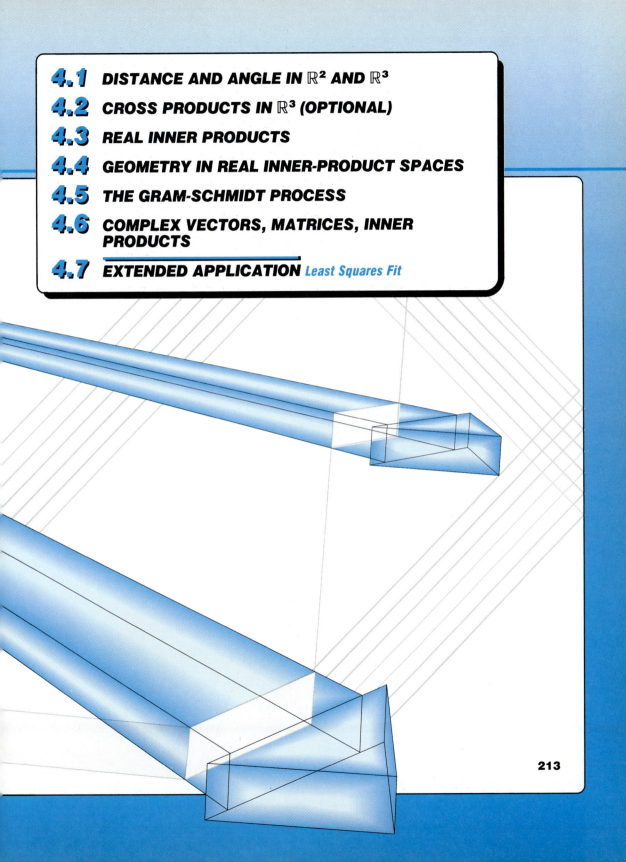

213

4.1 DISTANCE AND ANGLE IN $\mathbb{R}^2$ AND $\mathbb{R}^3$

Let us begin with the ordinary Cartesian coordinate system in 3-space. As usual, if the point P has coordinates (u_1, u_2, u_3) in the given system, the vector $\mathbf{u}$ from the origin $O: (0, 0, 0)$ to P, $\mathbf{u} = \overrightarrow{OP}$, is also written as $\mathbf{u} = (u_1, u_2, u_3)$. If $Q: (v_1, v_2, v_3)$ is any other point, the vector $\mathbf{v} = \overrightarrow{OQ} = (v_1, v_2, v_3)$, and the vector from P to Q is the vector (see Section 3.1)

$$\overrightarrow{PQ} = \mathbf{v} - \mathbf{u} = (v_1 - u_1, v_2 - u_2, v_3 - u_3).$$

Let l be the line in $\mathbb{R}^3$ through the given points P and Q. Then the vector $\mathbf{x}$ from the origin to any point $X: (x_1, x_2, x_3)$ of l is such that the vector from P to X is a scalar multiple of the vector from P to Q. Stated in vector language, $\mathbf{x} - \mathbf{u} = t(\mathbf{v} - \mathbf{u})$.

Thus, for any real number t,

$$\mathbf{x} = t\mathbf{v} + (1 - t)\mathbf{u}.$$

This is the **(parametric) vector equation** of the line determined by the points P and Q in $\mathbb{R}^3$. See Figure 4.1. If we write out the components one by one, we have the three *parametric* **scalar equations**

$$x_1 = (v_1 - u_1)t + u_1,$$
$$x_2 = (v_2 - u_2)t + u_2,$$
$$x_3 = (v_3 - u_3)t + u_3.$$

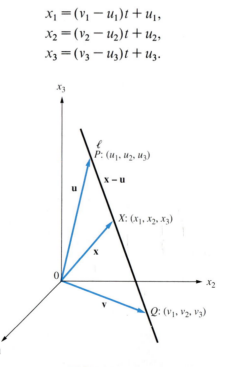

FIGURE 4.1

And if we solve these equations for the parameter t, we obtain the equations

$$t = \frac{x_1 - u_1}{v_1 - u_1} = \frac{x_2 - u_2}{v_2 - u_2} = \frac{x_3 - u_3}{v_3 - u_3}.$$

This last set of equations is called the **symmetric form** of the equation of the line in $\mathbb{R}^3$.

The components $v_1 - u_1$, $v_2 - u_2$, and $v_3 - u_3$ of the vector $\mathbf{v} - \mathbf{u}$ that appear in both the parametric form and the symmetric form of the equation of the line are called **direction numbers** of the line. Since any vector $\mathbf{w} = (\alpha, \beta, \gamma)$ along the line l will serve as the vector $\mathbf{v} - \mathbf{u}$, the components, α, β, and γ of $\mathbf{w}$ are also direction numbers of the line, and they are of course not unique but are proportional to $v_1 - u_1$, $v_2 - u_2$, and $v_3 - u_3$ because $\mathbf{w}$ must be a scalar multiple of $\mathbf{v} - \mathbf{u}$.

If a concept of length is available, as is usual in Euclidean space $\mathbb{R}^3$, then one often chooses $\mathbf{w}$ to have unit length, and then the components satisfy

$$\alpha^2 + \beta^2 + \gamma^2 = 1.$$

In that case they are called **direction cosines** for l. See Exercises 30 and 31.

Example 1   Find the equations for the line l in $\mathbb{R}^3$ determined by the two points $(1, -1, 2)$ and $(2, 0, 5)$. It makes little difference which point we use first, since the vector between them is $\mathbf{v} - \mathbf{u}$ or $\mathbf{u} - \mathbf{v} = -(\mathbf{v} - \mathbf{u})$. So let $\mathbf{u}$ be the vector from the origin to the point $(1, -1, 2)$; that is, $\mathbf{u} = (1, -1, 2)$; and let $\mathbf{v} = (2, 0, 5)$. Then $\mathbf{v} - \mathbf{u} = (1, 1, 3)$. The vector equation for the line l determined by the two points is $\mathbf{x} - \mathbf{u} = t(\mathbf{v} - \mathbf{u})$ or $\mathbf{x} = t\mathbf{v} + (1 - t)\mathbf{u}$, for any real number t. This becomes

$$\mathbf{x} = (x_1, x_2, x_3) = t(2, 0, 5) + (1 - t)(1, -1, 2) = (t + 1, t - 1, 3t + 2).$$

The scalar equations are, in *parametric form*,

$$x_1 = t + 1,$$
$$x_2 = t - 1,$$
$$x_3 = 3t + 2.$$

The *symmetric form* for the scalar equations is $\dfrac{x_1 - 1}{1} = \dfrac{x_2 + 1}{1} = \dfrac{x_3 - 2}{3}.$ ∎

We turn now to a discussion of the idea of the length or magnitude of a vector in $\mathbb{R}^3$. In Figure 4.2 (on the next page) we have sketched a typical vector $\mathbf{u} = (u_1, u_2, u_3)$ in $\mathbb{R}^3$. This sketch is, of course, a special case where the real numbers u_1, u_2, and u_3 are all positive.

In general the point P could be in any of the octants, but it is always located $|u_3|$ units from the x_1, x_2 plane, $|u_2|$ units from the x_1, x_3 plane, and $|u_1|$ units from the x_2, x_3 plane. Let E be the point in the x_1, x_2 plane directly below, as in Figure 4.2 (or *above* if $u_3 < 0$), the point P. That is, E has coordinates $(u_1, u_2, 0)$. Let the vector $\overrightarrow{OE}$ be denoted by $\mathbf{v}$. Now we wish to find the

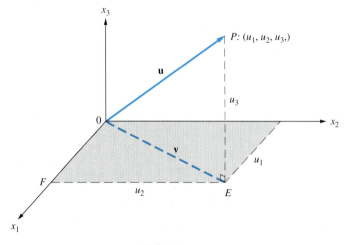

FIGURE 4.2

length of **u**, which we denote by $\|\mathbf{u}\|$. We'll obtain it using the Pythagorean theorem and the length $\|\mathbf{v}\|$ of **v**. We get

$$(\|\mathbf{u}\|)^2 = (\|\mathbf{v}\|)^2 + u_3^2, \tag{4.1}$$

because, as you can see in Figure 4.3 where the figure is reoriented, triangle *OEP* is a right triangle with the right angle at *E*. We have thus reduced the problem of finding the length of a vector in $\mathbb{R}^3$ to that of finding the length of the vector **v** that lies entirely in the x_1, x_2 plane, that is, essentially in $\mathbb{R}^2$.

Since triangle *OFE* (see Figure 4.3) is also a right triangle, we apply the Pythagorean theorem again to obtain the length of **v**, namely,

$$\|\mathbf{v}\|^2 = u_1^2 + u_2^2. \tag{4.2}$$

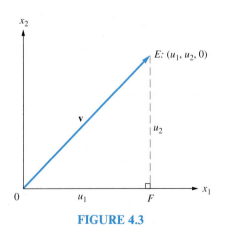

FIGURE 4.3

By substituting the value of $||\mathbf{v}||^2$ from equation (4.2) into equation (4.1) we obtain

$$||\mathbf{u}||^2 = u_1^2 + u_2^2 + u_3^2.$$

Therefore,

$$||\mathbf{u}|| = \sqrt{u_1^2 + u_2^2 + u_3^2}. \tag{4.3}$$

Even though our figures are drawn with u_1, u_2, and u_3 positive, the formula holds in all cases, as you can readily verify.

Example 2 Find the length of the vectors $\mathbf{u} = (2, -1, 3)$, and $\mathbf{w} = (1, 0, -2)$.

From (4.3) we readily calculate

$$||\mathbf{u}|| = \sqrt{(2)^2 + (-1)^2 + (3)^2} = \sqrt{4 + 1 + 9} = \sqrt{14};$$

and

$$||\mathbf{w}|| = \sqrt{(1)^2 + (0)^2 + (-2)^2} = \sqrt{1 + 0 + 4} = \sqrt{5}. \quad \blacksquare$$

Before we consider the question of angle in $\mathbb{R}^2$ and $\mathbb{R}^3$ let us make one or two other observations. First of all, you have probably already noticed that if $\mathbf{u}$ and $\mathbf{v}$ are any two geometric vectors, their linear span $L(\mathbf{u}, \mathbf{v})$ is either a plane or a line or the origin. If $\mathbf{u}$ and $\mathbf{v}$ are not collinear, they are linearly independent and therefore span a plane. See Figure 4.4, where they span the plane of this page.

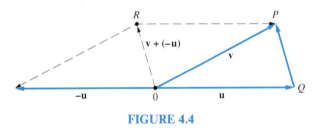

FIGURE 4.4

In this figure the third side of triangle OPQ formed by these vectors is the vector $\overrightarrow{QP} = \mathbf{v} - \mathbf{u}$. To see that this is so, use the parallelogram rule for adding two vectors and construct $\overrightarrow{OR} = \mathbf{v} + (-\mathbf{u})$ as in Figure 4.4. Clearly $\overrightarrow{OR} = \overrightarrow{QP}$, since both have the same magnitude and the same direction; therefore, $\overrightarrow{QP} = \mathbf{v} - \mathbf{u}$.

Now we first use these ideas to derive the distance formula in 3-space. Let $\mathbf{u}$ and $\mathbf{v}$ be two vectors in 3-space $\mathbb{R}^3$, and suppose $\mathbf{u} = (u_1, u_2, u_3)$ and $\mathbf{v} = (v_1, v_2, v_3)$. Then the distance between their respective terminal points Q and P (see Figure 4.4) is the length of the vector $\mathbf{v} - \mathbf{u}$. We have $\mathbf{v} - \mathbf{u} = (v_1 - u_1, v_2 - u_2, v_3 - u_3)$, so from (4.3) the distance is

$$d(\mathbf{u}, \mathbf{v}) = ||\mathbf{v} - \mathbf{u}|| = \sqrt{(v_1 - u_1)^2 + (v_2 - u_2)^2 + (v_3 - u_3)^2}.$$

This is the distance formula in 3-space $\mathbb{R}^3$. Its analog in the plane $\mathbb{R}^2$ is of course

$$d(\mathbf{u}, \mathbf{v}) = \sqrt{(v_1 - u_1)^2 + (v_2 - u_2)^2}.$$

We can introduce the concept of distance into the n-dimensional vector space $\mathbb{R}^n$ by generalizing these formulas as follows.

Definition Let $\mathbf{u}$ and $\mathbf{v}$ be any two vectors in the n-dimensional vector space $\mathbb{R}^n$. The distance $d(\mathbf{u}, \mathbf{v})$ between these vectors is defined to be the number

$$d(\mathbf{u}, \mathbf{v}) = \sqrt{(v_1 - u_1)^2 + (v_2 - u_2)^2 + \cdots + (v_n - u_n)^2}. \qquad \textbf{(4.4)}$$

Example 3 Let $\mathbf{u} = (1, 4, -3)$ and $\mathbf{v} = (-2, -1, 3)$ be two vectors in $\mathbb{R}^3$. Find the distance $d(\mathbf{u}, \mathbf{v})$ between them. From (4.4) then,

$$d(\mathbf{u}, \mathbf{v}) = \sqrt{(1 + 2)^2 + (4 + 1)^2 + (-3 - 3)^2} = \sqrt{9 + 25 + 36} = \sqrt{70}. \quad \blacksquare$$

When one introduces the concept of distance given by (4.4) or, equivalently, the definition of length given below into Cartesian n-space $\mathbb{R}^n$, then $\mathbb{R}^n$ is called *Euclidean n-space*.

Definition The vector space $\mathbb{R}^n$ with the length of a vector $\mathbf{v} = (v_1, v_2, \ldots, v_n)$ defined by the formula

$$\|\mathbf{v}\| = \sqrt{v_1^2 + v_2^2 + \cdots + v_n^2}$$

is called **Euclidean n-space $\mathbb{R}^n$**.

Example 4 Let $\mathbf{v} = (1, -1, 2, -3, 5)$ and $\mathbf{u} = (2, -3, 0, 4, 1)$ be two vectors in Euclidean 5-space, $\mathbb{R}^5$. Find the Euclidean "length" of $\mathbf{u} - \mathbf{v}$.

$$\begin{aligned}
\|\mathbf{u} - \mathbf{v}\| = \|\mathbf{v} - \mathbf{u}\| \\
= d(\mathbf{u}, \mathbf{v}) \\
= \sqrt{(1 - 2)^2 + (-1 + 3)^2 + (2 - 0)^2 + (-3 - 4)^2 + (5 - 1)^2} \\
= \sqrt{1 + 4 + 4 + 49 + 16} = \sqrt{74}. \quad \blacksquare
\end{aligned}$$

In Exercise 29 we ask you to use the distance formula in Euclidean 2-space $\mathbb{R}^2$ to derive the important law of cosines that you first encountered in your trigonometry class. Refer to Figures 4.5 and 4.6, where we have sketched two possible kinds of triangles. Their placement on the coordinate axes is highly arbitrary. We could just as well have placed the angle at B or C at the origin.

Using the usual convention for angles and sides as illustrated in Figures 4.5 and 4.6, we state the law of cosines as follows.

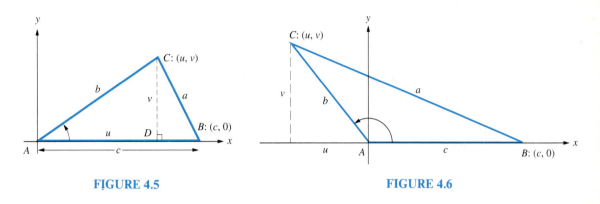

FIGURE 4.5   FIGURE 4.6

The Law of Cosines

Let ABC be a triangle with sides a, b, and c opposite angles A, B, and C, respectively. Then

$$a^2 = b^2 + c^2 - 2bc \cos A,$$
$$b^2 = a^2 + c^2 - 2ac \cos B,$$
$$c^2 = a^2 + b^2 - 2ab \cos C.$$

Let us now apply the law of cosines to calculate the cosine of the angle between two vectors in the plane $\mathbb{R}^2$. To this end suppose that we have the two points $P_1: (x_1, y_1)$ and $P_2: (x_2, y_2)$ as indicated in Figure 4.7. Consider the vectors $\mathbf{u} = \overrightarrow{OP_1}$ and $\mathbf{v} = \overrightarrow{OP_2}$. Also suppose that the angle P_1OP_2 is θ (radians) with $0 \le \theta \le \pi$. Then from the law of cosines we have

$$||\mathbf{v} - \mathbf{u}||^2 = ||\mathbf{u}||^2 + ||\mathbf{v}||^2 - 2||\mathbf{u}||\ ||\mathbf{v}|| \cos \theta. \tag{4.5}$$

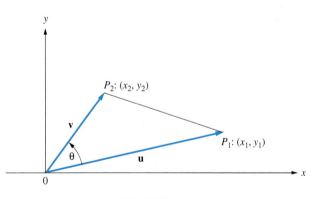

FIGURE 4.7

Applying the distance formula, we see that equation (4.5) becomes

$$(x_1 - x_2)^2 + (y_1 - y_2)^2 = (x_1^2 + y_1^2) + (x_2^2 + y_2^2) - 2\sqrt{x_1^2 + y_1^2}\sqrt{x_2^2 + y_2^2} \cos \theta.$$

Simplifying this equation, we get

$$-2x_1x_2 - 2y_1y_2 = -2\sqrt{x_1^2 + y_1^2}\sqrt{x_2^2 + y_2^2}\cos\theta;$$

and hence, *assuming that the denominator is not zero,*

$$\cos\theta = \frac{x_1x_2 + y_1y_2}{\sqrt{x_1^2 + y_1^2}\sqrt{x_2^2 + y_2^2}}. \tag{4.6}$$

Note that equation (4.6) also has the following form (see Figure 4.7):

$$\cos\theta = \frac{x_1x_2 + y_1y_2}{\|\mathbf{u}\| \cdot \|\mathbf{v}\|}. \tag{4.7}$$

Now do the same thing in 3-space $\mathbb{R}^3$, where the points are $P_1: (x_1, y_1, z_1)$ and $P_2: (x_2, y_2, z_2)$, the law of cosines is applied in the plane (subspace) spanned by the vectors $\mathbf{u}$ and $\mathbf{v}$, and the distance formula for $\mathbb{R}^3$ is used. Then formula (4.7) becomes the following (see Figure 4.8):

$$\cos\theta = \frac{x_1x_2 + y_1y_2 + z_1z_2}{\|\mathbf{u}\| \cdot \|\mathbf{v}\|}. \tag{4.8}$$

This formula for the cosine of the angle between two vectors in $\mathbb{R}^3$ is derived by repeating the steps that led to equations (4.5) and (4.6), only in this case in 3-space.

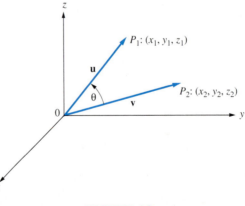

FIGURE 4.8

Notice that the numerators in both equations (4.7) and (4.8) involve a sum of products of the components of the vectors $\mathbf{u}$ and $\mathbf{v}$. This expression is not a vector but rather a *real* number, a scalar. It is, however, often called the *dot product* $\mathbf{u} \cdot \mathbf{v}$ of the two vectors $\mathbf{u}$ and $\mathbf{v}$. This is particularly true in classical physics and engineering. We call it the *inner product* (more specifically later, the *standard or Euclidean inner product*) of the two vectors. Formally, the definition in $\mathbb{R}^3$ is as follows.

> **Definition**   Let $\mathbf{u} = (x_1, y_1, z_1)$ and $\mathbf{v} = (x_2, y_2, z_2)$ be two vectors in $\mathbb{R}^3$. Then the **inner product** of the two vectors $\mathbf{u}$ and $\mathbf{v}$ is the scalar $<\mathbf{u}, \mathbf{v}> = x_1 x_2 + y_1 y_2 + z_1 z_2$.

Of course, if the two vectors $\mathbf{u}$ and $\mathbf{v}$ are in $\mathbb{R}^2$, this reduces to $<\mathbf{u}, \mathbf{v}> = x_1 x_2 + y_1 y_2$. The inner product is also often called the *scalar product* of the two vectors.

Example 5   Find the inner product of the vectors $\mathbf{u} = (1, -2)$ and $\mathbf{v} = (-3, 7)$ from $\mathbb{R}^2$. It is easy to calculate $<\mathbf{u}, \mathbf{v}> = (1)(-3) + (-2)(7) = -17$. ∎

Example 6   Find the inner product of the vectors $\mathbf{u} = (1, -2, 4)$ and $\mathbf{v} = (-3, 7, 5)$ from $\mathbb{R}^3$. It is easy to calculate $<\mathbf{u}, \mathbf{v}> = (1)(-3) + (-2)(7) + (4)(5) = 3$. ∎

Example 7   Find the inner product of the vectors $\mathbf{u} = (6, -4)$ and $\mathbf{v} = (-2, -3)$ from $\mathbb{R}^2$.

Notice that $<\mathbf{u}, \mathbf{v}> = (6)(-2) + (-4)(-3) = 0$, so if we used equation 4.7 we'd find that the angle between these two vectors has a cosine equal to 0. So $\theta = \pi/2$; thus the two vectors are perpendicular. You may want to sketch them. ∎

In general, whenever two vectors $\mathbf{u}$ and $\mathbf{v}$ have zero inner product, we say that they are *orthogonal.* The new word, as you'll see later, applies even when the vectors in question are not line segments in a Euclidean space. Formally then, for 2- and 3-space we have the following:

> Two vectors $\mathbf{u}$ and $\mathbf{v}$ from $\mathbb{R}^2$ or $\mathbb{R}^3$ are **orthogonal** if, and only if, $<\mathbf{u}, \mathbf{v}> = 0$.

Example 8   Are the two vectors $\mathbf{u} = (10, -2, 0)$ and $\mathbf{v} = (-1, -5, 7)$ from $\mathbb{R}^3$ orthogonal?

Since $<\mathbf{u}, \mathbf{v}> = (10)(-1) + (-2)(-5) + (0)(7) = 0$, these two vectors are indeed orthogonal. You may want to sketch them and see that they are perpendicular to each other in $\mathbb{R}^3$. ∎

We'll give some examples of orthogonal vectors that can hardly be called *perpendicular* later. However, it's not even easy to see what perpendicular would mean in $\mathbb{R}^4$. Let's extend the definition of inner product in $\mathbb{R}^3$ to $\mathbb{R}^4$ in the obvious way; namely:

> **Definition**   Let $\mathbf{u} = (x_1, y_1, z_1, w_1)$ and $\mathbf{v} = (x_2, y_2, z_2, w_2)$ be two vectors in $\mathbb{R}^4$; the **standard inner product** of $\mathbf{u}$ and $\mathbf{v}$ is the scalar $<\mathbf{u}, \mathbf{v}> = x_1 x_2 + y_1 y_2 + z_1 z_2 + w_1 w_2$.

Example 9   With this definition of Euclidean inner product in $\mathbb{R}^4$, the two vectors $\mathbf{u} = (1, -3, -1, 1)$ and $\mathbf{v} = (-1, -1, 1, -1)$ are orthogonal, since $<\mathbf{u}, \mathbf{v}> = (1)(-1) + (-3)(-1) + (-1)(1) + (1)(-1) = 0$. ■

We intend to generalize the concepts of length and angle to other vector spaces, forming other geometries. It's well known that Euclidean geometry is not valid in space exploration or even in transcontinental travel. One fairly simple, although surely not the only, extension of these basic ideas to more general geometries in a vector space applies the unifying concept of inner product. We therefore focus on the basic algebraic (arithmetic) properties of the standard inner product function considered here and state them as Theorem 4.1.

Theorem 4.1   If $\mathbf{u}$ and $\mathbf{v}$ are vectors in $\mathbb{R}^2$ or $\mathbb{R}^3$ and k_1, k_2 are real-number scalars, then

(i) $<\mathbf{u}, \mathbf{v}> = <\mathbf{v}, \mathbf{u}>$,
(ii) $<\mathbf{u} + \mathbf{v}, \mathbf{w}> = <\mathbf{u}, \mathbf{w}> + <\mathbf{v}, \mathbf{w}>$,
(iii) $<k_1\mathbf{u}, \mathbf{v}> = k_1<\mathbf{u}, \mathbf{v}>$,
(iv) $<k_1\mathbf{u} + k_2\mathbf{v}, \mathbf{w}> = k_1<\mathbf{u}, \mathbf{w}> + k_2<\mathbf{v}, \mathbf{w}>$,
(v) $<\mathbf{u}, \mathbf{u}> \geq 0$,
(vi) $<\mathbf{u}, \mathbf{u}> = 0$ if, and only if $\mathbf{u} = \mathbf{0}$.

Proof   The proof of this theorem is a straightforward application of the definition of inner product. We write out the details for part (iv) in $\mathbb{R}^3$ as a sample of what needs to be done here, the proof in $\mathbb{R}^2$ being completely analogous. The rest of the theorem is left for you to prove in the exercises.

Part (iv)   Let $\mathbf{u} = (x_1, y_1, z_1)$, $\mathbf{v} = (x_2, y_2, z_2)$, and $\mathbf{w} = (x_3, y_3, z_3)$. Then for any real numbers k_1 and k_2,

$$k_1\mathbf{u} + k_2\mathbf{v} = (k_1x_1 + k_2x_2, k_1y_1 + k_2y_2, k_1z_1 + k_2z_2),$$

so that

$$
\begin{aligned}
<k_1\mathbf{u} + k_2\mathbf{v}, \mathbf{w}> &= (k_1x_1 + k_2x_2)x_3 + (k_1y_1 + k_2y_2)y_3 + (k_1z_1 + k_2z_2)z_3, \\
&= (k_1x_1x_3 + k_2x_2x_3) + (k_1y_1y_3 + k_2y_2y_3) + (k_1z_1z_3 + k_2z_2z_3), \\
&= (k_1x_1x_3 + k_1y_1y_3 + k_1z_1z_3) + (k_2x_2x_3 + k_2y_2y_3 + k_2z_2z_3), \\
&= k_1(x_1x_3 + y_1y_3 + z_1z_3) + k_2(x_2x_3 + y_2y_3 + z_2z_3), \\
&= k_1<\mathbf{u}, \mathbf{w}> + k_2<\mathbf{v}, \mathbf{w}>.
\end{aligned}
$$

We also leave to you as an exercise the proof of the fact that parts (ii) and (iii) of this theorem are together equivalent to part (iv) (see Exercises 44 and 45). ❏

Example 10   Let $\mathbf{u} = (1, -1, 3)$, $\mathbf{v} = (2, 1, -4)$, and $\mathbf{w} = (3, -2, 1)$. Then

$$2\mathbf{u} + 3\mathbf{v} = 2(1, -1, 3) + 3(2, 1, -4) = (8, 1, -6),$$

so

$$<2\mathbf{u} + 3\mathbf{v}, \mathbf{w}> = <(8, 1, -6,), (3, -2, 1)>$$
$$= (8)(3) + (1)(-2) + (-6)(1) = 16.$$

Note that

$$2<\mathbf{u}, \mathbf{w}> + 3<\mathbf{v}, \mathbf{w}> = 2[(1)(3) + (-1)(-2) + (3)(1)]$$
$$+ 3[(2)(3) + (1)(-2) + (-4)(1)]$$
$$= 2[8] + 3[0] = 16,$$

as assured by the theorem.   ■

Notice now that with our definition of inner product for $\mathbf{u} = (x_1, y_1, z_1)$, we have

$$<\mathbf{u}, \mathbf{u}> = x_1^2 + y_1^2 + z_1^2 = ||\mathbf{u}||^2.$$

Therefore, parts (v) and (vi) of Theorem 4.1 say that the square of the length of any vector in $\mathbb{R}^3$ or $\mathbb{R}^2$ is nonnegative; in fact, *positive* unless $\mathbf{u}$ is the zero vector $\mathbf{0}$. Note also that for $\mathbf{u} = (x_1, y_1, z_1)$ and $\mathbf{v} = (x_2, y_2, z_2)$ the statement

$$||\mathbf{v} - \mathbf{u}||^2 = <\mathbf{v} - \mathbf{u}, \mathbf{v} - \mathbf{u}> = (x_2 - x_1)^2 + (y_2 - y_1)^2 + (z_2 - z_1)^2$$

is another version of the distance formula [see equation (4.4)]. By the same token, if we use the inner product notation in the expression for the cosine of the angle θ between the two vectors $\mathbf{u}$ and $\mathbf{v}$, we have the formula

$$\cos \theta = \frac{<\mathbf{u}, \mathbf{v}>}{<\mathbf{u}, \mathbf{u}>^{1/2} \cdot <\mathbf{v}, \mathbf{v}>^{1/2}}.$$

In the remaining sections of this chapter we show how these geometric ideas in Euclidean 2- and 3-space, $\mathbb{R}^2$ and $\mathbb{R}^3$, can be extended to other vector spaces such as $M_n(\mathbb{R})$ and P_n, etc. The ideas of distance and angle, expressed in terms of an inner product, can be generalized to nonstandard distances. We indicate how one might, thereby, look at other "geometries." In Section 4.6 we show how, with the proper modification, these concepts can also be extended to cover complex vector spaces.

One final comment concerning notation should be made. In Chapter 3 we discussed a number of vector spaces, including both of the vector spaces

$\overline{\mathbb{R}^n}$ and $\mathbb{R}^n$. In the one case we have *column* vectors and in the other *row* vectors. Notational confusion can result if the distinction between the two is not carefully observed. Analogous to what we said there about $\overline{\mathbb{R}^n}$, we note that one may view any physical vector $\mathbf{u}$ in $\mathbb{R}^3$ as a 1×3 matrix, that is, as an element in the space of *row matrices* $M_{1\times3}(\mathbb{R})$. From this point of view the Euclidean inner product in $\mathbb{R}^3$ of two vectors $\mathbf{u}$ and $\mathbf{v}$ is a matrix product $<\mathbf{u}, \mathbf{v}> = \mathbf{u}\mathbf{v}^T$. It is, however, a more common practice to make use of this idea in a special way. It is common to identify a physical vector with a *column* vector from $\mathbb{R}^3$, that is, to represent the physical vector not as a row but as a column. In that case the inner product would be $<\mathbf{u}, \mathbf{v}> = \mathbf{u}^T\mathbf{v}$. Please always be aware of the distinction between row and column vectors. You can verify the row vector notation easily as Exercise 19.

 ## Vocabulary

In this section we have considered several new concepts. These concepts will be the basis for major ideas to be used throughout the rest of the text, so be sure that you understand each of the ideas described by the terms in the following list. Review them until you feel secure enough to make some further abstractions based upon them.

vector equation of a line	parametric equations
symmetric form	direction numbers of a line
direction cosines	distance formula
Euclidean *n*-space, $\mathbb{R}^n$	Euclidean length
The law of cosines	inner product
dot product	scalar product
orthogonal vectors	

Exercises 4.1

In Exercises 1–3 write the parametric scalar equations for the line l determined by the given points.

1. $(1, 2, 1)$ and $(-1, -3, 0)$.   **2.** $(1, -2, -3)$ and $(3, -2, 1)$.   **3.** $(-2, 0, 1)$ and $(5, -4, 7)$.

4. Adapt the discussion of Example 1 to find the vector equation of a line l in 2-space $\mathbb{R}^2$ determined by the two points (u_1, u_2) and (v_1, v_2). What are the parametric scalar equations for this line?

5. Use the results of Exercise 4 to find the vector equation and the parametric scalar equations for the line l in $\mathbb{R}^2$ determined by the two points $(1, -2)$ and $(-3, 1)$.

In Exercises 6–11 find the Euclidean inner product $<\mathbf{u}, \mathbf{v}>$ of the two given vectors.

6. $\mathbf{u} = (1, -1)$, $\mathbf{v} = (-2, 1)$.   **7.** $\mathbf{u} = (2, -1)$, $\mathbf{v} = (3, 4)$.

8. $\mathbf{u} = (3, -2)$, $\mathbf{v} = (0, 0)$.   **9.** $\mathbf{u} = (1, -1, 1)$, $\mathbf{v} = (1, -2, 1)$.

10. $\mathbf{u} = (4, 1, -2)$, $\mathbf{v} = (1, -2, 1)$.   **11.** $\mathbf{u} = (1, 0, -1)$, $\mathbf{v} = (0, 0, 1)$.

12. Find the Euclidean lengths of the two vectors in Exercise 7.

13. Find the Euclidean lengths of the two vectors in Exercise 9.

14. Find the Euclidean lengths of the two vectors in Exercise 10.

15. Find the Euclidean lengths of the two vectors in Exercise 11.

16. Are the two vectors in Exercise 11 orthogonal?

17. Are the two vectors $\mathbf{u} = (1, 1, 1)$ and $\mathbf{v} = (1, -2, 1)$ orthogonal?

18. Are the two vectors $\mathbf{u} = (1, 1, 2)$ and $\mathbf{v} = (-1, -2, 1)$ orthogonal?

19. Let $\mathbf{u} = (u_1, u_2, u_3)$ and $\mathbf{v} = (v_1, v_2, v_3)$ be any two vectors in $\mathbb{R}^3$. Show that, considering each of these vectors as matrices of one row and three columns, we have $<\mathbf{u}, \mathbf{v}> = \mathbf{u}\mathbf{v}^T$.

20. Find the cosine of the angle between the two vectors in Exercise 6.

21. Find the cosine of the angle between the two vectors in Exercise 7.

22. Find the cosine of the angle between the two vectors in Exercise 8.

23. Find the cosine of the angle between the two vectors in Exercise 9.

24. Find the cosine of the angle between the two vectors $\mathbf{u} = (1, 1, 1)$ and $\mathbf{v} = (1, -2, 1)$.

25. Find the cosine of the angle between the two vectors $\mathbf{u} = (2, -4, 1)$ and $\mathbf{v} = (-1, -2, 4)$.

26. Use the distance formula to find the Euclidean distance between the two points $A: (2, -4, 1)$ and $B: (-1, -2, 4)$ in $\mathbb{R}^3$.

27. Use the distance formula to find the Euclidean distance between the two points $A: (1, -4, 1)$ and $B: (-1, -2, 3)$ in $\mathbb{R}^3$.

28. Use the distance formula to find the Euclidean distance between the two points $A: (2, 0, -3)$ and $B: (-1, -2, 0)$ in $\mathbb{R}^3$.

29. Use the distance formula in $\mathbb{R}^2$ to derive the law of cosines.

---

30. Show that if $\mathbf{u} = (u_1, u_2, u_3)$, then $\mathbf{u}$ can be expressed as $\|\mathbf{u}\|(\cos\theta_1, \cos\theta_2, \cos\theta_3)$, where θ_1, θ_2, and θ_3 are the angles between $\mathbf{u}$ and the x_1, x_2, and x_3 axes, respectively. See Figure 4.9.

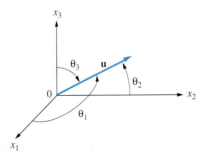

FIGURE 4.9

31. Show that the numbers $\cos\theta_1$, $\cos\theta_2$, and $\cos\theta_3$ in the preceding exercise satisfy the equation $(\cos\theta_1)^2 + (\cos\theta_2)^2 + (\cos\theta_3)^2 = 1$.

32. Suppose that $\mathbf{v}$ is a vector in $\mathbb{R}^2$ such that $<\mathbf{v}, \mathbf{x}> = 0$, for *every* vector $\mathbf{x}$ in $\mathbb{R}^2$. What can you say about $\mathbf{v}$? Give reasons!

33. Is there a vector $\mathbf{v}$ in $\mathbb{R}^2$ such that $<\mathbf{v}, \mathbf{x}> < 0$, for *every* vector $\mathbf{x}$ in $\mathbb{R}^2$? Give reasons!

34. Prove that $<\mathbf{0}, \mathbf{x}> = 0$, for *every* vector $\mathbf{x}$ in 2-space $\mathbb{R}^2$.

35. Prove that $<\mathbf{0}, \mathbf{x}> = 0$, for *every* vector $\mathbf{x}$ in 3-space $\mathbb{R}^3$.

36. Prove directly (i.e., using the usual rules of high school algebra) that

$$-1 \le \frac{x_1 x_2 + y_1 y_2}{\sqrt{x_1^2 + y_1^2}\sqrt{x_2^2 + y_2^2}} \le 1.$$

HINT: Consider the square of the fraction in the middle and compare the square of the numerator with the square of the denominator.

37. Given an example of two vectors $\mathbf{u} = (x_1, y_1)$ and $\mathbf{v} = (x_2, y_2)$ in Euclidean 2-space $\mathbb{R}^2$, for which the fraction in Exercise 36 is exactly equal to 1. Give another example where this fraction is exactly -1.

38. Carry out the details of the derivation for $\cos\theta$ in equation (4.8) for Euclidean 3-space $\mathbb{R}^3$.

39. Write out a formal proof for part (i) of Theorem 4.1 (page 222) in Euclidean 3-space $\mathbb{R}^3$.

40. Write out a formal proof for part (ii) of Theorem 4.1 in Euclidean 3-space $\mathbb{R}^3$.

41. Write out a formal proof for part (iii) of Theorem 4.1 in Euclidean 3-space $\mathbb{R}^3$.

42. Write out a formal proof for part (v) of Theorem 4.1 in Euclidean 3-space $\mathbb{R}^3$.

43. Write out a formal proof for part (vi) of Theorem 4.1 in Euclidean 3-space $\mathbb{R}^3$.

44. Show that in Euclidean 3-space $\mathbb{R}^3$ parts (ii) and (iii) of Theorem 4.1 imply part (iv).

45. Show that in Euclidean 3-space $\mathbb{R}^3$ part (iv) of Theorem 4.1 implies parts (ii) and (iii).

46. Use vectors in $\mathbb{R}^2$ to demonstrate that the triangle in 2-space whose vertices are

$$A: (1, 0); \quad B: \left(-\frac{1}{2}, \frac{\sqrt{3}}{2}\right); \text{ and } C: \left(-\frac{1}{2}, -\frac{\sqrt{3}}{2}\right)$$

is an equilateral triangle.

47. Use vectors in $\mathbb{R}^3$ to demonstrate that the triangle in 3-space whose vertices are $A: (1, -2, 3)$; $B: (4, 2, 15)$; and $C: (-3, 10, 0)$ is an isosceles right triangle.

48. Use vectors in $\mathbb{R}^2$ to demonstrate that a rectangle in the plane is the only parallelogram whose diagonals have equal length.

49. Use vectors in $\mathbb{R}^2$ to demonstrate that the diagonals of a rhombus (equilateral parallelogram) are perpendicular.

50. Show that in $\mathbb{R}^3$ we always have the inequality $||\mathbf{u} + \mathbf{v}|| \le ||\mathbf{u}|| + ||\mathbf{v}||$. When does equality hold?

4.2 CROSS PRODUCTS IN $\mathbb{R}^3$ (OPTIONAL)

In using vectors to describe the geometry of three-dimensional Euclidean space $\mathbb{R}^3$ and its application in physics and engineering, it is frequently useful to look at a vector perpendicular to the plane described by two other vectors. Consider the following example.

Example 1 Given the vectors $\mathbf{u}_1 = (1, -1, 1)$ and $\mathbf{u}_2 = (1, 0, 2)$ in $\mathbb{R}^3$. Find a vector that is perpendicular to the plane spanned by $\mathbf{u}_1$ and $\mathbf{u}_2$.

It is fairly easy to see that the two given vectors $\mathbf{u}_1$ and $\mathbf{u}_2$ are linearly independent and therefore span a two-dimensional subspace (i.e., a plane) in $\mathbb{R}^3$. Any vector $\mathbf{x} = (x_1, x_2, x_3)$ will lie in this plane if $\mathbf{x}$ is a linear combination of the vectors $\mathbf{u}_1$ and $\mathbf{u}_2$. That is, $\mathbf{x} = a_1\mathbf{u}_1 + a_2\mathbf{u}_2$. Thus,

$$\begin{aligned} a_1 + a_2 &= x_1, \\ -a_1 &= x_2, \\ a_1 + 2a_2 &= x_3. \end{aligned}$$

These three equations have solutions; i.e., the scalars a_1 and a_2 exist, provided

$$2x_1 + x_2 - x_3 = 0; \tag{4.9}$$

as you can easily compute. Equation (4.9) is the equation of the plane spanned by the two vectors $\mathbf{u}_1$ and $\mathbf{u}_2$. Now it is a fairly easy matter to show that any vector perpendicular (orthogonal) to this plane must have its inner product with all vectors in the plane equal to zero. In particular it must have zero inner product with all three of the vectors $\mathbf{u}_1$, $\mathbf{u}_2$, and $\mathbf{x}$. We could solve the problem one way by computing the inner products of an unknown vector $\mathbf{n} = (n_1, n_2, n_3)$ with $\mathbf{u}_1$ and $\mathbf{u}_2$; namely, $n_1 - n_2 + n_3$ and $n_1 + 2n_3$. Set both of these equal to zero and solve. (There will be infinitely many solutions, as you would expect both from the geometry and from the fact that we have but two homogeneous equations in the three unknowns.) One of these solutions will be the vector $\mathbf{n}^* = (-2, -1, 1)$.   ∎

In fact, the vector $\mathbf{n}^*$ was not pulled out of the air or selected from among the solutions to the given equations arbitrarily. Rather it was obtained from the vectors $\mathbf{u}_1$ and $\mathbf{u}_2$ as a sort of product which Hamilton defined in a series of papers about his quaternions between 1844 and 1855. It is called the **cross product**, $\mathbf{u}_1 \times \mathbf{u}_2$, of the vectors $\mathbf{u}_1$ and $\mathbf{u}_2$. Formally, we define the cross product as follows.

Definition   Let $\mathbf{u} = (u_1, u_2, u_3)$ and $\mathbf{v} = (v_1, v_2, v_3)$ be any two vectors in $\mathbb{R}^3$. The **cross product** of $\mathbf{u}$ and $\mathbf{v}$ is the vector

$$\mathbf{u} \times \mathbf{v} = (u_2 v_3 - u_3 v_2, -u_1 v_3 + u_3 v_1, u_1 v_2 - u_2 v_1).$$

You can readily verify that the vector $\mathbf{n}^*$ of Example 1 is the cross product of the two given vectors $\mathbf{u}_1$ and $\mathbf{u}_2$, and that it is indeed perpendicular (orthogonal) to the plane. This last fact is accomplished by checking that $<\mathbf{n}^*, a_1\mathbf{u}_1 + a_2\mathbf{u}_2> = 0$ for all scalars a_1 and a_2.

You may have observed that the components of the cross product $\mathbf{u}_1 \times \mathbf{u}_2$ of two vectors can be written as determinants of the following three 2×2 matrices:

$$\begin{bmatrix} u_2 & u_3 \\ v_2 & v_3 \end{bmatrix} ; \qquad -\begin{bmatrix} u_1 & u_3 \\ v_1 & v_3 \end{bmatrix} ; \qquad \text{and} \qquad \begin{bmatrix} u_1 & u_2 \\ v_1 & v_2 \end{bmatrix} .$$

In fact, if we let $\mathbf{e}_1 = (1, 0, 0)$, $\mathbf{e}_2 = (0, 1, 0)$, $\mathbf{e}_3 = (0, 0, 1)$ we see that the cross product $\mathbf{u}_1 \times \mathbf{u}_2$ can be viewed as the Laplace expansion across the first row of the "determinant" of the following matrix.

$$\begin{bmatrix} \mathbf{e}_1 & \mathbf{e}_2 & \mathbf{e}_3 \\ u_1 & u_2 & u_3 \\ v_1 & v_2 & v_3 \end{bmatrix} . \tag{4.10}$$

The matrix (4.10) is unusual in that its first row entries are vectors in $\mathbb{R}^3$ rather than real or complex numbers and its determinant is a vector rather than a number. However, this liberty with the definition of a determinant does, in fact, summarize the details of the cross product.

Theorem 4.2 Let $E = \{e_1, e_2, e_3\}$ be the standard basis for $\mathbb{R}^3$, and let $u = (u_1, u_2, u_3)$, and $v = (v_1, v_2, v_3)$ be vectors in $\mathbb{R}^3$. Then the cross product of u and v is the vector

$$u \times v = \det \begin{bmatrix} e_1 & e_2 & e_3 \\ u_1 & u_2 & u_3 \\ v_1 & v_2 & v_3 \end{bmatrix}. \tag{4.11}$$

This vector (4.11) is perpendicular to the plane (subspace) spanned by the vectors u and v.

Proof The discussion preceding this theorem proves all but the perpendicularity. To do this, we compute the scalar $<(u \times v), (a_1 u + a_2 v)>$

$$<(u \times v), (a_1 u + a_2 v)> = a_1 <u, (u \times v)> + a_2 <v, (u \times v)>,$$

from Theorem 4.1 (page 222). Then the expansion of this becomes

$$
\begin{aligned}
&<(u \times v), (a_1 u + a_2 v)> \\
&= a_1(u_1 u_2 v_3 - u_1 u_3 v_2 - u_2 u_1 v_3 + u_2 u_3 v_1 + u_3 u_1 v_2 - u_3 u_2 v_1) \\
&\quad + a_2(v_1 u_2 v_3 - v_1 u_3 v_2 - v_2 u_1 v_3 + v_2 u_3 v_1 + v_3 u_1 v_2 - v_3 u_2 v_1) \\
&= a_1(0) + a_2(0) = 0. \quad \blacksquare
\end{aligned}
$$

The following corollary follows immediately from the theorem.

Corollary $u \times v$ is orthogonal (perpendicular) to both u and v.

Example 2 Let $a = (2, -1, 4)$ and $b = (-3, 1, 2)$. Find $a \times b$.
From (4.11) we have

$$a \times b = \det \begin{bmatrix} e_1 & e_2 & e_3 \\ 2 & -1 & 4 \\ -3 & 1 & 2 \end{bmatrix}$$

$$= \det \begin{bmatrix} -1 & 4 \\ 1 & 2 \end{bmatrix} e_1 - \det \begin{bmatrix} 2 & 4 \\ -3 & 2 \end{bmatrix} e_2 + \det \begin{bmatrix} 2 & -1 \\ -3 & 1 \end{bmatrix} e_3$$

$$= -6 e_1 - 16 e_2 - 1 e_3 = (-6, -16, -1). \quad \blacksquare$$

Cross products have some very interesting algebraic properties which we list as Theorem 4.3. We ask you to verify these as exercises, as we just did in the proof of Theorem 4.2.

Theorem 4.3 Let $E = \{e_1, e_2, e_3\}$ be the standard basis for $\mathbb{R}^3$, and let $\mathbf{u} = (u_1, u_2, u_3)$, $\mathbf{v} = (v_1, v_2, v_3)$, and $\mathbf{w} = (w_1, w_2, w_3)$ be vectors in $\mathbb{R}^3$. Then

   (i) $\mathbf{u} \times \mathbf{v} = \mathbf{0}$ if, and only if, $\mathbf{u}$ and $\mathbf{v}$ are linearly dependent; in particular, $\mathbf{u} \times \mathbf{0} = \mathbf{0} \times \mathbf{u} = \mathbf{0}$.

  (ii) $\mathbf{u} \times \mathbf{v} = -\mathbf{v} \times \mathbf{u}$.

 (iii) $(\alpha \mathbf{u} \times \mathbf{v}) = \alpha(\mathbf{u} \times \mathbf{v})$ for any scalar α.

 (iv) $\mathbf{u} \times (\mathbf{v} + \mathbf{w}) = (\mathbf{u} \times \mathbf{v}) + (\mathbf{u} \times \mathbf{w})$.

  (v) $<(\mathbf{u} \times \mathbf{v}), \mathbf{w}> = <\mathbf{u}, (\mathbf{v} \times \mathbf{w})>$. This product is called the **scalar triple product** of $\mathbf{u}$, $\mathbf{v}$, and $\mathbf{w}$.

 (vi) $(\mathbf{u} \times \mathbf{v}) \times \mathbf{w} \neq \mathbf{u} \times (\mathbf{v} \times \mathbf{w})$. In general, the cross product is **not associative**.

(vii) $e_1 \times e_1 = e_2 \times e_2 = e_3 \times e_3 = \mathbf{0}$. See part (i).

(viii) $(\mathbf{u} \times \alpha \mathbf{v}) = \alpha(\mathbf{u} \times \mathbf{v})$ for any scalar α. See part (iii).

Let's take a look at the cross products of the standard basis vectors e_1, e_2, and e_3 for $\mathbb{R}^3$. It is easy to verify the following:

$$e_1 \times e_2 = e_3; \qquad e_3 \times e_1 = e_2; \qquad e_2 \times e_3 = e_1;$$
$$\text{whereas } e_2 \times e_1 = -e_3; \qquad e_1 \times e_3 = -e_2; \qquad \text{and} \qquad e_3 \times e_2 = -e_1.$$

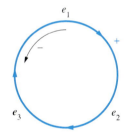

FIGURE 4.10

To remember these results, consider the circle in Figure 4.10, which is also reminiscent of Hamilton's guaternions. The cross product of two consecutive vectors in a *clockwise* direction is the third vector in that direction. By contrast, the cross product reading *counterclockwise* is the **negative** of the third vector.

Example 3 Find $(3e_1 - 2e_3) \times (4e_2 + e_3)$. Rather than writing out the two vectors as triples in $\mathbb{R}^3$, we use the results of Theorem 4.3 and Figure 4.10 to note that we have

$$(3e_1 - 2e_3) \times (4e_2 + e_3) = (3)(4)(e_1 \times e_2) \times (3)(1)(e_1 \times e_3)$$
$$- (2)(4)(e_3 \times e_2) - (2)(1)(e_3 \times e_3)$$
$$= 12e_3 + 3(-e_2) - 8(-e_1) - 2(\mathbf{0})$$
$$= 12e_3 - 3e_2 + 8e_1 = 8e_1 - 3e_2 + 12e_3$$
$$= (8, -3, 12). \quad \blacksquare$$

Let us now turn to some geometric interpretations of the cross product. First of all, although we have shown that the cross product of two linearly independent vectors is always orthogonal to the plane determined by them, there are always at least two such vectors, one in each direction. For example,

$e_1 \times e_2 = e_3$ is a unit vector orthogonal to the plane of e_1 and e_2, but so is $-e_3 = e_2 \times e_1$. Suppose that we know that the vectors n (for normal) and $-n$ are both orthogonal to the plane determined by the two vectors u and v. Which one is $u \times v$? The answer is given by the **right-hand rule**. (See Figure 4.11.) If the right hand is placed so that the index finger points in the direction of u while the middle finger points in the direction of v, then the thumb points in the direction of $u \times v$.

> The vector $n = u \times v$ is often said to be **normal** to the plane determined by the vectors u and v.

To use the word *normal* applied to the vector n in the statement above is quite common, but it should not be confused with a *normal vector* as part of an ortho*normal* basis described later in this chapter. For that reason we mostly refer to n as being *orthogonal* to the plane.

Having determined the direction of $u \times v$, let us now proceed to consider its magnitude. Let ϕ denote the angle between u and v (measured from u to v). Then by comparing their components using equation (4.11) it can be verified that

$$||u \times v||^2 = ||u||^2||v||^2 - <u, v>^2. \qquad \text{(Write out the details.)}$$

Then, since $<u, v>^2 = ||u||^2||v||^2 \cos^2 \phi$, we have

$$||u \times v||^2 = ||u||^2||v||^2 - ||u||^2||v||^2 \cos^2 \phi = ||u||^2||v||^2(1 - \cos^2 \phi),$$

or

> $$||u \times v||^2 = ||u||^2||v||^2 \sin^2 \phi. \qquad \textbf{(4.12)}$$

These geometric properties of $u \times v$ are often interpreted as indicated in Figure 4.12, where the vectors u and v are two adjacent sides of a parallelogram in $\mathbb{R}^3$. Then we have that the *area* of the parallelogram is $||u|| \cdot ||v|| \cdot \sin \phi = ||u \times v||$. Often this interpretation is used in elementary physics as the definition of the cross product.

> **Definition** $u \times v$ is the vector whose magnitude is the area of the parallelogram determined by u and v and whose direction is orthogonal to that parallelogram as determined by the right-hand rule.

In Example 1 we computed the equation of the plane in $\mathbb{R}^3$ formed by the two vectors $u_1 = (1, -1, 1)$ and $u_2 = (1, 0, 2)$ by setting up a linear combination of these two vectors and determining when the resulting system of linear equations had a solution. Compare equation (4.9) with the cross product $u_1 \times u_2$. Notice that the coefficients of the variables in equation (4.9) and the

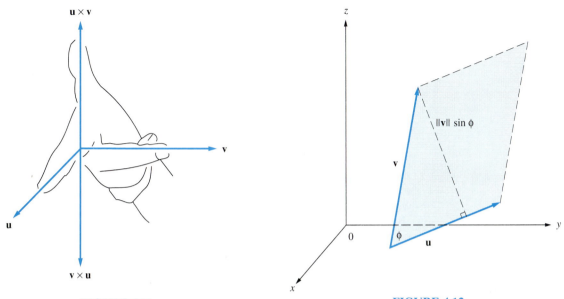

FIGURE 4.11   **FIGURE 4.12**

components of $\mathbf{u}_1 \times \mathbf{u}_2$ are numerically equal. In fact, equation (4.9) is the scalar form of the vector equation $-<\mathbf{u}_1 \times \mathbf{u}_2, \mathbf{x}> = 0$, where $\mathbf{x} = (x_1, x_2, x_3)$. The components of $\mathbf{n} = \mathbf{u}_1 \times \mathbf{u}_2$ are called the *direction numbers* of the plane. Since the plane determined by $\mathbf{u}_1$ and $\mathbf{u}_2$ must pass through the origin, we have equation (4.9). In general we have the following.

> The equation of a plane determined by two vectors **a** and **b** (and passing through the origin) is
> $$<(\mathbf{a} \times \mathbf{b}), \mathbf{x}> = 0,$$
> or, using the "dot product,"
> $$\mathbf{n} \cdot \mathbf{x} = 0, \text{ where } \mathbf{n} = (\mathbf{a} \times \mathbf{b}).$$

Example 4   Use the cross product to determine the direction numbers and write the equation of the plane passing through the origin O: $(0, 0, 0)$ and the points P_1: $(2, 3, -1)$ and P_2: $(1, -1, 2)$.

Let $\mathbf{a} = (2, 3, -1)$ be the vector from O to P_1, and let $\mathbf{b} = (1, -1, 2)$ be the vector from O to P_2. Then

$$\mathbf{a} \times \mathbf{b} = \det \begin{bmatrix} \mathbf{e}_1 & \mathbf{e}_2 & \mathbf{e}_3 \\ 2 & 3 & -1 \\ 1 & -1 & 2 \end{bmatrix} = (5, -5, -5)$$

is orthogonal to the desired plane. The direction numbers are therefore 5, -5, and -5; so the equation of the desired plane is $<(\mathbf{a} \times \mathbf{b}), \mathbf{x}> = 5x_1 - 5x_2 - 5x_3 = 0$. ∎

Let's turn to a more general plane. Suppose that we want to determine the equation of any plane Γ passing through a given point P and parallel to the plane spanned by two given linearly independent vectors **a** and **b**. Let the vector from the origin O to P be **c**. Then we seek the plane Γ which consists of all vectors $\mathbf{x} = \mathbf{z} + \mathbf{c}$, where **z** is any vector in the plane spanned by **a** and **b**. Thus $\mathbf{z} = s\mathbf{a} + t\mathbf{b}$, for arbitrary scalars s and t. (See Figure 4.13.) The vector parametric equation for the plane Γ is therefore

$$\mathbf{x} = s\mathbf{a} + t\mathbf{b} + \mathbf{c}. \tag{4.13}$$

Now, given this parametric form, how do we arrive at the nonparametric or standard form $ax_1 + bx_2 + cx_3 = p$ of the equation of this plane? First of all the vector $\mathbf{n} = \mathbf{a} \times \mathbf{b}$ is normal (orthogonal) to the desired plane Γ through P as well as to the plane spanned by **a** and **b**, since they are parallel. Then we note that the vector $\mathbf{x} - \mathbf{c}$ lies in the plane Γ and is therefore also orthogonal to $\mathbf{n} = \mathbf{a} \times \mathbf{b}$. Thus $0 = <(\mathbf{a} \times \mathbf{b}), \mathbf{x} - \mathbf{c}> = <(\mathbf{a} \times \mathbf{b}), \mathbf{x}> - <(\mathbf{a} \times \mathbf{b}), \mathbf{c}>$. So,

$$<(\mathbf{a} \times \mathbf{b}), \mathbf{x}> = <(\mathbf{a} \times \mathbf{b}), \mathbf{c}>. \tag{4.14}$$

Equation (4.14) can be written a little more simply as follows:

$$<\mathbf{n}, \mathbf{x}> = <\mathbf{n}, \mathbf{c}>.$$

If one prefers the "dot" for the inner product in $\mathbb{R}^3$, this equation reads $\mathbf{n} \cdot \mathbf{x} = \mathbf{n} \cdot \mathbf{c}$.

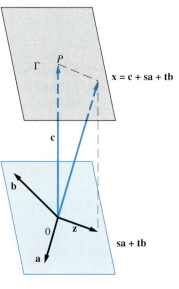

FIGURE 4.13

Example 5   Write the standard form (4.14) for the plane in Euclidean 3-space $\mathbb{R}^3$ determined by the three noncollinear points P_1: $(1, -2, 1)$, P_2: $(1, -3, -1)$, and P_3: $(2, -1, -1)$.

See Figure 4.14(a). Let **u** be the vector from the origin O to the point P_1, **v** the vector from O to the point P_2, and **w** the vector from O to the point P_3. Let $\mathbf{a} = \mathbf{u} - \mathbf{v} = (0, 1, 2)$, and $\mathbf{b} = \mathbf{u} - \mathbf{w} = (-1, -1, 2)$. The plane we seek is parallel to the plane spanned by **a** and **b** (because this plane goes through O; why?) and passing through any one of the given three points. See Figure 4.14(b). We select $\mathbf{c} = \mathbf{w}$. You can try the problem with $\mathbf{c} = \mathbf{u}$ or $\mathbf{c} = \mathbf{v}$ and see that the resulting equation is the same. Then, from equation (4.14), we have $\langle(\mathbf{a} \times \mathbf{b}), \mathbf{x}\rangle = \langle(\mathbf{a} \times \mathbf{b}), \mathbf{c}\rangle$ as the desired equation of the plane. Thus, since

$$\mathbf{a} \times \mathbf{b} = \det \begin{bmatrix} \mathbf{e}_1 & \mathbf{e}_2 & \mathbf{e}_3 \\ 0 & 1 & 2 \\ -1 & -1 & 2 \end{bmatrix} = (4, -2, 1),$$

we have $(4)x_1 - (2)x_2 + (1)x_3 = (4)(2) - (2)(-1) + (1)(-1) = 8 + 2 - 1 = 9$. So the equation of the plane is $4x_1 - 2x_2 + x_3 = 9$. See the graph of the plane in Figure 4.14(c). ▪

Notice that in Figure 4.14(c) the origin is inside the box and the xy plane has been rotated clockwise from the standard views as given in Figures 4.14(a) and (b). Notice also the position of the three axes.

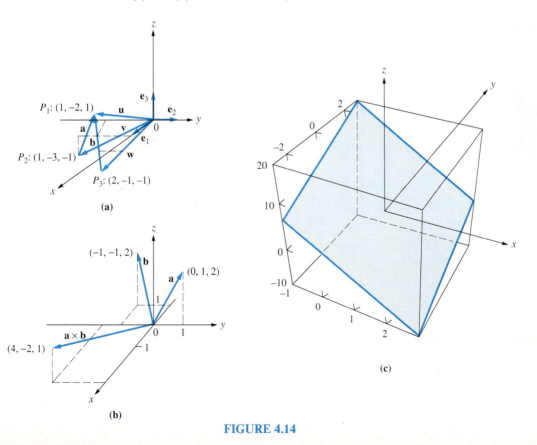

FIGURE 4.14

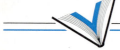

Vocabulary

In this section we have again met some new concepts. The vocabulary developed here is often used in applications of linear algebra to physics and engineering, so be sure that you understand each of the concepts described by the terms in the following list. Review these ideas until you feel secure enough to use them in other courses or applications.

cross product of two vectors
scalar triple product
vector equation of a plane
direction numbers of a plane

right-hand rule
not associative
normal vector to a plane

Exercises 4.2

Find the cross products for each of the following pairs of vectors.

1. $\mathbf{u} = (1, 0, -1)$; $\mathbf{v} = (0, -1, 1)$.

2. $\mathbf{u} = (1, 1, -1)$; $\mathbf{v} = (1, -1, 1)$.

3. $\mathbf{u} = (2, 2, -1)$; $\mathbf{v} = (2, -1, 1)$.

4. $\mathbf{u} = (1, 2, -1)$; $\mathbf{v} = (1, -2, 1)$.

5. $\mathbf{u} = (-3, 0, -1)$; $\mathbf{v} = (0, -2, 1)$.

6. $\mathbf{u} = (4, 1, -2)$; $\mathbf{v} = (1, -2, 1)$.

7. $\mathbf{u} = (1, 0, -3)$; $\mathbf{v} = (0, -1, 1)$.

8. $\mathbf{u} = (1, 1, -3)$; $\mathbf{v} = (1, -1, 1)$.

9. $\mathbf{u} = (-3, 0, -1)$; $\mathbf{v} = (0, -2, 2)$.

10. $\mathbf{u} = (4, 1, -2)$; $\mathbf{v} = (3, -2, 1)$.

11. $\mathbf{u} = (1, 0, -5)$; $\mathbf{v} = (0, -1, 1)$.

12. $\mathbf{u} = (0, 1, -1)$; $\mathbf{v} = (1, -4, 1)$.

13. $\mathbf{u} = (-3, 0, -2)$; $\mathbf{v} = (0, -2, 1)$.

14. $\mathbf{u} = (4, 1, -2)$; $\mathbf{v} = (0, -2, 4)$.

15. $\mathbf{u} = (4, 0, -2)$; $\mathbf{v} = (0, -1, 0)$.

16. $\mathbf{u} = (1, 0, -1)$; $\mathbf{v} = (4, -2, 1)$.

Frequently the standard basis for $\mathbb{R}^3$ *is written as* $\{\mathbf{i}, \mathbf{j}, \mathbf{k}\}$ *rather than as* $\{\mathbf{e}_1, \mathbf{e}_2, \mathbf{e}_3\}$, *with* $\mathbf{i} = \mathbf{e}_1$, *etc. The vector* $\mathbf{u} = (4, 1, -2) = 4\mathbf{i} + \mathbf{j} - 2\mathbf{k}$, *for example. Make this notational change and find the cross products* $\mathbf{u} \times \mathbf{v}$ *for each of the following vectors.*

17. $\mathbf{u} = \mathbf{i}$; $\mathbf{v} = -\mathbf{j}$.

18. $\mathbf{u} = \mathbf{i}$; $\mathbf{v} = -\mathbf{k}$.

19. $\mathbf{u} = 2\mathbf{i} - 3\mathbf{k}$; $\mathbf{v} = -2\mathbf{j} + \mathbf{k}$.

20. $\mathbf{u} = 2\mathbf{i} - 2\mathbf{j} - 3\mathbf{k}$; $\mathbf{v} = \mathbf{i} - 2\mathbf{j} + \mathbf{k}$.

21. $\mathbf{u} = \mathbf{i} + 2\mathbf{j} - 3\mathbf{k}$; $\mathbf{v} = 2\mathbf{i} - 2\mathbf{j} + \mathbf{k}$.

22. $\mathbf{u} = 3\mathbf{i} - 2\mathbf{j} - \mathbf{k}$; $\mathbf{v} = 4\mathbf{i} - 2\mathbf{j} + 5\mathbf{k}$.

23. $\mathbf{u} = 5\mathbf{i} + 3\mathbf{j} - 4\mathbf{k}$; $\mathbf{v} = 5\mathbf{i} - \mathbf{j} + 3\mathbf{k}$.

24. $\mathbf{u} = 3\mathbf{i} - \mathbf{j} - 6\mathbf{k}$; $\mathbf{v} = \mathbf{i} - 8\mathbf{j} + 2\mathbf{k}$.

25. Find two *unit* vectors orthogonal to the vectors $\mathbf{a} = (1, 1, -5)$ and $\mathbf{b} = (3, -1, 1)$.

26. Find two *unit* vectors orthogonal to the vectors $\mathbf{a} = (1, 3, -2)$ and $\mathbf{b} = (3, -2, 0)$.

27. Write the scalar equation of the plane spanned by $\mathbf{a} = (1, 1, -5)$ and $\mathbf{b} = (3, -1, 1)$.

28. Write the scalar equation of the plane spanned by $\mathbf{a} = (1, 3, -2)$ and $\mathbf{b} = (3, -2, 0)$.

29. Write the equation of the plane passing through the three points P_1: $(1, 3, -2)$, P_2: $(3, -2, 0)$, and P_3: $(1, 1, -5)$.

30. Write the equation of the plane passing through the three points P_1: $(1, 0, -2)$, P_2: $(1, -2, 0)$, and P_3: $(1, 0, -5)$.

31. Write the equation of the plane passing through the three points P_1: $(2, 2, -2)$, P_2: $(1, -2, 2)$, and P_3: $(1, 2, -5)$.

32. Find the area of the parallelogram whose two adjacent sides are the vectors $\mathbf{a} = (1, 1, -5)$ and $\mathbf{b} = (3, -1, 1)$.

33. Find the area of the parallelogram whose two adjacent sides are the vectors $\mathbf{a} = (1, 3, -2)$ and $\mathbf{b} = (3, -2, 0)$.

In Exercises 34–41 write out the details of the proof of Theorem 4.3 (page 229).

34. Part (i).   **35.** Part (ii).   **36.** Part (iii).

37. Part (iv).   **38.** Part (v).   **39.** Part (vi).

40. Part (vii).   **41.** Part (viii).

---

42. Show that the scalar triple product $< \mathbf{u}, (\mathbf{v} \times \mathbf{w}) >$ of part (v) of Theorem 4.3 can be written as det M, where M is the 3×3 matrix

$$M = \begin{bmatrix} u_1 & u_2 & u_3 \\ v_1 & v_2 & v_3 \\ w_1 & w_2 & w_3 \end{bmatrix}.$$

43. Show that the volume of the parallelepiped (box) in $\mathbb{R}^3$ determined by the (based at the origin) vectors $\mathbf{u}$, $\mathbf{v}$, and $\mathbf{w}$ is the value $V = |< \mathbf{u}, (\mathbf{v} \times \mathbf{w}) >| = |\det M|$, where M is as in Exercise 42 (or, if you wish, $|\mathbf{u} \cdot (\mathbf{v} \times \mathbf{w})|$). (Sketch a graph.)

44. Use Exercise 43 to find the volume of the parallelepiped determined by the origin and the three points P_1: $(2, 2, -2)$, P_2: $(1, -2, 2)$, and P_3: $(1, 2, -5)$.

45. Show that if two distinct planes $< \mathbf{n}, \mathbf{x} > = < \mathbf{n}, \mathbf{c} >$ and $< \mathbf{m}, \mathbf{x} > = < \mathbf{m}, \mathbf{c} >$ intersect, then their two normal vectors $\mathbf{n}$ and $\mathbf{m}$ cannot be parallel, so the vector $\mathbf{a} = \mathbf{n} \times \mathbf{m}$ is parallel to the line of their intersection. (Sketch a graph.)

46. Use the results of Exercise 45 to find the symmetric form of the equation (see Section 4.1) of the line of intersection of the two planes $2x_1 - 3x_2 + x_3 = 5$ and $x_1 + 4x_2 - 2x_3 = -2$.

47. Use the methods of Chapter 1 to find the parametric form of the equation of the line of intersection of the two planes in Exercise 46. Compare with Exercise 46.

48. Show that the torque about the origin of a rigid body resulting from a force vector $\mathbf{F}$ applied at the end point of the vector $\mathbf{x}$ on the body is completely determined by the vector $\mathbf{x} \times \mathbf{F}$. Show that the length of $\mathbf{x} \times \mathbf{F}$ is proportional to the magnitude of the torque and the direction of $\mathbf{x} \times \mathbf{F}$ is the positive axis of rotation: with your thumb along $\mathbf{x} \times \mathbf{F}$, the fingers of your right hand curl in the direction of the turning.

49. Find the torque, $\mathbf{x} \times \mathbf{F}$, about the origin if a force vector $\mathbf{F} = (1, 1, -1)$ is applied at the point $\mathbf{x} = (9, 0, 1)$.

4.3 REAL INNER PRODUCTS

In this section we carry on the program (announced at the end of Section 4.1) to view the concept of a real inner product in such a way that it can allow for geometric interpretations other than the conventional Euclidean plane geometry. We begin with the definition of the dot product (the "standard" or Euclidean inner product) in $\mathbb{R}^2$ and $\mathbb{R}^3$ in mind, but we extend these ideas to an arbitrary vector space with real-number scalars. In Section 4.6 we make the necessary modifications to this idea so as to make it apply when the scalars are real or complex numbers. Let us take the basic results of Theorem 4.1 (page 222) for the inner product in $\mathbb{R}^2$ and $\mathbb{R}^3$ as the starting point. We consider an arbitrary real vector space V and suppose that we can assign to each pair of vectors **u** and **v** from V a *unique real number,* which we denote by $<$**u**, **v**$>$, that satisfies the same properties as the dot product did in Theorem 4.1. Thus,

---

Definition Let V be any real vector space. A function that assigns to each pair of vectors **u** and **v** in V a unique real number, $<$**u**, **v**$>$, is called an **inner product** on V provided it has the following properties. For all vectors, **u**, **v**, **w**, . . . and *scalars a, b, c, . . .*

Property (1) $<$**u**, **v**$> = <$**v**, **u**$>$,
Property (2) $<a$**u** $+ b$**v**, **w**$> = a <$**u**, **w**$> + b <$**v**, **w**$>$,
Property (3) $<$**u**, **u**$> \geq 0$,
Property (4) $<$**u**, **u**$> = 0$ if, and only if, **u** $=$ **0**.

(4.15)

The real number $<$**u**, **v**$>$ assigned by the inner-product function on V is usually called the **inner product** of **u** and **v**, and the vector space V equipped with this function is called a real **inner-product space**. The fact that both $\mathbb{R}^2$ and $\mathbb{R}^3$ are inner-product spaces with respect to the "dot product" is the statement of Theorem 4.1. This was verified in its proof, most of which you provided in the exercises of Section 4.1.

Notice that our definition of a real inner product states that it is a *real-valued function.* Property (1) says that this is a *symmetric function,* while Property (2) requires it to be *linear with respect to its first argument.* This leads to the natural question: "Is the inner product also linear with respect to its second argument?" The answer is our first theorem.

Theorem 4.4 For all vectors, **u**, **v**, **w**, . . . in the inner-product space V and scalars *a, b, c, . . .* it is true that

$$<\text{**u**}, a\text{**v**} + b\text{**w**}> = a <\text{**u**}, \text{**v**}> + b <\text{**u**}, \text{**w**}>.$$

Proof To prove this, note that

$$
\begin{aligned}
<\text{**u**}, a\text{**v**} + b\text{**w**}> &= <a\text{**v**} + b\text{**w**}, \text{**u**}> && \text{by Property (1) of (4.15)}\\
&= a <\text{**v**}, \text{**u**}> + b <\text{**w**}, \text{**u**}> && \text{by Property (2)}\\
&= a <\text{**u**}, \text{**v**}> + b <\text{**u**}, \text{**w**}>. && \text{again by Property (1)}
\end{aligned}
$$

This proves the theorem. So the inner-product function is *linear with respect to each argument.* ❏

The most natural example of an inner-product function in Cartesian n-space $\mathbb{R}^n$ is the obvious extension of the dot product of Section 4.1. Let $V = \mathbb{R}^n$ and suppose that we have two arbitrary vectors **u** and **v** in $\mathbb{R}^n$. Then we can write

$$\text{**u**} = (u_1, u_2, \dots, u_n) \quad \text{and} \quad \text{**v**} = (v_1, v_2, \dots, v_n), \quad (4.16)$$

where all the numbers $u_1, u_2, \ldots, u_n$ and $v_1, v_2, \ldots, v_n$ are real numbers. Now consider the real number *defined* by

$$<\mathbf{u}, \mathbf{v}> = u_1v_1 + u_2v_2 + \cdots + u_nv_n = \sum_{i=1}^{n} u_iv_i. \qquad \textbf{(4.17)}$$

An argument similar to the one given to prove Theorem 4.1 (page 222) proves the following theorem. We leave the verification to you as an exercise.

Theorem 4.5   Let V be Cartesian n-space $\mathbb{R}^n$ and let $\mathbf{u}$ and $\mathbf{v}$ be any two vectors in V. Then the function $<\mathbf{u}, \mathbf{v}>$ defined in (4.17) is an inner product in $\mathbb{R}^n$.

The cases $n = 2$ and $n = 3$ of this theorem recover the definitions of the dot or scalar product in Section 4.1. As we remarked there, we call $\mathbb{R}^n$ with this particular inner product **Euclidean *n*-space.** Many authors call this inner product the **standard inner product** in $\mathbb{R}^n$, and use the title "Euclidean *n*-space" to apply to *any* vector space with *real-number scalars* on which an inner product is defined. We however reserve the term *Euclidean* for real vector spaces with this *(standard)* inner product. Other inner products are possible in $\mathbb{R}^n$ that do not give one the usual Euclidean length as the square root of the sum of the squares of the components, as we see later. One of these is the following.

Example 1   Let $\mathbf{u} = (u_1, u_2, u_3)$ and $\mathbf{v} = (v_1, v_2, v_3)$ be any two vectors in $\mathbb{R}^3$. The function

$$<\mathbf{u}, \mathbf{v}> = 2u_1v_1 + 3u_2v_2 + 4u_3v_3 \qquad \textbf{(4.18)}$$

defines an inner product (not the *standard* one) on $\mathbb{R}^3$.

Let's verify that this function is indeed an inner product on $\mathbb{R}^3$. We do this by showing that it satisfies the definition of an inner product given in (4.15). First of all, it is clear that $2u_1v_1 + 3u_2v_2 + 4u_3v_3$ is a real number, since $u_1, u_2, u_3, v_1, v_2,$ and v_3 are all real numbers. Notice also that

$$<\mathbf{u}, \mathbf{v}> = 2u_1v_1 + 3u_2v_2 + 4u_3v_3 = 2v_1u_1 + 3v_2u_2 + 4v_3u_3 = <\mathbf{v}, \mathbf{u}>$$

satisfying Property (1). For Property (2) let a and b be real numbers and suppose that $\mathbf{w} = (w_1, w_2, w_3)$ is also an arbitrary vector in $\mathbb{R}^3$. Then

$$a\mathbf{u} + b\mathbf{v} = a(u_1, u_2, u_3) + b(v_1, v_2, v_3) = (au_1 + bv_1, au_2 + bv_2, au_3 + bv_3).$$

So

$$
\begin{aligned}
<a\mathbf{u}+b\mathbf{v},\ \mathbf{w}> &= 2(au_1+bv_1)w_1+3(au_2+bv_2)w_2+4(au_3+bv_3)w_3 \\
&= 2au_1w_1+2bv_1w_1+3au_2w_2+3bv_2w_2+4au_3w_3+4bv_3w_3 \\
&= 2au_1w_1+3au_2w_2+4au_3w_3+2bv_1w_1+3bv_2w_2+4bv_3w_3 \\
&= a(2u_1w_1+3u_2w_2+4u_3w_3)+b(2v_1w_1+3v_2w_2+4v_3w_3) \\
&= a<\mathbf{u},\ \mathbf{w}>+b<\mathbf{v},\ \mathbf{w}>;\ \text{as desired.}
\end{aligned}
$$

To verify Properties (3) and (4) we note that $<\mathbf{u},\ \mathbf{u}>=2u_1^2+3u_2^2+4u_3^2 \geq 0$ for all real numbers u_1, u_2, and u_3. Furthermore, this can equal 0 if, and only if, u_1, u_2, and u_3 are all zero; that is, if $\mathbf{u}=\mathbf{0}$. This completes the demonstration that this function is, indeed, an inner product. That it is *not standard* is quite clear, and we look at such things as the nonstandard "length" that it defines later on.   ∎

Example 2   This is a generalization of Example 1. Let D be a diagonal $n \times n$ matrix with positive diagonal elements; $d_{ii}>0$. Let $\mathbf{u}$ and $\mathbf{v}$ be vectors in $\mathbb{R}^n$. Show that the function

$$
<\mathbf{u},\ \mathbf{v}> = \mathbf{u}\,D\mathbf{v}^T
$$

is an inner product on $\mathbb{R}^n$.

That $\mathbf{u}\,D\mathbf{v}^T$ is a real number can be seen by noticing that we have the product of the $1 \times n$ matrix $\mathbf{u}$ times the $n \times n$ matrix D times the $n \times 1$ matrix $\mathbf{v}^T$, resulting in a 1×1 matrix $\mathbf{u}\,D\mathbf{v}^T$. Thus, if $\mathbf{u}=(u_1,\ u_2,\ \ldots,\ u_n)$ and $\mathbf{v}=(v_1,\ v_2,\ \ldots,\ v_n)$ then

$$
\mathbf{u}\,D\mathbf{v}^T=(u_1,\ u_2,\ \ldots,\ u_n)
\begin{bmatrix}
d_{11} & 0 & \cdots & 0 \\
0 & d_{22} & \cdots & 0 \\
\vdots & \vdots & \cdots & \vdots \\
0 & 0 & \cdots & d_{nn}
\end{bmatrix}
\begin{bmatrix}
v_1 \\ v_2 \\ \vdots \\ v_n
\end{bmatrix}.
$$

So $\mathbf{u}\,D\mathbf{v}^T=u_1d_{11}v_1+u_2d_{22}v_2+\cdots+u_nd_{nn}v_n$. To verify that the conditions of the definition of an inner product are satisfied notice first that

$$
(\mathbf{u}\,D\mathbf{v}^T)^T=\mathbf{v}D^T\mathbf{u}^T=\mathbf{v}D\mathbf{u}^T=\mathbf{u}\,D\mathbf{v}^T
$$

because the transpose of a 1×1 matrix must equal itself. (You can also see this from the expanded expression given for $\mathbf{u}\,D\mathbf{v}^T$.) Making use of this fact, we can easily see that this function is symmetric, $<\mathbf{u},\ \mathbf{v}>=<\mathbf{v},\ \mathbf{u}>$. Furthermore, if $\mathbf{u}$, $\mathbf{v}$, and $\mathbf{w}$ are vectors in $\mathbb{R}^n$, and a and b are real numbers, the rules of matrix algebra give us

$$
\begin{aligned}
<a\mathbf{u}+b\mathbf{v},\ \mathbf{w}> &= (a\mathbf{u}+b\mathbf{v})D\mathbf{w}^T=(a\mathbf{u})D\mathbf{w}^T+(b\mathbf{v})D\mathbf{w}^T \\
&= a(\mathbf{u}\,D\mathbf{w}^T)+b(\mathbf{v}\,D\mathbf{w}^T) \\
&= a<\mathbf{u},\ \mathbf{w}>+b<\mathbf{v},\ \mathbf{w}>.
\end{aligned}
$$

Finally, $<\mathbf{u},\ \mathbf{u}> = \mathbf{u}\ D\mathbf{u}^T = u_1 d_{11} u_1 + u_2 d_{22} u_2 + \cdots + u_n d_{nn} u_n = d_{11} u_1^2 + d_{22} u_2^2 + \cdots + d_{nn} u_n^2 \geq 0$, because the entries in D are positive (each $d_{ii} > 0$). Also this sum can be equal to zero only when $\mathbf{u} = \mathbf{0}$.

In Chapter 6 we see how this example can be extended to allow D to be any symmetric positive definite matrix M (that is, M must have positive eigenvalues).

As a numerical example let's consider $\mathbb{R}^4$ with $D = \operatorname{diag}\ [1,\ 2,\ 1,\ 3]$, $\mathbf{u} = (1,\ 1,\ -1,\ 1)$, and $\mathbf{v} = (1,\ 2,\ -1,\ 2)$. Then

$$<\mathbf{u},\ \mathbf{v}> = \mathbf{u}\ D\mathbf{v}^T = (1,\ 1,\ -1,\ 1) \begin{bmatrix} 1 & 0 & 0 & 0 \\ 0 & 2 & 0 & 0 \\ 0 & 0 & 1 & 0 \\ 0 & 0 & 0 & 3 \end{bmatrix} \begin{bmatrix} 1 \\ 2 \\ -1 \\ 2 \end{bmatrix}$$

$$= 1 + 4 + 1 + 6 = 12. \quad \blacksquare$$

Inner products can also be defined on vector spaces other than $\mathbb{R}^n$. Two rather interesting cases are the next examples.

Example 3 Let us look at the vector space $M_n(\mathbb{R})$ of all real $n \times n$ matrices. The **trace** of any $n \times n$ matrix $C = (c_{ij})$ is defined to be the sum of the main diagonal elements:

$$\text{trace } C = \operatorname{tr} C = \sum_{i=1}^{n} c_{ii}.$$

The function

$$<A,\ B> = \operatorname{tr}(A^T B)$$

is a real inner product on $M_n(\mathbb{R})$, as you are asked to verify in Exercise 31. Therefore, if

$$A = \begin{bmatrix} 1 & 2 & 4 \\ 0 & 3 & 5 \\ 1 & 2 & 3 \end{bmatrix}, \quad \text{and} \quad B = \begin{bmatrix} 1 & -1 & 1 \\ 0 & 1 & -2 \\ 0 & 0 & 1 \end{bmatrix},$$

then

$$<A,\ B> = \operatorname{tr}(A^T B) = \operatorname{tr} \begin{bmatrix} 1 & -1 & 2 \\ 2 & 1 & -2 \\ 4 & 1 & -3 \end{bmatrix} = 1 + 1 + (-3) = -1. \quad \blacksquare$$

Example 4   (For students who have studied calculus) Consider the vector space $C[a, b]$ of all real-valued continuous functions defined on the interval $[a, b]$ of the real line. Let c and d be real numbers that satisfy $a \le c < d \le b$. Then for any two vectors (functions) f and g in $C[a, b]$ the function

$$<f, g> = \int_c^d f(x)\, g(x)\, dx$$

is an inner product on $C[a, b]$. This is called **the integral inner product** (from c to d).

To see that it is indeed an inner product note that the product of continuous functions is integrable and the definite integral of any real-valued continuous function is a real number. For the symmetric property of a real inner product we note that

$$<f, g> = \int_c^d f(x)\, g(x)\, dx = \int_c^d g(x)\, f(x)\, dx = <g, f>.$$

The linearity of the inner product also follows from properties of the definite integral. If r and s are any real numbers then

$$<rf + sg, h> = \int_c^d [rf(x) + sg(x)]h(x)\, dx = \int_c^d [rf(x)h(x) + sg(x)h(x)]\, dx$$

$$= r \int_c^d f(x)h(x)\, dx + s \int_c^d g(x)h(x)\, dx$$

$$= r<f, h> + s<g, h>.$$

Since the integral of a positive function is a positive number, and the integral of a nonnegative function is zero if and only if the function is the constant function 0, we have

$$<f, f> = \int_c^d f(x)^2\, dx \ge 0,$$

with equality only when f is the zero function.

In particular, consider the functions $f(x) = x$ and $g(x) = e^x$ and let $c = -1, d = 1$. Then we have

$$<f, g> = <x, e^x>$$

$$= \int_{-1}^1 xe^x\, dx = e^x[x - 1]_{-1}^1 = e[1 - 1] - e^{-1}[-1 - 1] = 2e^{-1}. \quad \blacksquare$$

Example 5   This is a special case of the previous example. We define an integral inner product on the vector space P_n of real polynomials as follows:

$$<p, q> = <p(x), q(x)> = \int_0^1 p(x)\, q(x)\, dx. \qquad \textbf{(4.19)}$$

The verification that this makes P_n a real inner-product space is left to you.

For $n = 1$ and $p(x) = 1 - x$ and $q(x) = 1 + x$, we have

$$<p(x), q(x)> = \int_0^1 (1-x)(1+x)\, dx = \int_0^1 (1-x^2)\, dx = \left[x - \frac{x^3}{3} \right]_0^1 = \frac{2}{3}.$$

As another example, suppose that $p(x) = x^2 - 2x + 3$ and $q(x) = 3x - 4$. These are both vectors in P_2. Then

$$<p(x), q(x)> = \int_0^1 (x^2 - 2x + 3)(3x - 4)dx$$

$$= \int_0^1 (3x^3 - 10x^2 + 17x - 12)dx$$

$$= \frac{3}{4}x^4 - \frac{10}{3}x^3 + \frac{17}{2}x^2 - 12x \Big|_0^1 = -\frac{73}{12}. \quad \blacksquare$$

Example 6   We can also define an inner product on P_n that more closely resembles the Euclidean inner product in $\mathbb{R}^{n+1}$, although this is not nearly so useful as the integral inner product in the study of function spaces. We define this inner product as follows: Let $p(x)$ and $q(x)$ be vectors (polynomials) in P_n. Then there are scalars (real numbers) $a_0, a_1, \ldots, a_n, b_0, b_1, \ldots, b_n$ so that

$$p(x) = a_0 + a_1 x + a_2 x^2 + \cdots + a_n x^n,$$

and

$$q(x) = b_0 + b_1 x + b_2 x^2 + \cdots + b_n x^n.$$

Consider the function

$$<p(x), q(x)> = \sum_{k=0}^n a_k b_k. \qquad \textbf{(4.20)}$$

Compare (4.20) with (4.17). We claim that this function is also an inner product on P_n. To see this, note first that $<p(x), q(x)>$ given by (4.20) is clearly a real number. Furthermore the commutativity of multiplication of real numbers tells us that, for *each* $k = 0, 1, \ldots, n$, $a_k b_k = b_k a_k$, so $< p(x), q(x) >$ will equal $< q(x), p(x) >$. This verifies Property (1) of the definition for this function.

Now let α and β be any two real numbers; then

$$\alpha p(x) = \alpha a_0 + \alpha a_1 x + \alpha a_2 x^2 + \cdots + \alpha a_n x^n,$$

and

$$\beta q(x) = \beta b_0 + \beta b_1 x + \beta b_2 x^2 + \cdots + \beta b_n x^n.$$

Given a third vector (polynomial) $r(x)$ from P_n, we write

$$r(x) = c_0 + c_1 x + c_2 x^2 + \cdots + c_n x^n.$$

Then we have, using (4.20), the following:

$$< \alpha p(x) + \beta q(x), r(x) > = \sum_{k=0}^{n} (\alpha a_k + \beta b_k) c_k = \sum_{k=0}^{n} (\alpha a_k c_k + \beta b_k c_k)$$

$$= \sum_{k=0}^{n} \alpha a_k c_k + \sum_{k=0}^{n} \beta b_k c_k$$

$$= \alpha \sum_{k=0}^{n} a_k c_k + \beta \sum_{k=0}^{n} b_k c_k$$

$$= \alpha < p(x), r(x) > + \beta < q(x), r(x) >.$$

This gives us the second property of the definition of an inner product.

Finally, we note that $< p(x), p(x) > = a_0^2 + a_1^2 + a_2^2 + \cdots + a_n^2 \geq 0$, being the sum of squares of real numbers. Furthermore, this sum can be equal to zero if, and only if, each a_i, $i = 0, 1, 2, \ldots, n$, is zero.

Thus (4.20) does indeed define an inner product on P_n. To see that it is different from the integral inner product one need only contrast the values of $< p(x), p(x) >$ under each. In the case $p(x) = 1 + x$, for the integral inner product

$$< p(x), p(x) > = \int_0^1 (1 + x)(1 + x)\, dx$$

$$= \int_0^1 (1 + x)^2\, dx = \frac{1}{3}(1 + x)^3 \Big|_0^1 = \frac{7}{3},$$

whereas, for the inner product of (4.20), $< p(x), p(x) > = (1)(1) + (1)(1) = 2$.

You may also wish to compare the value of $< p(x), q(x) >$ given in Example 5, for $p(x) = x^2 - 2x + 3$ and $q(x) = 3x - 4$, with the value $(1)(0) + (-2)(3) + (3)(-4) = -18$, given by this inner product (4.20). ∎

In the next section we describe how the concepts of *length, angle,* and *distance* can be defined on any vector space equipped with an inner product. To conclude this section, we examine a minor modification of Euclidean *n*-space that we use in our subsequent work.

From here on we let $\overline{\mathbb{R}^n}$ denote the space of n-dimensional column vectors **x**.

That is, $\overline{\mathbb{R}^n}$ is the space $M_{n\times 1}(\mathbb{R})$ of real $n \times 1$ matrices. If **v** is any vector in $\mathbb{R}^n$, then $\mathbf{v} \in M_{1\times n}(\mathbb{R})$, so $\mathbf{v}^T = \mathbf{k}$ is in $\overline{\mathbb{R}^n}$. Conversely if $\mathbf{k} \in \overline{\mathbb{R}^n}$, then $\mathbf{v} = \mathbf{k}^T \in \mathbb{R}^n$. We can make $\overline{\mathbb{R}^n}$ into a Euclidean space by defining an appropriate inner product.

Let **k** and **m** be vectors in $\overline{\mathbb{R}^n}$. Define the inner product $< \mathbf{k}, \mathbf{m} >$ by $< \mathbf{k}, \mathbf{m} > = \mathbf{k}^T\mathbf{m}$.

Therefore, if

$$\mathbf{k} = \begin{bmatrix} x_1 \\ x_2 \\ \vdots \\ x_n \end{bmatrix} \quad \text{and} \quad \mathbf{m} = \begin{bmatrix} y_1 \\ y_2 \\ \vdots \\ y_n \end{bmatrix}$$

then

$$< \mathbf{k}, \mathbf{m} > = (x_1, x_2, \ldots, x_n) \cdot \begin{bmatrix} y_1 \\ y_2 \\ \vdots \\ y_n \end{bmatrix}$$

$$= x_1 y_1 + x_2 y_2 + \cdots + x_n y_n = \sum_{k=1}^{n} x_k y_k.$$

You can easily see that this is indeed a Euclidean inner product, since it is quite like that of Example 2, with D being the identity matrix.

Example 7 Let

$$\mathbf{x} = \begin{bmatrix} 1 \\ 2 \\ 1 \\ -3 \end{bmatrix} \quad \text{and} \quad \mathbf{y} = \begin{bmatrix} 3 \\ -1 \\ 2 \\ -1 \end{bmatrix}.$$

Then

$$< \mathbf{x}, \mathbf{y} > = \mathbf{x}^T\mathbf{y} = (1, 2, 1, -3) \cdot \begin{bmatrix} 3 \\ -1 \\ 2 \\ -1 \end{bmatrix}$$

$$= (1)(3) + (2)(-1) + (1)(2) + (-3)(-1) = 6. \quad \blacksquare$$

Vocabulary

In this section we have again met several new concepts. The vocabulary developed here is used not only throughout the rest of the text but also in those applications of linear algebra which involve geometric ideas. Be sure that you understand each of the concepts described by the terms in the following list. Review these ideas until you feel secure enough to use them further.

real inner product	inner-product space
linear in each argument	symmetric function
Euclidean inner product	Euclidean space
standard inner product	Euclidean n-space $\mathbb{R}^n$
the "column" vector space $\overline{\mathbb{R}^n}$	trace of a matrix
integral inner product	trace inner product

Exercises 4.3

In Exercises 1–8 find the Euclidean inner product $<\mathbf{u}, \mathbf{v}>$ for the given vectors in the implied $\mathbb{R}^n$.

1. $\mathbf{u} = (1, -1, 2)$, $\mathbf{v} = (0, 2, 0)$.

2. $\mathbf{u} = (1, -1, 2, 1)$, $\mathbf{v} = (0, 0, 2, -1)$.

3. $\mathbf{u} = (1, -1, 2, 0, 1)$, $\mathbf{v} = (1, -2, 0, 2, -3)$.

4. $\mathbf{u} = (1, 0, -1, 0, 2)$, $\mathbf{v} = (0, 2, -3, 4, -3)$.

5. $\mathbf{u} = (-2, 1, -1, 2, 0, 1)$, $\mathbf{v} = (0, 0, 0, 0, 0, 0)$.

6. $\mathbf{u} = (-2, 1, -1, 2, 0, 1)$, $\mathbf{v} = (1, 0, 1, 0, 1, 0)$.

7. $\mathbf{u} = (1, 1, -1, 2, -4, 1)$, $\mathbf{v} = (-1, 1, -1, 1, -1, 1)$.

8. $\mathbf{u} = (-1, 1, -1, 1, -1, 1)$, $\mathbf{v} = (-1, 0, -2, 0, -3, 0)$.

In Exercises 9–12 use the inner product $<\mathbf{u}, \mathbf{v}> = \mathbf{u} \, D\mathbf{v}^T$ of Example 2 for the given vectors $\mathbf{u}$ and $\mathbf{v}$ from $\mathbb{R}^n$ and the given diagonal matrix D.

9. $\mathbf{u} = (1, -1, 2)$, $\mathbf{v} = (0, 2, 0)$, $D = 2I$.

10. $\mathbf{u} = (1, -1, 2, 1)$, $\mathbf{v} = (1, 0, 2, -1)$, $D = 3I$.

11. $\mathbf{u} = (1, -1, 2)$, $\mathbf{v} = (1, 2, -1)$, $D = \text{diag} [2, 3, 4]$.

12. $\mathbf{u} = (1, -1, 2, 1)$, $\mathbf{v} = (3, -1, 2, -1)$, $D = \text{diag} [2, 3, 4, 5]$.

In Exercises 13–15, use the trace inner product of Example 3 to find $<A, B>$ for the given matrices A and B.

13. $A = \begin{bmatrix} 0 & 1 \\ -1 & -1 \end{bmatrix}$, $B = \begin{bmatrix} 2 & 3 \\ 1 & -1 \end{bmatrix}$.

14. $A = \begin{bmatrix} 1 & -1 & 2 \\ 0 & 5 & -3 \\ 2 & 1 & -2 \end{bmatrix}$, $B = \begin{bmatrix} 2 & -1 & 5 \\ 4 & 3 & 1 \\ 2 & -1 & 0 \end{bmatrix}$.

15. $A = \begin{bmatrix} 1 & 0 & 0 & -1 \\ 0 & 1 & 1 & 0 \\ 1 & 1 & 0 & 1 \\ 0 & 1 & 1 & -1 \end{bmatrix}$, $B = \begin{bmatrix} 1 & -1 & 1 & -1 \\ 0 & 1 & -1 & 1 \\ 0 & 0 & 1 & -1 \\ 0 & 0 & 0 & 1 \end{bmatrix}$.

In Exercises 16–19 use the "Euclidean-like" inner product of Example 6 in P_n to find $<p(x), q(x)>$ for the given polynomials.

16. $p(x) = 1 - x$, $q(x) = 1 + x$. (Contrast with Example 5.)

17. $p(x) = x^2$, $q(x) = x - 3$.

18. $p(x) = 3x + 4$, $q(x) = 6x - 2$.

19. $p(x) = 3x^2 + 4x - 5$, $q(x) = 3x - 2$.

In Exercises 20–23 use the Euclidean inner product of Example 7 to find $< X, Y >$ in the vector space $\mathbb{R}^n = M_{n \times 1}(\mathbb{R})$ of column vectors.

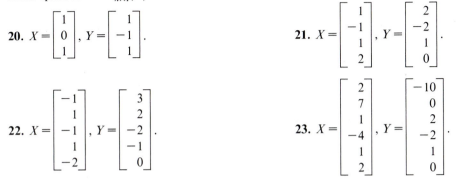

20. $X = \begin{bmatrix} 1 \\ 0 \\ 1 \end{bmatrix}$, $Y = \begin{bmatrix} 1 \\ -1 \\ 1 \end{bmatrix}$.

21. $X = \begin{bmatrix} 1 \\ -1 \\ 1 \\ 2 \end{bmatrix}$, $Y = \begin{bmatrix} 2 \\ -2 \\ 1 \\ 0 \end{bmatrix}$.

22. $X = \begin{bmatrix} -1 \\ 1 \\ -1 \\ 1 \\ -2 \end{bmatrix}$, $Y = \begin{bmatrix} 3 \\ 2 \\ -2 \\ -1 \\ 0 \end{bmatrix}$.

23. $X = \begin{bmatrix} 2 \\ 7 \\ 1 \\ -4 \\ 1 \\ 2 \end{bmatrix}$, $Y = \begin{bmatrix} -10 \\ 0 \\ 2 \\ -2 \\ 1 \\ 0 \end{bmatrix}$.

In Exercises 24–29 decide which of the given functions $Q(\mathbf{u}, \mathbf{v})$ on $\mathbb{R}^2$ are inner products and which are not. For those that are not inner products, give an example that violates one or more of the defining properties of an inner product. Let $\mathbf{u} = (u_1, u_2)$ and $\mathbf{v} = (v_1, v_2)$.

24. $Q(\mathbf{u}, \mathbf{v}) = u_1 v_2 + u_2 v_1$.

25. $Q(\mathbf{u}, \mathbf{v}) = u_1 u_2 v_1 v_2$.

26. $Q(\mathbf{u}, \mathbf{v}) = \mathbf{u} \, A\mathbf{v}^T$, where $A = \begin{bmatrix} -1 & -3 \\ 5 & 8 \end{bmatrix}$.

27. $Q(\mathbf{u}, \mathbf{v}) = \mathbf{u} \, A\mathbf{v}^T$, where $A = \begin{bmatrix} -1 & -3 \\ -3 & 8 \end{bmatrix}$.

28. $Q(\mathbf{u}, \mathbf{v}) = \mathbf{u} \, A\mathbf{v}^T$, where $A = \begin{bmatrix} 2 & 5 \\ 5 & 2 \end{bmatrix}$.

29. $Q(\mathbf{u}, \mathbf{v}) = \mathbf{u} \, A\mathbf{v}^T$, where $A = \begin{bmatrix} 2 & -3 \\ -3 & 3 \end{bmatrix}$.

---

30. Prove Theorem 4.5 (page 237).

31. Verify that the "trace inner product" of Example 3 satisfies the definition of an inner product in $M_n(\mathbb{R})$.

32. Which of the vectors, if any, in Exercises 1–8 satisfy the equation $< \mathbf{u}, \mathbf{v} > = 0$?

33. Which of the vectors (polynomials), if any, in Exercises 16–19 satisfy the equation $< \mathbf{u}, \mathbf{v} > = 0$?

34. (For students who have studied calculus) Show that, given the inner product of Example 4 on $C[-\infty, \infty]$, with $c = -\pi$ and $d = \pi$, the functions $\sin x$ and $\cos x$ satisfy $< \sin x, \cos x > = 0$. That is, they are orthogonal functions in this inner product.

35. Let $\mathbf{a} = (a_1, a_2)$ and $\mathbf{b} = (b_1, b_2)$ be vectors in ordinary Euclidean 2-space. Show that the area of the parallelogram that has one vertex at the origin and adjacent sides along the two vectors $\mathbf{a}$ and $\mathbf{b}$ is the number $|\det [\mathbf{a} : \mathbf{b}]|$; that is, the area equals the absolute value of the determinant of the 2×2 matrix whose column vectors are $\mathbf{a}$ and $\mathbf{b}$. HINT: sketch a figure first and note (4.12) in Section 4.2.

36. Let V be *any* real inner-product space. Prove that $< \mathbf{0}, \mathbf{v} > = 0$ for all vectors $\mathbf{v}$ in V. (HINT: Consider $< \mathbf{0} + \mathbf{0}, \mathbf{v} >$.)

37. Let V be *any* real inner-product space. Show that $< \mathbf{u}, a\mathbf{v} > = a < \mathbf{u}, \mathbf{v} >$ for all vectors $\mathbf{u}$ and $\mathbf{v}$ in V and any real number (scalar) a.

38. Let V be any real inner-product space. Show that $< a\mathbf{u}, b\mathbf{v} > = ab < \mathbf{u}, \mathbf{v} >$ for all vectors $\mathbf{u}$ and $\mathbf{v}$ in V and all real numbers (scalars) a and b.

39. Let V be any real inner-product space. Show that

$$< a_1\mathbf{u}_1 + a_2\mathbf{u}_2, a_3\mathbf{u}_3 + a_4\mathbf{u}_4 >$$
$$= a_1 a_3 < \mathbf{u}_1, \mathbf{u}_3 > + a_1 a_4 < \mathbf{u}_1, \mathbf{u}_4 >$$
$$+ a_2 a_3 < \mathbf{u}_2, \mathbf{u}_3 > + a_2 a_4 < \mathbf{u}_2, \mathbf{u}_4 >,$$

for all vectors $\mathbf{u}_1, \mathbf{u}_2, \mathbf{u}_3,$ and $\mathbf{u}_4$ in V, and all real numbers (scalars) $a_1, a_2, a_3,$ and a_4.

40. Let V be any real inner-product space. Show that

$$< \mathbf{u} + \mathbf{v}, \mathbf{u} + \mathbf{v} > = < \mathbf{u}, \mathbf{u} > + 2 < \mathbf{u}, \mathbf{v} >$$
$$+ < \mathbf{v}, \mathbf{v} >,$$

for all vectors $\mathbf{u}$ and $\mathbf{v}$ in V. HINT: Recall that $1\mathbf{u} = \mathbf{u}$.

In Exercises 41–50 students who have studied calculus should use the "integral" inner product of Example 4 to find $<f(x), g(x)>$ for the given functions in $C[a, b]$. Use the values $c = -1$ and $d = 1$, unless otherwise indicated.

41. $f(x) = x; g(x) = 1 + x$.

42. $f(x) = x^2; g(x) = 1 + x + x^2$.

43. $f(x) = 3x; g(x) = 1 + x^2$.

44. $f(x) = 2x - 6; g(x) = 4 + x^3$.

45. $f(x) = 1 - x - 2x^2; g(x) = 1 + x^3 + x^5$.

46. $f(x) = x; g(x) = e^{-x^2}$.

47. $f(x) = x - e^x; g(x) = e^{-x}$.

48. $f(x) = \sin^2 x; g(x) = \cos x, c = 0, d = \pi$.

49. $f(x) = x; g(x) = \cos x, c = -\pi, d = \pi$.

50. $f(x) = \tan x; g(x) = \sec^2 x, c = 0, d = \pi/4$.

---

51. Show that if $w(x) > 0$ for all x in $[a, b]$ then the following is an inner product (a **weighted integral inner product**) on $C[a, b]$:

$$<f, g> = \int_c^d f(x)\, g(x)\, w(x)\, dx,$$

for $a \le c < d \le b$. The function $w(x)$ is the "weight function" for this inner product.

52. Use the result of Exercise 51 to find the weighted integral inner product of the two functions in Exercise 41, if $w(x) = x^2$ ($c = -1, d = 1$).

53. Use the result of Exercise 51 to find the weighted integral inner product of the two functions in Exercise 42, if $w(x) = x^2$ ($c = -1, d = 1$).

54. Use the result of Exercise 51 to find the weighted integral inner product of the two functions in Exercise 45, if $w(x) = x^2$ ($c = -1, d = 1$).

55. Use the result of Exercise 51 to find the weighted integral inner product of the two functions in Exercise 47, if $w(x) = x^2$ ($c = -1, d = 1$).

4.4 GEOMETRY IN REAL INNER-PRODUCT SPACES

In Section 4.1 we developed the idea of the standard inner product (the "dot product") for ordinary Euclidean 2-space and 3-space from the notions of length and angle in Euclidean geometry. Now that we have generalized the concept of inner product, let us see how we can use this more general inner product to define the ideas of "length" and "angle" in *any* inner-product space. You might expect that different inner products give different lengths and angles.

Let V be any real vector space and suppose that V is an inner-product space with respect to a real inner product $<\mathbf{u}, \mathbf{v}>$. We begin with a generalization of the idea of length by defining the **norm** $||\mathbf{v}||$ of an arbitrary vector in V as follows:

$$||\mathbf{v}|| = <\mathbf{v}, \mathbf{v}>^{1/2}. \tag{4.21}$$

Since, in any real inner-product space V, $<\mathbf{v}, \mathbf{v}>$ is a nonnegative real number, it does indeed have a nonnegative real square root.

Example 1   If V is $\mathbb{R}^3$ equipped with the *standard* inner product, and $\mathbf{x} = (1, -1, 2)$, then $\|\mathbf{x}\|$ is its Euclidean length, $\sqrt{1^2 + (-1)^2 + 2^2} = \sqrt{6}$. On the other hand, suppose that the inner product in $\mathbb{R}^3$ is the following:

$$<\mathbf{u}, \mathbf{v}> = 2u_1 v_1 + 3u_2 v_2 + u_3 v_3, \text{ where } \mathbf{u} = (u_1, u_2, u_3) \text{ and } \mathbf{v} = (v_1, v_2, v_3).$$

In this case, the *norm* of $\mathbf{x}$ is

$$\|\mathbf{x}\| = <\mathbf{x}, \mathbf{x}>^{1/2} = [2(1)(1) + 3(-1)(-1) + (2)(2)]^{1/2} = \sqrt{9} = 3. \quad \blacksquare$$

Example 2   If V is $\mathbb{R}^4$ equipped with the *standard* inner product, and $\mathbf{x} = (1, -1, 2, 4)$, then the *norm* $\|\mathbf{x}\|$ of $\mathbf{x}$ is a sort of "Euclidean length," although we have never actually measured the length of a vector in 4-space. Nevertheless, in Euclidean 4-space

$$\|\mathbf{x}\| = <\mathbf{x}, \mathbf{x}>^{1/2} = \sqrt{1^2 + (-1)^2 + 2^2 + 4^2} = \sqrt{22}.$$

Now suppose that we equip $V = \mathbb{R}^4$ with the inner product of Example 2 of Section 4.3. That is, $<\mathbf{u}, \mathbf{v}> = \mathbf{u} D\mathbf{v}^T$, for $\mathbf{u} = (u_1, u_2, u_3, u_4)$ and $\mathbf{v} = (v_1, v_2, v_3, v_4)$ and $D = \text{diag}[1, 2, 1, 3]$. Then for our same vector $\mathbf{x} = (1, -1, 2, 4)$, its *norm* $\|\mathbf{x}\|$ is given by

$$\|\mathbf{x}\| = <\mathbf{x}, \mathbf{x}>^{1/2}$$

and

$$<\mathbf{x}, \mathbf{x}> = (1, -1, 2, 4) \cdot \begin{bmatrix} 1 & 0 & 0 & 0 \\ 0 & 2 & 0 & 0 \\ 0 & 0 & 1 & 0 \\ 0 & 0 & 0 & 3 \end{bmatrix} \begin{bmatrix} 1 \\ -1 \\ 2 \\ 4 \end{bmatrix}$$

$$= 1 + 2 + 4 + 48 = 55.$$

Hence $\|\mathbf{x}\| = \sqrt{55}$. Contrast this with its standard "Euclidean length." $\quad \blacksquare$

Perhaps better examples occur in vector spaces where even an extrapolated idea of "Euclidean length" has little meaning. For example, consider the polynomial spaces P_n.

Example 3   (For students who have studied calculus) Let P_n be the real inner-product space of all polynomials of degree n or less with the integral (from 0 to 1) inner product. Then the norm of a polynomial induced by this inner product is the following;

$$\|p(x)\| = <p, p>^{\frac{1}{2}} = \left(\int_0^1 p^2(x) \, dx \right)^{\frac{1}{2}}.$$

Let $p(x)$ be the polynomial x; then

$$\|p(x)\| = \left(\int_0^1 (x^2) \, dx \right)^{\frac{1}{2}} = \left(\frac{x^3}{3} \Big|_0^1 \right)^{\frac{1}{2}} = \left(\frac{1}{3} \right)^{\frac{1}{2}} = \frac{\sqrt{3}}{3}.$$

If $q(x) = 1 - x^3$, then

$$\|q(x)\| = \|1 - x^3\| = \left[\int_0^1 (1 - x^3)^2\, dx\right]^{\frac{1}{2}} = \left[\int_0^1 (1 - 2x^3 + x^6)\, dx\right]^{\frac{1}{2}}$$

$$= \left[x - \frac{2x^4}{4} + \frac{x^7}{7}\Big|_0^1\right]^{\frac{1}{2}} = \left[1 - \frac{1}{2} + \frac{1}{7}\right]^{\frac{1}{2}}$$

$$= \sqrt{\frac{9}{14}} = \frac{3}{\sqrt{14}} = \frac{3\sqrt{14}}{14}. \quad \blacksquare$$

These three examples illustrate that each inner-product function on a vector space V induces a norm for vectors in that space. This norm behaves in much the same way as we expect a measure of "size" or "magnitude" to behave. That is, the norm induced by an inner product has the properties given by Theorem 4.6 below and it also satisfies the inequalities of Theorem 4.7 and its corollary, which follow. These are precisely the properties that we require of any norm. The idea of *length* in ordinary Euclidean geometry is just one example of such a norm.

Theorem 4.6   Let V be a real inner-product space, let $\mathbf{u}$ and $\mathbf{v}$ be any vectors in V, and let a be any real number. Then

(i) $\|\mathbf{v}\| \geq 0$, with $\|\mathbf{v}\| = 0$, if and only if, $\mathbf{v} = \mathbf{0}$.
(ii) $\|a\mathbf{v}\| = |a|\|\mathbf{v}\|$.

Proof   Part (i) follows at once from Properties (3) and (4) of the definition of an inner product [see (4.15), page 236] and the definition of a norm. To prove part (ii), observe that

$$\|a\mathbf{v}\|^2 = <a\mathbf{v}, a\mathbf{v}> = a<\mathbf{v}, a\mathbf{v}>, \text{ by (4.15), Property (2)};$$
$$= a(a<\mathbf{v}, \mathbf{v}>), \text{ by Theorem 4.4};$$
$$= a^2 <\mathbf{v}, \mathbf{v}> = a^2\|\mathbf{v}\|^2.$$

So we've shown that, for any norm,

$$\|a\mathbf{v}\|^2 = a^2\|\mathbf{v}\|^2. \tag{4.22}$$

And upon taking the positive square roots of both sides of (4.22), and recalling that $\sqrt{a^2} = |a|$ for every real number a, we have

$$\|a\mathbf{v}\| = \sqrt{a^2}\|\mathbf{v}\| = |a|\,\|\mathbf{v}\|, \tag{4.23}$$

as desired.   ❏

In Section 4.1, equations (4.7) and (4.8) show that the angle θ between two nonzero vectors $\mathbf{u}$ and $\mathbf{v}$ in Euclidean 2-space $\mathbb{R}^2$ or 3-space $\mathbb{R}^3$ satisfy the equation

$$\cos \theta = \frac{<\mathbf{u}, \mathbf{v}>}{\|\mathbf{u}\|\,\|\mathbf{v}\|}. \tag{4.24}$$

It is always a fact that, for any real number θ, $|\cos \theta| \leq 1$. Therefore, equation (4.24) implies that

$$\left| \frac{<\mathbf{u}, \mathbf{v}>}{||\mathbf{u}|| \, ||\mathbf{v}||} \right| \leq 1, \qquad \mathbf{u} \neq \mathbf{0}, \mathbf{v} \neq \mathbf{0}. \tag{4.25}$$

So, multiplying both sides of the inequality (4.25) by the *positive* real number $||\mathbf{u}|| \, ||\mathbf{v}||$ gives us

$$|<\mathbf{u}, \mathbf{v}>| \leq ||\mathbf{u}|| \, ||\mathbf{v}||. \tag{4.26}$$

We have, then, that inequality (4.26) holds at least in the Euclidean inner-product spaces $\mathbb{R}^2$ or $\mathbb{R}^3$. The remarkable fact is that this inequality is true in *any* real inner-product space. It is precisely this celebrated *Cauchy-Schwarz inequality* that makes a geometry in an arbitrary inner-product space possible.

Theorem 4.7 (The Cauchy-Schwarz Inequality) Let V be any real inner-product space. Then for all vectors $\mathbf{u}$ and $\mathbf{v}$ in V it is true that

$$|<\mathbf{u}, \mathbf{v}>| \leq ||\mathbf{u}|| \, ||\mathbf{v}||. \tag{4.27}$$

Moreover, equality in (4.27) holds if, and only if, $\mathbf{u}$ and $\mathbf{v}$ are linearly dependent.

This inequality also holds when the scalars are complex numbers, so that this theorem is a special case of Theorem 4.13 in Section 4.6. One could refer to the more general proof given there, but we include the proof of Theorem 4.7 here for completeness.

Proof Let $\mathbf{u}$ and $\mathbf{v}$ be arbitrary vectors from the arbitrary *real* inner-product space V. If the vector $\mathbf{u} = \mathbf{0}$, then any vector $\mathbf{v}$ makes the set $\{\mathbf{u}, \mathbf{v}\}$ a linearly dependent set and both sides of (4.27) equal zero. (See Exercise 36 in Section 4.3.) So suppose that $\mathbf{u} \neq \mathbf{0}$. Then, of course, $<\mathbf{u}, \mathbf{u}> \, > 0$, which we use several times in what follows. Let x be a (variable) real number and consider the vector $\mathbf{s} = x\mathbf{u} + \mathbf{v}$. The properties of a real inner product assure us that $<\mathbf{s}, \mathbf{s}> \, \geq 0$ and that

$$0 \leq <\mathbf{s}, \mathbf{s}> \, = \, <x\mathbf{u} + \mathbf{v}, x\mathbf{u} + \mathbf{v}>$$
$$= x^2 <\mathbf{u}, \mathbf{u}> + 2x <\mathbf{u}, \mathbf{v}> + <\mathbf{v}, \mathbf{v}>. \tag{4.28}$$

This is a quadratic inequality, in x, of the form

$$ax^2 + 2bx + c \geq 0, \qquad a \neq 0.$$

By completing the square on the left, we have for every real number x,

$$a\left(x^2 + \frac{2b}{a}x + \frac{b^2}{a^2}\right) + \left(c - \frac{b^2}{a}\right) \geq 0,$$

or

$$a\left(x + \frac{b}{a}\right)^2 + \left(c - \frac{b^2}{a}\right) \geq 0. \tag{4.29}$$

In particular, the inequality (4.29) is true when $x = -b/a$. So substitute $-b/a$ for x and obtain

$$c - \frac{b^2}{a} \geq 0, \quad \text{and hence} \quad c \geq \frac{b^2}{a}. \tag{4.30}$$

Now replacing the values of a, b, and c from (4.28) into (4.30) we have the inequality

$$<\mathbf{v}, \mathbf{v}> \geq \frac{(<\mathbf{u}, \mathbf{v}>)^2}{(<\mathbf{u}, \mathbf{u}>)}.$$

But, as we remarked earlier, $<\mathbf{u}, \mathbf{u}>$ is positive when $\mathbf{u} \neq \mathbf{0}$, so we may multiply both sides of the above inequality by it to obtain

$$(<\mathbf{u}, \mathbf{v}>)^2 \leq (<\mathbf{u}, \mathbf{u}>)(<\mathbf{v}, \mathbf{v}>) = ||\mathbf{u}||^2||\mathbf{v}||^2,$$

from which the desired result follows by taking the positive square root of both sides.

Now when does one get equality in (4.27)? For both sides to be equal we must have in (4.28) $<\mathbf{s}, \mathbf{s}> = <x\mathbf{u} + \mathbf{v}, x\mathbf{u} + \mathbf{v}> = 0$, for some real number x. Hence the two sides of (4.27) are equal if, and only if, $x\mathbf{u} + \mathbf{v}$ is the zero vector, for some x. But $x\mathbf{u} + \mathbf{v} = \mathbf{0}$ if, and only if, $\mathbf{u}$ and $\mathbf{v}$ are linearly dependent. This completes the proof. ❏

As an important corollary of the Cauchy-Schwarz inequality we get the triangle inequality. This inequality together with the two properties (i) and (ii) in Theorem 4.6 (page 248) make up the usual definition of a (vector) **norm**.

Corollary (The Triangle Inequality)   Let $\mathbf{u}$ and $\mathbf{v}$ be any two vectors in a real inner-product space V. Then

$$||\mathbf{u} + \mathbf{v}|| \leq ||\mathbf{u}|| + ||\mathbf{v}||. \tag{4.31}$$

Proof   Remember that $||\mathbf{u} + \mathbf{v}||^2 = <\mathbf{u} + \mathbf{v}, \mathbf{u} + \mathbf{v}> = <\mathbf{u}, \mathbf{u}> + 2<\mathbf{u}, \mathbf{v}> + <\mathbf{v}, \mathbf{v}>$. Therefore,

$$||\mathbf{u} + \mathbf{v}||^2 = ||\mathbf{u}||^2 + 2<\mathbf{u}, \mathbf{v}> + ||\mathbf{v}||^2. \tag{4.32}$$

Now recall that, for all real numbers a, $a \leq |a|$. So, in particular, $<\mathbf{u}, \mathbf{v}> \leq |<\mathbf{u}, \mathbf{v}>|$; hence,

$$||\mathbf{u} + \mathbf{v}||^2 \leq ||\mathbf{u}||^2 + 2|<\mathbf{u}, \mathbf{v}>| + ||\mathbf{v}||^2. \tag{4.33}$$

By the Cauchy-Schwarz inequality (4.27), $|<\mathbf{u}, \mathbf{v}>| \leq ||\mathbf{u}|| \, ||\mathbf{v}||$, so we combine this with (4.33) to obtain

$$||\mathbf{u} + \mathbf{v}||^2 \leq ||\mathbf{u}||^2 + 2(||\mathbf{u}|| \, ||\mathbf{v}||) + ||\mathbf{v}||^2 = (||\mathbf{u}|| + ||\mathbf{v}||)^2. \tag{4.34}$$

Now take the positive square root of both sides of (4.34) to obtain (4.31), as desired. ❏

As we have remarked, the Cauchy-Schwarz and the triangle inequalities give a geometry in our inner-product space. In particular let's look at what the triangle inequality says about vectors in Euclidean 2-space $\mathbb{R}^2$. [You may want to entertain yourself with a picture of what this would look like if $\mathbb{R}^2$ were equipped with the inner product of Example 2 in Section 4.3 given $D = \text{diag} [2, 3]$. (See the Supplementary Exercises.)] In the ordinary Euclidean plane $\|\mathbf{u}\|$, $\|\mathbf{v}\|$ and $\|\mathbf{u} + \mathbf{v}\|$ would form the lengths of three sides of a triangle as depicted in Figure 4.15. Here, the triangle inequality gives us the familiar fact that *the length of a side of a triangle is at most as great as the sum of the lengths of the other two sides.* (Is this not also the case with any other inner product induced geometry?)

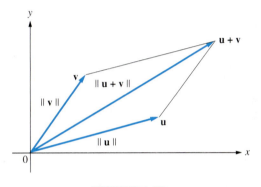

FIGURE 4.15

Example 4   Demonstrate the application of the triangle inequality and the Cauchy-Schwarz inequality to the vectors $\mathbf{v}_1 = (1, 2)$ and $\mathbf{v}_2 = (-3, -5)$ in $\mathbb{R}^2$ with the standard inner product. Here $\mathbf{v}_1 + \mathbf{v}_2 = (1, 2) + (-3, -5) = (-2, -3)$, and therefore, by definition,

$$\|\mathbf{v}_1 + \mathbf{v}_2\| = \sqrt{<\mathbf{v}_1 + \mathbf{v}_2, \mathbf{v}_1 + \mathbf{v}_2>} = \sqrt{(-2)^2 + (-3)^2} = \sqrt{13}.$$

Also

$$\|\mathbf{v}_1\| = \sqrt{<\mathbf{v}_1, \mathbf{v}_1>} = \sqrt{1^2 + 2^2} = \sqrt{5},$$

and

$$\|\mathbf{v}_2\| = \sqrt{<\mathbf{v}_2, \mathbf{v}_2>} = \sqrt{(-3)^2 + (-5)^2} = \sqrt{34}.$$

Observe that $\sqrt{13} \leq \sqrt{5} + \sqrt{34}$, satisfying the triangle inequality. Moreover, $<\mathbf{v}_1, \mathbf{v}_2> = (1)(-3) + (2)(-5) = -13$, so $|<\mathbf{v}_1, \mathbf{v}_2>| = 13$, and $13 \leq \sqrt{5} \cdot \sqrt{34}$ (verify this), satisfying the Cauchy-Schwarz inequality.   ∎

As we mentioned above, it is the Cauchy-Schwarz inequality that makes it possible for meaningful geometric concepts to be defined in an arbitrary real inner-product space V. In particular an "angle" concept can be defined.

Let **u** and **v** be two nonzero vectors from V; then we *define* the **angle** θ between **u and v** to be the angle such that

$$\cos \theta = \frac{<\mathbf{u}, \mathbf{v}>}{||\mathbf{u}|| \, ||\mathbf{v}||}, \qquad 0 \leq \theta \leq \pi.$$

This definition includes the usual definition in Euclidean 2- and 3-space. That it is a reasonable definition in an arbitrary inner-product space V follows from the fact that, for nonzero vectors **u** and **v**, the form (4.27) of the Cauchy-Schwarz inequality gives us the following, as you should verify. [Compare with (4.25).]

$$-1 \leq \frac{<\mathbf{u}, \mathbf{v}>}{||\mathbf{u}|| \, ||\mathbf{v}||} \leq 1.$$

Example 5   Find the cosine of the "angle" between the vectors $\mathbf{a} = (1, 0, 1, 1)$ and $\mathbf{b} = (2, 1, -1, 3)$ in both Euclidean 4-space $\mathbb{R}^4$ (standard inner product), and $\mathbb{R}^4$ with the inner product of Example 2 of Section 4.3; namely, $<\mathbf{a}, \mathbf{b}> = \mathbf{a} \, D \, \mathbf{b}^T$ for $D = \text{diag} \, [1, 2, 1, 3]$. The Euclidean "angle" θ satisfies

$$\cos \theta = \frac{<\mathbf{a}, \mathbf{b}>}{||\mathbf{a}|| \, ||\mathbf{b}||} = \frac{2 + 0 - 1 + 3}{\sqrt{1+0+1+1} \cdot \sqrt{4+1+1+9}}$$

$$= \frac{4}{\sqrt{3} \cdot \sqrt{15}} = 0.5963 \qquad \text{(approximately)}.$$

The nonstandard "angle" θ' satisfies

$$\cos \theta' = \frac{<\mathbf{a}, \mathbf{b}>}{||\mathbf{a}|| \, ||\mathbf{b}||}$$

$$= \frac{(1)(2) + (2)(0) + (1)(-1) + (3)(3)}{\sqrt{(1)(1)^2 + (2)(0)^2 + (1)(1)^2 + (3)(1)^2} \cdot \sqrt{(1)(2)^2 + (2)(1)^2 + (1)(-1)^2 + (3)(3)^2}}$$

$$= \frac{10}{\sqrt{5} \cdot \sqrt{34}} = 0.7670 \qquad \text{(approximately)}. \quad \blacksquare$$

We leave it to your imagination what an angle in $\mathbb{R}^4$ with either a Euclidean or a nonstandard inner product is. Let's look at one more example that does lead us to an idea that has wide application.

Example 6   Find the cosine of the angle between the two vectors $\mathbf{u} = (2, -1, 2, -1, 0)$ and $\mathbf{v} = (1, 3, 1, 1, 5)$ in Euclidean 5-space, $\mathbb{R}^5$ (with the standard inner product). From the definition given above for $\cos \theta$ we have

$$\cos \theta = \frac{<\mathbf{u}, \mathbf{v}>}{||\mathbf{u}|| \, ||\mathbf{v}||} = \frac{0}{||\mathbf{u}|| \, ||\mathbf{v}||} = 0. \quad \blacksquare$$

If we were in $\mathbb{R}^3$, we would say that **u** and **v** were "perpendicular," since $\cos^{-1} 0 = \pi/2$. But that idea doesn't seem to mean very much in $\mathbb{R}^5$. We choose a more general term.

> **Definition** Two vectors **u** and **v** in the real inner-product space V are called **orthogonal** if, and only if, $<$**u**, **v**$> = 0$. Moreover, a set $\{$**u**$_1$, **u**$_2$, . . . , **u**$_k\}$ of vectors is called an **orthogonal set** of vectors if, and only if, $<$**u**$_i$, **u**$_j> = 0$, for all $i \ne j$; $i, j = 1, 2, . . . , k$.

In the next several examples we illustrate why the term *orthogonal* is preferred over the word *perpendicular* for vectors whose inner product is zero.

Example 7 (For students who have studied calculus) Consider the inner-product space $C[a, b]$ of continuous real-valued functions with the integral inner product. In particular, consider the functions $\sin x$ and $\cos x$, and let $c = -\pi$, $d = \pi$. Then we have

$$< \sin x, \cos x > = \int_{-\pi}^{\pi} (\sin x)(\cos x)\, dx = \frac{1}{2}[\sin^2 x]_{-\pi}^{\pi}$$

$$= \frac{1}{2}[\sin^2 \pi - \sin^2 (-\pi)] = 0.$$

So sine and cosine are *orthogonal functions* (vectors) in this vector space. ■

Remember that an inner-product space depends on the choices of (1) a set of vectors, (2) a set of scalars, and (3) a specific inner product. A change in any one of these three results in a different inner-product space. For example, would the two functions *sine* and *cosine* be orthogonal if the integral were taken from 0 to 1?

Example 8 (For students who have studied calculus) Show that the two polynomials $p(x) = x - 1$ and $q(x) = 3x - 1$ are orthogonal in the inner-product space P_1 with the inner product of Example 5 in Section 4.3. We have

$$< p(x), q(x) > = \int_0^1 (x - 1)(3x - 1)\, dx = \int_0^1 (3x^2 - 4x + 1)\, dx$$

$$= [x^3 - 2x^2 + x]_0^1 = 0. \quad ■$$

Example 9 **Special Orthogonal Polynomials** (For students who have studied calculus) In advanced analysis, particularly in its applications, certain approximating orthogonal polynomials are used in various ways. For example, we define two of them here and give an inner product on $C[a, b]$ with respect to which each forms an orthogonal set of vectors.

Legendre Polynomials. For each nonnegative integer n we define the Legendre polynomial $P_n(x)$, of degree n, using Rodrigues' formula, by $P_0(x) = 0$, $P_1(x) = x$ and

$$P_n(x) = \frac{1}{2^n \, n!} \frac{d^n (x^2 - 1)^n}{dx^n} \qquad \text{for all } n > 1.$$

The first few of these Legendre polynomials are the first two given above and

$$P_2(x) = \frac{1}{2}(3x^2 - 1); \qquad\qquad P_3(x) = \frac{1}{2}(5x^3 - 3x).$$

$$P_4(x) = \frac{1}{8}(35x^4 - 30x^2 + 3); \qquad P_5(x) = \frac{1}{8}(63x^5 - 70x^3 + 15x).$$

The Legendre polynomials are orthogonal with respect to the integral inner product

$$< p(x),\, q(x) > = \int_{-1}^{1} p(x)q(x) \, dx.$$

Tchebycheff Polynomials. For each nonnegative integer n we define the Tchebycheff polynomial $T_n(x)$, of degree n, by $T_0(x) = 1$, $T_1(x) = x$, and $T_{n+1} = 2xT_n - T_{n-1}$; $n \geq 1$.

Tchebycheff polynomials have the interesting property that $T_n(\cos \theta) = \cos n\theta$. The first few of these, in addition to the two given above, are

$$T_2(x) = 2x^2 - 1, \qquad T_3(x) = 4x^3 - 3x, \ldots$$

These polynomials are orthogonal with respect to the weighted integral inner product

$$< p(x),\, q(x) > = \int_{-1}^{1} p(x)q(x)(1 - x^2)^{-1/2} \, dx. \quad \blacksquare$$

Any real vector space with an inner product has orthogonal vectors. We see later how these vectors make very useful basis vectors for the space. The standard basis vectors in $\mathbb{R}^n$ are always mutually orthogonal, as you can prove in the exercises. One more example.

Example 10   Show that the two matrices

$$A = \begin{bmatrix} 1 & 3 \\ -1 & 4 \end{bmatrix} \qquad \text{and} \qquad B = \begin{bmatrix} -5 & 2 \\ 5 & 1 \end{bmatrix}$$

are orthogonal in $M_2(\mathbb{R})$ with the trace inner product $< A,\, B > = \text{tr}(A^T B)$.
Observe that

$$A^T B = \begin{bmatrix} 1 & -1 \\ 3 & 4 \end{bmatrix} \cdot \begin{bmatrix} -5 & 2 \\ 5 & 1 \end{bmatrix} = \begin{bmatrix} -10 & 1 \\ 5 & 10 \end{bmatrix}.$$

Therefore, $< A,\, B > = \text{tr}(A^T B) = -10 + 10 = 0$. So A and B are orthogonal matrices under the given inner product. $\quad \blacksquare$

In our discussion above we have considered generalizations of both "angle" (or "direction") and "magnitude" for a vector in various guises. The concept of the norm $\|\mathbf{u}\|$ of a vector and the idea of orthogonality of two vectors in an inner-product space are the main ideas that we wish to use later. One additional definition will be useful.

> **Definition** A vector $\mathbf{v}$ in an inner-product space V is called a **unit vector** if, and only if, $\|\mathbf{v}\| = <\mathbf{v},\, \mathbf{v}>^{1/2} = 1$.

For example, the standard basis vectors $\mathbf{e}_i = (0,\, 0,\, \ldots ,\, 1,\, 0,\, \ldots ,\, 0)$, $i = 1,\, 2,\, \ldots ,\, n$, in $\mathbb{R}^n$ are also all *unit* vectors in Euclidean n-space.

Example 11 (For students who have studied calculus) Find the value of the scalar a so that the polynomial $p(x) = ax - 1$ is a unit vector in P_1 with the integral inner product.

We compute $\|p(x)\| = (< p(x),\, p(x) >)^{\frac{1}{2}}$.

$$\|p(x)\| = \|ax - 1\| = \left[\int_0^1 (ax - 1)^2\, dx\right]^{\frac{1}{2}} = \left[\int_0^1 (a^2 x^2 - 2ax + 1)\, dx\right]^{\frac{1}{2}}$$

$$= \left[\frac{a^2 x^3}{3} - ax^2 + x\Big|_0^1\right]^{\frac{1}{2}} = \left[\frac{a^2}{3} - a + 1\right]^{\frac{1}{2}}.$$

Therefore, since we want $\|p(x)\| = 1$, $a^2/3 - a + 1 = 1$, or $a^2/3 - a = 0$, from which we conclude that $a = 0$ or $a = 3$. Thus both of the polynomials $p_1(x) = -1$ and $p_2(x) = 3x - 1$ are unit vectors in P_1 with respect to this integral inner product. ∎

As we mentioned, the standard basis for $\mathbb{R}^n$ is made up of mutually orthogonal unit vectors. Such a basis is called an **orthonormal basis** and has many simplifying properties. Formally, for any real inner-product space V we have the following definition:

> **Definition** A basis $B = \{\mathbf{v}_1,\, \mathbf{v}_2,\, \ldots ,\, \mathbf{v}_n\}$ for a real inner-product space V is an **orthonormal basis** for V if both of the following conditions are satisfied:
>
> (i) Every $\mathbf{v}_i$ in B is a unit vector: $\|\mathbf{v}_i\| = 1$, $(<\mathbf{v}_i,\, \mathbf{v}_i> = 1)$ and
> (ii) The members of B are pairwise orthogonal: $<\mathbf{v}_i,\, \mathbf{v}_j> = 0$, for $i \neq j$.

In the next section we develop a method for converting a given basis for a real inner-product space into an orthonormal basis for that space. In Exercise 48, below, we ask you to prove that the *standard basis* $E = \{\mathbf{e}_1,\, \mathbf{e}_2,\, \ldots ,\, \mathbf{e}_n\}$ is an *orthonormal basis* for $\mathbb{R}^n$, for every positive integer n.

Vocabulary

In this section we have again met several new concepts. The vocabulary developed here is used not only throughout the rest of the text but also in those applications of linear algebra which involve geometric ideas. Be sure that you understand each of the concepts described by the terms in the following list. Review these ideas until you feel secure enough to use them further.

norm of a vector induced by an
    inner product
angle between two vectors in an
    inner-product space
Cauchy-Schwarz inequality
orthogonal vectors

unit vector
orthonormal basis
the triangle inequality
orthogonal polynomials
orthonormal set

Exercises 4.4

In Exercises 1–10 find the norm $\|\mathbf{u}\|$ (length) of the given vector in a Euclidean n-space $\mathbb{R}^n$ with the standard inner product.

1. $(1, -3)$.
2. $(0, 1)$.
3. $(1, 1, 1)$.
4. $(1, -2, -1)$.

5. $\left(\dfrac{1}{\sqrt{3}}, -\dfrac{1}{\sqrt{3}}, \dfrac{1}{\sqrt{3}}\right)$.
6. $\left(\dfrac{1}{\sqrt{2}}, 0, -\dfrac{1}{\sqrt{2}}\right)$.
7. $(1, -1, 1, -1, 0)$.

8. $(2, 0, -2, 0, 1, 5)$
9. $\left(\dfrac{1}{\sqrt{5}}, -\dfrac{1}{\sqrt{5}}, 0, -\dfrac{1}{\sqrt{5}}, \dfrac{1}{\sqrt{5}}, \dfrac{1}{\sqrt{5}}\right)$.
10. $(0, 0, 0, 0, 0)$.

11. Which of the vectors in Exercises 1–10 are "unit" vectors?

12. Find a unit vector that is a scalar multiple of the vector in Exercise 2.

13. Find a unit vector that is a scalar multiple of the vector in Exercise 3.

14. Find a unit vector that is a scalar multiple of the vector in Exercise 4.

15. Find a unit vector that is a scalar multiple of the vector in Exercise 7.

16. Find a unit vector that is a scalar multiple of the vector in Exercise 8.

In Exercises 17–21 show that the given vectors $\mathbf{u}$ and $\mathbf{v}$ are orthogonal with respect to the standard inner product in the appropriate Euclidean n-space $\mathbb{R}^n$.

17. $\mathbf{u} = (1, -2), \mathbf{v} = (2, 1)$.
18. $\mathbf{u} = (1, -3, 1), \mathbf{v} = (1, 1, 2)$.
19. $\mathbf{u} = (2, 3, 1), \mathbf{v} = (-1, 1, -1)$.
20. $\mathbf{u} = (0, 14, -2), \mathbf{v} = (19, 1, 7)$.
21. $\mathbf{u} = (1, -1, 1, -1), \mathbf{v} = (1, 1, 2, 2)$.

In Exercises 22–25 show that the given vectors $\mathbf{u}$ and $\mathbf{v}$ are not orthogonal with respect to the standard inner product but are orthogonal with respect to the inner product $<\mathbf{u}, \mathbf{v}> = \mathbf{u}\,D\mathbf{v}^T$ in Euclidean 3-space $\mathbb{R}^3$, given that $D = diag\,[1, 2, 3]$.

22. $\mathbf{u} = (2, 1, -2), \mathbf{v} = (3, 0, 1)$.
23. $\mathbf{u} = (3, -3, 1), \mathbf{v} = (2, 1, 0)$.
24. $\mathbf{u} = (3, 3, 1), \mathbf{v} = (-4, 1, 2)$.
25. $\mathbf{u} = (0, 3, 1), \mathbf{v} = (19, 1, -2)$.

---

26. Find *all* of the vectors $\mathbf{x} = (x, y, z)$ in Euclidean 3-space $\mathbb{R}^3$ that are orthogonal to the vector $\mathbf{y} = (1, 1, 1)$ with respect to the standard inner product in $\mathbb{R}^3$.

27. Find *all* of the vectors $\mathbf{x} = (x, y, z)$ in the inner-product space $\mathbb{R}^3$ that are orthogonal to the vector $\mathbf{y} = (1, 1, 1)$ with respect to the inner product $<\mathbf{u}, \mathbf{v}> = \mathbf{u}\,D\mathbf{v}^T$, given that $D = diag\,[1, 2, 3]$.

28. Find *all* of the vectors $\mathbf{x} = (x, y, z)$ in the inner-product space $\mathbb{R}^3$ that are orthogonal to the vector $\mathbf{y} = (1, 1, 1)$ with respect to the inner product $<\mathbf{u}, \mathbf{v}> = \mathbf{u}\, D\mathbf{v}^T$, given the $D = \text{diag}\, [1, 1, 4]$.

29. Find a vector (matrix) that is orthogonal to the matrix $\begin{bmatrix} -1 & -3 \\ 5 & 8 \end{bmatrix}$ in the inner-product space $M_2(\mathbb{R})$ with the trace inner product of Example 3 of Section 4.3.

In Exercises 30–37 find the norm $\|p(x)\| = <p(x), p(x)>^{\frac{1}{2}}$ of each of the given vectors (polynomials) in the inner-product space P_n with the integral (from 0 to 1) inner product of Example 5 of Section 4.3.

30. $p(x) = 1$.

31. $p(x) = x$.

32. $p(x) = 1 + x$.

33. $p(x) = 1 - x^2$.

34. $p(x) = 2x^2 - 1$.

35. $p(x) = 4x^3 - 3x$.

36. $p(x) = \frac{1}{2}(5x^3 - 3x)$.

37. $p(x) = \frac{1}{8}(35x^4 - 30x^2 + 3)$.

---

38. Find a and b so that the polynomial $q(x) = ax + b$ is orthogonal to $p(x) = 1$ and $\|q(x)\| = 1$, in P_1 with the integral (from 0 to 1) inner product of Example 5 of Section 4.3.

39. Find a polynomial $q(x)$ that is orthogonal to $p(x) = x$ and has unit norm, in P_2 with the integral (from 0 to 1) inner product of Example 5 of Section 4.3.

40. Find a polynomial $q(x)$ that is orthogonal to $p(x) = x^2$ and has unit norm, in P_2 with the integral (from 0 to 1) inner product of Example 5 of Section 4.3.

41. Verify that the two vectors $\mathbf{u}$ and $\mathbf{v}$ of Exercise 17 above satisfy the Cauchy-Schwarz inequality (the inner product of Exercise 17).

42. Verify that the two vectors $\mathbf{u}$ and $\mathbf{v}$ of Exercise 18 above satisfy the Cauchy-Schwarz inequality (the inner product of Exercise 18).

43. Verify that the two vectors (polynomials) of Exercises 33 and 34 above satisfy the Cauchy-Schwarz inequality (the integral inner product of those exercises).

44. Verify that the two vectors (polynomials) of Exercises 35 and 36 above satisfy the Cauchy-Schwarz inequality. (Use the integral inner product of those exercises.)

45. Show that the six Legendre polynomials, $P_0(x)$, $P_1(x), \ldots, P_5(x)$, given in Example 9 are indeed orthogonal.

46. Show that the four Tchebycheff polynomials given in Example 9 are indeed orthogonal.

47. Verify that the basis $E = \{(1, 0, 0, 0), (0, 1, 0, 0), (0, 0, 1, 0), (0, 0, 0, 1)\}$ is an orthonormal basis for $\mathbb{R}^4$ under the standard inner product.

48. Prove that the standard basis $E = \{\mathbf{e}_1, \mathbf{e}_2, \ldots, \mathbf{e}_n\}$ is an *orthonormal basis* for $\mathbb{R}^n$ (with the standard inner product), for every positive integer n.

49. Prove that the reordered basis

$$E' = \{\mathbf{e}_{i_1}, \mathbf{e}_{i_2}, \ldots, \mathbf{e}_{i_n}\}$$

(where the numbers $i_1, i_2, \ldots, i_n$ are the result of any permutation of the numbers $1, 2, \ldots, n$) is also an *orthonormal basis* for $\mathbb{R}^n$.

50. Let V be an arbitrary real inner-product space. Suppose that the vector $\mathbf{v} \in V$ has the property that $<\mathbf{v}, \mathbf{x}> = 0$ for every vector $\mathbf{x} \in V$. Prove that $\mathbf{v} = \mathbf{0}$.

51. Let V be an arbitrary real inner-product space. Suppose that the vector $\mathbf{v} \in V$ is a fixed vector in V. Show that the collection of all vectors $\mathbf{x} \in V$ that satisfy the property $<\mathbf{v}, \mathbf{x}> = 0$ is a subspace of V. It is called the **orthogonal complement** of $L(\mathbf{v})$. Interpret this geometrically in $\mathbb{R}^2$ and $\mathbb{R}^3$.

52. Interpret the triangle inequality in $\mathbb{R}^3$ for the vectors of Exercise 23.

53. Interpret the triangle inequality in $\mathbb{R}^3$ for the vectors of Exercise 25.

54. Interpret the triangle inequality in P_1 for the two vectors of Exercises 31 and 32.

55. Interpret the triangle inequality in P_2 for the two vectors of Exercises 33 and 34.

4.5 THE GRAM-SCHMIDT PROCESS

In this section we describe a method known as the *Gram-Schmidt process,* by means of which one can construct an orthonormal basis for any finite-dimensional inner-product space. This process gives a constructive proof for the fact that any finite-dimensional inner-product space has an orthonormal basis. We also see that once our orthonormal basis is established the given inner product takes on Euclidean-like characteristics.

Recall that if V is a real inner-product space, then a set of vectors $B = \{v_1, v_2, \ldots, v_n\}$ is an *orthonormal basis* for V provided that it is a basis whose elements have *both* of the following two properties.

1. Every one of the vectors $v_1, v_2, \ldots, v_n$ is a *unit* vector in V; that is, $\|v_i\| = <v_i, v_i>^{1/2} = 1$, for each $i = 1, 2, \ldots, n$.
2. Any two distinct vectors v_i and v_j from B are mutually orthogonal; that is, $<v_i, v_j> = 0$, whenever $i \neq j$, $i, j = 1, 2, \ldots, n$.

The usefulness of an orthonormal basis for a given inner-product space V becomes apparent from the following calculations. Suppose that $S = \{w_1, w_2, \ldots, w_n\}$ is *any* arbitrary basis for V. Let u and v be any two vectors from V. Since S is a basis, there exist unique scalars so that

$$u = a_1 w_1 + a_2 w_2 + \cdots + a_n w_n,$$

and

$$v = b_1 w_1 + b_2 w_2 + \cdots + b_n w_n.$$

Now, using the properties of an inner product we have the following messy calculation.

$$
\begin{aligned}
<u, v> &= <a_1 w_1 + a_2 w_2 + \cdots + a_n w_n, b_1 w_1 + b_2 w_2 + \cdots + b_n w_n> \\
&= a_1 b_1 <w_1, w_1> + a_1 b_2 <w_1, w_2> \\
&\quad + \cdots + a_1 b_n <w_1, w_n> \\
&\quad + a_2 b_1 <w_2, w_1> + a_2 b_2 <w_2, w_2> \\
&\quad + \cdots + a_2 b_n <w_2, w_n> \\
&\quad + \cdots + a_n b_1 <w_n, w_1> + a_n b_2 <w_n, w_2> \\
&\quad + \cdots + a_n b_n <w_n, w_n>.
\end{aligned}
\tag{4.35}
$$

Therefore, the value of $<u, v>$ for any two vectors in V is completely determined by the values of each of the inner products $<w_i, w_j>$ of the basis vectors, $i, j = 1, 2, \ldots, n$.

Example 1 The vectors $w_1 = (1, 1)$ and $w_2 = (1, -2)$ form a basis for Euclidean 2-space $\mathbb{R}^2$. (You can verify that.) Suppose that we have two vectors $u = (2, 3)$ and $v = (-1, 4)$. It takes but a little calculation to see that

$$u = \frac{7}{3} w_1 - \frac{1}{3} w_2 \quad \text{and} \quad v = \frac{2}{3} w_1 - \frac{5}{3} w_2.$$

Now, using the standard inner product in $\mathbb{R}^2$,

$$<\mathbf{u}, \mathbf{v}> = <\frac{7}{3}\mathbf{w}_1 - \frac{1}{3}\mathbf{w}_2, \frac{2}{3}\mathbf{w}_1 - \frac{5}{3}\mathbf{w}_2>$$

$$= \left(\frac{7}{3}\right)\left(\frac{2}{3}\right) <\mathbf{w}_1, \mathbf{w}_1> + \left(\frac{7}{3}\right)\left(-\frac{5}{3}\right) <\mathbf{w}_1, \mathbf{w}_2>$$

$$+ \left(\frac{2}{3}\right)\left(-\frac{1}{3}\right) <\mathbf{w}_2, \mathbf{w}_1> + \left(-\frac{1}{3}\right)\left(-\frac{5}{3}\right) <\mathbf{w}_2, \mathbf{w}_2>.$$

Since $<\mathbf{w}_1, \mathbf{w}_1> = 2$, $<\mathbf{w}_1, \mathbf{w}_2> = -1 = <\mathbf{w}_2, \mathbf{w}_1>$, and $<\mathbf{w}_2, \mathbf{w}_2> = 5$, this becomes

$$<\mathbf{u}, \mathbf{v}> = \left(\frac{7}{3}\right)\left(\frac{2}{3}\right)(2) + \left(\frac{7}{3}\right)\left(-\frac{5}{3}\right)(-1) + \left(\frac{2}{3}\right)\left(-\frac{1}{3}\right)(-1) + \left(-\frac{1}{3}\right)\left(-\frac{5}{3}\right)(5)$$

$$= \frac{28 + 35 + 2 + 25}{9} = \frac{90}{9} = 10. \quad \blacksquare$$

This very messy calculation (minimally messy because we have only dimension 2) is in marked contrast to the easy direct computation $<\mathbf{u}, \mathbf{v}> = (2)(-1) + (3)(4) = 10$. Actually the latter computation is based on the same principle as the former, except that here we have

$$\mathbf{u} = 2\mathbf{e}_1 + 3\mathbf{e}_2 \quad \text{and} \quad \mathbf{v} = -1\mathbf{e}_1 + 4\mathbf{e}_2.$$

And $\{\mathbf{e}_1 = (1, 0), \mathbf{e}_2 = (0, 1)\}$ is an *orthonormal basis* for $\mathbb{R}^2$.

If V is any real inner-product space and $B = \{\mathbf{w}_1, \mathbf{w}_2, \ldots, \mathbf{w}_n\}$ is an orthonormal basis for V, then since

$$<\mathbf{w}_i, \mathbf{w}_j> = \begin{cases} 0 & \text{if } i \neq j \\ 1 & \text{if } i = j \end{cases} \quad \textbf{(4.36)}$$

the messy equations (4.35) reduce to

$$<\mathbf{u}, \mathbf{v}> = a_1 b_1 + a_2 b_2 + \cdots + a_n b_n.$$

This resembles the standard inner product in $\mathbb{R}^n$, even though V may have a nonstandard inner product $<, >$ or may not be $\mathbb{R}^n$ at all. V was an *arbitrary* real inner-product space. What has happened is all of the mess has been absorbed by the orthonormal basis for V. We summarize this as the following theorem.

Theorem 4.8   Let $B = \{\mathbf{w}_1, \mathbf{w}_2, \ldots, \mathbf{w}_n\}$ be an orthonormal basis for the real inner-product space V. If $\mathbf{u} = a_1\mathbf{w}_1 + a_2\mathbf{w}_2 + \cdots + a_n\mathbf{w}_n$ and $\mathbf{v} = b_1\mathbf{w}_1 + b_2\mathbf{w}_2 + \cdots + b_n\mathbf{w}_n$ are any two vectors in V, then $<\mathbf{u}, \mathbf{v}> = a_1 b_1 + a_2 b_2 + \cdots + a_n b_n$.

A second theorem, with closely related results, tells us that the unique coefficients by which a given vector $\mathbf{u}$ is written as a linear combination of the vectors from an orthonormal basis for V are precisely the values of the inner products of $\mathbf{u}$ with the individual basis vectors. That is, if

$\mathbf{u} = a_1\mathbf{w}_1 + a_2\mathbf{w}_2 + \cdots + a_n\mathbf{w}_n$ as in the statement of Theorem 4.8, and $B = \{\mathbf{w}_1, \mathbf{w}_2, \ldots, \mathbf{w}_n\}$ is an *orthonormal* basis, then for each $i = 1, 2, \ldots, n$, $a_i = <\mathbf{u}, \mathbf{w}_i>$. You are asked to prove this as Exercise 25. The theorem is as follows.

Theorem 4.9   Let $B = \{\mathbf{w}_1, \mathbf{w}_2, \ldots, \mathbf{w}_n\}$ be an orthonormal basis for the real inner-product space V. If $\mathbf{u} = a_1\mathbf{w}_1 + a_2\mathbf{w}_2 + \cdots + a_n\mathbf{w}_n$ is any vector from V, then

$$\mathbf{u} = \sum_{i=1}^{n} <\mathbf{u}, \mathbf{w}_i> \mathbf{w}_i. \tag{4.37}$$

Example 2   Note that the set $\{\mathbf{v}_1, \mathbf{v}_2, \mathbf{v}_3\}$ given below is an orthonormal basis for $\mathbb{R}^3$ (verify that) with respect to the standard inner product.

$$\mathbf{v}_1 = \left(\frac{1}{\sqrt{2}}, 0, \frac{1}{\sqrt{2}}\right); \qquad \mathbf{v}_2 = \left(\frac{1}{\sqrt{3}}, \frac{1}{\sqrt{3}}, -\frac{1}{\sqrt{3}}\right); \qquad \mathbf{v}_3 = \left(-\frac{1}{\sqrt{6}}, \frac{2}{\sqrt{6}}, \frac{1}{\sqrt{6}}\right).$$

Then if $\mathbf{u} = (2, -1, 5)$ we can write $\mathbf{u}$ as a linear combination of $\mathbf{v}_1$, $\mathbf{v}_2$, and $\mathbf{v}_3$ by using Theorem 4.9. Since

$$<\mathbf{u}, \mathbf{v}_1> = \frac{7}{\sqrt{2}}, \qquad <\mathbf{u}, \mathbf{v}_2> = -\frac{4}{\sqrt{3}}, \qquad \text{and} \qquad <\mathbf{u}, \mathbf{v}_3> = \frac{1}{\sqrt{6}},$$

from (4.37) we get $$\mathbf{u} = \frac{7}{\sqrt{2}} \mathbf{v}_1 - \frac{4}{\sqrt{3}} \mathbf{v}_2 + \frac{1}{\sqrt{6}} \mathbf{v}_3.$$

Verify this by a direct calculation. ∎

Any finite-dimensional inner-product space must have a finite basis $S = \{\mathbf{s}_1, \mathbf{s}_2, \ldots, \mathbf{s}_n\}$, which could be arbitrary. We can, however, use the vectors in S to construct a new basis B for V which is an *orthonormal* basis. The process is named for the Danish actuary Jörgen Pederson Gram (1850–1916) and the German mathematician Erhardt Schmidt (1876–1959). This **Gram-Schmidt process** generalizes the idea of projecting a given vector in ordinary Euclidean space *along* a second given vector and then finding a new vector *orthogonal* to the second vector.

To see the main idea, let us first consider the case where we have two linearly independent vectors $\mathbf{u}$ and $\mathbf{v}$. They will span a two-dimensional subspace of V, so in Figure 4.16 we sketch a geometric view of Euclidean 2-space. Drop a perpendicular from the terminal point A of the vector $\mathbf{u}$ to the line along the vector $\mathbf{v}$. This determines the point B on that line. The vector from O to B is $\mathbf{w}_1$ and is called the **projection of u along v**, or the **projection of u on v**. Now the vector from B to A is $\mathbf{w}_2$ and is perpendicular to $\mathbf{w}_1$. This vector is called the **projection of u orthogonal to v**. Note first that

$$\mathbf{w}_2 = \mathbf{u} - \mathbf{w}_1 \tag{4.38}$$

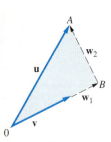

FIGURE 4.16

so $\mathbf{u} = \mathbf{w}_1 + \mathbf{w}_2$ is the sum of its two "components": its projections (1) *along (on)* $\mathbf{v}$, and (2) *orthogonal to* $\mathbf{v}$.

Let us find algebraic descriptions for these two projections (components of $\mathbf{u}$). First of all, it is evident that $\mathbf{w}_1$ is a scalar multiple of $\mathbf{v}$. That is, $\mathbf{w}_1 = a\mathbf{v}$, so that $\mathbf{u} = a\mathbf{v} + \mathbf{w}_2$. But $\mathbf{w}_2$ is orthogonal to $\mathbf{v}$, so $<\mathbf{w}_2, \mathbf{v}> = 0$; therefore,

$$<\mathbf{u}, \mathbf{v}> = <a\mathbf{v} + \mathbf{w}_2, \mathbf{v}> = a<\mathbf{v}, \mathbf{v}> + <\mathbf{w}_2, \mathbf{v}> = a<\mathbf{v}, \mathbf{v}>.$$

Since $\mathbf{v}$ is a nonzero vector, the number $<\mathbf{v}, \mathbf{v}>$ is positive, and we obtain $a = \dfrac{<\mathbf{u}, \mathbf{v}>}{<\mathbf{v}, \mathbf{v}>}$ so that

$$\mathbf{w}_1 = \text{Proj}_{\mathbf{v}}\, \mathbf{u} = \frac{<\mathbf{u}, \mathbf{v}>}{<\mathbf{v}, \mathbf{v}>}\mathbf{v}, \qquad (\mathbf{v} \neq \mathbf{0}). \qquad \textbf{(4.39)}$$

Now substitute this result (4.39) in equation (4.38) to get the formula for the **projection of u orthogonal to v:**

$$\mathbf{w}_2 = \mathbf{u} - \frac{<\mathbf{u}, \mathbf{v}>}{<\mathbf{v}, \mathbf{v}>}\mathbf{v}, \qquad (\mathbf{v} \neq \mathbf{0}). \qquad \textbf{(4.40)}$$

You can also verify that $\mathbf{w}_2$ is orthogonal to $\mathbf{v}$ by a direct calculation.

We remark here that, although we used a Euclidean geometry picture to illustrate and motivate this derivation, it is not necessary for the inner product to be the standard one. We take equations (4.39) and (4.40) as *definitions* of the two projections of $\mathbf{u}$ *along (on)* $\mathbf{v}$ and *orthogonal* to $\mathbf{v}$ for any inner product function on V.

Now let us simplify this process a little. If the vector $\mathbf{v}$ is a "unit vector," equations (4.39) and (4.40) can be simplified, since in that case $<\mathbf{v}, \mathbf{v}> = 1$. We state this formally as follows.

> **Definition**   Let V be a real inner-product space and let $\mathbf{u}$ and $\mathbf{v}$ be vectors in V. If $<\mathbf{v}, \mathbf{v}> = 1$ ($\mathbf{v}$ is a unit vector), then the projection of $\mathbf{u}$ along (or on) the unit vector $\mathbf{v}$ is
>
> $$\text{Proj}_{\mathbf{v}}\, \mathbf{u} = <\mathbf{u}, \mathbf{v}> \mathbf{v}, \qquad \textbf{(4.41)}$$
>
> and the projection of $\mathbf{u}$ orthogonal to $\mathbf{v}$ is
>
> $$\mathbf{u} - <\mathbf{u}, \mathbf{v}> \mathbf{v}.$$

Example 3   Let $\mathbf{u} = (1, 3, -2)$ and $\mathbf{v} = (1, 0, 1)$ be two vectors in Euclidean 3-space $\mathbb{R}^3$. Find the projections of $\mathbf{u}$ *along* and *orthogonal* to $\mathbf{v}$. Since $<\mathbf{v}, \mathbf{v}> = 2$, $\mathbf{v}$ is not a unit vector. We can use equations (4.39) and (4.40) directly, however, to obtain

$$\text{Proj}_{\mathbf{v}}\, \mathbf{u} = \frac{<\mathbf{u}, \mathbf{v}>}{<\mathbf{v}, \mathbf{v}>}\mathbf{v} = -\frac{1}{2}\,\mathbf{v} = -\frac{1}{2}(1, 0, 1) = \left(-\frac{1}{2}, 0, -\frac{1}{2}\right) = \mathbf{w}_1.$$

The projection of $\mathbf{u}$ *orthogonal* to $\mathbf{v}$ is

$$\mathbf{u} - \mathbf{w}_1 = (1, 3, -2) - \left(-\frac{1}{2}, 0, -\frac{1}{2}\right) = \left(\frac{3}{2}, 3, -\frac{3}{2}\right) = \mathbf{w}_2. \quad \blacksquare$$

Example 4   Construct an orthonormal basis for the linear span of the two vectors $\mathbf{u}$ and $\mathbf{v}$ of Example 3. Since the vectors $\mathbf{w}_1$ and $\mathbf{w}_2$ found in Example 3 are orthogonal, we can simply divide by their lengths to obtain the desired basis. An alternative order of computation, which is often more convenient for constructing an orthonormal basis for the linear span of two given vectors, is the following. We first "normalize" $\mathbf{v}$; that is, divide $\mathbf{v}$ by its norm, to obtain

$$\mathbf{v}_1 = \frac{\mathbf{v}}{||\mathbf{v}||} = \left(\frac{1}{\sqrt{2}}, 0, \frac{1}{\sqrt{2}}\right),$$

which is a unit vector in the same direction as $\mathbf{v}$. The projections of $\mathbf{u}$ along and orthogonal to $\mathbf{v}_1$ will be the same as the projections along and orthogonal to $\mathbf{v}$. They are

$$\text{Proj}_{\mathbf{v}_1}\mathbf{u} = <\mathbf{u}, \mathbf{v}_1> \mathbf{v}_1 = -\frac{1}{\sqrt{2}}\, \mathbf{v}_1 = \left(-\frac{1}{2}, 0, -\frac{1}{2}\right) = \mathbf{w}_1,$$

as above, and

$$\mathbf{w}_2 = \mathbf{u} - \mathbf{w}_1 = \left(\frac{3}{2}, 3, -\frac{3}{2}\right).$$

The vector

$$\mathbf{v}_2 = \frac{\mathbf{w}_2}{||\mathbf{w}_2||} = \frac{\mathbf{w}_2}{\sqrt{<\mathbf{w}_2, \mathbf{w}_2>}} = \frac{2}{3\sqrt{6}}\left(\frac{3}{2}, 3, -\frac{3}{2}\right) = \left(\frac{1}{\sqrt{6}}, \frac{2}{\sqrt{6}}, -\frac{1}{\sqrt{6}}\right)$$

is a unit vector that is orthogonal to $\mathbf{v}_1$, as you can readily verify. Therefore, the two vectors $\mathbf{v}_1$ and $\mathbf{v}_2$ form an orthonormal basis for the linear span, $\mathbf{L}(\{\mathbf{u}, \mathbf{v}\})$, of $\mathbf{u}$ and $\mathbf{v}$. $\blacksquare$

Let us recapitulate the method of Example 4. This gives the first steps in the Gram-Schmidt orthogonalization process.

Step 1   Select a nonzero vector $\mathbf{v}$, and divide $\mathbf{v}$ by its norm, $||\mathbf{v}|| = \sqrt{<\mathbf{v}, \mathbf{v}>}$ to obtain a unit vector $\mathbf{v}_1$ along $\mathbf{v}$.

Step 2   Take any other vector $\mathbf{u}$ that is linearly independent of $\mathbf{v}$, that is, is not a scalar multiple of $\mathbf{v}$. Then project $\mathbf{u}$ along (on) $\mathbf{v}_1$ to obtain the vector $\mathbf{w}_1$.

Step 3   Find the projection $\mathbf{w}_2 = \mathbf{u} - \mathbf{w}_1$ of $\mathbf{u}$ orthogonal to $\mathbf{v}_1$ (orthogonal to $\mathbf{v}$).

Step 4   Divide the orthogonal projection $\mathbf{w}_2$ by its norm $||\mathbf{w}_2|| = \sqrt{<\mathbf{w}_2, \mathbf{w}_2>}$ to obtain the unit vector $\mathbf{v}_2$ orthogonal to $\mathbf{v}_1$.

Thus the subspace $W = \mathbf{L}(\{\mathbf{u}, \mathbf{v}\})$ of V spanned by $\mathbf{u}$ and $\mathbf{v}$ has $\{\mathbf{v}_1, \mathbf{v}_2\}$ as an *orthonormal basis,* where

$$\mathbf{v}_1 = \frac{\mathbf{v}}{\|\mathbf{v}\|}, \quad \text{and} \quad \mathbf{v}_2 = \frac{\mathbf{u} - <\mathbf{u}, \mathbf{v}_1> \mathbf{v}_1}{\|\mathbf{u} - <\mathbf{u}, \mathbf{v}_1> \mathbf{v}_1\|}. \tag{4.42}$$

This process can be extended now to construct an orthonormal basis for any n-dimensional vector space V. We merely substitute for Steps 2 and 3 above the following two steps.

Step 2'   Select any vector $\mathbf{u}'$ that is linearly independent of the orthonormal set $\{\mathbf{v}_1, \mathbf{v}_2, \ldots, \mathbf{v}_k\}$; that is, $\mathbf{u}'$ is not in the subspace $W = \mathbf{L}(\{\mathbf{v}_1, \mathbf{v}_2, \ldots, \mathbf{v}_k\})$ spanned by this set. Then project $\mathbf{u}'$ on the subspace W to obtain the vector $\mathbf{w}_1$. Note that $\{\mathbf{v}_1, \mathbf{v}_2, \ldots, \mathbf{v}_k\}$ is an orthonormal basis for W.

Step 3'   Find the projection $\mathbf{w}_2 = \mathbf{u}' - \mathbf{w}_1$ of $\mathbf{u}'$ *orthogonal* to W (orthogonal to the linear span $\mathbf{L}(\{\mathbf{v}_1, \mathbf{v}_2, \ldots, \mathbf{v}_k\})$).

The projection of $\mathbf{u}'$ onto the subspace $W = \mathbf{L}(\{\mathbf{v}_1, \mathbf{v}_2, \ldots, \mathbf{v}_k\})$ is a vector $\mathbf{w}_1$ lying in W, and in fact

$$\mathbf{w}_1 = \sum_{i=1}^{k} <\mathbf{u}', \mathbf{v}_i> \mathbf{v}_i.$$

Figure 4.17 illustrates this when W is the two-dimensional subspace of Euclidean 3-space $\mathbb{R}^3$ spanned by the orthonormal basis $\{\mathbf{v}_1, \mathbf{v}_2\}$.

From this illustration, using Steps 1, 2', 3', and 4, we see that the third orthonormal basis vector is

$$\mathbf{v}_3 = \frac{\mathbf{u}' - <\mathbf{u}', \mathbf{v}_1> \mathbf{v}_1 - <\mathbf{u}', \mathbf{v}_2> \mathbf{v}_2}{\|\mathbf{u}' - <\mathbf{u}', \mathbf{v}_1> \mathbf{v}_1 - <\mathbf{u}', \mathbf{v}_2> \mathbf{v}_2\|}. \tag{4.43}$$

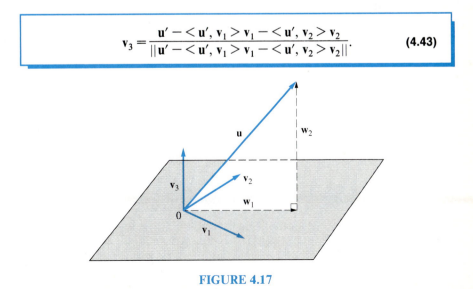

FIGURE 4.17

Example 5   Find a third basis vector, v_3, for $\mathbb{R}^3$ so that it together with the two orthonormal vectors v_1 and v_2 of Example 4 form an orthonormal basis for $\mathbb{R}^3$. Calculation shows that

$$W = \mathbf{L}\left(\left\{\left(\frac{1}{\sqrt{2}}, 0, \frac{1}{\sqrt{2}}\right), \left(\frac{1}{\sqrt{6}}, \frac{2}{\sqrt{6}}, -\frac{1}{\sqrt{6}}\right)\right\}\right)$$

is the set of all vectors (a, b, c) in $\mathbb{R}^2$ for which $a - b - c = 0$; i.e., the plane $x - y - z = 0$ in 3-space. We may select any vector u' that does not lie in W. For this example let $u' = e_1 = (1, 0, 0)$, a simple choice. Notice that although our u' is a "unit vector," it is not orthogonal to either v_1 or v_2. The projection of u' *onto* the subspace (plane) W is therefore

$$w_1 = <u', v_1> v_1 + <u', v_2> v_2$$

$$= \frac{1}{\sqrt{2}}\left(\frac{1}{\sqrt{2}}, 0, \frac{1}{\sqrt{2}}\right) + \frac{1}{\sqrt{6}}\left(\frac{1}{\sqrt{6}}, \frac{2}{\sqrt{6}}, -\frac{1}{\sqrt{6}}\right)$$

$$= \left(\frac{2}{3}, \frac{1}{3}, \frac{1}{3}\right).$$

So the projection of u' *orthogonal* to W is

$$w_2 = u' - w_1 = (1, 0, 0) - \left(\frac{2}{3}, \frac{1}{3}, \frac{1}{3}\right) = \left(\frac{1}{3}, -\frac{1}{3}, -\frac{1}{3}\right).$$

A simple calculation shows that w_2 is orthogonal to both v_1 and v_2. It is, however, not a unit vector. So we apply Step 4 to obtain the vector

$$v_3 = \frac{w_2}{\|w_2\|} = \frac{3}{\sqrt{3}} w_2 = \left(\frac{1}{\sqrt{3}}, -\frac{1}{\sqrt{3}}, -\frac{1}{\sqrt{3}}\right).$$

The desired orthonormal basis is therefore

$$B = \{v_1, v_2, v_3\} = \left\{\left(\frac{1}{\sqrt{2}}, 0, \frac{1}{\sqrt{2}}\right), \left(\frac{1}{\sqrt{6}}, \frac{2}{\sqrt{6}}, -\frac{1}{\sqrt{6}}\right), \left(\frac{1}{\sqrt{3}}, -\frac{1}{\sqrt{3}}, -\frac{1}{\sqrt{3}}\right)\right\}. \quad \blacksquare$$

Of course, a simpler method for Euclidean 3-space would be to select

$$w_2 = \frac{v_1 \times v_2}{\|v_1 \times v_2\|}.$$

However, the cross product will not generalize to other vector spaces or higher dimensions. The Gram-Schmidt method does generalize. We have the following theorem which inductively constructs an orthonormal basis for any finite-dimensional vector space V. Its proof essentially amounts to an application of the steps outlined.

> **Theorem 4.10**   Let V be a real inner-product space and let $S = \{v_1, v_2, \ldots, v_k\}$ be a set of mutually orthogonal unit vectors from V. Let W be the subspace of V spanned by the set S. Then any vector $\mathbf{u}$ in V can be written in the form $\mathbf{u} = \mathbf{w}_1 + \mathbf{w}_2$, where $\mathbf{w}_1$ is in W and $\mathbf{w}_2$ is orthogonal to the subspace W (i.e., to every vector in W). This is accomplished by setting
>
> $$\mathbf{w}_1 = <\mathbf{u}, \mathbf{v}_1> \mathbf{v}_1 + <\mathbf{u}, \mathbf{v}_2> \mathbf{v}_2 + \cdots + <\mathbf{u}, \mathbf{v}_k> \mathbf{v}_k,$$
>
> and
>
> $$\mathbf{w}_2 = \mathbf{u} - \mathbf{w}_1.$$
>
> If $\mathbf{w}_2 \neq \mathbf{0}$, let $\mathbf{v}_{k+1} = \dfrac{\mathbf{w}_2}{||\mathbf{w}_2||}$. Then the set $\{\mathbf{v}_1, \mathbf{v}_2, \ldots, \mathbf{v}_k, \mathbf{v}_{k+1}\}$ is orthonormal.

Proof   Clearly $\mathbf{w}_1$ is in $\mathbf{L}(S)$. For each $i = 1, 2, \ldots, k$,

$$<\mathbf{w}_2, \mathbf{v}_i> = <\mathbf{u} - \mathbf{w}_1, \mathbf{v}_i>$$
$$= <\mathbf{u}, \mathbf{v}_i> - <\mathbf{w}_1, \mathbf{v}_i> = <\mathbf{u}, \mathbf{v}_i> - \sum_{j=1}^{k} <\mathbf{u}, \mathbf{v}_j> <\mathbf{v}_j, \mathbf{v}_i>$$
$$= <\mathbf{u}, \mathbf{v}_i> - <\mathbf{u}, \mathbf{v}_i> = 0.$$

This is because the set S is an orthonormal set, so $<\mathbf{v}_i, \mathbf{v}_j> = 0$ when $i \neq j$. Therefore, $\mathbf{w}_2$ is orthogonal to every vector in $\mathbf{L}(S) = W$, as desired. The rest is an immediate consequence of our earlier remarks.   ◻

Throughout our discussion we have tacitly assumed that we were constructing an orthonormal basis for the vector space V. For an orthonormal set of vectors to be a basis for its linear span requires that the orthonormal set be linearly independent. It is; this is the claim of the following theorem, which we ask that you prove as Exercise 26.

> **Theorem 4.11**   If the set $S = \{v_1, v_2, \ldots, v_n\}$ is orthonormal, then it is linearly independent, and therefore is an orthonormal basis for $\mathbf{L}(S)$.

Example 6   Let $\mathbf{u} = (-1, 4)$ and $\mathbf{v} = (2, -3)$. Construct an orthonormal basis for $\mathbf{L}(\{\mathbf{u}, \mathbf{v}\})$ in the real inner-product space for which $<\mathbf{x}, \mathbf{y}> = x_1 y_1 + 3 x_2 y_2$. Let this basis include the appropriate scalar multiple of $\mathbf{v}$.

In this inner-product space $||\mathbf{v}|| = \sqrt{(2)^2 + 3(-3)^2} = \sqrt{31}$. The first basis vector is

$$\mathbf{v}_1 = \frac{\mathbf{v}}{||\mathbf{v}||} = \frac{1}{\sqrt{31}} (2, -3) = \left(\frac{2}{\sqrt{31}}, \frac{-3}{\sqrt{31}} \right).$$

Check to see that

$$<\mathbf{v}_1, \mathbf{v}_1> = \left(\frac{2}{\sqrt{31}}\right)^2 + 3\left(\frac{-3}{\sqrt{31}}\right)^2 = 1.$$

Now project $\mathbf{u}$ on $\mathbf{v}_1$. Thus

$$\mathbf{w}_1 = \text{Proj}_{\mathbf{v}_1}\mathbf{u} = <\mathbf{u}, \mathbf{v}_1> \mathbf{v}_1 = \left[(-1)\frac{2}{\sqrt{31}} + 3(4)\frac{-3}{\sqrt{31}}\right]\mathbf{v}_1 = \frac{-38}{\sqrt{31}}\mathbf{v}_1;$$

$$\mathbf{w}_1 = \left(-\frac{76}{31}, \frac{114}{31}\right).$$

So the projection of $\mathbf{u}$ *orthogonal* to $\mathbf{v}_1$ is

$$\mathbf{w}_2 = \mathbf{u} - \mathbf{w}_1 = \left(\frac{45}{31}, \frac{10}{31}\right).$$

Check to see that

$$<\mathbf{w}_2, \mathbf{v}_1> = \left(\frac{2}{\sqrt{31}}\right)\left(\frac{45}{31}\right) + 3\left(\frac{-3}{\sqrt{31}}\right)\left(\frac{10}{31}\right) = 0.$$

So $\mathbf{w}_2$ and $\mathbf{v}_1$ are indeed orthogonal. Finally we "normalize" $\mathbf{w}_2$, to obtain the vector

$$\mathbf{v}_2 = \frac{\mathbf{w}_2}{\|\mathbf{w}_2\|} = \frac{31}{5\sqrt{93}}\mathbf{w}_2 = \left(\frac{9}{\sqrt{93}}, \frac{2}{\sqrt{93}}\right).$$

The orthonormal basis for this two-dimensional inner-product space is $B = \{\mathbf{v}_1, \mathbf{v}_2\}$; that is,

$$B = \left\{\left(\frac{2}{\sqrt{31}}, \frac{-3}{\sqrt{31}}\right), \left(\frac{9}{\sqrt{93}}, \frac{2}{\sqrt{93}}\right)\right\}. \quad \blacksquare$$

Example 7   Write the vector $\mathbf{p} = (2, 3)$ in the inner-product space of Example 6 as a linear combination of the orthonormal basis vectors $\mathbf{v}_1$ and $\mathbf{v}_2$ found in that example.

From Theorem 4.9 (page 260) we have

$$\mathbf{p} = <\mathbf{p}, \mathbf{v}_1> \mathbf{v}_1 + <\mathbf{p}, \mathbf{v}_2> \mathbf{v}_2.$$

So

$$\mathbf{p} = \frac{-23}{\sqrt{31}}\mathbf{v}_1 + \frac{36}{\sqrt{93}}\mathbf{v}_2. \quad \blacksquare$$

As we learn in the next chapter, the vector $\mathbf{p}$ can be represented by these two components as the column vector

$$\left[\frac{-23}{\sqrt{31}}, \frac{36}{\sqrt{93}}\right]^T$$

in terms of this basis.

Example 8   (For students who have studied calculus) Using the integral (from 0 to 1) inner product in the space P_2 of polynomials of degree two or less, find an orthonormal basis that includes the appropriate scalar multiple of the polynomial x.

We already noticed in the exercises of Section 4.4 that $p(x) = x$ is not a "unit vector"; indeed

$$||x||^2 = \; < x, x > \; = \int_0^1 p(x)^2 \, dx = \int_0^1 x^2 \, dx = \left[\frac{x^3}{3} \right]_0^1 = \frac{1}{3}.$$

So the desired first basis vector is the vector $p_1(x) = x\sqrt{3}$. Now select any other vector in P_2 that is not a scalar multiple of this vector, say, $q(x) = 1$. Project $q(x)$ *along* and *orthogonal* to $p_1(x)$ as before.

$$w_1(x) = \; < q(x), p_1(x) > p_1(x) = \left(\int_0^1 q(x) p_1(x) \, dx \right) p_1(x)$$

$$= \left(\int_0^1 (1 x \sqrt{3}) \, dx \right) p_1(x) = \left(\sqrt{3} \int_0^1 x \, dx \right) p_1(x)$$

$$= \left(\frac{\sqrt{3}}{2} \right) (x \sqrt{3}) = \frac{3x}{2}$$

is the projection *along* $p_1(x)$. So

$$w_2(x) = 1 - \frac{3x}{2} = \frac{2 - 3x}{2}$$

is the projection *orthogonal* to $p_1(x)$. The desired second basis vector is

$$p_2(x) = \frac{w_2(x)}{||w_2(x)||}.$$

Since

$$||w_2(x)||^2 = \; < w_2(x), w_2(x) > \; = \int_0^1 \left(\frac{2 - 3x}{2} \right)^2 \, dx = \frac{1}{4},$$

$$p_2(x) = 2 \left(\frac{2 - 3x}{2} \right) = 2 - 3x.$$

Check to see that $< p_1(x), p_2(x) > \; = 0$ under this integral inner product, and that both $< p_1(x), p_1(x) > \; = \; < p_2(x), p_2(x) > \; = 1$, so we have an orthonormal basis for a two-dimensional subspace $W = \mathbf{L}(\{p_1(x), p_2(x)\})$ of P_2. ∎

Now to complete the basis to an orthonormal basis for P_2 itself, we select some polynomial from P_2 which is not in W. It should be clear that $h(x) = x^2$ will be such a polynomial. We project $h(x)$ *on* the subspace W to obtain the polynomial

$$\underline{w}_1(x) = \; < h(x), p_1(x) > p_1(x) + \; < h(x), p_2(x) > p_2(x).$$

Thus,

$$\underline{w}_1(x) = \left(\int_0^1 x^2 x\sqrt{3}\, dx \right)(x\sqrt{3}) + \left(\int_0^1 x^2\,(2-3x)\, dx \right)(2-3x)$$

$$= \frac{\sqrt{3}}{4}(x\sqrt{3}) + \left(\frac{-1}{12} \right)(2-3x) = x - \frac{1}{6}.$$

And the projection of $h(x)$ *orthogonal* to W is the vector (polynomial) $\underline{w}_2(x) = h(x) - \underline{w}_1(x)$;

$$\underline{w}_2(x) = x^2 - x + \frac{1}{6}.$$

The third vector in our orthonormal basis for P_2 is then

$$p_3(x) = \frac{\underline{w}_2(x)}{\|\underline{w}_2(x)\|} = 6\sqrt{5}\left(x^2 - x + \frac{1}{6} \right);$$

$$p_3(x) = 6\sqrt{5}x^2 - 6\sqrt{5}x + \sqrt{5}.$$

Vocabulary

In this section we have again met new concepts. The vocabulary developed here is used not only throughout the rest of the text but also in those applications of linear algebra that involve orthogonality. Be sure that you understand each of the concepts described by the terms in the following list. Review these ideas until you feel secure enough to use them further. Review also each of the steps in the Gram-Schmidt process.

orthonormal basis   orthonormal set
projection along (on) a vector   projection orthogonal to a vector
projection on a subspace   projection orthogonal to a subspace

Exercises 4.5

In Exercises 1–8 convert each of the given bases into an orthonormal basis with respect to the standard Euclidean inner product on the given space.

1. $\{(1, 1), (1, -1)\}$ in $\mathbb{R}^2$.   **2.** $\{(1, 1), (2, -3)\}$ in $\mathbb{R}^2$.

3. $\{(0, 3), (2, -1)\}$ in $\mathbb{R}^2$.   **4.** $\{(1, -2), (1, -3)\}$ in $\mathbb{R}^2$.

5. $\{(0, 1, 1), (1, -1, 0), (1, 0, -1)\}$ in $\mathbb{R}^3$.   **6.** $\{(1, 0, 1), (1, 1, 0), (0, 1, 1)\}$ in $\mathbb{R}^3$.

7. $\{(1, 1, 1), (1, -1, -1), (1, 3, 1)\}$ in $\mathbb{R}^3$.

8. $\{(1, 0, 1, 0), (1, 1, 0, 0), (0, 1, 1, 0), (0, 0, 1, 1)\}$ in $\mathbb{R}^4$.

9. Let V be Cartesian 2-space $\mathbb{R}^2$ with the inner product defined by

$$<\mathbf{x}, \mathbf{y}> = <(x_1, x_2), (y_1, y_2)>$$
$$= \frac{1}{2} x_1 y_1 + \frac{1}{3} x_2 y_2.$$

Convert the basis given in Exercise 1 into an orthonormal basis using this inner product.

10. Use the inner product of Exercise 9 to convert the basis given in Exercise 2 into an orthonormal basis.

11. Use the inner product of Exercise 9 to construct an orthonormal basis for $\mathbb{R}^2$ that includes the appropriate scalar multiple of the vector $\mathbf{u} = (2, -3)$.

12. Let V be Cartesian 3-space with the inner product defined by

$$<\mathbf{x}, \mathbf{y}> = <(x_1, x_2, x_3), (y_1, y_2, y_3)>$$
$$= x_1 y_1 + 2x_2 y_2 + \frac{1}{3} x_3 y_3.$$

Convert the basis given in Exercise 5 into an orthonormal basis using this inner product.

13. Use the inner product of Exercise 12 to convert the basis given in Exercise 6 into an orthonormal basis.

14. Use the inner product of Exercise 12 to construct an orthonormal basis for $\mathbb{R}^3$ that includes the appropriate scalar multiple of the vector $\mathbf{u} = (1, 2, -3)$.

15. Show that the standard basis of $M_2(\mathbb{R})$ is an orthonormal basis with respect to the trace inner product of Example 3 in Section 4.3. (See Exercise 11 in Section 3.4.)

16. Construct an orthonormal basis for $M_2(\mathbb{R})$, using the trace inner product, starting from the basis of matrices

17. Show that the standard basis $\{x^3, x^2, x, 1\}$ for P_3 is an orthonormal basis when using the "Euclidean-like" inner product of Example 6 in Section 4.3.

18. In the inner-product space P_2, using the inner product of Exercise 17, construct an orthonormal

basis that includes the "unit polynomial" $\frac{1}{\sqrt{2}} (1 - x^2)$.

19. Is the basis $\{x^3, x^2, x, 1\}$ for P_3 an orthonormal basis when using the integral, from -1 to 1, inner product of Example 9 of Section 4.4?

20. Use the integral inner product, from 0 to 1, in P_2 to construct an orthonormal basis starting with the polynomial $p(x) = 1$. Compare this with Example 8.

21. Use the integral inner product, from 0 to 1, in P_2 to construct an orthonormal basis starting with the polynomial $p(x) = 1 + x$. HINT: This problem requires very careful computation. Do not lose track of the process in the computational morass.

22. Do the first four of the Legendre polynomials $\{P_0(x), P_1(x), P_2(x), P_3(x)\}$ given in Example 9 of Section 4.4 form an orthonormal basis for the vector space P_3 under the given integral inner product? If not, what needs to be done so that they will?

23. Let V be an arbitrary real inner-product space. Show that if $\mathbf{u}$ and $\mathbf{v}$ are both nonzero and *orthogonal* in V, then the set $\{\mathbf{u}, \mathbf{v}\}$ is *linearly independent*.

24. Let V be an arbitrary real inner-product space. Show that if $\mathbf{u}_1$, $\mathbf{u}_2$, and $\mathbf{u}_3$ are three nonzero vectors in V for which $<\mathbf{u}_i, \mathbf{u}_j> = 0$, when $i \neq j$, then the set $\{\mathbf{u}_1, \mathbf{u}_2, \mathbf{u}_3\}$ is *linearly independent*.

25. Prove Theorem 4.9 (page 260).

26. Prove Theorem 4.11 (page 265). HINT: Assume $\alpha_1 \mathbf{v}_1 + \alpha_2 \mathbf{v}_2 + \alpha_3 \mathbf{v}_3 + \cdots + \alpha_n \mathbf{v}_n = \mathbf{0}$ and consider the inner product

$$<\mathbf{0}, \mathbf{v}_i>$$
$$= <\alpha_1 \mathbf{v}_1 + \alpha_2 \mathbf{v}_2 = \alpha_3 \mathbf{v}_3 + \cdots + \alpha_n \mathbf{v}_n, \mathbf{v}_i>,$$

for $i = 1, 2, \ldots, n$.

27. Let W be the subspace of $C[-\infty, \infty]$ spanned by the two functions $f(x) = \sin x$ and $g(x) = \cos x$. Use the integral inner product from $-\pi$ to π of Example 7 of Section 4.4 to find an orthonormal basis for $\mathbf{L}(\{\sin x, \cos x\})$.

28. Let W be the subspace of $C[-1, \infty]$ spanned by the two functions $f(x) = e^x$ and $g(x) = e^{-x}$. Use the integral inner product, from -1 to 1, of Ex-

ample 4 of Section 4.3 to find an orthonormal basis for $\mathbf{L}(\{e^x, e^{-x}\})$.

29. Prove the following "Pythagorean theorem" in any finite-dimensional real inner-product space V: Let W be a subspace of V and let $\mathbf{u}$ be any nonzero vector in V. If $\mathbf{w}_1$ is the projection of $\mathbf{u}$ on W and $\mathbf{w}_2$ is the projection of $\mathbf{u}$ orthogonal to W (see Theorem 4.10, page 265), then

$$\|\mathbf{u}\|^2 = \|\mathbf{w}_1\|^2 + \|\mathbf{w}_2\|^2.$$

30. Prove the following "norm approximation theorem" in any finite-dimensional real inner-product space V: Let W be a subspace of V and let $\mathbf{u}$ be any nonzero vector in V. If $\mathbf{w}_1$ is the projection of $\mathbf{u}$ on W and $\mathbf{v}$ is any vector in W, then

$$\|\mathbf{u} - \mathbf{v}\| \geq \|\mathbf{u} - \mathbf{w}_1\|.$$

31. Some people prefer to first obtain an orthogonal (rather than an orthonormal) set of vectors in using the Gram-Schmidt process for constructing an orthonormal basis. That is, they omit steps 3 and 3' and then normalize all the vectors at the end by dividing by their lengths. Write modified formulas (4.43) and (4.42) appropriate for this alternate method.

32. Use the method of Exercise 31 to convert the basis $G = \{(1, 2, 3), (1, 0, -1), (2, -2, 1)\}$ to an orthonormal basis for $\mathbb{R}^3$.

4.6 COMPLEX VECTORS, MATRICES, INNER PRODUCTS

Let V be a vector space over the complex numbers $\mathbb{C}$. If V is finite-dimensional, then in Chapter 5 we demonstrate that we may consider V to be essentially $\mathbb{C}^n$. For that reason, in this section, we mostly restrict our attention to the space $\mathbb{C}^n$.

The *standard basis* for $\mathbb{C}^n$, as a complex vector space, is the familiar standard basis

$$E = \{\mathbf{e}_1 = (1, 0, \ldots, 0), \mathbf{e}_2 = (0, 1, 0, \ldots, 0), \ldots, \mathbf{e}_n = (0, 0, \ldots, 1)\}.$$

We merely have a new and larger set of scalars, the complex numbers $\mathbb{C}$.

Example 1 In $\mathbb{C}^3$ if $\mathbf{u} = (1, 1 + i, i)$, then $\mathbf{u} = \mathbf{e}_1 + (1 + i)\mathbf{e}_2 + i\mathbf{e}_3$, and in $\mathbb{C}^5$, if $\mathbf{u} = (2, i, 1 + i, 1 - i, -3)$, then $\mathbf{u} = 2\mathbf{e}_1 + i\mathbf{e}_2 + (1 + i)\mathbf{e}_3 + (1 - i)\mathbf{e}_4 - 3\mathbf{e}_5$. ∎

Let $A = (z_{rs})$ be a matrix with complex number entries. Each element z_{rs} in A is a complex number $z_{rs} = x_{rs} + iy_{rs}$. See Appendix A.1. The complex matrix A can always be written in the form $A = B + iC$, where the matrices $B = (x_{rs})$ and $C = (y_{rs})$ are real matrices.

By $\bar{A}$ we mean the matrix all of whose entries are the complex conjugates $\bar{z}_{rs} = x_{rs} - iy_{rs}$ of the complex number entries $z_{rs} = x_{rs} + iy_{rs}$ in the matrix A.

Example 2 Suppose that

$$A = \begin{bmatrix} 1 & -i & 1 + i \\ 0 & 2 - i & 1 \\ -i & 1 + 3i & 2 - 5i \end{bmatrix}.$$

Then $A = B + iC$, where

$$B = \begin{bmatrix} 1 & 0 & 1 \\ 0 & 2 & 1 \\ 0 & 1 & 2 \end{bmatrix} \quad \text{and} \quad C = \begin{bmatrix} 0 & -1 & 1 \\ 0 & -1 & 0 \\ -1 & 3 & -5 \end{bmatrix};$$

and

$$\overline{A} = \begin{bmatrix} 1 & i & 1-i \\ 0 & 2+i & 1 \\ i & 1-3i & 2+5i \end{bmatrix} = B - iC, \qquad \text{same } B \text{ and } C. \quad \blacksquare$$

We now want to define an important matrix obtained from a given complex $m \times n$ matrix A. This matrix, denoted by A^*, is the transpose of $\overline{A}$; that is, A^* is the **conjugate transpose** of A and is most often called the **Hermitian adjoint** of A:

$$A^* = (\overline{A})^T.$$

For the matrix A of Example 2

$$A^* = \begin{bmatrix} 1 & 0 & i \\ i & 2+i & 1-3i \\ 1-i & 1 & 2+5i \end{bmatrix}.$$

It is left as an exercise for you to prove that, for every $m \times n$ complex matrix A, $A^* = (\overline{A^T})$ as well, since $(\overline{A})^T = (\overline{A^T})$. (See Exercise 37.)

Caution A^*, the Hermitian **adjoint** of a matrix A, is **not the same** as the classical adjoint, adj (A) discussed in Section 2.4. For this reason we prefer to call adj (A), the **adjugate** of A rather than the adjoint.

You may wish to contrast A^* of Example 2 with adj (A) for the given matrix A. Refer also to Example 1. For the vector **u** given there, considered as a row $(1 \times n)$ matrix, we have

$$\mathbf{u}^* = (1, 1+i, i)^* = \begin{bmatrix} 1 \\ 1-i \\ -i \end{bmatrix}.$$

Of course, since **u** is not square, it has no classical adjoint = adjugate.

It is fairly easy to see that the following properties hold for the Hermitian adjoint of any $m \times n$ complex matrix A; prove them as Exercises 38–40. In doing so, you want to remember that $\overline{AB} = (\overline{A}) \cdot (\overline{B})$, not $(\overline{B}) \cdot (\overline{A})$.

Property (1) $(A^*)^* = A.$

Property (2) $(AB)^* = B^*A^*.$ **(4.44)**

Property (3) $(A + B)^* = A^* + B^*.$

Now to the topic of this particular section, a (complex) **inner product** on any complex vector space V is a function that assigns to each pair of vectors **u** and **v** in V a *complex* number $<$ **u**, **v** $>$ in $\mathbb{C}$ and which has the following properties.

Property (1) $<$ **u**, **v** $> = \overline{<$ **v**, **u** $>}$.

Property (2) $<$ α**u** $+ \beta$**v**, **w** $> = \alpha <$ **u**, **w** $> + \beta <$ **v**, **w** $>$ for all complex scalars α and β.

Property (3) $<$ **u**, **u** $> \geq 0$; with equality if, and only if, **u** $= \mathbf{0}$.

Note that $<$ **u**, **u** $>$ is a nonnegative *real number*.

In the complex vector space $\mathbb{C}^n$, the *standard inner product* is defined as follows.

Definition   For two complex vectors **u** and **v**, from $\mathbb{C}^n$, with **u** $= (u_1, u_2, \ldots, u_n)$ and **v** $= (v_1, v_2, \ldots, v_n)$,

$$< \mathbf{u}, \mathbf{v} > = \sum_{i=1}^{n} u_i \cdot \overline{v_i}. \tag{4.45}$$

The complex vector space $\mathbb{C}^n$ on which any complex inner product is defined is often called a **unitary space.** If we consider row vectors in $\mathbb{C}^n$ to be $1 \times n$ matrices, the *standard* inner product can be expressed as a matrix product

$$< \mathbf{u}, \mathbf{v} > = \mathbf{u} \, \mathbf{v}^*.$$

Remember that the real inner-product space of n-tuple row vectors $\mathbb{R}^n$ is frequently called a *Euclidean space.* In such a space we have $<$ **u**, **v** $> = \mathbf{u} \, \mathbf{v}^T$. Take particular care to notice, however, that the usual "Euclidean product" or "standard inner product" of row vectors from $\mathbb{R}^n$,

$$\mathbf{u} \, \mathbf{v}^T = \sum_{i=1}^{n} u_i \cdot v_i$$

is *not* an inner product in a *complex* vector space. Just consider the following example.

Example 3   In $\mathbb{C}^2$ let **u** $= (1, i)$. Then

$$\mathbf{u} \, \mathbf{u}^T = (1, i) \begin{bmatrix} 1 \\ i \end{bmatrix} = (1)(1) + (i)(i) = 1 - 1 = 0.$$

This violates the nonnegative property (3), of the definition of an inner product. (See also Section 4.3.) As we will soon see, if this were an inner product in $\mathbb{C}^2$ we would, so to speak, have a *nonzero* vector **u** in $\mathbb{C}^2$ with zero "length" (norm). ■

The following example illustrates the concept of inner product in a *unitary* space.

Example 4 In the *unitary* space, $\mathbb{C}^3$, with the standard inner product, let $\mathbf{u} = (1, 1 + i, i)$ and $\mathbf{v} = (0, i, 2 + i)$. Then, by (4.45), $<\mathbf{u}, \mathbf{v}> = (1)(0) + (1 + i)(-i) + (i)(2 - i)$ $= 2 + i$. Notice also that

$$<\mathbf{v}, \mathbf{u}> = (0)(1) + (i)(1 - i) + (2 + i)(-i) = 2 - i = \overline{2 + i}.$$

Therefore,

$$\overline{<\mathbf{v}, \mathbf{u}>} = 2 + i = <\mathbf{u}, \mathbf{v}>,$$

verifying the first property of the definition above. ■

Now the **norm** $\|\mathbf{u}\|$ of a vector in a unitary space is defined in a similar way as in a real inner-product space; that is,

$$\|\mathbf{u}\| = <\mathbf{u}, \mathbf{u}>^{\frac{1}{2}} = (\mathbf{u}\,\mathbf{u}^*)^{\frac{1}{2}}.$$

Example 5 In the unitary space, $\mathbb{C}^3$ let $\mathbf{u} = (1, 1 + i, i)$. Then

$$\|\mathbf{u}\| = <\mathbf{u}, \mathbf{u}>^{\frac{1}{2}} = (1(1) + (1 + i)(1 - i) + (i)(-i))^{\frac{1}{2}} = (1 + 2 + 1)^{\frac{1}{2}} = 2.$$

> **Definition** In $\mathbb{C}^n$, if $\mathbf{u} = (u_1, u_2, \ldots, u_n)$, then $\|\mathbf{u}\|$ is the nonnegative real number
>
> $$\|\mathbf{u}\| = <\mathbf{u}, \mathbf{u}>^{\frac{1}{2}} = (\mathbf{u}\,\mathbf{u}^*)^{\frac{1}{2}} = \sqrt{u_1\,\overline{u_1} + u_2\,\overline{u_2} + \cdots + u_n\,\overline{u_n}}.$$

Notice that in the *one*-dimensional unitary space $\mathbb{C}$, $\|\mathbf{u}\|$ is the ordinary modulus (or absolute value) of the complex number **u**. ■

The standard inner product is not the only possible complex inner product that can be defined on $\mathbb{C}^n$, as you can see from the following example.

Example 6 Let A be the complex matrix $\begin{bmatrix} 2 & 0 \\ 0 & 4 \end{bmatrix}$. Notice that in this case $A^* = A$. The function $<\mathbf{u}, \mathbf{v}> = \mathbf{u}\,A\mathbf{v}^*$, for $\overline{\mathbf{u}}$ and $\overline{\mathbf{v}}$ in $\mathbb{C}^2$, is an inner product in $\mathbb{C}^2$.

That this is an inner product is verified by checking the properties one by one. First note that

$$\overline{<\mathbf{u}, \mathbf{v}>} = \overline{\mathbf{u}\,A\mathbf{v}^*} = \overline{\mathbf{u}}\,\overline{A}\,\overline{\mathbf{v}^*} = (\mathbf{u}^*)^T (A^*)^T \mathbf{v}^T = (\mathbf{v}A^*\,\mathbf{u}^*)^T$$
$$= (\mathbf{v}A\mathbf{u}^*)^T = (\mathbf{v}A\mathbf{u}^*) = <\mathbf{v}, \mathbf{u}>,$$

because $A = A^*$, and the transpose of the scalar $(\mathbf{v}A\mathbf{u}^*)$ is the scalar itself. So we have verified Property (1). For Property (2), write

$$<\alpha\mathbf{u} + \beta\mathbf{v}, \mathbf{w}> = (\alpha\mathbf{u} + \beta\mathbf{v})A\mathbf{w}^* = \alpha\mathbf{u}\,A\mathbf{w}^* + \beta\mathbf{v}\,A\mathbf{w}^*$$
$$= \alpha<\mathbf{u}, \mathbf{w}> + \beta<\mathbf{v}, \mathbf{w}>,$$

by the rules of matrix multiplication. Finally, if $\mathbf{u} = (u_1, u_2)$ is a vector in $\mathbb{C}^2$, then

$$<\mathbf{u}, \mathbf{u}> = \mathbf{u}\,A\mathbf{u}^* = 2|u_1|^2 + 4|u_2|^2,$$

since $u_1\overline{u_1} = |u_1|^2$ and $u_2\overline{u_2} = |u_2|^2$. This is always a nonnegative real number and can be zero if, and only if, both $u_1 = 0$ and $u_2 = 0$; and hence $\mathbf{u} = \mathbf{0}$. ∎

In Chapter 6 we see that this example can be generalized to allow a large class of matrices A that satisfy the relation $A^* = A$ and are appropriately nonnegative. Moreover, in Exercise 47 you will be asked to generalize this example to $\mathbb{C}^n$ and any real diagonal matrix D with positive entries. (See Example 2 in Section 4.3.)

Let us now examine the properties of a complex inner product a little more closely. We have the following theorem which extends similar results for real inner products. Contrast it with the appropriate theorems of the previous sections.

Theorem 4.12 Let V be any complex vector space on which an inner product is defined. Then

(i) $<\mathbf{u}, \alpha\mathbf{v} + \beta\mathbf{w}> = \overline{\alpha}<\mathbf{u}, \mathbf{v}> + \overline{\beta}<\mathbf{u}, \mathbf{w}>$.
(ii) $<\mathbf{u}, \mathbf{v}> = 0$ for every vector $\mathbf{v}$ in V if, and only if, $\mathbf{u} = \mathbf{0}$.
(iii) $<\mathbf{u}, <\mathbf{u}, \mathbf{v}>\mathbf{v}> = |<\mathbf{u}, \mathbf{v}>|^2$.

Proof This theorem follows directly from the definition of a complex inner product, so we leave the details of the proof to you as exercises. (See Exercises 41–43.) ❑

Orthogonal means the same in a complex inner-product space as it does in real inner-product space.

Definition $\mathbf{u}$ and $\mathbf{v}$ are orthogonal if, and only if, $<\mathbf{u}, \mathbf{v}> = 0$.

Therefore, the standard basis for $\mathbb{C}^n$ is an **orthonormal basis.** Other concepts described in the previous sections for real inner-product spaces will carry over to complex inner-product spaces as well. In particular, the Cauchy-Schwarz inequality (Theorem 4.7, page 249) holds in a complex inner-product space also.

Theorem 4.13 (Cauchy-Schwarz) Let V be any complex vector space on which an inner product is defined. Then

$$|<\mathbf{u}, \mathbf{v}>|^2 \le <\mathbf{u}, \mathbf{u}><\mathbf{v}, \mathbf{v}>,$$

for all $\mathbf{u}$ and $\mathbf{v}$ in V. Equality occurs if, and only if, $\mathbf{u}$ and $\mathbf{v}$ are linearly dependent over $\mathbb{C}$.

Proof This proof is similar to that for Theorem 4.7. Given $\mathbf{u}$ and $\mathbf{v}$ in V, if $\mathbf{v} = \mathbf{0}$, the assertion is trivially true [see Theorem 4.12(ii)], so suppose that $\mathbf{v} \ne \mathbf{0}$. Let t be any *real* number and consider the quadratic polynomial

$$p(t) \equiv <\mathbf{u}+t\mathbf{v}, \mathbf{u}+t\mathbf{v}> = <\mathbf{u}, \mathbf{u}>+t<\mathbf{v}, \mathbf{u}>+\bar{t}<\mathbf{u}, \mathbf{v}>+t\bar{t}<\mathbf{v}, \mathbf{v}>$$
$$= <\mathbf{u}, \mathbf{u}>+t<\mathbf{v}, \mathbf{u}>+t<\mathbf{u}, \mathbf{v}>+t^2<\mathbf{v}, \mathbf{v}>$$
$$= <\mathbf{u}, \mathbf{u}>+t(<\mathbf{v}, \mathbf{u}>+<\mathbf{u}, \mathbf{v}>)+t^2<\mathbf{v}, \mathbf{v}>$$
$$= <\mathbf{u}, \mathbf{u}>+2t\ \mathrm{Re}\ (<\mathbf{u}, \mathbf{v}>)+t^2<\mathbf{v}, \mathbf{v}>.$$

Here $\mathrm{Re}(<\mathbf{u}, \mathbf{v}>)$ denotes the real part of the complex number $<\mathbf{u}, \mathbf{v}>$.

This is a polynomial with real coefficients, and by Property (3) of the definition of an inner product, $p(t) \ge 0$, for every real number t. Since the graph of $y = p(t)$ lies *above* or *on* the t axis, roots of this $p(t)$ are either *nonreal* or else are *real and equal*. Hence its discriminant is either negative or zero. Thus,

$$(2\ \mathrm{Re}(<\mathbf{u}, \mathbf{v}>))^2 - 4(<\mathbf{u}, \mathbf{u}>)(<\mathbf{v}, \mathbf{v}>) \le 0,$$

or

$$4\ (\mathrm{Re}(<\mathbf{u}, \mathbf{v}>))^2 \le 4(<\mathbf{u}, \mathbf{u}>)(<\mathbf{v}, \mathbf{v}>).$$

Dividing by 4, we have

$$(\mathrm{Re}(<\mathbf{u}, \mathbf{v}>))^2 \le (<\mathbf{u}, \mathbf{u}>)(<\mathbf{v}, \mathbf{v}>).$$

This last inequality holds for any pair of vectors $\mathbf{u}$ and $\mathbf{v}$, so it is true when $\mathbf{v}$ is replaced by the vector $<\mathbf{u}, \mathbf{v}>\mathbf{v}$. The inequality then reads

$$(\mathrm{Re}(<\mathbf{u}, <\mathbf{u}, \mathbf{v}>\mathbf{v}>))^2 \le (<\mathbf{u}, \mathbf{u}>)(<<\mathbf{u}, \mathbf{v}>\mathbf{v}, <\mathbf{u}, \mathbf{v}>\mathbf{v}>),$$

or

$$(\mathrm{Re}(<\mathbf{u}, <\mathbf{u}, \mathbf{v}>\mathbf{v}>))^2 \le (<\mathbf{u}, \mathbf{u}>)(<\mathbf{v}, \mathbf{v}>)(<\mathbf{u}, \mathbf{v}>)(\overline{<\mathbf{u}, \mathbf{v}>})$$
$$= <\mathbf{u}, \mathbf{u}><\mathbf{v}, \mathbf{v}>|<\mathbf{u}, \mathbf{v}>|^2.$$

But from Theorem 4.12 (iii), $\mathrm{Re}(<\mathbf{u}, <\mathbf{u}, \mathbf{v}>\mathbf{v}>) = \mathrm{Re}|<\mathbf{u}, \mathbf{v}>|^2 = |<\mathbf{u}, \mathbf{v}>|^2$, so we have

$$(|<\mathbf{u}, \mathbf{v}>|^2)^2 \le <\mathbf{u}, \mathbf{u}><\mathbf{v}, \mathbf{v}>|<\mathbf{u}, \mathbf{v}>|^2.$$

If $|<\mathbf{u}, \mathbf{v}>| = 0$, the statement of the theorem is trivially true. Otherwise divide both sides of the last inequality by $|<\mathbf{u}, \mathbf{v}>|^2$ to obtain the inequality in the statement of the theorem. (Recall that $|<\mathbf{u}, \mathbf{v}>|^2$ is a *positive* real number in this case.)

For the equality case if $\mathbf{v} = \mathbf{0}$, equality clearly follows; moreover $\mathbf{u}$ and $\mathbf{v}$ are certainly linearly dependent. (See Theorem 4.12 (ii).) So suppose that $\mathbf{v} \neq \mathbf{0}$. In Exercise 54 we ask you to prove that for all vectors $\mathbf{u}$ and nonzero vectors $\mathbf{v}$ in V,

$$< \mathbf{u} - \frac{<\mathbf{u}, \mathbf{v}>}{<\mathbf{v}, \mathbf{v}>} \mathbf{v}, \mathbf{u} - \frac{<\mathbf{u}, \mathbf{v}>}{<\mathbf{v}, \mathbf{v}>} \mathbf{v}> = \frac{<\mathbf{u}, \mathbf{u}><\mathbf{v}, \mathbf{v}> - |<\mathbf{u}, \mathbf{v}>|^2}{<\mathbf{v}, \mathbf{v}>}. \qquad \textbf{(4.46)}$$

Therefore, $|<\mathbf{u}, \mathbf{v}>|^2 = <\mathbf{u}, \mathbf{u}><\mathbf{v}, \mathbf{v}>$ if, and only if,

$$\frac{<\mathbf{u}, \mathbf{u}><\mathbf{v}, \mathbf{v}> - |<\mathbf{u}, \mathbf{v}>|^2}{<\mathbf{v}, \mathbf{v}>} = 0.$$

Thus, from (4.46) equality holds if, and only if,

$$< \mathbf{u} - \frac{<\mathbf{u}, \mathbf{v}>}{<\mathbf{v}, \mathbf{v}>} \mathbf{v}, \mathbf{u} - \frac{<\mathbf{u}, \mathbf{v}>}{<\mathbf{v}, \mathbf{v}>} \mathbf{v}> = 0.$$

By Property (3) of the inner-product definition this is true if, and only if,

$$\mathbf{u} - \frac{<\mathbf{u}, \mathbf{v}>}{<\mathbf{v}, \mathbf{v}>} \mathbf{v} = \mathbf{0};$$

that is, the vectors $\mathbf{u}$ and $\mathbf{v}$ form a linearly dependent set. ◼

Corollary (Triangle Inequality) Let V be any complex inner-product space. Then for any $\mathbf{u}$ and $\mathbf{v}$ in V

$$||\mathbf{u} + \mathbf{v}|| \leq ||\mathbf{u}|| + ||\mathbf{v}||.$$

Proof Make the appropriate modification to the proof given in the real case and use the Cauchy-Schwarz inequality proved above. The details are left as Exercise 34. ◼

Orthogonal Complement

We conclude this section with a look at a particularly interesting subspace of either a complex (unitary), or a real inner-product space. Let V be a complex inner-product space.

> **Definition**   If W is any subset of V, the collection of vectors in V that are orthogonal to *every* vector in W is called the **orthogonal complement** of W, and is usually written $W^\perp$.

Example 7   Let $W = \{(1, i)\}$. Then the set $W^\perp = \{(z, w) : z, w \in \mathbb{C}$ and $\bar{z} + i\,\bar{w} = 0.\}$.

Thus, if $z = x + iy$ and $w = u + iv$, we have $\bar{z} = x - iy$ and $i\,\bar{w} = v + iu$, so $\bar{z} + i\,\bar{w} = (x + v) - i(y - u) = 0$. From this we conclude that $x + v = 0$ and $y - u = 0$. Therefore, $x = -v$ and $y = u$. Hence $w = u + iv = y - ix = -i(x + iy) = -iz$. So $W^\perp = \{(z, -iz) : z \in \mathbb{C}\}$. ■

Notice that if W is a subspace of V and $K = \{\mathbf{w}_1, \mathbf{w}_2, \ldots, \mathbf{w}_k\}$ is a basis for W, then every vector in $W^\perp$ is orthogonal to each of the basis vectors $\mathbf{w}_i$ because they are in W; so $K^\perp \supseteq W^\perp$. Conversely, since any vector in W is a linear combination of the basis vectors in K, any vector $\mathbf{v}$ in V that is in $K^\perp$ is orthogonal to all of the basis vectors, hence to every vector in W; so $W^\perp \supseteq K^\perp$. Thus we have equality of these two sets. We have proved part (i) of the following theorem.

> **Theorem 4.14**   Let V be a complex inner-product space, let W be a subspace of V, and let $W^\perp$ be the orthogonal complement of W. Then
>
> (i) If K is any basis for W, then $K^\perp = W^\perp$.
> (ii) $W^\perp$ is a subspace of V.
> (iii) $W \cap W^\perp = \{\mathbf{0}\}$.
> (iv) Every vector $\mathbf{x}$ in V can be written uniquely as $\mathbf{x} = \mathbf{w} + \mathbf{w}'$, where $\mathbf{w} \in W$ and $\mathbf{w}' \in W^\perp$.

Proof   For part (ii), let $\mathbf{a}$ and $\mathbf{b}$ be vectors in $W^\perp$ and let α and β be scalars (complex numbers). For any $\mathbf{w}$ in W, $<\alpha\mathbf{a} + \beta\mathbf{b}, \mathbf{w}> = \alpha<\mathbf{a}, \mathbf{w}> + \beta<\mathbf{b}, \mathbf{w}> = 0$. Therefore, the vector $\alpha\mathbf{a} + \beta\mathbf{b}$ is in $W^\perp$ whenever $\mathbf{a}$ and $\mathbf{b}$ are in $W^\perp$. Since $\mathbf{0} \in W^\perp$ (why?), $W^\perp$ is not empty. So $W^\perp$ is a subspace of V.

To prove part (iii), we suppose that $\mathbf{x} \in W \cap W^\perp$. Then $\mathbf{x}$ is orthogonal to itself, $<\mathbf{x}, \mathbf{x}> = 0$. But then $\mathbf{x} = \mathbf{0}$. (Why?)

For part (iv) first let $\mathbf{x} \in V$ be arbitrary and consult the proof of Theorem 4.10 in Section 4.5 (page 265). Let $\boldsymbol{\omega}$ be any vector in W. Project $\mathbf{x}$ onto $\boldsymbol{\omega}$, resulting in the vector $\mathbf{w}$ in W. The projection of $\mathbf{x}$ orthogonal to $\boldsymbol{\omega} = \mathbf{x} - \mathbf{w}$ is in $W^\perp$. Thus $\mathbf{x} = \mathbf{w} + (\mathbf{x} - \mathbf{w})$; $\mathbf{w} \in W$, $\mathbf{x} - \mathbf{w} \in W^\perp$. As an exercise, verify that the work done in Section 4.5 carries over for complex inner products.

For uniqueness suppose that $\mathbf{x} = \mathbf{w} + \mathbf{w}' = \mathbf{u} + \mathbf{u}'$, with $\mathbf{w}$ and $\mathbf{u}$ in W and $\mathbf{w}'$ and $\mathbf{u}'$ in $W^\perp$. Then $\mathbf{w} - \mathbf{u} = \mathbf{u}' - \mathbf{w}'$; but $\mathbf{w} - \mathbf{u} \in W$ whereas $\mathbf{u}' - \mathbf{w}' \in W^\perp$. (Why?) From part (iii) we conclude that both are $\mathbf{0}$, so $\mathbf{w} = \mathbf{u}$ and $\mathbf{w}' = \mathbf{u}'$, as required for the representation to be unique. ❑

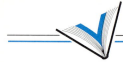

Vocabulary

In this section we have again met new concepts. If you are not familiar with the complex number system, study Appendix A.1 at the back of this text. Many of the concepts developed here will be used also in Section 6.6. Be sure that you understand each of them. Review these ideas until you feel secure enough to use them further.

standard basis for $\mathbb{C}^n$ conjugate
conjugate transpose   Hermitian adjoint
complex inner product   unitary space
orthogonal   orthogonal complement

Exercises 4.6

Find the conjugate transpose M^* for each of the complex matrices M in Exercises 1–6.

1. $M = \begin{bmatrix} 1 & 2i \\ -2i & 0 \end{bmatrix}$.
2. $M = \begin{bmatrix} 1-i & 2 \\ 0 & -2i \end{bmatrix}$.
3. $M = \begin{bmatrix} 2-i & 2 \\ -i & 1+i \end{bmatrix}$.

4. $M = \begin{bmatrix} 1 & 2i & 1+4i \\ 0 & 3i & 5 \\ 1-2i & 2 & 3+i \end{bmatrix}$.
5. $M = \begin{bmatrix} 1 & 2 & 4 \\ 0 & 3 & 5 \\ 1 & 2 & 3 \end{bmatrix}$.
6. $M = \begin{bmatrix} 2-i & 2 & 0 \\ -i & 1+i & 2-5i \\ 0 & 3+2i & 1-i \end{bmatrix}$.

Use the standard inner product in $\mathbb{C}^n$ to find $<\mathbf{u}, \mathbf{v}>$ for the following vectors $\mathbf{u}$ and $\mathbf{v}$.

7. $\mathbf{u} = (1, 2)$, $\mathbf{v} = (-1, 4)$.
8. $\mathbf{u} = (1, 2i)$, $\mathbf{v} = (-i, 4)$.

9. $\mathbf{u} = (i, 2i)$, $\mathbf{v} = (-i, 4i)$.
10. $\mathbf{u} = (1+i, 2i)$, $\mathbf{v} = (1-i, 1+4i)$.

11. $\mathbf{u} = (1-i, 1+2i)$, $\mathbf{v} = (1-3i, 2-3i)$.

12. $\mathbf{u} = (1-i, 1+2i, i)$, $\mathbf{v} = (1+i, 1-3i, 2-3i)$.

13. $\mathbf{u} = (1+3i, 1-2i, 2-i)$, $\mathbf{v} = (2+i, 1+3i, 2-5i)$.

Decide whether or not each of the following functions is an inner product in $\mathbb{C}^3$:

$\mathbf{u} = (u_1, u_2, u_3)$, $\mathbf{v} = (v_1, v_2, v_3)$, and $\lambda_1, \lambda_2,$ and λ_3 are arbitrary complex numbers.

14. $<\mathbf{u}, \mathbf{v}>_1 = u_1 v_1 + u_2 v_2 + i\, u_3 v_3$.
15. $<\mathbf{u}, \mathbf{v}>_2 = \lambda_1 u_1 \overline{v_1} + \lambda_2 u_2 \overline{v_2} + \lambda_3 u_3 \overline{v_3}$.

16. $<\mathbf{u}, \mathbf{v}>_3 = \overline{\lambda_1 u_1}\, v_1 + \overline{\lambda_2 u_2}\, v_2 + \overline{\lambda_3 u_3}\, v_3$.
17. $<\mathbf{u}, \mathbf{v}>_4 = |u_1||v_1| + |u_2||v_2| + |u_3||v_3|$.

18. $<\mathbf{u}, \mathbf{v}>_5 = |\lambda_1 u_1 v_1| + |\lambda_2 u_2 v_2| + |\lambda_3 u_3 v_3|$.

19. $<\mathbf{u}, \mathbf{v}>_6 = \mathbf{u}A\mathbf{v}^*$, where A is the 3×3 matrix $A = \begin{bmatrix} 1 & 0 & 0 \\ 0 & 3 & 0 \\ 0 & 0 & 5 \end{bmatrix}$.

20. Find $<\mathbf{u}, \mathbf{v}>_6$ (see Exercise 19) for the vectors $\mathbf{u}$ and $\mathbf{v}$ of Exercise 12.

21. Find $<\mathbf{u}, \mathbf{v}>_6$ (see Exercise 19) for the vectors $\mathbf{u}$ and $\mathbf{v}$ of Exercise 13.

22. Let $\lambda_1 = 2$, $\lambda_2 = -2$, and $\lambda_3 = 2i$. Find $<\mathbf{u}, \mathbf{v}>_2$ (see Exercise 15) for the vectors $\mathbf{u}$ and $\mathbf{v}$ of Exercise 12.

23. Let $\lambda_1 = 2$, $\lambda_2 = -2$, and $\lambda_3 = 2i$. Find $<\mathbf{u}, \mathbf{v}>_2$ (see Exercise 15) for the vectors $\mathbf{u}$ and $\mathbf{v}$ of Exercise 13.

For Exercises 24–30, find the standard norm $\|\mathbf{x}\|$ *for each of the vectors in Exercises 7–13.*

24. Exercise 7.   **25.** Exercise 8.   **26.** Exercise 9.   **27.** Exercise 10.

28. Exercise 11.   **29.** Exercise 12.   **30.** Exercise 13.

---

31. Using the inner product $<\mathbf{u},\ \mathbf{v}>_6$ (see Exercise 19), find $\|\mathbf{u}\|$ (the norm for that inner product) for the vector $\mathbf{u}$ of Exercise 12.

32. Using the inner product $<\mathbf{u},\ \mathbf{v}>_2$ of Exercise 15, find $\|\mathbf{u}\|$ (the norm for that inner product) for the vector $\mathbf{u}$ of Exercise 12.

33. Use the standard complex inner product in $\mathbb{C}^3$ to show that the vectors $\mathbf{u} = (1,\ i,\ -i)$ and $\mathbf{v} = (0, 1, 1)$ are orthogonal.

34. Write out the details of the proof of the triangle inequality (corollary to Theorem 4.13, page 275).

35. Prove or disprove that if $\lambda_1, \lambda_2, \ldots, \lambda_n$ are arbitrary complex numbers and $\mathbf{u}$ and $\mathbf{v}$ are vectors in $\mathbb{C}^n$, then $<\mathbf{u},\ \mathbf{v}>_\lambda = \lambda_1 u_1\ \overline{v_1} + \lambda_2 u_2\ \overline{v_2} + \cdots + \lambda_n u_n\ \overline{v_n}$ is an inner product in $\mathbb{C}^n$.

36. Prove or disprove that if A is an $n \times n$ complex matrix for which $A^* = A$, then the function $<\mathbf{u},\ \mathbf{v}>_A = \mathbf{u} A \mathbf{v}^*$ is an inner product in $\mathbb{C}^n$.

37. Prove that $\overline{A^T} = (\overline{A})^T$ for every complex matrix A.

38. Prove (4.44), Property (1). $(A^*)^* = A$.

39. Prove (4.44), Property (2). $(AB)^* = B^*A^*$.

40. Prove (4.44), Property (3). $(A + B)^* = A^* + B^*$.

41. Prove Theorem 4.12 (i), page 274.

42. Prove Theorem 4.12 (ii).

43. Prove Theorem 4.12 (iii).

44. Show that if V is an n-dimensional complex vector space and U is a subspace of V, then dim $U + $ dim $U^\perp = n$. (HINT: Pick a basis for U and extend to a basis for V with vectors that can be shown to form a basis for $U^\perp$.)

45. Let V be a complex vector space and let U and W be subspaces of V. We say that V is the **direct sum** of U and W, written $V = U \oplus W$ if, and only if, each vector $\mathbf{v}$ in V can be written as $\mathbf{v} = \mathbf{u} + \mathbf{w}$, for $\mathbf{u}$ in U and $\mathbf{w}$ in W, and $U \cap W = \{\mathbf{0}\}$. Use the last part of the proof of Theorem 4.14 (page 277) to prove that the representation of $\mathbf{v}$ as $\mathbf{u} + \mathbf{w}$ is unique.

46. Show that if V is a complex vector space and U is a subspace of V, then $V = U \oplus U^\perp$.

47. Prove the following theorem.

> **Theorem 4.15**   Let D be an $n \times n$ *real* diagonal matrix; $D = (d_{ij})$, with each $d_{ii} > 0$. Then the function $<\mathbf{u},\ \mathbf{v}> = \mathbf{u}\ D\mathbf{v}^*$ is an inner product on $\mathbb{C}^n$.

48. Use Theorem 4.15 and repeat Example 5 using the matrix $D = \text{diag}\ [2, 4, 6]$. That is, find the norm of the vector $\mathbf{u} = (1,\ 1 + i,\ i)$ with respect to this new inner-product function.

49. Use Theorem 4.15 and repeat Example 4 using the matrix $D = \text{diag}\ [2, 4, 6]$. That is, find $<\mathbf{u},\ \mathbf{v}>$ for $\mathbf{u} = (1,\ 1 + i,\ i)$ and $\mathbf{v} = (0,\ i,\ 2 + i)$ with respect to this new inner-product function.

50. Use Theorem 4.15 and repeat Example 5 using the matrix $D = \text{diag}\ [3, 1, 5]$. That is, find the norm of the vector $\mathbf{u} = (1,\ 1 + i,\ i)$ with respect to this new inner-product function.

51. Use Theorem 4.15 and repeat Example 4 using the matrix $D = \text{diag}\ [3, 1, 5]$. That is, find $<\mathbf{u},\ \mathbf{v}>$ for $\mathbf{u} = (1,\ 1 + i,\ i)$ and $\mathbf{v} = (0,\ i,\ 2 + i)$ with respect to this new inner-product function.

52. Use Theorem 4.15 and repeat Example 4 using the matrix $D = \text{diag}\ [1, 3, 9]$. That is, find $<\mathbf{u},\ \mathbf{v}>$ for $\mathbf{u} = (1,\ 1 + i,\ i)$ and $\mathbf{v} = (0,\ i,\ 2 + i)$ with respect to this new inner-product function.

53. Show that Theorem 4.15 fails when the matrix $D = \text{diag}\ [1, 0, 6]$. (HINT: Find an appropriate nonzero vector $\mathbf{u}$ for which $<\mathbf{u},\ \mathbf{u}> = 0$ with respect to this inner-product function.)

54. Let $\mathbf{u}$ and $\mathbf{v} \neq \mathbf{0}$ be any vectors in a complex inner-product space. Prove (4.46); that is,

$$<\mathbf{u} - \frac{<\mathbf{u},\ \mathbf{v}>}{<\mathbf{v},\ \mathbf{v}>}\ \mathbf{v},\ \mathbf{u} - \frac{<\mathbf{u},\ \mathbf{v}>}{<\mathbf{v},\ \mathbf{v}>}\ \mathbf{v}>$$
$$= \frac{<\mathbf{u},\ \mathbf{u}><\mathbf{v},\ \mathbf{v}> - |<\mathbf{u},\ \mathbf{v}>|^2}{<\mathbf{v},\ \mathbf{v}>}.$$

4.7 EXTENDED APPLICATION Least Squares Fit

In this section we look at an interesting application of matrix and vector methods in a real inner-product space. It is often desirable to construct a mathematical model that describes the results of an experiment. This can involve *fitting* an algebraic curve to the given data. Let's illustrate first with a spring-displacement example.

Example 1   Suppose that in an experiment in mechanics we displace a spring by attaching various weights to it. We then measure the amount of displacement (Figure 4.18). Suppose that the results of this experiment are those given in the following table.

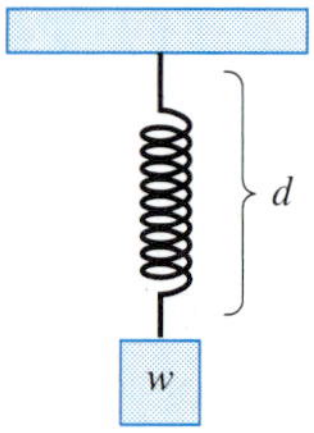

FIGURE 4.18

Weight/Displacement Table

Weight w, lb	Displacement d, in
2	4.18
3	6.31
4	8.26
5	11.10
10	20.82

We plot the data from this experiment in Figure 4.19. As indicated, the points appear to almost lie along the straight line sketched in the figure. That they do not actually form a line could be attributed to one or both of the following reasons: (1) experimental error in measuring the spring displacement, or (2) the spring displacement is not linear.

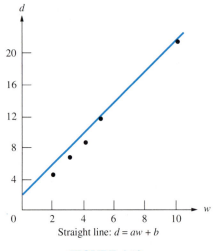

Straight line: $d = aw + b$

FIGURE 4.19

Let us, for the present, disregard the second possible reason. The data points appear to be *nearly* linear. Our job is, then, to determine the line that in some sense *best* fits the data. ∎

Other sorts of data points might be approximated by other types of curves; three popular possibilities are indicated in Figure 4.20. In each case the idea is to choose the curve that best fits the data points; that is, we want to find the coefficients in the equation of the curve that almost passes through the data points. Let's be more specific about what we mean by *best fits* by first concentrating on fitting a straight line to the given data points.

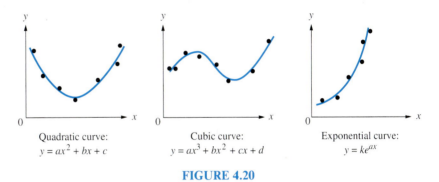

Quadratic curve:
$y = ax^2 + bx + c$

Cubic curve:
$y = ax^3 + bx^2 + cx + d$

Exponential curve:
$y = ke^{ax}$

FIGURE 4.20

Assume that we have n data points, experimentally determined. Write these as

$$(x_1, y_1), (x_2, y_2), \ldots, (x_n, y_n).$$

Suppose that we want to determine a straight line $y = ax + b$ passing through each of these points. If the points do indeed lie on a line, we have the following equations in the two unknown coefficients a and b.

$$y_1 = ax_1 + b,$$
$$y_2 = ax_2 + b,$$
$$\vdots \quad \vdots \quad \vdots \quad \vdots \quad \vdots$$
$$y_n = ax_n + b.$$

This is surely an overdetermined system of equations. In matrix notation this system is expressed as $Y = MS$, where

$$Y = \begin{bmatrix} y_1 \\ y_2 \\ \vdots \\ y_n \end{bmatrix}, \qquad M = \begin{bmatrix} x_1 & 1 \\ x_2 & 1 \\ \vdots & \vdots \\ x_n & 1 \end{bmatrix}, \qquad \text{and} \qquad S = \begin{bmatrix} a \\ b \end{bmatrix}.$$

If the given data points are genuinely collinear, then the system $Y = MS$ has a unique solution, since any two of them are enough to determine the line and the rank of M is two. If, however, the points are not collinear, Y will not be in the column space of M; the augmented matrix $[M : Y]$ will have rank three (see Theorem 3.14, page 204); and the system will be inconsistent. Because the data points (x_i, y_i) were obtained experimentally, this is what will most likely happen. We are then faced with finding the vector S that is best. What we want is the difference between the vector MS and the vector Y to be as small as possible. That is, we want the value of $\|Y - MS\|$ to be as small as possible.

Here we'll work with the space $\overline{\mathbb{R}^n}$ of column vectors equipped with the standard (Euclidean) inner product. (See Example 7 of Section 4.3.)

If $S' = \begin{bmatrix} a' \\ b' \end{bmatrix}$ is a vector so that $\|Y - MS'\|$ is a minimum, then the straight line

$$y = a'x + b'$$

is called the **least squares fit** to the data. This name is justified by the fact that when we have $\|Y - MS'\|$ take on a minimum value we are minimizing the inner product

$$
\begin{aligned}
\|Y - MS'\|^2 &= <Y - MS', \, Y - MS'> \\
&= (y_1 - a'x_1 - b')^2 + (y_2 - a'x_2 - b')^2 \\
&\quad + \cdots + (y_n - a'x_n - b')^2,
\end{aligned}
$$

as well.

In Figure 4.21 we indicate that the distance $d_1 = |y_1 - a'x_1 - b'|$ is the vertical distance from the data point (x_1, y_1) to the point (x_1, y'_1) on the desired line, $y = a'x + b'$. Thus the least squares fit will minimize the sum of the squares of these vertical errors $d_1^2 + d_2^2 + \cdots + d_n^2$.

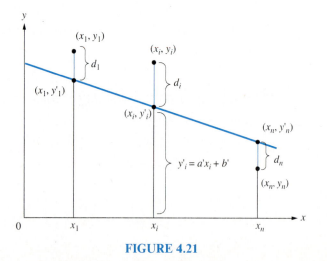

FIGURE 4.21

To calculate the vector S' that accomplished this minimization does not require calculus. It can be done with algebraic techniques. We note that if Y is a fixed vector in the inner-product space $\mathbb{R}^n$ and S varies over all vectors in the space $\mathbb{R}^2$, then the collection of all vectors MS forms a *subspace* of the vector space $\mathbb{R}^n$. This subspace is the column space W of the matrix M. That is, the vectors MS must lie in W and are therefore linear combinations of the column vectors of M.

Thus, for scalars a and b we have

$$MS = a \begin{bmatrix} x_1 \\ x_2 \\ \cdot \\ \cdot \\ \cdot \\ x_n \end{bmatrix} + b \begin{bmatrix} 1 \\ 1 \\ \cdot \\ \cdot \\ \cdot \\ 1 \end{bmatrix}.$$

In Exercise 30 of Section 4.5 we asked you to prove the "norm approximation theorem," which tells us that for $\|Y - MS'\|$ to be a minimum it must be orthogonal to this subspace W of $\mathbb{R}^n$. (You can do this proof as Exercise 19 below if you didn't do it in Section 4.5.) Therefore, the inner product $< MS, Y - MS' > = 0$, for all S. Thus we have

$$0 = < MS, Y - MS' > = (MS)^T(Y - MS')$$
$$= S^TM^T(Y - MS') = S^T(M^TY - M^TMS').$$

Remember that, in $\mathbb{R}^n$, the matrix form of the standard inner product $< A, B >$ is the scalar A^TB. We have just shown that the inner product $< S, M^TY - M^TMS' > = 0$ for *every* vector S in $\mathbb{R}^2$. But then the fixed vector $M^TY - M^TMS'$ must be the zero vector (see Exercise 50 in Section 4.4 or Exercise 20 below). It therefore follows that

$$M^TY = M^TMS'.$$

Since M^T is a $2 \times n$ matrix, the matrix M^TM is 2×2. So if M^TM is invertible, we can solve $M^TY = M^TMS'$ for the vector S', obtaining the solution

$$S' = (M^TM)^{-1}M^TY. \tag{4.47}$$

Except for when the n data points all lie in a vertical line ($x_i = y_i$ for each $i = 1, 2, \ldots, n$), the matrix M^TM is indeed invertible. We ask that you prove this as Exercise 18 below. The solution vector $S' = (a', b')^T$ is therefore given by (4.47). This gives us the coefficients of the line $y = a'x + b'$, which is the least squares fit to the given data points. Now try it out on some actual numbers.

Example 2   Find the straight line that is the least squares fit to the four points $(0, 2)$, $(1, 3)$, $(2, 5)$, and $(3, 7)$. Our matrix equation $Y = MS$ is as follows:

$$\begin{bmatrix} 2 \\ 3 \\ 5 \\ 7 \end{bmatrix} = \begin{bmatrix} 0 & 1 \\ 1 & 1 \\ 2 & 1 \\ 3 & 1 \end{bmatrix} \cdot \begin{bmatrix} a \\ b \end{bmatrix}.$$

Hence

$$M^T M = \begin{bmatrix} 14 & 6 \\ 6 & 4 \end{bmatrix}, \quad \text{and} \quad (M^T M)^{-1} = \begin{bmatrix} 0.2 & -0.3 \\ -0.3 & 0.7 \end{bmatrix}.$$

Therefore, from (4.47), we have the solution

$$S' = (M^T M)^{-1} M^T Y = \begin{bmatrix} 0.2 & -0.3 \\ -0.3 & 0.7 \end{bmatrix} \cdot \begin{bmatrix} 0 & 1 & 2 & 3 \\ 1 & 1 & 1 & 1 \end{bmatrix} \cdot \begin{bmatrix} 2 \\ 3 \\ 5 \\ 7 \end{bmatrix}$$

$$= \begin{bmatrix} 0.2 & -0.3 \\ -0.3 & 0.7 \end{bmatrix} \begin{bmatrix} 34 \\ 17 \end{bmatrix} = \begin{bmatrix} 1.7 \\ 1.7 \end{bmatrix}.$$

So the desired line is $y = 1.7x + 1.7$. ■

Example 3 Find the line that is the least squares fit to the data points of Example 1. See the table and Figures 4.18 and 4.19. From the table in Example 1 we obtain our matrix equation $Y = MS$ as follows:

$$\begin{bmatrix} 4.18 \\ 6.31 \\ 8.26 \\ 11.10 \\ 20.82 \end{bmatrix} = \begin{bmatrix} 2 & 1 \\ 3 & 1 \\ 4 & 1 \\ 5 & 1 \\ 10 & 1 \end{bmatrix} \begin{bmatrix} a \\ b \end{bmatrix}.$$

So that

$$M^T M = \begin{bmatrix} 154 & 24 \\ 24 & 5 \end{bmatrix}, \quad (M^T M)^{-1} = \frac{1}{194} \begin{bmatrix} 5 & -24 \\ -24 & 154 \end{bmatrix},$$

and

$$M^T Y = \begin{bmatrix} 324.03 \\ 50.67 \end{bmatrix}.$$

Thus, by (4.47),

$$S' = (M^T M)^{-1} M^T Y = \frac{1}{194} \begin{bmatrix} 404.07 \\ 26 \end{bmatrix} = \begin{bmatrix} 2.08 \\ 0.14 \end{bmatrix}.$$

So the desired line is

$$d = 2.08w + 0.14,$$

and, according to Hooke's law, the spring constant is 2.08 lb/in. ■

The method we have just described to find the straight line that is the least squares fit to a set of planar data points can be generalized to apply to any linear function of the form

$$y = a_1 x_1 + a_2 x_2 + \cdots + a_k x_k. \tag{4.48}$$

Suppose that we are given m data points $(x_{1i}, x_{2i}, \ldots, x_{ki}, y_i)$. Substitute them into this linear equation (4.48) to obtain a system of m linear equations in k variables. The resulting matrix equation $Y = MS$ is of the form

$$
\begin{bmatrix} y_1 \\ y_2 \\ \vdots \\ \\ y_m \end{bmatrix} = \begin{bmatrix} x_{11} & x_{21} & x_{31} & \cdots & x_{k1} \\ x_{12} & x_{22} & x_{32} & \cdots & x_{k2} \\ x_{13} & x_{23} & x_{33} & \cdots & x_{k3} \\ \vdots & \vdots & \vdots & \cdots & \vdots \\ \vdots & \vdots & \vdots & \cdots & \vdots \\ x_{1m} & x_{2m} & x_{3m} & \cdots & x_{km} \end{bmatrix} \begin{bmatrix} a_1 \\ a_2 \\ \vdots \\ \\ a_k \end{bmatrix}.
$$

Now the arguments used in the "line" case above carry over completely here. The minimizing vector S' is given by (4.47):

$$ S' = (M^T M)^{-1} M^T Y. $$

In particular, we can think of (4.48) as a polynomial of degree $k - 1$. We need merely set $t^{j-1} = x_j$ and $c_{j-1} = a_j$, for each $j = 1, 2, \ldots, k$. Then (4.48) would be of the form

$$ y = c_0 + c_1 t + c_2 t^2 + \cdots + c_{k-1} t^{k-1}. $$

Therefore, if we are given any polynomial of degree n

$$ y = a_n x^n + a_{n-1} x^{n-1} + \cdots + a_2 x^2 + a_1 x + a_0, $$

and m data points (x_i, y_i), $i = 1, 2, \ldots, m$, our matrix equation for this polynomial, $Y = MS$, is of the form

$$
\begin{bmatrix} y_1 \\ y_2 \\ \vdots \\ \\ y_m \end{bmatrix} = \begin{bmatrix} x_1^n & x_1^{n-1} & \cdots & x_1 & 1 \\ x_2^n & x_2^{n-1} & \cdots & x_2 & 1 \\ \vdots & \vdots & \vdots & \cdots & \vdots \\ \vdots & \vdots & \vdots & \cdots & \vdots \\ x_m^n & x_m^{n-1} & \cdots & x_m & 1 \end{bmatrix} \begin{bmatrix} a_n \\ a_{n-1} \\ \vdots \\ a_1 \\ a_0 \end{bmatrix}.
$$

The minimizing vector S' is still that given by (4.47); namely,

$$ S' = (M^T M)^{-1} M^T Y. $$

Example 4   Find the quadratic polynomial $y = ax^2 + bx + c$ that is the least squares fit to the data points $(-2, 3)$, $(-1, 1)$, $(0, 2)$, $(1, 3)$, and $(2, 5)$. (Refer back to Figure 4.20.) Making the appropriate substitution of these values for x and y, we obtain the system of equations $Y = MS$ of the form

$$
\begin{bmatrix} 3 \\ 1 \\ 2 \\ 3 \\ 5 \end{bmatrix} = \begin{bmatrix} 4 & -2 & 1 \\ 1 & -1 & 1 \\ 0 & 0 & 1 \\ 1 & 1 & 1 \\ 4 & 2 & 1 \end{bmatrix} \cdot \begin{bmatrix} a \\ b \\ c \end{bmatrix}.
$$

Therefore,

$$M^TM = \begin{bmatrix} 34 & 0 & 10 \\ 0 & 10 & 0 \\ 10 & 0 & 5 \end{bmatrix},$$

and its inverse

$$(M^TM)^{-1} = \begin{bmatrix} \dfrac{1}{14} & 0 & -\dfrac{1}{7} \\ 0 & \dfrac{1}{10} & 0 \\ -\dfrac{1}{7} & 0 & \dfrac{17}{35} \end{bmatrix}.$$

So computing $S' = (M^TM)^{-1}M^TY$, we have

$$S' = \begin{bmatrix} a \\ b \\ c \end{bmatrix} = \begin{bmatrix} \dfrac{1}{14} & 0 & -\dfrac{1}{7} \\ 0 & \dfrac{1}{10} & 0 \\ -\dfrac{1}{7} & 0 & \dfrac{17}{35} \end{bmatrix} \cdot \begin{bmatrix} 4 & 1 & 0 & 1 & 4 \\ -2 & -1 & 0 & 1 & 2 \\ 1 & 1 & 1 & 1 & 1 \end{bmatrix} \cdot \begin{bmatrix} 3 \\ 1 \\ 2 \\ 3 \\ 5 \end{bmatrix}$$

$$= \begin{bmatrix} \dfrac{1}{14} & 0 & -\dfrac{1}{7} \\ 0 & \dfrac{1}{10} & 0 \\ -\dfrac{1}{7} & 0 & \dfrac{17}{35} \end{bmatrix} \begin{bmatrix} 36 \\ 6 \\ 14 \end{bmatrix} = \begin{bmatrix} \dfrac{4}{7} \\ \dfrac{3}{5} \\ \dfrac{58}{35} \end{bmatrix}.$$

Therefore, the quadratic polynomial that *best* fits the data is the least squares fit

$$y = \frac{4}{7}x^2 + \frac{3}{5}x + \frac{58}{35} = \frac{1}{35}(20x^2 + 21x + 58). \quad \blacksquare$$

Example 5 It has been observed that in an egg-production program the number of low-cholesterol eggs laid per day by the flock seems to be related to the amounts of the two special feed mixes x_1 and x_2 used. This relationship is expressed by a linear equation of the form $y = a_1x_1 + a_2x_2$, where a_1 and a_2 are (hopefully) constants. To determine these constants, a research program is undertaken in which the amounts of the two feeds are varied. The results are given by the following table.

x_1	x_2	y
1	0	4
0	1	5
1	1	6
2	1	5
1	2	4

Find the best approximation to the constants a_1 and a_2 using the least squares fit. By substituting the given data points we obtain the matrix equation $Y = MS$ in the form

$$\begin{bmatrix} 4 \\ 5 \\ 6 \\ 5 \\ 4 \end{bmatrix} = \begin{bmatrix} 1 & 0 \\ 0 & 1 \\ 1 & 1 \\ 2 & 1 \\ 1 & 2 \end{bmatrix} \cdot \begin{bmatrix} a_1 \\ a_2 \end{bmatrix}.$$

Therefore,

$$M^T M = \begin{bmatrix} 7 & 5 \\ 5 & 7 \end{bmatrix}, \quad \text{and} \quad (M^T M)^{-1} = \begin{bmatrix} \dfrac{7}{24} & -\dfrac{5}{24} \\ -\dfrac{5}{24} & \dfrac{7}{24} \end{bmatrix}.$$

The least squares fit is $S' = (M^T M)^{-1} M^T Y$, computed as

$$S' = \begin{bmatrix} a_1 \\ a_2 \end{bmatrix} = \begin{bmatrix} \dfrac{7}{24} & -\dfrac{5}{24} \\ -\dfrac{5}{24} & \dfrac{7}{24} \end{bmatrix} \cdot \begin{bmatrix} 1 & 0 & 1 & 2 & 1 \\ 0 & 1 & 1 & 1 & 2 \end{bmatrix} \cdot \begin{bmatrix} 4 \\ 5 \\ 6 \\ 5 \\ 4 \end{bmatrix} = \begin{bmatrix} 2 \\ 2 \end{bmatrix}.$$

Our least squares fit equation is therefore $y = 2x_1 + 2x_2$. ∎

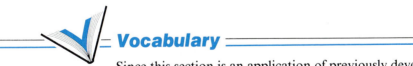

Vocabulary

Since this section is an application of previously developed material, we have not met a lot of new concepts. The procedure described here has only a few new terms other than those intrinsic to a particular use of the application. Be sure that you understand the following terms. Review the basic ideas of the technique described in this section until you feel secure enough to use it.

mathematical model   fitting a curve
data points   least squares fit

Exercises 4.7

In Exercises 1–6 find the line $y = ax + b$ that is the least squares fit to the given data points.

1. $(1, 1), (2, 2), (3, 4)$.

2. $(0, 3), (1, 5), (-1, 2)$.

3. $(0, 1.1), (1, 2.4), (2, 3.7)$.

4. $(0, -1.6), (1, 0.8), (2, 3.5), (3, 4.2)$.

5. $(0, -1.2), (1, -0.9), (2, 0.1), (3, 1.1), (4, 2.1)$.

6. $(1, -2.2), (2, -1.8), (3, -0.4), (4, 0.3), (5, 1.2)$.

In Exercises 7–12 find the quadratic polynomial $y = ax^2 + bx + c$ that is the least squares fit to the given data points. (Sketch the points.)

7. $(0, -1), (1, 2), (2, 4), (3, 7)$.

8. $(-2, 1), (-1, 3), (0, 5), (1, 4)$.

9. $(-2, -1), (-1, 2), (0, 4), (1, 1), (2, -3)$.

10. $(0, -1.6), (1, 0.8), (2, 3.5), (3, 4.2), (4, 0.3)$.

11. $(0, -1.2), (1, -0.9), (2, 0.1), (3, 1.1), (4, 2.1)$.

12. $(1, -2.2), (2, -1.8), (3, 0.4), (4, 0.3), (5, -1.2)$.

---

13. Find the cubic polynomial $y = ax^3 + bx^2 + cx + d$ that is the least squares fit to the data points $(1, -2.2), (2, -1.8), (3, 0.4), (4, 0.3), (5, 1.2), (6, 2.3)$. Sketch the points.

14. Find the cubic polynomial $y = ax^3 + bx^2 + cx + d$ that is the least squares fit to the data points $(0, -1.2), (1, -0.9), (2, 0.1), (3, 1.1), (4, 3.1), (5, 10.3), (6, 31.3)$. Sketch the points.

15. If $x_1, x_2, \ldots, x_n$ are linearly independent vectors in the real inner-product space V, then the $n \times n$ matrix $K = (k_{ij})$, where $k_{ij} = <x_i, x_j>$, $i, j = 1, 2, \ldots, n$, is called the **Gram matrix** associated with the vectors x_i. Let k_j denote the jth column vector of K. That is,
$$k_j = (k_{1j}, k_{2j}, \ldots, k_{nj})^T.$$
Suppose that $c_1 k_1 + c_2 k_2 + \cdots + c_n k_n = 0$.
Show that $\sum_{j=1}^{n} c_j <x_i, x_j> = 0$, $i = 1, 2, \ldots, n$.

16. Using the vectors $x_1, x_2, \ldots, x_n$, coefficients $c_1, c_2, \ldots, c_n$, and results of Exercise 15 suppose that $u = c_1 x_1 + c_2 x_2 + \cdots + c_n x_n$. Show that for each $i = 1, 2, \ldots, n$, $<x_i, u> = 0$. Explain why $u = 0$, and therefore the column vectors of the Gram matrix K are linearly independent.

17. Show that if A is any $m \times n$ matrix with linearly independent columns, the matrix $A^T A$ is nonsingular. HINT: Set $K = A^T A$ and refer to Exercises 15 and 16.

18. Show that the matrix M of the least squares method, as in (4.47), etc., has linearly independent columns if, and only if, at least two of the data point numbers $x_1, x_2, \ldots, x_n$ are distinct. Conclude from this that $M^T M$ is nonsingular if,

and only if, the given data points (x_i, y_i) do not lie in a vertical line in the xy plane.

19. Prove the "norm approximation theorem" in Exercise 30 of Section 4.5, if you haven't already done so.

20. Do Exercise 50 in Section 4.4 if you haven't already done so; namely, if in the inner-product space V a vector v is such that $<v, x> = 0$ for all $x \in V$, then $v = 0$.

21. The owner of a business finds that his sales for the first five months of the year are (in thousands of dollars) 3.8, 4.2, 4.9, 5.3, 5.1. Plot these points and fit a line, parabola, or cubic, as seems best to you, by the least squares method.

22. A strip of experimental metallic alloy is being tested. It is stretched to lengths of $\lambda = 63, 68$, and 72 inches by applied weights of $\omega = 1, 2$, and 3 tons. Assuming Hooke's law $\lambda = h + c\omega$, find this metal strip's normal length h by the least squares method.

23. Two radioactive chemicals in amounts x_1 and x_2 are being contained. We know the half-lives $\lambda_1 = 84$ years and $\lambda_2 = 115$ years, respectively, for each, but not the actual amounts of each present. We take radiation readings ρ, and anticipate that these readings will behave as the sum of two exponentials:
$$\rho = x_1 e^{-\mu_1 t} + x_2 e^{-\mu_2 t}$$
where $\mu_1 = (\ln 2)/84$, and $\mu_2 = (\ln 2)/115$. The actual readings for 7 time periods, given as (t, ρ) are as follows: $(0, 12.2), (1, 11.5), (2, 10.7), (3, 9.8), (4, 8.6), (5, 6.8), (6, 4.9)$. Find the amounts x_1 and x_2, by a least squares fit.

CHAPTER 4 SUMMARY

In this chapter, we studied the important concept of an inner-product space over the reals and saw that $\mathbb{R}^n$ is indeed an inner-product space with respect to the standard inner product. Many other diverse examples of inner-product spaces were also given. Further, the notion of an orthonormal basis for an inner-product space was discussed along with the Gram-Schmidt process for obtaining such an orthonormal basis for a given inner-product space. Moreover, the celebrated Cauchy-Schwarz inequality was proved, and as a corollary, the triangle inequality was derived, leading us to geometric ideas in the vector space. The concept of inner product was then extended to vector spaces with complex numbers as scalars, and many of the previous results were shown to have their analogs in complex vector spaces. The chapter concluded with an application to fitting a curve to data by the "least squares method."

CHAPTER 4 REVIEW EXERCISES

1. Find the standard inner product $<\mathbf{u}, \mathbf{v}>$ in Euclidean 3-space $\mathbb{R}^3$ for the vectors $\mathbf{u} = (1, 2, -1)$, and $\mathbf{v} = (3, 0, 2)$.

2. Find the nonstandard inner product $<\mathbf{u}, \mathbf{v}> = \mathbf{u}\, D\mathbf{v}^T$ in inner-product 3-space $\mathbb{R}^3$ for $D = \text{diag} [2, 3, 5]$ and $\mathbf{u} = (1, 2, -1)$, and $\mathbf{v} = (3, 0, 2)$.

3. Verify the Cauchy-Schwarz inequality for vectors $\mathbf{u} = (2, -3)$ and $\mathbf{v} = (-1, 5)$, using the standard inner product in $\mathbb{R}^2$.

4. Verify the triangle inequality for the two vectors in Exercise 3.

5. Convert the basis $H = \{(1, 1, 0), (0, -1, 1), (1, 0, -1)\}$ for $\mathbb{R}^3$ into an orthonormal basis, using the Gram-Schmidt process and the standard inner product in $\mathbb{R}^3$.

6. Let A be the complex matrix
$$A = \begin{bmatrix} 1+i & 3-i \\ 5 & 7-2i \end{bmatrix}.$$
Find its conjugate transpose ($=$ adjoint) A^*.

7. Find the standard inner product in $\mathbb{C}^2$, $<\mathbf{u}, \mathbf{v}> = \mathbf{u}\, \mathbf{v}^*$ of the two vectors $\mathbf{u} = (i, 1)$ and $\mathbf{v} = (1, i)$.

8. Verify the triangle inequality for the two vectors in Exercise 7.

9. Decide whether or not the function $f(\mathbf{u}, \mathbf{v}) = u_1 v_2 + 3u_2 v_1$ is an inner product on $\mathbb{C}^2$. If so, demonstrate it. If not, why not? Here $\mathbf{u} = (u_1, u_2)$, $\mathbf{v} = (v_1, v_2)$.

10. Show that the two polynomials $p(x) = 1 + x^2$ and $q(x) = -9 + 16x$ are orthogonal in P_2 with the integral, from 0 to 1, inner product.

CHAPTER 4 SUPPLEMENTARY EXERCISES AND PROJECTS

1. Show that $|a \cos \theta + b \sin \theta|^2 \le a^2 + b^2$. HINT: Use the Cauchy-Schwarz inequality with $\mathbf{u} = (a, b)$ and $\mathbf{v} = (\cos \theta, \sin \theta)$.

2. Extend the results of Exercise 1 to verify the inequality
$$\frac{1}{n} \left| \sum_{k=1}^{n} (a_k \cos kx + b_k \sin kx) \right|^2 \le \sum_{k=1}^{n} (a_k^2 + b_k^2).$$

3. An $n \times n$ matrix P is called an **orthogonal matrix** if $PP^T = I$. Show that the column vectors $\mathbf{p}_i$ of an orthogonal matrix P, $i = 1, 2, \ldots, n$, form an orthonormal set of vectors in the Euclidean vector space $\mathbb{R}^n$ of column vectors. HINT: Notice that the entry in *row i* and *column j* of PP^T is $< \mathbf{p}_i, \mathbf{p}_j, > = \mathbf{p}_i^T \mathbf{p}_j$.

4. Suppose that angle ABC is inscribed in a semicircle in the plane. Let the center of the semicircle be at the origin with the points A and C located at the opposite ends of the diameter. Let the vector from the origin to A be $-\mathbf{a}$ and from the origin to B be $\mathbf{b}$. Note that the vector from the origin to the point C must be $\mathbf{a}$, and prove that angle ABC is a right angle. HINT: Draw the figure and show that the two vectors $\mathbf{b} + \mathbf{a}$ and $\mathbf{b} - \mathbf{a}$ are orthogonal.

5. Let W denote the subspace of $\mathbb{C}^3$ that is spanned by the two vectors $\mathbf{u} = (1 + i, 1, i)$ and $\mathbf{v} = (-1, 0, 2 - i)$. Use the Gram-Schmidt process to find an orthonormal basis for W with respect to the standard inner product in $\mathbb{C}^3$; i.e., $< \mathbf{u}, \mathbf{v} > = \mathbf{u} \, \mathbf{v}^*$.

6. Extend the orthonormal basis of W in Exercise 5 to an orthonormal basis for all of $\mathbb{C}^3$.

7. Let A be any $n \times n$ complex matrix and let $< \mathbf{u}, \mathbf{v} > = \mathbf{u} \, \mathbf{v}^*$ be the standard inner product in $\mathbb{C}^n$. Prove that $< A\mathbf{u}, \mathbf{v} > = < \mathbf{u}, A^*\mathbf{v} >$, for all $\mathbf{u}$ and $\mathbf{v}$ in $\mathbb{C}^n$.

8. Refer to Exercise 7. An $n \times n$ complex matrix A is called a **unitary matrix** if $AA^* = A^*A = I$. If A is a unitary matrix prove that $< A\mathbf{u}, A\mathbf{v} > = < \mathbf{u}, \mathbf{v} >$, for all $\mathbf{u}$ and $\mathbf{v}$ in $\mathbb{C}^n$.

9. Let $B = \{p_0(x), p_1(x), \ldots, p_n(x)\}$ be an *orthonormal* basis for the inner-product space P_n of polynomials with the integral from -1 to 1 inner product. Show that the set of polynomials $H = \{q_0(x), q_1(x), \ldots, q_n(x)\}$ is also an orthonormal basis for P_n with the integral from a to b inner product where

$$q_i(x) = \sqrt{\frac{2}{b - a}} \, p_i(y), \text{ for } y = \frac{2x - (a + b)}{b - a}.$$

Project 1. **Orthogonal Bases for $\mathbb{R}^3$**

(See Osborne, Anthony, and Hans Liebeck, "Orthogonal bases for $\mathbb{R}^3$ with Integer Coordinates and Integer Lengths," American Mathematical Monthly, **96**, 1989, pp. 49–53.)

a. Show that the set of vectors

$$K = \left\{ \left(\frac{1}{\sqrt{2}}, \frac{1}{\sqrt{2}}, 0 \right), \left(\frac{1}{\sqrt{2}}, -\frac{1}{\sqrt{2}}, 0 \right), (0, 0, 1) \right\}$$

is an orthonormal basis for $\mathbb{R}^3$ with the standard inner product. Sketch this basis and the standard basis on a right-handed coordinate system.

b. Prove that the three vectors $\mathbf{a} = (r, s, t)$, $\mathbf{b} = (s, t, r)$, and $\mathbf{c} = (t, r, s)$, where r, s, and t are integers, form an orthogonal set of vectors in $\mathbb{R}^3$ if, and only if, $rs + st + tr = 0$.

c. Show that each of the three vectors $\mathbf{a}$, $\mathbf{b}$, $\mathbf{c}$ of part (b) has standard "length" $|r + s + t|$.

d. Show that for integers m, n, and k, with m and n relatively prime, the general solution to the equation $rs + st + tr = 0$ is given by $r = m(m+n)k$, $s = n(m+n)k$, and $t = -mnk$.

e. Use these results to construct two orthogonal bases B_1 and B_2 different from the standard basis E and the basis K of part (a) for $\mathbb{R}^3$.

f. Make the bases of part (e) into orthonormal bases for $\mathbb{R}^3$.

Project 2. Subspaces in an Inner-Product Space

Let V be a finite-dimensional real or complex inner-product space and let S and T be subspaces of V.

a. Show that the sum $S + T = \{\mathbf{s} + \mathbf{t} \,|\, \mathbf{s} \in S \text{ and } \mathbf{t} \in T\}$ of the two subspaces is a subspace of V, but that their union $S \cup T = \{\mathbf{x} \,|\, \mathbf{x} \in S \text{ or } \mathbf{x} \in T\}$ need not be a subspace.

b. Let V be Euclidean 3-space, $\mathbb{R}^3$, and let S be the xy plane and T the yz plane. Find both $S \cup T$ and $S + T$.

c. Repeat part (b) when V is $\mathbb{R}^4$, $S = L\{(1, 0, 0, 0), (0, 0, 1, 0)\}$ and $T = L\{(1, 1, 1, 1), (1, -1, 1, -1)\}$.

d. Demonstrate for the examples of parts (b) and (c) that $S \cup T \subset S + T$.

e. Prove that in general $S \cup T \subset S + T$.

f. Prove that $(S + T)^\perp = S^\perp \cap T^\perp$, where $\cap$ denotes the intersection of the two sets.

g. Prove that $S \cap T$ is a subspace of V and $(S \cap T)^\perp = S^\perp + T^\perp$.

Project 3. Bessel's Inequality

Let $B = \{\mathbf{v}_1, \mathbf{v}_2, \ldots, \mathbf{v}_n, \ldots\}$ be a collection (not necessarily finite) of orthogonal vectors in the real inner-product space V, and let $\mathbf{u}$ be any a vector in V.

a. Define the vector $\mathbf{w}_n = \mathbf{u} - \sum_{k=1}^{n} <\mathbf{u}, \mathbf{v}_k> \mathbf{v}_k$. Compute $\|\mathbf{w}_n\|^2$.

b. Suppose V is $\mathbb{R}^3$ and $B = E$, the standard basis (so $n = 3$). Let $\mathbf{u} = (1, 1, 1)$. Find $\mathbf{w}_3$ and compute $\|\mathbf{w}_3\|^2$.

c. Repeat (b) if V is $C[-\pi, \pi]$; see Example 7 in Section 4.4, $B = (1, \sin x, \cos x)$ and $\mathbf{u} = e^x$.

d. Prove that, in any real inner-product space V, **Bessel's inequality**

$$\sum_{k=1}^{n} |<\mathbf{u}, \mathbf{v}_k>|^2 \leq \|\mathbf{u}\|^2,$$

holds for each positive integer n.

e. (For students who have studied calculus) Show that the infinite series

$$\sum_{k=1}^{\infty} |<\mathbf{u}, \mathbf{v}_k>|^2$$

converges to a number $K \le ||\mathbf{u}||^2$.

f. (For students who have studied calculus) Show that the set of $n = 2k$ functions

$$B = \left\{ \frac{1}{\sqrt{\pi}} \cos x, \frac{1}{\sqrt{\pi}} \cos 2x, \ldots, \frac{1}{\sqrt{\pi}} \cos kx, \frac{1}{\sqrt{\pi}} \sin x, \frac{1}{\sqrt{\pi}} \sin 2x, \ldots, \frac{1}{\sqrt{\pi}} \sin kx \right\}$$

from $C[-\pi, \pi]$ is orthonormal under the integral inner product $\int_{-\pi}^{\pi} f(x)g(x)\, dx$. You

may find the following trigonometric identities useful:

$$\sin ax \cos bx = \frac{1}{2}[\sin(a+b)x + \sin(a-b)x];$$

$$\cos ax \cos bx = \frac{1}{2}[\cos(a+b)x + \cos(a-b)x];$$

$$\sin ax \sin bx = \frac{1}{2}[\cos(a-b)x - \cos(a+b)x].$$

g. (For students who have studied calculus) Let B be the set of $n = 2k$ functions in part (f) and let $\mathbf{u} = e^x$. Find $\mathbf{w}_n$ and $||\mathbf{u}||^2$. What is the K of part (e)?

h. (For students who have studied calculus) Do part (g) when B is the set of Legendre polynomials of Example 9 in Section 4.4.

Project 4. Fourier Series of Order n *(For students who have studied calculus)*

Let V be the inner-product space $C[a, b]$ of Example 4 in Section 4.3, and let $B = \{h_1(x), h_2(x), \ldots, h_n(x)\}$ be an orthonormal basis for the finite-dimensional subspace W of $C[a, b]$.

a. Let $g(x) = c_1 h_1(x) + c_2 h_2(x) + \cdots + c_n h_n(x)$ be any given function in W and construct the vector $\mathbf{c} = (c_1, c_2, \ldots, c_n)$ in $\mathbb{R}^n$. Demonstrate that for each function $f(x) \in C[a, b]$ if for each $i = 1, 2, \ldots, n$ we let $d_i = <f(x), h_i(x)>$, then for $\mathbf{d} = (d_1, d_2, \ldots, d_n)$ we have

$$||f(x) - g(x)||^2 = <\mathbf{c}, \mathbf{c}> - 2 <\mathbf{c}, \mathbf{d}> + ||f(x)||^2. \qquad (*)$$

b. Substitute $\mathbf{c} = \mathbf{c}' + \mathbf{d}$ into equation (*) of part (a) and demonstrate that

$$||f(x) - g(x)||^2 = <\mathbf{c}', \mathbf{c}'> - <\mathbf{d}, \mathbf{d}> + ||f(x)||^2. \qquad (**)$$

c. Use the results of part (b) to show that $||f(x) - g(x)||^2$ is a minimum when $\mathbf{c} = \mathbf{d}$, that is, when $c_i = <f(x), h_i(x)>$.

d. Show that if $q(x)$ is a vector (function) in W with the property that $||f(x) - g(x)||^2 = ||f(x) - q(x)||^2$, then $q(x) = g(x)$.

e. Use the results in parts (a)–(d) to prove the following.

> **Theorem** Let $f(x)$ be a vector (function) in the inner-product space $C[a, b]$ (with the integral inner product of Example 4, Section 4.3) and let W be a finite-dimensional subspace of $C[a, b]$. Then there is a unique function $g(x)$ in W so that $||f(x) - g(x)||^2$ is a minimum.

f. Refer to part (f) of Project 3. A **trigonometric polynomial** is a function of the form $p(x) = a_0 + a_1 \cos x + a_2 \cos 2x + \cdots + a_n \cos nx + b_1 \sin x + b_2 \sin 2x + \cdots + b_n \sin nx$. If a_n and b_n are not both zero, we say that $p(x)$ has **order n**. Show that the set $T_n[a, b]$ of all trigonometric polynomials of order less than or equal to n is a subspace of the inner-product space $C[a, b]$.

g. Use $T_n[a, b]$ of part (f) as the subspace W of this project. The minimizing function of part (e) is called the **Fourier series of order n** for a given function $f(x)$. *Show that the vector $\mathbf{c}$ in $\mathbb{R}^{2n+1}$, as first described in part (a), is $\mathbf{c} = (a_0, a_1, \ldots, a_n, b_1, \ldots, b_n)$ for this minimizing function. Show that $\mathbf{c}$ has coordinates (coefficients)*

$$a_0 = \int_{-\pi}^{\pi} f(x)\, dx, \qquad a_k = \frac{1}{\pi} \int_{-\pi}^{\pi} f(x) \cos kx\, dx, \qquad \text{for } k = 1, 2, \ldots, n$$

and

$$b_k = \frac{1}{\pi} \int_{-\pi}^{\pi} f(x) \sin kx\, dx, \qquad \text{for } k = 1, 2, \ldots, n.$$

h. Use the results of part (g) to find the Fourier series of order 2 for $f(x) = x$.

i. Use the results of part (g) to find the Fourier series of order for $f(x) = e^x$.

Project 5. Vector Norms

*Let V be a real or complex vector space. A function $|| \cdot ||$ from V to $\mathbb{R}$ is a (**vector) norm** if for all vectors $\mathbf{u}$ and $\mathbf{v}$ in V each of the following properties holds:*

(1) $||\mathbf{u}|| \geq 0$, with equality if, and only if, $\mathbf{u} = \mathbf{0}$.   Nonnegative property
(2) $||\alpha \mathbf{u}|| = |\alpha|\, ||\mathbf{u}||$, for all scalars α in $\mathbb{R}$ (or $\mathbb{C}$)   Homogeneous property
(3) $||\mathbf{u} + \mathbf{v}|| \leq ||\mathbf{u}|| + ||\mathbf{v}||$.   Triangle inequality

a. Quote the appropriate theorems to justify the statement that in an inner-product space V, $||\mathbf{u}|| = <\mathbf{u}, \mathbf{u}>^{1/2}$ is a vector norm.

b. Show that the function $||\mathbf{v}||_1 = \sum_{i=1}^{n} |v_i|$, where $\mathbf{v} = (v_1, v_2, \ldots, v_n)$, is a vector norm on $\mathbb{R}^n$.

c. Show that the function $\|\mathbf{v}\|_\infty = \displaystyle\max_{1 \le i \le n} |v_i|$ is a norm on the vector space $\mathbb{R}^n$. This norm is often called the **uniform norm**.

d. Show that the function $\|\mathbf{v}\|_p = \left(\displaystyle\sum_{i=1}^{n} |v_i|^p \right)^{1/p}$ is a norm on the vector space $\mathbb{R}^n$, for any integer $p \ge 1$. In particular note that the norm of part (a) is the case $p = 2$.

e. Show that when $p \ne 2$ in part (d) the norm $\|\mathbf{v}\|_p$ does not correspond to a Euclidean inner product.

f. Use the defining properties of a norm to show that in any vector space with a norm, $\|\mathbf{0}\| = 0$.

g. Show that in the vector space $\mathbb{R}^n$, $\|\mathbf{v}\|_\infty \le \|\mathbf{v}\|_2$.

h. Show that in vector space $\mathbb{R}^2$, $\|\mathbf{v}\|_2 \le \|\mathbf{v}\|_1$. HINT: Use the triangle inequality.

i. Show that in any vector space with a norm, $\|\mathbf{u} + \mathbf{v}\| \ge \big| \|\mathbf{u}\| - \|\mathbf{v}\| \big|$.

j. (For students who have studied calculus) Show that the function $\|f(x)\| = \displaystyle\int_a^b |f(x)|\, dx$ is a norm on $C[a, b]$.

k. Show that the function $\|f(x)\| = \displaystyle\max_{1 \le i \le n} |f(x)|$ is a norm on $C[a, b]$.

Project 6. Least Squares Solution to an Overdetermined System of Equations

A continuation of Project 5 in Chapter 3. Recall that A is an $m \times n$ matrix of rank n, with $m > n$, and $A\mathbf{x} = \mathbf{b}$ is a system of linear equations, and $\mathbf{x}' = [(A^TA)^{-1}A^T]\mathbf{b}$. See that project.

a. Show that $(A\mathbf{x}' - \mathbf{b}) = \mathbf{z}$ is a vector orthogonal to the column space of the matrix A; that is, $\mathbf{z}$ is orthogonal to every linear combination of the column vectors $\mathbf{a}_i$ of the matrix A.

b. In case $n = 3$, the ith equations in the two systems of equations $A\mathbf{x} = \mathbf{b}$ and $A\mathbf{x} = \mathbf{b}'$ are parallel planes. Discuss. Generalize this idea to any n. Show that the "distance" between the two "planes" is

$$d_i = \frac{|b_i' - b_i|}{\sqrt{\displaystyle\sum_{i=1}^{n} a_{ij}^2}}.$$

HINT: Draw the case $n = 2$ and generalize.

c. Show d_i is the distance from the vector $\mathbf{b}'$ to the "plane" corresponding to the ith equation in $A\mathbf{x} = \mathbf{b}$.

d. Normalize the system $A\mathbf{x} = \mathbf{b}$ by multiplying both sides by the diagonal matrix $D = (d_{ij})$, where

$$d_{ii} = \frac{1}{\sqrt{\sum_{i=1}^{n} a_{ij}^2}}, \text{ and } d_{ij} = 0 \text{ for } i \neq j.$$

Show that $\mathbf{x}'$ doesn't change for the new system $DA\mathbf{x} = D\mathbf{b}$.

e. Show that in the normalized system of part (d) the distance of part (b) becomes simply

$$d_i = |b_i' - b_i| \text{ so that } ||\mathbf{b}' - \mathbf{b}|| = \sqrt{\sum_{i=1}^{n} d_i^2}.$$

So since the least squares solution $\mathbf{x}'$ is the vector that minimizes $||\mathbf{b}' - \mathbf{b}||$ we have that this least squares solution to an overdetermined system of equations is the vector such that the sum of the squares of the distance from the vector to each subset defined by the linear equations is a minimum.

f. Sketch a graph of the equations and find the vector (point) $\mathbf{x}'$, which is the least squares solution for the overdetermined system of equations

$$x_1 - 2x_2 = 1,$$
$$2x_1 + 3x_2 = -1,$$
$$x_1 + 5x_2 = 2.$$

5 LINEAR TRANSFORMATIONS

In this chapter we study a particularly important specific class of functions, called linear transformations, from one vector space U to another, V. We consider important subsets connected with these functions, such as the range of a given transformation in V and the kernel (or null space) of the transformation as a subset of its domain U. It turns out that the kernel is, in fact, a subspace of the domain U and the range is a subspace of V. The dimensions of these subspaces, the nullity and the rank of a linear transformation, are then considered, and the so-called Sylvester's law of nullity relating these numbers is proved.

We then see that matrices not only may be considered as vectors in a finite-dimensional vector space of matrices, but they also can be used to represent linear transformations from one finite-dimensional vector space to another. This representation is carefully defined and illustrated by many examples.

The function algebra of linear transformations is then examined. We begin with the definition of addition and multiplication (i.e., composition) of two linear transformations and consider as well the transformation that results from multiplying a given linear transformation by a scalar. We then take note of the natural correspondence between the resulting transformation algebra and the algebra of matrices with which we have worked in previous chapters. Invertible linear transformations are then singled out and criteria for a linear transformation to be invertible are found. Then the connection with nonsingular matrices is established.

We also look at the process whereby one may change the basis (or coordinate system) in a given vector space to another basis and show how this leads, in a natural way, to the concept of similarity of matrices. Finally we turn to an even closer tie with geometry by considering orthogonal (inner-product preserving) linear transformations for inner-product spaces and establishing some criteria for orthogonality.

5.1 *DEFINITIONS AND EXAMPLES*

After spending the preceding two chapters discussing the internal structure of vector spaces, we turn now to mappings or functions from one vector space to another. Recall from previous mathematics courses that the concept of a *function* is of primary importance. In Appendix A.2 we review some of this material. Remember that a function or a mapping or a transformation is made up of three things: two sets, the *domain* and the *codomain,* and a *rule of correspondence* between them. (See Figure 5.1.) Thus:

> **Definition** If U is a set and V is a set (in our case both will be vector spaces), then a **function** f from U into V is a rule that associates with each member u of U a unique member of V, denoted by $f(u)$.

The set of all elements in V that are of the form $f(u)$ is called the **range** (or **image**) of the function f. Observe that the range of any function f is always contained in the codomain of the function.

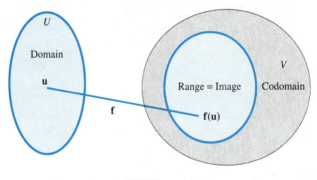

FIGURE 5.1

When the range of f is *exactly equal* to the codomain V of f, we say that the function f is **onto** V. In other words, a function f from U is *onto* V if every element v in V is of the form $v = f(u)$ for some u in U. Recall that some functions are one-to-one. A function f from U into V is one-to-one if distinct elements x and y in U have distinct function values $f(x)$ and $f(y)$ in V. A useful way of putting this is the following "contrapositive" formulation:

The function $f: U \to V$ is one-to-one if $f(x) = f(y)$ in V always implies that $x = y$, each x and y in U.

In calculus you dealt largely with a one-dimensional situation in which $U = V = \mathbb{R}$, the set of real numbers. We need now to lift our thoughts out of the one-dimensional world and consider multidimensional functions. When dealing with a multidimensional domain, the synonyms **transformation** and **mapping** are most often used for *function*.

Example 1 Let U and V both be the Euclidean plane ($\mathbb{R}^2$) and let T be the transformation (function) that sends each point P: (a, b) in the plane to the point $T(P)$: $(2a, 2b)$. Thus T stretches everything in the plane by a factor of 2. See Figure 5.2 for a diagram of this transformation. Note that the square region $ABCD$ in U is transformed into the square region $T(A)T(B)T(C)T(D)$ by the transformation T. In vector notation $U = V = \mathbb{R}^2$, so we have $T(\mathbf{u}) = 2\mathbf{u}$ for each $\mathbf{u}$ in U. ■

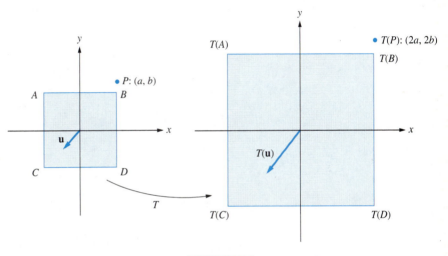

FIGURE 5.2

Example 2 Let U be 3-space $\mathbb{R}^3$ and let V be the two-dimensional xy plane, $V = \mathbb{R}^2$. Consider T to be the transformation that projects every point in U into the xy plane. We have indicated this projection T in Figure 5.3.

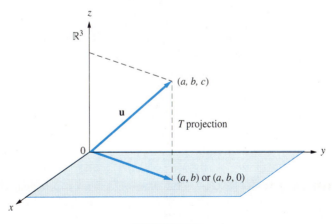

FIGURE 5.3

Note that $T(a, b, c) = (a, b)$ for each vector $\mathbf{u} = (a, b, c)$ in $\mathbb{R}^3$. Thus each vector $\mathbf{u}$ in $\mathbb{R}^3$ is essentially projected orthogonally onto $\mathbb{R}^2$ considered as a subspace of $\mathbb{R}^3$. See Section 4.5. Hence, this could be viewed as

$$T(\mathbf{u}) = <\mathbf{u}, \mathbf{e}_1> \mathbf{e}_1 + <\mathbf{u}, \mathbf{e}_2> \mathbf{e}_2,$$

where $\mathbf{e}_1$ and $\mathbf{e}_2$ are the first two of the standard basis vectors in $\mathbb{R}^3$.

The transformations of Examples 1 and 2 have special properties that distinguish them from other sorts of transformations (functions) from one vector space to another. In both of these examples it is true that, *for all vectors* $\mathbf{u}$ *and* $\mathbf{v}$ *in U and all scalars* α

$$T(\mathbf{u} + \mathbf{v}) = T(\mathbf{u}) + T(\mathbf{v}) \qquad \text{and} \qquad T(\alpha \mathbf{u}) = \alpha T(\mathbf{u}). \qquad \textbf{(5.1)}$$

You can quickly verify that the transformation of Example 1 does satisfy this. (See also Example 3.) In the case of the transformation given in Example 2, let us verify it here. Note that if $\mathbf{u} = (u_1, u_2, u_3)$ and $\mathbf{v} = (v_1, v_2, v_3)$ are any two vectors in $\mathbb{R}^3$ and α is any scalar in $\mathbb{R}$, then it is true that $\mathbf{u} + \mathbf{v} = (u_1 + v_1, u_2 + v_2, u_3 + v_3)$ and $\alpha \mathbf{u} = (\alpha u_1, \alpha u_2, \alpha u_3)$. Therefore, when T is applied to these two vectors, we have

$$T(\mathbf{u} + \mathbf{v}) = T(u_1 + v_1, u_2 + v_2, u_3 + v_3) = (u_1 + v_1, u_2 + v_2)$$
$$= (u_1, u_2) + (v_1, v_2) = T(\mathbf{u}) + T(\mathbf{v})$$

and

$$T(\alpha \mathbf{u}) = T(\alpha u_1, \alpha u_2, \alpha u_3) = (\alpha u_1, \alpha u_2) = \alpha(u_1, u_2) = \alpha T(\mathbf{u}),$$

as desired. Thus both of the equations (5.1) are true for this transformation T, as we asserted. ∎

Not every transformation (function) from one vector space to another satisfies the conditions (5.1). For example, suppose that we let T be the transformation from the Euclidean plane $\mathbb{R}^2$ to itself that associates with each point (a, b) the point $(a + 1, b + 1)$. This is the familiar translation of axes where the origin is moved to $(1, 1)$ and the point $P: (-1, -1)$ in the domain becomes the origin $(0', 0')$ in the range. See Figure 5.4.

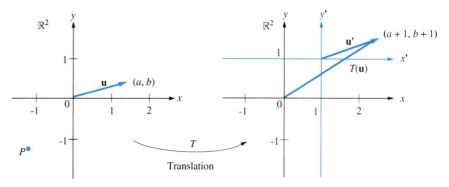

FIGURE 5.4

Notice that if $\mathbf{u} = (u_1, u_2)$ and $\mathbf{v} = (v_1, v_2)$ are any two vectors in $\mathbb{R}^2$ and α is any scalar in $\mathbb{R}$, then $\mathbf{u} + \mathbf{v} = (u_1 + v_1, u_2 + v_2)$ and $\alpha\mathbf{u} = (\alpha u_1, \alpha u_2)$. Therefore, when this translation T is applied to these two vectors, we have

$$T(\mathbf{u} + \mathbf{v}) = T(u_1 + v_1, u_2 + v_2) = (u_1 + v_1 + 1, u_2 + v_2 + 1)$$

but, on the other hand,

$$T(\mathbf{u}) + T(\mathbf{v}) = (u_1 + 1, u_2 + 1) + (v_1 + 1, v_2 + 1)$$
$$= (u_1 + v_1 + 2, u_2 + v_2 + 2) \neq T(\mathbf{u} + \mathbf{v}).$$

So (5.1) *fails* to hold for this transformation.

Notice also that $T(\alpha\mathbf{u}) = T(\alpha u_1, \alpha u_2) = (\alpha u_1 + 1, \alpha u_2 + 1)$ is *not* the same as

$$\alpha T(\mathbf{u}) = \alpha(u_1 + 1, u_2 + 1) = (\alpha u_1 + \alpha, \alpha u_2 + \alpha).$$

Those transformations between vector spaces U and V that *do satisfy* both properties of (5.1) are called **linear transformations.** The transformation of Figure 5.4 is *not linear* because it fails to satisfy these conditions. Had it satisfied but one of these conditions it would still not be linear; *both* are required. Even though translations and their generalization to affine transformations of $\mathbb{R}^n$ are *not linear* they are nevertheless important in mathematics. However, to discuss them here would take us beyond our present objective. We refer the interested reader to Project 7 at the end of this chapter. We concentrate on those transformations that satisfy both of the properties in (5.1). We state the following formal definition:

Definition Let U and V be any two vector spaces. A function $T: U \rightarrow V$, from U into V is called a **linear transformation** if, and only if, for all vectors $\mathbf{u}$ and $\mathbf{v}$ in U and all scalars (real numbers, complex numbers) α, T satisfies:

  (i) $T(\mathbf{u} + \mathbf{v}) = T(\mathbf{u}) + T(\mathbf{v})$,   and   (ii) $T(\alpha\mathbf{u}) = \alpha T(\mathbf{u})$.   **(5.2)**

We devote most of the rest of this section to examples of linear transformations on various vector spaces with which we are familiar. In the rest of the chapter we look at properties of transformations in general and see how matrix algebra can be used to calculate them. Note that the first property of the definition of a linear transformation (5.2(i)) states that we obtain the same result if we first add two vectors in the domain U and then transform the resultant to the codomain V by way of the transformation, as if we had first transformed the vectors separately and then added their images as vectors in V. How would you verbally express what the second property (5.2(ii)) says?

Example 3   Let $U = V = \mathbb{R}^2$ be Euclidean 2-space (i.e., the coordinate plane); and consider the transformation T given by $T(\mathbf{u}) = T(u_1, u_2) = (2u_1, -u_2)$. See Figure 5.5.

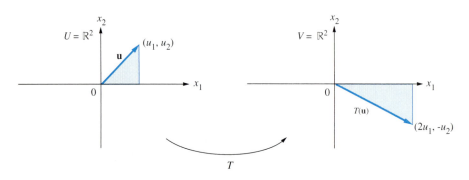

FIGURE 5.5

Observe that this T is linear; for if $\mathbf{u} = (u_1, u_2)$ and $\mathbf{v} = (v_1, v_2)$ are any two vectors in $\mathbb{R}^2$ and α is any scalar in $\mathbb{R}$, then $\mathbf{u} + \mathbf{v} = (u_1 + v_1, u_2 + v_2)$ and $\alpha\mathbf{u} = (\alpha u_1, \alpha u_2)$. Therefore, when this transformation T is applied to these two vectors we have

$$T(\mathbf{u} + \mathbf{v}) = T(u_1 + v_1, u_2 + v_2) = (2(u_1 + v_1), -(u_2 + v_2))$$
$$= (2u_1 + 2v_1, -u_2 - v_2)$$
$$= (2u_1, -u_2) + (2v_1, -v_2) = T(\mathbf{u}) + T(\mathbf{v}),$$

and, for any scalar α in $\mathbb{R}$,

$$T(\alpha\mathbf{u}) = T(\alpha u_1, \alpha u_2) = (2\alpha u_1, -\alpha u_2) = \alpha(2u_1, -u_2) = \alpha T(\mathbf{u}).$$

Since this transformation T satisfies *both* of the properties listed in the definition (5.2), it is a *linear transformation* from $\mathbb{R}^2$ to $\mathbb{R}^2$. ∎

> **Notational Remark**   In the above example, and in others before it, we wrote $T(\mathbf{u})$ as $T(u_1, u_2)$ rather than the perhaps more correct $T((u_1, u_2))$. Generally we continue to use this sort of notational simplification in what follows.

It is not difficult to show that the two-piece definition of a linear transformation given in (5.2) is equivalent to the following one-equation alternative formulation. We ask that you write out the proof as Exercise 22. Since the theorem claims to be an equivalent form, you need to prove both that (5.2(i) and (ii)) imply (5.3) below and, conversely, that (5.3) implies (5.2(i) and (ii)).

> **Theorem 5.1**   A function $T: U \to V$ from the vector space U into the vector space V is a **linear transformation** if, and only if, T satisfies the following property. For all vectors **u** and **v** in U and all scalars κ and μ,
>
> $$T(\kappa\mathbf{u} + \mu\mathbf{v}) = \kappa T(\mathbf{u}) + \mu T(\mathbf{v}). \tag{5.3}$$

Since this theorem gives us but one equation to check rather than two, we often use this criterion for determining whether a given vector function T is a linear transformation.

Example 4   Let $U = \mathbb{R}^2$ (a plane as in Example 3) and let $V = \mathbb{R}^3$ (3-space; see Example 2); define $T: U \to V$ by $T(x_1, x_2) = (x_1, -x_2, x_1 - x_2)$. Is T a linear transformation?

It is not difficult to verify (5.3) for this transformation. For all vectors $\mathbf{u} = (u_1, u_2)$ and $\mathbf{v} = (v_1, v_2)$, in $\mathbb{R}^2$, and all scalars κ and μ, we have

$$
\begin{aligned}
T(\kappa\mathbf{u} + \mu\mathbf{v}) &= T(\kappa(u_1, u_2) + \mu(v_1, v_2)) = T((\kappa u_1, \kappa u_2) + (\mu v_1, \mu v_2)) \\
&= T(\kappa u_1 + \mu v_1, \kappa u_2 + \mu v_2) \\
&= (\kappa u_1 + \mu v_1, -\kappa u_2 - \mu v_2, \kappa u_1 + \mu v_1 - \kappa u_2 - \mu v_2) \\
&= (\kappa u_1, -\kappa u_2, \kappa u_1 - \kappa u_2) + (\mu v_1, -\mu v_2, \mu v_1 - \mu v_2) \\
&= \kappa T(u_1, u_2) + \mu T(v_1, v_2) \\
&= \kappa T(\mathbf{u}) + \mu T(\mathbf{v}).
\end{aligned}
$$

So T is indeed a linear transformation. ∎

Example 5   **The Identity Transformation** The function $I: U \to U$ defined by $I(\mathbf{x}) = \mathbf{x}$ for all vectors **x** in U is readily seen to satisfy the criteria for a linear transformation. It is called the **identity transformation** on U. ∎

Example 6   **The Zero Transformation** Another special linear transformation is the transformation $O: U \to V$, defined by $O(\mathbf{x}) = \mathbf{0}$, for all **x** in U. This is again readily seen to satisfy the criteria for being a linear transformation. This transformation "collapses" every vector in U into the zero vector and is called the **zero transformation.** ∎

In addition to these two special transformations, there is a large class of transformations that will figure prominently in the rest of the text. These are the matrix transformations.

Example 7   **Matrix Transformations** This example will be exploited in subsequent sections. Let A be any $m \times n$ real (or complex) matrix. We define a function T_A on the vector space of n-column vectors, $\mathbb{R}^n = M_{n \times 1}(\mathbb{R})$ (or $\mathbb{C}^n = M_{n \times 1}(\mathbb{C})$) to the corresponding column space of m-column vectors $\mathbb{R}^m = M_{m \times 1}(\mathbb{R})$ (or $\mathbb{C}^m = M_{m \times 1}(\mathbb{C})$) by the matrix equation

$$T_A(X) = AX;$$

that is, the transformation is: *multiply X on the left by A.* Since it is easily seen, by the properties of matrix algebra in Chapter 1, that $A(cX + dY) = AcX + AdY = cAX + dAY$, we have

$$T_A(X) = AX \text{ is a linear transformation.}$$

In particular, if

$$A = \begin{bmatrix} 1 & 2 & 0 & 3 \\ 0 & 1 & 8 & 1 \\ -1 & 0 & 5 & 0 \end{bmatrix} \quad \text{and} \quad X = \begin{bmatrix} x_1 \\ x_2 \\ x_3 \\ x_4 \end{bmatrix}$$

then we have

$$T_A(X) = AX = \begin{bmatrix} 1 & 2 & 0 & 3 \\ 0 & 1 & 8 & 1 \\ -1 & 0 & 5 & 0 \end{bmatrix} \begin{bmatrix} x_1 \\ x_2 \\ x_3 \\ x_4 \end{bmatrix} = \begin{bmatrix} x_1 + 2x_2 + 3x_4 \\ x_2 + 8x_3 + x_4 \\ -x_1 + 5x_3 \end{bmatrix}.$$

So, for example,

$$T_A \left(\begin{bmatrix} 1 \\ 0 \\ -1 \\ 2 \end{bmatrix} \right) = \begin{bmatrix} 7 \\ -6 \\ -6 \end{bmatrix}.$$

We call this class of transformations **matrix transformations** or, often, *multiplication by the matrix A.* ∎

Example 8 As a special case of Example 7, let us model Euclidean 2-space by $\overline{\mathbb{R}^2}$ and let θ be a fixed angle (real number). Suppose A is the matrix $\begin{bmatrix} \cos\theta & -\sin\theta \\ \sin\theta & \cos\theta \end{bmatrix}$.

Let us look at the matrix transformation T_A. Suppose that $\mathbf{x} = \begin{bmatrix} x \\ y \end{bmatrix}$. Then

$$T_A(\mathbf{x}) = A\mathbf{x} = \begin{bmatrix} \cos\theta & -\sin\theta \\ \sin\theta & \cos\theta \end{bmatrix} \cdot \begin{bmatrix} x \\ y \end{bmatrix} = \begin{bmatrix} x\cos\theta - y\sin\theta \\ x\sin\theta + y\cos\theta \end{bmatrix} = \begin{bmatrix} x' \\ y' \end{bmatrix}.$$

If we identify $\overline{\mathbb{R}^2}$ with the plane in the usual way, we see that this transformation, which we henceforth denote by T_θ, is a rotation of every vector in the plane through an angle θ. Thus $T_\theta(\mathbf{u})$ is a vector that results from rotating $\mathbf{u}$ about the origin through an angle θ. The transformation T_θ is a rotation of the axes of the plane, since of course the standard basis vectors are also rotated. You will recognize the familiar rotation of axes and change of coordinates equations in the matrix equality above. See Figure 5.6. ∎

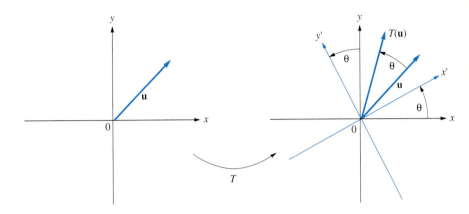

FIGURE 5.6 Rotation through an angle θ

Example 9 Consider the function $T: \mathbb{R}^3 \to P_2$ from Cartesian 3-space to the space of polynomials of degree two or less with real coefficients given by the formula

$$T(a, b, c) = ax^2 + bx + c.$$

It is easy to verify that this function is also a linear transformation. Do this as Exercise 21. ∎

Example 10 **A Vector Function That Is Not a Linear Transformation** Consider the vector function $T: \mathbb{R}^3 \to \mathbb{R}^3$, defined by the rule $T(a, b, c) = (a, 1, 1)$.

This function is *not* a linear transformation. It surely is a function from $\mathbb{R}^3 \to \mathbb{R}^3$, because each vector $\mathbf{x}$ in $\mathbb{R}^3$ has a unique vector $T(\mathbf{x})$ assigned to it by T. It fails, however, to satisfy the criteria for a linear transformation. To see this, note that if $\mathbf{u} = (x, y, z)$ then, for any scalar $k \neq 1$, $k\mathbf{u} = (kx, ky, kz)$. But $T(k\mathbf{u}) = (kx, 1, 1)$, while $kT(\mathbf{u}) = (kx, k, k)$. The vectors $T(k\mathbf{u})$ and $kT(\mathbf{u})$ are *not equal* (since $k \neq 1$), so T is not linear. ∎

Example 11 **The Derivative Is a Linear Transformation** (For students who have studied calculus) Consider the differentiation function D that assigns to each differentiable function $f(x)$ in the function space $C_d[a, b]$ its derivative $f'(x)$. That is, $D(f) = f' = df/dx$. D *is a linear transformation* from the space of all differentiable functions (on $[a, b]$) to the space of all functions (on $[a, b]$). Use the properties of derivatives of functions that you learned in calculus to verify that this is true. (See Exercise 46.) ∎

Example 12 **The Integral Is a Linear Transformation** (For students who have studied calculus) Remember the properties of the integral of the sum of two integrable functions and of a constant times a function that you learned in calculus.

These tell us that the function L from the space of all real-valued functions integrable on an interval $[a, b]$ to the space of real numbers $\mathbb{R}$ defined by

$$L(f) = \int_a^b f(x)\,dx$$

is a linear transformation. Verify this as Exercise 51. In particular note that if we let $a = 0$ and $b = 1$, we have the particular transformation

$$L_1(f) = \int_0^1 f(x)\,dx, \text{ and } L_1(x^2) = \frac{1}{3}, \text{ and } L_1(\cos x) = \sin 1, \text{ etc.} \quad \blacksquare$$

Vocabulary

In this section we have again encountered new concepts and terms. Since the vocabulary developed in this section is basic to the study of linear transformations and is used repeatedly, be sure that you understand each of the concepts described by the terms in the following list. Review these ideas until you feel secure enough to use them in the rest of this course and in your later use of linear algebra. You also may want to review Appendix A.2.

domain
codomain
range
image
function
mapping
one-to-one function
onto function
transformation

linear transformation
identity linear transformation
zero linear transformation
matrix transformation

Exercises 5.1

In Exercises 1–10 determine whether or not the given function $T: \mathbb{R}^2 \to \mathbb{R}^2$ is a linear transformation. Here $\mathbf{x} = (x_1, x_2)$ is an arbitrary vector in $\mathbb{R}^2$.

1. $T(x_1, x_2) = (x_1, -x_2)$.

2. $T(x_1, x_2) = (-x_2, -x_1)$.

3. $T(x_1, x_2) = (x_1, 1 - x_2)$.

4. $T(x_1, x_2) = (x_1^2, x_2^2)$.

5. $T(x_1, x_2) = (2x_1, 3x_2)$.

6. $T(x_1, x_2) = (x_1 + x_2, x_1 - x_2)$.

7. $T(x_1, x_2) = (2x_1 + 1, 3x_2 - 1)$.

8. $T(x_1, x_2) = (2x_1 + 3x_2, 3x_1 - 2x_2)$.

9. $T(x_1, x_2) = (2x_1, 0)$.

10. $T(x_1, x_2) = (x_1 + 3x_2, x_2)$ (a shear).

In Exercises 11–16 determine whether or not the given function $T: \mathbb{R}^3 \to \mathbb{R}^3$ is a linear transformation. Here $\mathbf{x} = (x_1, x_2, x_3)$ is an arbitrary vector in $\mathbb{R}^3$.

11. $T(x_1, x_2, x_3) = (x_1, -x_2, 0)$.

12. $T(x_1, x_2, x_3) = (x_3 - x_2, 0, x_3 - x_1)$.

13. $T(x_1, x_2, x_3) = (x_1, 1 - x_2, x_3)$.

14. $T(x_1, x_2, x_3) = (x_1 - x_2, x_2 + x_3, x_3 - x_1)$.

15. $T(x_1, x_2, x_3) = (2x_1, 3x_2, 4x_3)$.

16. $T(x_1, x_2, x_3) = (x_1^2 + x_2, x_1^2 - x_2, x_3^2 - x_2)$.

---

17. Show that the function T in Example 1 is a linear transformation.

18. Use (5.2) to show that the function T in Example 4 is a linear transformation.

19. Show that the function T in Example 5 is a linear transformation.

20. Show that the function T in Example 6 is a linear transformation.

21. Show that the function $T: \mathbb{R}^3 \to P_2$, of Example 9 is a linear transformation.

22. Prove that the definition of a linear transformation given by Theorem 5.1 (page 303) is equivalent to the definition (5.2) first given. HINT: In your proof that the second definition implies the first, note that $T(\mathbf{x} + \mathbf{y}) = T(1\mathbf{x} + 1\mathbf{y})$.

23. Show that if $T: \mathbb{R}^1 \to \mathbb{R}^1$ is any *linear* transformation, then for each x in $\mathbb{R}^1$ (i.e., real number) $T(x) = mx$, where $m = T(1)$.

24. Use the matrix A of Example 7 to compute $T_A(X)$

for $X = \begin{bmatrix} 1 \\ 1 \\ 0 \\ 2 \end{bmatrix}$.

25. Use the matrix A of Example 7 to compute $T_A(X)$

for $X = \begin{bmatrix} 1 \\ -1 \\ 3 \\ -2 \end{bmatrix}$.

26. Verify that if A is *any* $m \times n$ real matrix, the transformation $H: \mathbb{R}^m \to \mathbb{R}^n$ given by $H(\mathbf{x}) = \mathbf{x}A$, $\mathbf{x}$ any vector in $\mathbb{R}^m$, is always a linear transformation. (Writers who use this kind of matrix transformation usually use a different function notation as well. They write $T(\mathbf{x})$ as $\mathbf{x}T$; the name of the function is written on the right of its argument.)

27. Use the matrix A of Example 7 to compute $H(\mathbf{x}) = \mathbf{x}A$ for $\mathbf{x} = (1, -1, 5)$.

28. Use the matrix A of Example 7 to compute $H(\mathbf{x}) = \mathbf{x}A$ for $\mathbf{x} = (2, 3, -1)$.

29. Use the matrix $A = \begin{bmatrix} 1 & -1 & 2 \\ 0 & -1 & -2 \\ 2 & -1 & 0 \end{bmatrix}$ to compute

$H(\mathbf{x}) = \mathbf{x}A$ for $\mathbf{x} = (1, -1, 5)$.

30. Use the matrix $A = \begin{bmatrix} 1 & -1 & 2 \\ 0 & -1 & -2 \\ 2 & -1 & 0 \end{bmatrix}$ to compute

$H(\mathbf{x}) = \mathbf{x}A$ for $\mathbf{x} = (2, -3, 1)$.

31. Verify that if A is a *fixed* $n \times n$ real matrix, then the function $K_A: M_n \to M_n$ given by $K_A(X) = AX - XA$ is a linear transformation (operator) on the vector space $M_n = M_n(\mathbb{R})$ of $n \times n$ real matrices.

32. Use the matrix $A = \begin{bmatrix} 1 & -1 & 2 \\ 0 & -1 & -2 \\ 2 & -1 & 0 \end{bmatrix}$ to compute

$K_A(X) = AX - XA$ for $X = \begin{bmatrix} 2 & 2 & 4 \\ 0 & 1 & -1 \\ 1 & 2 & 3 \end{bmatrix}$.

33. Use the matrix $A = \begin{bmatrix} 1 & -1 & 2 \\ 0 & -1 & -2 \\ 2 & -1 & 0 \end{bmatrix}$ to compute

$K_A(X) = AX - XA$ for $X = \begin{bmatrix} 0 & -2 & 3 \\ 0 & 1 & -1 \\ -1 & 0 & 3 \end{bmatrix}$.

34. Verify that if A is *any* $m \times n$ complex matrix, then the transformation $T_A: \mathbb{C}^n \to \mathbb{C}^m$ defined by $T_A(\mathbf{x}) = A\mathbf{x}$ is a linear transformation on $\mathbb{C}^n$.

35. Use the matrix $A = \begin{bmatrix} 1 & 0 & i \\ 0 & i & 1 \\ -i & 0 & 1 \end{bmatrix}$ to compute

$T_A(\mathbf{x})$ for $\mathbf{x} = \begin{bmatrix} 1 \\ i \\ 0 \end{bmatrix}$.

36. Use the matrix $A = \begin{bmatrix} 1 & 0 & i \\ 0 & i & 1 \\ -i & 0 & 1 \end{bmatrix}$ to compute

$T_A(\mathbf{x})$ for $\mathbf{x} = \begin{bmatrix} i \\ -1 \\ -2 \end{bmatrix}$.

37. Show that the linear equation $y = m\mathbf{x} + \mathbf{b}$, $\mathbf{b} \neq \mathbf{0}$, does *not* give a linear transformation when the variables are from $\mathbb{R}^2$; that is, for $\mathbf{x} = (x_1, x_2)$ and $\mathbf{b} = (b_1, b_2)$ in $\mathbb{R}^2$ and m a scalar, show that $m\mathbf{x} + \mathbf{b}$ is not a linear transformation on $\mathbb{R}^2$ if $\mathbf{b} \neq \mathbf{0}$.

38. Verify that if A is a *fixed* $n \times n$ complex matrix, then the function $K_A: M_n(\mathbb{C}) \to M_n(\mathbb{C})$ given by $K_A(X) = AX - XA$ is a linear transformation (operator) on the vector space of all $n \times n$ complex matrices. (See Exercise 31.)

39. Use the matrix $A = \begin{bmatrix} 1 & 0 & i \\ 0 & i & 1 \\ -i & 0 & 1 \end{bmatrix}$ to compute

$K_A(X)$ for $X = \begin{bmatrix} 1 & -i & -2 \\ 1 & 0 & i \\ 0 & 0 & i \end{bmatrix}$.

40. Is the function $C: M_n(\mathbb{C}) \to M_n(\mathbb{C})$, given by $C(X) = X^* (= \overline{X}^T)$ for every X in $M_n(\mathbb{C})$, a linear transformation? Why, or why not?

41. Let $B = \{\mathbf{u}_1, \mathbf{u}_2, \ldots, \mathbf{u}_n\}$ be a basis for the *real* vector space U, and let $T: U \to U$ be a linear transformation (operator) such that $T(\mathbf{u}_i) = \mathbf{u}_i$ for each of the vectors in the basis B. Show that T is the identity transformation I; $T(\mathbf{x}) = I(\mathbf{x}) = \mathbf{x}$, on U.

42. Let $B = \{\mathbf{u}_1, \mathbf{u}_2, \ldots, \mathbf{u}_n\}$ be a basis for the *real* vector space U, and let $T: U \to V$ be a linear transformation. Show that for each $\mathbf{x}$ in U, the vector $T(\mathbf{x})$ in V is a linear combination of the vectors $T(\mathbf{u}_i)$, $i = 1, 2, \ldots, n$.

43. Let $B = \{\mathbf{u}_1, \mathbf{u}_2, \ldots, \mathbf{u}_n\}$ be a basis for the *real* vector space U, and let $T: U \to V$ be a linear transformation. Does the set $T(B) = \{T(\mathbf{u}_1), T(\mathbf{u}_2), \ldots, T(\mathbf{u}_n)\}$ necessarily span V? Give reasons.

44. Let $B = \{\mathbf{u}_1, \mathbf{u}_2, \ldots, \mathbf{u}_n\}$ be a basis for the *real* vector space U, and let $T: U \to V$ be a

linear transformation. Show that the set $T(B) = \{T(\mathbf{u}_1), T(\mathbf{u}_2), \ldots, T(\mathbf{u}_n)\}$ spans a subspace W of V. (This subspace is called the *range* of T; see Section 5.2.)

45. With the sets and transformation of Exercise 44, is the set $T(B)$ a basis for W? Give reasons.

Exercises 46–55 are for students who have studied calculus.

46. Prove that the differential operator $D: C_d[a, b] \to F[a, b]$, given by $D(f) = df/dx$, is a linear transformation from the space of real-valued differentiable functions defined on $[a, b]$ into $F[a, b]$, the space of all real-valued functions defined on $[a, b]$.

47. Let D^2 and D^3 denote the second and third derivatives, respectively, and let $C_{3d}[a, b]$ be the space of functions with continuous first, second, and third derivatives. Show that the transformation $T: C_{3d}[a, b] \to F[a, b]$ given by $T(f) = (D^3 + 2D^2 - 3D)(f)$ is a linear transformation.

48. Use the linear transformation T of Exercise 47 to find $T(5x^3 - 1)$.

49. Use the linear transformation T of Exercise 47 to find $T(\sin x)$.

50. Let D^k denote the kth derivative, k a positive integer. Show that the operator

$$T = D^n + a_{n-1}D^{n-1} + \cdots + a_1 D,$$

from the vector (function) space $C_{nd}[a, b]$ into the space $F[a, b]$ is linear for any given real or complex number coefficients a_i.

51. Show that the integral operator $L(f) = \int_a^b f(x)\, dx$ of Example 12 is linear.

52. Find $L_1(5x^3 - 1)$, where $L_1(f) = \int_0^1 f(x)\, dx$ is the integral linear operator of Example 12.

53. Find $L_1(\sin x)$, where L_1 is the integral linear operator of Example 12. (See Exercise 52.)

54. Find $L_1(\cos 2x)$, where L_1 is the integral linear operator of Example 12.

55. Find $L_1(e^x)$, where L_1 is the integral linear operator of Example 12.

5.2 THE RANGE AND KERNEL OF A LINEAR TRANSFORMATION

In this section we single out two important subspaces of the domain and the codomain of a linear transformation and briefly discuss their properties. Before actually defining these, however, let us show that a linear transformation always takes the zero vector to the zero vector in the following sense:

> **Theorem 5.2** If $T: U \to V$ is any linear transformation from the vector space U into the vector space V, then $T(\mathbf{0_U}) = \mathbf{0_V}$.

We used subscripts in the statement of the theorem to distinguish between the zero vectors of U and V, but we drop this cumbersome notation, since the context will make it clear which space we're talking about.

Proof Note first that in U since $\mathbf{0} + \mathbf{0} = \mathbf{0}$ we have

$$T(\mathbf{0} + \mathbf{0}) = T(\mathbf{0}). \tag{5.4}$$

But T is a linear transformation, so applying that to (5.4) we have, in V, the equation

$$T(\mathbf{0}) + T(\mathbf{0}) = T(\mathbf{0}). \tag{5.5}$$

From (5.5) we conclude that $T(\mathbf{0})$ must be the zero vector in V. (Why?) ❑

Theorem 5.2 shows us that a given linear transformation $T: U \to V$ always maps the zero vector of U onto the zero vector of V. It may well happen that for this transformation T there is some nonzero vector $\mathbf{x}$ in U for which we also have $T(\mathbf{x}) = \mathbf{0}$. In Example 6 of the previous section we saw the extreme case where the zero transformation O mapped *every* vector in U to the zero vector of V. What can we say in general about those vectors $\mathbf{x}$ in U which satisfy the equation $T(\mathbf{x}) = \mathbf{0}$? It turns out that they form a *subspace* of the domain U of the transformation T. Before proving this fact, however, we first introduce some terminology with which to describe the set of all such vectors.

> **Definition** Let $T: U \to V$ be a linear transformation of the vector space U into the vector space V. The collection of all those vectors $\mathbf{x}$ in U such that $T(\mathbf{x}) = \mathbf{0}$ (the zero vector of V) is called the **kernel of T** and is denoted by ker T. The kernel of T is also often called the **null-space** of T.

Example 1   Let $U = V = \mathbb{R}^2$ be Euclidean 2-space and let $T: U \to V$ be defined by the equation

$$T(x_1, x_2) = (x_1, 0). \qquad \textbf{(5.6)}$$

We leave it to you to verify that this transformation, a *projection* of $\mathbb{R}^2$ *onto* the x_1 axis, is in fact a linear transformation. What is its kernel? To find the answer, we wish to know which vectors project onto the zero vector. You can verify that

$$K = \text{kernel of } T = \{\mathbf{u} = (0, x_2), x_2 \in \mathbb{R}\}.$$

In other words, the kernel of this projection is the set of all vectors that lie along the x_2 axis. Notice also that, as we asserted above, this collection is indeed a subspace of $\mathbb{R}^2$. ■

We prove that this is always the case for every linear transformation. We have the following theorem.

Theorem 5.3   The kernel K of any linear transformation $T: U \to V$ is a subspace of the domain U of the transformation.

Proof   Because of Theorem 3.3 (page 179), you will recall that it is sufficient to prove (i) that the kernel K is not empty, (ii) that it is closed with respect to vector addition, and (iii) that it is closed with respect to scalar multiplication. We proceed to do each of these. First note that, by Theorem 5.2, $T(\mathbf{0}) = \mathbf{0}$, so the zero vector of U is always in the kernel K of T. Therefore, $K = \ker T$ is not empty. Next suppose that the vectors $\mathbf{x}$ and $\mathbf{y}$ are both in K. Then by definition of K, both $T(\mathbf{x}) = \mathbf{0}$ and $T(\mathbf{y}) = \mathbf{0}$. Then, by definition of a linear transformation, we have

$$T(\mathbf{x} + \mathbf{y}) = T(\mathbf{x}) + T(\mathbf{y}) = \mathbf{0} + \mathbf{0} = \mathbf{0}.$$

So (ii) is true; that is, $T(\mathbf{x} + \mathbf{y}) = \mathbf{0}$, so $\mathbf{x} + \mathbf{y}$ is in $\ker T$ (the kernel K of T). Now suppose that κ is any scalar. Since $T(\mathbf{x}) = \mathbf{0}$, and T is linear, we have

$$T(\kappa \mathbf{x}) = \kappa T(\mathbf{x}) = \kappa \mathbf{0} = \mathbf{0}.$$

Therefore, $\kappa \mathbf{x}$ is in $K = \ker T$; thus, part (iii). Therefore, by Theorem 3.3, K is a subspace of U. This completes the proof of the theorem. ❏

Example 2   Let $T: \mathbb{R}^3 \to \mathbb{R}^2$ be the following linear transformation. Find the kernel K of T.

$$T(x, y, z) = (x - y, x - z) \qquad \textbf{(5.7)}$$

We leave it for you to verify that equation (5.7) does indeed define a *linear* transformation of $\mathbb{R}^3$ into $\mathbb{R}^2$. Its kernel K consists of all those vectors $\mathbf{u} = (u_1, u_2, u_3)$ in $\mathbb{R}^3$ which have the property that $T(\mathbf{u}) = \mathbf{0} = (0, 0)$ in $\mathbb{R}^2$. That is, $T(\mathbf{u}) = (u_1 - u_2, u_1 - u_3) = (0, 0)$. Hence it must be true that $u_1 - u_2 = 0$ and $u_1 - u_3 = 0$. This is a homogeneous system of two equations in three unknowns that has the general solution $u_1 = u_2 = u_3 = t$. Therefore, each vector $\mathbf{x} = (t, t, t)$ in the kernel of T is a scalar multiple of the vector $\mathbf{v} = (1, 1, 1)$. This vector forms a basis for the kernel of the transformation T of (5.7); $K = \{t\mathbf{v} \mid t \text{ any scalar}\}$. Therefore, the kernel K is a one-dimensional subspace of $\mathbb{R}^3$. ∎

We remind you that for any linear transformation $T: U \to V$, the *kernel* of T is a subspace of the *domain U* of the transformation.

Let us now shift our attention and focus on the *codomain V* of a given transformation $T: U \to V$. To illustrate this, we consider another example.

Example 3   Let $T: \mathbb{R}^3 \to \mathbb{R}^4$ be the following linear transformation:

$$T(x, y, z) = (x + y, 0, x + z, 0).$$

Describe all of those vectors $\mathbf{v}$ in $\mathbb{R}^4$ which are images $T(\mathbf{u})$ of vectors $\mathbf{u}$ in $\mathbb{R}^3$ under this transformation T.

Once again we leave it to you to verify that this T is indeed a *linear* transformation. From its defining equation we see that any vector in $\mathbb{R}^4$ that has a nonzero second or fourth component, such as $(1, -1, 1, 1)$ *cannot* possibly be an image $T(\mathbf{u})$ of any vector $\mathbf{u}$ in $\mathbb{R}^3$ under this transformation. So, just which vectors of the form $(a, 0, b, 0)$ are actually images? To answer this question we must find numbers a and b such that $a = x + y$ and $b = x + z$. This gives us an underdetermined system of two equations in the three unknowns x, y, and z. This system has solutions for every a and b. One such solution is $x = b - 1$, $y = a - b + 1$, $z = 1$, as you can verify. In this case the vector $\mathbf{u} = (b - 1, a - b + 1, 1)$ is such that $T(\mathbf{u}) = (a, 0, b, 0)$.

Now for any two real numbers a and b, you can see that

$$(a, 0, b, 0) = a(1, 0, 0, 0) + b(0, 0, 1, 0). \tag{5.8}$$

In fact, for any real number t the vector $\mathbf{u}_t = (b - t, a - b + t, t)$ has the property that $T(\mathbf{u}_t) = (a, 0, b, 0)$. At any rate, those vectors in $\mathbb{R}^4$ which are images under T of vectors in $\mathbb{R}^3$ all are of the form (5.8). Therefore, they form a two-dimensional subspace of $\mathbb{R}^4$, namely, the linear span of the two vectors $\mathbf{e}_1 = (1, 0, 0, 0)$ and $\mathbf{e}_3 = (0, 0, 1, 0)$. ∎

Let us give this collection of images a name. The name that we choose is the *range* of T, the same that is used for any function. We restate the definition as it applies here to linear transformations of vector spaces as follows.

Definition   Let $T: U \to V$ be a linear transformation of the vector space U into the vector space V. The collection $T(U)$ of all those vectors $\mathbf{y}$ in V for which there exists some vector $\mathbf{x}$ in U so that $T(\mathbf{x}) = \mathbf{y}$ is called the **range** of T. It is also often called the **image** of T.

In other words, to be in the *range* of T, a vector in V must be of the form $T(\mathbf{x})$, where $\mathbf{x}$ is some vector in U. See Figure 5.7.

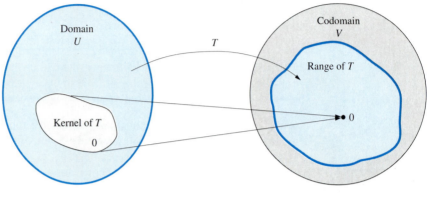

FIGURE 5.7

The *kernel of T* is a *subspace* of the *domain* U of the transformation T. The *range of T* is a subset of the *codomain* of any function T, but in the particular case where T is a linear transformation between vector spaces we have more. The range is a vector space also, not merely a piece of the codomain vector space. We ask you (Exercise 27) to prove this as Theorem 5.4.

> **Theorem 5.4**   The range of a linear transformation $T: U \to V$ is a subspace of V.

The next two examples illustrate the range of a transformation.

Example 4   Consider again the projection linear transformation $T: \mathbb{R}^2 \to \mathbb{R}^2$ of Example 1. In view of (5.6), we see that the range of T is the set of all vectors $(x_1, 0)$, where x_1 is any real number. Thus the range of T is precisely the set of all points on the x_1 axis.

Note that the x_1 axis is a subspace of $\mathbb{R}^2$; whereas any other horizontal line, say one that is one unit above the axis (the set $H = \{(x_1, 1), x_1 \in \mathbb{R}\}$, that is), is merely a *subset* of the codomain $\mathbb{R}^2$ but is not a *subspace*. (Why?) ∎

Example 5   Find the range of the transformation T of Example 2.

Here we have $T: \mathbb{R}^3 \to \mathbb{R}^2$, with (a, b) in the codomain $\mathbb{R}^2$ being the image $T(\mathbf{u})$ of a vector $\mathbf{u} = (x, y, z)$ in $\mathbb{R}^3$ when $a = x - y$ and $b = x - z$. Again this gives us an underdetermined system of linear equations. The augmented matrix for this system is

$$\begin{bmatrix} 1 & -1 & 0 & a \\ 1 & 0 & -1 & b \end{bmatrix},$$

which is row equivalent to

$$\begin{bmatrix} 1 & 0 & -1 & b \\ 0 & 1 & -1 & b-a \end{bmatrix}.$$

Therefore, the system always has solutions. So every (a, b) in $\mathbb{R}^2$ is the image $T(\mathbf{u})$ of some $\mathbf{u}$ in $\mathbb{R}^3$ of the form $\mathbf{u} = (b+t, b-a+t, t)$, for all real numbers t. Thus all of $\mathbb{R}^2$ is the range in this case.   ∎

Now refer back to Example 3. We found the range of the given transformation. Its kernel is readily seen to consist of all of these vectors in $\mathbb{R}^3$ of the form $\mathbf{u} = (x, -x, -x)$. You can easily compute that $T(x, -x, -x) = (0, 0, 0, 0)$. Conversely, any vector $\mathbf{u}$ of which $T(\mathbf{u}) = \mathbf{0}$ is of the form $x(1, -1, -1)$. Therefore, the kernel of this transformation is spanned by the vector $(1, -1, -1)$ and is as a result one-dimensional.

It is instructive to examine the results of Examples 1–5 a little. We summarize them in the following table.

Examples	Linear Transformation	Kernel	Range
1 and 4	$T(x_1, x_2) = (x_1, 0)$	$\{x_2(0, 1)\} \subset \mathbb{R}^2$	$\{x_1(1, 0)\} \subset \mathbb{R}^2$
2 and 5	$T(x, y, z) = (x-y, x-z)$	$\{x(1, 1, 1)\} \subset \mathbb{R}^3$	$\{a(1, 0) + b(0, 1)\} \subseteq \mathbb{R}^2$
3	$T(x, y, z) = (x+y, 0, x+z, 0)$	$\{x(1, -1, -1)\} \subset \mathbb{R}^3$	$\{a(1, 0, 0, 0) + b(0, 0, 1, 0)\} \subset \mathbb{R}^4$

Note that in each case given in this table the dimension of the *kernel* **plus** the dimension of the *range* of the linear transformation T is **equal to** the dimension of its *domain*. This is always the case. We state the following theorem and then give some further illustration of the ideas involved before giving its proof.

Theorem 5.5   (Sylvester's Law of Nullity) Let $T: U \rightarrow V$ be a linear transformation. Then

(Dimension of kernel of T) + (dimension of range of T)
= dimension of domain U of T.

First consider the following example.

Example 6   Let $T: \mathbb{R}^4 \rightarrow \mathbb{R}^3$ be given by the formula

$$T(x_1, x_2, x_3, x_4) = (0, x_1 - x_2, x_3 - x_4).$$

Then $T(x_1, x_2, x_3, x_4) = \mathbf{0} = (0, 0, 0)$ if, and only if, $x_1 - x_2 = 0$ and $x_3 - x_4 = 0$, that is, when $x_1 = x_2$ and $x_3 = x_4$. Thus vectors $\mathbf{u} = (a, a, b, b)$ form the kernel, ker T, of T. Each of these is a linear combination of the two vectors $v_1 = (1, 1, 0, 0)$ and $v_2 = (0, 0, 1, 1)$, as you can verify. So the dimension of ker T is 2. Those vectors w in the codomain $\mathbb{R}^3$ which are images $T(x)$ of vectors x in $\mathbb{R}^4$ must have zero first component. However, given any vector of the form $z = (0, s, t)$ in $\mathbb{R}^3$, you can verify that z is the image of at least the vector $(s, 0, t, 0)$ in $\mathbb{R}^4$. Hence z lies in the range of T; thus the dimension of the range of T is also 2. Note that range T has $\{(0, 1, 0)\}, (0, 0, 1)\}$ as a basis. Therefore, as indicated in Theorem 5.5,

Dimension of ker T + dimension of range T = dimension of $\mathbb{R}^4$ = 4. ■

Example 7 Let $A = (a_{ij})$ be an $m \times n$ matrix, $i = 1, 2, \ldots, m; j = 1, 2, \ldots, n$. Let us consider the matrix transformation of $T_A: \mathbb{R}^n \to \mathbb{R}^m$ given by $T_A(x) = Ax$. (See Example 7 of Section 5.1.) The kernel of T_A is the collection of all those column vectors

$$x = \begin{bmatrix} x_1 \\ x_2 \\ \vdots \\ x_n \end{bmatrix}$$

that are solutions to the equation $T_A(x) = Ax = 0$, that is, to the homogeneous system of equations

$$
\begin{aligned}
a_{11}x_1 + a_{12}x_2 + \cdots + a_{1n}x_n &= 0, \\
a_{21}x_1 + a_{22}x_2 + \cdots + a_{2n}x_n &= 0, \\
\vdots \qquad \vdots \qquad\qquad \vdots \quad &\;\; \vdots \\
a_{m1}x_1 + a_{m2}x_2 + \cdots + a_{mn}x_n &= 0.
\end{aligned}
$$

This is the *solution* space of the system of homogeneous equations $Ax = 0$. This solution space is also often called the **null space** *of the matrix A.* We discuss the connection between the *kernel* of a matrix transformation and the *solution* space or *null* space of the matrix further in the next section, but we surely have the following in the case of matrix transformation:

> The **kernel** of the transformation T_A = the **null space** of A = the **solution space** of $Ax = 0$.

We also take note of the fact that the *range* of the transformation T_A consists of all of those column vectors ($m \times 1$) matrices

$$b = \begin{bmatrix} b_1 \\ b_2 \\ \vdots \\ b_m \end{bmatrix}$$

in $\mathbb{R}^m$ which are such that $T_A(x) = b$, for some x in $\mathbb{R}^n$, that is, those b for which the system of equations $Ax = b$ has a solution. Thus the range of T_A is the set of all $m \times 1$ matrices for which there exists a column vector (an $n \times 1$ matrix) $\mathbf{x}$ so that $A\mathbf{x} = \mathbf{b}$. As we learned in Section 3.5, then, the range of T_A is precisely the collection of all column vectors that lie in the *column space* of the matrix A.

We also noted in Chapter 3 that the dimension of the column space of a matrix is the *rank* of that matrix. This terminology is applied to matrix transformations, and as we shall shortly see, to all linear transformations as well. Let's precisely define the words. ∎

Definition   The **rank** of any linear transformation $T: U \rightarrow V$ is the dimension of its range. The **nullity** of T is the dimension of its null space (kernel).

Accordingly, we can formulate the statement of Theorem 5.5 in the following form. This form is most often known as **Sylvester's law of nullity.**

(The nullity of T) + (the rank of T)
= the dimension of the domain of T.

Now for any linear transformation $T: U \rightarrow V$ from a finite-dimensional vector space U into the vector space V, consider the following.

Let $B = \{\mathbf{u}_1, \mathbf{u}_2, \ldots, \mathbf{u}_n\}$ be a **basis** for the domain U of T. The vectors $T(\mathbf{u}_1), T(\mathbf{u}_2), \ldots, T(\mathbf{u}_n)$ in V **span** (generate) $T(U) =$ the *range* of T in V.

To see why this is so, suppose that $\mathbf{w}$ is any vector in $T(U)$ (the range of $T \subseteq V$). Then there must be at least one vector $\mathbf{x}$ in U such that $\mathbf{w} = T(\mathbf{x})$. But, since the set B is a basis for U, there must exist scalars $\alpha_1, \alpha_2, \ldots, \alpha_n$ so that

$$\mathbf{x} = \alpha_1 \mathbf{u}_1 + \alpha_2 \mathbf{u}_2 + \cdots + \alpha_n \mathbf{u}_n. \tag{5.9}$$

Hence since T is a linear transformation,

$$\mathbf{w} = T(\mathbf{x}) = \alpha_1 T(\mathbf{u}_1) + \alpha_2 T(\mathbf{u}_2) + \cdots + \alpha_n T(\mathbf{u}_n). \tag{5.10}$$

This verifies our claim above that the vectors $T(\mathbf{u}_1), T(\mathbf{u}_2), \ldots, T(\mathbf{u}_n)$ span the range of T. Note that we have proved the following fact also.

The action of the transformation T is completely determined by its action on a given **basis** for its domain U.

The vectors $T(\mathbf{u}_i)$, $i = 1, 2, \ldots, n$, span the range of T and completely determine the action of T on U. However, they do not necessarily form a linearly independent set, so they may not be a basis for the *range* of T (in V).

For example, consider the set of vectors

$$T(\mathbf{e}_1) = T(1, 0, 0, 0) = (0, 1, 0),$$
$$T(\mathbf{e}_2) = T(0, 1, 0, 0) = (0, -1, 0),$$
$$T(\mathbf{e}_3) = T(0, 0, 1, 0) = (0, 0, 1),$$
$$T(\mathbf{e}_4) = T(0, 0, 0, 1) = (0, 0, -1).$$

which are images of the standard basis E for $\mathbb{R}^4$ under the transformation T of Example 6. While these four vectors *span* the range of T in $\mathbb{R}^3$, they are not a *basis* for any subspace of $\mathbb{R}^3$, because they are linearly *dependent*.

To select a basis from among the above set is, however, not difficult. For a systematic approach to such a task simply think of these vectors as the rows of a matrix. Then change the problem to that of finding a basis for the row space of this resulting matrix. We do this by row reduction, as we did in Chapter 3. We justify this process more in the next section. In any event, here we have

$$\begin{bmatrix} 0 & 1 & 0 \\ 0 & -1 & 0 \\ 0 & 0 & 1 \\ 0 & 0 & -1 \end{bmatrix} \rightarrow \begin{bmatrix} 0 & 1 & 0 \\ 0 & 0 & 1 \\ 0 & 0 & 0 \\ 0 & 0 & 0 \end{bmatrix}.$$

Now, upon changing the problem back to a vector problem, we find that the range of this transformation T is the two-dimensional subspace of $\mathbb{R}^3$, which is spanned by the two vectors $\mathbf{e}'_2 = (0, 1, 0)$ and $\mathbf{e}'_3 = (0, 0, 1)$. You can check this directly. So the *rank* of T is two. Therefore, by Theorem 5.5 the *nullity* is also two, as we already noted in Example 6.

Example 8 Let $T_A : \mathbb{R}^4 \to \mathbb{R}^3$ be multiplication by the matrix

$$A = \begin{bmatrix} 1 & 1 & -1 & 2 \\ 2 & 1 & 4 & 3 \\ 5 & 6 & 1 & 9 \end{bmatrix}.$$

(See Example 7.) Find a basis for the range of T_A and find a basis for the kernel of T_A. Remember that $T_A(\mathbf{x}) = A\mathbf{x}$ for each column vector (4×1 matrix) in $\mathbb{R}^4$. Now, making use of what we have learned, let us see what T_A does to a basis for $\mathbb{R}^4$. We have

$$T_A(\mathbf{e}_1) = A \begin{bmatrix} 1 \\ 0 \\ 0 \\ 0 \end{bmatrix} = \begin{bmatrix} 1 \\ 2 \\ 5 \end{bmatrix}, \qquad T_A(\mathbf{e}_2) = A \begin{bmatrix} 0 \\ 1 \\ 0 \\ 0 \end{bmatrix} = \begin{bmatrix} 1 \\ 1 \\ 6 \end{bmatrix},$$

$$T_A(\mathbf{e}_3) = A \begin{bmatrix} 0 \\ 0 \\ 1 \\ 0 \end{bmatrix} = \begin{bmatrix} -1 \\ 4 \\ 1 \end{bmatrix}, \qquad T_A(\mathbf{e}_4) = A \begin{bmatrix} 0 \\ 0 \\ 0 \\ 1 \end{bmatrix} = \begin{bmatrix} 2 \\ 3 \\ 9 \end{bmatrix}.$$

Note the interesting fact that the resulting vectors $T_A(\mathbf{e}_1)$, $T_A(\mathbf{e}_2)$, $T_A(\mathbf{e}_3)$, and $T_A(\mathbf{e}_4)$ are precisely the columns of the matrix A. In the next section we show that this always happens. For now we proceed with the calculations in this example. To find a basis for the range of T_A we find a basis for the column space of A, as we noted above. As discussed in Section 3.5, this is most easily accomplished by row reducing A^T to its reduced row echelon form N. Thus

$$A^T = \begin{bmatrix} 1 & 2 & 5 \\ 1 & 1 & 6 \\ -1 & 4 & 1 \\ 2 & 3 & 9 \end{bmatrix} \rightarrow N = \begin{bmatrix} 1 & 0 & 0 \\ 0 & 1 & 0 \\ 0 & 0 & 1 \\ 0 & 0 & 0 \end{bmatrix}.$$

Thus, we see from N that the range of T_A has dimension three and has the three standard basis vectors for $\mathbb{R}^3$ as a basis.

To find a basis for the kernel of T, as we noted above, we seek those vectors $\mathbf{x}$ in $\mathbb{R}^4$ which satisfy $A\mathbf{x} = \mathbf{0}$. The augmented matrix for this system is, of course, $[A: \mathbf{0}]$, but as we noted in Chapter 1 row reduction of this matrix will not affect the last column of zeros, so it is only necessary to row reduce A itself. You can compute that this A is row equivalent to

$$C = \begin{bmatrix} 1 & 0 & 0 & \dfrac{11}{6} \\ 0 & 1 & 0 & 0 \\ 0 & 0 & 1 & -\dfrac{1}{6} \end{bmatrix}.$$

Therefore, if
$$\begin{cases} x_1 = -\dfrac{11}{6}x_4, \\ x_2 = 0, \\ x_3 = \dfrac{1}{6}x_4, \end{cases} \quad \text{then} \quad \mathbf{x} = \begin{bmatrix} x_1 \\ x_2 \\ x_3 \\ x_4 \end{bmatrix}$$

is in the kernel of T_A. So, if we select $x_4 = 6$, the column vector

$$\mathbf{v} = \begin{bmatrix} -11 \\ 0 \\ 1 \\ 6 \end{bmatrix}$$

spans the kernel of T_A. Note that here the dimension of the kernel is *one*, the dimension of the range is *three*, and the dimension of $\mathbb{R}^4$ is *four*, concurring with Theorem 5.5.   ■

Let us now proceed with our proof of Theorem 5.5. It is an application of our discussion but requires some detailed argument.

Proof of Theorem 5.5   Let $T: U \rightarrow V$ be a linear transformation from the n-dimensional vector space U to the (m-dimensional) vector space V. First consider the case where the kernel, ker $T = \{\mathbf{0}\}$. Suppose that $B = \{\mathbf{u}_1, \mathbf{u}_2, \ldots, \mathbf{u}_n\}$

is a **basis** for the domain U of T. As we noted in the discussion just preceding Example 8, the set $T(B) = \{T(\mathbf{u}_1), T(\mathbf{u}_2), \ldots, \mathbf{T}(\mathbf{u}_n)\}$ *spans* the range $T(U)$ of T in V. Since ker $T = \{\mathbf{0}\}$, the set $T(B)$ is linearly independent. Otherwise one of these vectors, say $T(\mathbf{u}_1)$, is a linear combination of the rest of them. That is,

$$T(\mathbf{u}_1) = c_2 T(\mathbf{u}_2) + c_3 T(\mathbf{u}_3) + \cdots + c_n T(\mathbf{u}_n)$$
$$= T(c_2 \mathbf{u}_2 + c_3 \mathbf{u}_3 + \cdots + c_n \mathbf{u}_n) = T(\mathbf{v}),$$

where $\mathbf{v} = c_2 \mathbf{u}_2 + c_3 \mathbf{u}_3 + \cdots + c_n \mathbf{u}_n$.

And, therefore, $T(\mathbf{u}_1 - \mathbf{v}) = \mathbf{0}$, so $\mathbf{u}_1 - \mathbf{v}$ is the kernel of T. But ker $T = \{\mathbf{0}\}$, so $\mathbf{u}_1 - \mathbf{v} = \mathbf{0}$, or $\mathbf{u}_1 = \mathbf{v}$. But this says that

$$\mathbf{u}_1 = c_2 \mathbf{u}_2 + c_3 \mathbf{u}_3 + \cdots + c_n \mathbf{u}_n.$$

This is a contradiction of the fact that B is a basis, and the vectors in B are linearly *independent*. Thus $T(B)$ must be linearly independent also, and hence is a basis for the range of T. So the dimension of ker T is *zero* and the dimension of the range is n; thus

$$\dim \ker T + \dim \text{range } T = \dim U,$$

as desired.

Now let us assume that ker $T \neq \{\mathbf{0}\}$. Then dim ker $T = r \geq 1$. (Why?) We have $1 \leq r \leq n$. In the other extreme case $r = n$; then ker T is all of U. (Why?) Hence $T(\mathbf{u}) = \mathbf{0}$, for *all* vectors $\mathbf{u} \in U$. This implies that the range of T is just the zero vector, and thus range $T = \{\mathbf{0}\}$. In this case dim ker $T = n$, dim range $T = 0$, and again their sum is dim U, as desired.

Finally the only case left is that in which $1 \leq r < n$. (Remember that r is dimension of ker T.) Now let $S = \{\mathbf{u}_1, \mathbf{u}_2, \ldots, \mathbf{u}_r\}$ be a basis for the kernel, ker T, of T. By Theorem 3.9 (page 192) we can extend S to a basis B for all of U by augmenting S with some $n - r$ linearly independent vectors $\mathbf{u}_{r+1}, \mathbf{u}_{r+2}, \ldots, \mathbf{u}_n$. We shall show that the set of vectors $H = \{T(\mathbf{u}_{r+1}), T(\mathbf{u}_{r+2}), \ldots, T(\mathbf{u}_n)\}$ is a basis for the range of T in V. To do that we must show that it both (1) *spans* the range and (2) is *linearly independent*. First to show (1) that H spans the range of T in V, let $\mathbf{v}$ be any vector in the range, so $\mathbf{v} = T(\mathbf{w})$ for some vector $\mathbf{w}$ in U. Since $\mathbf{w} \in U$, $\mathbf{w}$ is a linear combination of the basis vectors in B; that is,

$$\mathbf{w} = k_1 \mathbf{u}_1 + k_2 \mathbf{u}_2 + \cdots + k_r \mathbf{u}_r + k_{r+1} \mathbf{u}_{r+1} + \cdots + k_n \mathbf{u}_n.$$

Therefore,

$$\mathbf{v} = T(\mathbf{w}) = T(k_1 \mathbf{u}_1 + k_2 \mathbf{u}_2 + \cdots + k_r \mathbf{u}_r + k_{r+1} \mathbf{u}_{r+1} + \cdots + k_n \mathbf{u}_n)$$
$$= k_1 T(\mathbf{u}_1) + k_2 T(\mathbf{u}_2) + \cdots + k_r T(\mathbf{u}_r) + k_{r+1} T(\mathbf{u}_{r+1})$$
$$+ \cdots + k_n T(\mathbf{u}_n)$$
$$= k_1 \mathbf{0} + k_2 \mathbf{0} + \cdots + k_r \mathbf{0} + k_{r+1} T(\mathbf{u}_{r+1}) + \cdots + k_n T(\mathbf{u}_n)$$
$$= \mathbf{0} + k_{r+1} T(\mathbf{u}_{r+1}) + \cdots + k_n T(\mathbf{u}_n)$$
$$= k_{r+1} T(\mathbf{u}_{r+1}) + \cdots + k_n T(\mathbf{u}_n),$$

because the vectors $\mathbf{u}_1, \mathbf{u}_2, \ldots, \mathbf{u}_r$ all lie in the kernel of T. Thus the set H *spans* the range of T.

Moreover the set H is linearly independent; for if

$$\mathbf{0} = c_{r+1}T(\mathbf{u}_{r+1}) + c_{r+2}T(\mathbf{u}_{r+2}) + \cdots + c_n T(\mathbf{u}_n), \qquad (*)$$

then

$$\mathbf{0} = T(c_{r+1}\mathbf{u}_{r+1} + c_{r+2}\mathbf{u}_{r+2} + \cdots + c_n\mathbf{u}_n) = T(\mathbf{z}).$$

Therefore, the vector

$$\mathbf{z} = c_{r+1}\mathbf{u}_{r+1} + c_{r+2}\mathbf{u}_{r+2} + \cdots + c_n\mathbf{u}_n$$

must be in ker T. But the set $S = \{\mathbf{u}_1, \mathbf{u}_2, \ldots, \mathbf{u}_r\}$ is a basis for ker T so $\mathbf{z}$ is also a linear combination of the vectors in S,

$$\mathbf{z} = c_1\mathbf{u}_1 + c_2\mathbf{u}_2 + \cdots + c_r\mathbf{u}_r.$$

Equating these two representations of $\mathbf{z}$ we have

$$\mathbf{z} = c_1\mathbf{u}_1 + c_2\mathbf{u}_2 + \cdots + c_r\mathbf{u}_r = c_{r+1}\mathbf{u}_{r+1} + c_{r+2}\mathbf{u}_{r+2} + \cdots + c_n\mathbf{u}_n.$$

Therefore,

$$-c_1\mathbf{u}_1 - c_2\mathbf{u}_2 - \cdots - c_r\mathbf{u}_r + c_{r+1}\mathbf{u}_{r+1} + c_{r+2}\mathbf{u}_{r+2} + \cdots + c_n\mathbf{u}_n = \mathbf{0}.$$

But the vectors $\mathbf{u}_1, \mathbf{u}_2, \ldots, \mathbf{u}_r, \mathbf{u}_{r+1}, \ldots, \mathbf{u}_n$ form a basis for U and are linearly independent. So each of the coefficients c_i, $i = 1, 2, \ldots, n$ must be zero. In particular the coefficients $c_{r+1}, \ldots, c_n$ in the equation (*) above are all *forced to be zero*. Hence H is linearly independent, as claimed. Since H is a basis for the range of T, the dimension of range T is $n - r$, and once again we have

(dimension of the kernel of T) + (dimension of the range of T)

= dimension of the domain of T.

This completes the proof of the theorem.   ❏

Vocabulary

In this section we have encountered some new concepts and terms. The vocabulary developed in this section is basic to the study of linear transformations and is used repeatedly. Be sure that you understand each of the concepts described by the terms in the following list. Review these ideas until you feel secure enough to use them in the rest of this course and in your future use of linear algebra. You also may want to review Appendix A.2.

kernel of T

null space of T

nullity of T

range of T

rank of T

Sylvester's law of nullity

Exercises 5.2

In Exercises 1–6 $T_A: \overline{\mathbb{R}^2} \to \overline{\mathbb{R}^2}$ is defined by $T_A(\mathbf{x}) = A\mathbf{x}$, where $A = \begin{bmatrix} -1 & 3 \\ 2 & -6 \end{bmatrix}$.

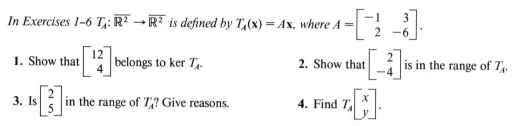

1. Show that $\begin{bmatrix} 12 \\ 4 \end{bmatrix}$ belongs to ker T_A.

2. Show that $\begin{bmatrix} 2 \\ -4 \end{bmatrix}$ is in the range of T_A.

3. Is $\begin{bmatrix} 2 \\ 5 \end{bmatrix}$ in the range of T_A? Give reasons.

4. Find $T_A \begin{bmatrix} x \\ y \end{bmatrix}$.

5. Find a basis for the range of T_A.

6. Find a basis for the kernel of T_A.

7. Find a basis for the range of the linear transformation T of Example 3 in Section 5.1; namely, $T(x, y) = (2x, -y)$.

8. Find a basis for the kernel of the linear transformation T of Example 3 in Section 5.1. (See Exercise 7.) Confirm Sylvester's law of nullity (Theorem 5.5, page 313) for this example.

9. Find a basis for the range of the linear transformation T of Example 1 in Section 5.1; namely,

$$T(x, y) = (2x, 2y).$$

10. Find a basis for the kernel of the linear transformation T of Example 1 in Section 5.1. (See Exercise 9.) Confirm Sylvester's law of nullity (Theorem 5.5) for this example.

11. Find a basis for the range of the linear transformation T of Example 2 in Section 5.1; namely,

$$T(x, y, z) = (x, y).$$

12. Find a basis for the kernel of the linear transformation T of Example 2 in Section 5.1. (See Exercise 11.) Confirm Sylvester's law of nullity for this example.

13. Find a basis for the range of the linear transformation T of Example 4 in Section 5.1; namely,

$$T(x, y) = (x, -y, x - y).$$

14. Find a basis for the kernel of the linear transformation T of Example 4 in Section 5.1. (See Exercise 13.) Confirm Sylvester's law of nullity for this example.

15. Find a basis for the range and for the kernel of the identity linear transformation I of Example 5 in Section 5.1, when $U = \mathbb{R}^3$; namely, $I(\mathbf{x}) = \mathbf{x}$. Confirm Sylvester's law of nullity for this transformation.

16. Find a basis for the range and for the kernel of the zero linear transformation O of Example 6 in Section 5.1, when $U = \mathbb{R}^3$; namely, $O(\mathbf{x}) = \mathbf{0}$. Confirm Sylvester's law of nullity for this transformation.

17. Find a basis for the range and for the kernel of the linear transformation T of Exercise 8 in Section 5.1; namely,

$$T(x_1, x_2) = (2x_1 + 3x_2, 3x_1 - 2x_2).$$

Confirm Sylvester's law of nullity for this transformation.

18. Find a basis for the range and for the kernel of the linear transformation T of Exercise 10 in Section 5.1; namely, $T(x_1, x_2) = (x_1 + 3x_2, x_2)$. Confirm Sylvester's law of nullity for this transformation.

19. Find a basis for the range and for the kernel of the linear transformation T of Exercise 11 in Section 5.1; namely, $T(x_1, x_2, x_3) = (x_1, -x_2, 0)$. Confirm Sylvester's law of nullity for this transformation.

20. Find a basis for the range and for the kernel of the linear transformation T of Exercise 12 in Section 5.1; namely,

$$T(x_1, x_2, x_3) = (x_3 - x_2, 0, x_3 - x_1).$$

Confirm Sylvester's law of nullity for this transformation.

21. Find a basis for the range and for the kernel of the linear transformation T of Exercise 14 in Section 5.1; namely,

$$T(x_1, x_2, x_3) = (x_1 - x_2, x_2 + x_3, x_3 - x_1).$$

Confirm Sylvester's law of nullity for this transformation.

22. Suppose that U is a vector space of dimension 3 with basis $B = \{\mathbf{u}_1, \mathbf{u}_2, \mathbf{u}_3\}$. Now suppose $T: U \rightarrow U$ is a linear transformation with $T(\mathbf{u}_1) = \mathbf{u}_3$, $T(\mathbf{u}_2) = \mathbf{u}_2$, $T(\mathbf{u}_3) = \mathbf{u}_1$. Prove that the kernel of $T = \{\mathbf{0}\}$.

23. Find a basis for the range and for the kernel of the linear transformation T of Example 7 in Section 5.1; namely, the matrix transformation $T_A(x_1, x_2, x_3, x_4)^T = A(x_1, x_2, x_3, x_4)^T$. Confirm Sylvester's law of nullity for this transformation. Recall that

$$A = \begin{bmatrix} 1 & 2 & 0 & 3 \\ 0 & 1 & 8 & 1 \\ -1 & 0 & 5 & 0 \end{bmatrix}.$$

24. Find a basis for the range and for the kernel of the linear transformation H of Exercises 29–30 in Section 5.1; namely, the matrix transformation $H(x_1, x_2, x_3) = (x_1, x_2, x_3)A$. Confirm Sylvester's law of nullity for this transformation. Recall that

$$A = \begin{bmatrix} 1 & -1 & 2 \\ 0 & -1 & -2 \\ 2 & -1 & 0 \end{bmatrix}.$$

25. Find a basis for the range and for the kernel of the linear transformation $K_A: M_n \rightarrow M_n$ of Exercises 31–33 in Section 5.1; namely, $K_A(X) = AX - XA$, where A is the matrix just given in Exercise 24. Confirm Sylvester's law of nullity for this transformation.

26. Find a basis for the range and for the kernel of the linear transformation $T_A: \mathbb{C}^n \rightarrow \mathbb{C}^m$ of Exercises 34–36 in Section 5.1; namely, $T_A(\mathbf{x}) = A\mathbf{x}$, where A is the following matrix:

$$\begin{bmatrix} 1 & 0 & i \\ 0 & i & 1 \\ -i & 0 & 1 \end{bmatrix}.$$

Confirm Sylvester's law of nullity for this transformation.

27. Prove Theorem 5.4 (page 312). Use either Theorem 3.3 (page 179) or Exercise 61 of Section 3.3.

28. Let $D: P_n \rightarrow P_n$ be the differentiation linear operator (see Example 11 of Section 5.1) which is easily defined formally, even if you have not studied calculus, by

$$D(a_0 + a_1 x + a_2 x^2 + \cdots + a_{n-1} x^{n-1} + a_n x^n)$$
$$= a_1 + 2a_2 x + \cdots + (n-1) a_{n-1} x^{n-2}$$
$$+ na_n x^{n-1}.$$

Find a basis for the range of this linear transformation.

29. Find a basis for the kernel of the linear differentiation operator of Exercise 28. Confirm Sylvester's law of nullity for this transformation.

30. Use Sylvester's law of nullity (Theorem 5.5) to prove that if A is any $n \times n$ matrix, the dimension of the solution space $S(A)$ of A (solutions to $A\mathbf{x} = \mathbf{0}$) is $n - rank\ A$. Discuss why this is the dimension of the kernel of the matrix transformation $T_A(\mathbf{x}) = A\mathbf{x}$. This is called the *nullity of the matrix A*. Thus *nullity of* $A = n - rank\ A$.

31. Suppose that $T: U \rightarrow V$ is a linear transformation from the vector space U into the vector space V. Let $B = \{\mathbf{u}_1, \mathbf{u}_2, \ldots, \mathbf{u}_n\}$ be a basis for the domain U of T. Prove that the set of vectors $\{T(\mathbf{u}_1), T(\mathbf{u}_2), \ldots, T(\mathbf{u}_n)\}$ in V is a basis for $T(U) = the\ range\ of\ T$ in V if, and only if, $\ker T = \{\mathbf{0}\}$.

32. Prove the following theorem:

Theorem The following are equivalent statements:

(1) The kernel of the linear transformation $T: U \rightarrow V$ is equal to $\{\mathbf{0}\}$.
(2) T is one-to-one.
(3) The vectors $\{T(\mathbf{u}_i) \mid i = 1, 2, \ldots, n\}$ as in Exercise 31 above are a basis for the range of T in V.

---

33. Prove that if $T: U \rightarrow U$ is a linear transformation, T is one-to-one if, and only if, T is *onto* U; that is, the range of T in U is all of U. Show that this statement is false if the domain and codomain of T are not the same set. HINT: See Exercise 32, part (1), and use Theorem 5.5.

34. Suppose that U is an n-dimensional vector space and that $T: U \rightarrow U$ is a linear transformation. Suppose that T takes a basis $B = \{\mathbf{u}_1, \mathbf{u}_2, \ldots, \mathbf{u}_n\}$ of U to a basis $B' = \{\mathbf{u}'_1, \mathbf{u}'_2, \ldots, \mathbf{u}'_n\}$ of U (possibly the same basis). Find the kernel of T and the range of T.

5.3 MATRIX REPRESENTATIONS

In this section we show that every vector in a finite-dimensional vector space U can be represented by a one-column matrix, and every linear transformation T from U to any finite-dimensional vector space V can be regarded as a matrix transformation T_A for some appropriate matrix A. That is, we see how all such linear transformations on *finite-dimensional vector spaces* can be represented as matrix transformations of the type considered in Sections 5.1 and 5.2.

In particular, let U be an n-dimensional vector space and V be m-dimensional and let $T: U \rightarrow V$ be a linear transformation. We show how one can realize T as the matrix transformation $T_A: \overline{\mathbb{R}^n} \rightarrow \overline{\mathbb{R}^m}$, defined by

$$T_A(\mathbf{x}) = A\mathbf{x}, \tag{5.11}$$

for a suitable $m \times n$ matrix A. (If the scalars are from $\mathbb{C}$, the map will be $T_A: \overline{\mathbb{C}^n} \rightarrow \overline{\mathbb{C}^m}$.) This will justify thinking of the range of a transformation as the column space of a matrix and thinking of the null space or kernel of a *matrix* and a *transformation* as essentially the same thing.

First of all let us show that each vector $\mathbf{u}$ in an n-dimensional vector space U can be represented by a column vector ($n \times 1$ matrix) in $\overline{\mathbb{R}^n}$ or $\overline{\mathbb{C}^n}$. To this end suppose that $S = \{\mathbf{u}_1, \mathbf{u}_2, \ldots, \mathbf{u}_n\}$ is a *given ordered basis* for U. Then, since S is a basis, there exist scalars $a_1, a_2, \ldots, a_n$ so that $\mathbf{u} = a_1\mathbf{u}_1 + a_2\mathbf{u}_2 + \cdots + a_n\mathbf{u}_n$. We have the following lemma, which we ask you to prove as Exercise 41, stating that the scalars that appear in this representation of $\mathbf{u}$ are *uniquely determined by the basis S*.

Lemma   If $\mathbf{u}$ is any vector from the vector space U with ordered basis $S = \{\mathbf{u}_1, \mathbf{u}_2, \ldots, \mathbf{u}_n\}$, then the representation $\mathbf{u} = a_1\mathbf{u}_1 + a_2\mathbf{u}_2 + \cdots + a_n\mathbf{u}_n$ of $\mathbf{u}$ as a linear combination of these basis vectors is unique.

These scalars, which are called the **coordinates** of $\mathbf{u}$ with respect to the basis S, are therefore **unique**. They depend only upon the vector $\mathbf{u}$ and the ordered basis S. Thus we may represent the vector $\mathbf{u}$ by a column vector ($n \times 1$ matrix) consisting of these unique scalars; write

$$[\mathbf{u}]_S = \begin{bmatrix} a_1 \\ a_2 \\ \vdots \\ a_n \end{bmatrix}. \tag{5.12}$$

We call the representation (5.12) of the vector $\mathbf{u}$ by the matrix (column vector) $[\mathbf{u}]_S$, the **matrix representation** of $\mathbf{u}$, or the **coordination** of $\mathbf{u}$ with respect to S, or more simply, the **S-coordinatization** of $\mathbf{u}$.

Example 1   Let $U = \mathbb{R}^2$ be Euclidean 2-space, and let $S = E = \{e_1 = (1, 0), e_2 = (0, 1)\}$ be the standard basis for $\mathbb{R}^2$. Then the E-coordinatization, or *standard-coordinatization of the vector* $\mathbf{u} = (3, -1) = 3\mathbf{e}_1 - 1\mathbf{e}_2$, will clearly be the column vector $\mathbf{u}^T = [\mathbf{u}]_E = \begin{bmatrix} 3 \\ -1 \end{bmatrix}$. More generally the E-coordinatization or standard coordinatization of any vector $\mathbf{x} = (x_1, x_2) \in \mathbb{R}^2$ will be $\mathbf{x}^T$; i.e.,

$$[\mathbf{x}]_E = \begin{bmatrix} x_1 \\ x_2 \end{bmatrix}. \quad \blacksquare$$

It is not difficult to generalize this example to see that if $U = \mathbb{R}^n$, we have the following.

> The standard coordinatization of the (row) vector $\mathbf{x} = (x_1, x_2 \ldots, x_n)$ in $\mathbb{R}^n$ is the (column) vector $\mathbf{x}^T$ in $\mathbb{R}^n$.

Example 2   Let $U = P_2$, the space of real polymomials of degree two or less. Consider the standard basis for P_2, namely, $E = \{1, x, x^2\}$. Then if $p(x) = ax^2 + bx + c$ is any polynomial (vector) in P_2, the standard coordinatization of $p(x)$ is the matrix (column vector)

$$[p(x)]_E = \begin{bmatrix} c \\ b \\ a \end{bmatrix}.$$

In particular, if $g(x) = x^2 - 1$, then

$$[g(x)]_E = \begin{bmatrix} -1 \\ 0 \\ 1 \end{bmatrix}, \quad \text{and} \quad [x]_E = \begin{bmatrix} 0 \\ 1 \\ 0 \end{bmatrix}. \quad \blacksquare$$

Example 3   Let $U = \mathbb{R}^3$, and let $B = \{\mathbf{u}_1 = (1, 1, 1), \mathbf{u}_2 = (0, 1, 1), \mathbf{u}_3 = (1, 0, -1)\}$ be a basis for U. You should verify that B is a basis. Let $\mathbf{v} = (2, -3, 4)$. The standard coordinatization for $\mathbf{v}$ is, of course, $\mathbf{v}^T = \begin{bmatrix} 2 \\ -3 \\ 4 \end{bmatrix}$. Contrast this with the B-coordinatization of $\mathbf{v}$, which is $[\mathbf{v}]_B = \begin{bmatrix} 9 \\ -12 \\ -7 \end{bmatrix}$. We obtain this from the fact that

$$\mathbf{v} = (2, -3, 4) = 9(1, 1, 1) - 12(0, 1, 1) - 7(1, 0, -1)$$
$$= 9\mathbf{u}_1 - 12\mathbf{u}_2 - 7\mathbf{u}_3.$$

Check this computation. This example reminds us that the coordinatization of a vector depends not merely on the vector itself but also on the basis for the vector space. ∎

Example 4   In $\mathbb{C}^3$ with the standard ordered basis E, if $\mathbf{u} = (1, 1 + i, i)$, then since

$$\mathbf{u} = \mathbf{e}_1 + (1 + i)\mathbf{e}_2 + i\mathbf{e}_3,$$

therefore,

$$[\mathbf{u}]_E = \begin{bmatrix} 1 \\ 1 + i \\ i \end{bmatrix} = \mathbf{u}^T.$$

Contrast this standard coordinatization with the coordinatization with respect to the basis $S = \{\mathbf{v}_1 = (1, i, 0), \mathbf{v}_2 = (0, i, -1), \mathbf{v}_3 = (0, 0, i)\}$ for $\mathbb{C}^3$. (Verify that S is a basis.) To find the S-coordinatization of $\mathbf{u}$ set

$$\mathbf{u} = (1, 1 + i, i) = c_1\mathbf{v}_1 + c_2\mathbf{v}_2 + c_3\mathbf{v}_3 = c_1(1, i, 0) + c_2(0, i, -1) + c_3(0, 0, i).$$

Solving the resulting system of equations for c_1, c_2, and c_3, we get $c_1 = 1$, $c_2 = -i$, and $c_3 = 0$. Hence,

$$[\mathbf{u}]_S = \begin{bmatrix} 1 \\ -i \\ 0 \end{bmatrix}.$$

Carry out the computations. ∎

Example 5   Let $\mathbb{P}_4$ be the vector space of all polynomials of degree four or less with *complex* coefficients. A "standard" ordered basis for $\mathbb{P}_4$ is $E = \{1, x, x^2, x^3, x^4\}$. So the vector (polynomial)

$$p(x) = x^4 + ix^3 - 2x^2 + (1 - i)x - (1 + 2i)$$

has as its standard coordinatization the column vector

$$[p(x)]_E = \begin{bmatrix} -(1 + 2i) \\ (1 - i) \\ -2 \\ i \\ 1 \end{bmatrix}. \quad \blacksquare$$

We have the following theorem that says that this correspondence between a given vector $\mathbf{u} \in U$ and its matrix representation in $\mathbb{R}^n$ (or $\overline{\mathbb{C}^n}$), *with respect to a given ordered basis B* (that is, the B-coordinatization of $\mathbf{u}$), is a very nice one-to-one operations-preserving linear map. Such a map is called an **isomorphism**, and we say that the vector spaces U and $\mathbb{R}^n$ (or $\overline{\mathbb{C}^n}$, depending on the scalars used) are **isomorphic**.

> **Theorem 5.6** Let U be an n-dimensional real (or complex) vector space with a given ordered basis B. The correspondence $\Phi_B: U \to \overline{\mathbb{R}^n}$ (or $\overline{\mathbb{C}^n}$) given by $\Phi_B(\mathbf{x}) = [\mathbf{x}]_B$, for all $\mathbf{x} \in U$; that is, the B-coordinatization of $\mathbf{x}$ has the following properties:
>
> (i) Φ_B is *one-to-one* and *onto* $\overline{\mathbb{R}^n}$ (or $\overline{\mathbb{C}^n}$).
> (ii) $\Phi_B(\mathbf{u} + \mathbf{w}) = \Phi_B(\mathbf{u}) + \Phi_B(\mathbf{w})$, for *all* $\mathbf{u}$, $\mathbf{w}$ in U.
> (iii) $\Phi_B(k\mathbf{u}) = k\Phi_B(\mathbf{u})$, *for all* $\mathbf{u}$ in U and any scalar k.
> (iv) $\Phi_B(\mathbf{0}) = \mathbf{0}$.
> (v) $\ker \Phi_B = \{\mathbf{0}\}$.
> (vi) $\operatorname{range} \Phi_B = \mathbb{R}^n$ (or $\mathbb{C}^n$).

We leave most of the details of the proof of this theorem as exercises for you. Parts (ii) and (iii) say essentially that Φ_B is a linear transformation, and part (i) is equivalent to parts (v) and (vi) together. We prove part (ii) to show you how the proofs go. You should write out the rest of the proofs. See Exercises 51–55.

Proof of part (ii) Let $B = \{\mathbf{v}_1, \mathbf{v}_2, \ldots, \mathbf{v}_n\}$ be a basis for the vector space U. Then for all vectors $\mathbf{u}$ and $\mathbf{w}$ in U there exist unique scalars (see the lemma at the beginning of this section) α_i and β_i, $i = 1, 2, \ldots, n$, such that

$$\mathbf{u} = \alpha_1 \mathbf{v}_1 + \alpha_2 \mathbf{v}_2 + \cdots + \alpha_n \mathbf{v}_n,$$
$$\mathbf{w} = \beta_1 \mathbf{v}_1 + \beta_2 \mathbf{v}_2 + \cdots + \beta_n \mathbf{v}_n.$$

Therefore,

$$\mathbf{u} + \mathbf{w} = (\alpha_1 + \beta_1)\mathbf{v}_1 + (\alpha_2 + \beta_2)\mathbf{v}_2 + \cdots + (\alpha_n + \beta_n)\mathbf{v}_n.$$

So that the B-coordinatizations for the vectors involved are as follows:

$$[\mathbf{u}]_B = \begin{bmatrix} \alpha_1 \\ \alpha_2 \\ \vdots \\ \alpha_n \end{bmatrix}, \quad [\mathbf{w}]_B = \begin{bmatrix} \beta_1 \\ \beta_2 \\ \vdots \\ \beta_n \end{bmatrix}, \quad \text{and} \quad [\mathbf{u} + \mathbf{w}]_B = \begin{bmatrix} \alpha_1 + \beta_1 \\ \alpha_2 + \beta_2 \\ \vdots \\ \alpha_n + \beta_n \end{bmatrix}.$$

Properties of matrix addition applied to these $n \times 1$ matrices now allow us to conclude that

$$\Phi_B(\mathbf{u} + \mathbf{w}) = [\mathbf{u} + \mathbf{w}]_B = [\mathbf{u}]_B + [\mathbf{w}]_B = \Phi_B(\mathbf{u}) + \Phi_B(\mathbf{w}),$$

as desired. ∎

We proceed now to show that not only can vectors be represented by matrices (in this case one-column matrices) but linear transformations on finite-dimensional vector spaces also have matrix representations. Again this representation is basis-dependent. The computations become somewhat involved. One should take care to not get bogged down in the computational details and miss the overall concept. To see what happens, let U be a given n-dimensional vector space with basis $G = \{\mathbf{u}_1, \mathbf{u}_2, \ldots, \mathbf{u}_n\}$. Also let V be an m-dimensional vector space with a given basis $H = \{\mathbf{v}_1, \mathbf{v}_2, \ldots, \mathbf{v}_m\}$. And let $T: U \to V$ be a given linear transformation from U into V. Now any vector $\mathbf{u}$ in U has a *unique* representation as a linear combination of the basis vectors in G; thus

$$\mathbf{u} = \alpha_1 \mathbf{u}_1 + \alpha_2 \mathbf{u}_2 + \cdots + \alpha_n \mathbf{u}_n. \tag{5.13}$$

Since T is a linear transformation, we may also write

$$\begin{aligned} T(\mathbf{u}) &= T(\alpha_1 \mathbf{u}_1 + \alpha_2 \mathbf{u}_2 + \cdots + \alpha_n \mathbf{u}_n) \\ &= \alpha_1 T(\mathbf{u}_1) + \alpha_2 T(\mathbf{u}_2) + \cdots + \alpha_n T(\mathbf{u}_n). \end{aligned} \tag{5.14}$$

Each of the vectors $T(\mathbf{u}_i)$, for $i = 1, 2, \ldots, n$, is in the codomain V of T. Since H is a basis for V, each of these vectors has a *unique* representation as a linear combination of these basis vectors. Therefore, for $i = 1, 2, \ldots, n$,

$$T(\mathbf{u}_i) = a_{1i} \mathbf{v}_1 + a_{2i} \mathbf{v}_2 + \cdots + a_{mi} \mathbf{v}_m = \sum_{j=1}^{m} a_{ji} \mathbf{v}_j.$$

Let us write these equations in expanded form.

$$\begin{aligned} T(\mathbf{u}_1) &= a_{11} \mathbf{v}_1 + a_{21} \mathbf{v}_2 + \cdots + a_{m1} \mathbf{v}_m, \\ T(\mathbf{u}_2) &= a_{12} \mathbf{v}_1 + a_{22} \mathbf{v}_2 + \cdots + a_{m2} \mathbf{v}_m, \\ &\ \ \vdots \qquad\ \ \vdots \qquad\ \ \vdots \qquad\qquad \vdots \\ T(\mathbf{u}_n) &= a_{1n} \mathbf{v}_1 + a_{2n} \mathbf{v}_2 + \cdots + a_{mn} \mathbf{v}_m. \end{aligned} \tag{5.15}$$

We now form the $m \times n$ matrix A whose ith column vector is the H-coordinatization of the vector $T(\mathbf{u}_i)$; that is, the *columns* of the matrix A are the matrix representations of the vectors $T(\mathbf{u}_i)$ with respect to the basis H for V. Notice that A is the transpose of the coefficient matrix for the linear system in (5.15).

$$A = \begin{bmatrix} a_{11} & a_{12} & a_{13} & \cdots & a_{1n} \\ a_{21} & a_{22} & a_{23} & \cdots & a_{2n} \\ a_{31} & a_{32} & a_{33} & \cdots & a_{3n} \\ \vdots & \vdots & \vdots & \cdots & \vdots \\ a_{m1} & a_{m2} & a_{m3} & \cdots & a_{mn} \end{bmatrix}. \tag{5.16}$$
$$\quad [T(\mathbf{u}_1)]_H \ \ [T(\mathbf{u}_2)]_H \qquad\qquad [T(\mathbf{u}_n)]_H$$

> **Definition** We define the **matrix representation** of the linear transformation $T: U \to V$ with respect to the bases G for U and H for V to be the matrix in (5.16) and write this in symbols as
>
> $$A = [T]_H^G.$$

We now proceed to show that if $\mathbf{u}$ is any vector in U, then the transformed vector $T(\mathbf{u})$ in V can be computed as a matrix transformation as follows:

$$[T(\mathbf{u})]_H = [T]_H^G[\mathbf{u}]_G. \qquad \textbf{(5.17)}$$

That is, the representation of the transformed vector $T(\mathbf{u})$ in V with respect to the basis H of V (the H-coordinatization of $T(\mathbf{u})$) is obtained by matrix multiplication.

Multiply the G-coordinatization of the vector $\mathbf{u}$ in U by the matrix representation of the transformation $T: U \rightarrow V$ with respect to the bases G for U and H for V.

This is indicated symbolically in (5.17). To see that this actually gives us the vector $T(\mathbf{u})$, let $\mathbf{u} \in U$ and use (5.13) to obtain

$$[\mathbf{u}]_G = \begin{bmatrix} \alpha_1 \\ \alpha_2 \\ \vdots \\ \alpha_n \end{bmatrix}.$$

From (5.14) and (5.15) it can be verified that (check this)

$$[T(\mathbf{u})]_H = \begin{bmatrix} \beta_1 \\ \beta_2 \\ \vdots \\ \beta_n \end{bmatrix}.$$

where

$$
\begin{aligned}
\beta_1 &= a_{11}\alpha_1 + a_{12}\alpha_2 + \cdots + a_{1n}\alpha_n, \\
\beta_2 &= a_{21}\alpha_1 + a_{22}\alpha_2 + \cdots + a_{2n}\alpha_n, \\
&\;\vdots \qquad\quad \vdots \qquad\quad \vdots \qquad\qquad\quad \vdots \\
\beta_m &= a_{m1}\alpha_1 + a_{m2}\alpha_2 + \cdots + a_{mn}\alpha_n.
\end{aligned}
\qquad \textbf{(5.18)}
$$

Or, using (5.16) we write this system of equations in matrix form as

$$[T(\mathbf{u})]_H = A[\mathbf{u}]_G.$$

Example 6   Let $U = V = \mathbb{R}^2$ be Euclidean 2-space. Find the matrix representation of the linear transformation of Example 1 of Section 5.1 relative to the standard basis E for $\mathbb{R}^2$:

$$E = \{\mathbf{e}_1 = (1, 0), \mathbf{e}_2 = (0, 1)\}.$$

Illustrate (5.17) for this representation. Recall that this transformation T was given by

$$T(x_1, x_2) = (2x_1, 2x_2).$$

To compute the matrix A in (5.16) which represents T we find the two vectors $T(\mathbf{e}_1)$ and $T(\mathbf{e}_2)$ and represent them with respect to the basis E for the co-domain. Then write these as the column vectors of A. Clearly $T(\mathbf{e}_1) = T(1, 0) = (2, 0)$ and $T(\mathbf{e}_2) = T(0, 1) = (0, 2)$. So the E-coordinatizations are $[T(\mathbf{e}_1)]_E = (2, 0)^T$ and $[T(\mathbf{e}_2)]_E = (0, 2)^T$; thus,

$$[T]_E^E = \begin{bmatrix} 2 & 0 \\ 0 & 2 \end{bmatrix}.$$

When the domain and codomain are the same vector space, and the same basis B is used in both, we simplify the notation by writing only the subscript. So $[T]_B^B$ would appear more simply as $[T]_B$. Here then we'll write $A = [T]_E$. To illustrate (5.17) write the standard coordinatization of an arbitrary vector $\mathbf{x} = (x_1, x_2)$ in $\mathbb{R}^2$ as $[\mathbf{x}]_E = \begin{bmatrix} x_1 \\ x_2 \end{bmatrix}$. Then note that

$$[T(\mathbf{x})]_E = \begin{bmatrix} 2x_1 \\ 2x_2 \end{bmatrix} = \begin{bmatrix} 2 & 0 \\ 0 & 2 \end{bmatrix} \cdot \begin{bmatrix} x_1 \\ x_2 \end{bmatrix} = [T]_E [\mathbf{x}]_E. \quad \blacksquare$$

This is a simple example because of the dimensions of the vector spaces and the simplicity of the bases. Yet it illustrates the principles involved without the cumbersome computations that might otherwise be involved. Another such example is the following.

Example 7   Find the matrix representation of the linear transformation $T(x_1, x_2) = (x_1, -x_2, x_1 - x_2)$ of Example 4 in Section 5.1, with respect to the standard bases for $\mathbb{R}^2$ and $\mathbb{R}^3$, and illustrate (5.17) for this transformation.

As in Example 6, we compute $T(\mathbf{e}_1)$ and $T(\mathbf{e}_2)$ and then coordinatize these with respect to the standard basis $E = \{(1, 0, 0), (0, 1, 0), (0, 0, 1)\}$ for $\mathbb{R}^3$. We use these as the column vectors of the matrix representation for T. Thus,

$$T(\mathbf{e}_1) = T(1, 0) = (1, 0, 1) \quad \text{and} \quad T(\mathbf{e}_2) = T(0, 1) = (0, -1, -1). \quad \textbf{(5.19)}$$

Therefore, we have the matrix

$$A = [T]_E = \begin{bmatrix} 1 & 0 \\ 0 & -1 \\ 1 & -1 \end{bmatrix}.$$

$$\underbrace{\qquad}_{[T(\mathbf{e}_1)]_E} \quad \underbrace{\qquad}_{[T(\mathbf{e}_2)]_E}$$

To illustrate (5.17), we note that if $\mathbf{x} = (x_1, x_2)$ is any vector in $\mathbb{R}^2$ then $[\mathbf{x}]_E = \begin{bmatrix} x_1 \\ x_2 \end{bmatrix}$, so

$$[T]_E[\mathbf{x}]_E = \begin{bmatrix} 1 & 0 \\ 0 & -1 \\ 1 & -1 \end{bmatrix} \begin{bmatrix} x_1 \\ x_2 \end{bmatrix} = \begin{bmatrix} x_1 \\ -x_2 \\ x_1 - x_2 \end{bmatrix} = [T(\mathbf{x})]_E. \quad \blacksquare$$

From now on we call the matrix representation of a linear transformation T with respect to the standard basis for both domain and codomain, as in Examples 6 and 7, the **standard matrix representation** of T. Often we denote this simply by $[T]$, leaving off the subscript E.

Now let us refer to the comments we made in the previous section about the kernel and range of a linear transformation being comparable with the null space and the column space of a matrix, respectively. Because of Theorem 5.6 (page 325), our derivation above, and the connection between the algebra of linear transformations and matrix algebra that will be described by the theorems in the next section, we are able to use the matrix representation $[T]_H^G$ of a given transformation T to find the kernel and range of the linear transformation. In fact most often we use the standard matrix representation $[T]$ of the transformation.

We can find the *range* of T by writing the *transpose* of its standard matrix representation $[T] = A$, and row reducing that transpose. Since, as we discussed in Section 3.5, the column space of a matrix is equal to the row space of its transpose, we determine a basis for the row space of $A^T = ([T])^T$, and these vectors will then span (generate) the column space of $[T]$. Because of Theorem 5.6 these column vectors will be the standard coordinatizations of a basis for the range of T.

In Example 7 above we note that

$$([T])^T = \begin{bmatrix} 1 & 0 & 1 \\ 0 & -1 & -1 \end{bmatrix}$$

has linearly independent rows, so a basis for the column space of $[T]$ consists of the two column vectors $\begin{bmatrix} 1 \\ 0 \\ 1 \end{bmatrix}$ and $\begin{bmatrix} 0 \\ -1 \\ -1 \end{bmatrix}$. Or, if we wish, we have the slightly simpler basis consisting of the two vectors $\begin{bmatrix} 1 \\ 0 \\ 1 \end{bmatrix}$ and $\begin{bmatrix} 0 \\ 1 \\ 1 \end{bmatrix}$. The latter two are the standard coordinatizations of the vectors $(1, 0, 1)$ and $(0, 1, 1)$ from the codomain $\mathbb{R}^3$. So the range of the transformation T is the two-dimensional subspace of $\mathbb{R}^3$ which these two vectors span (generate); namely, the set

$$\text{Range } T = \{(a, b, a + b): a, b \in \mathbb{R}\}.$$

The range is two-dimensional, so the rank of T is *two*. The nullity of T must be zero, by Sylvester's law of nullity. Thus ker $T = \{0\}$.

As another example of this process, refer to Example 8 in the preceding section. Let us look at one more example.

Example 8 Let $F: P_2 \rightarrow P_2$ be the linear transformation

$$F(c + bx + ax^2) = (c - a) + (a + b)x + (2c - b - 3a)x^2.$$

The standard ordered basis for P_2 is $\{1, x, x^2\}$, so the standard matrix representation of F is found by computing $F(1) = 1 + 2x^2$; $F(x) = x - x^2$; and $F(x^2) = -1 + x - 3x^2$. The standard coordinatizations (matrix representations) for these vectors are the following:

$$[F(1)]_E = \begin{bmatrix} 1 \\ 0 \\ 2 \end{bmatrix}, \qquad [F(x)]_E = \begin{bmatrix} 0 \\ 1 \\ -1 \end{bmatrix}, \qquad \text{and} \qquad [F(x^2)]_E = \begin{bmatrix} -1 \\ 1 \\ -3 \end{bmatrix}.$$

So

$$[F]_E = \begin{bmatrix} 1 & 0 & -1 \\ 0 & 1 & 1 \\ 2 & -1 & -3 \end{bmatrix} = [F].$$

If we row reduce $[F]$, we can find the solution space for the system of homogeneous equations $[F]\mathbf{x} = \mathbf{0}$ by the methods of Chapter 1. This gives us solutions of the form $\mathbf{x} = (t, -t, t)^T$. So a *standard coordinatization* for a basis for ker F is the set $S = \{(1, -1, 1)^T\}$. This set consists of the standard coordinatization (as a matrix) representing the polynomial (vector) $1 - x + x^2$. Therefore, a basis for the kernel of F is the single vector set $\overline{S} = \{1 - x + x^2\}$, and the nullity of F is *one*. We know that the rank of F is therefore *two*. (Why?)

A basis for the range of $F =$ the *column* space of $[F]$ is obtained by *row* reducing $([F])^T = \begin{bmatrix} 1 & 0 & 2 \\ 0 & 1 & -1 \\ -1 & 1 & -3 \end{bmatrix}$. The row echelon form for this matrix is

$N = \begin{bmatrix} 1 & 0 & 2 \\ 0 & 1 & -1 \\ 0 & 0 & 0 \end{bmatrix}$, so the two column vectors $\begin{bmatrix} 1 \\ 0 \\ 2 \end{bmatrix}$ and $\begin{bmatrix} 0 \\ 1 \\ -1 \end{bmatrix}$ span (generate)

the column space of $[F]$.

These are the coordinatizations as matrices of the polynomials $1 + 2x^2$ and $x - x^2$, respectively. Hence a basis for the range of F in P_2 is the set $B = \{(1 + 2x^2), (x - x^2)\}$. This verifies that the rank of F is *two*. Thus range F is the set of all polynomials of the form

$$p(x) = c(1 + 2x^2) + d(x - x^2) = c + dx + (2c - d)x^2; \; c, d \in \mathbb{R}. \quad \blacksquare$$

Example 9 (For those who have studied calculus) Consider the integral linear transformation defined on P_1 by $L(p(x)) = \int p(x)\, dx \in P_2$, with the constant of integration equal to zero. Find the standard matrix representation of L.

The standard ordered basis for P_1 is $\{1, x\}$; so we compute the action of L on these two vectors: $L(1) = x$, and $L(x) = x^2/2$. The standard ordered basis for P_2 is $E = \{1, x, x^2\}$; so the (standard) E-coordinatizations for the two vectors $L(1)$ and $L(x)$ are

$$[L(1)]_E = \begin{bmatrix} 0 \\ 1 \\ 0 \end{bmatrix} \quad \text{and} \quad [L(x)]_E = \begin{bmatrix} 0 \\ 0 \\ \frac{1}{2} \end{bmatrix}.$$

So the standard matrix representation of L has these two vectors as its columns:

$$[L]_E = [L] = \begin{bmatrix} 0 & 0 \\ 1 & 0 \\ 0 & \frac{1}{2} \end{bmatrix}. \quad \blacksquare$$

To see that this will actually accomplish the transformation $L: P_1 \rightarrow P_2$ we note that any vector in P_1 is a linear polynomial of the form $p(x) = a + bx$ whose standard coordinatization as a matrix is

$$[p(x)]_E = \begin{bmatrix} a \\ b \end{bmatrix}.$$

So the standard coordinatization in P_2 of $L(p(x))$ is

$$L(p(x)) = \int p(x)\, dx = ax + \frac{bx^2}{2} \leftrightarrow \begin{bmatrix} 0 \\ a \\ \frac{b}{2} \end{bmatrix} = [L]_E [p(x)]_E = \begin{bmatrix} 0 & 0 \\ 1 & 0 \\ 0 & \frac{1}{2} \end{bmatrix} \begin{bmatrix} a \\ b \end{bmatrix}.$$

From these examples we see that a linear transformation can just as well be calculated as a matrix transformation. We have used the symbol $\leftrightarrow$ here to denote the correspondence Φ_E of Theorem 5.6 (page 325) to keep our notation from becoming unnecessarily cluttered. On the right side of $\leftrightarrow$ we have the matrix transformation $L_A: \mathbb{R}^2 \rightarrow \mathbb{R}^3$, accomplishing the transformation $L: P_1 \rightarrow P_2$ with which we began on the left. In Exercise 49 we ask you to find a basis for the kernel and for the range of this transformation.

Example 10 (For those who have studied calculus) Given the integral transformation of Example 9. You can verify that the ordered sets $G = \{1 + x, 1 - x\}$ and $K = \{1 - x, 1 + x, 1 - x^2\}$ are bases for P_1 and P_2, respectively. This will require you to do some calculation—see Exercise 39 for a suggested method. Find the matrix representation $[L]_K^G$ for this integral transformation.

We proceed as we did in Example 8 to find the action of L on the basis vectors for the domain P_1. This time, however, we use the G-basis. Thus,

$$L(1 + x) = x + \frac{x^2}{2} \quad \text{and} \quad L(1 - x) = x - \frac{x^2}{2}.$$

Our task here is now a bit harder; we must represent each of these vectors with respect to the K-basis for P_2. You can compute that

$$L(1 + x) = x + \frac{x^2}{2} = a_{11}(1 - x) + a_{21}(1 + x) + a_{31}(1 - x^2)$$

$$= \frac{-1}{4}(1 - x) + \frac{3}{4}(1 + x) + \frac{-1}{2}(1 - x^2)$$

and

$$L(1 - x) = x - \frac{x^2}{2} = a_{12}(1 - x) + a_{22}(1 + x) + a_{32}(1 - x^2)$$

$$= \frac{-3}{4}(1 - x) + \frac{1}{4}(1 + x) + \frac{1}{2}(1 - x^2).$$

Hence the matrix representation for L with respect to the G- and K-bases is

$$[L]_K^G = \begin{bmatrix} \dfrac{-1}{4} & \dfrac{-3}{4} \\[2mm] \dfrac{3}{4} & \dfrac{1}{4} \\[2mm] \dfrac{-1}{2} & \dfrac{1}{2} \end{bmatrix}.$$

The simple calculation of $L(a + bx)$ done at the end of Example 9 as a matrix computation can also be accomplished with this matrix. However, it is necessary to write $a + bx$ in terms of the G-basis for P_1. Since $a + bx = \dfrac{a + b}{2}(1 + x) + \dfrac{a - b}{2}(1 - x)$, therefore,

$$[a + bx]_G = \begin{bmatrix} \dfrac{a + b}{2} \\[2mm] \dfrac{a - b}{2} \end{bmatrix}.$$

So

$$[L(a + bx)]_K = [L]_K^G [a + bx]_G = \begin{bmatrix} \dfrac{-1}{4} & \dfrac{-3}{4} \\[2mm] \dfrac{3}{4} & \dfrac{1}{4} \\[2mm] \dfrac{-1}{2} & \dfrac{1}{2} \end{bmatrix} \begin{bmatrix} \dfrac{a + b}{2} \\[2mm] \dfrac{a - b}{2} \end{bmatrix} = \begin{bmatrix} \dfrac{-2a + b}{4} \\[2mm] \dfrac{2a + b}{4} \\[2mm] \dfrac{-2b}{4} \end{bmatrix}.$$

Thus,

$$[L(a+bx)]_K \leftrightarrow \frac{-2a+b}{4}(1-x) + \frac{2a+b}{4}(1+x) + \frac{-2b}{4}(1-x^2)$$

$$= ax + \frac{bx^2}{2} = L(a+bx). \quad \blacksquare$$

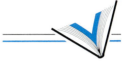

Vocabulary

In this section we have learned more new concepts and terms. This vocabulary is basic to the study of linear transformations, so be sure that you understand each of the concepts described by the terms in the following list. Review these ideas until you feel secure enough to use them in the rest of this course and in your later use of linear algebra.

coordinates of a vector with respect   standard coordinatization
   to a basis   matrix representation
the *S*-coordinatization of a vector   isomorphic
standard matrix representation   isomorphism

Exercises 5.3

In Exercises 1–10 represent each of the given vectors as a column vector $[\bullet]_E$ in $\overline{\mathbb{R}^n}$ or $\overline{\mathbb{C}^n}$ with respect to the "standard" basis for the given vector space. That is, find their standard coordinatization.

1. $(1, -2) \in \mathbb{R}^2$.

2. $(1, 2, 3) \in \mathbb{R}^3$.

3. $1 + x + x^3$ in P_3.

4. $(1, 2, 3, -1) \in \mathbb{R}^4$.

5. $(1, 1 - i, 1 + i) \in \mathbb{C}^3$.

6. $1 - 2x + 3x^2 - x^3$ in P_3.

7. $(-i, 1, 1 - 2i, 3 + 5i) \in \mathbb{C}^4$.

8. $(1 + i) - 2ix + (1 - 3i)x^2 - x^3$ in P_3.

9. $\begin{bmatrix} -1 & -3 \\ 5 & 8 \end{bmatrix}$ in $M_2(\mathbb{R})$.

10. $\begin{bmatrix} -1 & -3i \\ 1 + 5i & 8 - i \end{bmatrix}$ in $M_2(\mathbb{C})$.

In Exercises 11–16 find the standard matrix representation $[T]_E$ for each of the given linear transformations $T: \mathbb{R}^3 \to \mathbb{R}^3$.

11. $T(x_1, x_2, x_3) = (x_1 + x_2, x_3 - x_1, x_2 - x_3)$.

12. $T(x_1, x_2, x_3) = (x_1 - 2x_2, 2x_3 - x_1, 2x_2 - 3x_3)$.

13. $T(x_1, x_2, x_3) = (x_1 + x_2 + x_3, -x_1, x_2 - x_3)$.

14. $T(x_1, x_2, x_3) = (x_1 - 2x_2 - 2x_3, -x_1 + 2x_2 - 3x_3, 0)$.

15. $T(x_1, x_2, x_3) = (x_1, x_2, 0)$.

16. $T(x_1, x_2, x_3) = (x_1, x_2, x_3)$.

In Exercises 17–22 let S be the basis for $\mathbb{R}^3$ of Example 3. Represent each of the given vectors as a column vector $[\mathbf{u}]_S$ in $\overline{\mathbb{R}^3}$ with respect to this basis. That is, find the unique S-coordinatization of the given vector $\mathbf{u}$.

17. $\mathbf{u} = (1, 1, 1)$.

18. $\mathbf{u} = (-1, 0, 1)$.

19. $\mathbf{u} = (1, 2, 3)$.

20. $\mathbf{u} = (1, 0, 0)$.

21. $\mathbf{u} = (0, 0, 1)$.

22. $\mathbf{u} = (1, -1, 0)$.

In Exercises 23–28 let S be the basis for $\mathbb{R}^3$ *of Example 3. Find the matrix representation* $[T]_S$ *for each of the given linear transformations* $T: \mathbb{R}^3 \to \mathbb{R}^3$.

23. $T(x_1, x_2, x_3) = (x_1, x_2, x_3)$.

24. $T(x_1, x_2, x_3) = (x_1, x_2, 0)$.

25. $T(x_1, x_2, x_3) = (0, 0, 0)$.

26. $T(x_1, x_2, x_3) = (0, -x_1, 0)$.

27. $T(x_1, x_2, x_3) = (x_1 + x_2 + x_3, -x_1, x_2 - x_3)$.

28. $T(x_1, x_2, x_3) = (x_1 + x_2, x_3 - x_1, x_2 - x_3)$.

---

29. Let $H: P_1 \to P_2$ be the linear transformation $H(p(x)) = xp(x)$. Use the standard ordered bases, $\{1, x\}$ for P_1 and $\{1, x, x^2\}$ for P_2, to represent H as a matrix transformation; find $[H]$.

30. Repeat Exercise 29 if $H: P_2 \to P_3$.

31. Use the matrix $[H]$ of Exercise 29 to find $H(1 + 4x)$. Remember to first write the standard matrix representation $[1 + 4x]_E$ for the polynomial $1 + 4x$.

32. Find $H(1 - x + 2x^2)$ using the results of Exercise 30. (See Exercise 31.)

33. Let $D: P_2 \to P_1$ be the following differentiation linear transformation: $D(a + bx + cx^2) = b + 2cx$. Use the standard ordered bases, $\{1, x\}$ for P_1 and $\{1, x, x^2\}$ for P_2, to represent D as a matrix transformation.

34. Repeat Exercise 33 if $D: P_3 \to P_2$; i.e.,

$$D(a_0 + a_1 x + a_2 x^2 + a_3 x^3)$$
$$= a_1 + 2a_2 x + 3a_3 x^2.$$

35. Suppose $H: P_1 \to P_2$ is the linear transformation $H(p(x)) = xp(x)$ of Exercise 29. Use the ordered bases, $G = \{1 - x, 1 + x\}$ for P_1 and $K = \{1 - x, 1 + x, 1 - x^2\}$ for P_2, to represent H as a matrix transformation $[H]_K^G$.

36. Use the basis G of Exercise 35 to find the matrix representation $[p(x)]_G$ in $\mathbb{R}^2$ for the polynomial $p(x) = 1 + 4x$.

37. Use the matrix obtained in Exercise 35 to compute $[H(1 + 4x)]_K$ in $\mathbb{R}^3$. Remember to use the results of Exercise 35, that is, $[H(1 + 4x)]_K = [H]_K^G [1 + 4x]_G$.

38. Compare the results of Exercises 37 and 31. Show that both matrices (column vectors in $\mathbb{R}^3$) represent the same polynomial in P_2.

39. Justify for the following method for determining whether a given set $S = \{v_1, v_2, \ldots, v_n\}$ of vectors from an n-dimensional vector space V is a *basis* for V.

Step 1 Coordinatize each v_i with respect to the standard basis E of V, that is, find $[v_i]_E$.

Step 2 From the matrix M whose column vectors $m_i = [v_i]_E$, $i = 1, 2, \ldots, n$.

Step 3 Find the rank of M by row reduction, or by row reducing M^T. If it turns out that M has rank n, then S is a basis.

40. Let V be the ordinary x, y plane. Consider the transformation $R_\theta: V \to V$ defined by

$$R_\theta(x, y) = (x \cos \theta + y \sin \theta, -x \sin \theta + y \cos \theta).$$

a. Show that R_θ is a linear transformation.

b. Demonstrate geometrically that R_θ is a rotation of the x, y plane through an angle $-\theta$.

c. Find the standard matrix representation $A = [R_\theta]_E$.

d. Show that $\det A = 1$.

41. Prove the lemma that if u is a vector in the vector space U and $S = \{u_1, u_2, \ldots, u_n\}$ is an ordered basis for U then the representation of u as a linear combination of the vectors of S is unique. HINT: Suppose that there are two ways to represent u as a linear combination of the vectors u_i and subtract the equations involved.

---

42. Use the results of Exercise 11 to find bases for the range and for the kernel of the linear transformation given there.

43. Use the results of Exercise 12 to find bases for the range and for the kernel of the linear transformation given there.

44. Use the results of Exercise 13 to find bases for the range and for the kernel of the linear transformation given there.

45. Use the results of Exercise 14 to find bases for the range and for the kernel of the linear transformation given there.

46. Use the results of Exercise 15 to find bases for the range and for the kernel of the linear transformation given there.

47. Use the results of Exercise 16 to find bases for the range and for the kernel of the linear transformation given there.

48. Use the results of Exercise 24 to find bases for the range and for the kernel of the linear transformation given there.

49. Find bases for the range and for the kernel of the integral transformation of Example 9.

50. Verify that the rank and the nullity of the integral transformation of Examples 9 and 10 remain the same when one uses the matrix representation $[L]_K^G$ of Example 10. Compare the bases for each with the answers to Exercise 49.

---

51. Prove part (iv) of Theorem 5.6 (page 325). HINT: See the lemma about unique representation. What is the representation of the zero vector in any basis?

52. Prove part (iii) of Theorem 5.6.

53. Prove part (v) of Theorem 5.6.

54. Prove part (vi) of Theorem 5.6.

55. Prove part (i) of Theorem 5.6. HINT: Use parts (v) and (vi) and see Exercise 32 of Section 5.2.

5.4 *THE ALGEBRA OF LINEAR TRANSFORMATIONS*

In this section we consider the function algebra formed by linear transformations. We relate this to the matrix algebra of Chapter 1 via the matrix representation of a transformation from a finite-dimensional vector space U into a finite-dimensional vector space V as described in the previous section. Since linear transformations are functions, it is not surprising to learn that they possess many of the same algebraic properties as ordinary real or complex valued functions. Here, however, the functions are vector functions rather than number functions. Specifically we here define the *sum, product by a scalar,* and *product* (composition) of linear transformations. It is remarkable to find that linear transformations of vector spaces themselves form a vector space of transformations.

Sum of Two Transformations

Let $T_1: U \to V$ and $T_2: U \to V$ be linear transformations from the vector space U into the vector space V. Then the **sum** transformation $T_1 + T_2$ of the two transformations is defined as follows.

> **Definition** $(T_1 + T_2): U \to V$ is the transformation such that, for all $\mathbf{u} \in U$,
> $$(T_1 + T_2)(\mathbf{u}) = T_1(\mathbf{u}) + T_2(\mathbf{u}) \text{ in } V. \tag{5.20}$$

That is, the sum transformation $T_1 + T_2$ is the transformation that maps the vector $\mathbf{u}$ from the space U to the vector in V that is the sum of the two vectors $T_1(\mathbf{u})$ and $T_2(\mathbf{u})$ to which $\mathbf{u}$ was mapped by the two functions T_1 and T_2 individually. See the examples that follow. First let us also define the product of a transformation by a scalar.

Product of a Scalar and a Transformation

Let $T: U \rightarrow V$ be any linear transformation from the vector space U into the vector space V, and let α be any scalar. Then the **scalar product** αT of α and T is defined as follows.

> **Definition** $\alpha T: U \rightarrow V$ is the transformation such that, for all $\mathbf{u} \in U$,
> $$(\alpha T)(\mathbf{u}) = \alpha(T(\mathbf{u})) \text{ in } V. \tag{5.21}$$

Example 1   Let $U = V = \mathbb{R}^2$ be 2-space and let T_1 and T_2 be defined by the equations

$$T_1(x_1, x_2) = (x_2, x_1) \qquad \text{and} \qquad T_2(x_1, x_2) = (2x_1, 2x_2).$$

Then the sum transformation $T_1 + T_2$ is given by (5.20) and results in

$$\begin{aligned}
(T_1 + T_2)(x_1, x_2) &= T_1(x_1, x_2) + T_2(x_1, x_2) \\
&= (x_2, x_1) + (2x_1, 2x_2) \\
&= (x_2 + 2x_1, x_1 + 2x_2).
\end{aligned}$$

So, in particular,

$$(T_1 + T_2)(2, -3) = (-3 + 4, 2 - 6) = (1, -4).$$

If α is any scalar, then the scalar product αT_1 is the transformation given by (5.21), namely,

$$(\alpha T_1)(x_1, x_2) = \alpha(T_1(x_1, x_2)) = \alpha(x_2, x_1) = (\alpha x_2, \alpha x_1).$$

So, for the particular vector $(2, -3)$ of $\mathbb{R}^2$, we have

$$4T_1(2, -3) = 4(-3, 2) = (-12, 8). \quad \blacksquare$$

Example 2   Let $U = P_2$ and $V = P_3$. Suppose that L_1 and L_2 are defined by the equations

$$L_1(a + bx + cx^2) = x(a + bx + cx^2)$$

and

$$L_2(a + bx + cx^2) = cx + bx^2 + ax^3.$$

We leave it as an exercise for you to show that both L_1 and L_2 are linear transformations. Given that, we have

$$\begin{aligned}
(L_1 + L_2)(a + bx + cx^2) &= L_1(a + bx + cx^2) + L_2(a + bx + cx^2) \\
&= ax + bx^2 + cx^3 + cx + bx^2 + ax^3 \\
&= (a + c)x + 2bx^2 + (a + c)x^3.
\end{aligned}$$

So for the particular polynomials $1 - x^2$ and $1 + x^2$, we see that $(L_1 + L_2)(1 - x^2) = 0$, while $(L_1 + L_2)(1 + x^2) = 2x + 2x^3$. Note also that for any scalar α,

$$\alpha L_1(a + bx + cx^2) = \alpha ax + \alpha bx^2 + \alpha cx^3. \quad \blacksquare$$

We leave it as exercises (see Exercises 41 and 42) for you to check that the sum of two transformations and the product of a scalar and a transformation are indeed always still linear transformations as defined in Section 5.1. Let us now turn to the composition of two linear transformations.

Composition (Product) of Two Transformations

Since linear transformations are functions, we might expect that the composition of two of them would also give a linear transformation. Granted that the composition of two transformations is indeed a vector function, we leave it to you to check the linearity of that composite function. Let

$$T_1 \colon U \to V \qquad \text{and} \qquad T_2 \colon V \to W$$

be linear transformations with vector space domains and codomains as indicated. The definition of the **composition (= product)** $T_2 T_1$ of T_1 and T_2 is as follows. See Figure 5.8.

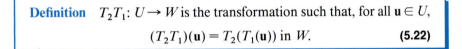

> **Definition** $T_2 T_1 \colon U \to W$ is the transformation such that, for all $\mathbf{u} \in U$,
>
> $$(T_2 T_1)(\mathbf{u}) = T_2(T_1(\mathbf{u})) \text{ in } W. \qquad \textbf{(5.22)}$$

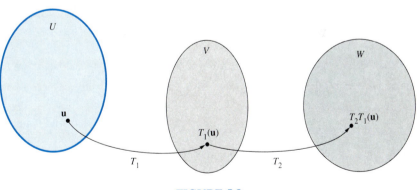

FIGURE 5.8

It is extremely important to emphasize here that *order is very essential* in this definition (5.22). Indeed, we *cannot* be certain that the composition (product) $T_1 T_2$ (in the opposite order) even makes sense.

Since we are defining products of transformations to be the same as their composition as functions, the product T_2T_1 is sometimes denoted by the function composition notation $T_2 \circ T_1$. Our preference will be for the **product notation** because we want to exploit the correspondence between linear transformations and matrices described in the previous section. You are asked in Exercise 43 to prove, without reference to the matrix representations of the transformations, that the product of two linear transformations is always a linear transformation.

Example 3 Let T_1 and T_2 be the linear transformations from $\mathbb{R}^2$ to $\mathbb{R}^2$ given in Example 1. In this particular case both of the composite functions *(products)* T_1T_2 and T_2T_1 are defined. Indeed, we have the highly unusual case where $T_1T_2 = T_2T_1$:

$$T_1T_2\,(x_1, x_2) = T_1(2x_1, 2x_2) = (2x_2, 2x_1),$$

and $\qquad T_2T_1\,(x_1, x_2) = T_2(x_2, x_1) = (2x_2, 2x_1).$ ∎

On the other hand, the product L_1L_2 of the linear transformations L_1 and L_2 of Example 2 is not even defined because the domain of L_1 is not the range of L_2. For example, what would L_1 do with $L_2(1 - x^2)$?

But consider the following example where both composite transformations are defined; however, they are *not the same* transformation.

Example 4 Let $U = V = \mathbb{R}^2$. It can be shown that the functions T_1 and T_2 defined by

$$T_1(x_1, x_2) = (x_1 + 2x_2, -x_2) \qquad \text{and} \qquad T_2(x_1, x_2) = (-x_1, 4x_1 + x_2)$$

are indeed linear transformations. Then (verify this)

$$T_1T_2(x_1, x_2) = T_1(-x_1, 4x_1 + x_2) = (7x_1 + 2x_2, -4x_1 - x_2),$$

while

$$T_2T_1(x_1, x_2) = T_2(x_1 + 2x_2, -x_2) = (-x_1 - 2x_2, 4x_1 + 7x_2).$$

So here both are defined, but

$$T_1T_2 \neq T_2T_1. \quad ∎$$

This example shows that for linear transformations, as well as for matrices, the **commutative law for multiplication (function composition) does not hold.** In fact, it is true that

> Linear transformations obey the same laws of algebra as do matrices.

Indeed we have the following useful lemma. In stating it we have simplified matters by only discussing the case where the domain and the codomain of the linear transformation T are the same vector space U, so that we rule out in advance the problem that the product may not exist. When the domain and the codomain are the same vector space U, one usually says that $T: U \rightarrow U$ *is a linear transformation* **on** *the vector space U, or that T: U → U is a linear* **operator** *on U.*

Lemma Let $\mathbb{A}$ denote the set of all linear transformations (operators) $T: U \rightarrow U$ on the n-dimensional vector space U. With the definitions of addition, and scalar multiplication given above, $\mathbb{A}$ is a vector space. Furthermore with multiplication of transformations defined as composition of maps, $\mathbb{A}$ also satisfies the same properties as the matrix algebra $M_n(\mathbb{R})$ discussed in Chapter 1.

Proof The easiest way to prove this is to let B be any ordered basis for U, to define $\Omega_B : \mathbb{A} \rightarrow M_n(\mathbb{R})$ by $\Omega_B(T) = [T]_B$, and then show that Ω_B is indeed an *isomorphism* in that it satisfies the properties of Theorem 5.6 (page 325). We ask you to do this in Exercises 35–40.

Nevertheless, most of the details for a more direct proof of the lemma are contained in our discussion above. Equations (5.20) and (5.21) are the important facts for verifying that $\mathbb{A}$ is indeed a vector space. The zero transformation is the transformation $O(\mathbf{x}) = \mathbf{0}$, for all $\mathbf{x}$ in U. The rest of the details you can easily supply. You may wish to refer to Theorems 1.1 and 1.2 (pages 32 and 34) where these properties are proved for $M_n(\mathbb{R})$. Equation (5.22) defines multiplication in $\mathbb{A}$, and the remaining details are fairly straightforward. For example, if $T: U \rightarrow U$, $G: U \rightarrow U$, and $F: U \rightarrow U$ are all linear operators from $\mathbb{A}$, then for each $\mathbf{x}$ in U we have

$$(T(F + G))(\mathbf{x}) = T(F(\mathbf{x}) + G(\mathbf{x})) = T(F(\mathbf{x})) + T(G(\mathbf{x}))$$
$$= (TF)(\mathbf{x}) + (TG)(\mathbf{x}) = (TF + TG)(\mathbf{x}).$$

Thus, the *left distributive law* holds in $\mathbb{A}$. This sort of proof, while not nearly so elegant, does check all of the important properties of the algebra $\mathbb{A}$. We ask as exercises that you show that the commutative law for addition, the associative laws for addition and multiplication, and the other distributive law hold in $\mathbb{A}$, together with the laws for scalar multiplication. In Exercise 36 we ask that you prove that one can identify the composition of two linear transformations with the product of their matrix representations. ❑

Continuing the comparison with matrices, we note that some, but not all, linear transformations have *inverses* in the same sense that functions have inverses. Refer to Appendix A.2. For linear transformations we have the following definition:

Definition A linear transformation $T: U \rightarrow V$ is **invertible** if there exists a linear transformation $L: V \rightarrow U$ such that $LT(\mathbf{u}) = \mathbf{u}$, for all $\mathbf{u}$ in U and $TL(\mathbf{v}) = \mathbf{v}$, for all $\mathbf{v}$ in V.

That is, the linear transformation $T: U \rightarrow V$ is an invertible linear transformation if it is invertible as a function. In this case its function inverse is a linear transformation (see Exercise 60) going the other direction from V to U. That means that the inverse linear transformation L must have the property that

LT is the identity transformation I_U on U: $LT(\mathbf{u}) = \mathbf{u}$, for all $\mathbf{u}$ in U **(5.23)**

and

TL is the identity transformation I_V on V: $TL(\mathbf{v}) = \mathbf{v}$, for all $\mathbf{v}$ in V,   **(5.24)**

as illustrated in Figure 5.9. In this case we say that L is the **inverse** of T (and vice versa).

You might expect that not all linear transformations have inverses. Even if T is a linear operator, it may not be invertible. For example, if $T: \mathbb{R}^2 \rightarrow \mathbb{R}^2$ is the projection linear transformation

$$T(x_1, x_2) = (x_1, 0) \tag{5.25}$$

then no linear transformation L that satisfies properties (5.23) and (5.24) can possibly exist. To prove this, first observe that by (5.25),

$$T(0, 1) = (0, 0) \quad \text{and} \quad T(0, 0) = (0, 0),$$

and hence $T(0, 1) = T(0, 0)$. Now, if T were to have an inverse L, then by applying L to both sides of this last equation, we would get

$$LT(0, 1) = L(0, 0),$$

or

$$(0, 1) = (0, 0)$$

(by (5.23)). This is clearly impossible. This shows that no function L satisfying (5.23) and (5.24) can possibly exist for this T; so T has *no* inverse.

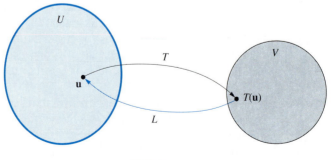

FIGURE 5.9

When a linear transformation $T: U \rightarrow U$ *on* U has an inverse $L: U \rightarrow U$, equations (5.23) and (5.24) can be seen to yield

$$LT = TL = I_U = \text{the identity map on } U.$$

In Exercise 57 we ask you to prove that such an L, which is the inverse of T, is indeed *unique*. We denote the *unique* inverse of the linear transformation T by T^{-1}. Thus the above equation may now be written as follows:

$$T^{-1}T = TT^{-1} = I_U = \text{the identity map on } U.$$

Note, too, that in this case the inverse transformation $T^{-1}: U \rightarrow U$ is indeed a linear transformation in its own right (see Exercise 60).

When does T^{-1} exist? The following theorem answers this question when T is a given linear transformation (operator) on a given vector space.

Theorem 5.7   Let $T: U \rightarrow U$ be a linear transformation and let $B = \{\mathbf{u}_1, \mathbf{u}_2, \ldots, \mathbf{u}_n\}$ be a basis for U. Then T is invertible and T^{-1} exists if, and only if, any (and hence all) of the following equivalent conditions hold:

(i) T is one-to-one
(ii) The kernel of $T = \{\mathbf{0}\}$ in U.
(iii) The range of $T = U$; that is, T is *onto* U.
(iv) The set $T(B) = \{T(\mathbf{u}_1), T(\mathbf{u}_2), \ldots, T(\mathbf{u}_n)\}$ is a basis for U.

As in the case of matrices, whenever T_1 and T_2 are both invertible linear operators on the vector space U, then $T_1 T_2$ is also invertible and

$$(T_1 T_2)^{-1} = T_2^{-1} T_1^{-1}. \tag{5.26}$$

In other words, the inverse of a product is equal to the product of the inverses in the reverse order. Moreover,

$$(T^{-1})^{-1} = T \tag{5.27}$$

for any invertible linear transformation $T: U \rightarrow U$; that is, the inverse of the inverse of T is T itself.

Example 5   Let $U = P_2$ be the space of all polynomials of degree 2 or less. Let $T: P_2 \rightarrow P_2$ be the following linear transformation (verify that it is indeed a linear transformation):

$$T(ax^2 + bx + c) = (a - b)x^2 + (a + c)x + (a + b).$$

Then the inverse transformation T^{-1} exists and is given by

$$T^{-1}(ax^2 + bx + c) = \frac{1}{2}(a + c)x^2 + \frac{1}{2}(c - a)x + \frac{1}{2}(2b - a - c), \tag{5.28}$$

as can be verified by calculating

$$TT^{-1}(ax^2 + bx + c) = T\left(\frac{1}{2}(a + c)x^2 + \frac{1}{2}(c - a)x + \frac{1}{2}(2b - a - c)\right)$$

$$= \left[\frac{1}{2}(a + c) - \frac{1}{2}(c - a)\right]x^2$$

$$+ \left[\frac{1}{2}(a + c) + \frac{1}{2}(2b - a - c)\right]x + \left[\frac{1}{2}(a + c) + \frac{1}{2}(c - a)\right]$$

$$= ax^2 + bx + c. \quad \blacksquare$$

Rather than a complicated functional solution to the equation $TT^{-1}(\mathbf{u}) = \mathbf{u}$, perhaps the easiest way to determine T^{-1}, if it exists, is to represent T by a matrix and then find the inverse of the matrix, if that inverse exists. This is the substance of the next theorem. Its proof follows from the same ideas as those used to prove Theorem 5.6 (page 325), except that we would here be concerned with the mapping Ω_B described in the proof of our lemma (preceding Theorem 5.7, page 341) from the algebra $\mathbb{A}$ of all linear transformations (operators) $T: U \to U$ on the n-dimensional vector space U, to the algebra $M_n(\mathbb{R})$ of $n \times n$ matrices. We omit the details of the proof but suggest that you work them out using the mapping Ω_B.

> **Theorem 5.8**   The linear transformation $T: U \to U$ is invertible if, and only if, any matrix representation A of T is a nonsingular (invertible) matrix. Moreover, A^{-1} represents T^{-1}.

Example 6   Let $T: \mathbb{R}^2 \to \mathbb{R}^2$ be the linear transformation that maps each vector in the plane to its symmetric image about the x axis (about the vector $\mathbf{e}_1 = (1, 0)$). See Figure 5.10.

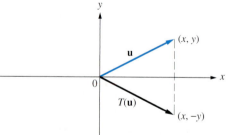

FIGURE 5.10

To find the standard matrix representation of T we note that $T(\mathbf{e}_1) = \mathbf{e}_1 = (1, 0)$, while $T(\mathbf{e}_2) = -\mathbf{e}_2 = (0, -1)$. Hence the standard matrix representation of T is the matrix

$$A = [T]_E = \begin{bmatrix} 1 & 0 \\ 0 & -1 \end{bmatrix}.$$

It is easy enough to verify that A^{-1} exists and equals the matrix A itself; therefore, $T^{-1} = T$. $\blacksquare$

As a less trivial illustration of Theorem 5.8, note that, with respect to the standard ordered basis $E = \{1, x, x^2\}$ for P_2, the transformation of Example 5 can be represented by the matrix (verify this)

$$A = [T]_E = \begin{bmatrix} 0 & 1 & 1 \\ 1 & 0 & 1 \\ 0 & -1 & 1 \end{bmatrix}.$$

Then, using the matrix-inversion algorithm (or your favorite computer program), we find that

$$A^{-1} = [T^{-1}]_E = \begin{bmatrix} -\dfrac{1}{2} & 1 & -\dfrac{1}{2} \\ \dfrac{1}{2} & 0 & -\dfrac{1}{2} \\ \dfrac{1}{2} & 0 & \dfrac{1}{2} \end{bmatrix}.$$

Hence by Theorem 5.8, T^{-1} exists and, moreover, A^{-1} is the standard matrix representation of T^{-1}. Thus, since the E-coordinatization of the polynomial $ax^2 + bx + c$ is $\begin{bmatrix} c \\ b \\ a \end{bmatrix}$ we have

$$[T^{-1}(ax^2 + bx + c)]_E = A^{-1} \begin{bmatrix} c \\ b \\ a \end{bmatrix} = \begin{bmatrix} -\dfrac{1}{2}a + b - \dfrac{1}{2}c \\ \dfrac{1}{2}(c - a) \\ \dfrac{1}{2}(a + c) \end{bmatrix}.$$

Compare this with equation (5.28).

Vocabulary

In this section we have encountered new concepts and terms. This vocabulary is basic to the study of linear transformations, so be sure that you understand each of the concepts described by the terms in the following list. Review these ideas until you feel secure enough to use them in the rest of this course and in your later use of linear algebra.

sum of two linear transformations
composition of transformations
linear operator
algebra of linear transformations

scalar product kT
product of two linear
   transformations
invertible linear transformations

Exercises 5.4

Exercises 1–10 involve the linear transformations T_1 and T_2 of Example 4; namely,

$$T_1(x_1, x_2) = (x_1 + 2x_2, - x_2) \quad and \quad T_2(x_1, x_2) = (-x_1, 4x_1 + x_2).$$

1. Compute directly the formula for the transformation $L = \frac{1}{2}T_1T_2 - 2T_2T_1$.

2. Find the standard matrix representation $A = [T_1]_E$ for T_1.

3. Find the standard matrix representation $B = [T_2]_E$ for T_2.

4. Find the standard matrix representation $C = [T_1 + T_2]_E$ for $T_1 + T_2$.

5. Find the standard matrix representation $G = [T_1T_2]_E$ for T_1T_2.

6. Compute the transformation T_1^{-1} directly.

7. Use the results of Exercise 6 to find the standard matrix representation $[T_1^{-1}]_E$ for T_1^{-1}.

8. Compare $[T_1^{-1}]_E$ in Exercise 7 with $A^{-1} = [T_1]_E^{-1}$ computed from the results of Exercise 2.

9. Compute the transformation T_2^{-1} directly.

10. Use the results of Exercise 9 to find the standard matrix representation $[T_2^{-1}]_E$ for T_2^{-1}.

11. Compare $[T_2^{-1}]_E$ in Exercise 10 with $B^{-1} = [T_2]_E^{-1}$ computed from the results of Exercise 3.

12. Demonstrate that the standard matrix representation for the transformation $(T_1T_2)^{-1}$ is the matrix $[T_2^{-1}]_E [T_1^{-1}]_E$.

13. Use the results of Exercise 1 to find the standard matrix representation $[L]_E$ for the linear transformation L defined there. Compare it with the matrix result $\frac{1}{2}AB - 2BA$.

14. Using the matrix representation, find a basis for the kernel of the transformation L of Exercise 1.

15. Using the matrix representation, find a basis for the range of the transformation L of Exercise 1.

In Exercises 16–25 let T_1 and T_2 be linear operators on $\mathbb{R}^3$ defined as follows:

$$T_1(x_1, x_2, x_3) = (x_3, x_1, x_3), \quad and \quad T_2(x_1, x_2, x_3) = (x_1 - x_3, x_2 - x_3, x_3 - 2x_2).$$

16. Find the standard matrix representation for T_1.

17. Find the standard matrix representation for T_2.

18. Find the standard matrix representation for $T_1 + T_2$.

19. Find the standard matrix representation for $3T_1 - 2T_2$.

20. Find the standard matrix representation for T_1T_2.

21. Find the standard matrix representation for T_2T_1.

22. Find the standard matrix representation $[T_1^{-1}]$ for T_1^{-1}, if T_1^{-1} exists.

23. Find the standard matrix representation $[T_2^{-1}]$ for T_2^{-1}, if T_2^{-1} exists.

24. Find a basis for the kernel of the transformation T_1 of Exercise 16.

25. Find a basis for the range of the transformation T_1 of Exercise 16.

26. Let $T_3: \mathbb{R}^3 \rightarrow \mathbb{R}^3$ be the linear operator $T_3(x_1, x_2, x_3) = (x_1 + x_3, x_2 + x_3, x_3 - x_2)$. Find the standard matrix representation $[T_3]$ for T_3.

27. Let T_3 be the linear transformation of Exercise 26. Find the standard matrix representation $[T_3^{-1}]$ for T_3^{-1}; then write T_3^{-1} as a function, $T_3^{-1}(x_1, x_2, x_3) = (\,\cdot\,)$.

28. With $T_2(x_1, x_2, x_3) = (x_1 - x_3, x_2 - x_3, x_3 - 2x_2)$, and T_3 as above, find the standard matrix representation for $(T_2 T_3)^{-1}$. Show that this is $[T_3^{-1}][T_2^{-1}]$.

---

29. Let $T: \mathbb{R}^3 \rightarrow \mathbb{R}^2$ be the linear transformation $T(x_1, x_2, x_3) = (x_1 - x_2 + x_3, x_3 - 2x_1)$. Write the standard matrix representation $[T]$ for T. Does T^{-1} exist? Give reasons.

30. Let $F: \mathbb{R}^2 \rightarrow \mathbb{R}^3$ be the linear transformation $F(x_1, x_2) = (x_1 - x_2, 2x_1 - 2x_2, 2x_1 - x_2)$. Write the standard matrix representation $[F]$ for F. Does F^{-1} exist? Give reasons.

31. With T and F as in Exercises 29 and 30, find the standard matrix representations $[TF]$ and $[FT]$. Note that $[TF] = I_2$.

32. Using the results of Exercise 31, write the two transformations TF and FT as functions.

33. Refer to Exercise 32. Find the kernel and the range of the transformation FT. Discuss why FT is not the identity transformation on $\mathbb{R}^2$.

34. Verify that the transformations T_1 and T_2 of Example 4 are indeed linear.

35. Verify that the map $\Omega_B: \mathbb{A} \rightarrow M_n(\mathbb{R})$ defined in the lemma of this section by $\Omega_B(T) = [T]_B$ has the property that for T and S any linear operators on the finite-dimensional vector space U with basis B, $\Omega_B(S + T) = [S]_B + [T]_B$. NOTE: By definition, $\Omega_B(S + T) = [S + T]_B$.

36. Verify that the map $\Omega_B: \mathbb{A} \rightarrow M_n(\mathbb{R})$ defined in the lemma of this section by $\Omega_B(T) = [T]_B$ has the property that for T and S any linear operators on the finite-dimensional vector space U with basis B, $\Omega_B(ST) = [S]_B[T]_B$.

37. Verify that the map $\Omega_B: \mathbb{A} \rightarrow M_n(\mathbb{R})$ defined in the lemma of this section by $\Omega_B(T) = [T]_B$ has the property that, for T any linear operator on the finite-dimensional vector space U with basis B and any scalar k from $\mathbb{R}$, $\Omega_B(kT) = k[T]_B$.

38. Verify that the map $\Omega_B: \mathbb{A} \rightarrow M_n(\mathbb{R})$ defined in the lemma of this section by $\Omega_B(T) = [T]_B$ has the property that ker $\Omega_B = O$; the zero transformation $O(\mathbf{x}) = \mathbf{0}$, for all $\mathbf{x}$ in U.

39. Show that the map $\Omega_B: \mathbb{A} \rightarrow M_n(\mathbb{R})$ is *one-to-one*. (See Exercise 38.)

40. Show that the map $\Omega_B: \mathbb{A} \rightarrow M_n(\mathbb{R})$ is *onto* $M_n(\mathbb{R})$. (Show range $\Omega_B = M_n(\mathbb{R})$.)

41. Show that the sum of two linear transformations as defined in (5.20) is always a linear transformation.

42. Show that the product of a scalar and a linear transformation as defined in (5.21) is always a linear transformation.

43. Show that the product (composition) of two linear transformations, when defined, is a linear transformation.

44. Show that the algebra $\mathbb{A}$ of all linear operators on the vector space U satisfies the *commutative* law for operator addition: $S + T = T + S$.

45. Show that the algebra $\mathbb{A}$ of all linear operators on the vector space U satisfies the *associative* law for addition: $S + (T + F) = (S + T) + F$.

46. Show that the algebra $\mathbb{A}$ of all linear operators on the vector space U satisfies the *associative* law for multiplication: $S(TF) = (ST)F$.

47. Show that the algebra $\mathbb{A}$ of all linear operators on a vector space U has both an *additive identity* O (see Exercise 38) and a *multiplicative identity* I (defined by $I(\mathbf{x}) = \mathbf{x}$, all $\mathbf{x}$ in U).

48. Show that the algebra $\mathbb{A}$ of all linear operators on the vector space U satisfies the *right distributive* law: $(T + F)S = (TS) + (FS)$.

49. Show that the algebra $\mathbb{A}$ of all linear operators on the vector space U satisfies the following law for multiplication by a scalar: $k(T + S) = kT + kS$.

50. Show that the algebra $\mathbb{A}$ of all linear operators on the vector space U satisfies the following law for multiplication by scalars:

$$(k + m)\,(T) = kT + mT.$$

51. Show that the algebra $\mathbb{A}$ of all linear operators on the vector space U satisfies the following law for multiplication by scalars: $(km)(T) = k(mT)$.

52. Show that the algebra $\mathbb{A}$ of all linear operators on the vector space U satisfies the following law for multiplication by scalars:

$$k(ST) = (kS)T = S(kT).$$

In Exercises 53–56 prove Theorem 5.7 (page 341). HINT: Use Sylvester's law of nullity.

53. Prove part (i).   **54.** Prove part (ii).   **55.** Prove part (iii).   **56.** Prove part (iv).

---

57. Prove that if $T: V \rightarrow V$ is an invertible linear transformation, then its inverse T^{-1} is unique.

58. Prove the statements (5.26) and (5.27) which follow Theorem 5.7, namely,

$$(T_1 T_2)^{-1} = T_2^{-1} T_1^{-1}, \quad \text{and} \quad (T^{-1})^{-1} = T.$$

59. Prove Theorem 5.8 (page 342). HINT: Start with an arbitrary basis $B = \{\mathbf{u}_1, \mathbf{u}_2 \ldots, \mathbf{u}_n\}$ for U; write $A = [T]_B$.

60. Prove that if $T: U \rightarrow U$ is an invertible linear transformation, then $T^{-1}: U \rightarrow U$ is also a linear transformation.

5.5 CHANGE OF BASIS AND SIMILARITY

In this section we compare the matrix representation of a given linear transformation (operator) T on the vector space U relative to one basis with the matrix representation of T relative to a different basis for U. These will be different matrices, yet they represent the same linear operator. How are they alike? This leads us to the important concept of *similarity* of matrices. Let us begin with an example.

Example 1   Let $T: \mathbb{R}^3 \rightarrow \mathbb{R}^3$ be the linear operator on $\mathbb{R}^3$ defined by

$$T(x_1, x_2, x_3) = (x_1 + x_2, x_2 + x_3, x_1 + x_3).$$

Check to see that the mapping T is indeed a linear transformation. Then, with respect to the standard basis for $\mathbb{R}^3$, namely,

$$E = \{\mathbf{e}_1 = (1, 0, 0), \mathbf{e}_2 = (0, 1, 0), \mathbf{e}_3 = (0, 0, 1)\},$$

the transformation T has the *standard matrix representation*

$$M = [T] = \begin{bmatrix} 1 & 1 & 0 \\ 0 & 1 & 1 \\ 1 & 0 & 1 \end{bmatrix}.$$
$$\;[T(\mathbf{e}_1)] \quad [T(\mathbf{e}_2)] \quad [T(\mathbf{e}_3)]$$

The columns of M are the matrix representations of the action of T on the basis vectors expressed in that ordered basis.

Now it is not difficult for you to verify that the set

$$G = \{\mathbf{u}_1 = (1, -1, 0), \mathbf{u}_2 = (1, 0, -1), \mathbf{u}_3 = (0, 1, 1)\}$$

is also a basis for $\mathbb{R}^3$. Relative to this basis the matrix representation for T is a different matrix. To compute it we must find $T(\mathbf{u}_1)$, $T(\mathbf{u}_2)$, and $T(\mathbf{u}_3)$ and express each of these as a linear combination of $\mathbf{u}_1$, $\mathbf{u}_2$, and $\mathbf{u}_3$. Thus,

$$T(\mathbf{u}_1) = (0, -1, 1) = 1\mathbf{u}_1 + (-1)\mathbf{u}_2 + 0\mathbf{u}_3,$$
$$T(\mathbf{u}_2) = (1, -1, 0) = 1\mathbf{u}_1 + 0\mathbf{u}_2 + 0\mathbf{u}_3,$$
$$T(\mathbf{u}_3) = (1, 2, 1) = 0\mathbf{u}_1 + 1\mathbf{u}_2 + 2\mathbf{u}_3.$$

You can verify these equations by the methods of the previous chapters. That is, compute the coefficients of $\mathbf{u}_1$, $\mathbf{u}_2$, and $\mathbf{u}_3$ by solving the resulting three systems of equations. (See Section 1.6.) Therefore,

$$[T(\mathbf{u}_1)]_G = \begin{bmatrix} 1 \\ -1 \\ 0 \end{bmatrix}, \qquad [T(\mathbf{u}_2)]_G = \begin{bmatrix} 1 \\ 0 \\ 0 \end{bmatrix}, \qquad \text{and} \qquad [T(\mathbf{u}_3)]_G = \begin{bmatrix} 0 \\ 1 \\ 2 \end{bmatrix}.$$

So the *matrix representation of T relative to the G-basis* is

$$K = [T]_G = \begin{bmatrix} 1 & 1 & 0 \\ -1 & 0 & 1 \\ 0 & 0 & 2 \end{bmatrix}. \quad \blacksquare$$

$$[T(\mathbf{u}_1)]_G \quad [T(\mathbf{u}_2)]_G \quad [T(\mathbf{u}_3)]_G$$

How are the matrices M and K related, as matrices? The answer is provided by the following theorem, whose proof we simply sketch.

Theorem 5.9 Let $T: U \rightarrow U$ be a linear operator on the finite-dimensional vector space U. Let U have bases G and H. Let $A = [T]_G$ be the matrix representation of T relative to the G-basis and let $B = [T]_H$ be the matrix representation of T relative to the H-basis. Then there exists a nonsingular transition matrix S such that

$$A = SBS^{-1}. \tag{5.29}$$

Proof Outline The **transition matrix** S is the matrix representation of the identity linear operator $I(\mathbf{u}) = \mathbf{u}$ on U, which transforms each vector $\mathbf{u}$ expressed as a linear combination of the H-basis vectors into the same vector $\mathbf{u}$ expressed as a linear combination of the G-basis vectors. That is, $S = [I]_G^H$. (See Section 5.3.) Since S represents the identity transformation, it is clearly invertible so $S^{-1} = [I]_H^G$ represents the change of basis from the G-basis to the H-basis. Then we must have

$$A[\mathbf{u}]_G = [T]_G[\mathbf{u}]_G = [I]_G^H[T]_H[I]_H^G[\mathbf{u}]_G = SBS^{-1}[\mathbf{u}]_G,$$

for every vector $\mathbf{u}$ in U. Hence $A = SBS^{-1}$, as desired. ◘

In the case of Example 1 we notice that the coordinatizations of the G-basis vectors $\mathbf{u}_1$, $\mathbf{u}_2$, and $\mathbf{u}_3$ in terms of the standard E-basis are

$$[\mathbf{u}_1]_E = \begin{bmatrix} 1 \\ -1 \\ 0 \end{bmatrix} = \mathbf{u}_1^T, \qquad [\mathbf{u}_2]_E = \begin{bmatrix} 1 \\ 0 \\ -1 \end{bmatrix} = \mathbf{u}_2^T, \qquad \text{and} \qquad [\mathbf{u}_3]_E = \begin{bmatrix} 0 \\ 1 \\ 1 \end{bmatrix} = \mathbf{u}_3^T.$$

Therefore, the transition matrix or **change of basis matrix** $S = [I]_E^G$ **from the** G-basis **to the** *standard E-basis* is

$$S = \begin{bmatrix} 1 & 1 & 0 \\ -1 & 0 & 1 \\ 0 & -1 & 1 \end{bmatrix}.$$

> In $\mathbb{R}^n$, the **change of basis matrix** *from* a given basis $B = \{v_1, v_2, \dots, v_n\}$ *to* the **standard** basis E is always easy to write down. Simply write the given ordered basis vectors as the *columns* of S: $S = [v_1{}^T : v_2{}^T : \dots : v_n{}^T]$ as a partitioned matrix.

The reason for this is that, for any u in $\mathbb{R}^n$, the standard representation for u is u^T in $\mathbb{R}^n$.

The matrix S is nonsingular and its inverse

$$S^{-1} = [v_1{}^T : v_2{}^T : \dots : v_n{}^T]^{-1}$$

is the **transition** matrix **from the** standard basis E **to the basis** G, or the *change of basis matrix from E to G*. You can verify that in our example above,

$$S^{-1} = \begin{bmatrix} \frac{1}{2} & -\frac{1}{2} & \frac{1}{2} \\ \frac{1}{2} & \frac{1}{2} & -\frac{1}{2} \\ \frac{1}{2} & \frac{1}{2} & \frac{1}{2} \end{bmatrix}.$$

You can also verify that the columns of S^{-1} are the coordinatizations of the standard basis vectors e_1, e_2, and e_3 in terms of the G-basis vectors for $\mathbb{R}^3$. That is,

$$e_1 = \frac{1}{2}u_1 + \frac{1}{2}u_2 + \frac{1}{2}u_3,$$

$$e_2 = -\frac{1}{2}u_1 + \frac{1}{2}u_2 + \frac{1}{2}u_3,$$

$$e_3 = \frac{1}{2}u_1 - \frac{1}{2}u_2 + \frac{1}{2}u_3.$$

These coefficients can be determined by solving the resulting systems of equations, or in what amounts to the same thing, by calculating the inverse for the matrix S.

You should now use matrix multiplication to see that it is indeed the case that in Example 1 above the matrix $M = SKS^{-1}$. That is, show that

$$M = \begin{bmatrix} 1 & 1 & 0 \\ 0 & 1 & 1 \\ 1 & 0 & 1 \end{bmatrix} = \begin{bmatrix} 1 & 1 & 0 \\ -1 & 0 & 1 \\ 0 & -1 & 1 \end{bmatrix} \begin{bmatrix} 1 & 1 & 0 \\ -1 & 0 & 1 \\ 0 & 0 & 2 \end{bmatrix} \begin{bmatrix} \frac{1}{2} & -\frac{1}{2} & \frac{1}{2} \\ \frac{1}{2} & \frac{1}{2} & -\frac{1}{2} \\ \frac{1}{2} & \frac{1}{2} & \frac{1}{2} \end{bmatrix} = SKS^{-1}.$$

In $\mathbb{R}^n$, the **change of basis matrix from** the standard basis E **to** another basis $B = \{v_1, v_2, \ldots, v_n\}$ is found in the same manner. One may either solve the resulting system of equations for the coefficients expressing each e_i as a linear combination of $v_1, v_2, \ldots, v_n$, or one may first find the transition matrix $[v_1^T: v_2^T: \ldots : v_n^T]$ *from* B *to* the standard basis and then find the inverse matrix: $[v_1^T: v_2^T: \ldots : v_n^T]^{-1}$.

Example 2   Consider the vector space P_3 of polynomials of degree three or less with real coefficients. The "standard ordered basis" for P_3 is the set of polynomials

$$E = \{1, x, x^2, x^3\}.$$

Now look at the set (basis)

$$G = \{p_1(x) = 1 - x, p_2(x) = 1 + x, p_3(x) = 1 + x + x^2, p_4(x) = 1 - x^3\}.$$

Each of the polynomials in G has a standard matrix representation as a vector in $\mathbb{R}^4$ as follows:

$$[p_1(x)] = \begin{bmatrix} 1 \\ -1 \\ 0 \\ 0 \end{bmatrix}, \quad [p_2(x)] = \begin{bmatrix} 1 \\ 1 \\ 0 \\ 0 \end{bmatrix}, \quad [p_3(x)] = \begin{bmatrix} 1 \\ 1 \\ 1 \\ 0 \end{bmatrix}, \quad \text{and} \quad [p_4(x)] = \begin{bmatrix} 1 \\ 0 \\ 0 \\ -1 \end{bmatrix}.$$

Therefore, the *change of basis matrix S from G to E* has these vectors as its column vectors. The fact that S is nonsingular verifies that G is also a basis for P_3. (Why?) Thus,

$$S = \begin{bmatrix} 1 & 1 & 1 & 1 \\ -1 & 1 & 1 & 0 \\ 0 & 0 & 1 & 0 \\ 0 & 0 & 0 & -1 \end{bmatrix}.$$

The *change of basis matrix from E to G* is the inverse S^{-1} of S, namely:

$$S^{-1} = \begin{bmatrix} \frac{1}{2} & -\frac{1}{2} & 0 & \frac{1}{2} \\ \frac{1}{2} & \frac{1}{2} & -1 & \frac{1}{2} \\ 0 & 0 & 1 & 0 \\ 0 & 0 & 0 & -1 \end{bmatrix}.$$

Now suppose that $T: P_3 \rightarrow P_3$ is the following linear transformation (verify that it is, in fact, a linear transformation)

$$T(a + bx + cx^2 + dx^3) = (a + b)x + (c + d)x^3.$$

We can represent T *relative to the standard basis and also relative to the basis* G. Since

$$T(1) = x, \qquad T(x) = x, \qquad T(x^2) = x^3, \qquad T(x^3) = x^3,$$

the *standard representation* $A = [T]$ for T is easy to write down, using the matrix representations for each of these vectors as the columns. Indeed,

$$A = \begin{bmatrix} 0 & 0 & 0 & 0 \\ 1 & 1 & 0 & 0 \\ 0 & 0 & 0 & 0 \\ 0 & 0 & 1 & 1 \end{bmatrix}.$$

Since we know from our theorem that $A = SBS^{-1}$, for $B = [T]_G$, we can use matrix algebra to solve for the matrix B:

$$[T]_G = B = S^{-1}AS$$

$$= \begin{bmatrix} \frac{1}{2} & -\frac{1}{2} & 0 & \frac{1}{2} \\ \frac{1}{2} & \frac{1}{2} & -1 & \frac{1}{2} \\ 0 & 0 & 1 & 0 \\ 0 & 0 & 0 & -1 \end{bmatrix} \begin{bmatrix} 0 & 0 & 0 & 0 \\ 1 & 1 & 0 & 0 \\ 0 & 0 & 0 & 0 \\ 0 & 0 & 1 & 1 \end{bmatrix} \begin{bmatrix} 1 & 1 & 1 & 1 \\ -1 & 1 & 1 & 0 \\ 0 & 0 & 1 & 0 \\ 0 & 0 & 0 & -1 \end{bmatrix}$$

$$= \begin{bmatrix} 0 & -1 & -\frac{1}{2} & -1 \\ 0 & 1 & \frac{3}{2} & 0 \\ 0 & 0 & 0 & 0 \\ 0 & 0 & -1 & 1 \end{bmatrix}.$$

Thus, if $h(x) = a + bx + cx^2 + dx^3 = c_1p_1(x) + c_2p_2(x) + c_3p_3(x) + c_4p_4(x)$ is any polynomial in P_3, then $[T(h(x))]_G = B[h(x)]_G =$

$$\begin{bmatrix} 0 & -1 & -\frac{1}{2} & -1 \\ 0 & 1 & \frac{3}{2} & 0 \\ 0 & 0 & 0 & 0 \\ 0 & 0 & -1 & 1 \end{bmatrix} \begin{bmatrix} c_1 \\ c_2 \\ c_3 \\ c_4 \end{bmatrix} = \begin{bmatrix} -c_2 - \frac{1}{2}c_3 - c_4 \\ c_2 + \frac{3}{2}c_3 \\ 0 \\ -c_3 + c_4 \end{bmatrix}. \quad \blacksquare$$

In an arbitrary vector space V without an easily accessible standard basis, the computational task of finding the matrix representation for a transformation and the change of basis matrices is slightly more complicated. See pages (483–486) in the Appendix for algorithms that accomplish this.

The relationship $A = SBS^{-1}$ between two different representations A and B for the same linear transformation T given by (5.29) is an interesting relationship between matrices and is worth looking at in its own right.

A given $n \times n$ matrix A is said to be **similar** to the $n \times n$ matrix B if there exists an $n \times n$ nonsingular matrix S so that

$$A = SBS^{-1}. \qquad (5.30)$$

The matrix M of Example 1 is *similar* to the matrix K of that example, and the matrix A of Example 2 is *similar* to the matrix B of that example.

Example 3 Suppose that we have two matrices

$$A = \begin{bmatrix} 1 & 0 \\ 0 & -1 \end{bmatrix} \text{ and } B = \begin{bmatrix} -1 & 0 \\ 0 & 1 \end{bmatrix}.$$

Is A similar to B?

In order for A to be *similar* to B there must exist at least one 2×2 *non-singular* matrix S such that $A = SBS^{-1}$. How does one find such an S or know if one can exist? Since these two matrices are quire simple, we try a simple computational approach. Suppose such an S exists and multiply equation (5.30) by it on the right to obtain $AS = (SBS^{-1})S = SB(S^{-1}S) = SB$. So if a nonsingular such S exists, then it satisfies $AS = SB$. Suppose then that

$$S = \begin{bmatrix} a & b \\ c & d \end{bmatrix}, \qquad (5.31)$$

where a, b, c, and d are unknown scalars. We have then that

$$\begin{bmatrix} 1 & 0 \\ 0 & -1 \end{bmatrix} \begin{bmatrix} a & b \\ c & d \end{bmatrix} = \begin{bmatrix} a & b \\ c & d \end{bmatrix} \begin{bmatrix} -1 & 0 \\ 0 & 1 \end{bmatrix},$$

or

$$\begin{bmatrix} a & b \\ -c & -d \end{bmatrix} = \begin{bmatrix} -a & b \\ -c & d \end{bmatrix}. \qquad (5.32)$$

For the matrix equation in (5.32) to hold it is necessary that $a = -a$ and $d = -d$; therefore, we must have $a = d = 0$. Substituting these in the matrix S in (5.31) we see that

$$S = \begin{bmatrix} 0 & b \\ c & 0 \end{bmatrix}. \qquad (5.33)$$

So there are infinitely many choices of S for which $AS = SB$, since b and c can be chosen arbitrarily. But remember that S must be *nonsingular* so both $b \neq 0$ and $c \neq 0$ (why?). One possible choice is $b = c = 1$. Then S^{-1} can easily be computed by hand. Remember that any nonzero values for b and c (not

necessarily equal) can have been chosen. In our particular case we have se-
lected the matrix $S = \begin{bmatrix} 0 & 1 \\ 1 & 0 \end{bmatrix}$. You should have no difficulty computing the

inverse $S^{-1} = \begin{bmatrix} 0 & 1 \\ 1 & 0 \end{bmatrix}$. This is our candidate for a solution to equation

(5.30). Substitute and see that

$$SBS^{-1} = \begin{bmatrix} 0 & 1 \\ 1 & 0 \end{bmatrix} \begin{bmatrix} -1 & 0 \\ 0 & 1 \end{bmatrix} \begin{bmatrix} 0 & 1 \\ 1 & 0 \end{bmatrix} = \begin{bmatrix} 1 & 0 \\ 0 & -1 \end{bmatrix} = A.$$

So A is *similar* to B. ■

A remark is in order here that does *not* have general application to problems of this sort. But it does relate to our work with elementary matrices in
Chapters 1 and 2. Note that the matrix S in this example is an elementary
matrix. You may have noticed that one *row* interchange followed by one *column* interchange transforms the matrix B of this example into the matrix A.
The elementary matrix that interchanges rows *one* and *two* was our choice for
the matrix S. To multiply B on the *left* by S accomplishes this row operation.
To multiply it on the *right* by the same elementary matrix—notice that here
$S = S^{-1}$, so we have indeed done that—interchanges columns *one and two*.
We have not studied column operations in this text, but this remark at this
point may be interesting to you.

Example 4 Is the matrix $A = \begin{bmatrix} 2 & 0 \\ 0 & 2 \end{bmatrix}$ similar to the matrix $B = \begin{bmatrix} 2 & 1 \\ 0 & 2 \end{bmatrix}$?

Let us proceed as in Example 3, and try to find a nonsingular matrix S so
that $A = SBS^{-1}$. If such a matrix exists, it will satisfy $AS = SB$. So we look for
an S such that

$$\begin{bmatrix} 2 & 0 \\ 0 & 2 \end{bmatrix} \begin{bmatrix} a & b \\ c & d \end{bmatrix} = \begin{bmatrix} a & b \\ c & d \end{bmatrix} \begin{bmatrix} 2 & 1 \\ 0 & 2 \end{bmatrix}.$$

Therefore, if S exists

$$\begin{bmatrix} 2a & 2b \\ 2c & 2d \end{bmatrix} = \begin{bmatrix} 2a & a + 2b \\ 2c & c + 2d \end{bmatrix}.$$

For this to be true it is necessary that $2b = a + 2b$, so $a = 0$; and

$2d = c + 2d$, so $c = 0$. But then the matrix $S = \begin{bmatrix} 0 & b \\ 0 & d \end{bmatrix}$. Such a matrix is *sin-*

gular and, therefore, we cannot get from the equation $AS = SB$ to the desired
$A = SBS^{-1}$. The matrix S^{-1} does not exist; hence A and B are *not similar*. In
fact, it is not sufficient for similarity that we merely have just $AS = SB$. ■

This example is also a manifestation of a property that we ask you to prove as an easy matrix algebra exercise:

> If $K = \lambda I$ is a scalar matrix, then the only matrix similar to K is K itself. **(5.34)**

The relation of similarity of matrices has the following attractive and important features.

1. Reflectivity

> Every $n \times n$ matrix A is similar to itself. **(5.35)**

Proof Note that $A = IAI^{-1}$, where I is the identity matrix. ❏

2. Symmetry

> If A is similar to B, then B is similar to A. **(5.36)**

Proof Suppose that A is similar to B. Then there exists a nonsingular matrix S such that $A = SBS^{-1}$. Solving this simple matrix equation for B (multiply by S on the right and S^{-1} on the left) we get $B = S^{-1}AS = (S^{-1})A(S^{-1})^{-1}$, or $B = PAP^{-1}$, where P is the nonsingular matrix S^{-1}. So B is similar to A. ❏

3. Transitivity

> If A is similar to B and B is similar to C, then A is similar to C. **(5.37)**

Proof Since we are assuming that A is similar to B *and* B is similar to C, there must exist nonsingular matrices P and Q such that $A = PBP^{-1}$ and $B = QCQ^{-1}$. Hence (give the reasons)

$$A = P(QCQ^{-1})P^{-1} = (PQ)C(Q^{-1}P^{-1}) = (PQ)C(PQ)^{-1}.$$

That is, $A = SCS^{-1}$, where S is the nonsingular matrix (PQ). Thus A is similar to C. ❏

Similarity of matrices is another example of the important mathematical concept that we have used before and now formally define.

> **Definition** An **equivalence relation** on a set of objects (in this case matrices) is any relation that is **reflexive, symmetric,** and **transitive. (5.38)**

We have just shown that similarity of (square) matrices is an equivalence relation on the set of square matrices. You can also readily verify that *row equivalence* of matrices, as discussed in Chapter 1, is a different equivalence relation on sets of $m \times n$ matrices. Recall too that equivalence relations were discussed in Section 1.7 in connection with communication networks. We remarked in Section 1.7 that a major reason why an equivalence relation on a given set has significance is that it divides up the objects of that set, or *partitions* the set in question, into mutually disjoint subsets known as **equivalence classes.** Each equivalence class is such that every element in the set belongs to *exactly one* of the equivalence classes. See Supplementary Project 2 in Chapter 1. This partitioning property when applied to *matrix similarity* gives us the proof of the following theorem.

> **Theorem 5.10** All matrices similar to a given $n \times n$ matrix A belong to a class called a **similarity class.** Every $n \times n$ matrix belongs to exactly one similarity class, and no two distinct similarity classes have any elements (i.e., matrices) in common.

We stated in (5.34) above that the scalar matrix κI is the only matrix in its entire similarity class, whereas other matrices have many matrices in their individual similarity class. Each linear transformation T gives rise to a similarity class of matrix representations of T. In Chapter 6 we learn a method for finding a "simplest" (in a certain sense) member of the similarity class containing a given matrix; thus, we find the "canonical representative" of the associated transformation. However, here we can do this only for some special types of matrices. More general results must be postponed to more advanced courses.

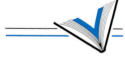 ## Vocabulary

In this section we again learned several new concepts. This related terminology is basic to the study of linear transformations, so be sure that you understand each of the concepts described by the terms in the following list. Review these ideas until you feel secure enough to use them in the rest of this course and in your future use of linear algebra.

transition matrix	change of basis matrix
similar matrices	equivalence relation
equivalence class	reflexive
symmetric	transitive
partition	similarity class

Exercises 5.5

In Exercises 1–5 determine whether or not the two given matrices are similar.

1. $A = \begin{bmatrix} 0 & 1 \\ 0 & 0 \end{bmatrix}$, $B = \begin{bmatrix} 0 & 0 \\ 1 & 0 \end{bmatrix}$. **2.** $A = \begin{bmatrix} 1 & 1 \\ 0 & 1 \end{bmatrix}$, $B = \begin{bmatrix} 1 & 0 \\ 1 & 1 \end{bmatrix}$. **3.** $A = \begin{bmatrix} 1 & 0 \\ 0 & 0 \end{bmatrix}$, $B = \begin{bmatrix} 0 & 0 \\ 0 & 1 \end{bmatrix}$.

4. $A = \begin{bmatrix} 1 & 0 & 0 \\ 0 & 1 & 0 \\ 0 & 0 & -1 \end{bmatrix}$, $B = \begin{bmatrix} -1 & 0 & 0 \\ 0 & 1 & 0 \\ 0 & 0 & 1 \end{bmatrix}$. **5.** $A = \begin{bmatrix} 1 & 0 & 0 \\ 0 & -1 & 0 \\ 0 & 0 & -1 \end{bmatrix}$, $B = \begin{bmatrix} -1 & 0 & 0 \\ 0 & -1 & 0 \\ 0 & 0 & 1 \end{bmatrix}$.

6. Show directly that the only matrix similar to the 3×3 identity matrix I_3 is I_3 itself. That is, consider $A = SI_3S^{-1}$.

7. Use Exercise 6 as a hint to prove our assertion (5.34) that the only matrix similar to a scalar matrix $K = \kappa I$ is the matrix K itself.

8. Carry out the computations necessary to find the coefficients c_1, c_2, c_3, and c_4 in Example 2, and find $[T(h(x))]_G$.

9. Show that if A and B are two similar $n \times n$ matrices then $\det A = \det B$. HINT: What can you say about the determinant of a product, and the determinant of S^{-1} relative to $\det S$?

10. Is the converse of Exercise 9 true? If so, prove it; if not, give a counterexample.

11. Can a singular matrix be similar to a nonsingular matrix? Give reasons.

12. Prove, or disprove by finding a counterexample. *If A and B are similar matrices, then either both are singular or both are nonsingular.*

13. Prove that *similar matrices have the same rank.*

14. Prove that if A and B are similar matrices, then for every positive integer n, A^n and B^n are similar.

15. Give an example of two row equivalent matrices that are *not* similar. See Example 4.

16. Find the rank and the nullity of the linear transformation $T: \mathbb{R}^3 \to \mathbb{R}^3$ found in Example 1: $T(x_1, x_2, x_3) = (x_1 + x_2, x_2 + x_3, x_1 + x_3)$. HINT: Use the standard matrix representation $M = [T]$.

17. Use the matrix representation $K = [T]_G$ to find the rank and the nullity of the linear transformation $T: \mathbb{R}^3 \to \mathbb{R}^3$ of Example 1. (Compare with Exercise 16.)

18. Use the standard matrix representation $A = [T]$ to find the rank and the nullity of the linear transformation $T: P_3 \to P_3$ from Example 2: $T(a + bx + cx^2 + dx^3) = (a + b)x + (c + d)x^3$.

19. Use the matrix representation $B = [T]_G$ to find the rank and the nullity of the linear transformation $T: P_3 \to P_3$ of Example 2. (Compare with Exercise 18.)

Exercises 20–30 refer to the linear transformation $T: \mathbb{R}^3 \to \mathbb{R}^3$ given by $T(x_1, x_2, x_3) = (x_2 - x_1, 0, x_3 - x_1)$. (Verify that it is a linear transformation.)

20. Find the standard matrix representation $[T]$ of T.

21. Find the rank and the nullity of T.

22. Consider the basis $B = \{\mathbf{u}_1 = (1, 0, 1), \mathbf{u}_2 = (0, -1, -1), \mathbf{u}_3 = (1, 1, 0)\}$ for $\mathbb{R}^3$. Write down the change of basis matrix S from the basis B to the standard basis E for $\mathbb{R}^3$. Verify that B is indeed a basis by showing that S is nonsingular.

23. Find the change of basis matrix S^{-1} (see Exercise 22) from the standard basis E to the basis B for $\mathbb{R}^3$.

24. Find the matrix representation $[T]_B$ of T relative to the basis B of Exercise 22.

25. Show that the matrices $[T]$ and $[T]_B$ of Exercises 20 and 24 are indeed similar.

26. Let $H = \{\mathbf{v}_1 = (1, 1, -1), \mathbf{v}_2 = (-1, 1, -1), \mathbf{v}_3 = (0, 1, 1)\}$ be another basis for $\mathbb{R}^3$. Write down the change of basis matrix P from the basis H to the standard basis E for $\mathbb{R}^3$. Verify that H is indeed a basis by showing that P is nonsingular.

27. Find the change of basis matrix P^{-1} (see Exercise 26) from the standard basis E to the basis H for $\mathbb{R}^3$.

28. Find the matrix representation $[T]_H$ of T relative to the basis H of Exercise 26.

29. Show, by calculation, that the matrices $[T]$ and $[T]_H$ of Exercises 20 and 28 are similar.

30. Show that the matrices $[T]_B$ and $[T]_H$ of Exercises 24 and 28 are indeed similar. What is the nonsingular matrix used?

Exercises 31–41 refer to the linear transformation (verify that it is) $T: P_2 \rightarrow P_2$ given by $T(a + bx + cx^2) = ax - (a + b + c)x^2$.

31. Find the standard matrix representation $[T]$ of T.

32. Find the rank and the nullity of T.

33. Consider the basis $B = \{p_1(x) = 2x, p_2(x) = -1, p_3(x) = x + x^2\}$ for P_2. Write down the change of basis matrix S from the basis B to the standard basis E for P_2. Verify that B is indeed a basis by showing that the S is nonsingular.

34. Find the change of basis matrix S^{-1} (see Exercise 33) from the standard basis E to the basis B for P_2.

35. Find the matrix representation $[T]_B$ of T relative to the basis B of Exercise 33.

36. Show that the matrices $[T]$ and $[T]_B$ of Exercises 31 and 35 are indeed similar.

37. Let $H = \{q_1(x) = 2x^2, q_2(x) = -1 + x^2, q_3(x) = x - x^2\}$ be another basis for P_2. Write down the change of basis matrix R from the basis H to the standard basis E for P_2. Verify that H is indeed a basis by showing that R is nonsingular.

38. Find the change of basis matrix R^{-1} (see Exercise 37) from the standard basis E to the basis H for P_2.

39. Find the matrix representation $[T]_H$ of T relative to the basis H of Exercise 37.

40. Show, by calculation, that the matrices $[T]$ and $[T]_H$ of Exercises 31 and 39 are similar.

41. Show that the matrices $[T]_B$ and $[T]_H$ of Exercises 35 and 39 are indeed similar. What is the nonsingular matrix used?

Exercises 42–49 refer to the linear transformation $T: M_2(\mathbb{C}) \rightarrow M_2(\mathbb{C})$ given by $T(A) = A^T$ (the transpose of A).

42. Show directly that this transformation is a linear transformation on $M_2(\mathbb{C})$.

43. Show that the set of matrices

$$E = \left\{ E_{11} = \begin{bmatrix} 1 & 0 \\ 0 & 0 \end{bmatrix}, E_{12} = \begin{bmatrix} 0 & 1 \\ 0 & 0 \end{bmatrix}, \right.$$

$$\left. E_{21} = \begin{bmatrix} 0 & 0 \\ 1 & 0 \end{bmatrix}, E_{22} = \begin{bmatrix} 0 & 0 \\ 0 & 1 \end{bmatrix} \right\}$$

is a basis for $M_2(\mathbb{C})$. This basis is called the *standard* basis for $M_2(\mathbb{C})$.

44. Find the standard matrix representation $[T]$ of $T(A) = A^T$. HINT: $[T]$ is a 4×4 matrix.

45. Find the rank and the nullity of T.

46. Consider the basis

$$B = \left\{ J_1 = I = \begin{bmatrix} 1 & 0 \\ 0 & 1 \end{bmatrix}, J_2 = \begin{bmatrix} 0 & 1 \\ -1 & 0 \end{bmatrix}, \right.$$

$$\left. J_3 = \begin{bmatrix} 0 & -1 \\ 0 & 1 \end{bmatrix}, J_4 = \begin{bmatrix} -1 & 0 \\ 1 & 0 \end{bmatrix} \right\} \text{ for } M_2(\mathbb{C}).$$

Write down the change of basis matrix S from the basis B to the standard basis E for $M_2(\mathbb{C})$. Verify that B is indeed a basis by showing that S is nonsingular.

47. Find the change of basis matrix S^{-1} (see Exercise 46) from the standard basis E to the basis B for $M_2(\mathbb{C})$.

48. Find the matrix representation $[T]_B$ of $T(A) = A^T$ relative to the basis B of Exercise 46.

49. Show that the matrices $[T]$ and $[T]_B$ of Exercises 44 and 48 are indeed similar. What is the nonsingular matrix used?

50. Prove the following theorem that we have tacitly used in the exercises above.

Theorem   Let $G = \{\mathbf{v}_1, \mathbf{v}_2, \ldots, \mathbf{v}_n\}$ be a set of n vectors from $\mathbb{R}^n$. The set G is a basis for $\mathbb{R}^n$ if, and only if, the $n \times n$ matrix $S = [\mathbf{v}_1^T : \mathbf{v}_2^T : \ldots : \mathbf{v}_n^T]$ is nonsingular.

---

51. Prove that if A is any 2×2 matrix with real or complex entries, then A is similar to its transpose, A^T.

52. Let A and B be $n \times n$ real matrices. Suppose that $p(x) = c_k x^k + c_{k-1} x^{k-1} + \cdots + c_1 x + c_0$ is a polynomial with real coefficients. Define

$p(A) = c_k A^k + c_{k-1} A^{k-1} + \cdots + c_1 A + c_0 I$ (and similarly for $p(B)$, where I is the $n \times n$ identity matrix). Prove that if A and B are similar, then so are the two matrices $p(A)$ and $p(B)$. HINT: Use the definition of similarity and see Exercise 14.

53. With $p(A)$ as defined in Exercise 52, prove that if A is a diagonal matrix, then so is $p(A)$.

54. With $p(A)$ as defined in Exercise 52, prove that if A is an upper- (or lower-) triangular matrix, then so is $p(A)$.

55. With $p(A)$ as defined in Exercise 52, suppose that A is an $n \times n$ diagonal or upper-triangular or lower-triangular matrix with diagonal entries a_{ii}. Prove that the entries on the main diagonal of the matrix $p(A)$ are the real numbers $p(a_{ii})$, $i = 1, 2, \ldots, n$.

56. Is every $n \times n$ matrix that is similar to a diagonal matrix, diag $[d_1, d_2, \ldots, d_n]$, necessarily a diagonal matrix itself? See Exercise 7. Why or why not?

5.6 ORTHOGONAL LINEAR TRANSFORMATIONS

We remarked earlier that it is often true that applications of linear algebra have geometric content. In particular a change of basis is related to changing a coordinate system. Often one is concerned with only those linear transformations that will preserve the geometry of a given space. Of particular interest are those linear transformations or operators on a vector space that preserve the length of vectors and angles between vectors. Remember that these ideas are most meaningful in inner-product spaces. (You should review the definitions and concepts of Chapter 4.) Let us first look at a simple example in the space $\mathbb{R}^2$ of ordered pairs of real numbers which, as we have previously noted, can be identified with the ordinary Euclidean geometric plane. Recall Example 8 in Section 5.1 and suppose that we rotate the plane about the origin through an angle of 30°. Consider the following example.

Example 1 Let $T_{30°}: \mathbb{R}^2 \to \mathbb{R}^2$ be the rotation of the plane through 30°. In Figure 5.11 we see that the point $P: (a, b)$ becomes transformed by $T_{30°}$ to the point $Q: (a', b')$.

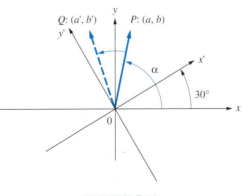

FIGURE 5.11

If $\mathbf{u} = \overrightarrow{OP}$ is the vector from the origin to P and $\mathbf{u}$ makes an angle α with the positive x axis, then $T_{30°}(\mathbf{u}) = \overrightarrow{OQ}$ will make an angle of $\alpha + 30°$ with the same axis. Set $\|\mathbf{u}\| = \|\overrightarrow{OP}\|$ and note that this number, say $\|\mathbf{u}\| = r$, is not changed by the rotation. Therefore, we have that

$$r = \|\mathbf{u}\| = \|\overrightarrow{OP}\| = \|\overrightarrow{OQ}\| = \|T_{30°}(\mathbf{u})\|.$$

From elementary trigonometry we also note that

$$a = r\cos\alpha \qquad \text{while} \qquad b = r\sin\alpha.$$

It follows that

$$a' = r\cos(\alpha + 30°), \qquad \text{while} \qquad b' = r\sin(\alpha + 30°).$$

Then, from the appropriate trigonometric identities we compute

$$
\begin{aligned}
a' &= r\cos(\alpha + 30°) \\
&= r(\cos\alpha\cos 30° - \sin\alpha\sin 30°) \\
&= (r\cos\alpha)\left[\frac{\sqrt{3}}{2}\right] - (r\sin\alpha)\left[\frac{1}{2}\right] \\
&= \frac{\sqrt{3}}{2}a - \frac{1}{2}b.
\end{aligned}
$$

Similarly,

$$
\begin{aligned}
b' &= r\sin(\alpha + 30°) \\
&= r(\sin\alpha\cos 30° + \cos\alpha\sin 30°) \\
&= r\sin\alpha\left[\frac{\sqrt{3}}{2}\right] + r\cos\alpha\left[\frac{1}{2}\right] \\
&= \frac{\sqrt{3}}{2}b + \frac{1}{2}a = \frac{1}{2}a + \frac{\sqrt{3}}{2}b.
\end{aligned}
$$

Therefore, we have a formula that defines $T_{30°}(\mathbf{x})$ as follows: For each $\mathbf{x} = (a, b)$ in $\mathbb{R}^2$,

$$T_{30°}(a, b) = \left[\frac{\sqrt{3}}{2}a - \frac{1}{2}b, \frac{1}{2}a + \frac{\sqrt{3}}{2}b\right].$$

The standard matrix representation for this transformation is the matrix

$$A = [T_{30°}] = \begin{bmatrix} \dfrac{\sqrt{3}}{2} & -\dfrac{1}{2} \\ \dfrac{1}{2} & \dfrac{\sqrt{3}}{2} \end{bmatrix}. \quad \blacksquare$$

Compare this with Example 8 in Section 5.1 and with Exercise 22 of this section. From the geometry of the situation we see that this rotation does indeed preserve the length of a vector and the angle between two vectors in the plane. Let us verify that it also *preserves the standard inner product* between any two vectors in $\mathbb{R}^2$.

Suppose that $\mathbf{u} = (u_1, u_2)$ and $\mathbf{v} = (v_1, v_2)$ are two vectors in $\mathbb{R}^2$. The standard inner product in $\mathbb{R}^2$ gives the number $<\mathbf{u}, \mathbf{v}> = u_1 v_1 + u_2 v_2$. Now let us transform the vectors $\mathbf{u}$ and $\mathbf{v}$ by $T_{30°}$. You can see that

$$T_{30°}(\mathbf{u}) = \left(\frac{\sqrt{3}}{2}u_1 - \frac{1}{2}u_2, \frac{1}{2}u_1 + \frac{\sqrt{3}}{2}u_2\right),$$

and

$$T_{30°}(\mathbf{v}) = \left(\frac{\sqrt{3}}{2}v_1 - \frac{1}{2}v_2, \frac{1}{2}v_1 + \frac{\sqrt{3}}{2}v_2\right).$$

Therefore,

$$
\begin{aligned}
<T_{30°}(\mathbf{u}), T_{30°}(\mathbf{v})> &= \left(\frac{\sqrt{3}}{2}u_1 - \frac{1}{2}u_2\right)\left(\frac{\sqrt{3}}{2}v_1 - \frac{1}{2}v_2\right) \\
&\quad + \left(\frac{1}{2}u_1 + \frac{\sqrt{3}}{2}u_2\right)\left(\frac{1}{2}v_1 + \frac{\sqrt{3}}{2}v_2\right) \\
&= \frac{3}{4}u_1 v_1 - \frac{\sqrt{3}}{4}u_1 v_2 - \frac{\sqrt{3}}{4}u_2 v_1 + \frac{1}{4}u_2 v_2 + \frac{1}{4}u_1 v_1 \\
&\quad + \frac{\sqrt{3}}{4}u_1 v_2 + \frac{\sqrt{3}}{4}u_2 v_1 + \frac{3}{4}u_2 v_2 \\
&= u_1 v_1 + u_2 v_2 = <\mathbf{u}, \mathbf{v}>.
\end{aligned}
$$

Thus, $<T_{30°}(\mathbf{u}), T_{30°}(\mathbf{v})> = <\mathbf{u}, \mathbf{v}>$.

This example is an example of an important particular kind of linear transformation. For reasons that will become clearer as we proceed, such an inner-product preserving transformation is called an **orthogonal linear transformation**. Formally, we have the following definition.

> **Definition** Let U be a real inner-product space. A linear transformation $T: U \rightarrow U$ is called an **orthogonal linear transformation** (operator) if, and only if, for each $\mathbf{u}$ and $\mathbf{v}$ in U, $<T(\mathbf{u}), T(\mathbf{v})> = <\mathbf{u}, \mathbf{v}>$.

That is, $T: U \rightarrow U$ is orthogonal if and only if T "preserves" inner products.

Example 2   Let $T: \mathbb{R}^2 \rightarrow \mathbb{R}^2$ be the linear transformation:

$$T(x_1, x_2) = (-x_2, x_1).$$

Now let $\mathbf{u} = (u_1, u_2)$ and $\mathbf{v} = (v_1, v_2)$ be two arbitrary vectors in $\mathbb{R}^2$ considered as Euclidean 2-space with the standard inner product. Let us compute the inner product of the two vectors $T(\mathbf{u})$ and $T(\mathbf{v})$. Since $T(\mathbf{u}) = T(u_1, u_2) = (-u_2, u_1)$ and $T(\mathbf{v}) = T(v_1, v_2) = (-v_2, v_1)$, we have

$$<T(\mathbf{u}), T(\mathbf{v})> = (-u_2)(-v_2) + u_1 v_1 = u_2 v_2 + u_1 v_1 = u_1 v_1 + u_2 v_2 = <\mathbf{u}, \mathbf{v}>.$$

So T is an orthogonal linear transformation. In particular, if $\mathbf{u} = \mathbf{v}$, the above equation yields

$$< T(\mathbf{u}),\, T(\mathbf{u}) > = < \mathbf{u},\, \mathbf{u} >.$$

Because of this, for all $\mathbf{u}$ in $\mathbb{R}^2$, we readily compute that

$$\|T(\mathbf{u})\| = \|\mathbf{u}\|. \quad \blacksquare$$

Hence T is a "norm-preserving" linear transformation. This fact is not restricted to our particular example. In fact, we have the following property of all orthogonal linear transformations.

Theorem 5.11 Let U be any inner-product space, and let $T: U \rightarrow U$ be a linear operator on U. The transformation T is an orthogonal linear operator if, and only if, T preserves the inner-product induced vector norm in the sense that $\|T(\mathbf{u})\| = \|\mathbf{u}\|$, for every $\mathbf{u}$ in U.

Proof First observe that if T is an orthogonal linear transformation, then surely for each u in U,

$$\|\mathbf{u}\|^2 = < \mathbf{u},\, \mathbf{u} > = < T(\mathbf{u}),\, T(\mathbf{u}) > = \|T(\mathbf{u})\|^2.$$

Therefore, $\|T(\mathbf{u})\| = \|\mathbf{u}\|$, so an orthogonal linear transformation always preserves the norm (length) that it defines.

The converse states that if T preserves the norm induced by the inner product, then it is an orthogonal linear transformation. You are asked to prove this as Exercise 35. ▢

Even more is true about orthogonal linear transformations. In Section 4.3 we noted that if θ is the angle between two nonzero vectors $\mathbf{v}_1$ and $\mathbf{v}_2$ in the inner-product space U, then

$$\cos\theta = \frac{< \mathbf{v}_1,\, \mathbf{v}_2 >}{\|\mathbf{v}_1\|\,\|\mathbf{v}_2\|}.$$

Since this expression involves just inner products and lengths, we can conclude:

Corollary 1 An orthogonal linear transformation preserves angles between vectors.

An orthogonal linear transformation is also often called an **isometry.** This term comes from the Greek and means "of equal measure or dimensions." The measure intended in the present situation is the norm (or length), which, of course, is preserved under an orthogonal linear transformation.

It is often useful to note that it is sufficient to verify that inner products are preserved just for all pairs of members of a basis for V in order to demonstrate that T is an orthogonal linear transformation. This is the content of the following corollary. The proof depends on the fact that the action of a transformation is completely determined by its action on a basis and is left to you as Exercise 36.

Corollary 2   Let $B = \{\mathbf{v}_1, \mathbf{v}_2, \ldots, \mathbf{v}_n\}$ be a basis for the real inner-product space V. Then the linear transformation $T: V \rightarrow V$ is an orthogonal linear transformation if, and only if, for each pair of (not necessarily distinct) vectors $\mathbf{v}_i$ and $\mathbf{v}_j$ from B, $<T(\mathbf{v}_i), T(\mathbf{v}_j)> = <\mathbf{v}_i, \mathbf{v}_j>$.

Let us return briefly to Example 2 and compute the matrix representation for the given linear transformation T.

Example 3   Find the matrix representation of the linear transformation $T(x, y) = (-y, x)$, of Example 2, with respect to the standard (orthonormal) basis for $\mathbb{R}^2$.
   It is clear that $T(\mathbf{e}_1) = (0, 1)$ and $T(\mathbf{e}_2) = (-1, 0)$ so

$$B = [T]_E = \begin{bmatrix} 0 & -1 \\ 1 & 0 \end{bmatrix}$$
$$\qquad [T(\mathbf{e}_1)] \quad [T(\mathbf{e}_2)]$$

is the standard matrix representation of T. Note also that

$$B^{-1} = \begin{bmatrix} 0 & 1 \\ -1 & 0 \end{bmatrix} = B^T.$$

It is also true that, for the matrix A of Example 1,

$$A^{-1} = \begin{bmatrix} \dfrac{\sqrt{3}}{2} & \dfrac{1}{2} \\ \dfrac{-1}{2} & \dfrac{\sqrt{3}}{2} \end{bmatrix} = A^T. \quad \blacksquare$$

The fact that in each of these cases the inverse and the transpose of these matrices A and B are the same is of interest to us, and motivates the following definition.

Definition   An $n \times n$ real matrix P is called an **orthogonal matrix** if, and only if, $PP^T = P^TP = I_n$.

The matrices A and B of Example 3 are, then, both *orthogonal matrices.*

Example 4 The following matrices are all orthogonal matrices. You may check, by matrix multiplication in each case, to see that $PP^T = P^TP = I$.

$$P = I_n; \qquad P = \begin{bmatrix} 0 & 0 & 1 \\ 0 & 1 & 0 \\ 1 & 0 & 0 \end{bmatrix} = J_3; \qquad P = J_n; \qquad P = \begin{bmatrix} 1 & 0 & 0 & 0 \\ 0 & 0 & 1 & 0 \\ 0 & 1 & 0 & 0 \\ 0 & 0 & 0 & 1 \end{bmatrix};$$

$$P = \begin{bmatrix} 0 & -1 \\ -1 & 0 \end{bmatrix} = -J_2; \qquad P = -I_4; \qquad P = \begin{bmatrix} \dfrac{1}{\sqrt{2}} & -\dfrac{1}{\sqrt{3}} & \dfrac{1}{\sqrt{6}} \\[2mm] 0 & \dfrac{1}{\sqrt{3}} & \dfrac{2}{\sqrt{6}} \\[2mm] \dfrac{1}{\sqrt{2}} & \dfrac{1}{\sqrt{3}} & -\dfrac{1}{\sqrt{6}} \end{bmatrix}. \qquad \blacksquare$$

We now suspect that any orthogonal linear transformation may be represented by an orthogonal matrix.

> However, it is *not true* that every matrix representation of an orthogonal linear transformation is an orthogonal matrix.

The facts are given by the following theorem. You are asked to supply the proof as Exercise 37. Also verify this result for the special situations in Examples 1, 2, and 3.

> **Theorem 5.12** Let $T: U \to U$ be a linear transformation of the real inner-product space U, and let $B = \{\mathbf{u}_1, \mathbf{u}_2, \ldots, \mathbf{u}_n\}$ be an **orthonormal** basis for U. Then T is an orthogonal linear transformation if, and only if, the matrix representation $[T]_B$ of T is an orthogonal matrix.

We remarked earlier that the choice of the word *orthogonal* to describe a linear transformation that preserves inner products would become apparent. Look at the column vectors of the orthogonal matrices in Examples 3 and 4 above. As you can easily check, they are mutually orthonormal. Of course, that is what $PP^T = I_n$ says when viewed as a statement about the columns of P. The following theorem characterizes (real) orthogonal matrices.

> **Theorem 5.13** An $n \times n$ real matrix P is orthogonal if, and only if, any one (hence all) of the following equivalent conditions is satisfied:
>
> (i) $PP^T = P^TP = I_n$.
> (ii) $P^{-1} = P^T$.
> (iii) The column vectors of P are mutually orthonormal.
> (iv) The row vectors of P are mutually orthonormal.

The proof of this theorem and some other interesting properties of orthogonal matrices are left to you as exercises. In particular, see Exercises 38–40.

Applications of the ideas developed in this chapter play an important role in Chapter 6.

Vocabulary

In this section we learned three important new concepts. They are basic to the study of both linear transformations and matrix theory, so be sure that you understand each of them. Review these ideas until you feel secure enough to use them in the rest of this course and in your future use of linear algebra.

orthogonal linear transformation   isometry
orthogonal matrix

Exercises 5.6

In Exercises 1–9 determine which of the given matrices are orthogonal.

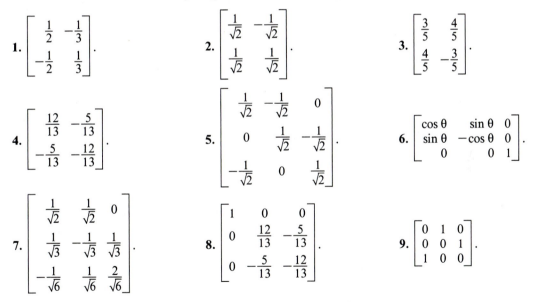

1. $\begin{bmatrix} \frac{1}{2} & -\frac{1}{3} \\ -\frac{1}{2} & \frac{1}{3} \end{bmatrix}$.

2. $\begin{bmatrix} \frac{1}{\sqrt{2}} & -\frac{1}{\sqrt{2}} \\ \frac{1}{\sqrt{2}} & \frac{1}{\sqrt{2}} \end{bmatrix}$.

3. $\begin{bmatrix} \frac{3}{5} & \frac{4}{5} \\ \frac{4}{5} & -\frac{3}{5} \end{bmatrix}$.

4. $\begin{bmatrix} \frac{12}{13} & -\frac{5}{13} \\ -\frac{5}{13} & -\frac{12}{13} \end{bmatrix}$.

5. $\begin{bmatrix} \frac{1}{\sqrt{2}} & -\frac{1}{\sqrt{2}} & 0 \\ 0 & \frac{1}{\sqrt{2}} & -\frac{1}{\sqrt{2}} \\ -\frac{1}{\sqrt{2}} & 0 & \frac{1}{\sqrt{2}} \end{bmatrix}$.

6. $\begin{bmatrix} \cos\theta & \sin\theta & 0 \\ \sin\theta & -\cos\theta & 0 \\ 0 & 0 & 1 \end{bmatrix}$.

7. $\begin{bmatrix} \frac{1}{\sqrt{2}} & \frac{1}{\sqrt{2}} & 0 \\ \frac{1}{\sqrt{3}} & -\frac{1}{\sqrt{3}} & \frac{1}{\sqrt{3}} \\ -\frac{1}{\sqrt{6}} & \frac{1}{\sqrt{6}} & \frac{2}{\sqrt{6}} \end{bmatrix}$.

8. $\begin{bmatrix} 1 & 0 & 0 \\ 0 & \frac{12}{13} & -\frac{5}{13} \\ 0 & -\frac{5}{13} & -\frac{12}{13} \end{bmatrix}$.

9. $\begin{bmatrix} 0 & 1 & 0 \\ 0 & 0 & 1 \\ 1 & 0 & 0 \end{bmatrix}$.

In Exercises 10–13 determine which of the given linear operators T, on $\mathbb{R}^n$ equipped with the standard inner product, are orthogonal.

10. $T(x_1, x_2) = \left(\dfrac{3x_1}{5} + \dfrac{4x_2}{5}, \dfrac{4x_1}{5} - \dfrac{3x_2}{5} \right)$.

11. $T(x_1, x_2, x_3) = \left(x_3, \dfrac{x_1\sqrt{3}}{2} + \dfrac{x_2}{2}, \dfrac{x_1}{2} - \dfrac{x_2\sqrt{3}}{2} \right)$.

12. $T(x_1, x_2, x_3) = \left(\dfrac{3x_1}{5} - \dfrac{4x_2}{5}, x_3, \dfrac{4x_2}{5} + \dfrac{3x_3}{5} \right)$.

13. $T(x_1, x_2, x_3) = \left(\dfrac{5x_1}{13} + \dfrac{12x_3}{13}, \dfrac{12x_2}{13} - \dfrac{5x_3}{13}, x_1 \right)$.

14. Write the standard matrix representation $P = [T]$ for the linear transformation of Exercise 10.

15. Write the standard matrix representation $P = [T]$ for the linear transformation of Exercise 11.

16. Write the standard matrix representation $P = [T]$ for the linear transformation of Exercise 12.

17. Write the standard matrix representation $P = [T]$ for the linear transformation of Exercise 13.

In Exercises 18–21 (if you have studied calculus) let $T: P_1 \rightarrow P_1$ be the linear transformation on the vector space of real linear polynomials equipped with the integral inner product,

$$< p, q > = \int_0^1 p(x)\, q(x)\, dx,$$

defined by $T(ax + b) = xb\sqrt{3} + \dfrac{a\sqrt{3}}{3}.$

18. Is the "standard" basis $E = \{1, x\}$ for P_1 orthonormal with respect to the integral inner product?

19. Is the basis $G = \{p(x) = 1, q(x) = \sqrt{3}(2x - 1)\}$ for P_1 orthonormal with respect to the integral inner product?

20. Represent the given transformation T by a matrix $Q = [T]_G$ relative to the basis G of Exercise 19.

21. Is the matrix Q of Exercise 20 an orthogonal matrix?

22. Show that every 2×2 orthogonal matrix P must have either form A or form B, where

$$A = \begin{bmatrix} \cos \phi & -\sin \phi \\ \sin \phi & \cos \phi \end{bmatrix},$$

$$B = \begin{bmatrix} \cos \phi & \sin \phi \\ \sin \phi & -\cos \phi \end{bmatrix},$$

for some ϕ, $0 \leq \phi \leq \pi$. HINT: Write an arbitrary 2×2 matrix and use the fact that it is orthogonal to arrive at some identities that must be satisfied. Remember that $\cos^2 x + \sin^2 x = 1$.

23. Show geometrically that the matrix form A in Exercise 22 represents, with respect to the standard basis for the plane $\mathbb{R}^2$, a rotation about the origin through an angle ϕ. HINT: See Example 1.

24. Show geometrically that the matrix form B in Exercise 22 represents, with respect to the standard basis for the plane $\mathbb{R}^2$, a reflection of the plane through a vector $\mathbf{w}$ whose angle of inclination with $\mathbf{e}_1 = (1, 0)$ is $\phi/2$. HINT: Draw a figure and use the trigonometric identities for $\cos (x + y)$ and $\sin (x + y)$.

25. Use the results of Exercises 22–24 to show that an orthogonal matrix P represents a rotation of the plane if, and only if, $\det P = 1$.

26. Use the results of Exercises 22–24 to show that an orthogonal matrix P represents a reflection of the plane through a line along a given vector $\mathbf{w}$ if, and only if, $\det P = -1$.

27. Prove that if P is any $n \times n$ orthogonal matrix then $\det P = 1$ or -1.

In Exercises 28–31, let V be a real inner-product space and let k be a fixed scalar such that $k > 1$. Let $T: V \rightarrow V$ be defined by $T(\mathbf{x}) = k\mathbf{x}$, for all $\mathbf{x} \in V$. REMARK: Such linear transformations are called magnifications.

28. Show that T is a *linear* transformation.

29. Show that for all nonzero vectors $\mathbf{u}$ and $\mathbf{v}$ in V,
$$\frac{< T(\mathbf{u}),\, T(\mathbf{v}) >}{\| T(\mathbf{u}) \| \, \| T(\mathbf{v}) \|} = \frac{< \mathbf{u},\, \mathbf{v} >}{\| \mathbf{u} \| \, \| \mathbf{v} \|}.$$

30. Use the results of Exercise 29 to show that this transformation T preserves angles between vectors. Show that T *does not* preserve lengths (norms) or inner products.

31. Is this linear transformation T an orthogonal linear operator on V? Give reasons.

In Exercises 32–34 let $L: \mathbb{R}^2 \rightarrow \mathbb{R}^2$ be the function $L(x_1, x_2) = (x_2, -x_1).$

32. Show that L is a linear transformation.

33. Is L orthogonal? Give reasons.

34. Does L preserve the length of a vector? the angle between vectors? inner products? Explain.

35. Finish the proof of Theorem 5.11 (page 360) by showing that if $T: U \to U$ has the property that $\|T(\mathbf{u})\| = \|\mathbf{u}\|$ for all $\mathbf{u} \in U$, then T is orthogonal. HINT: Consider $\|T(\mathbf{u} + \mathbf{v})\|^2 = \|\mathbf{u} + \mathbf{v}\|^2$.

36. Prove Corollary 2 to Theorem 5.11.

37. Prove Theorem 5.12 (page 362). HINT: Use Theorem 5.6 (page 325), and see the proofs of Theorems 5.7 and 5.8 (pages 341 and 342).

38. Prove that parts (i) and (ii) of Theorem 5.13 (page 362) are equivalent. (Remember that part (i) is our definition of an orthogonal matrix.)

39. Prove that parts (i) and (iii) of Theorem 5.13 are equivalent.

40. Prove that parts (i) and (iv) of Theorem 5.13 are equivalent.

41. Suppose V is a real inner-product space and let $F : V \to V$ be a linear transformation with the property that whenever two vectors $\mathbf{u}$ and $\mathbf{v}$ are orthogonal, $< \mathbf{u}, \mathbf{v} > = 0$, then $F(\mathbf{u})$ and $F(\mathbf{v})$ are also orthogonal: $< F(\mathbf{u}), F(\mathbf{v}) > = 0$. Does this force F to be an isometry (orthogonal linear operator)? Give reasons.

42. Let Q be an $n \times n$ orthogonal matrix and let X and Y be any $n \times 1$ matrices (column vectors). Prove that $X^T Y = (QX)^T (QY)$. Conversely, is any $n \times n$ matrix Q with this property orthogonal? Explain.

43. Let P and Q be $n \times n$ orthogonal matrices. Prove that PQ is also orthogonal.

44. Let the complex vector space $\mathbb{C}^n$ have the standard inner product $< \mathbf{u}, \mathbf{v} > = \mathbf{u}\,\mathbf{v}^*$ and let $L: \mathbb{C}^n \to \mathbb{C}^n$ be a linear operator on $\mathbb{C}^n$ with the property that $< L(\mathbf{u}), L(\mathbf{v}) > = < \mathbf{u}, \mathbf{v} >$. Show that if A is the standard matrix representation $[L]$ of L, then $AA^* = A^*A = I_n$. Such a matrix is called a *unitary* matrix.

In Exercises 45–48 we use the definition of a unitary matrix given in Exercise 44.

45. Prove that the column vectors of an $n \times n$ unitary matrix A are mutually orthonormal in the unitary (inner-product) space $\overline{\mathbb{C}^n}$.

46. Prove that the row vectors of an $n \times n$ unitary matrix A are mutually orthonormal in the unitary (inner-product) space $\mathbb{C}^n$.

47. Let $\mathbf{x}$ be any vector in the unitary space $\mathbb{C}^n$, and let A be a unitary matrix. Prove that if $\mathbf{y} = A\mathbf{x}^T$ then $\|\mathbf{y}\| = \|\mathbf{x}\|$; that is, show that $\mathbf{y}^*\mathbf{y} = \mathbf{x}\mathbf{x}^*$.

48. Let K and M be $n \times n$ unitary matrices. Prove that KM is also unitary.

49. Suppose that A is an orthogonal matrix such that $A^2 = I_n$. Prove that $A = A^T$.

50. Find all of the $n \times n$ real orthogonal matrices that are also diagonal.

51. Can you describe all of the $n \times n$ complex diagonal matrices A that have the property that $A^{-1} = A^T$?

CHAPTER 5 SUMMARY

In this chapter, we studied the important concepts of a linear transformation $T: U \to V$ and its associated kernel, ker T, and range, range T. We noted that the dimensions of these subspaces are the nullity and the rank of T, respectively. We saw that it is always true that (nullity of T) + (rank of T) = dimension of U. Diverse examples were given to illustrate these concepts. Next we studied the fundamental notion of the matrix representation of a given linear transformation, together with many illustrative examples. The algebra of linear transformations was then studied and its similarities and differences with ordinary algebra were noted. One striking difference between the ordinary algebra of the real and complex numbers and this algebra of transformations was the failure of the commutative law for multiplication: $T_1 T_2$ is in general different from $T_2 T_1$. We also found that many nonzero linear transformations have no multiplicative inverse. On the other hand, we saw that the algebra of linear transformations is isomorphic (i.e., abstractly identical) to the algebra of matrices that we encountered earlier. Finally the important concepts of similarity of matrices and of orthogonal linear transformations were discussed, with many examples.

CHAPTER 5 REVIEW EXERCISES

1. Show that $T: \mathbb{R}^2 \to \mathbb{R}^3$ given by $T(x, y) = (-x, y, y - x)$ is a linear transformation.

2. Find a basis for the kernel, ker T, of the transformation T of Exercise 1.

3. Find a basis for the range, range T, of the transformation T of Exercise 1.

4. Use the results of Exercises 1, 2, and 3 to find the rank and nullity of the given transformation T and to verify Sylvester's law of nullity.

5. Find the standard matrix representation $[T]$ of the transformation T in Exercise 1. (The standard bases for $\mathbb{R}^2$ and $\mathbb{R}^3$)

6. Let $T: \mathbb{R}^3 \to \mathbb{R}^3$ be given by the formula $T(x_1, x_2, x_3) = (x_1 + x_2, x_2 + x_3, x_3 + x_1)$. Verify that this T is a linear transformation and write its standard matrix representation $[T]$.

7. Is the transformation of Exercise 6 invertible? If it is, find $[T]^{-1}$.

8. Use the result of Exercise 7 to write the formula for $T^{-1}(x_1, x_2, x_3)$.

9. Are the two matrices A and B similar,

   where $A = \begin{bmatrix} -1 & 0 & 0 \\ 0 & 1 & 0 \\ 0 & 0 & 1 \end{bmatrix}$ and

   $B = \begin{bmatrix} 1 & 0 & 0 \\ 0 & 1 & 0 \\ 1 & 0 & -1 \end{bmatrix}$? Give reasons.

10. Which of the following two matrices, if any, are orthogonal? Give reasons.

    $K = \begin{bmatrix} 0 & 0 & 1 \\ 0 & 1 & 0 \\ 1 & 0 & 0 \end{bmatrix}$ and $M = \begin{bmatrix} 0 & -1 & 0 \\ 0 & 0 & -1 \\ -1 & 0 & 0 \end{bmatrix}$.

CHAPTER 5 SUPPLEMENTARY EXERCISES AND PROJECTS

1. Let A be a diagonal matrix $A = \text{diag}[a_1, a_2, \ldots, a_n]$, and let $B = SAS^{-1}$ for some nonsingular matrix S. Denote the column vectors of S by $\mathbf{s}_i$, $i = 1, 2, \ldots, n$. Show that $B\mathbf{s}_i = a_i\mathbf{s}_i$ for each $i = 1, 2, \ldots, n$.

2. Show that the set $\mathbb{L}$ of all linear transformations $T: U \to V$ from a given vector space U to a vector space V (same scalar field) is a vector space over the same scalar field. HINT: Consider what modifications are necessary in the proof of the lemma in Section 5.4 about the algebra $\mathbb{A}$ of linear operators on U.

3. Suppose that A is an $n \times n$ upper- (or lower-) triangular matrix that is also real orthogonal. Prove that every entry on the main diagonal of A is equal to 1 or -1.

4. Suppose that A is an $n \times n$ upper (or lower) triangular *complex* matrix that also satisfies $A^{-1} = A^T$. Is it true that every entry on the main diagonal of A is equal to 1 or -1? Give reasons.

5. In quantum mechanics the Pauli theory of electron spin makes use of three linear operators, T_1, T_2, and T_3, on $\mathbb{C}^2$. Their matrix representations with respect to the basis used are the following:

$$[T_1] = \begin{bmatrix} 0 & 1 \\ 1 & 0 \end{bmatrix} = J; \quad [T_2] = \begin{bmatrix} 0 & -i \\ i & 0 \end{bmatrix} = A; \quad \text{and} \quad [T_3] = \begin{bmatrix} 1 & 0 \\ 0 & -1 \end{bmatrix} = B.$$

Show that each of these is its own inverse and is therefore nonsingular.

6. Make a multiplication table including the identity matrix I_2 and the three matrices of Exercise 5. Observe that the product of any two of these matrices is a scalar times one of the others.

Project 1. (Numerical)

Consider a linear transformation $L: U \to V$ represented, with respect to the ordered bases $G = \{u_1, u_2, u_3, u_4\}$ for U and $H = \{v_1, v_2, v_3, v_4, v_5\}$ for V, by the matrix

$$A = \begin{bmatrix} 1 & 1 & 1 & -2 \\ 2 & 1 & 0 & 3 \\ 4 & 2 & 2 & 5 \\ 0 & 1 & 0 & -6 \\ -1 & 0 & -1 & -4 \end{bmatrix} = [L]_H^G.$$

a. What is the rank of L?

b. What is the nullity of L?

c. If $x \in U$ has G-coordinate representation $[x]_G = (1, 0, 0, 1)^T$, what is the H-coordinatization for $L(x)$; i.e., what is $[L(x)]_H$?

d. What is the G-coordinatization for a basis for ker L?

e. What is the H-coordinatization for a basis for range L?

f. Repeat part (c) for the vector $[x]_G = (-1, 3, -2, 5)^T$.

g. Suppose that $U = P_3$ and $G = \{1, x, x^2, x^3\}$, the standard basis, while $V = P_4$ with H the standard basis $\{1, x, x^2, x^3, x^4\}$. What is a basis (of polynomials) for the kernel of L?

h. With U, V, G, and H as in part (g), what is a basis for the range of L?

i. With U, V, G, and H, as in part (g), find $L(1 - 2x^2 + 3x^3)$.

Project 2. (Computational)

Suppose that $U = M_2(\mathbb{R})$ and $V = M_3(\mathbb{R})$ and $F: M_2(\mathbb{R}) \to M_3(\mathbb{R})$ is the mapping

$$F: \begin{bmatrix} a & b \\ c & d \end{bmatrix} \to \begin{bmatrix} a & b & 0 \\ 0 & c & d \\ a-b & 0 & b-d \end{bmatrix}.$$

a. Verify that F is a linear transformation.

b. What is the standard matrix representation of F?

c. What is the rank of F?

d. What is a basis for range F?

e. What is a basis for ker F?

f. What is the nullity of F?

Project 3. (Computational) Function Space

Let U be the subspace of $C[a, b]$, the space of continuous functions defined on the closed interval $[a, b]$, that is spanned by the vectors (functions) $f_1(x) = 1$, $f_2(x) = e^x$, and $f_3(x) = e^{-x}$. Let D be the differentiation linear operator on U.

a. Show that the set $B = \{f_1(x), f_2(x), f_3(x)\}$ is linearly independent, hence a basis for U. HINT: Consider $x = 0, 1, -1$, and show that $c_1 f_1 + c_2 f_2 + c_3 f_3 = 0$ forces $c_1 = c_2 = c_3 = 0$.

b. Find the B-coordinatization for the function $f_4(x) = \cosh x = \dfrac{e^x + e^{-x}}{2}$.

c. Find the B-coordinatization for the function $f_5(x) = \sinh x = \dfrac{e^x - e^{-x}}{2}$.

d. Show that the set $G = \{1, \cosh x, \sinh x\}$ is linearly independent; hence that it is also a basis for U.

e. Find the transition (change-of-basis) matrix S representing the change of coordinates from the B-basis to the G-basis.

f. Find the matrix representation $K = [D]_B$ of the differentiation operator with respect to the basis B.

g. Find the matrix representation $M = [D]_G$ of the differentiation operator with respect to the basis G.

h. Show that M and K are similar matrices.

Project 4. (Theoretical) Idempotent and Nilpotent Transformations

A linear transformation (operator) $L: U \to U$ is called idempotent whenever $L^2 = L$; that is, $L(L(\mathbf{x})) = L(\mathbf{x})$, for all $\mathbf{x}$ in the domain of L. The transformation $L: U \to U$ is called nilpotent of index k whenever $L^k(\mathbf{x}) = \mathbf{0}$ for all $\mathbf{x}$ in U, and $L^{k-1}(\mathbf{x}) \neq \mathbf{0}$ for some $\mathbf{x}$ in U, k an integer ≥ 1. We write simply $L^k = 0$, $L^{k-1} \neq 0$. (Notice that the zero transformation is nilpotent of index 1.)

a. Show that if L is idempotent, and U has dimension n, there exists a basis $B = \{\mathbf{u}_1, \mathbf{u}_2, \ldots, \mathbf{u}_n\}$ for U such that $L(\mathbf{u}_i) = \mathbf{u}_i$ for $i = 1, 2, \ldots, r = $ rank of L and $L(\mathbf{u}_i) = \mathbf{0}$ for the rest of the vectors ($i = r + 1, \ldots, n$) in the basis B.

b. Describe the matrix representation of the idempotent linear operator L of part (a) with respect to the basis B described there.

c. Must any matrix representation A of an idempotent linear operator be an idempotent matrix ($A^2 = A$) no matter what the basis for U? Prove your assertion.

d. Give an example of a nilpotent linear transformation $L: \mathbb{R}^3 \to \mathbb{R}^3$ of index 3 (i.e., $L^3 = 0$, $L^2 \neq 0$) and a nonzero vector $\mathbf{u}$ which is such that $L^2(\mathbf{u}) \neq \mathbf{0}$, but $L^3(\mathbf{u}) = \mathbf{0}$.

e. Let L be a nilpotent transformation of index $k > 1$ on U and suppose $\mathbf{u} \in U$ is such that $L^{k-1}(\mathbf{u}) \neq \mathbf{0}$. Show that the set $\{\mathbf{u}, L(\mathbf{u}), \dots, L^{k-1}(\mathbf{u})\}$ is linearly independent.

f. Let U have dimension n and suppose that $L: U \to U$ is *nilpotent of index n*. Show that there exists a vector $\mathbf{u}$ in U such that the set $B = \{\mathbf{u}, L(\mathbf{u}), \dots, L^{n-1}(\mathbf{u})\}$ is a basis for U.

g. Describe the matrix $M = [L]_B$ which represents the nilpotent linear transformation L of part (f) with respect to the basis B of that exercise.

h. Must *any* matrix representation A of a nilpotent linear operator be a nilpotent matrix no matter what the basis for U? Proof?

Project 5. *(Theoretical) Dual Space*

*We are concerned with a linear transformation from a given vector space U to its set of scalars K ($\mathbb{R}$ or $\mathbb{C}$). Such a linear transformation is called a **linear functional**. The collection of all linear functionals (together with the same set of scalars K) on the given vector space U is denoted by U* and is called the **dual space** of U.*

a. Show that the dual space U^* of U is a vector space.

b. Let $B = \{\mathbf{u}_1, \mathbf{u}_2, \dots, \mathbf{u}_n\}$ be a basis for U and for each $j = 1, 2, \dots, n$ define the function

$$f_j(\mathbf{u}_i) = \begin{cases} 1 & \text{if } i = j \\ 0 & \text{if } i \neq j \end{cases} \qquad \text{for } i = 1, 2, \dots, n.$$

Explain why this completely defines f_j as a linear functional on U.

c. Show that the set $B^* = \{f_1, f_2, \dots, f_n\}$ is a basis for the dual space U^*. B is called the **dual basis.**

d. Show that U and U^* have the same dimension.

e. Let E be the standard basis for $\mathbb{R}^2$. Describe the dual space $\mathbb{R}^{2*}$ and its dual basis E^*.

f. Describe the dual basis G^* for the basis $G = \{(1, 1), (1, -1)\}$ of $\mathbb{R}^2$.

g. Let E be the standard basis for $\mathbb{R}^3$. Describe the dual space $\mathbb{R}^{3*}$ and its dual basis E^*.

h. Describe the dual basis G^* for the basis $G = \{(0, 1, 1), (1, -1, 0), (1, 1, 1)\}$ of $\mathbb{R}^3$.

i. Let U be a finite-dimensional vector space with set of scalars K (K is $\mathbb{R}$ or $\mathbb{C}$) and let $\mathbf{u}$ be a fixed nonzero vector in U. Define the function $T_\mathbf{u}: U^* \to K$ by $T_\mathbf{u}(f) = f(\mathbf{u})$ for all f in U^*. Show that $T_\mathbf{u}$ is a linear functional on U^*.

j. Denote the set of all the linear functionals $T_\mathbf{u}$ as $\mathbf{u}$ runs over U as described in part (i) by U^{**}. Show that U^{**} is a vector space with set of scalars K. U^{**} is called the **second dual** of U.

k. Define the mapping $\psi: U \to U^{**}$ by $\psi(\mathbf{u}) = T_\mathbf{u}$. Show that ψ is an isomorphism and, in contrast with the isomorphism ϕ_B of Theorem 5.6 (page 325), is independent of any basis for U.

Project 6. (Theoretical) Adjoints

Here we are concerned with a function $*: M_n(K) \to M_n(K)$*, where K is the set of scalars (K is* $\mathbb{R}$ *or* $\mathbb{C}$*). This function, called an* **adjoint** *(or sometimes an* **involution***), has the following properties: for all A, B* $\in M_n(K)$*,*

$$\text{(i) } (A^*)^* = A; \qquad \text{(ii) } (A + B)^* = A^* + B^*; \qquad \text{(iii) } (AB)^* = B^*A^*.$$

Notice that we do not require that for all scalars $(cA)^* = cA^*$*. For some adjoint functions this is true while for others it is not.*

a. Is the matrix transpose A^T an adjoint function on $M_n(\mathbb{R})$? Explain. Does $(cA)^T = cA^T$?

b. Is the matrix inverse A^{-1} an adjoint function on $M_n(\mathbb{R})$? Explain (watch out!). Does $(cA)^{-1} = cA^{-1}$ when A is invertible?

c. Is the conjugate transpose A^* of a matrix A, defined in Section 4.6, an adjoint function on $M_n(\mathbb{C})$? Explain. Does $(cA)^* = cA^*$?

d. Is adj: $A \to \text{adj}(A)$, where adj A is the adjugate of a matrix A (classical adjoint) described in Chapter 2, an adjoint function on $M_n(\mathbb{R})$? Explain.

e. Is det: $A \to \det(A)$ (see Chapter 2) an adjoint function on $M_n(\mathbb{R})$? Explain.

f. Is tr: $A \to \text{tr}(A)$, where tr(A) is the sum of the diagonal elements of A, an adjoint function on $M_n(\mathbb{R})$? Explain.

Project 7. (Geometry) Affine Transformations

In Section 5.1, Figure 5.4, we considered a change of coordinate system for the plane, resulting from translating the origin from (0, 0) to a new point that had coordinates (1, 1) in the old system. We noted that this is not a linear transformation. Nevertheless, translations are used in mathematics; we encounter them in our discussion of quadratic forms in the next chapter. Translation, as well as all linear transformations, are special cases of a more general class of transformation that we define here. In keeping with our agreement in this chapter, all our vectors will be column vectors, so we identify the geometric plane with $\overline{\mathbb{R}^2}$*, and 3-space with* $\overline{\mathbb{R}^3}$*, etc.*

Definition   A transformation L from $\overline{\mathbb{R}^n}$ to $\overline{\mathbb{R}^n}$ is called an **affine transformation** if there exists an $n \times n$ matrix A and a fixed vector $\mathbf{k}$ in $\overline{\mathbb{R}^n}$, so that

$$L(\mathbf{x}) = A\mathbf{x} + \mathbf{k}.$$

---

a. Show that the translation of Figure 5.4 in Section 5.1 is an affine transformation of $\overline{\mathbb{R}^2}$, where A is the identity matrix I_2 and $\mathbf{k} = \begin{bmatrix} 1 \\ 1 \end{bmatrix}$.

b. Demonstrate that a translation $L(\mathbf{x}) = \mathbf{x} + \mathbf{k}$ on Euclidean 2-space $\overline{\mathbb{R}^2}$ preserves distances between points. That is, let $P: (a, b)$ and $Q: (c, d)$ be two points in $\overline{\mathbb{R}^2}$, and let $\mathbf{k} = \begin{bmatrix} h \\ k \end{bmatrix}$. Then for $\mathbf{x} = \overrightarrow{OP} = \begin{bmatrix} a \\ b \end{bmatrix}$ and $\mathbf{y} = \overrightarrow{OQ} = \begin{bmatrix} c \\ d \end{bmatrix}$, show that $\| \overrightarrow{PQ} \| = \| \overrightarrow{P'Q'} \|$.

HINT: $\| \mathbf{x} + \mathbf{k} - (\mathbf{y} + \mathbf{k}) \| = \| \mathbf{x} - \mathbf{y} \|$. (See Figure 5.12.)

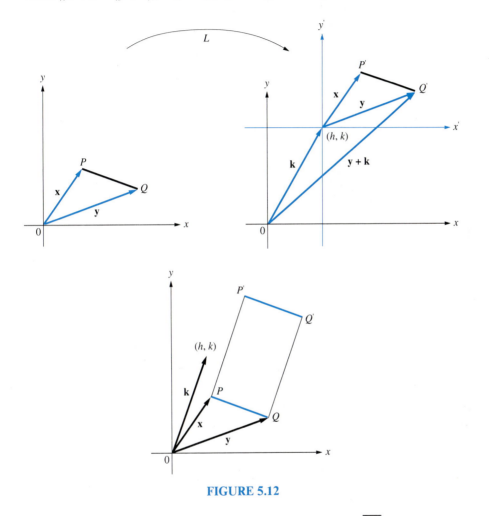

FIGURE 5.12

c. Show that a translation $L(\mathbf{x}) = \mathbf{x} + \mathbf{k}$ on Euclidean n-space $\overline{\mathbb{R}^n}$ preserves distances between points.

d. Find the equation of the image of the line $y = 3x + 2$ in Euclidean 2-space $\overline{\mathbb{R}^2}$ under

the translation $L(\mathbf{x}) = \mathbf{x} + \mathbf{k}$ given that $\mathbf{k} = \begin{bmatrix} 2 \\ 3 \end{bmatrix}$. HINT: Note that the vector $\mathbf{x}$ to any

point on the given line is $\mathbf{x} = \begin{bmatrix} x \\ y \end{bmatrix} = \begin{bmatrix} x \\ 3x + 2 \end{bmatrix}$. Find the relationship between the

coordinates of the transformed vector that will yield the translated equation.

e. Show that a translation $L(\mathbf{x}) = \mathbf{x} + \mathbf{k}$ on Euclidean n-space $\overline{\mathbb{R}^n}$ preserves angles between vectors only in a certain sense. Discuss that sense. Show that in $\overline{\mathbb{R}^2}$ the angle between $\mathbf{x} + \mathbf{k}$ and $\mathbf{y} + \mathbf{k}$ is not the same as between $\mathbf{x}$ and $\mathbf{y}$.

f. Show that an affine transformation $L(\mathbf{x}) = A\mathbf{x} + \mathbf{k}$ is linear if, and only if, $\mathbf{k} = \mathbf{0}$.

g. Let $L: \overline{\mathbb{R}^2} \rightarrow \overline{\mathbb{R}^2}$ be the affine transformation $L(\mathbf{x}) = A\mathbf{x} + \mathbf{k}$ defined by $A = \begin{bmatrix} 2 & -1 \\ 1 & -2 \end{bmatrix}$ and $\mathbf{k} = \begin{bmatrix} 2 \\ -3 \end{bmatrix}$. Does L preserve lengths? Why or why not?

h. Answer the question of part (g) if $A = \begin{bmatrix} \dfrac{1}{\sqrt{2}} & -\dfrac{1}{\sqrt{2}} \\ \dfrac{1}{\sqrt{2}} & \dfrac{1}{\sqrt{2}} \end{bmatrix}$ and $\mathbf{k} = \begin{bmatrix} 2 \\ -3 \end{bmatrix}$.

i. Use the affine transformation of part (g) to show that the image of the square $OABC$ with vertices O: $(0, 0)$, A: $(1, 0)$, B: $(1, 1)$, and C: $(0, 1)$ is the parallelogram $O'A'B'C'$ of Figure 5.13.

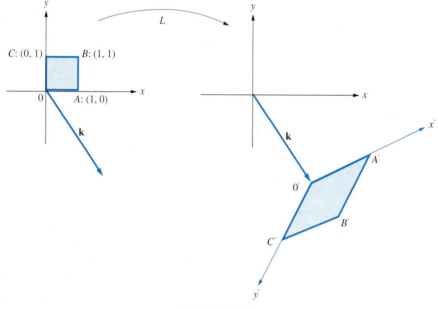

FIGURE 5.13

j. Use the affine transformation of part (g) to find the image of triangle ABC, where A: $(-1, 1)$, B: $(2, -4)$ and C: $(3, 1)$. Sketch a figure.

k. Use the affine transformation of part (h) to find the image of the square $OABC$ of part (i). Sketch a figure.

l. Use the affine transformation of part (h) to find the image of triangle *ABC* of part (j). Sketch a figure.

m. Let $L(\mathbf{x}) = A\mathbf{x} + \mathbf{k}$ be the affine transformation $L: \overline{\mathbb{R}^3} \to \overline{\mathbb{R}^3}$ determined by the matrix

$$A = \begin{bmatrix} \dfrac{1}{\sqrt{2}} & \dfrac{1}{\sqrt{2}} & 0 \\ -\dfrac{1}{\sqrt{2}} & \dfrac{1}{\sqrt{2}} & 0 \\ 0 & 0 & 1 \end{bmatrix} \text{ and the vector } \mathbf{k} = \begin{bmatrix} 1 \\ 1 \\ 2 \end{bmatrix}. \text{ Determine the image of the cube whose}$$

vertices are O: (0, 0, 0), A: (0, 0, 1), B: (0, 1, 0), C: (0, 1, 1), D: (1, 1, 0), E: (1, 0, 0), F: (1, 0, 1), and G: (1, 1, 1). Sketch a graph.

n. Does the affine transformation of part (m) preserve lengths in $\overline{\mathbb{R}^3}$?

o. (See Michael Barnsley, *Fractals Everywhere,* p. 105.) Let $L(\mathbf{x}) = A\mathbf{x} + \mathbf{k}$ be the affine

transformation $L: \overline{\mathbb{R}^3} \to \overline{\mathbb{R}^3}$ determined by the matrix $A = \begin{bmatrix} 0.2 & 0.2 & 0 \\ 0.2 & 0.2 & 0 \\ 0 & 0 & 0.3 \end{bmatrix}$ and the

vector $\mathbf{k} = \begin{bmatrix} 0 \\ 0.8 \\ 0 \end{bmatrix}$. Determine the image of the cube of part (m).

p. Does the affine transformation of part (o) preserve lengths in $\overline{\mathbb{R}^3}$?

q. Can you generalize the results of parts (g) and (h) to a theorem for $\overline{\mathbb{R}^2}$? How about in $\overline{\mathbb{R}^n}$?

6
EIGENVALUES, EIGENVECTORS, AND SOME APPLICATIONS

In this chapter we study the very important and widely applied notion of the eigenvalues and eigenvectors associated with a linear transformation on a vector space or of a square matrix. We see that these define subspaces of the domain vector space called eigenspaces. Allied with the eigenvalue concept are the characteristic polynomial and characteristic equation of a square matrix, which we discuss. These ideas are then related to the concept of matrix similarity discussed in the preceding chapter. It is shown, for example, that similar matrices always have the same eigenvalues, and corresponding eigenspaces, but not conversely.

Criteria for a given square matrix to be similar to a diagonal matrix are then discussed in detail and a method for diagonalizing a matrix, when it is possible, is given. The connection between eigenvalues and a diagonal "canonical similarity form" for a given matrix is discussed for certain classes of square matrices. In particular, real symmetric matrices and their complex counterparts, Hermitian matrices, are introduced and are shown always to be diagonalizable.

We then exploit these ideas in studying certain classical geometric concepts. The application of eigenvalues, eigenvectors, and diagonalization to a given quadratic form leads us to a very useful tool with which to discuss the conic sections and quadric surfaces. For centuries these curves and solids have been an important source of mathematical models for scientific description. We also apply eigenvalues and eigenvectors to describe Markov chains, which are so useful in the behavioral sciences.

An exploration of matrices with complex number entries lead us to see that unitary, Hermitian, skew-Hermitian, and normal matrices yield generalizations of the earlier concepts developed for real matrices. In this connection we discuss both Schur's theorem and the spectral theorem. Finally, we examine an application to the solution of systems of ordinary differential equations.

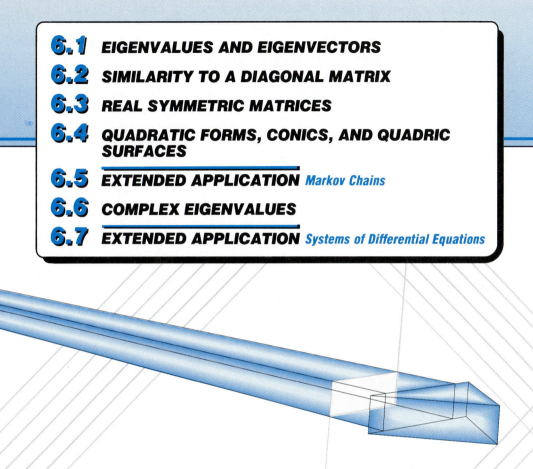

6.1 EIGENVALUES AND EIGENVECTORS

In the preceding chapter we learned that any vector $\mathbf{x}$ from a given finite-dimensional vector space V, with ordered basis B, can be represented by the column vector $[\mathbf{x}]_B$. This representation is a vector in one of the two vector spaces $\mathbb{R}^n$ or $\mathbb{C}^n$ of all $n \times 1$ *column vectors.* For that reason we, for the most part, are concerned here with vectors in $\mathbb{R}^n$ or $\mathbb{C}^n$. Unless otherwise specified, all vectors $\mathbf{x}$ are real or complex *column* vectors: $\mathbf{x} = (x_1, x_2, \ldots, x_n)^T$. A linear operator, $T\colon V \to V$, on V can then be represented, with respect to the basis B, by an $n \times n$ matrix $A = [T]_B$.

It is common to apply linear algebra to situations where a linear transformation T is given and one is concerned with those vectors that are rather "stable" vectors under T. That is, one is concerned with those vectors $\mathbf{x}$ in the domain of T that are such that $T(\mathbf{x}) = a\mathbf{x}$ for some scalar a. Let us begin with an example from biology that illustrates this concept.

Example 1   Consider an ecosystem containing m species of life. Define the population vector $\mathbf{n}(t)$ of the ecosystem to be the column vector with m entries whose ith entry $n_i(t)$ is the population of the ith species at time t. We assume that there is a matrix $A(t)$, called the change (or transition) matrix of the system from time t to time $t + 1$, such that

$$\mathbf{n}(t + 1) = A(t)\mathbf{n}(t).$$

The entries in the matrix $A(t)$ are the numbers that reflect the possibility that an individual member of the given species will survive to time $t + 1$ or affect the survival of members of other species, or that new individuals will come into existence during this time interval.

In a stable ecosystem the value of $\mathbf{n}(t + 1)$ is proportional to the value of $\mathbf{n}(t)$; that is, at time $t + 1$ the species in the system will still be in the same proportion to each other as they were at time t. This gives rise to the problem: For what values of $\mathbf{n}(t)$ and what proportionally constant λ is true that

$$\mathbf{n}(t + 1) = A(t)\mathbf{n}(t) = \lambda\mathbf{n}(t)? \quad \blacksquare$$

The general form of this problem is interesting in its own right. It is applied so often in contexts as wide-ranging as biology, economics, engineering, physics, chemistry, etc., that we give the following definition.

Definition   Let A be an $n \times n$ matrix. A scalar λ is called an **eigenvalue** of the matrix A corresponding to a nonzero **eigenvector** $\mathbf{x}$ if

$$A\mathbf{x} = \lambda\mathbf{x}. \tag{6.1}$$

It is clear that the zero vector **0** would satisfy equation (6.1) for any scalar λ, so the zero vector is *not allowed* to be an eigenvector.

Example 2 Given the matrix $A = \begin{bmatrix} 1 & 0 & 1 \\ 1 & 2 & 0 \\ 3 & 0 & 1 \end{bmatrix}$. Show that $\lambda = 2$ is an *eigenvalue* of the

matrix A with $\mathbf{x} = \begin{bmatrix} 0 \\ 1 \\ 0 \end{bmatrix}$ as a corresponding *eigenvector* of A.

Direct computation shows that the definition (6.1) is satisfied. We have

$$A\mathbf{x} = \begin{bmatrix} 1 & 0 & 1 \\ 1 & 2 & 0 \\ 3 & 0 & 1 \end{bmatrix} \begin{bmatrix} 0 \\ 1 \\ 0 \end{bmatrix} = \begin{bmatrix} 0 \\ 2 \\ 0 \end{bmatrix} = 2\mathbf{x}. \quad \blacksquare$$

To find the eigenvalues and the corresponding eigenvectors for any given $n \times n$ matrix A we use matrix algebra and first note that we may write equation (6.1) in the equivalent matrix form as

$$\lambda I\mathbf{x} - A\mathbf{x} = \mathbf{0}.$$

Then factor out the vector (matrix) $\mathbf{x}$ on the right. Therefore,

> If A is any $n \times n$ matrix, then λ is an eigenvalue corresponding to a non-zero eigenvector $\mathbf{x}$ if, and only if,
>
> $$(\lambda I - A)\mathbf{x} = \mathbf{0}. \tag{6.2}$$

Equation (6.2) is the matrix equation of a homogeneous system of linear equations. Such a system will have a nontrivial solution $\mathbf{x}$ if, and only if, the determinant of the coefficient matrix $\lambda I - A$ is zero. We have proved the following theorem.

> **Theorem 6.1** The scalar λ is an eigenvalue for the $n \times n$ matrix A if, and only if,
>
> $$\det(\lambda I - A) = 0. \tag{6.3}$$

Note that $p(x) = \det(xI - A)$ is a polynomial of degree n in the variable x. Equation (6.3), $p(\lambda) = 0$, is called the **characteristic equation** of the matrix A. An *eigenvalue* λ is a zero or root of this equation, so it is frequently called a **characteristic value** of A. Other names for eigenvalues that are sometimes used are *proper value* and *latent root*. See the following examples.

Example 3   Find all of the eigenvalues for the matrix $A = \begin{bmatrix} 1 & -2 \\ 0 & 3 \end{bmatrix}$. Note that

$$\lambda I - A = \begin{bmatrix} \lambda & 0 \\ 0 & \lambda \end{bmatrix} - \begin{bmatrix} 1 & -2 \\ 0 & 3 \end{bmatrix} = \begin{bmatrix} \lambda - 1 & 2 \\ 0 & \lambda - 3 \end{bmatrix}.$$

Therefore, the *characteristic equation* for A is

$$\det \begin{bmatrix} \lambda - 1 & 2 \\ 0 & \lambda - 3 \end{bmatrix} = (\lambda - 1)(\lambda - 3) = 0.$$

As you can readily see, its roots are 1 and 3. Therefore, the eigenvalues of the matrix A are 1 and 3. The corresponding eigenvectors $\mathbf{x}$ for A are found by substituting these values for λ separately into equation (6.2) and solving the resulting homogeneous systems of equations. Thus, for $\lambda = 3$,

$$(3I - A)\mathbf{x} = \begin{bmatrix} 3 - 1 & 2 \\ 0 & 3 - 3 \end{bmatrix} \begin{bmatrix} x_1 \\ x_2 \end{bmatrix} = \begin{bmatrix} 2 & 2 \\ 0 & 0 \end{bmatrix} \begin{bmatrix} x_1 \\ x_2 \end{bmatrix} = \begin{bmatrix} 0 \\ 0 \end{bmatrix}.$$

From our previous work with homogeneous systems of equations one can conclude easily that all solutions to this system have the form $\mathbf{x} = \begin{bmatrix} t \\ -t \end{bmatrix}$, hence are proportional to the vector

$$\mathbf{x}_0 = \begin{bmatrix} 1 \\ -1 \end{bmatrix}.$$

Thus each of these vectors $\mathbf{x}$ is an eigenvector corresponding to the eigenvalue $\lambda = 3$ for the matrix A. That is, $A\mathbf{x} = 3\mathbf{x}$ for each of them. You can very readily calculate, for example, that

$$A\mathbf{x}_0 = \begin{bmatrix} 1 & -2 \\ 0 & 3 \end{bmatrix} \begin{bmatrix} 1 \\ -1 \end{bmatrix} = \begin{bmatrix} 3 \\ -3 \end{bmatrix} = 3\mathbf{x}_0.$$

A similar calculation shows that for the eigenvalue $\lambda = 1$ we have the homogeneous system

$$\begin{bmatrix} 0 & 2 \\ 0 & -2 \end{bmatrix} \begin{bmatrix} x_1 \\ x_2 \end{bmatrix} = \begin{bmatrix} 0 \\ 0 \end{bmatrix}.$$

All solutions to this system are of the form $\mathbf{x} = \begin{bmatrix} t \\ 0 \end{bmatrix}$ and therefore are proportional to

$$\mathbf{x}_1 = \begin{bmatrix} 1 \\ 0 \end{bmatrix}. \quad \blacksquare$$

If A is an $n \times n$ matrix and λ_0 is an eigenvalue for A, then the homogeneous system of linear equations $(\lambda_0 I - A)\mathbf{x} = \mathbf{0}$ *always* has nontrivial solutions. (Why do we know that?) The vector space, which we studied previously and called the solution space of the coefficient matrix $(\lambda_0 I - A)$, consisting of all of these solutions, is called the **eigenspace of A corresponding to** λ_0. Thus,

---

Definition   The **eigenspace** of an $n \times n$ matrix A, corresponding to an eigenvalue λ_0, is the solution space of the matrix $(\lambda_0 I - A)$.

In Examples 2 and 3 the eigenspaces were each one-dimensional. Each eigenspace was spanned (generated) by the given particular eigenvector $\mathbf{x}_0$ or $\mathbf{x}_1$. Consider, however, the following example.

Example 4   Find a basis for each eigenspace of the matrix $A = \begin{bmatrix} 2 & -1 & 0 \\ -1 & 2 & 0 \\ 0 & 0 & 3 \end{bmatrix}$. For this matrix the characteristic equation (verify this) is

$$\det(\lambda I - A) = (\lambda - 3)^2(\lambda - 1) = 0.$$

So the eigenvalues (characteristic roots) for A are $\lambda_1 = 3$ and $\lambda_2 = 1$. For the eigenvalue $\lambda_1 = 3$ we solve the matrix equation

$$(3I - A)\mathbf{x} = \mathbf{0}.$$

This is

$$\begin{bmatrix} 1 & 1 & 0 \\ 1 & 1 & 0 \\ 0 & 0 & 0 \end{bmatrix}\begin{bmatrix} x_1 \\ x_2 \\ x_3 \end{bmatrix} = \begin{bmatrix} 0 \\ 0 \\ 0 \end{bmatrix}.$$

Solving this homogeneous system of equations, we see that x_3 can be arbitrary and that $x_1 = -x_2$, so let $x_3 = t$, $x_2 = s$ and then $x_1 = -s$. So each eigenvector is of the form $\begin{bmatrix} -s \\ s \\ t \end{bmatrix}$, where s and t are arbitrary. Hence the solution space for $3I - A$ is two-dimensional. A basis for this space, which is the eigenspace of A corresponding to $\lambda_1 = 3$ is therefore $\left\{ \mathbf{u} = \begin{bmatrix} -1 \\ 1 \\ 0 \end{bmatrix}, \mathbf{v} = \begin{bmatrix} 0 \\ 0 \\ 1 \end{bmatrix} \right\}$. It consists of the two linearly independent eigenvectors $\mathbf{u}$ and $\mathbf{v}$, so as we noted this eigenspace is two-dimensional. We obtained this particular basis by setting $s = 1$, $t = 0$, for $\mathbf{u}$ and then $s = 0$, $t = 1$ to obtain $\mathbf{v}$. This gave an obvious linear independence. It is also fairly obvious that any eigenvector corresponding to $\lambda_1 = 3$ is a linear combination of these two vectors of the form

$$\begin{bmatrix} -s \\ s \\ t \end{bmatrix} = s\begin{bmatrix} -1 \\ 1 \\ 0 \end{bmatrix} + t\begin{bmatrix} 0 \\ 0 \\ 1 \end{bmatrix} = s\mathbf{u} + t\mathbf{v}.$$

For the second eigenvalue $\lambda_2 = 1$ we solve the system $(1I - A)\mathbf{x} = \mathbf{0}$ and obtain eigenvectors

$$\mathbf{x} = \begin{bmatrix} t \\ t \\ 0 \end{bmatrix} = t \begin{bmatrix} 1 \\ 1 \\ 0 \end{bmatrix}.$$

Therefore, the vector $\mathbf{w} = \begin{bmatrix} 1 \\ 1 \\ 0 \end{bmatrix}$ will form a basis for the eigenspace of A corresponding to the eigenvalue $\lambda_2 = 1$. ∎

So far we have discussed eigenvalues and eigenvectors for an $n \times n$ **matrix** A. Suppose that, as we remarked at the beginning of the section, we have a **linear transformation** $T: V \rightarrow V$; that is, suppose that we have a linear operator T on the finite-dimensional vector space V. Then eigenvalues and eigenvectors of this transformation can also be defined. This is done in a rather natural way. We have the following definition.

Definition   Let $T: V \rightarrow V$ be a linear operator on V. The **eigenvectors** of T are all nonzero vector solutions to the equation

$$T(\mathbf{x}) = \lambda \mathbf{x},$$

where λ is a scalar called an **eigenvalue** of the transformation T.

We shall shortly prove that the eigenvalues and eigenvectors of a given linear transformation $T: V \rightarrow V$ can be computed from *any* matrix representation of that transformation with respect to a basis for the vector space V. It can also be shown that the eigenvectors of T corresponding to a given eigenvalue λ_0 plus the zero vector (which cannot be an eigenvector, by definition) form a subspace of V. This subspace is the same as the kernel of the associated linear transformation $L_0 = (\lambda_0 I - T)$, where I is the identity linear operator on V, defined by $I(\mathbf{x}) = \mathbf{x}$ for all $\mathbf{x} \in V$. We prove

---

Lemma   Let λ_0 be an eigenvalue of the linear transformation $T: V \rightarrow V$ and suppose that $\mathbf{x}$ and $\mathbf{y}$ are eigenvectors of T corresponding to λ_0. Then any nonzero linear combination $\mathbf{z} = \alpha \mathbf{x} + \beta \mathbf{y}$ of the eigenvectors $\mathbf{x}$ and $\mathbf{y}$ is an eigenvector of T corresponding to λ_0. Therefore, the eigenvectors of T corresponding to λ_0 (together with $\mathbf{0}$) form a subspace of V.

Proof   Clearly we have the following computation:

$$T(\mathbf{z}) = T(\alpha \mathbf{x} + \beta \mathbf{y}) = \alpha T(\mathbf{x}) + \beta T(\mathbf{y}) = \alpha \lambda_0 \mathbf{x} + \beta \lambda_0 \mathbf{y} = \lambda_0 (\alpha \mathbf{x} + \beta \mathbf{y}) = \lambda_0 \mathbf{z}.$$

So unless $\mathbf{z} = \mathbf{0}$, $\mathbf{z}$ is an eigenvector of T. Clearly, $T(\mathbf{0}) = \lambda_0 \mathbf{0}$. Therefore, we have a subspace of V. We leave it to you to show the similar calculation required to demonstrate that this subspace is precisely the *kernel* of the linear transformation $L_0 = (\lambda_0 I - T)$. It is called the **eigenspace of** T corresponding to λ_0. ❑

Example 5 Let $V = \mathbb{R}^2$ and let $\mathbf{w}$ be a fixed vector in V. Let $T: \mathbb{R}^2 \to \mathbb{R}^2$ be projection onto (along) the vector $\mathbf{w}$. That is, for each vector $\mathbf{x}$ in the plane, $T(\mathbf{x}) = \text{proj}_{\mathbf{w}} \mathbf{x}$. We can find eigenvalues and eigenvectors for T geometrically, since we seek solutions to the equation

$$T(\mathbf{x}) = \lambda \mathbf{x}.$$

Notice that every vector parallel to $\mathbf{w}$ is of the form $\mathbf{v} = \alpha \mathbf{w}$, where α is a scalar. Since T projects $\mathbf{v}$ onto $\mathbf{w}$, in this particular case we would have

$$T(\mathbf{v}) = T(\alpha \mathbf{w}) = \alpha T(\mathbf{w}) = \alpha \mathbf{w} = \mathbf{v} = (1)\mathbf{v}.$$

So $\mathbf{v} = \alpha \mathbf{w}$ is an eigenvector of T corresponding to the eigenvalue $\lambda_0 = 1$.

Now notice that if the nonzero vector $\mathbf{v}$ is *orthogonal* to $\mathbf{w}$, then $T(\mathbf{v}) = \mathbf{0}$; that is, T projects it onto $\mathbf{0}$. Thus we have

$$T(\mathbf{v}) = \mathbf{0} = 0\mathbf{v}.$$

So these nonzero vectors *orthogonal* to $\mathbf{w}$ are eigenvectors of T corresponding to the eigenvalue $\lambda_1 = 0$.

If $\mathbf{v}$ is any other vector in $V = \mathbb{R}^2$, neither parallel nor orthogonal to the given vector $\mathbf{w}$, then the projection $T(\mathbf{v})$ onto $\mathbf{w}$ is not a scalar multiple of the given vector $\mathbf{v}$. See Figure 6.1. Such a vector's projection is not parallel to $\mathbf{v}$ at all, so it cannot be an eigenvector (characteristic vector) of T.

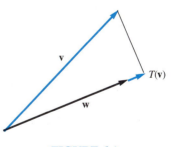

FIGURE 6.1

Therefore, the eigenvalues for this transformation (projection along $\mathbf{w}$) in $\mathbb{R}^2$ are $\lambda_0 = 1$ and $\lambda_1 = 0$, with corresponding eigenvectors respectively parallel and orthogonal to the given vector $\mathbf{w}$ onto which the projection goes. ■

For another linear transformation $T: V \to V$, the geometric determination of the eigenvalues and corresponding eigenvectors for a given linear transformation can be more difficult. It is easier to use matrix ideas. Generally it is best for one to select a matrix representation $A = [T]_B$ for T, with respect to any given basis B for the vector space V, and then calculate the characteristic equation $\det(\lambda I - A) = 0$ for this matrix representation. The eigenvalues and corresponding eigenvectors for A will be the same as those for T. The fact that the eigenvalues are independent of the choice of basis is a result of our work in Chapter 5. Let us summarize this in the following important theorem.

Theorem 6.2   Let $T: V \to V$ be a linear operator on the finite-dimensional vector space V. If H and H' are bases for V and if A and A' are the matrix representations of T relative to H and H', respectively, then the eigenvalues of A and A' are equal. That is,

Similar matrices have the same eigenvalues.

Proof   Since $A = [T]_H$ and $A' = [T]_{H'}$ are both matrix representations of the transformation T, they are similar matrices. Therefore, for some invertible matrix S, we have

$$A = S^{-1}A'S.$$

Hence,

$$\lambda I - A = \lambda I - S^{-1}A'S = \lambda S^{-1}S - S^{-1}A'S$$
$$= \lambda S^{-1}IS - S^{-1}A'S = S^{-1}(\lambda I - A')S.$$

Note that matrix multiplication is not commutative, so we have factored S^{-1} out on the left and S out on the right. From this equation we deduce that A and A' *have the same characteristic equation.* This follows from

$$0 = \det(\lambda I - A) = \det(\lambda I - S^{-1}A'S) = \det(S^{-1}(\lambda I - A')S)$$
$$= \det(S^{-1})\det(\lambda I - A')\det S = \det(S^{-1}) \cdot \det S \cdot \det(\lambda I - A')$$
$$= \det(S^{-1}S) \cdot \det(\lambda I - A') = \det I \cdot \det(\lambda I - A')$$
$$= 1 \cdot \det(\lambda I - A'). \quad \blacksquare$$

We are able to move the factors around here because determinants are real (or complex) numbers and their multiplication is commutative. Therefore, since the characteristic equations of A and A' are the same, A and A' have the same eigenvalues. *Furthermore, if* $\mathbf{x}$ *is an eigenvector of A corresponding to the eigenvalue λ, then $S\mathbf{x}$ is an eigenvector of A' corresponding to λ.* You can readily verify this fact.

Example 6   Let T be the transformation on the vector space P_2 of real polynomials of degree two or less defined by the formula

$$T(c + bx + ax^2) = (5c - a) + 2bx + (2a - c)x^2.$$

First let H be the standard basis $E = \{1, x, x^2\}$ for P_2. The standard matrix representation for T is (verify this)

$$[T]_E = A = \begin{bmatrix} 5 & 0 & -1 \\ 0 & 2 & 0 \\ -1 & 0 & 2 \end{bmatrix}.$$

The characteristic equation $0 = \det(\lambda I - A)$ for A, and therefore for T, is, in factored form,

$$(\lambda - 2)(\lambda^2 - 7\lambda + 9) = 0.$$

The roots of this equation, hence the eigenvalues of A and T, are

$$\lambda_1 = 2, \qquad \lambda_2 = \frac{7}{2} + \frac{\sqrt{13}}{2}, \qquad \text{and} \qquad \lambda_3 = \frac{7}{2} - \frac{\sqrt{13}}{2}.$$

Corresponding *standard matrix representations* for the eigenvectors of T are the following vectors (verify this):
For $\lambda_1 = 2$:

$$[\mathbf{v}_1] = \begin{bmatrix} 0 \\ t \\ 0 \end{bmatrix} \leftrightarrow \mathbf{v}_1 = tx. \qquad \text{(Note that } T(\mathbf{v}_1) = T(tx) = 2tx = 2\mathbf{v}_1.)$$

For $\lambda_2 = \dfrac{7}{2} + \dfrac{\sqrt{13}}{2}$,

$$[\mathbf{v}_2] = \begin{bmatrix} \dfrac{-3 - \sqrt{13}}{2}t \\ 0 \\ t \end{bmatrix} \leftrightarrow \mathbf{v}_2 = t\left(\frac{-3 - \sqrt{13}}{2} + 0x + x^2\right).$$

For $\lambda_3 = \dfrac{7}{2} - \dfrac{\sqrt{13}}{2}$,

$$[\mathbf{v}_3] = \begin{bmatrix} \dfrac{-3 + \sqrt{13}}{2}t \\ 0 \\ t \end{bmatrix} \leftrightarrow \mathbf{v}_3 = t\left(\frac{-3 + \sqrt{13}}{2} + 0x + x^2\right). \quad \blacksquare$$

As you can see, in general, finding the eigenvalues of the transformation T or of any $n \times n$ matrix A is a problem in finding the roots of a polynomial equation $p(x) = 0$. This is not always a simple task, and the techniques can be quite involved. One must either use the classical ideas for solving polynomial equations, some of which you may have learned in college algebra, or use appropriate approximation techniques and a computer. Often some (or all) of the roots of the characteristic equation $\det(\lambda I - A) = 0$ for a given real matrix A are *not real* numbers. However, we defer discussion of complex eigenvalues until later (Section 6.6).

We conclude this section with several general observations about the characteristic equation. First of all, quite naturally

> The polynomial $p_A(\lambda) = \det(\lambda I - A)$ of a matrix (or transformation) is called its **characteristic polynomial.**

All of the remarks that follow are based on the theory of polynomial equations which classically form a large part of any college algebra or precalculus mathematics course. First of all, as a result of applying the fundamental theorem of algebra, we know that the characteristic polynomial $p_A(\lambda)$ can be written as the product of linear factors over the complex numbers in the following form:

$$p_A(\lambda) = (\lambda - \lambda_1)^{m_1}(\lambda - \lambda_2)^{m_2} \cdots (\lambda - \lambda_k)^{m_k}.$$

Then $\lambda_1, \lambda, \ldots, \lambda_k$ are the distinct eigenvalues of A, and the positive integers $m_1, m_2, \ldots, m_k$ are their respective multiplicities as zeros of the characteristic polynomial.

If A is $n \times n$, then $m_1 + m_2 + \cdots + m_k = n$. The integer m_i is called the *algebraic multiplicity* of the eigenvalue λ_i. The number of linearly independent eigenvectors of A associated with λ_i is the dimension of the eigenspace associated with λ_i and is often called the *geometric multiplicity* of the eigenvalue λ_i. It can be shown that the geometric multiplicity of an eigenvalue is less than or equal to the algebraic multiplicity. When it is strictly less than, the matrix is sometimes called *defective*.

It is also true that any eigenvalue λ_i is a zero of $p_A(\lambda) = \det(\lambda I - A)$ if, and only if, it is also a zero of the polynomial

$$(\lambda_1 - \lambda)^{m_1}(\lambda_2 - \lambda)^{m_2} \cdots (\lambda_k - \lambda)^{m_k} = (-1)^n p_A(\lambda) = \det(A - \lambda I) = P(\lambda).$$

Now $P(0) = \det A = \lambda_1^{m_1}\lambda_2^{m_2} \cdots \lambda_k^{m_k}$. So we have proved that

> The determinant of A is the product of its eigenvalues.

Thus, if we list all n of the *not necessarily distinct* eigenvalues of the $n \times n$ matrix A as $\lambda_1, \lambda, \ldots, \lambda_n$, then we have

> $$\det A = \prod_{i=1}^{n} \lambda_i.$$

Of course, the rest of the coefficients of the characteristic polynomial, as with all polynomials, are (symmetric) functions of the roots, hence of the eigenvalues. See Exercise 50 for the 3×3 case.

One final observation: If A is an $n \times n$ *real* matrix, then only real numbers would appear in the computation of the determinant of $\lambda I - A$, so the characteristic polynomial, $p_A(\lambda)$, of A has *real* coefficients. Thus, *if the complex number λ_0 is an eigenvalue* (zero of $p_A(\lambda)$), *then its complex conjugate $\bar{\lambda}_0$ is also an eigenvalue.* You can make additional observations from your knowledge of polynomials.

In Theorem 6.2 (page 382) we proved that similar matrices have the same eigenvalues since they have the same characteristic polynomials. *The converse is false,* however. The two matrices

$$I = \begin{bmatrix} 1 & 0 \\ 0 & 1 \end{bmatrix} \quad \text{and} \quad K = \begin{bmatrix} 1 & 1 \\ 0 & 1 \end{bmatrix}$$

have the same eigenvalues, yet they are not similar. (Why not?)

The matrix K is defective. Both have eigenvalue $\lambda = 1$, whose algebraic multiplicity in each case is two. However, the matrix K has but one linearly independent eigenvector, $\mathbf{x} = \begin{bmatrix} 1 \\ 0 \end{bmatrix} = \mathbf{e}_1$, associated with $\lambda = 1$. The first matrix, I, on the other hand, clearly has both $\mathbf{e}_1 = \begin{bmatrix} 1 \\ 0 \end{bmatrix}$ and $\mathbf{e}_2 = \begin{bmatrix} 0 \\ 1 \end{bmatrix}$ as eigenvectors. Verify this.

Vocabulary

In this section we have encountered some important and interesting new concepts and definitions. Be sure that you understand all of these concepts and definitions. We list below the new terms to add to your linear algebra vocabulary. Be sure that you review them sufficiently until you feel at home with all of them.

eigenvalue of matrix A (of a linear transformation T)

eigenvector of a matrix A (of a linear transformation T)

characteristic equation

algebraic multiplicity

eigenspace

characteristic polynomial

geometric multiplicity

Exercises 6.1

In Exercises 1–9 for each given matrix A

(a) *Write the characteristic polynomial $p_A(\lambda) = \det (\lambda I - A)$.*
(b) *Find the eigenvalues of A.*
(c) *Find an eigenvector of A corresponding to each eigenvalue.*

1. $\begin{bmatrix} -1 & 2 \\ 2 & -1 \end{bmatrix}.$
2. $\begin{bmatrix} 0 & -1 \\ -1 & 0 \end{bmatrix}.$
3. $\begin{bmatrix} 1 & 1 \\ 1 & 1 \end{bmatrix}.$
4. $\begin{bmatrix} -3 & 2 \\ -3 & 2 \end{bmatrix}.$
5. $\begin{bmatrix} \frac{1}{2} & \frac{1}{2} \\ \frac{1}{2} & -\frac{1}{2} \end{bmatrix}.$

6. $\begin{bmatrix} 1 & 2 & 4 \\ 0 & 3 & 5 \\ 0 & 0 & 2 \end{bmatrix}$.
7. $\begin{bmatrix} 1 & -1 & 0 \\ 0 & 1 & -1 \\ 0 & 0 & 1 \end{bmatrix}$.
8. $\begin{bmatrix} 1 & 0 & 3 \\ 2 & 3 & 2 \\ 3 & 0 & 1 \end{bmatrix}$.
9. $\begin{bmatrix} 2 & 0 & 0 & 0 \\ 0 & -1 & 0 & 5 \\ 0 & 4 & 6 & 4 \\ 0 & 5 & 0 & -1 \end{bmatrix}$.

10. Find bases for each of the eigenspaces of the matrix of Exercise 3.

11. Find bases for each of the eigenspaces of the matrix of Exercise 4.

12. Find bases for each of the eigenspaces of the matrix of Exercise 6.

13. Find bases for each of the eigenspaces of the matrix of Exercise 7.

14. Find bases for each of the eigenspaces of the matrix of Exercise 8.

15. Find bases for each of the eigenspaces of the matrix of Exercise 9.

In Exercises 16–21 a linear transformation is described geometrically. Sketch the appropriate figure and determine from the geometry of the situation the eigenvalues and eigenvectors associated with each. Do this intuitively rather than first determining the matrix of the transformation.

16. $T: \mathbb{R}^2 \to \mathbb{R}^2$ is a reflection through a line along a fixed vector $\mathbf{w} \in \mathbb{R}^2$.

17. $T: \mathbb{R}^2 \to \mathbb{R}^2$ is a projection orthogonal to a fixed vector $\mathbf{w} \in \mathbb{R}^2$.

18. $T: \mathbb{R}^3 \to \mathbb{R}^3$ is a reflection through a line along a fixed vector $\mathbf{w} \in \mathbb{R}^3$.

19. $T: \mathbb{R}^3 \to \mathbb{R}^3$ is a projection along a fixed vector $\mathbf{w} \in \mathbb{R}^3$.

20. $T: \mathbb{R}^3 \to \mathbb{R}^3$ is a projection orthogonal to a fixed vector $\mathbf{w} \in \mathbb{R}^3$.

21. $T: \mathbb{R}^2 \to \mathbb{R}^2$ is a counterclockwise rotation of $90°$ followed by a reflection in the y axis.

In Exercises 22–27 find the eigenspace associated with each eigenvalue of the given matrix.

22. $\begin{bmatrix} 4 & 1 & 0 \\ 1 & 4 & 0 \\ 0 & -3 & 0 \end{bmatrix}$.
23. $\begin{bmatrix} 5 & 0 & 1 \\ 1 & 1 & 0 \\ -7 & 1 & 0 \end{bmatrix}$.
24. $\begin{bmatrix} -1 & 0 & 0 \\ 0 & 3 & 2 \\ 0 & 2 & 3 \end{bmatrix}$.

25. $\begin{bmatrix} 2 & 0 & -2 \\ -1 & 2 & -1 \\ -2 & 0 & 2 \end{bmatrix}$.
26. $\begin{bmatrix} 0 & 0 & 2 & 0 \\ 1 & 0 & 1 & 0 \\ 0 & 1 & -2 & 0 \\ 0 & 0 & 0 & 1 \end{bmatrix}$.
27. $\begin{bmatrix} 3 & 1 & 0 & 0 \\ 0 & 1 & 0 & 0 \\ 0 & 0 & \frac{1}{2} & \frac{1}{2} \\ 0 & 0 & \frac{1}{2} & -\frac{1}{2} \end{bmatrix}$.

In Exercises 28–31 let $T: \mathbb{R}^3 \to \mathbb{R}^3$ be the linear operator defined by

$$T(x_1, x_2, x_3) = (x_1 - x_3, 3x_2, x_3 - x_1).$$

28. Write the standard matrix representation $A = [T]$.

29. Write the characteristic equation of T (of A).

30. Find the eigenvalues of T.

31. Find bases for each of the eigenspaces of T. Watch out! Each of these is a subspace of $\mathbb{R}^3$, not of $\overline{\mathbb{R}^3}$.

In Exercises 32–35 let $T: P_2 \to P_2$ be the linear operator defined by

$$T(c + bx + ax^2) = (4c + a) + (b - 2c)x + (a - 2c)x^2.$$

32. Write the standard matrix representation $A = [T]$. (The standard basis is $\{1, x, x^2\}$.)

33. Write the characteristic equation of T (of A).

34. Find the eigenvalues of T.

35. Find bases for each of the eigenspaces of T. Watch out! Each of these is a subspace of P_2, not of $\mathbb{R}^3$.

In Exercises 36–39 let $T: M_2(\mathbb{R}) \to M_2(\mathbb{R})$ *be the linear operator defined by*

$$T\left(\begin{bmatrix} a & b \\ c & d \end{bmatrix}\right) = \begin{bmatrix} 10a - 9b & 4a - 2b \\ -2c + 5d & c + 2d \end{bmatrix}.$$

36. Write the standard matrix representation $A = [T]$.

37. Write the characteristic equation of T (of A).

38. Find the eigenvalues of T.

39. Find bases for each of the eigenspaces of T. Watch out! Each of these is a subspace of $M_2(\mathbb{R})$, not of $\mathbb{R}^4$.

40. Prove that the two matrices

$$A = \begin{bmatrix} 1 & 0 & 0 \\ 0 & -1 & 1 \\ 0 & 0 & -1 \end{bmatrix} \quad \text{and} \quad B = \begin{bmatrix} 1 & 0 & 0 \\ 0 & -1 & 0 \\ 0 & 0 & -1 \end{bmatrix}$$

are not similar, even though they have the same eigenvalues. HINT: Show that if $AS = SB$, then S is singular.

41. Show that the real matrix $\begin{bmatrix} -1 & 2 \\ -2 & 1 \end{bmatrix}$ has no real eigenvalues. However, considered as a matrix in $M_2(\mathbb{C})$ it has complex eigenvalues and eigenvectors. What are they?

42. Repeat Exercise 41 for the matrix

$$R = \begin{bmatrix} \dfrac{1}{\sqrt{2}} & \dfrac{1}{\sqrt{2}} \\ -\dfrac{1}{\sqrt{2}} & \dfrac{1}{\sqrt{2}} \end{bmatrix}$$

which represents a 45° rotation of the plane. Interpret this geometrically.

43. Prove that $\lambda = 0$ is an eigenvalue of a matrix A if, and only if, A is singular.

44. Prove, by induction, that if λ is an eigenvalue of the matrix A, then λ^n is an eigenvalue of A^n, for every positive integer n.

45. Let A have eigenvalues $\lambda_1 = 2$ and $\lambda_2 = -1$. What are the eigenvalues of A^3?

46. Recall that a matrix is *idempotent* if $A^2 = A$. Show that the eigenvalues of any idempotent matrix must be either 0 or 1.

47. Recall that a matrix is *nilpotent* if $A^k = 0$, for some positive integer k. Show that all of the eigenvalues of a nilpotent matrix must be 0.

48. Prove the following theorem.

Theorem Let A be any $n \times n$ matrix. The following are equivalent statements:

(a) λ is an eigenvalue of A.
(b) The homogeneous system of linear equations $(\lambda I - A)X = 0$ has a nontrivial solution.
(c) The dimension of the null space of the linear operator $L(\mathbf{x}) = (\lambda I - A)\mathbf{x}$ on $\mathbb{R}^n$ is not zero.
(d) The matrix $(\lambda I - A)$ is singular.
(e) $\det(\lambda I - A) = 0$.

49. Show that the characteristic equation for the 3×3 matrix $A = (a_{ij})$ can be written in the form:

$$p_A(\lambda) = \lambda^3 - (\text{trace}(A))\lambda^2 + (A_{11} + A_{22} + A_{33})\lambda - \det(A) = 0.$$

Recall that trace (A) is the sum of the diagonal elements, and A_{ii} is the cofactor (Chapter 2) of the diagonal element a_{ii} of A; $i = 1, 2, 3$.

50. Show that if $\lambda_1, \lambda_2, \lambda_3$ are the three, not necessarily distinct, eigenvalues of the 3×3 matrix A, then the characteristic *polynomial* of A can be written as

$$p_A(x) = x^3 - (\lambda_1 + \lambda_2 + \lambda_3)x^2 + (\lambda_1\lambda_2 + \lambda_1\lambda_3 + \lambda_2\lambda_3)x - \lambda_1\lambda_2\lambda_3.$$

Exercises 51–55 are concerned with the $n \times n$ matrix J, which has ones on the sinistral diagonal (running from the upper right corner to the lower left corner) and zeros elsewhere. Find the eigenvalues of J in each case.

51. $J_2 = \begin{bmatrix} 0 & 1 \\ 1 & 0 \end{bmatrix}$.

52. $J_3 = \begin{bmatrix} 0 & 0 & 1 \\ 0 & 1 & 0 \\ 1 & 0 & 0 \end{bmatrix}$.

53. $J_4 = \begin{bmatrix} 0 & 0 & 0 & 1 \\ 0 & 0 & 1 & 0 \\ 0 & 1 & 0 & 0 \\ 1 & 0 & 0 & 0 \end{bmatrix}$.

54. Write the characteristic polynomial and find the eigenvalues of J_5.

55. What do you suppose the eigenvalues of J_n are?

56. Let $T: V \to V$ be a linear operator on the n-dimensional vector space V. Suppose that B is a basis for V and that A is the $n \times n$ matrix $[T]_B$. Suppose also that λ_0 is an eigenvalue of T. Prove that the two vector spaces (eigenspaces) (1) the *null space of the matrix* $(\lambda_0 I - A)$, and (2) the *kernel of the transformation* $(\lambda_0 I - T)$ are isomorphic.

57. If you have learned to use the software available in your college (department) computer center,

use it to find the eigenvalues for the matrices of Exercises 1–9.

58. If you have learned to use some appropriate computer software, use a computer to find the eigenvalues for the matrix

$$K = \begin{bmatrix} 1 & 0 & 0 & 0 & 0 & -3 \\ 2 & 2 & 0 & 0 & 0 & -1 \\ 3 & 0 & 1 & 0 & 0 & 4 \\ 4 & 0 & 0 & -1 & 0 & 3 \\ -1 & 0 & 0 & 0 & 1 & 2 \\ -3 & 0 & 0 & 0 & 0 & 1 \end{bmatrix}.$$

59. Use a computer to find bases for the eigenspaces for the matrix K of Exercise 58.

6.2 SIMILARITY TO A DIAGONAL MATRIX

If $T: V \to V$ is a linear transformation from the finite-dimensional vector space V into itself (linear operator on V), we have seen that T can be represented relative to a given ordered basis B for V as multiplication by a matrix $A = [T]_B$. The choice of the basis for V determines the matrix A. A change in basis will change the matrix representation for T. We have also shown that all such matrix representations of T are similar; that is, $A' = P^{-1}AP$. Moreover, similar matrices have the same eigenvalues. In this section we are interested in determining under what conditions a basis B for V can be chosen so that the matrix $A = [T]_B$ is a diagonal matrix.

> **Definition** An $n \times n$ matrix A is said to be **diagonalizable** if there exists an invertible matrix P such that the matrix $P^{-1}AP = D$ is a diagonal matrix. That is, $D = (d_{ij})$, where $d_{ij} = 0$ when $i \neq j$.

It is *not true* that every $n \times n$ matrix A is similar to a diagonal matrix. That is, not every $n \times n$ matrix can be diagonalized. This is shown by Example 2 below. However, if A *can* be diagonalized, then it must have n linearly independent eigenvectors, as we prove in the following theorem.

> **Theorem 6.3** An $n \times n$ matrix A is similar to a diagonal matrix D if, and only if, A has n linearly independent eigenvectors. Furthermore, in this case, the diagonal elements of D are the eigenvalues of A as well as those of D.

Proof Assume that $P^{-1}AP = D = (d_{ij})$ is a diagonal matrix. Then we have $AP = PD$ or

$$AP = PD = \begin{bmatrix} p_{11} & p_{12} & p_{13} & \cdots & p_{1n} \\ p_{21} & p_{22} & p_{23} & \cdots & p_{2n} \\ p_{31} & p_{32} & p_{33} & \cdots & p_{3n} \\ \vdots & \vdots & \vdots & \cdots & \vdots \\ p_{n1} & p_{n2} & p_{n3} & \cdots & p_{nn} \end{bmatrix} \cdot \begin{bmatrix} d_{11} & 0 & 0 & \cdots & 0 \\ 0 & d_{22} & 0 & \cdots & 0 \\ 0 & 0 & d_{33} & \cdots & 0 \\ \vdots & \vdots & \vdots & \cdots & \vdots \\ 0 & 0 & 0 & \cdots & d_{nn} \end{bmatrix}$$

$$= \begin{bmatrix} d_{11}p_{11} & d_{22}p_{12} & \cdots & d_{nn}p_{1n} \\ d_{11}p_{21} & d_{22}p_{22} & \cdots & d_{nn}p_{2n} \\ \vdots & \vdots & \cdots & \vdots \\ d_{11}p_{n1} & d_{22}p_{n2} & \cdots & d_{nn}p_{nn} \end{bmatrix}.$$

Let us denote the column vectors of the matrix P by $\mathbf{p}_i$ for each $i = 1, 2, \ldots, n$. Then the above matrix equation can be restated in the following form:

$$A\mathbf{p}_i = d_{ii}\mathbf{p}_i, \qquad \text{for each } i = 1, 2, \ldots, n. \tag{6.4}$$

Since none of the column vectors of a nonsingular matrix is the *zero* vector, we see from (6.4) that the vectors $\mathbf{p}_i$ are each eigenvectors of the matrix A corresponding to the eigenvalue d_{ii}. Furthermore, since P is invertible, its column vectors must be linearly independent, as desired.

Conversely, if the matrix A has n linearly independent eigenvectors $\mathbf{p}_1$, $\mathbf{p}_2, \ldots, \mathbf{p}_n$ corresponding to the eigenvalues $\lambda_1, \lambda_2, \ldots, \lambda_n$, then let the matrix P be constructed with the eigenvectors $\mathbf{p}_i$ as column vectors: $P = [\mathbf{p}_1 : \mathbf{p}_2 : \ldots : \mathbf{p}_n]$. Since its column vectors are linearly independent, this matrix P is surely invertible. Thus,

$$AP = \left[A\mathbf{p}_1 : A\mathbf{p}_2 : \dots : A\mathbf{p}_n \right] = \left[\lambda_1 \mathbf{p}_1 : \lambda_2 \mathbf{p}_2 : \dots : \lambda_n \mathbf{p}_n \right]$$

$$= P \begin{bmatrix} \lambda_1 & 0 & 0 & \dots & 0 \\ 0 & \lambda_2 & 0 & \dots & 0 \\ 0 & 0 & \lambda_3 & \dots & 0 \\ \vdots & \vdots & \vdots & \vdots & \vdots \\ 0 & 0 & 0 & \dots & \lambda_n \end{bmatrix} = PD.$$

Therefore, multiplying both extreme sides by P^{-1} on the left, we have $P^{-1}AP = D$; that is, A is similar to a diagonal matrix. This completes the proof of Theorem 6.3. ❑

The proof gives a method for constructing the matrix P that will diagonalize the matrix A, should that be possible.

Example 1   Diagonalize, if possible, the matrix $A = \begin{bmatrix} 1 & -2 \\ 0 & 3 \end{bmatrix}$.

From Example 3 of Section 6.1 we know that the eigenvalues and corresponding eigenvectors of this matrix A are

For the eigenvalue $\lambda = 1$, $\mathbf{x}_1 = \begin{bmatrix} 1 \\ 0 \end{bmatrix}$ and for $\lambda = 3$, $\mathbf{x}_0 = \begin{bmatrix} 1 \\ -1 \end{bmatrix}$.

Let these be the respective column vectors $\mathbf{x}_1 = \mathbf{p}_1$ and $\mathbf{x}_0 = \mathbf{p}_2$ of the matrix P given by the proof of the theorem.

$$P = \begin{bmatrix} 1 & 1 \\ 0 & -1 \end{bmatrix}.$$

We compute

$$P^{-1} = \begin{bmatrix} 1 & 1 \\ 0 & -1 \end{bmatrix}.$$

Thus,

$$P^{-1}AP = \begin{bmatrix} 1 & 1 \\ 0 & -1 \end{bmatrix} \begin{bmatrix} 1 & -2 \\ 0 & 3 \end{bmatrix} \begin{bmatrix} 1 & 1 \\ 0 & -1 \end{bmatrix} = \begin{bmatrix} 1 & 0 \\ 0 & 3 \end{bmatrix} = D.$$ ■

Contrast this with the following example.

Example 2   Show that the matrix $A = \begin{bmatrix} 1 & -2 \\ 2 & -3 \end{bmatrix}$ is not diagonalizable.

The characteristic equation for A is

$$\det (\lambda I - A) = \det \begin{bmatrix} \lambda - 1 & 2 \\ -2 & \lambda + 3 \end{bmatrix} = (\lambda + 1)^2 = 0.$$

Therefore, A has a single eigenvalue $\lambda_0 = -1$. This fact by itself is not sufficient to show that A is not similar to a diagonal matrix since the diagonal matrix $-I_2$ has the same property and is already diagonal. In the case of the matrix A, however, the eigenspace of A associated with $\lambda_0 = -1$ has dimension *one*. This fact is determined, in the usual way, by solving the system of equations $A\mathbf{x} = -\mathbf{x}$ or the corresponding homogeneous system $(-1I - A)\mathbf{x} = \mathbf{0}$.

The coefficient matrix for the homogeneous system $(-1I - A)\mathbf{x} = \mathbf{0}$ is the matrix

$$(-1I - A) = \begin{bmatrix} -2 & 2 \\ -2 & 2 \end{bmatrix}.$$

So the solutions to the system have the form $x_1 = x_2 = t$, and each eigenvector of A has the form

$$t\mathbf{u} = t\begin{bmatrix} 1 \\ 1 \end{bmatrix}, t \neq 0.$$

Since the single vector $\mathbf{u}$ constitutes a basis for the only eigenspace of A, A fails to have *two* linearly independent eigenvectors and cannot, by Theorem 6.3, be similar to a diagonal matrix.   ■

Let us warn you again against making the *inductive* leap to the statement that A was not similar to a diagonal matrix because it had a single eigenvalue. Rather the reason A was not diagonalizable was because the sum of the dimensions of all its eigenspaces was smaller than $n = 2$. It is easy to see that the $n \times n$ matrix $2I$ (any n) has the single eigenvalue $\lambda = 2$, yet is similar to (indeed already is) a diagonal matrix. However, its eigenspaces have dimensions that total to n, with the standard basis for $\mathbb{R}^n$ as its set of linearly independent eigenvectors. Consider also the following example that illustrates each of the steps in the diagonalizing process.

Example 3   Diagonalize, if possible, the following matrix:

$$A = \begin{bmatrix} 2 & 1 & 1 \\ 1 & 2 & 1 \\ 1 & 1 & 2 \end{bmatrix}.$$

Step 1   Find the eigenvalues as roots of the following characteristic equation: $\det(\lambda I - A) = 0$. Here we see that

$$\det(\lambda I - A) = \det \begin{bmatrix} \lambda - 2 & -1 & -1 \\ -1 & \lambda - 2 & -1 \\ -1 & -1 & \lambda - 2 \end{bmatrix}$$

$$= \lambda^3 - 6\lambda^2 + 9\lambda - 4 = (\lambda - 1)^2(\lambda - 4).$$

Therefore, A has eigenvalues $\lambda_1 = 1$ and $\lambda_2 = 4$. Note that there are but *two* distinct eigenvalues of A, yet A is a 3×3 matrix. To determine whether or not A is similar to a diagonal matrix, we proceed to the next step.

Step 2 Find the eigenvectors and eigenspaces corresponding to each eigenvalue. For the matrix of this example, and for $\lambda_1 = 1$, solve the homogeneous system of the equations $((1)I - A)x = 0$: that is, solve

$$\begin{bmatrix} -1 & -1 & -1 \\ -1 & -1 & -1 \\ -1 & -1 & -1 \end{bmatrix} \cdot \begin{bmatrix} x_1 \\ x_2 \\ x_3 \end{bmatrix} = \begin{bmatrix} 0 \\ 0 \\ 0 \end{bmatrix}.$$

The *augmented* matrix for this system of equations reduces to the matrix (verify this)

$$\begin{bmatrix} 1 & 1 & 1 & 0 \\ 0 & 0 & 0 & 0 \\ 0 & 0 & 0 & 0 \end{bmatrix}.$$

Therefore, the solutions to the system are of the form $x_1 = -x_2 - x_3$, $x_2 = s$, $x_3 = t$. *Two* linearly independent eigenvectors correspond to $\lambda_1 = 1$. These can be found most easily by setting $t = 0$, $s = 1$, and then $t = 1$, $s = 0$ to obtain the two vectors

$$\mathbf{u} = \begin{bmatrix} -1 \\ 1 \\ 0 \end{bmatrix} \quad \text{and} \quad \mathbf{v} = \begin{bmatrix} -1 \\ 0 \\ 1 \end{bmatrix}.$$

Any eigenvector $\mathbf{x}_1 = \begin{bmatrix} -s-t \\ s \\ t \end{bmatrix}$ is expressible as a linear combination of $\mathbf{u}$ and $\mathbf{v}$; that is,

$$\mathbf{x}_1 = s\mathbf{u} + t\mathbf{v}.$$

For $\lambda_2 = 4$, solve the homogeneous system of equations $(4I - A)x = 0$, whose augmented matrix is

$$\begin{bmatrix} 2 & -1 & -1 & 0 \\ -1 & 2 & -1 & 0 \\ -1 & -1 & 2 & 0 \end{bmatrix}.$$

This matrix is row equivalent to the matrix (verify this)

$$\begin{bmatrix} 1 & 0 & -1 & 0 \\ 0 & 1 & -1 & 0 \\ 0 & 0 & 0 & 0 \end{bmatrix}.$$

Therefore, the solutions to the system are of the form $x_1 = x_2 = x_3 = t$, so every eigenvector corresponding to the eigenvalue $\lambda_2 = 4$ is of the form $\begin{bmatrix} t \\ t \\ t \end{bmatrix}$. Setting $t = 1$, we obtain $\mathbf{w} = \begin{bmatrix} 1 \\ 1 \\ 1 \end{bmatrix}$ as the single basis eigenvector for the eigenspace of A corresponding to the eigenvalue $\lambda_2 = 4$.

Step 3   Compare the number of linearly independent eigenvectors of A with the size of A. If A is $n \times n$, are there n linearly independent eigenvectors? If so, we may complete the computation of the nonsingular matrix P.

Since our matrix A in this example does indeed have three eigenvectors **u**, **v**, and **w**, even though there were but *two* distinct eigenvalues, we need now to verify that they are linearly independent. If they are, then the matrix $P = [\mathbf{u}: \mathbf{v}: \mathbf{w}]$ whose columns are these eigenvectors will be such that

$$P^{-1}AP = \begin{bmatrix} 1 & 0 & 0 \\ 0 & 1 & 0 \\ 0 & 0 & 4 \end{bmatrix} = \begin{bmatrix} \lambda_1 & 0 & 0 \\ 0 & \lambda_1 & 0 \\ 0 & 0 & \lambda_2 \end{bmatrix}.$$

We already know that **u** and **v** are linearly independent; we chose them to be such. To see that all three are, we must show that **w** is not a linear combination of the other two, that is, that **w** is not in the space spanned by **u** and **v**. This can be accomplished in a number of different ways, as discussed in earlier chapters. Perhaps the easiest way in this case is to set the matrix $P = [\mathbf{u}: \mathbf{v}: \mathbf{w}]$ and note that it has nonzero determinant. Thus,

$$P = \begin{bmatrix} -1 & -1 & 1 \\ 1 & 0 & 1 \\ 0 & 1 & 1 \end{bmatrix}.$$
$$\quad\ \mathbf{u} \quad \mathbf{v} \quad \mathbf{w}$$

Then you can calculate that $P^{-1}AP$ is as indicated. This completes the diagonalization of A.   ∎

We conclude this section with two theorems, the first of which disposes of the problem mentioned in step 3 above. We prove the following theorem, which tells us that it is not necessary to verify each time that (as in that example) the vector **w** is linearly independent of the vectors **u** and **v**. It will always be the case that eigenvectors belonging to different eigenvalues are linearly independent.

Theorem 6.4   Let A be an $n \times n$ matrix. The distinct eigenvectors of A corresponding to distinct eigenvalues of A are linearly independent.

Proof   Let A have distinct eigenvalues $\lambda_1, \lambda_2, \ldots, \lambda_m$, where $m \leq n$ and $\lambda_i \neq \lambda_j$ for $i \neq j$. Let $\mathbf{u}_1, \mathbf{u}_2, \ldots, \mathbf{u}_m$ be eigenvectors of A corresponding, respectively, to these eigenvalues. Thus, $A\mathbf{u}_i = \lambda_i\mathbf{u}_i$, for each $i = 1, 2, \ldots, m$. Note that this may not include all of the eigenvectors of A, since a given eigenvalue may determine an eigenspace of A whose dimension is more than one. However, since $\mathbf{u}_i$ can be any nonzero vector from the eigenspace of A corresponding to the eigenvalue λ_i, these m eigenvectors will be sufficient for the proof.

Since $\mathbf{u}_1 \neq \mathbf{0}$, the set $\{\mathbf{u}_1\}$ consisting of the single vector $\mathbf{u}_1$ is linearly independent. Let k be the *largest* of the integers $1, 2, \ldots, m$ such that the set $\{\mathbf{u}_1, \mathbf{u}_2, \ldots, \mathbf{u}_k\}$ is linearly independent. Thus $1 \leq k \leq m$. Now if $k \neq m$, then $k < m$, and the set of vectors $\{\mathbf{u}_1, \mathbf{u}_2, \ldots, \mathbf{u}_k, \mathbf{u}_{k+1}\}$ is linearly dependent because of our choice of k. Hence there exist scalars, $a_1, a_2, \ldots, a_{k+1}$, not all zero, such that

$$a_1\mathbf{u}_1 + a_2\mathbf{u}_2 + \cdots + a_k\mathbf{u}_k + a_{k+1}\mathbf{u}_{k+1} = \mathbf{0}. \tag{6.5}$$

This being the case, it must be true that

$$\begin{aligned}\mathbf{0} = A\mathbf{0} &= A(a_1\mathbf{u}_1 + a_2\mathbf{u}_2 + \cdots + a_k\mathbf{u}_k + a_{k+1}\mathbf{u}_{k+1}) \\ &= a_1 A\mathbf{u}_1 + a_2 A\mathbf{u}_2 + \cdots + a_k A\mathbf{u}_k + a_{k+1} A\mathbf{u}_{k+1}\end{aligned}$$

or

$$\mathbf{0} = a_1\lambda_1\mathbf{u}_1 + a_2\lambda_2\mathbf{u}_2 + \cdots + a_k\lambda_k\mathbf{u}_k + a_{k+1}\lambda_{k+1}\mathbf{u}_{k+1}. \tag{6.6}$$

Now, multiply both sides of equation (6.5) by λ_{k+1} to get

$$\mathbf{0} = a_1\lambda_{k+1}\mathbf{u}_1 + a_2\lambda_{k+1}\mathbf{u}_2 + \cdots + a_k\lambda_{k+1}\mathbf{u}_k + a_{k+1}\lambda_{k+1}\mathbf{u}_{k+1}. \tag{6.7}$$

Subtracting equation (6.7) from equation (6.6) results in the equation

$$\mathbf{0} = a_1(\lambda_1 - \lambda_{k+1})\mathbf{u}_1 + a_2(\lambda_2 - \lambda_{k+1})\mathbf{u}_2 + \cdots + a_k(\lambda_k - \lambda_{k+1})\mathbf{u}_k + \mathbf{0};$$

but the set of vectors $\{\mathbf{u}_1, \mathbf{u}_2, \ldots, \mathbf{u}_k\}$ is linearly independent. So this equation forces each of the coefficients to be zero; that is, for each $i = 1, 2, \ldots, k$.

$$a_i(\lambda_i - \lambda_{k+1}) = 0.$$

However, the eigenvalues are all distinct so $(\lambda_i - \lambda_{k+1}) \neq 0$. This means that

$$a_1 = a_2 = \cdots = a_k = 0.$$

But this fact together with equation (6.5) means that $a_{k+1} = 0$ also. (Why?) We have contradicted the fact that the coefficients a_i in (6.5) were *not* all zero. Hence it must be true that k is *not* less than m, and therefore $k = m$. Thus the eigenvectors $\mathbf{u}_1, \mathbf{u}_2, \ldots, \mathbf{u}_m$ are linearly independent, as asserted by the theorem. ◘

As an immediate consequence of this theorem and Theorem 6.3 (page 389), it follows that if the $n \times n$ matrix A has n *distinct* eigenvalues, then it must have n linearly independent eigenvectors corresponding to those eigenvalues. It must therefore be diagonalizable. We saw in Example 3, and in the matrix just preceding that example, that while distinct eigenvalues are sufficient for A to be diagonalizable, this is not a necessary condition. We state all of this formally as the following theorem.

> **Theorem 6.5** A sufficient, but not a necessary, condition for the $n \times n$ matrix A to be similar to a diagonal matrix is that A have n distinct eigenvalues.

Warning If the eigenvalues of an $n \times n$ matrix A are *not* distinct (as in Examples 2 and 3 above), then A may or may not be similar to a diagonal matrix, and further analysis is needed to settle this.

Examples 2 and 3 above explain what can happen if there is repetition in the set of eigenvalues of A.

Example 4 Find, if possible, a diagonal matrix similar to the matrix

$$A = \begin{bmatrix} -1 & -2 & 3 \\ 0 & 1 & 5 \\ 0 & 0 & 2 \end{bmatrix}.$$

The characteristic polynomial for A is

$$p_A(\lambda) = \det(\lambda I - A) = (\lambda + 1)(\lambda - 1)(\lambda - 2).$$

Therefore, A has the three distinct eigenvalues -1, $+1$, and 2. Hence A is similar to the diagonal matrix

$$D = \begin{bmatrix} -1 & 0 & 0 \\ 0 & 1 & 0 \\ 0 & 0 & 2 \end{bmatrix}.$$

To find a matrix P so that $D = P^{-1}AP$, we must find eigenvectors of A corresponding to these distinct eigenvalues. As illustrated in step 2 of Example 3, these can be the following. (Verify this.)

For $\lambda_1 = -1$:

$$\mathbf{p}_1 = \begin{bmatrix} 1 \\ 0 \\ 0 \end{bmatrix}.$$

For $\lambda_2 = 1$:

$$\mathbf{p}_2 = \begin{bmatrix} 1 \\ -1 \\ 0 \end{bmatrix}.$$

For $\lambda_3 = 2$:

$$\mathbf{p}_3 = \begin{bmatrix} -7 \\ 15 \\ 3 \end{bmatrix}.$$

Thus we have the following diagonalizing matrices:

$$P = \begin{bmatrix} 1 & 1 & -7 \\ 0 & -1 & 15 \\ 0 & 0 & 3 \end{bmatrix}, \quad \text{and} \quad P^{-1} = \begin{bmatrix} 1 & 1 & -\dfrac{8}{3} \\ 0 & -1 & 5 \\ 0 & 0 & \dfrac{1}{3} \end{bmatrix}. \quad \blacksquare$$

Example 5 Let T be the linear transformation in $\mathbb{R}^2$ that reflects every point in $\mathbb{R}^2$ through the line $y = x$. Does there exist a basis for $\mathbb{R}^2$ such that the matrix representation for T is a diagonal matrix? This reflection is sketched in Figure 6.2.

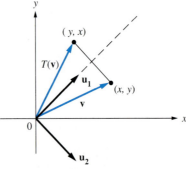

FIGURE 6.2

This reflection T takes the point $(1, 1)$ to itself and the point $(1, -1)$ to its negative, so the answer to the question is *yes*. Here is how it works out by matrix algebra. Since T takes each vector $\mathbf{v} = (x, y)$ in $\mathbb{R}^2$ to the vector $T(\mathbf{v}) = (y, x)$, the standard matrix representation $[T]_E$ for T is (as you can easily verify) the matrix

$$A = \begin{bmatrix} 0 & 1 \\ 1 & 0 \end{bmatrix}.$$

Now solve the characteristic equation $\det (\lambda I - A) = \lambda^2 - 1 = 0$ to find the eigenvalues for T. These are $\lambda_1 = 1$ and $\lambda_2 = -1$. The two corresponding eigenspaces are found to be spanned, respectively, by the vectors (verify this)

$$\mathbf{u}_1 = \begin{bmatrix} 1 \\ 1 \end{bmatrix} \quad \text{and} \quad \mathbf{u}_2 = \begin{bmatrix} 1 \\ -1 \end{bmatrix}.$$

Note these in Figure 6.2. Hence the set of vectors $\{(1, 1), (1, -1)\}$ is a basis for $\mathbb{R}^2$, relative to which the matrix representation for T is the diagonal matrix

$$D = \begin{bmatrix} 1 & 0 \\ 0 & -1 \end{bmatrix} = \begin{bmatrix} \dfrac{1}{2} & \dfrac{1}{2} \\ \dfrac{1}{2} & -\dfrac{1}{2} \end{bmatrix} \cdot \begin{bmatrix} 0 & 1 \\ 1 & 0 \end{bmatrix} \cdot \underbrace{\begin{bmatrix} 1 & 1 \\ 1 & -1 \end{bmatrix}}_{\mathbf{u}_1 \quad \mathbf{u}_2} = P^{-1}AP. \quad \blacksquare$$

Example 6   Consider the transformation $T: \mathbb{R}^2 \to \mathbb{R}^2$ given by the equation

$$T(x, y) = (y - x, x - 3y).$$

The matrix representing T with respect to the standard basis for $\mathbb{R}^2$ is (verify this)

$$A = \begin{bmatrix} -1 & 1 \\ 1 & -3 \end{bmatrix}.$$

The characteristic polynomial $p_A(\lambda)$ for A, and hence for T, is $\lambda^2 + 4\lambda + 2$ (check this also). Therefore, T has eigenvalues $\lambda_1 = -2 + \sqrt{2}$ and $\lambda_2 = -2 - \sqrt{2}$. Also, a basis for the eigenspaces of T consists of the two linearly independent eigenvectors

$$\mathbf{v}_1 = \begin{bmatrix} 1 + \sqrt{2} \\ 1 \end{bmatrix} \quad \text{and} \quad \mathbf{v}_2 = \begin{bmatrix} 1 - \sqrt{2} \\ 1 \end{bmatrix}.$$

Relative to the basis $\{(1 + \sqrt{2}, 1), (1 - \sqrt{2}, 1)\}$ of $\mathbb{R}^2$, or $\{\mathbf{v}_1, \mathbf{v}_2\}$ for $\overline{\mathbb{R}^2}$, the transformation T can be represented by the diagonal matrix

$$D = \begin{bmatrix} -2 + \sqrt{2} & 0 \\ 0 & -2 - \sqrt{2} \end{bmatrix}$$

$$= \begin{bmatrix} \dfrac{\sqrt{2}}{4} & \dfrac{2 - \sqrt{2}}{4} \\ -\dfrac{\sqrt{2}}{4} & \dfrac{2 + \sqrt{2}}{4} \end{bmatrix} \cdot \begin{bmatrix} -1 & 1 \\ 1 & -3 \end{bmatrix} \cdot \underbrace{\begin{bmatrix} 1 + \sqrt{2} & 1 - \sqrt{2} \\ 1 & 1 \end{bmatrix}}_{\mathbf{v}_1 \quad \mathbf{v}_2} = P^{-1}AP. \quad \blacksquare$$

Example 7   Consider the linear transformation $T: \mathbb{R}^3 \to \mathbb{R}^3$ given by the following formula:

$$T(x_1, x_2, x_3) = (x_1 - 4x_2 + 2x_3, 3x_1 - 4x_2, 3x_1 - x_2 - 3x_3).$$

The standard matrix representation for T is

$$A = \begin{bmatrix} 1 & -4 & 2 \\ 3 & -4 & 0 \\ 3 & -1 & -3 \end{bmatrix}.$$

The characteristic polynomial $\det(\lambda I - A) = \lambda^3 + 6\lambda^2 + 11\lambda + 6 = (\lambda + 1)(\lambda + 2)(\lambda + 3)$ is the characteristic polynomial for T as well. The eigenvalues of T are therefore $\lambda_1 = -1, \lambda_2 = -2$, and $\lambda_3 = -3$. A basis for each eigenspace of T corresponding to each of these respective eigenvalues comes from solving the three homogeneous systems of equations

$$(-I - A)\mathbf{x} = \mathbf{0};$$
$$(-2I - A)\mathbf{x} = \mathbf{0};$$
$$(-3I - A)\mathbf{x} = \mathbf{0}.$$

Respective basis vectors are the eigenvectors

$$\mathbf{u}_1 = \begin{bmatrix} 1 \\ 1 \\ 1 \end{bmatrix}, \qquad \mathbf{u}_2 = \begin{bmatrix} 2 \\ 3 \\ 3 \end{bmatrix}, \qquad \text{and} \qquad \mathbf{u}_3 = \begin{bmatrix} 1 \\ 3 \\ 4 \end{bmatrix}.$$

These three linearly independent eigenvectors span $\overline{\mathbb{R}^3}$. Relative to the ordered basis $\{\mathbf{u}_1, \mathbf{u}_2, \mathbf{u}_3\}$ for $\overline{\mathbb{R}^3}$, or $\{(1, 1, 1), (2, 3, 3), (1, 3, 4)\}$ for $\mathbb{R}^3$, the transformation T is represented by the diagonal matrix

$$D = \begin{bmatrix} -1 & 0 & 0 \\ 0 & -2 & 0 \\ 0 & 0 & -3 \end{bmatrix}$$

$$= \begin{bmatrix} 3 & -5 & 3 \\ -1 & 3 & -2 \\ 0 & -1 & 1 \end{bmatrix} \begin{bmatrix} 1 & -4 & 2 \\ 3 & -4 & 0 \\ 3 & -1 & -3 \end{bmatrix} \begin{bmatrix} 1 & 2 & 1 \\ 1 & 3 & 3 \\ 1 & 3 & 4 \end{bmatrix} = P^{-1}AP. \quad \blacksquare$$

Vocabulary

Some very important facts pertaining to the diagonalization of a matrix were discussed in this section. The linkage between this concept and that of the eigenvectors of a given matrix was exploited. Also a connection between the eigenvalues being distinct and the matrix being diagonalizable has been established. You should study the concepts we list here until you feel confident that you understand them thoroughly.

diagonalizable matrix A

criterion for similarity to a diagonal matrix

connection between eigenvalues and diagonalization

Exercises 6.2

In Exercises 1–9, find a nonsingular matrix P so that for each given matrix A, $P^{-1}AP$ is a diagonal matrix.

1. $A = \begin{bmatrix} 2 & 3 \\ 1 & 0 \end{bmatrix}.$

2. $A = \begin{bmatrix} 0 & 2 \\ 3 & 0 \end{bmatrix}.$

3. $A = \begin{bmatrix} \frac{1}{2} & \frac{1}{2} \\ \frac{1}{2} & \frac{1}{2} \end{bmatrix}.$

4. $A = \begin{bmatrix} -\frac{1}{\sqrt{2}} & \frac{1}{\sqrt{2}} \\ \frac{1}{\sqrt{2}} & \frac{1}{\sqrt{2}} \end{bmatrix}.$

5. $A = \begin{bmatrix} 0 & 0 \\ 0 & 0 \end{bmatrix}.$

6. $A = \begin{bmatrix} 2 & 0 & 0 \\ 0 & 1 & 1 \\ 0 & 1 & 1 \end{bmatrix}.$

7. $A = \begin{bmatrix} 1 & 2 & 0 \\ 2 & 1 & 0 \\ 0 & 0 & -2 \end{bmatrix}.$

8. $A = \begin{bmatrix} 1 & 0 & 3 \\ 2 & 1 & 2 \\ 3 & 0 & 1 \end{bmatrix}.$

9. $A = \begin{bmatrix} 8 & 9 & 9 \\ 3 & 2 & 3 \\ -9 & -9 & -10 \end{bmatrix}.$

10. Show that the matrix $A = \begin{bmatrix} 2 & 1 \\ 0 & 2 \end{bmatrix}$ is not diago-

nalizable, while the matrix $B = \begin{bmatrix} 2 & 1 \\ 1 & 2 \end{bmatrix}$ is. What

diagonal matrix is similar to B?

11. Show that the matrix $C = \begin{bmatrix} 1 & -1 \\ 3 & -2 \end{bmatrix}$ is not simi-

lar to any diagonal matrix with real entries.

In Exercises 12–22, decide which of the given matrices is similar to a diagonal matrix and which is not. If the matrix is diagonalizable, find both the diagonal matrix and the nonsingular matrix P that diagonalizes it.

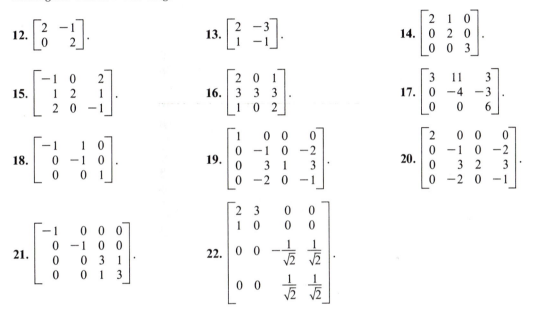

12. $\begin{bmatrix} 2 & -1 \\ 0 & 2 \end{bmatrix}$.

13. $\begin{bmatrix} 2 & -3 \\ 1 & -1 \end{bmatrix}$.

14. $\begin{bmatrix} 2 & 1 & 0 \\ 0 & 2 & 0 \\ 0 & 0 & 3 \end{bmatrix}$.

15. $\begin{bmatrix} -1 & 0 & 2 \\ 1 & 2 & 1 \\ 2 & 0 & -1 \end{bmatrix}$.

16. $\begin{bmatrix} 2 & 0 & 1 \\ 3 & 3 & 3 \\ 1 & 0 & 2 \end{bmatrix}$.

17. $\begin{bmatrix} 3 & 11 & 3 \\ 0 & -4 & -3 \\ 0 & 0 & 6 \end{bmatrix}$.

18. $\begin{bmatrix} -1 & 1 & 0 \\ 0 & -1 & 0 \\ 0 & 0 & 1 \end{bmatrix}$.

19. $\begin{bmatrix} 1 & 0 & 0 & 0 \\ 0 & -1 & 0 & -2 \\ 0 & 3 & 1 & 3 \\ 0 & -2 & 0 & -1 \end{bmatrix}$.

20. $\begin{bmatrix} 2 & 0 & 0 & 0 \\ 0 & -1 & 0 & -2 \\ 0 & 3 & 2 & 3 \\ 0 & -2 & 0 & -1 \end{bmatrix}$.

21. $\begin{bmatrix} -1 & 0 & 0 & 0 \\ 0 & -1 & 0 & 0 \\ 0 & 0 & 3 & 1 \\ 0 & 0 & 1 & 3 \end{bmatrix}$.

22. $\begin{bmatrix} 2 & 3 & 0 & 0 \\ 1 & 0 & 0 & 0 \\ 0 & 0 & -\frac{1}{\sqrt{2}} & \frac{1}{\sqrt{2}} \\ 0 & 0 & \frac{1}{\sqrt{2}} & \frac{1}{\sqrt{2}} \end{bmatrix}$.

In Exercises 23–27, T is a linear operator on $\mathbb{R}^2$ or $\mathbb{R}^3$. Determine whether or not there is an ordered basis for $\mathbb{R}^2$ or $\mathbb{R}^3$ such that the matrix representation of T relative to that basis is a diagonal matrix. If there is, find such a basis and the diagonal matrix representation of T.

23. $T(x_1, x_2) = (x_1, 0)$.

24. $T(x_1, x_2) = (x_1 + 3x_2, x_2)$.

25. $T(x_1, x_2) = (3x_1 + x_2, x_1 + 3x_2)$.

26. $T(x_1, x_2) = (x_2 - x_1, x_1 - 3x_2)$.

27. $T(x_1, x_2, x_3) = (2x_3 - x_1, x_1 + 2x_2 + x_3, 2x_1 - x_3)$.

In Exercises 28 and 29, T is the linear operator on P_2, the space of real polynomials, defined by the formula

$$T(c + bx + ax^2)$$
$$= (2a + 5c) + (3b - a - c)x + (5a + 2c)x^2.$$

28. Write the standard matrix representation of T (with respect to the basis $\{1, x, x^2\}$).

29. Find a basis B for P_2 such that the matrix representation $[T]_B$ of T with respect to that basis is a diagonal matrix. State both the basis and the matrix.

30. Prove, by induction, that for any $n \times n$ matrix A and any nonsingular matrix P, $(P^{-1}AP)^k = P^{-1}A^kP$, for every positive integer k.

In Exercises 31–34, find A^9 for the given matrix A. HINT: In each case first find a diagonal matrix similar to A and use the result of Exercise 30.

31. $A = \begin{bmatrix} 2 & 0 \\ -1 & 1 \end{bmatrix}$.

32. $A = \begin{bmatrix} 4 & 1 \\ 1 & 4 \end{bmatrix}$.

33. $A = \begin{bmatrix} 1 & 0 & -1 \\ 2 & -1 & 2 \\ -1 & 0 & 1 \end{bmatrix}$.

34. $A = \begin{bmatrix} 2 & -1 & 1 \\ 3 & -2 & 3 \\ 3 & -1 & 0 \end{bmatrix}$.

35. Suppose that the $n \times n$ matrix A has characteristic polynomial $p_A(\lambda) = (\lambda - a)^n$ and that A is similar to a diagonal matrix. What can you say about the elements of A? HINT: If $D = aI$, what matrices can be similar to D? (See Exercise 7 in Section 5.5.)

36. Suppose that the $n \times n$ matrix A has characteristic polynomial $p_A(\lambda) = (\lambda - 1)^k(\lambda + 1)^r$, with $r + k = n$, and that A is similar to a diagonal matrix. Prove that $A^{-1} = A$. HINT: Choose the matrix P so that the diagonal matrix D is partitioned with I and $-I$ in diagonal blocks.

37. Use the diagonalization $D = P^{-1}AP$ of the matrix A in Exercise 1 to compute A^{-1} from D^{-1}.

38. Use the diagonalization $P^{-1}AP$ of the matrix A in Exercise 3 to compute A^{-1}.

39. Use the diagonalization $P^{-1}AP$ of the matrix A in Exercise 4 to compute A^{-1}.

40. Use the diagonalization $P^{-1}AP$ of the matrix A in Exercise 7 to compute A^{-1}.

41. Use the diagonalization $P^{-1}AP$ of the matrix A in Exercise 9 to compute A^{-1}.

42. Use the diagonalization $P^{-1}AP$ of the matrix A in Exercise 21 to compute A^{-1}.

43. Suppose that $A = (a_{ij})$ is an upper-triangular matrix with distinct diagonal entries. Prove that A is similar to a diagonal matrix.

44. Matrix Polynomials Let $p(t) = a_m t^m + a_{m-1}t^{m-1} + \cdots + a_1 t + a_0$. If A is an $n \times n$ matrix, then by $p(A)$ we mean the matrix

$$a_m A^m + a_{m-1}A^{m-1} + \cdots + a_1 A + a_0 I.$$

Find $p(A)$ when $p(t) = t^2 + 2t - 3$ for the matrix A of Exercise 1.

45. Repeat Exercise 44 for the matrix A of Exercise 7.

46. Repeat Exercise 44 for the matrix A of Exercise 9.

47. Repeat Exercise 44 for the matrix A of Exercise 21.

48. Repeat Exercise 44 for the matrix A of Exercise 22.

49. Prove that if $p(t)$ is a given polynomial and λ is an eigenvalue of the $n \times n$ matrix A, and $\mathbf{x}$ is an associated eigenvector, then $p(\lambda)$ is an eigenvalue of the matrix $p(A)$ and $\mathbf{x}$ is an eigenvector of $p(A)$ associated with $p(\lambda)$. HINT: Write out $p(A)\mathbf{x}$.

50. The Cayley-Hamilton Theorem If $p_A(t)$ is the characteristic polynomial of the $n \times n$ matrix A, then $p_A(A) = 0$. Prove this theorem for the case when A is a 2×2 diagonal matrix.

51. Verify the Cayley-Hamilton theorem for the matrix $A = \begin{bmatrix} -1 & 2 \\ 0 & 3 \end{bmatrix}$.

52. Use the results of Exercise 51 to show that $A(A - 2I_2) = 3I_2$, for the given matrix A. Conclude that A is invertible, and find A^{-1}.

53. Prove the Cayley-Hamilton theorem for a 3×3 diagonal matrix D.

54. Prove the Cayley-Hamilton theorem for *any* $n \times n$ diagonal matrix D.

6.3 *REAL SYMMETRIC MATRICES*

In Theorems 6.3, 6.4, and 6.5 (pages 389, 393, and 395) we found that an $n \times n$ matrix A is surely similar to a diagonal matrix when it has n distinct eigenvalues. On the other hand, when a matrix A has any eigenvalue whose algebraic multiplicity (as a zero of its characteristic polynomial $p_A(\lambda) =$ det $(\lambda I - A)$) is greater than one, this matrix may, or may not, be similar to a diagonal matrix. In general A will be diagonalizable when there are "enough" linearly independent eigenvectors, as noted in Theorem 6.3.

There are, fortunately, a number of theorems which tell us that certain special types of matrices are always diagonalizable. In this section we look at one of these special classes, while in Section 6.6 we go into the idea in even greater detail. Also in this section all the matrices and eigenvalues will be real, whereas in Section 6.6 we discuss complex matrices and eigenvalues. Section 6.6 subsumes all that is done here for the *special* case of real symmetric matrices, and so it could be read instead of this section. However, since the real symmetric matrices are interesting and applicable in their own right, we take the time to consider them separately. See, for example, the next two sections. We make the following formal definition.

Definition   A real matrix A is (real) **symmetric** if $A = A^T$.

It should be fairly obvious that only a square, $n \times n$, matrix can be symmetric.

Example 1   The matrices

$$M = \begin{bmatrix} 7 & 2 & 1 \\ 2 & 7 & -1 \\ 1 & -1 & 4 \end{bmatrix}, N = \begin{bmatrix} -8 & 5 & 4 \\ 5 & 3 & 1 \\ 4 & 1 & 0 \end{bmatrix} \text{ and } K = \begin{bmatrix} 1 & -7 & 0 & -3 \\ -7 & 0 & 2 & -1 \\ 0 & 2 & -1 & 5 \\ -3 & -1 & 5 & 11 \end{bmatrix}$$

are real symmetric matrices because $M = M^T$, $N = N^T$, and $K = K^T$. On the other hand, the matrix

$$H = \begin{bmatrix} 1 & 2 & -4 \\ 0 & 3 & 5 \\ 1 & -2 & 3 \end{bmatrix}$$

is not a symmetric matrix, since $H^T \neq H$.   ∎

The following theorem is stated here without proof; it follows from Theorem 6.16 in Section 6.6 (page 443). See Exercise 35 for the 2×2 case.

> **Theorem 6.6**   Let A be any real symmetric matrix. Then there exists a real nonsingular matrix P such that $P^{-1}AP$ is a diagonal matrix. This implies that all of the eigenvalues of A are real.

According to this theorem, every real symmetric matrix A must have only real eigenvalues, since these are the diagonal entries of the diagonal real matrix $P^{-1}AP$. Therefore, the characteristic polynomial $p_A(\lambda) = \det(\lambda I - A)$, which is a polynomial with real coefficients, must have only real roots. To say this, however, is not to say that these real roots are rational numbers, nor that they are easy to compute. You should not be misled by the relative ease of calculation in the examples and exercises of this section. In the first place, to find real roots of real polynomials is often not an easy task, and in most cases requires the aid of computers. Furthermore, it may be necessary to approximate the eigenvalues of a given real symmetric matrix by other numerical methods. You are referred to other sources for a discussion of these methods.

Example 2   The matrix M of Example 1 is symmetric. Therefore, according to Theorem 6.6 it has real eigenvalues. Its characteristic equation is

$$\det(\lambda I - M) = \det \begin{bmatrix} \lambda - 7 & -2 & -1 \\ -2 & \lambda - 7 & 1 \\ -1 & 1 & \lambda - 4 \end{bmatrix} = \lambda^3 - 18\lambda^2 + 99\lambda - 162$$

$$= (\lambda - 3)(\lambda - 6)(\lambda - 9) = 0.$$

So M has the three distinct eigenvalues $\lambda_1 = 3$, $\lambda_2 = 6$, and $\lambda_3 = 9$. Each of the three eigenspaces corresponding to these eigenvalues is one-dimensional. According to Theorem 6.6 there is a real matrix P such that $P^{-1}MP$ is the diagonal matrix diag [3, 6, 9]. We find the matrix P, as we did in the previous section, by finding a basis for $\overline{\mathbb{R}^3}$ consisting of the linearly independent eigenvectors of M. These eigenvectors can be chosen as

$$\mathbf{p}_1 = \begin{bmatrix} -1 \\ 1 \\ 2 \end{bmatrix}, \qquad \mathbf{p}_2 = \begin{bmatrix} 1 \\ -1 \\ 1 \end{bmatrix}, \qquad \text{and} \qquad \mathbf{p}_3 = \begin{bmatrix} 1 \\ 1 \\ 0 \end{bmatrix}.$$

So the matrix

$$P = \begin{bmatrix} -1 & 1 & 1 \\ 1 & -1 & 1 \\ 2 & 1 & 0 \end{bmatrix}$$

is the desired matrix, as you can readily verify. A similar computation gives the matrix

$$Q = \begin{bmatrix} 1 & 1 & -3 \\ -4 & 2 & 1 \\ 7 & 1 & 1 \end{bmatrix}$$
$$\quad \mathbf{q}_1 \;\; \mathbf{q}_2 \;\; \mathbf{q}_3$$

as a matrix of eigenvectors of the real symmetric matrix N of Example 1 (verify this). ■

Note that for each of the two real symmetric matrices M and N the eigenvectors are mutually orthogonal. That is, for M

$$< \mathbf{p}_1, \mathbf{p}_2 > = < \mathbf{p}_1, \mathbf{p}_3 > = < \mathbf{p}_2, \mathbf{p}_3 > = 0$$

and for N

$$< \mathbf{q}_1, \mathbf{q}_2 > = < \mathbf{q}_1, \mathbf{q}_3 > = < \mathbf{q}_2, \mathbf{q}_3 > = 0.$$

This is not a coincidence. Indeed we prove the following theorem.

Theorem 6.7   If A is any real symmetric matrix, then eigenvectors corresponding to distinct eigenvalues of A are mutually orthogonal.

Proof   Let λ_1 and λ_2 be *distinct* $(\lambda_1 \neq \lambda_2)$ eigenvalues of a real *symmetric* matrix A and let $\mathbf{v}_1$ and $\mathbf{v}_2$ be the corresponding eigenvectors considered as $n \times 1$ matrices. Then the matrix

$$\mathbf{v}_1^T A \mathbf{v}_2$$

is also a real symmetric matrix, because it is 1×1. Therefore, since $A^T = A$, we have

$$\mathbf{v}_1^T A \mathbf{v}_2 = (\mathbf{v}_1^T A \mathbf{v}_2)^T = \mathbf{v}_2^T A^T \mathbf{v}_1 = \mathbf{v}_2^T A \mathbf{v}_1 \tag{6.8}$$

by the properties of the transpose. Also $A\mathbf{v}_1 = \lambda_1 \mathbf{v}_1$ and $A\mathbf{v}_2 = \lambda_2 \mathbf{v}_2$, so that equation (6.8) becomes

$$\mathbf{v}_1^T \lambda_2 \mathbf{v}_2 = \mathbf{v}_2^T \lambda_1 \mathbf{v}_1, \quad \text{or} \quad \lambda_2 \mathbf{v}_1^T \mathbf{v}_2 = \lambda_1 \mathbf{v}_2^T \mathbf{v}_1. \tag{6.9}$$

But the product $\mathbf{v}_1^T \mathbf{v}_2$ is also a 1×1 matrix, so it is symmetric; hence

$$\mathbf{v}_1^T \mathbf{v}_2 = (\mathbf{v}_1^T \mathbf{v}_2)^T = \mathbf{v}_2^T \mathbf{v}_1.$$

Substituting this in equation (6.9) gives us $\lambda_2 \mathbf{v}_1^T \mathbf{v}_2 = \lambda_1 \mathbf{v}_1^T \mathbf{v}_2$, which gives us the fact that

$$(\lambda_1 - \lambda_2)\mathbf{v}_1^T \mathbf{v}_2 = 0.$$

Since the eigenvalues were distinct $(\lambda_1 - \lambda_2) \neq 0$. Thus the 1×1 matrix $\mathbf{v}_1^T \mathbf{v}_2 = [0]$ or the inner product $< \mathbf{v}_1, \mathbf{v}_2 > = \mathbf{v}_1^T \mathbf{v}_2 = 0$. The two eigenvectors $\mathbf{v}_1$ and $\mathbf{v}_2$ are therefore orthogonal. This completes the proof of the theorem.   ❏

Recall from Chapter 5 that a matrix P is said to be *orthogonal* if, and only if, P is nonsingular and $P^{-1} = P^T$; that is, if, and only if, $P^T P = I$.

Definition   An $n \times n$ matrix A is said to be **orthogonally diagonalizable** if it is possible to find an orthogonal matrix P such that $P^T A P = P^{-1} A P = D$, a diagonal matrix.

Notice that the two matrices P and Q of Example 2 are *not* orthogonal matrices, even though their column vectors are mutually orthogonal. They can, however, easily be transformed into orthogonal matrices by *normalizing* each column vector; that is, by dividing each column vector by its *norm* (= length). Since these were bases for eigenspaces, that merely replaces each basis vector with a "unit" basis vector. Consider the process illustrated in the following example.

Example 3   Find *orthogonal* matrices P_1 and Q_1 which orthogonally diagonalize the matrices M and N, respectively, of Examples 1 and 2.

The vectors

$$\mathbf{u}_1 = \frac{\mathbf{p}_1}{\|\mathbf{p}_1\|} = \begin{bmatrix} -\dfrac{1}{\sqrt{6}} \\ \dfrac{1}{\sqrt{6}} \\ \dfrac{2}{\sqrt{6}} \end{bmatrix}, \mathbf{u}_2 = \frac{\mathbf{p}_2}{\|\mathbf{p}_2\|} = \begin{bmatrix} \dfrac{1}{\sqrt{3}} \\ -\dfrac{1}{\sqrt{3}} \\ \dfrac{1}{\sqrt{3}} \end{bmatrix}, \text{and } \mathbf{u}_3 = \frac{\mathbf{p}_3}{\|\mathbf{p}_3\|} = \begin{bmatrix} \dfrac{1}{\sqrt{2}} \\ \dfrac{1}{\sqrt{2}} \\ 0 \end{bmatrix}$$

are a set of orthonormal eigenvectors for the matrix M. Therefore, the matrix

$$P_1 = \begin{bmatrix} -\dfrac{1}{\sqrt{6}} & \dfrac{1}{\sqrt{3}} & \dfrac{1}{\sqrt{2}} \\ \dfrac{1}{\sqrt{6}} & -\dfrac{1}{\sqrt{3}} & \dfrac{1}{\sqrt{2}} \\ \dfrac{2}{\sqrt{6}} & \dfrac{1}{\sqrt{3}} & 0 \end{bmatrix}$$

is orthogonal and is such that $P_1^T A P_1$ is the diagonal matrix $D = \text{diag} [3, 6, 9]$ we had before (verify this).

Similarly, you should also verify that the matrix N is orthogonally diagonalized by the matrix

$$Q_1 = \begin{bmatrix} \dfrac{1}{\sqrt{66}} & \dfrac{1}{\sqrt{6}} & -\dfrac{3}{\sqrt{11}} \\ -\dfrac{4}{\sqrt{66}} & \dfrac{2}{\sqrt{6}} & \dfrac{1}{\sqrt{11}} \\ \dfrac{7}{\sqrt{66}} & \dfrac{1}{\sqrt{6}} & \dfrac{1}{\sqrt{11}} \end{bmatrix}. \quad \blacksquare$$

The two matrices described so far have both had $n(n = 3$ here) distinct eigenvectors. As you are aware, this is not always true for a given $n \times n$ matrix, not even for a real symmetric matrix. Nevertheless, because of Theorem 6.6 (page 402), it is always true that

> Any $n \times n$ real symmetric matrix A will be diagonalizable, in fact, orthogonally diagonalizable.

Therefore, every $n \times n$ real symmetric matrix A must have n linearly independent eigenvectors. The following theorem, whose proof we omit, assures us of this.

> **Theorem 6.8**   If an eigenvalue λ of an $n \times n$ real symmetric matrix A has algebraic multiplicity k, then the eigenspace of A corresponding to λ is k-dimensional. **That is, for a *real symmetric matrix,* the algebraic multiplicity of an eigenvalue is equal to its geometric multiplicity.**

Using Theorems 6.6–6.8, we arrive at the following algorithm for orthogonally diagonalizing a real symmetric matrix A:

Step 1   Find a basis of eigenvectors for each eigenspace of A.

Step 2   Apply the Gram-Schmidt process (Section 4.5), where necessary, to each of the basis vectors of step 1 to obtain an orthonormal basis for each eigenspace.

Step 3   Form the orthogonal matrix P whose columns are the orthonormal basis vectors obtained in step 2.

Example 4   Find a matrix P that orthogonally diagonalizes the matrix

$$A = \begin{bmatrix} -1 & 2 & 2 \\ 2 & -1 & 2 \\ 2 & 2 & -1 \end{bmatrix}.$$

The characteristic equation for the matrix A is

$$\det(\lambda I - A) = \det \begin{bmatrix} \lambda+1 & -2 & -2 \\ -2 & \lambda+1 & -2 \\ -2 & -2 & \lambda+1 \end{bmatrix} = (\lambda+3)^2(\lambda-3) = 0.$$

Therefore, A has two eigenspaces. One of these is a subspace S_1 of dimension two corresponding to the eigenvalue $\lambda_1 = -3$. It is spanned by the two linearly independent eigenvectors

$$\mathbf{u}_1 = \begin{bmatrix} -1 \\ 0 \\ 1 \end{bmatrix} \quad \text{and} \quad \mathbf{u}_2 = \begin{bmatrix} -1 \\ 1 \\ 0 \end{bmatrix},$$

as you can compute. The second eigenspace S_2 is one-dimensional and corresponds to the eigenvalue $\lambda_2 = 3$. It is spanned by the eigenvector

$$\mathbf{u}_3 = \begin{bmatrix} 1 \\ 1 \\ 1 \end{bmatrix}.$$

Note that $\mathbf{u}_3$ is orthogonal to both $\mathbf{u}_1$ and $\mathbf{u}_2$, as guaranteed by Theorem 6.7 (page 403), but that $<\mathbf{u}_1, \mathbf{u}_2> = 1$. Thus $\mathbf{u}_1$ and $\mathbf{u}_2$ are *not* orthogonal. By applying the Gram-Schmidt process we obtain the following new orthonormal basis for S_1:

$$\mathbf{v}_1 = \frac{\mathbf{u}_1}{\|\mathbf{u}_1\|} = \begin{bmatrix} -\dfrac{1}{\sqrt{2}} \\ 0 \\ \dfrac{1}{\sqrt{2}} \end{bmatrix} \quad \text{and} \quad \mathbf{v}_2 = \frac{\mathbf{u}_2 - <\mathbf{u}_2, \mathbf{v}_1> \mathbf{v}_1}{\|\mathbf{u}_2 - <\mathbf{u}_2, \mathbf{v}_1> \mathbf{v}_1\|} = \begin{bmatrix} -\dfrac{1}{\sqrt{6}} \\ \dfrac{2}{\sqrt{6}} \\ -\dfrac{1}{\sqrt{6}} \end{bmatrix}.$$

Then divide $\mathbf{u}_3$ by its norm $\|\mathbf{u}_3\|$ to obtain for S_2 the orthonormal basis

$$\mathbf{v}_3 = \frac{\mathbf{u}_3}{\|\mathbf{u}_3\|} = \begin{bmatrix} \dfrac{1}{\sqrt{3}} \\ \dfrac{1}{\sqrt{3}} \\ \dfrac{1}{\sqrt{3}} \end{bmatrix}.$$

Therefore, the orthogonal matrix

$$P = \begin{bmatrix} -\dfrac{1}{\sqrt{2}} & -\dfrac{1}{\sqrt{6}} & \dfrac{1}{\sqrt{3}} \\ 0 & \dfrac{2}{\sqrt{6}} & \dfrac{1}{\sqrt{3}} \\ \dfrac{1}{\sqrt{2}} & -\dfrac{1}{\sqrt{6}} & \dfrac{1}{\sqrt{3}} \end{bmatrix}$$

is such that

$$P^T A P = \text{diag} [-3, -3, 3] = \begin{bmatrix} -3 & 0 & 0 \\ 0 & -3 & 0 \\ 0 & 0 & 3 \end{bmatrix}. \quad \blacksquare$$

Vocabulary

The concepts of a real symmetric matrix, the diagonalization of such a matrix, and the nature of its eigenvalues were discussed in this section. Study these concepts carefully and, in particular, emphasize the algorithm used to orthogonally diagonalize a real symmetric matrix. The following terms should be studied until you feel confident that you understand them thoroughly.

real symmetric matrix A
diagonalization of a real symmetric matrix

orthogonal matrix
real eigenvalues
orthogonally diagonalize

Exercises 6.3

In Exercises 1–12, find the eigenvalues, and then the corresponding eigenspaces, for each of the given matrices.

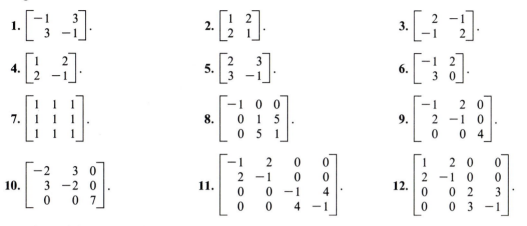

1. $\begin{bmatrix} -1 & 3 \\ 3 & -1 \end{bmatrix}$.

2. $\begin{bmatrix} 1 & 2 \\ 2 & 1 \end{bmatrix}$.

3. $\begin{bmatrix} 2 & -1 \\ -1 & 2 \end{bmatrix}$.

4. $\begin{bmatrix} 1 & 2 \\ 2 & -1 \end{bmatrix}$.

5. $\begin{bmatrix} 2 & 3 \\ 3 & -1 \end{bmatrix}$.

6. $\begin{bmatrix} -1 & 2 \\ 3 & 0 \end{bmatrix}$.

7. $\begin{bmatrix} 1 & 1 & 1 \\ 1 & 1 & 1 \\ 1 & 1 & 1 \end{bmatrix}$.

8. $\begin{bmatrix} -1 & 0 & 0 \\ 0 & 1 & 5 \\ 0 & 5 & 1 \end{bmatrix}$.

9. $\begin{bmatrix} -1 & 2 & 0 \\ 2 & -1 & 0 \\ 0 & 0 & 4 \end{bmatrix}$.

10. $\begin{bmatrix} -2 & 3 & 0 \\ 3 & -2 & 0 \\ 0 & 0 & 7 \end{bmatrix}$.

11. $\begin{bmatrix} -1 & 2 & 0 & 0 \\ 2 & -1 & 0 & 0 \\ 0 & 0 & -1 & 4 \\ 0 & 0 & 4 & -1 \end{bmatrix}$.

12. $\begin{bmatrix} 1 & 2 & 0 & 0 \\ 2 & -1 & 0 & 0 \\ 0 & 0 & 2 & 3 \\ 0 & 0 & 3 & -1 \end{bmatrix}$.

13. Find the eigenvalues of the real symmetric matrix

$$A = \begin{bmatrix} a & b \\ b & a \end{bmatrix}, \quad \text{where } b \neq 0.$$

14. Find an orthogonal matrix P so that $P^T A P$ is a diagonal matrix, for the matrix A of Exercise 13.

15. Find an orthogonal matrix P so that $P^T A P$ is a diagonal matrix, where A is the matrix of Exercise 1.

16. Find an orthogonal matrix P so that $P^T A P$ is a diagonal matrix, where A is the matrix of Exercise 2.

17. Find an orthogonal matrix P so that $P^T A P$ is a diagonal matrix, where A is the matrix of Exercise 3.

18. Find an orthogonal matrix P so that $P^T A P$ is a diagonal matrix, for the matrix of Exercise 4.

19. Find an orthogonal matrix P so that $P^T A P$ is a diagonal matrix, where A is the matrix of Exercise 5.

20. Show that an *orthogonal* matrix P so that $P^T A P$ is a diagonal matrix, where A is the matrix of Exercise 6, doesn't exist. Find a *nonsingular* matrix S so that $S^{-1} A S$ is a diagonal matrix.

21. Find an orthogonal matrix P so that $P^T A P$ is a diagonal matrix, where A is the matrix of Exercise 7.

22. Find an orthogonal matrix P so that $P^T A P$ is a diagonal matrix, where A is the matrix of Exercise 8.

23. Find an orthogonal matrix P so that $P^T A P$ is a diagonal matrix, where A is the matrix of Exercise 9.

24. Find an orthogonal matrix P so that $P^T A P$ is a diagonal matrix, where A is the matrix of Exercise 10.

25. Find an orthogonal matrix P so that P^TAP is a diagonal matrix, where A is the matrix of Exercise 11.

26. Find an orthogonal matrix P so that P^TAP is a diagonal matrix, where A is the matrix of Exercise 12.

27. Show that if A is any $n \times n$ real matrix, then the matrices A^TA, AA^T, and $A + A^T$ are all real *symmetric* matrices.

28. A real matrix A is called **skew-symmetric** if $A^T = -A$. Write an example of a 3×3 skew-symmetric matrix.

29. Prove that the diagonal entries of any skew-symmetric (see Exercise 28) matrix are always zero.

30. Show that for any $n \times n$ real matrix A, $A - A^T$ is skew-symmetric.

31. Combine the results of Exercise 27 and 30 to show that every $n \times n$ real matrix A is the sum of a symmetric and a skew-symmetric matrix.

32. Show that the eigenvalues of a 2×2 real skew-symmetric matrix cannot be real numbers.

33. Use the results of Exercise 30 in Section 6.2 to find A^{10} for the matrix A of Exercise 3.

34. Use the results of Exercise 30 in Section 6.2 to find A^{10} for the matrix A of Exercise 7.

---

35. Verify Theorem 6.6 (page 402) for the case of 2×2 real symmetric matrices. Note that Exercise 14 is a special case.

36. Two $n \times n$ matrices A and B are called **orthogonally similar** if there exists an orthogonal matrix P so that $B = P^TAP$. Why is every real symmetric matrix orthogonally similar to a diagonal matrix?

37. Show that if A and B are orthogonally similar (see Exercise 36), but not necessarily diagonal, matrices then A is symmetric if, and only if, B is symmetric.

38. If you have a computer available with the appropriate software, use them to determine if the matrix K of Exercise 58 of Section 6.1 is similar to a diagonal matrix.

39. If you have a computer available with the appropriate software, use them to determine the diagonal matrix to which the following real symmetric matrix is similar.

$$M = \begin{bmatrix} 1 & 2 & 4 & 0 & 3 \\ 2 & 4 & 0 & 3 & 1 \\ 4 & 0 & 3 & 1 & -1 \\ 0 & 3 & 1 & -1 & -2 \\ 3 & 1 & -1 & -2 & 1 \end{bmatrix}.$$

40. If you have a computer available with the appropriate software, use them to determine the diagonal matrix to which the following real symmetric matrix is similar.

$$K = \begin{bmatrix} 3 & 1 & -1 & -2 & 3 & -3 \\ 1 & -1 & -2 & 3 & -3 & 1 \\ -1 & -2 & 3 & -3 & 1 & -4 \\ -2 & 3 & -3 & 1 & -4 & -2 \\ 3 & -3 & 1 & -4 & -2 & 2 \\ -3 & 1 & -4 & -2 & 2 & 0 \end{bmatrix}.$$

41. If you have a computer available with the appropriate software, use them to determine bases for the eigenspaces of the symmetric matrix M in Exercise 39.

42. If you have a computer available with the appropriate software, use them to determine bases for the eigenspaces of the symmetric matrix K in Exercise 40.

6.4 QUADRATIC FORMS, CONICS, AND QUADRIC SURFACES

Real symmetric matrices can be applied quite naturally to the study of analytic geometry and, thereby, to many mathematical models. In this section we consider such an application. We study quadratic forms and their two- and three-dimensional graphs. These are the classical curves called *conic sections* and the three-dimensional surfaces called *quadric surfaces*.

Conic Sections

The most general *linear equation in two variables x and y* is, of course,

$$ax + by + c = 0.$$

Its graph is a line in 2-space. The most general *quadratic equation in two variables* has the form

$$ax^2 + 2bxy + cy^2 + dx + ey + f = 0 \qquad \text{(6.10)}$$

where at least one of the real numbers a, b, c is not zero. The reason for using $2b$ in (6.10) rather than simply a single letter, say β, will appear shortly. It loses no generality; our b is just $\beta/2$. However, $2b$ will shortly be seen to be more convenient.

> **Definition**   The expression
> $$ax^2 + 2bxy + cy^2 \qquad \text{(6.11)}$$
> is called the **quadratic form** associated with the general quadratic equation of (6.10).

Example 1   Consider the quadratic equation

$$3x^2 + 5xy - 2y^2 - 7x + 8y - 3 = 0.$$

This is in the form (6.10) with $a = 3$, $b = 5/2$, $c = -2$, $d = -7$, $e = 8$, and $f = -3$. The *quadratic form* associated with it is the expression

$$3x^2 + 5xy - 2y^2. \quad \blacksquare$$

Example 2   Some quadratic equations and their associated quadratic forms are the following.

Quadratic Equation	Quadratic Form
$x^2 + 3xy - y^2 - 4x + 4y - 6 = 0$	$x^2 + 3xy - y^2$
$x^2 + y^2 = 9$	$x^2 + y^2$
$xy - 6 = 0$	xy

$\blacksquare$

In plane analytic geometry it is shown that the graph of the general quadratic equation is one of the four conic sections: a circle, an ellipse, a parabola, or a hyperbola, or else one of the degenerate cases of these conics.

Example 3   The graph of the quadratic equation $x^2 + xy + y^2 - 1 = 0$ is the ellipse sketched in Figure 6.3. Note that this ellipse has its axes rotated about the origin 45° from the usual position of the coordinate axes. Its equation does not have one of the standard forms for the equation of an ellipse with its center at the origin and its axes along the coordinate axes, namely,

$$\frac{x^2}{u^2} + \frac{y^2}{v^2} = 1. \quad \blacksquare$$

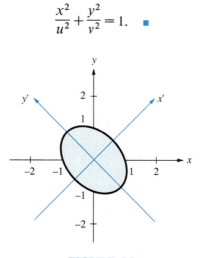

FIGURE 6.3

We develop a method here, using real symmetric matrices, whereby the rotation of axes can be performed. To do this we represent the general quadratic equation of (6.10) in matrix notation as follows: Given the equation $ax^2 + 2bxy + cy^2 + dx + ey + f = 0$, let

$$X = \begin{bmatrix} x \\ y \end{bmatrix}, \quad A = \begin{bmatrix} a & b \\ b & c \end{bmatrix}, \quad N = [d, e], \quad \text{and} \quad f = [f]. \quad \textbf{(6.12)}$$

Then equation (6.10) becomes (verify this)

$$X^T A X + N X + f = 0 \quad \textbf{(6.13)}$$

and the associated quadratic form $ax^2 + 2bxy + cy^2$ has the matrix expression

$$X^T A X.$$

> The symmetric matrix A is called the **matrix of the quadratic form.**

Example 4   The ellipse of Example 3 is represented by the matrix equation $X^T A X - 1 = 0$, where

$$A = \begin{bmatrix} 1 & \frac{1}{2} \\ \frac{1}{2} & 1 \end{bmatrix}, \quad \text{and} \quad 1 \text{ is the } 1 \times 1 \text{ matrix } [1]. \quad \blacksquare$$

We adopt the convention that 1×1 matrices $[a]$ will not be distinguished from their single entry a. To rotate the axes and eliminate the xy term from the quadratic equation $x^2 + xy + y^2 - 1 = 0$, we proceed as follows: Consider equation (6.13).

Step 1 Find a matrix P that orthogonally diagonalizes the real symmetric matrix A.

Step 2 Interchange the columns of P if necessary, so that det $P = 1$, assuring that P represents a rotation of the plane. (See Example 1 of Section 5.6 and Exercises 22–26 of that section.)

Step 3 Introduce a transformation of coordinates

$$X = PX', \quad \text{where } X' = \begin{bmatrix} x' \\ y' \end{bmatrix}. \tag{6.14}$$

Step 4 Substitute (6.14) into (6.13) and obtain the transformed equation

$$(PX')^T A(PX') + N(PX') + f = 0, \text{ or } X'^T P^T A P X' + N(PX') + f = 0.$$

Since P orthogonally diagonalizes A, we get $P^T A P = D = \begin{bmatrix} \lambda_1 & 0 \\ 0 & \lambda_2 \end{bmatrix}$, where λ_1 and λ_2 are the eigenvalues of A. Thus the quadratic equation becomes

$$X'^T D X' + N(PX') + f = 0,$$

the scalar form of which is the equation

$$\lambda_1 x'^2 + \lambda_2 y'^2 + d' x' + e' y' + f = 0,$$

which contains no *cross-product* term $x' y'$.

Note that this is a new linear algebra proof that the graph of a quadratic equation is a conic section. We have discussed the main ideas contained in the following theorem and how its proof should go.

Theorem 6.9 Let the matrix form of the general quadratic equation in two variables be

$$X^T A X + K X + f = 0.$$

Let the matrix P orthogonally diagonalize A, with det $P = 1$. Then the substitution $X = PX'$ rotates the axes so that the equation of the rotated conic contains no $x' y'$ term and is

$$X'^T P^T A P X' + K(PX') + f = 0, \quad \text{or} \quad X'^T D X' + K' X' + f = 0.$$

Example 5 Rotate the axes to eliminate the xy term from the equation $x^2 + xy + y^2 - 1 = 0$ of the ellipse of Example 3. From Example 4 we have the equation $X^T A X = 1$, where $A = \begin{bmatrix} 1 & \frac{1}{2} \\ \frac{1}{2} & 1 \end{bmatrix}$. Since the characteristic equation of A is $\det(\lambda I - A) = \lambda^2 - 2\lambda + \frac{3}{4} = (\lambda - \frac{3}{2})(\lambda - \frac{1}{2}) = 0$, we consider the two eigenvalues $\lambda_1 = \frac{3}{2}$ and $\lambda_2 = \frac{1}{2}$. Associated orthonormal eigenvectors are, respectively,

$$\mathbf{p}_1 = \begin{bmatrix} \dfrac{1}{\sqrt{2}} \\ \dfrac{1}{\sqrt{2}} \end{bmatrix} \quad \text{and} \quad \mathbf{p}_2 = \begin{bmatrix} -\dfrac{1}{\sqrt{2}} \\ \dfrac{1}{\sqrt{2}} \end{bmatrix}.$$

So that

$$P = \begin{bmatrix} \dfrac{1}{\sqrt{2}} & -\dfrac{1}{\sqrt{2}} \\ \dfrac{1}{\sqrt{2}} & \dfrac{1}{\sqrt{2}} \end{bmatrix}$$

is the matrix that orthogonally diagonalizes A. Therefore, $D = P^T A P = \begin{bmatrix} \frac{3}{2} & 0 \\ 0 & \frac{1}{2} \end{bmatrix}$. Since $\det P = 1$, this is the desired rotation matrix, so we substitute $X = PX'$ into the equation $X^T A X = 1$ to get $X'^T (P^T A P) X' = 1$, or $X'^T D X' = 1$, or $\frac{3}{2} x'^2 + \frac{1}{2} y'^2 = 1$. In standard form this is

$$\frac{x'^2}{\dfrac{2}{3}} + \frac{y'^2}{2} = 1.$$

This is the equation of an ellipse with major axis of length $2\sqrt{2}$ along the y' axis, and the minor axis of length $2\sqrt{\frac{2}{3}} = \frac{2\sqrt{6}}{3}$ along the x' axis. Refer once more to Figure 6.3.

Note that this particular matrix P and, in all cases, the matrix P has the form

$$\begin{bmatrix} \cos\phi & -\sin\phi \\ \sin\phi & \cos\phi \end{bmatrix}. \tag{6.15}$$

(Refer again to Section 5.6. See particularly Exercise 22.) ∎

Example 6 Describe and sketch the graph of the conic section whose matrix equation is

$$x^2 + 2xy + y^2 - x + y = 0. \tag{6.16}$$

The associated quadratic form $x^2 + 2xy + y^2$ determines the symmetric matrix

$$A = \begin{bmatrix} 1 & 1 \\ 1 & 1 \end{bmatrix}.$$

The matrix form of equation (6.16) is therefore

$$X^T A X + [-1, 1]X = 0.$$

The matrix A has eigenvalues $\lambda_1 = 0$ and $\lambda_2 = 2$, with associated orthonormal eigenvectors

$$\mathbf{p}_1 = \begin{bmatrix} -\dfrac{1}{\sqrt{2}} \\ \dfrac{1}{\sqrt{2}} \end{bmatrix} \quad \text{and} \quad \mathbf{p}_2 = \begin{bmatrix} \dfrac{1}{\sqrt{2}} \\ \dfrac{1}{\sqrt{2}} \end{bmatrix}.$$

Since

$$\det \begin{bmatrix} -\dfrac{1}{\sqrt{2}} & \dfrac{1}{\sqrt{2}} \\ \dfrac{1}{\sqrt{2}} & \dfrac{1}{\sqrt{2}} \end{bmatrix} = -1,$$

we interchange the columns $\mathbf{p}_1$ and $\mathbf{p}_2$, so that

$$P = \begin{bmatrix} \dfrac{1}{\sqrt{2}} & -\dfrac{1}{\sqrt{2}} \\ \dfrac{1}{\sqrt{2}} & \dfrac{1}{\sqrt{2}} \end{bmatrix} \quad \det P = 1, \quad \text{and} \quad P^T A P = D = \begin{bmatrix} 2 & 0 \\ 0 & 0 \end{bmatrix}.$$

Performing the linear transformation $X = PX'$ we obtain the matrix form of (6.16):

$$X'^T D X' + [-1, 1]PX' = [x', y'] \begin{bmatrix} 2 & 0 \\ 0 & 0 \end{bmatrix} \begin{bmatrix} x' \\ y' \end{bmatrix}$$

$$+ [-1, 1] \begin{bmatrix} \dfrac{1}{\sqrt{2}} & -\dfrac{1}{\sqrt{2}} \\ \dfrac{1}{\sqrt{2}} & \dfrac{1}{\sqrt{2}} \end{bmatrix} \begin{bmatrix} x' \\ y' \end{bmatrix} = 0.$$

This is

$$2x'^2 + \frac{2}{\sqrt{2}} y' = 0. \tag{6.17}$$

Equation (6.17) yields the standard form of the equation of a parabola, namely,

$$x'^2 = -\frac{1}{\sqrt{2}} y' \qquad \text{or} \qquad y' = -\sqrt{2}\,(x')^2.$$

The vertex of this parabola is at the origin and it opens along the negative (rotated) y' axis. Since the transformation matrix P again has the form (6.15), we obtain $\sin \phi = 1/\sqrt{2}$ and $\cos \phi = 1/\sqrt{2}$. Thus $\phi = 45°$. The parabola is sketched in Figure 6.4. ■

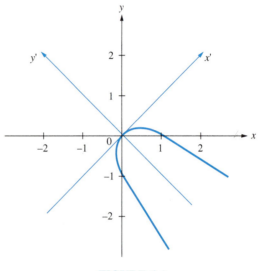

FIGURE 6.4

Sometimes it is necessary to translate the origin to a different point before the equation is actually in the standard form for a conic.

> A translation of the origin is, of course, not a linear transformation in 2-space $\mathbb{R}^2$.

(Why is this true? You may wish to refer to Project 7 at the end of Chapter 5.) The translation can, however, be either done *prior* to the rotation which *is* a linear transformation of $\mathbb{R}^2$, and the new plane considered as $\mathbb{R}^2$, or it can be done *after* a rotation in $\mathbb{R}^2$, as we do in the following example.

Example 7   Identify the conic whose equation is $2x^2 - 4xy - y^2 - 6x - 8y = -13$. The matrix A of the associated quadratic form is

$$A = \begin{bmatrix} 2 & -2 \\ -2 & -1 \end{bmatrix}.$$

You can verify that it has eigenvalues $\lambda_1 = 3$ and $\lambda_2 = -2$, with corresponding associated orthonormal eigenvectors

$$\mathbf{p}_1 = \begin{bmatrix} \dfrac{2}{\sqrt{5}} \\ -\dfrac{1}{\sqrt{5}} \end{bmatrix} \quad \text{and} \quad \mathbf{p}_2 = \begin{bmatrix} \dfrac{1}{\sqrt{5}} \\ \dfrac{2}{\sqrt{5}} \end{bmatrix}.$$

The matrix P that orthogonally diagonalizes A is

$$P = \begin{bmatrix} \dfrac{2}{\sqrt{5}} & \dfrac{1}{\sqrt{5}} \\ -\dfrac{1}{\sqrt{5}} & \dfrac{2}{\sqrt{5}} \end{bmatrix}, \quad \text{so } P^T A P = \begin{bmatrix} 3 & 0 \\ 0 & -2 \end{bmatrix}.$$

Therefore, the *rotated* form of the given equation is

$$X'^T P^T A P X' + K P X' = -13, \text{ or } 3x' - 2y'^2 - \frac{4}{\sqrt{5}}x' - \frac{22}{\sqrt{5}}y' = -13. \quad \blacksquare$$

This equation is not yet in standard form. It is easier to recognize the conic and sketch it if the equation is put in standard form, so we translate the axes to a new origin. To do this, complete the square (or see Project 7 in Chapter 5) as follows:

$$3\left[x'^2 - \frac{4}{3\sqrt{5}}x' \right] - 2\left[y'^2 + \frac{11}{\sqrt{5}}y' \right] = -13.$$

So

$$3\left[x'^2 - \frac{4}{3\sqrt{5}}x' + \left[-\frac{2}{3\sqrt{5}} \right]^2 \right] - 2\left[y'^2 + \frac{11}{\sqrt{5}}y' + \left[\frac{11}{2\sqrt{5}} \right]^2 \right]$$

$$= -13 + 3\left[-\frac{2}{3\sqrt{5}} \right]^2 - 2\left[\frac{11}{2\sqrt{5}} \right]^2.$$

This equation simplifies to

$$3\left[x' - \frac{2}{3\sqrt{5}} \right]^2 - 2\left[y' + \frac{11}{2\sqrt{5}} \right]^2 = -\frac{149}{6},$$

which is equivalent to

$$\frac{\left[y' + \dfrac{11}{2\sqrt{5}} \right]^2}{149/12} - \frac{\left[x' - \dfrac{2}{3\sqrt{5}} \right]^2}{149/18} = 1.$$

This is the *standard form* for the equation of a hyperbola with center at $\left[\dfrac{2}{3\sqrt{5}}, -\dfrac{11}{2\sqrt{5}}\right]$ in the $x'y'$ coordinate system, whose transverse axis lies along the y' axis, and whose conjugate axis lies along the x' axis. The x' and y' axes are rotated through an angle ϕ such that $\cos\phi = 2/\sqrt{5}$ and $\sin\phi = -1/\sqrt{5}$ [see (6.15)], that is, approximately $-26.57°$.

Quadric Surfaces

Let us now extend these same ideas to $\mathbb{R}^3$. After discussing the three-dimensional manifestation, we give an indication as to how these same concepts are extended to $\mathbb{R}^n$ in general.

Of course, the general linear equation in three variables

$$ax + by + cz = d$$

is the equation of a *plane* in $\mathbb{R}^3$. Consider the *general quadratic equation* in three variables:

$$
\begin{aligned}
&a_{11}x_1^2 + a_{22}x_2^2 + a_{33}x_3^2 + 2a_{12}x_1x_2 + 2a_{13}x_1x_3 \\
&\quad + 2a_{23}x_2x_3 + bx_1 + cx_2 + dx_3 + f = 0,
\end{aligned}
\tag{6.18}
$$

where not all of the coefficients a_{ij} are zero. The *associated quadratic form* is

$$a_{11}x_1^2 + a_{22}x_2^2 + a_{33}x_3^2 + 2a_{12}x_1x_2 + 2a_{13}x_1x_3 + 2a_{23}x_2x_3.$$

The matrix expression of this quadratic form is

$$
X^TAX = [x_1, x_2, x_3]\underbrace{\begin{bmatrix} a_{11} & a_{12} & a_{13} \\ a_{12} & a_{22} & a_{23} \\ a_{13} & a_{23} & a_{33} \end{bmatrix}}_{A}\underbrace{\begin{bmatrix} x_1 \\ x_2 \\ x_3 \end{bmatrix}}_{X}.
\tag{6.19}
$$
$$\underbrace{}_{X^T}$$

The symmetric matrix A is the **matrix of the quadratic form** just as in the two-variable case considered above. It is clear that the quadratic (6.18) can be written also in matrix form as

$$X^TAX + KX + f = 0, \qquad \text{where } K = (b, c, d).$$

Example 8   Given the quadratic equation

$$3x_1^2 + 7x_3^2 + 2x_1x_2 - 3x_1x_3 + 4x_2x_3 - 3x_1 = 4.$$

The matrix form (6.19) is

$$
[x_1, x_2, x_3]\begin{bmatrix} 3 & 1 & -\dfrac{3}{2} \\ 1 & 0 & 2 \\ -\dfrac{3}{2} & 2 & 7 \end{bmatrix}\begin{bmatrix} x_1 \\ x_2 \\ x_3 \end{bmatrix} + [-3, 0, 0]\begin{bmatrix} x_1 \\ x_2 \\ x_3 \end{bmatrix} - 4 = 0. \quad \blacksquare
\tag{6.20}
$$

The graphs of quadratic equations in *three* variables are called **quadrics** or **quadric surfaces.** In Figures 6.5 through 6.10 we have sketched the usual nondegenerate quadrics and have given one of the standard forms of their equations in the three variables x, y, and z.

Each of these quadrics could, of course, be oriented along different axes with corresponding variations in their equations. The presence of one or more *cross product terms* $x_i x_j$ (don't confuse this with the cross product of two vectors) of a nondegenerate quadric indicates that the quadric has been

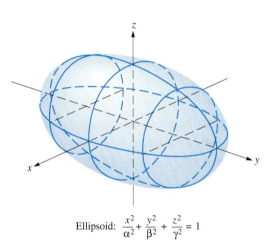

Ellipsoid: $\dfrac{x^2}{\alpha^2} + \dfrac{y^2}{\beta^2} + \dfrac{z^2}{\gamma^2} = 1$

FIGURE 6.5

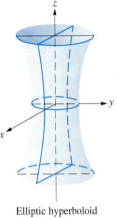

Elliptic hyperboloid
of one sheet:

$$\frac{x^2}{\alpha^2} + \frac{y^2}{\beta^2} - \frac{z^2}{\gamma^2} = 1$$

FIGURE 6.6

Elliptic cone:

$$\frac{x^2}{\alpha^2} + \frac{y^2}{\beta^2} = \frac{z^2}{\gamma^2}$$

FIGURE 6.7

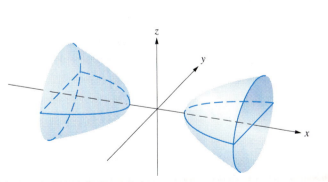

Elliptic hyperboloid of two sheets:

$$\frac{x^2}{\alpha^2} - \frac{y^2}{\beta^2} - \frac{z^2}{\gamma^2} = 1$$

FIGURE 6.8

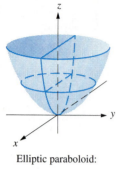

Elliptic paraboloid:

$$\frac{x^2}{\alpha^2} + \frac{y^2}{\beta^2} = z$$

FIGURE 6.9

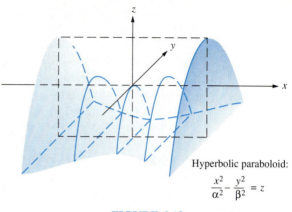

Hyperbolic paraboloid:

$$\frac{x^2}{\alpha^2} - \frac{y^2}{\beta^2} = z$$

FIGURE 6.10

rotated out of its standard position with respect to the standard axes, while the presence of both x_i^2 and x_i terms generally indicates that the "center" of the quadric surface is located at some point other than the origin. This would be a translation of the standard form. Remember that our reference point is the standard basis for $\mathbb{R}^3$. (And our figures use the right-handed coordinate system.)

Example 9 Name and sketch the quadric surface whose matrix equation is

$$X^T A X = [x, y, z] \begin{bmatrix} 4 & 0 & 0 \\ 0 & -9 & 0 \\ 0 & 0 & 8 \end{bmatrix} \begin{bmatrix} x \\ y \\ z \end{bmatrix} = 72.$$

By matrix multiplication we arrive at the scalar quadratic equation $4x^2 - 9y^2 + 8z^2 = 72$. Upon dividing by 72, we arrive at the standard form

$$\frac{x^2}{18} - \frac{y^2}{8} + \frac{z^2}{9} = 1$$

which is one of the standard forms for the equation of an elliptic hyperboloid of one sheet. This one has its axis oriented along the y axis, and its graph is sketched in Figure 6.11. ∎

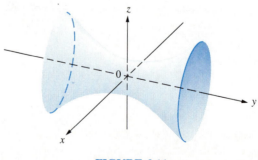

FIGURE 6.11

Example 10   Name and sketch the graph of the quadric surface whose equation is

$$x^2 - 4xz + 2y^2 + 4z^2 - 3x + 2y - 4z = 5.$$

The matrix form for this equation is (verify this)

$$X^T A X + K X = [x, y, z] \begin{bmatrix} 1 & 0 & -2 \\ 0 & 2 & 0 \\ -2 & 0 & 4 \end{bmatrix} \begin{bmatrix} x \\ y \\ z \end{bmatrix} + [-3, 2, -4] \begin{bmatrix} x \\ y \\ z \end{bmatrix} = 5.$$

We find that the eigenvalues of the real symmetric matrix

$$A = \begin{bmatrix} 1 & 0 & -2 \\ 0 & 2 & 0 \\ -2 & 0 & 4 \end{bmatrix}$$

are $\lambda_1 = 0$, $\lambda_2 = 2$, and $\lambda_3 = 5$. The corresponding orthonormal eigenvectors are

$$\mathbf{p}_1 = \begin{bmatrix} \frac{2}{\sqrt{5}} \\ 0 \\ \frac{1}{\sqrt{5}} \end{bmatrix}, \qquad \mathbf{p}_2 = \begin{bmatrix} 0 \\ 1 \\ 0 \end{bmatrix}, \qquad \text{and} \qquad \mathbf{p}_3 = \begin{bmatrix} -\frac{1}{\sqrt{5}} \\ 0 \\ \frac{2}{\sqrt{5}} \end{bmatrix}.$$

The determinant of the matrix

$$P = \begin{bmatrix} \frac{2}{\sqrt{5}} & 0 & -\frac{1}{\sqrt{5}} \\ 0 & 1 & 0 \\ \frac{1}{\sqrt{5}} & 0 & \frac{2}{\sqrt{5}} \end{bmatrix}$$

is 1; hence a rotation is involved. Furthermore $P^T A P$ is a diagonal matrix:

$$D = P^T A P = \begin{bmatrix} 0 & 0 & 0 \\ 0 & 2 & 0 \\ 0 & 0 & 5 \end{bmatrix}.$$

Therefore, making the orthogonal coordinate transformation $PX' = X$ we effect a change of basis for $\mathbb{R}^3$ from the standard basis to the basis of eigenvectors $G = \{\mathbf{p}_1^T, \mathbf{p}_2^T, \mathbf{p}_3^T\}$. We then have the new transformed equation $X'^T P^T A P X' + K P X' = 5$ or

$$[x', y', z'] \begin{bmatrix} 0 & 0 & 0 \\ 0 & 2 & 0 \\ 0 & 0 & 5 \end{bmatrix} \begin{bmatrix} x' \\ y' \\ z' \end{bmatrix} + [-3, 2, -4] \begin{bmatrix} \frac{2}{\sqrt{5}} & 0 & -\frac{1}{\sqrt{5}} \\ 0 & 1 & 0 \\ \frac{1}{\sqrt{5}} & 0 & \frac{2}{\sqrt{5}} \end{bmatrix} \begin{bmatrix} x' \\ y' \\ z' \end{bmatrix} = 5.$$

Multiplying this out, we arrive at the equation

$$2y'^2 + 5z'^2 - 2\sqrt{5}x' + 2y' - \sqrt{5}z' = 5.$$

The presence of first-degree terms in y' and z' indicates that the center of this surface is not at the origin, so we must translate the axes. Remember that, unlike the rotation, this is not a linear transformation of $\mathbb{R}^3$. We do the translation by completing the square in the equation. Thus (verify this)

$$2\left[y'^2 + y' + \frac{1}{4}\right] + 5\left[z'^2 - \frac{1}{\sqrt{5}}z' + \frac{1}{20}\right] = 2\sqrt{5}x' + \frac{23}{4}.$$

$$2\left[y' + \frac{1}{2}\right]^2 + 5\left[z' - \frac{1}{2\sqrt{5}}\right]^2 = 2\sqrt{5}\left[x' + \frac{23}{8\sqrt{5}}\right].$$

This equation may also be written as

$$\frac{(y' + 1/2)^2}{\sqrt{5}} + \frac{(z' - 1/2\sqrt{5})^2}{2\sqrt{5}/5} - \frac{(x' + 23/8\sqrt{5})^2}{1}$$

which is the equation of an elliptic paraboloid, somewhat as in Figure 6.9, but oriented so that its axis of symmetry is the positive x'' axis (the translated x' axis). See Figure 6.12. ∎

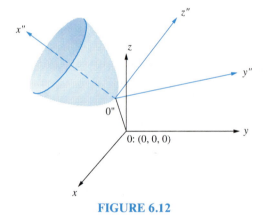

FIGURE 6.12

Although the problem of translating the axes to establish a new center introduces a slight complication, the results demonstrated in Examples 7 and 10 are three-dimensional illustrations of the following important geometric theorem. Its proof is left to you as Exercises 55 and 56. In doing so, note that it is a geometric extension of Theorem 6.6 (page 402).

Theorem 6.10. The Principal Axis Theorem   Let A be an $n \times n$ real symmetric matrix. The quadratic form $X^T A X$ is equivalent to the sum of squares

$$\lambda_1 y_1^2 + \lambda_2 y_2^2 + \cdots + \lambda_n y_n^2.$$

Here the λ_i are the eigenvalues of A and $X = PY$, where P is the orthogonal matrix that diagonalizes A.

Vocabulary

In this section real symmetric matrices were applied to the study of quadratic forms and their two- and three-dimensional graphs, giving rise to such classical curves as conic sections and quadric surfaces. Be sure that you digest these ideas and that you study the following terms until you feel confident that you understand them thoroughly.

quadratic form
conic sections
rotation of axes
names of the various quadric
   surfaces

names and standard forms of the
   conic sections
matrix of a quadratic form
quadric surfaces
standard form of equations

Exercises 6.4

In Exercises 1–18, write the quadratic form and its resulting symmetric matrix for each of the given quadratic equations.

1. $xy = 3$.

2. $8xy = 5$.

3. $7xy = -2$.

4. $2x^2 - 4y^2 = 7$.

5. $53x^2 - 72xy + 32y^2 = 80$.

6. $49x^2 + 64y^2 = 25$.

7. $16x^2 - 24xy + 9y^2 - 60x - 80y + 100 = 0$.

8. $64x^2 - 240xy + 225y^2 + 1020x - 544y = 0$.

9. $2x^2 - 3xy + 4y^2 - 7x + 2y + 7 = 0$.

10. $9x^2 + 4y^2 - 36x - 24y + 36 = 0$.

11. $x^2 + y^2 + z^2 = 16$.

12. $xy + yz = 9$.

13. $x^2 - y^2 - z^2 = 16$.

14. $9y^2 + z^2 = 9x$.

15. $4x^2 + 4y^2 + 4z^2 + 4xy + 4xz + 4yz = 7$.

16. $2xy - 6xz = 1$.

17. $2x^2 - 3y^2 + 6z^2 - 2xy + 4yz + 2xz = 18$.

18. $144x^2 + 100y^2 + 81z^2 - 216xz - 540x - 720z = 0$.

19. Express the quadratic equation in Exercise 1 in matrix form.

20. Express the quadratic equation in Exercise 2 in matrix form.

21. Express the quadratic equation in Exercise 3 in matrix form.

22. Express the quadratic equation in Exercise 4 in matrix form.

23. Express the quadratic equation in Exercise 5 in matrix form.

24. Express the quadratic equation in Exercise 6 in matrix form.

25. Express the quadratic equation in Exercise 7 in matrix form.

26. Express the quadratic equation in Exercise 8 in matrix form.

27. Express the quadratic equation in Exercise 9 in matrix form.

28. Express the quadratic equation in Exercise 10 in matrix form.

29. Express the quadratic equation in Exercise 11 in matrix form.

30. Express the quadratic equation in Exercise 12 in matrix form.

31. Express the quadratic equation in Exercise 13 in matrix form.

32. Express the quadratic equation in Exercise 14 in matrix form.

33. Express the quadratic equation in Exercise 15 in matrix form.

34. Express the quadratic equation in Exercise 16 in matrix form.

35. Express the quadratic equation in Exercise 17 in matrix form.

36. Express the quadratic equation in Exercise 18 in matrix form.

37. Name the conic section that is the graph of the equation in Exercise 1. Sketch it.

38. Name the conic section that is the graph of the equation in Exercise 2. Sketch it.

39. Name the conic section that is the graph of the equation in Exercise 3. Sketch it.

40. Name the conic section that is the graph of the equation in Exercise 4. Sketch it.

41. Name the conic section that is the graph of the equation in Exercise 5. Sketch it.

42. Name the conic section that is the graph of the equation in Exercise 6. Sketch it.

43. Name the conic section that is the graph of the equation in Exercise 7. Sketch it.

44. Name the conic section that is the graph of the equation in Exercise 8. Sketch it.

45. Name the conic section that is the graph of the equation in Exercise 9. Sketch it.

46. Name the conic section that is the graph of the equation in Exercise 10. Sketch it.

47. Name the quadric surface that is the graph of the equation in Exercise 11. Discuss its graph.

48. Name the quadric surface that is the graph of the equation in Exercise 12. Discuss its graph.

49. Name the quadric surface that is the graph of the equation in Exercise 13. Discuss its graph.

50. Name the quadric surface that is the graph of the equation in Exercise 14. Discuss its graph.

51. Name the quadric surface that is the graph of the equation in Exercise 15. Discuss its graph.

52. Name the quadric surface that is the graph of the equation in Exercise 16. Discuss its graph.

53. Name the quadric surface that is the graph of the equation in Exercise 17. Discuss its graph.

54. Name the quadric surface that is the graph of the equation in Exercise 18. Discuss its graph.

55. Prove the principal axis theorem (Theorem 6.10, page 420) for $\mathbb{R}^3$.

56. Sketch the proof of the principal axis theorem (Theorem 6.10) for $\mathbb{R}^n$.

In Exercises 57–60, consider coefficients a, b, c, d, e, and f of the general quadratic equation in two variables of equation (6.10).

57. Show that for any rotation of the plane the discriminant $b^2 - ac$ is invariant. HINT: Remember that a rotation is accomplished by an orthogonal matrix of the form (6.15).

58. Show that the number $R = \det \begin{bmatrix} a & b & d \\ b & c & e \\ d & e & f \end{bmatrix}$ is invariant under any rotation of the plane.

59. Show that the discriminant $b^2 - ac$ is negative, zero, or positive when the conic is, respectively, an ellipse, parabola, or hyperbola.

60. Show that if the number R of Exercise 58 is zero, the conic degenerates into a point or into one or two straight lines; these are called *degenerate conics*.

It is possible that the general quadratic equation cannot be satisfied by any real numbers at all; for example, $x^2 + y^2 + z^2 + 1 = 0$. In this case the equation has no graph in $\mathbb{R}^2$ or $\mathbb{R}^3$, and is called an imaginary conic or an imaginary quadric surface. Determine whether the following equations represent a degenerate or an imaginary conic, and if degenerate, sketch its graph.

61. $x^2 - y^2 = 0$.

62. $x^2 + y^2 + 3 = 0$.

63. $9x^2 + 25y^2 = 0$.

64. $16x^2 - 9y^2 = 0$

65. $9x^2 + 12xy + 4y^2 = 52$.

66. $x^2 + 2y^2 + z^2 + 9 = 0$.

67. $x^2 - 2xy + y^2 = 0$.

68. $x^2 + 2xy + y^2 = 0$.

69. $4x^2 + y^2 + z^2 = 0$.

6.5 *EXTENDED APPLICATION* Markov Chains

In this section we look at another application of linear algebra. Our presentation requires only a minimal understanding of the concepts of probability, such as you would study in a college algebra, finite mathematics, or elementary probability course. The proofs of the theorems, however, require quite a bit more mathematics and are not given here.

Consider a finite sequence of events (or experiments) in which the outcome of a given event depends only upon the outcome of the event that immediately precedes it in the sequence, but not on the other events of the sequence. This gives rise to the study of what are called *Markov processes* or *chains*. We give more precise definitions after looking at an example.

Example 1 Suppose that a maze is constructed in the form illustrated in Figure 6.13. The maze consists of three chambers (numbered 1, 2, and 3 for convenience). Each is painted a different color, as indicated in the figure. The experimenter places a mouse into one of these chambers and then, at periodic intervals, observes where the mouse is. Since the mouse is not under constant observation, it is not possible to determine its exact movements. Therefore, the movements are stated as probabilities. Some notation is helpful at this point.

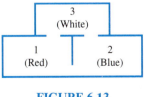

FIGURE 6.13

The situation or condition in a Markov process at which the experiment begins (here the chamber in which the mouse is first placed) is called the **initial state** of the process. If the initial state is chosen by a chance device that selects condition (chamber) j with a probability p_j^0, the initial probability distribution is given by a column vector

$$\mathbf{p}^0 = \begin{bmatrix} p_1^0 \\ p_2^0 \\ \vdots \\ p_n^0 \end{bmatrix}.$$

Then in our example, if the mouse is *always* placed initially in chamber 1, the vector $\mathbf{p}^0$ will be $\mathbf{p}^0 = (1, 0, 0)^T$. On the other hand, if the mouse had an equal chance of being placed in any one of the three chambers, the initial probability distribution would be

$$\mathbf{p}^0 = \begin{bmatrix} \dfrac{1}{3} \\ \dfrac{1}{3} \\ \dfrac{1}{3} \end{bmatrix}.$$

Now denote by p_{ij} the probability that the mouse will move *from* chamber j *to* chamber i. Because the mouse has an affinity for certain colors, the probability that it moves from chamber 2 to chamber 3, for instance, might be $p_{32} = \frac{1}{2}$, while the probability that it moves from chamber 2 to chamber 1 might be $p_{12} = \frac{1}{4}$. Suppose also that it remains in chamber 2 with probability $\frac{1}{4}$, so $p_{22} = \frac{1}{4}$. We can display all of these so-called **transition probabilities** in a matrix $A = (p_{ij})$, which is called the **transition matrix** *for the Markov chain.* This is not the same as the *change of basis matrix,* which was also called a transition matrix in Section 5.5. The context will rule out any confusion in this double use of the same name. A possible transition matrix for this particular example could be the following:

$$P = \begin{array}{c} \begin{array}{ccc} 1 & \; 2 & \; 3 \end{array} \\ \begin{bmatrix} 0 & \dfrac{1}{4} & \dfrac{1}{6} \\ \dfrac{1}{3} & \dfrac{1}{4} & \dfrac{1}{2} \\ \dfrac{2}{3} & \dfrac{1}{2} & \dfrac{1}{3} \end{bmatrix} \end{array}. \quad \blacksquare$$

Let us now make the promised formal definitions.

> **Definition** A **Markov chain** is a sequence of n events in which each event has m possible outcomes $a_1, a_2, \ldots, a_m$ called **states** and the probability p_{ij} that a particular state occurs depends only on the outcome of the preceding event.

We assume that these probabilities are nonnegative numbers p_{ij} between 0 and 1, which represent the probability that the outcome (state) a_i of a given event occurs provided that outcome a_j occurred on the preceding event. Thus $p_{1j} + p_{2j} + \cdots + p_{mj} = 1$, for each $j = 1, 2, \ldots, m$.

> **Definition** A column vector **p** is called a **probability vector** if it has nonnegative components whose sum is 1.

The vectors $\mathbf{p}^0$ in Example 1 are probability vectors, as are the columns of the matrix P in that example.

> **Definition** A matrix A is called the **transition matrix for a Markov chain** if it is square and its column vectors are probability vectors; that is, the entries in each column of A are *nonnegative real numbers whose sum is* 1.

The matrix P of Example 1 is an example.

Example 2 Determine which of the following are transition matrices for a Markov chain:

$$J = \begin{bmatrix} 0 & 1 \\ 1 & 0 \end{bmatrix}; \ B = \begin{bmatrix} \dfrac{1}{2} & \dfrac{1}{2} \\ \dfrac{1}{4} & \dfrac{3}{4} \end{bmatrix}; \ I_3; \ C = \begin{bmatrix} 0.1 & 0.3 & 0.6 \\ 0.7 & 0.4 & 0.2 \\ 0.2 & 0.3 & 0.2 \end{bmatrix}; \ D = \begin{bmatrix} 1 & 2 & 4 \\ 0 & 3 & 5 \\ 1 & 2 & 3 \end{bmatrix}.$$

Note that the column sum for each column in the matrices J, C, and I_3 is one, while for the matrices B and D it is not. Thus J, C, and I_3 are transition matrices for a Markov chain, and the matrices B and D are not. Is B^T a transition matrix for a Markov chain?

Let us return to Example 1 and the mouse. Denote by the number $p_j^{(n)}$ the probability that after n observations the mouse is in chamber j. Then the probability vector

$$\mathbf{p}^{(n)} = (p_1^{(n)}, p_2^{(n)}, p_3^{(n)})^T$$

represents the distribution for this situation. It can be shown that for $j = 1, 2, 3$ these probabilities satisfy the equation

$$p_j^{(n)} = p_{j1}\, p_1^{(n-1)} + p_{j2}\, p_2^{(n-1)} + p_{j3}\, p_3^{(n-1)}.$$

Thus the probability that the mouse is in chamber j after n steps is the sum of three terms composed of the probability of being in any of the three chambers after $n - 1$ steps multiplied by the probability of moving from that chamber to chamber j at the nth step. Note that these equations can be written in matrix form as

$$\mathbf{p}^{(n)} = P\mathbf{p}^{(n-1)}$$

where P is the transition matrix of the chain. (Verify this.) Let us iterate this process by letting n take on the values $1, 2, \ldots$. We get

$$\mathbf{p}^{(1)} = P\mathbf{p}^{(0)},$$
$$\mathbf{p}^{(2)} = P\mathbf{p}^{(1)} = PP\mathbf{p}^{(0)} = P^2\mathbf{p}^{(0)},$$
$$\mathbf{p}^{(3)} = P\mathbf{p}^{(2)} = P^3\mathbf{p}^{(0)}$$

etc. ∎

So for all Markov chains we have the following theorem.

> **Theorem 6.11** Let P be the transition matrix for a Markov chain. The probability distribution $\mathbf{p}^{(k)}$ after k steps is given by $\mathbf{p}^{(k)} = P^k\mathbf{p}^{(0)}$, where $\mathbf{p}^{(0)}$ is the initial condition.

Example 3 Let us refer back to the mouse of Example 1 and its maze in Figure 6.13. This time we assume that the mouse has an equal chance of being placed initially in any one of the three chambers. Therefore, $\mathbf{p}^{(0)} = (\frac{1}{3}, \frac{1}{3}, \frac{1}{3})^T$. Suppose further that the transition matrix for this chain is

$$P = \begin{bmatrix} 0.1 & 0.3 & 0.5 \\ 0.7 & 0.4 & 0.2 \\ 0.2 & 0.3 & 0.3 \end{bmatrix}.$$

Then, after three observations, the position of the mouse is given by $P^3\mathbf{p}^{(0)}$ as follows:

$$P^3\mathbf{p}^{(0)} = \begin{bmatrix} 0.1 & 0.3 & 0.5 \\ 0.7 & 0.4 & 0.2 \\ 0.2 & 0.3 & 0.3 \end{bmatrix} \begin{bmatrix} 0.1 & 0.3 & 0.5 \\ 0.7 & 0.4 & 0.2 \\ 0.2 & 0.3 & 0.3 \end{bmatrix} \begin{bmatrix} 0.1 & 0.3 & 0.5 \\ 0.7 & 0.4 & 0.2 \\ 0.2 & 0.3 & 0.3 \end{bmatrix} \begin{bmatrix} \frac{1}{3} \\ \frac{1}{3} \\ \frac{1}{3} \end{bmatrix}$$

$$= \begin{bmatrix} 0.294 & 0.294 & 0.298 \\ 0.438 & 0.436 & 0.428 \\ 0.268 & 0.270 & 0.274 \end{bmatrix} \begin{bmatrix} \frac{1}{3} \\ \frac{1}{3} \\ \frac{1}{3} \end{bmatrix} = \begin{bmatrix} 0.295 \\ 0.434 \\ 0.271 \end{bmatrix} = \mathbf{p}^{(3)}.$$

This last vector expresses the probability of the mouse's position in each chamber after three observations.

Suppose that we were to do the mouse-maze experiment by placing the mouse in the *first* chamber every time. In this case the initial state is given by the vector $\mathbf{p}^{(0)} = (1, 0, 0)^T$. Therefore, after three observations, the position of the mouse is given by

$$P^3 \mathbf{p}^{(0)} = \begin{bmatrix} 0.294 & 0.294 & 0.298 \\ 0.438 & 0.436 & 0.428 \\ 0.268 & 0.270 & 0.274 \end{bmatrix} \begin{bmatrix} 1 \\ 0 \\ 0 \end{bmatrix} = \begin{bmatrix} 0.294 \\ 0.438 \\ 0.268 \end{bmatrix} = \mathbf{p}^{(3)}.$$

Note that $\mathbf{p}^{(3)}$ is the first column of the matrix P^3. This is not a coincidence since the entries $p_{ij}^{(k)}$ in the matrix P^k of a Markov chain give the probabilities of passing from state a_j to a_i after k stages (for all i and j). ∎

Example 4   Consider the following transition matrix and some of its successive powers:

$$R = \begin{bmatrix} 0.5 & 0.7 \\ 0.5 & 0.3 \end{bmatrix}; \qquad R^2 = \begin{bmatrix} 0.6 & 0.56 \\ 0.4 & 0.44 \end{bmatrix};$$

$$R^3 = \begin{bmatrix} 0.58 & 0.588 \\ 0.42 & 0.412 \end{bmatrix}; \qquad R^4 = \begin{bmatrix} 0.584 & 0.5824 \\ 0.416 & 0.4176 \end{bmatrix}.$$

If we continue with the powers of R, we see that they approach the matrix

$$R' = \begin{bmatrix} 0.5867 & 0.5867 \\ 0.4167 & 0.4167 \end{bmatrix} = \begin{bmatrix} \dfrac{7}{12} & \dfrac{7}{12} \\ \dfrac{5}{12} & \dfrac{5}{12} \end{bmatrix} \qquad \text{(approximately).} \quad ∎$$

It is interesting to note that the vector $\mathbf{u} = \begin{bmatrix} \frac{7}{12} \\ \frac{5}{12} \end{bmatrix}$ is an *eigenvector* for the matrix R corresponding to the *eigenvalue* 1. That is, it satisfies $R\mathbf{u} = \mathbf{u}$. In the theory of Markov chains, such a vector is called a *fixed point of the matrix R*.

> **Definition**   A probability vector $\mathbf{u}$ is a **fixed point** of a given transition matrix A if, and only if, $A\mathbf{u} = \mathbf{u}$.

In Example 4 we noticed that, as n increased, the powers R^n of the matrix R there became more and more like a matrix W whose column vectors were equal to the fixed-point vector for R. This is not a coincidence; for certain transition matrices such is always the case. These are the so-called *regular transition matrices*.

> **Definition**   A transition matrix A of a Markov chain is called **regular** if some power A^n of A has only positive entries.

The matrix R of Example 4 is regular since $R = R^1$ has only positive entries. Of course, any transition matrix for a Markov chain has only *nonnegative* entries. But no zeros are allowed in the above power A^n of a regular transition matrix. The identity matrix I_n is a transition matrix for a Markov chain that is *not* regular. (Why?) We have the following theorem, whose proof requires the theory of limits.[1]

> **Theorem 6.12**   If A is a regular transition matrix of a Markov chain, then
>
> (a) A has a unique fixed-point probability vector **w** whose components are positive.
> (b) The powers A^n of A (where n is a positive integer) approach a matrix W whose column vectors are each equal to **w**.

In many experiments the researcher hopes that, no matter how the process begins, it will settle down to some predictable stable behavior. Such is not always the case; but, as the above theorem states, when a regular Markov chain is involved, stable long-range behavior is predictable. The mouse-maze experiment of Example 1 is just such a case. While it might appear at first that the mouse exhibited a preference for some chambers because of having been released in a particular place, after a long number of observations the transition probabilities will stabilize at values independent of the particular chamber into which the mouse is put first. This is because the transition matrix is regular. The following is a classical example.

Example 5   **The Spread of a Rumor** Assume that a given piece of information is passed to individual a_1, who in turn passes it on to individual a_2, who passes it on to a_3, and so on. Each time the information is passed to a new individual, we assume that there is a probability p that the information will be reversed. Thus if a_i receives the message as true, there is a probability p that it will be passed to individual a_{i+1} as false; hence the probability that it is passed on as true is $1 - p$.

---

[1] See R. A. Horn and C. R. Johnson, *Matrix Analysis,* Cambridge University Press, New York, 1985, pp. 495–507.

This gives us an example of a regular Markov chain with the transition matrix

$$P = \begin{array}{c} \\ \text{False} \\ \text{True} \end{array} \begin{array}{cc} \text{False} & \text{True} \\ \begin{bmatrix} 1-p & p \\ p & 1-p \end{bmatrix} \end{array}.$$

We are, of course, interested in successive powers of the matrix P, since the probability distribution for the nth individual will be $P^n \mathbf{u}$, where $\mathbf{u}$ is the initial distribution.

We show that the fixed-point probability vector $\mathbf{w}$ for P is $\mathbf{w} = \begin{bmatrix} \frac{1}{2} \\ \frac{1}{2} \end{bmatrix}$ by

demonstrating a general formula for the fixed point of any positive 2×2 transition matrix for a Markov chain. Let a and b be real numbers with $0 < a < 1$ and $0 < b < 1$. Then any positive 2×2 transition matrix will have the form

$$P = \begin{bmatrix} 1-a & b \\ a & 1-b \end{bmatrix}.$$

You can verify that 1 is an eigenvalue for P. The corresponding eigenvector $\mathbf{w}$ (the fixed point) is obtained by solving $P\mathbf{w} = \mathbf{w}$, or the homogeneous system $(I - P)\mathbf{w} = 0$. This is easy since

$$(I - P) = \begin{bmatrix} a & -b \\ -a & b \end{bmatrix} \text{ row reduces to } \begin{bmatrix} a & -b \\ 0 & 0 \end{bmatrix},$$

whence $\mathbf{w} = (w_1, w_2)^T$ is such that $aw_1 - bw_2 = 0$. Now since $\mathbf{w}$ is also supposed to be a probability vector, we want $w_1 + w_2 = 1$. Therefore, we solve these two equations to obtain the particular eigenvector (fixed point)

$$\mathbf{w} = \begin{bmatrix} \dfrac{b}{a+b} \\ \dfrac{a}{a+b} \end{bmatrix}.$$

So, in the case of Example 5, where $a = b = p$, $\mathbf{w} = \begin{bmatrix} \frac{1}{2} \\ \frac{1}{2} \end{bmatrix}$, as desired. ∎

Let us now turn our attention to a different type of Markov chain. We modify the mouse-maze example as follows.

Example 6 Suppose that a mouse is put into the maze of Figure 6.14. Suppose also that the mouse moves from compartment to compartment, but when he arrives in compartment 4 he remains in it, since there is food there, and does not move on to any other compartment. Suppose that the transition matrix for this Markov chain is the matrix

$$T = \begin{array}{c} \begin{array}{cccc} 1 & 2 & 3 & 4 \end{array} \\ \begin{bmatrix} 0.3 & 0.2 & 0 & 0 \\ 0.3 & 0.3 & 0.3 & 0 \\ 0 & 0.2 & 0.3 & 0 \\ 0.4 & 0.3 & 0.4 & 1 \end{bmatrix} \end{array}.$$

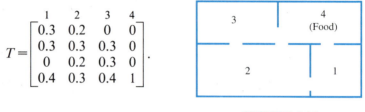

FIGURE 6.14

The last column of T indicates that, if the mouse is in chamber (state) 4, he will remain there. ∎

> **Definition**   A state in a Markov chain is called an **absorbing state** if it is impossible to leave it.
>   A Markov chain is called an **absorbing chain** if it has at least one absorbing state and it is possible to reach some absorbing state from every nonabsorbing state in the chain (not necessarily in one step).

Example 7   The transition matrix

$$A = \begin{array}{c} \begin{array}{ccc} a_1 & a_2 & a_3 \end{array} \\ \begin{array}{c} a_1 \\ a_2 \\ a_3 \end{array} \begin{bmatrix} 1 & 0 & 0 \\ 0 & \dfrac{1}{3} & \dfrac{1}{2} \\ 0 & \dfrac{2}{3} & \dfrac{1}{2} \end{bmatrix} \end{array}$$

fails to be the transition matrix for an absorbing Markov chain because, even though state a_1 is an absorbing state, it is not possible to reach it from the nonabsorbing state a_2. From a_2 one can move only to a_3 or remain in a_2, and from a_3 one can only move to a_2 or remain in a_3. ∎

Let us examine a third example. This is an example of an absorbing chain of the type often referred to as the *random walk with absorbing barriers.*

Example 8   Consider a game with two players, each starting with two marbles. On each play of the game player A has probability p of winning a marble from his opponent, player B, and a probability $1 - p$ of losing a marble to him. The game ends when one of the players has gambled away all of his marbles.

As a Markov chain, this game has five states $a_0, a_1, a_2, a_3,$ and a_4 corresponding to player A having 0, 1, 2, 3, or 4 marbles. States a_0 and a_4 are absorbing states. (Why?) State a_0 corresponds to player A losing all of his marbles, so player B wins. State a_4 corresponds to player A winning all of the marbles, and the game. The transition matrix for this game is

$$
T = \begin{array}{c} \\ a_0 \\ a_1 \\ a_2 \\ a_3 \\ a_4 \end{array}
\begin{array}{ccccc} a_0 & a_1 & a_2 & a_3 & a_4 \end{array}
\left[
\begin{array}{ccccc}
1 & 1-p & 0 & 0 & 0 \\
0 & 0 & 1-p & 0 & 0 \\
0 & p & 0 & 1-p & 0 \\
0 & 0 & p & 0 & 0 \\
0 & 0 & 0 & p & 1
\end{array}
\right].
$$

Note that, for example, $p_{24} = 0$: it is impossible for player A to have four marbles and after one play of the game to have two marbles. He may win or lose only one marble at a time. (Note that in the matrix $T = (t_{ij})$, $t_{11} = p_{00} = 1$, and in general $t_{ij} = p_{i-1,\,j-1}$, the probability that player A has $j-1$ marbles and after one play of the game has $i-1$ marbles.) The matrix T is a transition matrix for an absorbing Markov chain. ∎

This example raises three questions typical of such chains:

1. What is the probability that player A will win? In general, what is the probability that the process will end in a given absorbing state?
2. On the average, how long will it require for the game to end, that is, for the process to be absorbed?
3. On the average, how many times will player A hold 1, 2, or 3 marbles? In general, how many times will the process be in each nonabsorbing state?

To answer these questions, we develop a general technique for handling absorbing Markov chains.

First, we renumber the states of an absorbing Markov chain so that the absorbing states are the first ones. If there are r absorbing states and $s = n - r$ nonabsorbing states, we list them so that $a_0, a_1, \ldots, a_{r-1}$ are the absorbing states and $a_r, a_{r+1}, \ldots, a_{n-1}$ are the nonabsorbing states. Therefore, the transition matrix can be written in block form (see Section 1.6) as

$$
T = \begin{bmatrix} I_r & R \\ 0 & S \end{bmatrix},
$$

where 0 is an $s \times r$ zero matrix, R is an $r \times s$ matrix, and S is an $s \times s$ matrix. This is the *canonical form for the transition matrix of an absorbing Markov chain with r absorbing states and s nonabsorbing states.*

It is intuitively plausible that, in any absorbing Markov chain, the probability that the process will eventually be absorbed is one. We have already shown that the entries p_{ij} of the matrix T^n give the probability of being in state i after n steps in the chain, if we began in state j. It is an exercise in block multiplication to show that

$$
T^n = \begin{bmatrix} I_r & Q \\ 0 & S^n \end{bmatrix},
$$

where we shall not bother to compute the $r \times s$ matrix Q. The form of T^n indicates that the matrix S^n gives the probabilities of being in each of the s nonabsorbing states of the process after n steps, from each possible nonabsorbing initial state. But these probabilities must approach zero as n increases; that is, S^n approaches the $s \times s$ zero matrix O. In such a case the matrix $I - S$ can be shown to be invertible. We consider its inverse.

> **Definition**   The matrix $F = (I - S)^{-1}$ is called the **fundamental matrix** for an absorbing Markov chain.

The transition matrix for the marble game of Example 8 is

$$\begin{bmatrix} I_2 & R \\ 0 & S \end{bmatrix},$$

where

$$R = \begin{bmatrix} 1-p & 0 & 0 \\ 0 & 0 & p \end{bmatrix} \quad \text{and} \quad S = \begin{bmatrix} 0 & 1-p & 0 \\ p & 0 & 1-p \\ 0 & p & 0 \end{bmatrix}.$$

Therefore, the fundamental matrix for this example is the *inverse of*

$$\begin{bmatrix} 1 & 0 & 0 \\ 0 & 1 & 0 \\ 0 & 0 & 1 \end{bmatrix} - \begin{bmatrix} 0 & 1-p & 0 \\ p & 0 & 1-p \\ 0 & p & 0 \end{bmatrix} = \begin{bmatrix} 1 & p-1 & 0 \\ -p & 1 & p-1 \\ 0 & -p & 1 \end{bmatrix}.$$

Set $q = 1 - p$ and the fundamental matrix F is the following (verify this):

$$F = \frac{1}{1-2pq} \begin{bmatrix} 1-pq & q & q^2 \\ p & 1 & q \\ p^2 & p & 1-pq \end{bmatrix}.$$

To see how the fundamental matrix F helps to answer the questions posed above we state and illustrate three basic principles. Our first principle provides the answer to the third question.

> **Principle 1**   The entries in the fundamental matrix $F = (I - S)^{-1}$ of an absorbing Markov chain give the average (mean) number of times in each nonabsorbing state of the chain for each possible nonabsorbing initial state.

Example 8 (Continued) We have already calculated F for the marble game. Since the game begins in state a_2 (each player has two marbles), the second column of F

$$\frac{q}{1-2pq}, \qquad \frac{1}{1-2pq}, \qquad \frac{p}{1-2pq},$$

gives the average number of times that player A will have one marble, two marbles, and three marbles, respectively. Now if we add these numbers together we obtain the total average number of times that player A can play the game; that is,

$$\frac{p+q+1}{1-2pq} = \frac{2}{1-2pq}. \qquad \blacksquare$$

This is the answer to the third question. The following principle provides the general answer to question 2.

Principle 2 The vector ($1 \times s$ matrix) $\mathbf{n} = \mathbf{c}F$, where F is the fundamental matrix of an absorbing Markov chain with r absorbing states and $s = n - r$ nonabsorbing states and $\mathbf{c}$ is the vector $\mathbf{c} = (1, 1, \ldots, 1)$ in $\mathbb{R}^s$, has as its entries the average (mean) number of steps before being absorbed, for each possible nonabsorbing initial state.

Notice that each entry n_i in $\mathbf{n}$ is simply the sum of the entries in column i of F. To illustrate we have the following example.

Example 9 Let us return to the mouse-maze chain of Example 6. We have the matrix

$$T = \begin{bmatrix} I_1 & R \\ 0 & S \end{bmatrix}, \qquad \text{where } S = \begin{bmatrix} 0.3 & 0.2 & 0 \\ 0.3 & 0.3 & 0.3 \\ 0 & 0.2 & 0.3 \end{bmatrix}.$$

Hence,

$$I - S = \begin{bmatrix} 0.7 & -0.2 & 0 \\ -0.3 & 0.7 & -0.3 \\ 0 & -0.2 & 0.7 \end{bmatrix},$$

and the fundamental matrix is

$$F = (I - S)^{-1} = \begin{bmatrix} 1.66 & 0.54 & 0.23 \\ 0.81 & 1.89 & 0.81 \\ 0.23 & 0.54 & 1.66 \end{bmatrix}.$$

Therefore, our vector $\mathbf{n} = (1, 1, 1)F = (2.7, 2.97, 2.7)$. The mouse will always reach the food after three steps in the chain. The average number of observations necessary to find it in chamber 4 with the food is 2.7 if it is initially placed in chamber 1 or chamber 3, and 2.97 if it is initially placed in chamber 2. $\blacksquare$

The answer to the first question is our third principle.

Principle 3   Let b_{ij} be the probability that an absorbing Markov chain with r absorbing and $n - r = s$ nonabsorbing states will be absorbed (end) in state a_i, if it begins in the nonabsorbing state a_j. Set $B = (b_{ij})$. Then $B = RF$, where F is the fundamental matrix of the chain and R is the $r \times s$ matrix in the upper right block of the canonical form T of the transition matrix for the chain.

Example 10   For the mouse-maze chain of Example 6, the canonical form of the transition matrix is

$$T = \begin{bmatrix} 1 & 0.4 & 0.3 & 0.4 \\ 0 & 0.3 & 0.2 & 0 \\ 0 & 0.3 & 0.3 & 0.3 \\ 0 & 0 & 0.2 & 0.3 \end{bmatrix},$$

whence $R = [0.4, 0.3, 0.4]$. We have already found the fundamental matrix F in Example 9; thus $B = RF$, or

$$B = [0.4, 0.3, 0.4] \begin{bmatrix} 1.66 & 0.54 & 0.23 \\ 0.81 & 1.89 & 0.81 \\ 0.23 & 0.54 & 1.66 \end{bmatrix} = [0.999, 0.999, 0.999].$$

The probability that the mouse will end up in chamber 4 with the food when it begins initially in any one of the other three chambers is 0.999, or nearly one. ∎

Example 11   For the marble game of Example 8 we have the following (remember that $q = 1 - p$):

$$B = \begin{bmatrix} q & 0 & 0 \\ 0 & 0 & p \end{bmatrix} \cdot \frac{1}{1 - 2pq} \begin{bmatrix} 1 - pq & q & q^2 \\ p & 1 & q \\ p^2 & p & 1 - pq \end{bmatrix}$$

$$= \frac{1}{1 - 2pq} \begin{bmatrix} q - pq^2 & q^2 & q^3 \\ p^3 & p^2 & p - p^2 q \end{bmatrix}.$$

Hence the probability that player A (who began with two marbles) will win, is $\dfrac{p^2}{1 - 2pq}$, and the probability that he will lose, and player B will win, is $\dfrac{q^2}{1 - 2pq}$. By substituting in $q = 1 - p$ we confirm our intuition, namely,

$$\frac{q^2}{1 - 2pq} + \frac{p^2}{1 - 2pq} = \frac{(1 - p)^2 + p^2}{1 - 2p(1 - p)} = \frac{1 - 2p + 2p^2}{1 - 2p + 2p^2} = 1. \quad ∎$$

Vocabulary

In this section we have introduced an application of linear algebra to Markov chains. We list below the new specialized terms you will need to add to your vocabulary, as applied to Markov chains. Be sure that you review these sufficiently until you feel at home with each of them.

Markov process or chain
transition matrix for a Markov
   chain
fixed point
regular Markov chain
nonabsorbing state

fundamental matrix
states
probability vector
probability distribution
absorbing state
absorbing chain

Exercises 6.5

In Exercises 1–15, decide if the given matrix is a transition matrix from a regular Markov chain, or if it is a transition matrix from an absorbing Markov chain, or if it is neither of these. Write the canonical form for each that is a transition matrix from an absorbing Markov chain.

1. $\begin{bmatrix} \frac{1}{3} & 0 \\ \frac{2}{3} & 1 \end{bmatrix}$.

2. $\begin{bmatrix} \frac{1}{3} & 1 \\ \frac{2}{3} & 0 \end{bmatrix}$.

3. $\begin{bmatrix} 1 & 0.3 \\ 0 & 0.7 \end{bmatrix}$.

4. $\begin{bmatrix} 1 & 0 \\ 0 & 1 \end{bmatrix}$.

5. $\begin{bmatrix} 0 & 1 \\ 1 & 0 \end{bmatrix}$.

6. $\begin{bmatrix} \frac{1}{2} & \frac{2}{3} & 0 \\ \frac{1}{2} & 0 & 0 \\ 0 & \frac{1}{3} & 1 \end{bmatrix}$.

7. $\begin{bmatrix} 0 & \frac{1}{2} & 0 & \frac{1}{4} \\ 0 & \frac{1}{2} & 0 & \frac{3}{4} \\ \frac{2}{3} & 0 & 1 & 0 \\ \frac{1}{3} & 0 & 0 & 0 \end{bmatrix}$.

8. $\begin{bmatrix} \frac{1}{2} & 0 & \frac{1}{3} \\ \frac{1}{2} & \frac{1}{2} & \frac{1}{3} \\ 0 & \frac{1}{2} & \frac{1}{3} \end{bmatrix}$.

9. $\begin{bmatrix} \frac{1}{2} & 0 & \frac{1}{3} & \frac{1}{6} \\ \frac{1}{2} & \frac{1}{2} & \frac{1}{3} & \frac{1}{6} \\ 0 & \frac{1}{2} & \frac{1}{3} & \frac{2}{3} \\ 0 & 1 & 0 & 0 \end{bmatrix}$.

10. $\begin{bmatrix} 0.7 & 0.4 & 0 \\ 0.1 & 0.3 & 0 \\ 0.2 & 0.3 & 1 \end{bmatrix}$.

11. $\begin{bmatrix} 0.2 & 0.3 & 0.1 \\ 0.4 & 0.3 & 0.5 \\ 0.4 & 0.4 & 0.4 \end{bmatrix}$.

12. $\begin{bmatrix} 1 & 0.2 & 0.3 & 1 \\ 0.1 & 0.4 & 1 & 0.3 \\ 0.5 & 1 & 0.4 & 0.4 \\ 0.4 & 0 & 0 & 0 \end{bmatrix}$.

$$13.\begin{bmatrix} 0 & 1 & 0.1 & 0 & 0.2 \\ 0.1 & 0 & 0.2 & 0 & 0.3 \\ 0.2 & 0 & 0.3 & 1 & 0.4 \\ 0.3 & 0 & 0.4 & 0 & 0 \\ 0.4 & 0 & 0 & 0 & 0.1 \end{bmatrix}.$$

$$14.\begin{bmatrix} \frac{1}{3} & 0 & 1 & 0 & 0 \\ \frac{1}{3} & \frac{1}{5} & 0 & \frac{1}{2} & 1 \\ \frac{1}{3} & \frac{1}{5} & 0 & 0 & 0 \\ 0 & \frac{2}{5} & 0 & \frac{1}{2} & 0 \\ 0 & \frac{1}{5} & 0 & 0 & 0 \end{bmatrix}.$$

$$15.\begin{bmatrix} 1 & 0 & 0 & 0.1 & 0.2 \\ 0 & 1 & 0 & 0.1 & 0.2 \\ 0 & 0 & 1 & 0.2 & 0.2 \\ 0 & 0 & 0 & 0.2 & 0.2 \\ 0 & 0 & 0 & 0.4 & 0.2 \end{bmatrix}.$$

16. Find the fixed-point probability vector for the matrix of Exercise 8, if it has one.

17. Find the fixed-point probability vector for the matrix of Exercise 9, if it has one.

18. Find the fixed-point probability vector for the matrix of Exercise 10, if it has one.

19. Find the fixed-point probability vector for the matrix of Exercise 11, if it has one.

---

20. Find the fixed-point probability vector for the matrix

$$\begin{bmatrix} p & q & 1-p-q \\ q & 1-p-q & p \\ 1-p-q & p & q \end{bmatrix},$$

where $0 < p < \frac{1}{2}$, and $0 < q < \frac{1}{2}$.

21. Find the fixed-point probability vector for the matrix

$$\begin{bmatrix} 1-a & b & 0 & 0 \\ 0 & 0 & 1-c & d \\ a & 1-b & 0 & 0 \\ 0 & 0 & c & 1-d \end{bmatrix},$$

where a, b, c, and d are real numbers in the open interval $(0, 1)$.

In Exercises 22–30, find the matrix W approached by the powers A^n of the given matrix A as n becomes large.

$$22.\begin{bmatrix} \frac{1}{2} & \frac{1}{3} \\ \frac{1}{2} & \frac{2}{3} \end{bmatrix}.$$

$$23.\begin{bmatrix} \frac{5}{8} & \frac{3}{8} \\ \frac{3}{8} & \frac{5}{8} \end{bmatrix}.$$

$$24.\begin{bmatrix} 0.8 & 0.6 \\ 0.2 & 0.4 \end{bmatrix}.$$

$$25.\begin{bmatrix} 0.8 & 0.2 \\ 0.2 & 0.8 \end{bmatrix}.$$

$$26.\begin{bmatrix} 0.6 & 1 \\ 0.4 & 0 \end{bmatrix}.$$

$$27.\begin{bmatrix} \frac{5}{8} & 1 \\ \frac{3}{8} & 0 \end{bmatrix}.$$

$$28.\begin{bmatrix} 0 & 0 & 0.5 \\ 1 & 0 & 0.5 \\ 0 & 1 & 0 \end{bmatrix}.$$

$$29.\begin{bmatrix} \frac{1}{4} & \frac{3}{5} & 0 \\ \frac{1}{2} & \frac{2}{5} & \frac{2}{3} \\ \frac{1}{4} & 0 & \frac{1}{3} \end{bmatrix}.$$

$$30.\begin{bmatrix} 0 & 1 & 0.1 & 0 & 0.2 \\ 0.1 & 0 & 0.2 & 0 & 0.3 \\ 0.2 & 0 & 0.3 & 0 & 0.4 \\ 0.3 & 0 & 0.4 & 1 & 0 \\ 0.4 & 0 & 0 & 0 & 0.1 \end{bmatrix}.$$

31. Find the fundamental matrix F for the absorbing Markov chain whose transition matrix is the matrix of Exercise 1.

32. Find the fundamental matrix F for the absorbing Markov chain whose transition matrix is the matrix of Exercise 6.

33. Find the fundamental matrix F for the absorbing Markov chain whose transition matrix is the matrix of Exercise 7.

34. Find the fundamental matrix F for the absorbing Markov chain whose transition matrix is the matrix of Exercise 10.

35. Find the fundamental matrix F for the absorbing Markov chain whose transition matrix is the matrix of Exercise 13.

36. Find the fundamental matrix F for the absorbing Markov chain whose transition matrix is the matrix of Exercise 15.

37. Find the vector $\mathbf{n} = \mathbf{c}\,F$ for the absorbing Markov chain whose fundamental matrix F is the matrix you found in Exercise 31.

38. Find the vector $\mathbf{n} = \mathbf{c}\,F$ for the absorbing Markov chain whose fundamental matrix F is the matrix you found in Exercise 32.

39. Find the vector $\mathbf{n} = \mathbf{c}\,F$ for the absorbing Markov chain whose fundamental matrix F is the matrix you found in Exercise 33.

40. Find the vector $\mathbf{n} = \mathbf{c}\,F$ for the absorbing Markov chain whose fundamental matrix F is the matrix you found in Exercise 34.

41. Find the vector $\mathbf{n} = \mathbf{c}\,F$ for the absorbing Markov chain whose fundamental matrix F is the matrix you found in Exercise 35.

42. Find the vector $\mathbf{n} = \mathbf{c}\,F$ for the absorbing Markov chain whose fundamental matrix F is the matrix you found in Exercise 36.

---

43. Show that 1 is an eigenvalue for a 3×3 transition matrix

$$P = \begin{bmatrix} 1-a-b & c & e \\ a & 1-c-d & f \\ b & d & 1-e-f \end{bmatrix}.$$

44. Find the fixed-point vector $\mathbf{w}$ for the matrix of Exercise 43.

45. Work out the probabilities for the marble game of Example 8 if $p = q = \frac{1}{2}$.

46. A certain sales representative visits three cities: Springfield, Portland, and Middletown. To avoid being bored, he doesn't cover them in the same order each time; rather he conducts a Markov chain of transitions from one city to another. His matrix is as follows:

$$T = \begin{array}{c c} & \begin{array}{c c c} S & P & M \end{array} \\ & \begin{bmatrix} 0 & 0.5 & 0.3 \\ 0.6 & 0 & 0.7 \\ 0.4 & 0.5 & 0 \end{bmatrix} \end{array}.$$

Assuming that he starts at Springfield, what is the probability he will return to Springfield after three transitions?

47. What are the various probabilities for the salesperson in Exercise 46 to be in each city after four transitions?

48. Suppose that the quality of the air in Midtown is checked every day to see whether it is within acceptable standards (*C*—clean) or not (*D*—dirty) with respect to a certain particulate. Assume that if a given day is C, the probability it will be C the next day is 2/5, while if it is D the probability it will be C the next day is 1/5. What is the probability that it will be D three days after if on day *one* it is D?

49. Immediately after inauguration the president is faced with a serious economic slump and must choose a set of economic advisers from two conflicting philosophical persuasions. The policies of group I will produce a boom year within the next year with probability 0.7, but the probability of a recession the year following this boom is 0.6. The policies of group II will bring a boom year following a recession with probability 0.5 and will follow a boom year with one of recession with probability 0.3. The president's concern is to have a boom year three years hence, during the next presidential campaign. Which philosophy (if consistently followed) will give the greatest likelihood of producing this favorable turn of events? HINT: First set up the transition matrices for each of the two groups.

50. In the United Kingdom, next-generation descendants of members of the Labour Party vote Labour with probability 0.5, vote Conservative with probability 0.4, and vote Liberal with probability 0.1. The probabilities for descendants of Conservative Party members are 0.7 Conservative, 0.2 Labour, and 0.1 Liberal. For descendants of Liberal Party members the probabilities are 0.2 Conservative, 0.4 Labour, and 0.4 Liberal. Given these statistics, what is the probability that the *granddaughter* of a Liberal party member will vote Labour? What is the long-range voting pattern for offspring of each party affiliation?

51. Apply the spread-of-a-rumor example (Example 5) to the following case. The governor tells a person a_1 his intentions with respect to speculation that he will resign and seek a U.S. Senate seat at the next election. The first hearer a_1 relays the news to person a_2, who relays it to person a_3, etc. Let the probability that any one person will reverse the message when he passes it on be $p = 0.2$. What is the probability for the nth person to be told that the governor will indeed seek the senate seat? HINT: Use a two-state Markov chain.

In Exercises 52–55, suppose that we play a coin-flipping game using a fair coin. Player A has $3 and player B has $2. The coin is flipped. If it is a head, A pays B one dollar; if it is a tail, B pays A one dollar.

52. Model this game as an absorbing Markov chain. Write the transition matrix in canonical form.

53. Find the fundamental matrix F, the vector $\mathbf{n}$, and the matrix B for this model.

54. How long is the game likely to last?

55. What are the probabilities of winning for each player?

In Exercises 56 and 57 we consider an annual sales convention traditionally held in Boston. While attending the convention, Barry Brown has a choice of four moderately expensive hotels. Hotels M and N are, however, less expensive than the more luxurious hotels K and L. The following array lists the probabilities that if Mr. Brown stays in one of the hotels the first year he will go to a given hotel the next year.

$$
\begin{array}{c c c c c}
 & M & N & K & L \\
M & \frac{1}{3} & 0 & 0 & \frac{1}{3} \\
N & 0 & 1 & \frac{1}{2} & \frac{1}{3} \\
K & \frac{2}{3} & 0 & 0 & \frac{1}{3} \\
L & 0 & 0 & \frac{1}{2} & 0
\end{array}
$$

56. If Mr. Brown stays in Hotel M the first year, what is the probability he will stay in a luxury hotel (K or L) during the following year?

57. Find the canonical form for the matrix of this Markov chain, and then find the matrices F and B and the vector $\mathbf{n}$, and interpret them.

In Exercises 58–60, we consider a certain sociological experiment in which it was discovered that the educational level of children depended upon the educational level of their parents. The people involved were classified into three groups: those with no higher than one year of secondary school (S) (grades 9–12), those who had at least one term of college (C), and those who had at least one college degree (D). If a parent is in one of these groups, the probabilities that a child will belong to any one of the groups is given by the following table. (HINT: In these exercises you will first want to turn the table into a transition matrix for a Markov chain.)

		Child		
		S	C	D
Parent	S	0.7	0.2	0.1
	C	0.4	0.4	0.2
	D	0.1	0.2	0.7

58. What is the probability that the grandchild of a parent in group S will graduate from college?

59. Suppose that a different experimenter finds that, if the parent has some college education (C) the child *always* attends college. Modify the transition matrix accordingly.

60. Given the situation in Exercise 59, what is the average number of generations before the offspring of each class will attend some college?

6.6 *COMPLEX EIGENVALUES*

In Section 4.6 we discussed complex matrices and vectors. We have already noted in the first part of this chapter that a real matrix can have complex numbers as eigenvalues. In this section we look a little closer at complex matrices and their eigenvalues and eigenvectors. In particular we generalize Theorem 6.7 (page 403).

Certain special types of matrices (and therefore linear transformations on unitary or Euclidean space) are of interest. Most of these ideas we have already encountered, particularly in the real number case. Complex analogs are given here. A matrix A is called by the indicated name when it satisfies the indicated condition.

Euclidean (Real) Space		Unitary (Complex) Space	
Real symmetric:	$A^T = A$	Hermitian:	$A^* = A$
Orthogonal:	$A^T = A^{-1}$	Unitary:	$A^* = A^{-1}$
Skew-symmetric:	$A^T = -A$	Skew-Hermitian:	$A^* = -A$
Real normal:	$AA^T = A^T A$	Normal:	$AA^* = A^*A$

Since each matrix defines a linear transformation, and each linear operator T on a unitary space has a matrix representation, linear transformations are called by the same names as are the matrices, when they satisfy the specified condition.

Example 1 The matrix $M = \begin{bmatrix} 5 & 3-2i \\ 3+2i & 1 \end{bmatrix}$ is Hermitian, since

$$M^* = \begin{bmatrix} \overline{5} & \overline{3+2i} \\ \overline{3-2i} & \overline{1} \end{bmatrix} = \begin{bmatrix} 5 & 3-2i \\ 3+2i & 1 \end{bmatrix} = M.$$

Is it normal? The answer is, *every Hermitian matrix is normal* since $MM^* = M^2 = M^*M$. ∎

Of course, every unitary matrix U is also normal, because $U^*U = I = UU^*$. On the other hand, the matrix $K = \begin{bmatrix} 1 & -1 \\ 1 & 1 \end{bmatrix}$ is normal (even real normal), as you can verify. Yet K is not unitary; neither is it Hermitian since it is not symmetric. Of course, every *real symmetric* matrix is *Hermitian*; this is because A^* then is precisely A^T. We ask you to prove as Exercise 19 that every skew-Hermitian matrix is also normal.

Notice that in the unitary space $\overline{\mathbb{C}^n}$, $\|\mathbf{x}\|^2 = <\mathbf{x}, \mathbf{x}> = \mathbf{x}^*\mathbf{x}$ for all $\mathbf{x}$ in $\overline{\mathbb{C}^n}$. Since $\mathbf{x}$ is a column vector, $\mathbf{x}^*\mathbf{x}$ is a 1×1 matrix. In this vector space the *standard* inner product $<\mathbf{u}, \mathbf{v}>$ is the matrix product $\mathbf{v}^*\mathbf{u}$, as you can readily verify.

We saw in Section 6.3 that real symmetric matrices are especially nice in that they can be orthogonally diagonalized. In a similar way Hermitian matrices are especially well behaved. It is left as Exercise 20 for you to show that the following properties of the Hermitian adjoint * hold in $M_n(\mathbb{C})$.

Lemma 6.13   Let A, B be matrices in $M_n(\mathbb{C})$, then

(i) $(A^*)^* = A$.
(ii) $(A + B)^* = A^* + B^*$.
(iii) $(cA)^* = \bar{c}\, A^*$ for any scalar c.
(iv) $(AB)^* = B^*A^*$.

We can then prove the following important theorem.

Theorem 6.14   The eigenvalues of an Hermitian matrix are all real numbers; and furthermore, eigenvectors corresponding to distinct eigenvalues are orthogonal.

Proof   If A is Hermitian, $A^* = A$. Let λ be an eigenvalue of A and consider the eigenvector $\mathbf{x}$ from $\overline{\mathbb{C}^n}$ corresponding to λ: $A\mathbf{x} = \lambda\mathbf{x} = A^*\mathbf{x}$. Now consider the matrix $\alpha = \mathbf{x}^*A\mathbf{x}$. Since α is a 1×1 matrix (a scalar in $\mathbb{C}$), it follows that

$$\alpha^* = (\mathbf{x}^*A\mathbf{x})^* = \mathbf{x}^*A\mathbf{x} = \alpha.$$

So α is a real number. Since $A\mathbf{x} = \lambda\mathbf{x}$, we have

$$\alpha = \mathbf{x}^*A\mathbf{x} = \mathbf{x}^*\lambda\mathbf{x} = \lambda\mathbf{x}^*\mathbf{x} = \lambda||\mathbf{x}||^2.$$

Therefore, the eigenvalue

$$\lambda = \frac{\alpha}{||\mathbf{x}||^2}$$

is a real number, as desired.   ◘

The proof of the second part of the theorem duplicates the proof of Theorem 6.7 for real symmetric matrices except that $\mathbf{x}^*$ is used in place of $\mathbf{x}^T$. The details are left to you as Exercise 21. Theorem 6.7 is, of course, a special case of Theorem 6.14.

Theorem 6.6 (page 402) for real symmetric matrices also has a more general statement. It is called the *spectral theorem* for complex matrices. Before stating and proving it, we prove a more general theorem due to Schur.

Recall our definition of unitary matrices given above. It follows from that definition (see Exercise 22) that a matrix U is **unitary** if, and only if, its column vectors $\{\mathbf{u}_1, \mathbf{u}_2, \ldots, \mathbf{u}_n\}$ determine an orthonormal set of vectors in $\overline{\mathbb{C}^n}$.

Example 2   The matrix $U = \dfrac{1}{\sqrt{3}}\begin{bmatrix} 1-i & -1 \\ 1 & 1+i \end{bmatrix}$ is unitary.

You can readily verify this by showing that $UU^* = U^*U = I$.   ■

Unitary matrices are the complex analogs of *orthogonal* matrices in the real case. Now we turn to Schur's triangularization theorem.

> **Theorem 6.15 (Schur's Theorem)**   For every $n \times n$ complex matrix A, there exists a unitary matrix U such that $M = U^*AU$ is upper-triangular. Furthermore, the diagonal entries of M are the eigenvalues of A.

Outline of the Proof   The proof is essentially constructive and proceeds by induction on the size of the matrix. The theorem is obvious if $n = 1$. Suppose then that $n > 1$ and let us assume that the theorem is true for all complex $(n-1) \times (n-1)$ matrices. Let λ_1 be an eigenvalue of A and let $\mathbf{v}_1$ be a corresponding eigenvector of unit length; that is, $< \mathbf{v}_1, \mathbf{v}_1 > = 1$, for the complex inner product on $\overline{\mathbb{C}^n}$. We construct a unitary matrix $U^{(1)}$ so that its first column vector $\mathbf{u}_1 = \mathbf{v}_1$. This is accomplished in the following way. We complete $\{\mathbf{v}_1\}$ to a basis $\{\mathbf{v}_1, \mathbf{y}_2, \ldots, \mathbf{y}_n\}$. Then use the Gram-Schmidt process to construct an orthonormal basis $\{\mathbf{u}_1 = \mathbf{v}_1, \mathbf{u}_2, \ldots, \mathbf{u}_n\}$ for $\overline{\mathbb{C}^n}$. (This is most easily accomplished if we use the standard inner product in $\overline{\mathbb{C}^n}$.) Now, let the kth column vector of $U^{(1)}$ be $\mathbf{u}_k$, $k = 1, 2, \ldots, n$. It can be directly verified that $U^{(1)}$ is unitary. So let us compute the first column of the matrix $U^{(1)*}AU^{(1)}$:

$$(U^{(1)*}AU^{(1)})_1 = U^{(1)*}A(U^{(1)})_1 = U^{(1)*}A\mathbf{u}_1 = U^{(1)*}\lambda_1\mathbf{u}_1 = \lambda_1 U^{(1)*}\mathbf{u}_1$$
$$= \lambda_1 U^{(1)*}(U^{(1)})_1 = \lambda_1(U^{(1)*}U^{(1)})_1 = \lambda_1\mathbf{e}_1.$$

Therefore,

$$U^{(1)*}AU^{(1)} = \begin{bmatrix} \lambda_1 & Z \\ 0 & \\ \vdots & A_1 \\ 0 & \end{bmatrix},$$

where Z is a $1 \times (n-1)$ submatrix and A_1 is an $(n-1) \times (n-1)$ submatrix. Therefore, by our induction assumption there is an $(n-1) \times (n-1)$ unitary matrix K so that K^*A_1K is upper-triangular with the eigenvalues of A_1 as its diagonal elements. Let the matrix $U^{(2)}$ be constructed as the block matrix

$$\begin{bmatrix} 1 & O \\ 0 & \\ \vdots & K \\ 0 & \end{bmatrix},$$

and let $U = U^{(1)}U^{(2)}$. Then

$$U*AU = (U^{(1)}U^{(2)})*A(U^{(1)}U^{(2)}) = U^{(2)}*(U^{(1)}*AU^{(1)})U^{(2)} = \begin{bmatrix} \lambda_1 & & & Y \\ 0 & & & \\ \vdots & & K*A_1K & \\ 0 & & & \end{bmatrix}$$

is upper-triangular, as desired. You should verify that the diagonal elements are indeed the eigenvalues of A. ❏

Example 3 We illustrate the method of proof of Schur's theorem by applying it to the following matrix.

$$A = \begin{bmatrix} 1 & 0 & i \\ 1 & i & 1 \\ i & 0 & 1 \end{bmatrix}.$$

You can readily find, using the techniques of the first two sections of this chapter, that $\lambda = i$ is an eigenvalue and $x = (0, 1, 0)^T = e_2$ is an eigenvector for this matrix. The basis consisting of $\{e_2, e_1, e_3\}$ is already an orthonormal basis for $\overline{C^3}$; although it is not the standard ordered basis. Therefore, the matrix $U^{(1)} = [e_2 : e_1 : e_3]$ is unitary, and we have

$$U^{(1)}*AU^{(1)} = \begin{bmatrix} 0 & 1 & 0 \\ 1 & 0 & 0 \\ 0 & 0 & 1 \end{bmatrix} \begin{bmatrix} 1 & 0 & i \\ 1 & i & 1 \\ i & 0 & 1 \end{bmatrix} \begin{bmatrix} 0 & 1 & 0 \\ 1 & 0 & 0 \\ 0 & 0 & 1 \end{bmatrix} = \begin{bmatrix} i & 1 & 1 \\ 0 & 1 & i \\ 0 & i & 1 \end{bmatrix}.$$

Here $Z = [1, 1]$ and $A_1 = \begin{bmatrix} 1 & i \\ i & 1 \end{bmatrix}$.

Now we focus on the 2×2 submatrix A_1. It has eigenvalue $\lambda_1 = 1 + i$ with corresponding eigenvector $v_1 = (1, 1)^T$ (or also $\lambda_2 = 1 - i$, with eigenvec-

tor $v_2 = (1, -1)^T$). The unit eigenvector $u_1 = \dfrac{v_1}{\|v_1\|} = \begin{bmatrix} \dfrac{1}{\sqrt{2}} \\ \dfrac{1}{\sqrt{2}} \end{bmatrix}$ can be completed

to a basis for $\overline{C^2}$ with the other eigenvector $u_2 = \begin{bmatrix} \dfrac{1}{\sqrt{2}} \\ -\dfrac{1}{\sqrt{2}} \end{bmatrix}$. So the matrix

$$K = \begin{bmatrix} \dfrac{1}{\sqrt{2}} & \dfrac{1}{\sqrt{2}} \\ \dfrac{1}{\sqrt{2}} & -\dfrac{1}{\sqrt{2}} \end{bmatrix}$$ is unitary and $K*A_1K$ is triangular. In fact, in this example,

it is diagonal:

$$K^*A_1K = \begin{bmatrix} \dfrac{1}{\sqrt{2}} & \dfrac{1}{\sqrt{2}} \\ \dfrac{1}{\sqrt{2}} & -\dfrac{1}{\sqrt{2}} \end{bmatrix} \begin{bmatrix} 1 & i \\ i & 1 \end{bmatrix} \begin{bmatrix} \dfrac{1}{\sqrt{2}} & \dfrac{1}{\sqrt{2}} \\ \dfrac{1}{\sqrt{2}} & -\dfrac{1}{\sqrt{2}} \end{bmatrix} = \begin{bmatrix} 1+i & 0 \\ 0 & 1-i \end{bmatrix}.$$

The matrix $U^{(2)}$ is therefore the matrix

$$\begin{bmatrix} 1 & 0 & 0 \\ 0 & \dfrac{1}{\sqrt{2}} & \dfrac{1}{\sqrt{2}} \\ 0 & \dfrac{1}{\sqrt{2}} & -\dfrac{1}{\sqrt{2}} \end{bmatrix},$$

and the triangularizing unitary matrix $U = U^{(1)}U^{(2)}$ is

$$\begin{bmatrix} 0 & \dfrac{1}{\sqrt{2}} & \dfrac{1}{\sqrt{2}} \\ 1 & 0 & 0 \\ 0 & \dfrac{1}{\sqrt{2}} & -\dfrac{1}{\sqrt{2}} \end{bmatrix}.$$

We have

$$U^*AU = \begin{bmatrix} 0 & 1 & 0 \\ \dfrac{1}{\sqrt{2}} & 0 & \dfrac{1}{\sqrt{2}} \\ \dfrac{1}{\sqrt{2}} & 0 & -\dfrac{1}{\sqrt{2}} \end{bmatrix} \begin{bmatrix} 1 & 0 & i \\ 1 & i & 1 \\ i & 0 & 1 \end{bmatrix} \begin{bmatrix} 0 & \dfrac{1}{\sqrt{2}} & \dfrac{1}{\sqrt{2}} \\ 1 & 0 & 0 \\ 0 & \dfrac{1}{\sqrt{2}} & -\dfrac{1}{\sqrt{2}} \end{bmatrix}$$

$$= \begin{bmatrix} i & \sqrt{2} & 0 \\ 0 & 1+i & 0 \\ 0 & 0 & 1-i \end{bmatrix} = M$$

is the desired upper-triangular matrix. ∎

If the given complex matrix A is normal, then it is always the case that the upper-triangular matrix M of Schur's theorem is a diagonal matrix. This is the statement of the following theorem.

Theorem 6.16 If A is an $n \times n$ normal matrix, then there exists a unitary matrix U such that U^*AU is a diagonal matrix.

Proof   By Schur's theorem there is a unitary matrix U such that U^*AU is an upper-triangular matrix M. Since A is normal and U is unitary, we have

$$M^*M = (U^*AU)^*(U^*AU) = U^*A^*UU^*AU = U^*A^*AU$$
$$= U^*AA^*U = U^*AUU^*A^*U = MM^*.$$

The proof is completed by showing that an upper-triangular normal matrix must be diagonal. We do this by equating the diagonal elements of MM^* and M^*M.

The fact that the entry in the $(1, 1)$ position of M^*M is the same as that of MM^* means that

$$\overline{m_{11}}\, m_{11} = m_{11}\, \overline{m_{11}} + \sum_{j=2}^{n} m_{1j}\, \overline{m_{1j}} = |m_{11}|^2 + \sum_{j=2}^{n} |m_{1j}|^2.$$

This means that $\sum_{j=2}^{n} |m_{1j}|^2 = 0$. A sum of nonnegative terms equaling zero means that each of the individual terms is zero, so we conclude that the rest of the entries in row 1 are zeros; $m_{1j} = 0$, for $j = 2, 3, \ldots, n$. We then proceed to row two and equate the diagonal elements there. A similar argument shows that $m_{2j} = 0$ for $j = 3, 4, \ldots, n$. In this way we successively calculate that all of the entries to the right of the diagonal element in each row of M are zero; $m_{ij} = 0$ for $j > i$, each $i = 1, 2, \ldots, n$. Since M is upper-triangular, the entries to the left of the diagonal in each row were already zero; $m_{ij} = 0$ for $j < i$, each $i = 1, 2, \ldots, n$. Thus M is a diagonal matrix, as desired.   ◘

Example 4   Demonstrate that the matrix $A = \begin{bmatrix} 2 & 0 & 0 \\ 0 & 2 & i \\ 0 & -i & 0 \end{bmatrix}$ is unitarily similar to a diagonal matrix by finding an appropriate unitary matrix U.

To do this, proceed as in Example 3. It is fairly easy to calculate that the eigenvalues of A are $\lambda_1 = 2$, $\lambda_2 = 1 + \sqrt{2}$, and $\lambda_3 = 1 - \sqrt{2}$, since $\det(\lambda I - A) = (\lambda - 2)(\lambda^2 - 2\lambda - 1)$. Notice also that $A^* = A$; A is Hermitian. Since $\mathbf{e}_1$ is an eigenvector of A corresponding to $\lambda_1 = 2$, the 3×3 identity matrix I can serve as the matrix $U^{(1)}$ in our algorithmic method. We need then to focus on the 2×2 submatrix

$$A_1 = \begin{bmatrix} 2 & i \\ -i & 0 \end{bmatrix}.$$

The eigenvalues of A_1 are $\lambda_2 = 1 + \sqrt{2}$, and $\lambda_3 = 1 - \sqrt{2}$, with corresponding eigenvectors

$$\mathbf{x}_2 = \begin{bmatrix} (1 + \sqrt{2})i \\ 1 \end{bmatrix} \quad \text{and} \quad \mathbf{x}_3 = \begin{bmatrix} (1 - \sqrt{2})i \\ 1 \end{bmatrix}.$$

While $\mathbf{x}_2$ and $\mathbf{x}_3$ are mutually orthogonal, $\mathbf{x}_3{}^*\mathbf{x}_2 = 0$, they are *not* unit vectors. Let

$$\alpha = (4 + 2\sqrt{2})^{1/2} = \|\mathbf{x}_2\| = (\mathbf{x}_2{}^*\mathbf{x}_2)^{1/2} \approx 2.6131$$

and

$$\beta = (4 - 2\sqrt{2})^{1/2} = \|\mathbf{x}_3\| = (\mathbf{x}_3{}^*\mathbf{x}_3)^{1/2} \approx 1.0824.$$

The desired unitary matrix K is then the matrix

$$\begin{bmatrix} \dfrac{(1 + \sqrt{2})i}{\alpha} & \dfrac{(1 - \sqrt{2})i}{\beta} \\ \dfrac{1}{\alpha} & \dfrac{1}{\beta} \end{bmatrix}.$$

The matrix $U^{(2)}$ is

$$\begin{bmatrix} 1 & 0 & 0 \\ 0 & \dfrac{(1 + \sqrt{2})i}{\alpha} & \dfrac{(1 - \sqrt{2})i}{\beta} \\ 0 & \dfrac{1}{\alpha} & \dfrac{1}{\beta} \end{bmatrix}.$$

This is also the unitary matrix $U = U^{(1)}U^{(2)} = IU^{(2)}$ which diagonalizes A. Thus,

$$U^*AU = \begin{bmatrix} 1 & 0 & 0 \\ 0 & -\dfrac{(1 + \sqrt{2})i}{\alpha} & \dfrac{1}{\alpha} \\ 0 & -\dfrac{(1 - \sqrt{2})i}{\beta} & \dfrac{1}{\beta} \end{bmatrix} \begin{bmatrix} 2 & 0 & 0 \\ 0 & 2 & i \\ 0 & -i & 0 \end{bmatrix} \begin{bmatrix} 1 & 0 & 0 \\ 0 & \dfrac{(1 + \sqrt{2})i}{\alpha} & \dfrac{(1 - \sqrt{2})i}{\beta} \\ 0 & \dfrac{1}{\alpha} & \dfrac{1}{\beta} \end{bmatrix}$$

$$= \begin{bmatrix} 2 & 0 & 0 \\ 0 & 1 + \sqrt{2} & 0 \\ 0 & 0 & 1 - \sqrt{2} \end{bmatrix}. \quad \blacksquare$$

Notice that both of the two matrices designated A_1 in this example and in Example 3 had mutually orthogonal eigenvectors. In each case the matrices are normal matrices; that is, $A_1A_1{}^* = A_1{}^*A_1$. In Exercise 31 we ask you to prove the important spectral theorem as a corollary to Theorems 6.15 and 6.16 (pages 441 and 443).

Corollary (The Spectral Theorem) An $n \times n$ matrix A is normal if, and only if, it has a complete set of orthonormal eigenvectors.

As a final comment, let us return to a remark we made in Chapter 4 that weighted inner products can be defined using an Hermitian matrix with positive eigenvalues. See, in particular, Example 6 in Section 4.6. From Theorem 6.14 (page 440) we know that the eigenvalues of an Hermitian matrix are always real numbers, so if they are also required to be positive we have the following theorem.

Theorem 6.17   If A is an $n \times n$ Hermitian matrix with positive eigenvalues, then the function

$$< \mathbf{u}, \mathbf{v} > = \mathbf{v}^* A \mathbf{u}$$

is an inner product on $\overline{\mathbb{C}^n}$.

Proof   The proof is a straightforward verification of the properties of an inner product. (See Section 4.6.) The first two properties are easy calculations, and we leave them to you as Exercise 32. To prove the third property, note that if $\mathbf{u} = (u_1, u_2, \ldots, u_n)^T$ is a vector in $\overline{\mathbb{C}^n}$ and A is Hermitian, then there is some unitary matrix U so that U^*AU is a diagonal matrix. Note $U^*AU = D = \mathrm{diag}\,[\lambda_1, \lambda_2, \ldots, \lambda_n]$, so $A = UDU^*$. By hypothesis each $\lambda_i > 0$, for each $i = 1, 2, \ldots, n$. The vector $U^*\mathbf{u}$ is in $\overline{\mathbb{C}^n}$. Write $U^*\mathbf{u} = (v_1, v_2 \ldots, v_n)^T$. So we have

$$< \mathbf{u}, \mathbf{u} > = \mathbf{u}^* A \mathbf{u} = \mathbf{u}^* U D U^* \mathbf{u} = (U^* \mathbf{u})^* D (U^* \mathbf{u})$$
$$= \lambda_1 \overline{v_1}\, v_1 + \lambda_2 \overline{v_2}\, v_2 + \cdots + \lambda_n \overline{v_n}\, v_n$$
$$= \lambda_1 |v_1|^2 + \lambda_2 |v_2|^2 + \cdots + \lambda_n |v_n|^2 \geq 0.$$

Equality can occur if, and only if, each $|v_i| = 0$; which happens if, and only if, $U^*\mathbf{u} = \mathbf{0}$. But since U is unitary, $UU^* = I$, so $UU^*\mathbf{u} = U\mathbf{0} = \mathbf{0}$, or $I\mathbf{u} = \mathbf{0}$. Thus $U^*\mathbf{u} = \mathbf{0}$, if and only if $\mathbf{u} = \mathbf{0}$.   ∎

Example 5   Let $\mathbf{u} = \begin{bmatrix} 1 \\ i \\ 1 \end{bmatrix}$ and $\mathbf{v} = \begin{bmatrix} i \\ 0 \\ -1 \end{bmatrix}$ and let A be the matrix of Example 4. Find the inner product $< \mathbf{u}, \mathbf{v} > = \mathbf{v}^* A \mathbf{u}$ under the inner product induced by A on $\overline{\mathbb{C}^n}$. Also find $\|\mathbf{u}\|$ and $\|\mathbf{v}\|$.

We have $< \mathbf{u}, \mathbf{v} > = \mathbf{v}^* A \mathbf{u} = [-i,\ 0,\ -1] \begin{bmatrix} 2 & 0 & 0 \\ 0 & 2 & i \\ 0 & -i & 0 \end{bmatrix} \begin{bmatrix} 1 \\ i \\ 1 \end{bmatrix} = -2i - 1.$

For the "lengths" of $\mathbf{u}$ and $\mathbf{v}$, we have

$$< \mathbf{u}, \mathbf{u} > = \mathbf{u}^* A \mathbf{u} = [1,\ -i,\ 1] \begin{bmatrix} 2 & 0 & 0 \\ 0 & 2 & i \\ 0 & -i & 0 \end{bmatrix} \begin{bmatrix} 1 \\ i \\ 1 \end{bmatrix} = 6,$$

so $\|\mathbf{u}\| = \sqrt{6}$; while

$$<\mathbf{v}, \mathbf{v}> = \mathbf{v}^* A\mathbf{v} = [-i, 0, -1]\begin{bmatrix} 2 & 0 & 0 \\ 0 & 2 & i \\ 0 & -i & 0 \end{bmatrix}\begin{bmatrix} i \\ 0 \\ -1 \end{bmatrix} = 2,$$

so $\|\mathbf{v}\| = \sqrt{2}$ in this inner-product space. ∎

Vocabulary

In this section we have looked at the complex number analogs of earlier linear algebra concepts that were previously discussed only for real numbers. We list below the new specialized terms you will need to add to your vocabulary. Be sure that you review these until you feel comfortable with them.

Hermitian matrix   unitary matrix
skew-Hermitian matrix   normal matrix
spectral theorem   Schur's theorem
diagonalization of a normal matrix

Exercises 6.6

1. Verify that the matrix $A_1 = \begin{bmatrix} 1 & i \\ i & 1 \end{bmatrix}$ of Example 3 is normal, but is not Hermitian, nor

   is it skew-Hermitian.

2. Verify directly by computation that the matrix $A_1 = \begin{bmatrix} 2 & i \\ -i & 0 \end{bmatrix}$ of Example 4 is

   normal.

3. Is the matrix $A = \begin{bmatrix} 1 & 0 & i \\ 1 & i & 1 \\ i & 0 & 1 \end{bmatrix}$ of Example 3 normal? Is it Hermitian? Is it skew-

   Hermitian?

4. Show that the matrix A of Example 3 is similar to a diagonal matrix.

5. Is the matrix $A = \begin{bmatrix} \dfrac{i}{\sqrt{2}} & \dfrac{i}{\sqrt{2}} \\ \dfrac{i}{\sqrt{2}} & \dfrac{-i}{\sqrt{2}} \end{bmatrix}$ normal? Is it Hermitian? Is it skew-Hermitian?

6. Is the matrix $A = \begin{bmatrix} 1 & 2-i & i \\ 2+i & 2 & -1 \\ -i & -1 & 3 \end{bmatrix}$ normal? Is it Hermitian? Is it skew-Hermitian?

In Exercises 7–12, find a unitary matrix U so that U*MU is a diagonal matrix for the given matrix M.

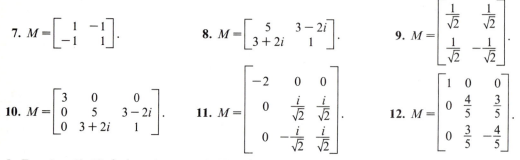

7. $M = \begin{bmatrix} 1 & -1 \\ -1 & 1 \end{bmatrix}$.

8. $M = \begin{bmatrix} 5 & 3-2i \\ 3+2i & 1 \end{bmatrix}$.

9. $M = \begin{bmatrix} \frac{1}{\sqrt{2}} & \frac{1}{\sqrt{2}} \\ \frac{1}{\sqrt{2}} & -\frac{1}{\sqrt{2}} \end{bmatrix}$.

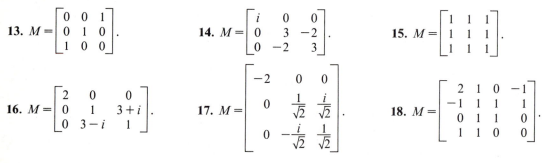

10. $M = \begin{bmatrix} 3 & 0 & 0 \\ 0 & 5 & 3-2i \\ 0 & 3+2i & 1 \end{bmatrix}$.

11. $M = \begin{bmatrix} -2 & 0 & 0 \\ 0 & \frac{i}{\sqrt{2}} & \frac{i}{\sqrt{2}} \\ 0 & -\frac{i}{\sqrt{2}} & \frac{i}{\sqrt{2}} \end{bmatrix}$.

12. $M = \begin{bmatrix} 1 & 0 & 0 \\ 0 & \frac{4}{5} & \frac{3}{5} \\ 0 & \frac{3}{5} & -\frac{4}{5} \end{bmatrix}$.

In Exercises 13–18, find a unitary matrix U so that U*MU is an upper-triangular matrix for the given matrix M.

13. $M = \begin{bmatrix} 0 & 0 & 1 \\ 0 & 1 & 0 \\ 1 & 0 & 0 \end{bmatrix}$.

14. $M = \begin{bmatrix} i & 0 & 0 \\ 0 & 3 & -2 \\ 0 & -2 & 3 \end{bmatrix}$.

15. $M = \begin{bmatrix} 1 & 1 & 1 \\ 1 & 1 & 1 \\ 1 & 1 & 1 \end{bmatrix}$.

16. $M = \begin{bmatrix} 2 & 0 & 0 \\ 0 & 1 & 3+i \\ 0 & 3-i & 1 \end{bmatrix}$.

17. $M = \begin{bmatrix} -2 & 0 & 0 \\ 0 & \frac{1}{\sqrt{2}} & \frac{i}{\sqrt{2}} \\ 0 & -\frac{i}{\sqrt{2}} & \frac{1}{\sqrt{2}} \end{bmatrix}$.

18. $M = \begin{bmatrix} 2 & 1 & 0 & -1 \\ -1 & 1 & 1 & 1 \\ 0 & 1 & 1 & 0 \\ 1 & 1 & 0 & 0 \end{bmatrix}$.

19. Prove that every skew-Hermitian ($A^* = -A$) matrix is normal.

20. Prove Lemma 6.13 (page 440) for A and B matrices in $M_n(\mathbb{C})$. Specifically prove the following:
    (i) $(A^*)^* = A$.   (ii) $(A + B)^* = A^* + B^*$.
    (iii) $(cA)^* = \bar{c}\, A^*$ for any scalar c.
    (iv) $(AB)^* = B^*A^*$.

21. Prove directly the second part of Theorem 6.14 (page 440): The eigenvectors of an Hermitian matrix corresponding to distinct eigenvalues are orthogonal. Don't use the spectral theorem.

In Exercises 22–25, prove the stated property of a unitary matrix $U \in Mn(\mathbb{C})$.

22. U is **unitary** if, and only if, its column vectors $\{\mathbf{u}_1, \mathbf{u}_2, \ldots, \mathbf{u}_n\}$ determine an orthonormal set of vectors in $\overline{\mathbb{C}^n}$.

23. $\|U\mathbf{x}\| = \|\mathbf{x}\|$ for all $\mathbf{x}$ in $\overline{\mathbb{C}^n}$ (standard inner product: $\|\mathbf{x}\|^2 = \mathbf{x}^*\mathbf{x}$).

24. $\|\mathbf{x}U\| = \|\mathbf{x}\|$ for all $\mathbf{x}$ in $\mathbb{C}^n$ (standard inner product: $\|\mathbf{x}\|^2 = \mathbf{x}\mathbf{x}^*$).

25. If λ is an eigenvalue of U, then $|\lambda| = 1$.

26. Prove that if A is a normal $n \times n$ matrix and $B = U^*AU$, where U is unitary, then the matrix B is normal.

27. Show that the diagonal elements of an Hermitian matrix must be real.

28. Show that the diagonal elements of a skew-Hermitian matrix must be pure imaginary.

29. Prove that if N is a normal matrix then $\|N\mathbf{x}\| = \|N^*\mathbf{x}\|$ for all $\mathbf{x}$ in $\overline{\mathbb{C}^n}$ (standard inner product: $\|\mathbf{x}\| = (\mathbf{x}^*\mathbf{x})^{1/2}$).

30. Prove that if A is Hermitian and $A^2 = 0$, then $A = 0$.

31. Prove the spectral theorem; that is, an $n \times n$ matrix A is normal if, and only if, it has a complete set of *orthonormal* eigenvectors. HINT: Use Theorems 6.15 and 6.16 (pages 441 and 443).

32. Complete the proof of Theorem 6.17 (page 446).

33. Use the matrix $A = \begin{bmatrix} 1 & 0 & 0 \\ 0 & 5 & 3-2i \\ 0 & 3+2i & 1 \end{bmatrix}$ to find

$<\mathbf{u}, \mathbf{v}>$, $\|\mathbf{u}\|$, and $\|\mathbf{v}\|$ for the vectors of Example 5.

34. Show that if A is Hermitian with positive eigenvalues, then the function $<\mathbf{u}, \mathbf{v}> = \mathbf{u}A\mathbf{v}^*$ is an inner product in $\mathbb{C}^n$, the space of n-dimensional complex *row* vectors.

35. Use the matrix of Exercise 33 to find $<\mathbf{u}, \mathbf{v}>$, $\|\mathbf{u}\|$, and $\|\mathbf{v}\|$ for the vectors $\mathbf{u} = (1, i, -i)$ and $\mathbf{v} = (1, 1+i, 1-i)$ in the inner-product space of Exercise 34.

6.7 EXTENDED APPLICATION *Systems of Differential Equations*

This section is addressed only to those who have studied sufficient calculus to have encountered the concept of a differential equation. In Exercise 41 of Section 3.1 you were asked to verify that the collection $U = D[a, b]$ of all real-valued differentiable functions defined on the closed interval $[a, b]$ is a real vector space. Here let V be the subspace of U consisting of all these functions whose first derivative is also differentiable. Then it can be verified that the transformation $D: V \to U$ that maps each twice-differentiable function f in V into its derivative, namely, $D(f) = f' = df/dx$, is a linear transformation (see Exercise 19).

In many applications of mathematics to physics, chemistry, economics, and so forth, the ideas considered are expressible in the form of an equation involving functions and their derivatives. Such equations are called **differential equations**. One of the simplest differential equations is the equation

$$y' = ay, \tag{6.21}$$

where $y' = \dfrac{dy}{dx}$ is the derivative of the function $y = f(x)$ and a is a constant.

Expressed in linear operator terms, the differential equation of (6.21) becomes

$$D(y) = ay.$$

In other words, if a nonzero y that satisfies (6.21) exists, then a is an *eigenvalue* of the linear transformation D (derivative), and the function y is an *eigenvector* of D corresponding to a. It can be verified that the eigenspace corresponding to the eigenvalue a is spanned by the function $y = e^{ax}$. Thus every solution to (6.21) is a scalar multiple of this one and has the form

$$y = ce^{ax}, \tag{6.22}$$

where c is an arbitrary real number (scalar). This solution (6.22) is called the **general solution** to the differential equation (6.21).

If a particular problem generating the differential equation imposes additional conditions that allow us to identify a particular vector in the eigenspace of solutions, we call that vector a **particular solution** of the differential equation.

Example 1   Solve the differential equation

$$y' = 3y \qquad\qquad (6.23)$$

subject to the *initial condition* $y(0) = 5$.

The general solution to equation (6.23) is $y = ce^{3x}$. The fact that $y(0) = 5$ gives us

$$5 = ce^{3(0)} = ce^0 = c,$$

so the particular solution we seek to (6.23) is $y = 5e^{3x}$.   ∎

While there are many ways in which linear algebra aids in the solution of more complicated differential equations than the simple one of (6.21), we indicate only one such idea here. You are referred to texts on differential equations for an exposition of the other areas. Since we have the tools for solving systems of linear equations, we consider systems of linear differential equations of the form

$$y'_1 = a_{11}y_1 + a_{12}y_2 + \cdots + a_{1n}y_n,$$
$$y'_2 = a_{21}y_1 + a_{22}y_2 + \cdots + a_{2n}y_n,$$
$$\cdot \qquad \cdot \qquad\qquad \cdot \qquad \cdot$$
$$y'_i = a_{i1}y_1 + a_{i2}y_2 + \cdots + a_{in}y_n, \qquad\qquad (6.24)$$
$$\cdot \qquad \cdot \qquad\qquad \cdot \qquad \cdot$$
$$y'_n = a_{n1}y_1 + a_{n2}y_2 + \cdots + a_{nn}y_n,$$

where $y_i = f_i(x)$ are functions to be determined and each a_{ij} is a given real number (constant), $i, j = 1, 2, \ldots, n$. Let $A = (a_{ij})$ be the $n \times n$ coefficient matrix of the system (6.24) and let the two column vectors ($n \times 1$ matrices) Y and Y' be as follows:

$$Y = \begin{bmatrix} y_1 \\ y_2 \\ \vdots \\ y_n \end{bmatrix} \quad \text{and} \quad Y' = \begin{bmatrix} y'_1 \\ y'_2 \\ \vdots \\ y'_n \end{bmatrix},$$

where $y'_i = dy_i/dx = f'_i(x)$ is the derivative of the function $y_i = f_i(x)$. Then the system (6.24) can be written as the matrix equation

$$Y' = AY.$$

Example 2   Suppose that we have the system of linear differential equations

$$y'_1 = 2y_1,$$
$$y'_2 = -y_2, \qquad\qquad (6.25)$$
$$y'_3 = 7y_3.$$

Write the system in matrix form, solve it, and find the particular solution that satisfies the given initial conditions

$$y_1(0) = 1, \qquad y_2(0) = -1, \qquad \text{and} \qquad y_3(0) = 4.$$

The matrix form of (6.25) is

$$Y' = \begin{bmatrix} 2 & 0 & 0 \\ 0 & -1 & 0 \\ 0 & 0 & 7 \end{bmatrix} Y.$$

Since the system (6.25) consists entirely of equations of the form (6.21), which involves only a single function in each equation, we can solve each of them as we did in Example 1. Thus the general solution to this system is

$$y_1 = c_1 e^{2x}, \qquad y_2 = c_2 e^{-x}, \qquad y_3 = c_3 e^{7x}.$$

The particular solution that satisfies the given initial conditions is obtained as follows:

$$\begin{aligned} y_1(0) &= 1 &&= c_1 e^{2(0)} = c_1, \\ y_2(0) &= -1 &&= c_2 e^{-(0)} = c_2, \\ y_3(0) &= 4 &&= c_3 e^{7(0)} = c_3. \end{aligned}$$

Therefore, the particular solution to the system (6.25) is

$$Y = \begin{bmatrix} e^{2x} \\ -e^{-x} \\ 4e^{7x} \end{bmatrix}. \quad \blacksquare$$

Now the system (6.25) was easily solved. This was because each equation involved only one function; the matrix A of the system was diagonal. To handle a more complicated system $Y' = AY$ when A is *not* diagonal is, of course, more difficult. It should nevertheless remind you of our previous work in this chapter. We shall try to replace the more complicated system with another system of equations whose matrix of coefficients *is* diagonal. That is, we shall try to change the variables so as to obtain a diagonal coefficient matrix. The changed system will result from replacing each function $y_i = f_i(x)$ with a linear combination of new functions $z_i = z_i(x)$ as in the form

$$\begin{aligned} y_1 &= p_{11}z_1 + p_{12}z_2 + \cdots + p_{1n}z_n, \\ y_2 &= p_{21}z_1 + p_{22}z_2 + \cdots + p_{2n}z_n, \\ &\ \ \vdots \qquad\qquad\qquad\qquad\ \ \vdots \\ y_i &= p_{i1}z_1 + p_{i2}z_2 + \cdots + p_{in}z_n, \\ &\ \ \vdots \qquad\qquad\qquad\qquad\ \ \vdots \\ y_n &= p_{n1}z_1 + p_{n2}z_2 + \cdots + p_{nn}z_n. \end{aligned} \qquad \textbf{(6.26)}$$

In matrix form this is

$$Y = PZ.$$

For this substitution we wish to determine the entries p_{ij} of P in such a way that the new system of differential equations in the new but unknown functions $z_i = z_i(x)$, $i = 1, 2, \ldots, n$ has a diagonal matrix of coefficients. It is easy to see that by differentiating each of the equations in (6.26) with respect to x we obtain the system of differential equations

$$Y' = PZ'$$

since the entries p_{ij} in P are constants. Substituting this result in the original system of differential equations, we have

$$Y' = AY$$
$$PZ' = APZ.$$

If the matrix P is nonsingular, we may multiply both sides of this last equation above by P^{-1} to obtain the following new system of differential equations:

$$Z' = P^{-1}APZ = MZ.$$

If $M = P^{-1}AP = D$ is a diagonal matrix, we will have reached our goal. Therefore, it is clear that the matrix P of our substitution (6.26) should be the matrix P that diagonalizes the matrix A. Hence, if A is diagonalizable, we can easily solve the new system and ultimately the original system. Consider the following example, where we outline the procedure.

Example 3   Solve the system of differential equations

$$y'_1 = 2y_1 + 3y_2,$$
$$y'_2 = 7y_1 - 2y_2.$$

Step 1   Write the system in matrix form as $Y' = AY$, where

$$A = \begin{bmatrix} 2 & 3 \\ 7 & -2 \end{bmatrix}.$$

Step 2   Find a matrix P that diagonalizes A (if possible).

The eigenvalues of A are the roots of its characteristic equation $\lambda^2 - 25 = 0$. Thus we can find the eigenvectors that correspond to the eigenvalues $\lambda_1 = 5$ and $\lambda_2 = -5$. We have done this earlier in the chapter. Two such eigenvectors are

For $\lambda_1 = 5$: $\mathbf{p}_1 = \begin{bmatrix} 1 \\ 1 \end{bmatrix}$ and   for $\lambda_2 = -5$: $\mathbf{p}_2 = \begin{bmatrix} 3 \\ -7 \end{bmatrix}.$

So A is diagonalized by the matrix

$$P = \begin{bmatrix} 1 & 3 \\ 1 & -7 \end{bmatrix}.$$

Step 3   Make the substitutions $Y = PZ$ and $Y' = PZ'$ to obtain a new diagonal system $Z' = DZ$, where $D = P^{-1}AP$. You can verify that in this example

$$D = P^{-1}AP = \begin{bmatrix} 5 & 0 \\ 0 & -5 \end{bmatrix}, \quad \text{so} \quad Z' = \begin{bmatrix} 5 & 0 \\ 0 & -5 \end{bmatrix}Z.$$

Step 4   Solve the system $Z' = DZ$. In this example the new system is $z'_1 = 5z_1$, $z'_2 = -5z_2$ and its general solution is $z_1 = c_1 e^{5x}$, $z_2 = c_2 e^{-5x}$.

Step 5   Determine the solutions Y from the substitution $Y = PZ$. In our example

$$Y = PZ = \begin{bmatrix} 1 & 3 \\ 1 & -7 \end{bmatrix}\begin{bmatrix} c_1 e^{5x} \\ c_2 e^{-5x} \end{bmatrix} = \begin{bmatrix} c_1 e^{5x} + 3c_2 e^{-5x} \\ c_1 e^{5x} - 7c_2 e^{-5x} \end{bmatrix},$$

or $y_1 = c_1 e^{5x} + 3c_2 e^{-5x}, \qquad y_2 = c_1 e^{5x} - 7c_2 e^{-5x}.$

To find a particular solution one would now use whatever initial conditions are imposed by the problem to evaluate the constants c_1 and c_2.   ∎

In this discussion we have assumed that the matrix A of the given system of differential equations is diagonalizable. When this is not the case, other more complicated methods are needed. Of course, we always have Schur's theorem and A can be made upper-triangular. But we leave this complication, as applied to differential equations, for the specific texts in that subject.

Vocabulary

In this section we have shown how to apply the concepts of linear algebra to certain systems of differential equations. We list below the new specialized terms to add to your vocabulary. Be sure that you review these until you feel comfortable with them.

differential equation   particular solution
general solution   initial conditions

Exercises 6.7

In Exercises 1–6, find the general solution to each of the following systems of differential equations.

1. $y'_1 = 3y_1 - y_2,$
       $y'_2 = 2y_1.$

2. $y'_1 = 3y_1 + y_2,$
       $y'_2 = y_1 + 3y_2.$

3. $y'_1 = y_1 + 4y_2,$
       $y'_2 = 2y_1 + 3y_2.$

4. $y'_1 = \frac{1}{2}y_1 + 3y_2,$
       $y'_2 = 3y_1 - y_2.$

5. $y'_1 = -y_1 + 2y_2 + 2y_3,$
       $y'_2 = 2y_1 - y_2 + 2y_3,$
       $y'_3 = 2y_1 + 2y_2 - y_3.$

6. $y'_1 = y'_2 = y'_3 = y_1 + y_2 + y_3.$
       HINT: Write three equations.

In Exercises 7–12, find the particular solution to each of the systems of differential equations in Exercises 1–6 respectively that satisfies the given initial conditions.

7. For the system of Exercise 1: $y_1(0) = 1$, $y_2(0) = -1$.

8. For the system of Exercise 2: $y_1(0) = -2$, $y_2(0) = 3$.

9. For the system of Exercise 3: $y_1(0) = -1$, $y_2(0) = 5$.

10. For the system of Exercise 4: $y_1(0) = 4$, $y_2(0) = -2$.

11. For the system of Exercise 5: $y_1(0) = y_3(0) = 1$, $y_2(0) = 0$.

12. For the system of Exercise 6: $y_1(0) = 1$, $y_2(0) = 2$, $y_3(0) = -5$.

---

13. Find the general solution to the system of differential equations

$$\begin{aligned} y'_1 &= 8y_1 + 9y_2 + 9y_3, \\ y'_2 &= 3y_1 + 2y_2 + 3y_3, \\ y'_3 &= -9y_1 - 9y_2 - 10y_3. \end{aligned}$$

14. Find the particular solution to the system in Exercise 13 that satisfies the initial conditions

$$y_1(0) = y_2(0) = 1, \qquad y_3(0) = 0.$$

15. Find the general solution to the system of differential equations

$$\begin{aligned} y'_1 &= 2y_1 - y_2, \\ y'_2 &= -y_1 + 2y_2, \\ y'_3 &= \qquad\quad y_3 - y_4, \\ y'_4 &= \qquad\quad -y_3 + y_4. \end{aligned}$$

16. Solve the *second-order* differential equation $y'' - y' = 6y$. HINT: Set $y_1 = y$, $y_2 = y'$; then

note that $y'_2 = 6y_1 + y_2$ and $y'_1 = y_2$. Solve the system.

17. Use the method of Exercise 16 to solve

$$y'' + 2y' - 8y = 0.$$

18. Solve the differential equation

$$y''' + y'' - 17y' + 15y = 0.$$

HINT: Expand the method of Exercise 16.

19. Prove that the transformation $D: V \to U$ that maps each twice-differentiable function f into its derivative $D(f) = f'$ is a linear transformation.

20. Prove that every solution of the *first-order* differential equation $y' = ay$ (where a is a constant) is of the form $y = ce^{ax}$, by letting $y = f(x)$ be any solution and showing that it is a scalar multiple of $f_0(x) = e^{ax}$. HINT: Show that $f(x)e^{-ax}$ is always a constant.

CHAPTER 6 SUMMARY

In this chapter, we discussed some very important concepts associated with matrices and linear transformations. These included the eigenvalues and corresponding eigenvectors, the associated eigenspaces, the characteristic polynomial and characteristic equation for a given matrix or transformation. We also developed some criteria for a given $n \times n$ matrix to be similar to a diagonal matrix. Important classes of matrices were introduced, such as the real symmetric matrices, unitary, Hermitian, and normal matrices. Again diagonalization of such matrices was considered. In particular, we saw that any real symmetric matrix is similar to a diagonal matrix whose diagonal elements are its eigenvalues. Moreover every complex $n \times n$ matrix was shown to be similar to an upper-triangular matrix with its eigenvalues on the main diagonal. For normal matrices we even have the sharper result that it is similar to a diagonal matrix. These ideas were applied to well-known standard mathematical models such as the conic sections and quadric surfaces and their associated quadratic forms culminating in the principal axis theorem. Further applications of these matrix methods to Markov chains as well as to systems of differential equations concluded the chapter.

CHAPTER 6 REVIEW EXERCISES

1. Show that the vector $\mathbf{v} = (1, -1)$ is an eigenvector of the linear transformation T: $\mathbb{R}^2 \to \mathbb{R}^2$ given the formula
$$T(x, y) = (-18x - 20y, 15x + 17y).$$

2. What is the eigenvalue of the transformation T in Exercise 1 to which the given eigenvector $\mathbf{v} = (1, -1)$ corresponds?

3. Write the standard matrix representation for the transformation T of Exercise 1 and find all of its eigenvalues. Also find bases for the corresponding eigenspaces.

4. Let $A = \begin{bmatrix} 2 & 0 & 0 \\ -1 & -1 & 6 \\ 0 & 8 & 1 \end{bmatrix}$.

Find the eigenvalues and corresponding eigenvectors for A.

5. Is the matrix A in Exercise 4 similar to a diagonal matrix? If so, to what diagonal matrix is it similar?

6. Use the Cayley-Hamilton theorem (See Exercise 50 of Section 6.2) to find A^{-1} for the matrix A of Exercise 4, if it exists. Check your result.

7. Find the eigenvalues and corresponding eigenvectors for the matrix
$$K = \begin{bmatrix} -1 & -1 & -1 \\ -1 & -1 & -1 \\ -1 & -1 & -1 \end{bmatrix}.$$

8. Is it possible to find an orthogonal matrix P such that $P^T K P$ is a diagonal matrix where K is the matrix of Exercise 7? If so, find such a matrix P.

9. Find the standard form and sketch the graph of the conic section whose equation is $9 - 4xy = 0$.

10. Find the standard form and sketch the graph of the conic section whose equation is $x^2 - 2xy + 3y^2 = 1$.

11. Find a unitary matrix U such that $U^* M U$ is an upper-triangle matrix where
$$M = \begin{bmatrix} -i & 0 & 0 \\ 0 & -3 & 2 \\ 0 & 2 & -3 \end{bmatrix}.$$

12. Use the techniques of Section 6.7 to find the general solution to the system of differential equations:
$$y'_1 = 2y_1,$$
$$y'_2 = 3y_1 - y_2.$$

CHAPTER 6 SUPPLEMENTARY EXERCISES AND PROJECTS

1. Let A be a 2×2 matrix with trace u and determinant v. Find the characteristic polynomial of A.

2. Recall that an $n \times n$ matrix A is *nilpotent* if, and only if, $A^k = O$, for some positive integer k. Prove that all of the eigenvalues of a nilpotent matrix are zero.

3. Use the Cayley-Hamilton theorem to prove that if all the eigenvalues of an $n \times n$ matrix A are zero, then A is nilpotent. This is the converse of Exercise 2.

4. Prove that if an $n \times n$ matrix A is similar to a diagonal matrix, then so is A^T.

5. Prove that if an $n \times n$ matrix A is a real normal matrix, then so is A^T.

6. Prove that if an $n \times n$ matrix A is a skew-symmetric matrix, then so is A^T.

7. Prove that if an $n \times n$ matrix A is an orthogonal matrix, then so is A^T.

8. Recall that an $n \times n$ matrix A is *idempotent* if, and only if, $A^2 = A$. Prove that if A is a nonsingular idempotent matrix, then so is A^{-1}.

9. Prove that all of the eigenvalues of an idempotent matrix are either *zero* or *one*.

10. Show that the eigenvalues of the matrix $\begin{bmatrix} a & 0 & b \\ c & d & c \\ b & 0 & a \end{bmatrix}$ are $a - b$, $a + b$, and d.

Project 1. (Numerical)

Consider the matrix $A = \begin{bmatrix} 1 & 0 & 0 \\ -1 & 3 & 0 \\ 0 & 4 & -6 \end{bmatrix}$.

a. Find the eigenvalues and corresponding eigenvectors for A.

b. Is A similar to a diagonal matrix D? If so, find such a diagonal matrix D, and show that A and D are similar. If not, why not?

c. Recall that the characteristic polynomial of A is $p_A(\lambda) = \det(\lambda I - A)$. Find $p_A(A)$.

d. Use the result of part (c) to show that $(A - I)(A - 3I)(A + 6I) = O$ (the 3×3 zero matrix).

e. Use the results of parts (c) and (d) to show that A^{-1} exists and is, in fact, equal to a certain quadratic matrix polynomial in A. What is this polynomial?

f. Use the result of part (e) to find A^{-1}.

g. Find the eigenvalues and corresponding eigenvectors for the 4×4 matrix

$$M = \begin{bmatrix} 1 & 0 & 0 & -2 \\ 2 & -1 & 0 & -1 \\ -1 & 0 & 3 & 2 \\ -2 & 0 & 0 & 1 \end{bmatrix}.$$

h. Is M similar to a diagonal matrix D'? If so, find such a diagonal matrix D', and demonstrate by matrix multiplication that M and D' are similar. If M is not similar to a diagonal matrix demonstrate why it is not.

i. Recall that the characteristic polynomial of M is $p_M(\lambda) = \det(\lambda I - M)$. Find $p_M(M)$.

j. Use the result of part (i) to show that $(M + I)^2(M - 3I)^2 = O$ (the 4×4 zero matrix).

k. Show that the matrix $(M^2 - 2M - 3I)$ is nilpotent of index 2.

l. Find M^{-1} as a polynomial in M, if it exists.

Project 2. *(Theoretical) Matrix Congruence*

*The $n \times n$ matrix A is said to be **congruent** to the $n \times n$ matrix B if there exists a nonsingular matrix P so that $B = P^T A P$.*

a. Is every $n \times n$ real symmetric matrix congruent to a diagonal matrix? Why?

b. Show that every $n \times n$ matrix A is congruent to itself; that is, congruence is a *reflexive* relation on $M_n(\mathbb{R})$.

c. Show that if A is congruent to B then B is congruent to A; that is, congruence is a *symmetric* relation on $M_n(\mathbb{R})$.

d. Use the results of parts (b) and (c), together with a demonstration that congruence is a *transitive* relation on $M_n(\mathbb{R})$, to prove that congruence is an *equivalence relation* on $M_n(\mathbb{R})$. That is, prove that if A is congruent to B and B is congruent to C, then A is congruent to C.

e. Show that every $n \times n$ *nonsingular real symmetric matrix* A is congruent to its inverse A^{-1}.

f. Show that congruent matrices are equivalent (see Project 4 in Chapter 3), but $n \times n$ equivalent matrices need not be congruent.

g. Prove that two $n \times n$ real symmetric matrices A and B are congruent if, and only if, they have the same number of positive eigenvalues and the same number of negative eigenvalues. (An eigenvalue λ with algebraic multiplicity m is counted m times.)

h. If an $n \times n$ real symmetric matrix A has p positive eigenvalues and q negative eigenvalues, the number $s = p - q$ is called the **signature** of A. Prove that two $n \times n$ real symmetric matrices A and B are congruent if, and only if, they have the same rank and the same signature.

Project 3. *(Theoretical) Hermitian Congruence*

*The $n \times n$ complex matrix A is said to be **Hermitely congruent** to the $n \times n$ matrix B if there exists a nonsingular matrix P so that $B = P^* A P$.*

a. Prove that Hermitian congruence is a *reflexive* relation on $M_n(\mathbb{C})$. (See Project 2.)

b. Prove that Hermitian congruence is a *symmetric* relation on $M_n(\mathbb{C})$.

c. Prove that Hermitian congruence is a *transitive* relation on $M_n(\mathbb{C})$. Hence, Hermitian congruence is an *equivalence relation* on $M_n(\mathbb{C})$.

d. Prove that every $n \times n$ Hermitian matrix A of rank r is Hermitely congruent to the $n \times n$ (block) matrix

$$B = \begin{bmatrix} I_p & 0 & 0 \\ 0 & -I_{r-p} & 0 \\ 0 & 0 & 0 \end{bmatrix},$$

where the integer p, called the **index** of A, is uniquely determined by A.

e. Prove that two $n \times n$ Hermitian matrices A and B are Hermitely congruent if, and only if, they have the same rank and the same index.

Project 4. *(Computational) Differential Equations*

An nth-order differential equation may be regarded as a system of first-order differential equations by making the identification $y = y_1$, $y' = y_2$, $y'' = y_3$, and so on. Solve the differential equations of this project by making this identification and solving the resulting linear system of first-order equations.

a. Show that $y''' - 4y'' + y' + 6y = 0$ can be replaced by the following system of linear differential equations:

$$\begin{aligned} y'_1 &= & y_2, \\ y'_2 &= & y_3, \\ y'_3 &= -6y_1 - y_2 + 4y_3. \end{aligned}$$

b. Solve the system of first-order differential equations of part (a), and demonstrate that this gives a solution to the given third-order differential equation.

c. Use the method of this project to solve the differential equation $y'' - 2y' - 3y = 0$.

d. Use the method of this project to solve the differential equation $y'' - 3y' + 2y = 0$.

e. Use the method of this project to solve the differential equation $y''' - 7y' + 6y = 0$.

f. Generalize the results of Section 6.7 to obtain the following theorem:

Theorem   Let A be an $n \times n$ diagonalizable matrix, and let P be a nonsingular matrix such that $P^{-1}AP = D = \text{diag}[\lambda_1, \lambda_2, \ldots, \lambda_n]$. Then the system of first-order differential equations $Y' = AY$ has the general solution

$$Y = P \, \text{diag}\,[e^{\lambda_1 x}, e^{\lambda_2 x}, \ldots, e^{\lambda_n x}] \cdot (c_1, c_2, \ldots, c_n)^T,$$

where $c_1, \ldots, c_n$ are constants.

---

g. Let k_1, k_2, and k_3 be real numbers. Show that the matrix

$$C(p) = \begin{bmatrix} 0 & 1 & 0 \\ 0 & 0 & 1 \\ k_3 & k_2 & k_1 \end{bmatrix}$$

has characteristic polynomial $p(x) = x^3 - k_1 x^2 - k_2 x - k_3$. $C(p)$ is called the **companion matrix** for $p(x)$.

h. Refer to part (g) and describe an $n \times n$ companion matrix whose characteristic polynomial is $p(x) = x^n - k_1 x^{n-1} - k_2 x^{n-2} - \cdots - k_{n-1}x - k_n$.

i. Show that the nth-order differential equation

$$y^{(n)} + k_{n-1}y^{(n-1)} + k_{n-2}y^{(n-2)} + \cdots + k_1 y' + k_0 y = 0$$

has as its associated matrix (see Section 6.7 and parts (a) and (b)) the companion matrix whose characteristic polynomial is

$$p(x) = x^n + k_{n-1}x^{n-1} + k_{n-2}x^{n-2} + \cdots + k_1 x + k_0.$$

j. Use part (i) and those parts preceding it to show that if the polynomial
$$p(x) = x^n + k_{n-1}x^{n-1} + k_{n-2}x^{n-2} + \cdots + k_1x + k_0$$
has n distinct real zeroes $\lambda_1, \lambda_2, \ldots, \lambda_n$, then the nth-order differential equation
$$y^{(n)} + k_{n-1}y^{(n-1)} + k_{n-2}y^{(n-2)} + \cdots + k_1y' + k_0y = 0$$
has the set $\{e^{\lambda_1 x}, e^{\lambda_2 x}, \ldots, e^{\lambda_n x}\}$ as a basis for its solution space.

k. Use the result of part (j) to solve the 6th-order differential equation
$$y^{(6)} - 14y^{(4)} + 49y'' - 36y = 0.$$
HINT: $x^6 - 14x^4 + 49x^2 - 36 = (x^2 - 9)(x^2 - 4)(x^2 - 1)$.

Project 5. *(Theoretical) Some Interesting Matrices*

a. Let A and B be $n \times n$ matrices with the property that $AB = BA$ (A and B commute). Suppose that λ is an eigenvector of A with a one-dimensional eigenspace spanned by the eigenvector $\mathbf{x}$. Show that there exists a scalar μ such that $B\mathbf{x} = \mu\mathbf{x}$; that is, μ is an eigenvalue of B with associated eigenvector $\mathbf{x}$. HINT: first show that $B\mathbf{x}$ is an eigenvector of A associated with the eigenvalue λ.

b. Given the conditions of part (a), show that $\lambda + \mu$ is an eigenvalue of $A + B$ and $\lambda\mu$ is an eigenvalue of AB.

c. An $n \times n$ upper-triangular Toeplitz matrix is a matrix of the form

$$T = \begin{bmatrix} a_0 & a_1 & a_2 & \cdots & a_{n-2} & a_{n-1} \\ 0 & a_0 & a_1 & a_2 & \cdots & a_{n-2} \\ \vdots & \vdots & \vdots & \vdots & \vdots\vdots\vdots & \vdots \\ & & & & \cdots & a_2 \\ 0 & 0 & 0 & & \cdots & a_1 \\ 0 & 0 & 0 & & \cdots & a_0 \end{bmatrix} \qquad \text{Let } N = \begin{bmatrix} 0 & 1 & 0 & \cdots & 0 \\ 0 & 0 & 1 & \cdots & 0 \\ \vdots & \vdots & \vdots & \vdots\vdots\vdots & \vdots \\ 0 & 0 & 0 & \cdots & 1 \\ 0 & 0 & 0 & \cdots & 0 \end{bmatrix}$$

be the $n \times n$ nilpotent matrix with *ones* on the "superdiagonal" and *zeros* elsewhere: $N = (\eta_{ij})$; where $\eta_{ij} = 1$, for $j = i + 1$ and $\eta_{ij} = 0$, for $j \neq i + 1$. Write T as a polynomial in N.

d. Use the result of part (c) to prove that $n \times n$ upper-triangular Toeplitz matrices commute with each other.

e. (William Watkins) Let A and B be $n \times n$ real matrices and further suppose S is an $n \times n$ nonsingular complex matrix such that $S^{-1}BS = A$. Show that there exist real matrices P and Q such that $(P + iQ)A = B(P + iQ)$ so that $PA = BP$ and $QA = BQ$. Conclude that if either P or Q is nonsingular, A and B are similar over the reals.

f. (William Watkins) Use the information in part (e) to show that if both P and A are singular real matrices there is a real number r so that $P + rQ$ is a nonsingular real matrix and hence A and B are still similar over the reals. HINT: Show that there is a real number r that is not a zero of the polynomial $p(x) = \det(P + xQ)$.

APPENDICES

The Appendices to the text are to help you as you study the material. Refer to them freely as you work your way through the text.

Appendices A.1 and A.2 review ideas that should have been part of your prerequisite background but, perhaps, were not; or, perhaps these ideas were covered in a prerequisite course that you took some time ago and have forgotten. Since we make free use of these notions in the text, we give brief summaries of the main points here. However, these are merely thumbnail sketches, as it were. If you need more information, you are referred to more elaborate treatments in standard textbooks. Most standard college algebra, trigonometry, and precalculus texts discuss the complex numbers. Many of these also discuss functions and sets.

Appendix A.3 gathers together several of the linear algebra algorithms discussed in the text. They are presented in simplified form for convenient reference. You should consult the appropriate sections of the text for the full explanations of the procedures that are listed in this appendix. In many cases your college computer center, your personal computer, or your calculator will use algorithms different from those given in the text for actual numerical computations. You should always consult the appropriate manuals for these technological helps. These manuals should provide you with a discussion of the methods that are used and their specific limitations. Our algorithms are for hand calculation and do not necessarily yield the result in the shortest possible time.

We have listed several books, papers, and software programs in **Appendix A.4. Appendix A.5** contains a listing of the Greek alphabet and frequently used symbols.

The listing of Appendices is:

A.1 *THE COMPLEX NUMBERS*

As you are aware, the quadratic polynomial $y = x^2 + 1$ has a graph which is always above the x-axis and does not cross it. The quadratic equation $x^2 + 1 = 0$ has no real number solutions; the number $x^2 + 1$ is always positive. On the other hand, the convenience of having solutions to *all* quadratic equations motivated mathematicians to extend the real number system $\mathbb{R}$ to a larger system $\mathbb{C}$ in order to include such solutions. We will look at this larger number system: the complex numbers, $\mathbb{C}$.

> **Definition**   A **complex number** is any number of the form $a + bi$, where a and b are real numbers and i satisfies the equation $i^2 = -1$.

Note that this new number i, which we introduce here, is *not* a real number; every real number a satisfies $a^2 \geq 0$. We agree to identify the real number a with the complex number $a + 0i$ so that every real number may be viewed also as a complex number. The new number i is identified with the complex number $0 + 1i$, and more generally, we identify bi with $0 + bi$, where b is a real number. A mathematically more rigorous development of the complex number system is possible, but we'll stay with this more informal approach here.

In the symbol for the complex number $z = a + bi$, the real number a is called the *real part* of z, or Re(z), and the real coefficient b of bi is called the *imaginary part* of z, or Im(z). Thus, in the complex number $z = 3 - 5i$, Re(z) = 3 and Im(z) = -5.

The acceptance of complex numbers as numbers was aided by noting that, just as the real numbers can be put into one-to-one correspondence with the points on "the real line," complex numbers can be made to correspond one-to-one with points in the plane. That is, the number $a + bi$ corresponds to the point (a, b) of the "complex plane." See Figure A1.1 where we have sketched the points corresponding to the five complex numbers $0 + i$, $3 - 2i$, $-3 + 5i$, $4 + 0i$, and $-4 - 3i$. The horizontal axis is called the *real axis,* and the vertical axis is called the *imaginary axis* in the scheme often referred to as the *Gaussian plane* or an *Argand diagram.*

The algebraic rules for adding and subtracting complex numbers, then, are the same as if we were adding or subtracting the vectors in the plane from the origin to the point representing the complex number. Thus, for each pair $a + bi$ and $c + di$ of complex numbers:

> $$(a + bi) + (c + di) = (a + c) + (b + d)i$$
> $$(a + b) - (c + di) = (a - c) + (b - d)i.$$

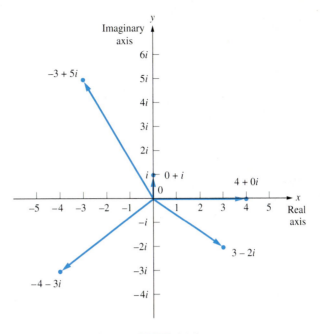

FIGURE A1.1

The definition for the multiplication of complex numbers is motivated by the ordinary rules for real number algebra and the fact that $i^2 = -1$. We, therefore, define multiplication of the two complex numbers $a + bi$ and $c + di$ as follows.

$$(a + bi)(c + di) = (ac - bd) + (ad + bc)i$$

Notice that if we approach this as a real number algebra product of two binomials and if i is treated as a formal symbol, the product is just what we would expect to get. Remembering that $i^2 = -1$, the term resulting from multiplying (bi) and (di) is $(bd)i^2 = (bd)(-1) = -bd$. This is part of the real part of the product.

Dividing complex numbers is less obvious. Let us first consider inverting a nonzero complex number. In general, we note that for $(a + bi)^{-1}$, $a + bi \neq 0$, to be written in the prescribed form for a complex number we need to do a little work, as follows.

$$\frac{1}{a + bi} = \frac{(a - bi)}{(a + bi)(a - bi)} = \frac{(a - bi)}{(a^2 + b^2)} = \left(\frac{a}{a^2 + b^2}\right) + i\left(\frac{-b}{a^2 + b^2}\right)$$

Consequently, to divide the complex number $a + bi$ by the nonzero complex number $c + di$, one adopts the same approach we just used to find $\dfrac{1}{a + bi}$. That is, multiply both the numerator and the denominator of the fraction $\dfrac{a + bi}{c + di}$ by the number $c - di$. This number is called the *conjugate* of the denominator. Thus,

$$\frac{a + bi}{c + di} = \frac{(a + bi)(c - di)}{(c + di)(c - di)} = \frac{(ac + bd) + (bc - ad)i}{c^2 + d^2}$$

$$= \left(\frac{ac + bd}{c^2 + d^2}\right) + i\left(\frac{bc - ad}{c^2 + d^2}\right).$$

Example 1 Let $z_1 = 2 + 3i$ and $z_2 = -3 + 4i$. Then we have the following:

$$z_1 + z_2 = (2 + 3i) + (-3 + 4i) = [2 + (-3)] + (3 + 4)i = -1 + 7i;$$
$$z_1 - z_2 = (2 + 3i) - (-3 + 4i) = [2 - (-3)] + (3 - 4)i = 5 - i;$$
$$z_1 z_2 = (2 + 3i)(-3 + 4i)$$
$$= [2(-3) - (3)(4)] + [2(4) + 3(-3)]i = -18 - i;$$
$$\frac{z_1}{z_2} = \frac{(2 + 3i)}{(-3 + 4i)} = \frac{(2 + 3i)(-3 - 4i)}{(-3 + 4i)(-3 - 4i)} = \frac{6 - 17i}{(-3)^2 + (4)^2}$$
$$= \frac{6 - 17i}{25} = \frac{6}{25} - \frac{17}{25}i. \quad\blacksquare$$

In the process of actually dividing two complex numbers, we used the idea of the conjugate of a complex number. This important concept has many applications, so we define it formally and then look at some of its properties.

> **Definition** The **conjugate** of the complex number $z = x + yi$ is the complex number $\bar{z} = x - yi$.

The overbar notation is used for the process of replacing a complex number by its conjugate—called *conjugation*. Therefore, $\overline{a + bi} = a - bi$, $\overline{1 + 3i} = 1 - 3i$, and $\overline{7 - 4i} = 7 - (-4i) = 7 + 4i$, etc. The basic facts about conjugates are the content of the following lemma.

> **Lemma** If z is any complex number, then
>
> (i) $\bar{\bar{z}} = z$.   (ii) $\overline{z_1 + z_2} = \bar{z_1} + \bar{z_2}$.   (iii) $\overline{z_1 - z_2} = \bar{z_1} - \bar{z_2}$.
>
> (iv) $\overline{z_1 z_2} = \bar{z_1} \cdot \bar{z_2}$.   (v) $\overline{\left(\dfrac{z_1}{z_2}\right)} = \dfrac{\bar{z_1}}{\bar{z_2}}$, if $z_2 \neq 0$.

Proof Let $z = x + yi$ be any complex number in $\mathbb{C}$, then $\bar{z} = x - yi$. From this we easily get (i): $\bar{\bar{z}} = \overline{x - yi} = x + yi = z$.

For (ii), let $z_1 = x_1 + y_1 i$ and $z_2 = x_2 + y_2 i$. So $z_1 + z_2 = (x_1 + x_2) + (y_1 + y_2)i$, and its conjugate is

$$\overline{z_1 + z_2} = (x_1 + x_2) - (y_1 + y_2)i = (x_1 - y_1 i) + (x_2 - y_2 i) = \overline{z_1} + \overline{z_2}.$$

Part (iii) follows immediately from (ii), just use $-z_2$ for z_2.

To do (iv), we note that

$$z_1 z_2 = (x_1 x_2 - y_1 y_2) + (x_1 y_2 + y_1 x_2)i.$$

Hence,

$$\overline{z_1 z_2} = (x_1 x_2 - y_1 y_2) - (x_1 y_2 + y_1 x_2)i = (x_1 - y_1 i)(x_2 - y_2 i) = \overline{z_1} \cdot \overline{z_2}.$$

By using the fact that $\left(\dfrac{1}{z}\right) = \left(\dfrac{x}{x^2 + y^2}\right) + i\left(\dfrac{y}{x^2 + y^2}\right) = \dfrac{1}{x - yi} = \dfrac{1}{\bar{z}}$, part (v) follows from part (iv). ◘

We illustrate with the following example.

Example 2 Let $z_1 = 5 - 4i$ and $z_2 = -3 + 14i$. Then we have the following.

$$\overline{\overline{z_1}} = \overline{\overline{5 - 4i}} = \overline{5 + 4i} = 5 - 4i = z_1$$
$$\overline{z_1 + z_2} = \overline{2 + 10i} = 2 - 10i = (5 + 4i) + (-3 - 14i) = \overline{z_1} + \overline{z_2}$$
$$\overline{z_1 - z_2} = \overline{8 - 18i} = 8 + 18i = (5 + 4i) - (-3 - 14i) = \overline{z_1} - \overline{z_2}$$
$$\overline{z_1 z_2} = \overline{41 + 82i} = 41 - 82i = (5 + 4i)(-3 - 14i) = \overline{z_1} \cdot \overline{z_2}$$

Finally, we verify that

$$\frac{z_1}{z_2} = \frac{5 - 4i}{-3 + 14i} = \frac{(5 - 4i)(-3 - 14i)}{(-3 + 14i)(-3 - 14i)} = \frac{-71 - 58i}{205},$$

so

$$\left(\frac{\overline{z_1}}{z_2}\right) = \frac{-71 + 58i}{205} = \frac{5 + 4i}{-3 - 14i} = \frac{\overline{z_1}}{\overline{z_2}}. \quad ■$$

One further concept that we need to consider is that of the *absolute value*, or *norm*, of the complex number z. This coincides with the ordinary absolute value of a real number when z is real (i.e., $z = x + 0i$) and with the concept of the Euclidean norm for the vector (x, y) when we consider $\mathbb{C}$, the complex plane, to be $\mathbb{R}^2$.

Definition The absolute value $|z|$ of the complex number $z = x + yi$ is defined to be the real number $\sqrt{x^2 + y^2} = |x + yi|$.

You can quickly verify that, if z is real, $z = x + 0i$, then

$$|z| = \sqrt{x^2 + 0^2} = \sqrt{x^2} = \begin{cases} x & \text{when } x \geq 0 \\ -x & \text{when } x < 0 \end{cases} = |x|, \quad \text{as we remarked above.}$$

See Chapter 4 and the discussion on Euclidean norms for the correspondence between the absolute value of a complex number and vector norms.

Example 3   If $z = 3 - 4i$, then $|z| = |3 - 4i| = \sqrt{(3)^2 + (-4)^2} = \sqrt{9 + 16} = \sqrt{25} = 5$. Whereas if $z = -1 + 2i$, then $|z| = |-1 + 2i| = \sqrt{(-1)^2 + (2)^2} = \sqrt{1 + 4} = \sqrt{5}$. ∎

A.2 SETS AND FUNCTIONS

Sets

The notation and terminology of set theory is fundamental in present-day mathematics. Having been developed in the nineteenth century principally by George Boole (1815–1864) and Georg F. L. P. Cantor (1845–1918), its effect on both the methods and the content of twentieth-century mathematics has been profound. We shall sketch here a brief and very informal outline of the basic vocabulary of set theory as an aid to understanding the definitions and exposition of the text. You are referred to more detailed treatments for a complete examination of the subject.

In modern mathematics, "set" is most often an undefined concept, a "primitive term" in the logical construct. We do not attempt to define the word but rather use it as a basis for other mathematical definitions. As a primitive term, however, there needs to be general agreement about what a set is. In mathematics, **set** means approximately the same thing as it does in ordinary usage in English; such as a "set of dishes." That is, a set is a collection of objects under some unifying property: a set of golf clubs, a set of table flatware, and so on. Besides *collection*, some synonyms for set are *category, heap, batch, bunch, herd* (of cattle), *flock* (of sheep), *gaggle* (of geese), *clique, covey*, ... etc. The individual objects that make up the set are called its **elements** or its **members.**

In this text we follow a fairly widespread convention that sets are denoted by uppercase Latin letters A, B, C, U, V, and so on, whereas their elements are denoted by lowercase letters a, b, x, u, y, z, etc. The sentence

"The object x belongs to (is an element or member of) the set S,"

is symbolized as

$$x \in S.$$

The negation of this statement is written $x \notin S$; and read "x does not belong to (is not an element of) S."

Some examples of sets for you to consider follow.

Examples of Sets

1. The set of all students in this university.
2. The set of all known planets in our solar system.
3. The set of all red-headed women students in this class.
4. The set of all letters in the English alphabet.
5. The set of all former presidents of the United States.
6. The set of all months of the year.
7. The set $\mathbb{Z}$ of all integers.
8. The set $\mathbb{Q}$ of all rational numbers.
9. The set $\mathbb{R}$ of all real numbers.
10. The set $\mathbb{C}$ of all complex numbers.
11. The set $\mathbb{R}^n$ of all n-tuples $(x_1, x_2, \ldots, x_n)$ of real numbers, n a positive integer.
12. The set of all the letters in the word "Mississippi."
13. The set $[0, 1]$ of all the real numbers $x \in \mathbb{R}$ which satisfy the inequality $0 \leq x \leq 1$; this set is called the *closed* unit interval.
14. The set $(0, 1)$ of all the real numbers $x \in \mathbb{R}$ which satisfy the inequality $0 < x < 1$; this set is called the *open* unit interval.

In describing a set, the standard notation is to use braces { } in one of two ways. The first method, sometimes called the *roster notation,* is to include all of the members of the set between the braces—not parentheses nor brackets. For example, one writes

$$A = \{a, b, c\}$$

for the set consisting of the first three letters of the English alphabet. The roster notation for set number 2 in our list above is $P = \{$Mercury, Venus, Earth, Mars, Jupiter, Saturn, Uranus, Neptune, Pluto$\}$. This method is used most often when there are but a small number of elements in the set, although ellipsis (three dots: ...) sometimes are used to indicate that there are items omitted from the list. For example, the set of all former presidents of the United States (number 5 above) might be written a little ambiguously as

$$\{\text{Washington, Adams, Jefferson, Madison, Monroe}, \ldots\}.$$

The second way of describing a set using the braces notation is more flexible and most widely used. Since a set is a collection of things which satisfy some defining property, say P, one may use the **set-builder notation:**

$$S = \{x: x \text{ has property } P\}.$$

This symbol is read "S is the set of all elements x such that x has property P." The letter x, used in this generic way to represent any element of S, is often called a **variable.**

The set of all former presidents of the United States would then appear as

$$\{p : p \text{ is a former president of the United States}\}.$$

Using set-builder notation, the set of positive integer $\mathbb{Z}^+$ can then be *defined* as

$$\mathbb{Z}^+ = \{x : x \in \mathbb{Z} \text{ and } x \geq 1\}.$$

The **closed unit interval** is *defined* as the following set:

$$[0, 1] = \{x : x \in \mathbb{R} \text{ and } 0 \leq x \leq 1\};$$

and the **closed interval** $[a, b]$ for $a, b \in \mathbb{R}$, $a \leq b$, by

$$[a, b] = \{x : x \in \mathbb{R} \text{ and } a \leq x \leq b\}.$$

The **open interval** (a, b) for $a, b \in \mathbb{R}$, $a < b$, is *defined* by

$$(a, b) = \{x : x \in \mathbb{R} \text{ and } a < x < b\}.$$

Notice that $a \notin (a, b)$, and $b \notin (a, b)$, while both a and b *are* elements of $[a, b]$. Every real number x which is an element of (a, b) is also an element of $[a, b]$, but not conversely. This is one example of the concept of a *subset* of a given set. More precisely,

---

Definition   A set B is a **subset** of a given set A, written $B \subseteq A$, if, and only if, $x \in B$ implies $x \in A$.

Two sets are equal, $A = B$, provided both contain precisely the same elements. Therefore,

---

$A = B$ if, and only if, $B \subseteq A$ and $A \subseteq B$.

For example, the sets $\{2, 4, 6, 8\}$ and $\{4, 8, 6, 2\}$ are equal, whereas the sets $\{1, 3, 5, 7\}$ and $\{3, 5, 7\}$ are not equal, although $\{3, 5, 7\} \subseteq \{1, 3, 5, 7\}$. It is trivially true that if $A = B$, then surely $A \subseteq B$.

Turning the subset symbol around, we have $A \supseteq B$, which is often described by saying that A is a **superset** of B. By definition, $A \supseteq B$ means $B \subseteq A$. If we wish to call attention to the condition that $B \subseteq A$ but they are *not equal*, then we can write $B \subset A$ and call B a **proper subset** of A.

There are some consequences of these definitions that seem a little bit strange when one first encounters them, but in fact need to be considered. *First*, every set is a subset of itself; $A \subseteq A$, for every set A. *Second*, it is sometimes convenient to think of the set with no elements, called **the empty set,** $\varnothing$. Then for every set A, $\varnothing \subseteq A$.

One should be very careful to avoid one of the perils of set theory and set notations. At all times one must distinguish between the *element* x and the *set* $\{x\}$, whose only member is x. These are two different objects. It is something like distinguishing between a peanut and a peanut can with only one peanut in it—the can is not the peanut; or, between a given student and a course in which that student is the only registrant—the course is not the student.

It is very common to use pictorial representations for sets. These representations are called *Venn diagrams.* For example, if A and B are sets, both subsets of some "universal" set U, and $B \subset A$, then we can picture this by the Venn diagram of Figure A2.1.

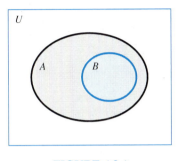

FIGURE A2.1

Given any two sets S and T, we can consider new sets formed from them. In each instance we give the formal definition of the created set, and then sketch a Venn diagram representing the concept. In each case the defined set is shaded in the Venn diagram.

The **complement of T in S,** denoted by $S \backslash T$ (or sometimes $S - T$), is defined as follows.

Definition	$S \backslash T = \{x : x \in S \text{ and } x \notin T\}$

One representation of this concept is the two overlapping sets in Figure A2.2. The gray portion of S is $S \backslash T$.

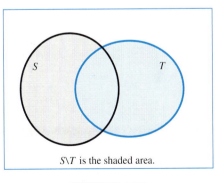

$S \backslash T$ is the shaded area.

FIGURE A2.2

For example, if $S = 2\mathbb{Z} = \{2n: n \in \mathbb{Z}\}$, the even integers, and $T = 3\mathbb{Z} = \{3n: n \in \mathbb{Z}\}$, the multiples of three, then $S \backslash T$ is the set of all those even integers that are not multiples of three; consequently, not multiples of six: $S \backslash T = \{n \in \mathbb{Z}: n = 2k, \text{ but } n \neq 6m; m, k \in \mathbb{Z}\}$.

The **union** of S and T is denoted by $S \cup T$ (see Figure A2.3). This union is defined as follows.

Definition	$S \cup T = \{x: x \in S \text{ or } x \in T\}$

Here "or" means "either $x \in S$ or $x \in T$ or x is in *both* S and T."

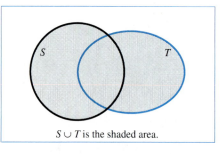

$S \cup T$ is the shaded area.

FIGURE A2.3

For example, if as above $S = 2\mathbb{Z} = \{2n: n \in \mathbb{Z}\}$ and $T = 3\mathbb{Z} = \{3n: n \in \mathbb{Z}\}$, then $S \cup T$ is the set of all those integers that are either even *or* are multiples of three *or* both. Thus,

$$S \cup T = \{n \in \mathbb{Z}: n = 2k, \text{ or } n = 3k, k \in \mathbb{Z}\}.$$

The **intersection** of S and T is denoted by $S \cap T$ (see Figure A2.4). This intersection is defined as follows.

Definition	$S \cap T = \{x: x \in S \text{ and } x \in T\}$

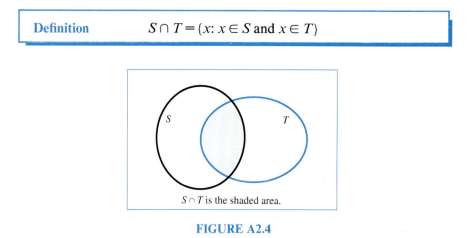

$S \cap T$ is the shaded area.

FIGURE A2.4

For example, if $S = 2\mathbb{Z} = \{2n: n \in \mathbb{Z}\}$ and $T = 3\mathbb{Z} = \{3n: n \in \mathbb{Z}\}$ as above, then $S \cap T$ is the set of all those even integers that are also multiples of three; thus, are multiples of six:

$$S \cap T = \{6n: n \in \mathbb{Z}\} = 6\mathbb{Z}.$$

The **Cartesian product** of S and T is denoted by $S \times T$ (see Figure A2.5). This Cartesian product is defined as follows.

Definition $S \times T = \{(s, t): s \in S \text{ and } t \in T\}$

The idea from which the Cartesian product springs is the usual Cartesian coordinate system with which you are familiar. A finite situation is sketched in Figure A2.5 where we suppose that S is represented horizontally by the indicated dots and T vertically by the indicated dots. Then $S \times T$, the set of ordered pairs (s, t), can be represented by the indicated points in the plane. As a second example, consider as before the sets $S = 2\mathbb{Z} = \{2n: n \in \mathbb{Z}\}$ and $T = 3\mathbb{Z} = \{3n: n \in \mathbb{Z}\}$. Then,

$$S \times T = 2\mathbb{Z} \times 3\mathbb{Z} = \{(2k, 3m): m, k \in \mathbb{Z}\}.$$

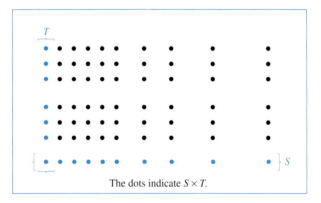

The dots indicate $S \times T$.

FIGURE A2.5

Example   Suppose that we have the sets $S = \{a, b, c, d\}$, $T = \{b, d, f, h\}$, and $R = \{a, e, i\}$. Then the following are true.

Complement: $T \backslash S = \{f, h\}$; $S \backslash T = \{a, c\}$; $S \backslash R = \{b, c, d\}$; $R \backslash S = \{e, i\}$.

Union: $S \cup T = \{a, b, c, d, f, h\} = T \cup S$; $S \cup R = \{a, b, c, d, e, i\}$; $R \cup T = \{a, b, d, e, f, h, i\}$.

Intersection: $S \cap T = \{b, d\} = T \cap S$; $S \cap R = \{a\}$; $R \cap T = \{ \} = \varnothing$.

Cartesian product: $S \times T = \{(a, b), (a, d), (a, f), (a, h), (b, b), (b, d),$
$(b, f), (b, h), (c, b), (c, d), (c, f), (c, h),$
$(d, b), (d, d), (d, f), (d, h)\};$

$R \times T = \{(a, b), (a, d), (a, f), (a, h), (e, b), (e, d),$
$(e, f), (e, h), (i, b), (i, d), (i, f), (i, h)\};$
whereas

$T \times R = \{(b, a), (b, e), (b, i), (d, a), (d, e), (d, i),$
$(f, a), (f, e), (f, i), (h, a), (h, e),$
$(h, i)\}.$ ∎

If two sets, such as T and R in the above example, are such that their intersection is the empty set, $\varnothing$, we say that they are disjoint.

Definition A and B are **disjoint** if, and only if, $A \cap B = \varnothing$.

In the exercises in the text and in the exposition itself you will find many examples of sets.

Functions

The concept of *function* is all pervasive in mathematics. It is as fundamental a mathematical idea as the concepts of *number* and *shape.* Because the concept has developed, sometimes independently, in every branch of mathematics it has several widely used synonyms. What are called *functions* in calculus are called *mappings* in algebra and *transformations* in geometry. In actual fact, these words are used interchangeably by most mathematicians, no matter what branch of mathematics they may be working in.

Like many other mathematical concepts, the idea of a function or transformation or mapping has undergone considerable evolution, beginning with Descartes who, in 1637, used it to mean some positive integral power x^n of a variable x. Our major encounter with the function concept will be in looking at maps from one vector space to another. We also consider collections of certain types of functions as vector spaces (the individual functions are the vectors).

A formal definition of *function* can be made using only the ideas of set theory, a function being a specific subset of the Cartesian product of the domain and the codomain. However, the intuitive idea is more easily understood and, for that matter, more easily used. In the intuitive definition every function consists of three parts or constituents; without them there is no function. The three parts are two sets, and a rule connecting them.

Definition A **function** f from a given set D, called its **domain,** into a second set S, called its **codomain,** is a rule which assigns to each $x \in D$ a unique element $f(x)$ of S.

Any calculator user has had experience with the function concept and "function keys" on the calculator. The domain of a given calculator function, say the ⌊sin⌋ key, is some set of numbers, in this case all of $\mathbb{R}$. The rule is contained in the circuitry of the calculator, and the codomain is the set of numbers that can appear after the function key is pressed. One enters x, pushes ⌊sin⌋ and $\sin(x)$ is read.

Another example of the function concept is the fact that the area of a circle is a "function" of the length of the radius of the circle. The domain of this function is all of the positive real numbers (nonnegative real numbers if you allow for a circle of zero radius) $\mathbb{R}^+$, its codomain is the same set, and the rule is given by the formula $A = \pi r^2$. In this statement r is a (variable) generic member of the domain and A is the member of the codomain associated with it by the rule, "multiply r by itself and then by the number π."

There are several ways to symbolize the function concept. The most explicit is to name the function, say f, and write

$$f: D \rightarrow S$$

indicating that f "maps" elements of the *domain D* into the *codomain S.* The trigonometric function *sine* would appear in this notation as

$$\sin: \mathbb{R} \rightarrow \mathbb{R}.$$

You are probably more familiar with another notation for this function; namely, the *equation*

$$y = \sin x.$$

This latter symbol is the statement, "the element (number) y in the codomain of the sine function is associated with the element x from the domain by the function *sine*." The general notation is $y = f(x)$, or sometimes $y = xf$. We call $f(x)$—most precisely read as "f at x" or "f of x"—the **image of x under f.** Both the arrow notation and the $f(x)$ notation are often used in the same discourse. When the mapping idea is stressed, the arrow notation is most often employed, and the equation $y = f(x)$ is replaced with the following arrow scheme:

$$x \rightarrow f(x) \qquad \text{or} \qquad x \overset{f}{\rightarrow} y.$$

It is important to observe that the collection of all of the elements $f(x)$ in the codomain which the function f assigns to members of the domain is a *subset* of the codomain and may be a much smaller set than the stated codomain. This set is called the *range* or the *image* of the function or transformation of map f. See Figure A2.6 on the next page.

Definition If $f: D \rightarrow S$ is a function from D into S, then the **range (= image) of f** is the subset $R = \{f(x): x \in D\} \subseteq S$.

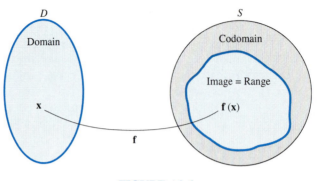

FIGURE A2.6

The *range* of the sine function is not all of $\mathbb{R}$, but rather the closed interval

$$[-1, 1] = \{x: -1 \leq x \leq 1\}.$$

In general, we say that f is a function from the domain **into** the codomain. It is **onto** its range (image) when the domain is D and the range and the codomain S coincide. In that special case we generally write the following:

$$f: D \underset{\text{onto}}{\rightarrow} S.$$

Definition A function $f: D \rightarrow S$ is **onto** S if, and only if, every $s \in S$ is of the form $s = f(x)$ for some $x \in D$.

The cube function, $s = x^3$ is *onto* the real numbers because each real number s is the cube of some real number $x = \sqrt[3]{s}$. On the other hand, the square function $y = x^2$ is *not onto* the reals because no negative real number is the square of any real number. The *range* of the square function is the non-negative reals $\mathbb{R}^+$.

Another important concept is that of a *one-to-one* function. The cube function is also one-to-one in that each real number is the cube of precisely one real number. The square function doesn't have this property: $4 = 2^2$, and $4 = (-2)^2$.

Definition A function $f: D \rightarrow S$ is **one-to-one** if, and only if, whenever $f(x) = f(y)$ in S, then $x = y$ in D.

In a one-to-one function no two different elements of the domain can have the same image in the range. The permutations discussed in Section 2.1 of the text are one-to-one functions from the finite set $S = \{1, 2, \ldots, n\}$ onto itself. For example, consider the permutation π of the finite set $S = \{1, 2, 3, 4, 5\}$. Here is a function (mapping) table.

$$\boxed{\begin{array}{c} \boldsymbol{x \in S \rightarrow \pi(x) \in S} \\ \hline 1 \rightarrow 2 \\ 2 \rightarrow 4 \\ 3 \rightarrow 5 \\ 4 \rightarrow 1 \\ 5 \rightarrow 3 \end{array}}$$

The determinant function is a function whose domain is the set of $n \times n$ matrices and whose codomain is the set of scalars. For real matrices, det: $M_n(\mathbb{R}) \rightarrow \mathbb{R}$. This function is onto $\mathbb{R}$ but is not one-to-one. Why?

The inner product $<\mathbf{u}, \mathbf{v}>$ of two vectors of a vector space V is a function whose domain is the set $V \times V$ and whose codomain is the scalar field. When V is a real vector space,

$$\text{inner product: } V \times V \rightarrow \mathbb{R}.$$

Again, this is a function *onto* $\mathbb{R}$, but it is also not one-to-one.

Linear transformations are special functions whose domain is one vector space U and whose codomain is another, possibly the same, vector space V; $L: U \rightarrow V$. Linear transformations in general need not be either *one-to-one* or *onto*. Nevertheless, some special linear transformations may have one or the other or both of these properties.

We list some examples of other functions of a specific nature that are worth noting.

The identity map.   This function has for its domain and its codomain (in this case, its *codomain* = its *range*) the same set and sends each element of the domain to itself: $\iota: D \underset{\text{onto}}{\rightarrow} D$ is defined by $\iota(x) = x$, for all $x \in D$. This function is both *one-to-one* and *onto D*.

The zero map.   This function sends every element of the domain D to the distinguished zero element of the codomain S: $O(x) = 0$, for all x in D.

Constant functions.   The zero map is a special case of a constant function (map) where every element of the domain D is sent to the same element of the codomain S. The range is a single element set $\{c\} \subseteq S$: $f_c(x) = c$, for every $x \in D$. If D has more than one element, these functions are clearly not one-to-one. And unless S has just one element, they are not onto.

Continuous functions, *differentiable* functions, and *integrable* functions are defined in calculus courses. A proper treatment of these requires more space than we can allow here. You should carefully review your calculus text. Notice, in particular, the following.

1. Every continuous function is integrable on its domain.
2. Every differentiable function is continuous, but not conversely. For example, the absolute value function $f(x) = |x|$ is continuous at zero, but not differentiable there.
3. Polynomial functions are everywhere differentiable.

Inverse Functions

It is not infrequent that we want to know which elements of the domain of a function were associated, by the function, with given elements of the co-domain. Such questions as, "What angle has a cosine equal to one-half?" or "What is the fourth root of 16?", etc., are often of interest.

Let $f: D \rightarrow S$ be a given function. Let $A \subseteq S$ be a particular subset of S. We define the **inverse image** of A under f as follows.

Definition	$f^{-1}(A) = \{x \in D : f(x) \in A\}$

That is, $f^{-1}(A)$ is the collection of all those elements of D which are mapped into A by the function f. (If A is not a subset of the range of f, then $f^{-1}(A)$ could be empty.) Suppose that A consists of a single element $s \in S$, then the definition specializes to

$$f^{-1}(s) = \{x \in D : f(x) = s\}.$$

For example, $\cos^{-1} \frac{1}{2}$ is an infinite set consisting of all those numbers whose cosine is $\frac{1}{2}$. Among its elements are $\frac{\pi}{3}$, $-\frac{\pi}{3}$, $\frac{7\pi}{3}$, etc. In fact,

$$\cos^{-1} \frac{1}{2} = \left\{ \pm \frac{\pi}{3} + 2k\pi, \; k \in \mathbb{Z} \right\}.$$

The fact that the inverse image of an element of S under f is more than a single element of the set D means that the function f was not one-to-one. Why? This problem causes some difficulty for many students of trigonometry and other mathematics classes. When one depresses the $\boxed{\cos^{-1}}$ key on one's calculator a single unique answer is expected (and obtained), not infinitely many. It is for this reason that the *arc cosine* is defined very carefully so as to be the inverse function of a special piece, or branch, of the cosine function. That is, the *domain* of the *cosine* function is first restricted to the interval $[0, \pi]$ on which the cosine function is one-to-one, then its inverse *is* a function.

If $f: D \rightarrow S$ is a one-to-one function, then $f^{-1}(s)$ is a single element of D for each s in the range A of f, so that f^{-1} is also a function with domain S and codomain D. We call it the **inverse function** of f. When f is not one-to-one, as in the case of the trigonometric functions and the square function, restrictions are placed on its domain *before* an inverse function can be defined.

A.3 ALGORITHMS

We have gathered together here several of the algorithms discussed in the text. They are presented in simplified form for convenient reference. You should consult the appropriate sections of the text for the full explanations of the procedures that we list here. Note that most of these are for use with hand calculation and, as such, give precise results. However, they are not the most numerically simple, and they may not be practical in large or complicated numerical problems.

In many cases your college computer center, your personal computer, or your calculator may use algorithms different from these for their computations. You should always consult the appropriate manuals which are available for the technological help you are using. These manuals ought to provide you with a discussion of the methods that are being used, together with their specific limitations. For a discussion of useful numerical linear algebra techniques and the many interesting problems connected with them, you are referred to more advanced texts.

The Gaussian Reduction of a System of Equations (See Section 1.1)

Gaussian Elimination Algorithm   Given is a system of m linear equations in n variables such as (1.1) in Section 1.1.

Step 1   Scan the system of equations and select one of them, equation i, from the system/subsystem being considered for which the *leftmost* variable $x_{\mu(j)}$ has a nonzero coefficient $a_{i\mu(j)}$. (The number $\mu(j)$ will be 1 when the process begins and, in general, is the smallest variable number with a nonzero coefficient remaining in the subsystem.) Call $x_{\mu(j)}$ the **pivot variable,** call equation i the **pivot equation,** and call $a_{i\mu(j)}x_{\mu(j)}$ in this equation the **pivot term.**

Step 2   Exchange equations so that the pivot equation becomes the first equation in the system/subsystem. This uses the first elementary operation.

Step 3   Use the pivot term to eliminate the pivot variable from all equations below the pivot equation in the system/subsystem. That is, replace each following equation with its sum with the appropriate multiple of the pivot equation making use of elementary operations 2 and 3.

Step 4   Save the pivot equation and consider only the subsystem formed from the remaining equations. None of these contain the variable $x_{\mu(j)}$. If the subsystem has a "bad equation" of the form $0x_k + 0x_{k+1} + \cdots + 0x_n = b \neq 0$, the original system was inconsistent and has no solution. (Move the bad equation to the bottom of the system and delete it from the subsystem if you desire to continue.) If there are any nontrivial equations remaining in the subsystem return to Step 1 with the subsystem assuming the role of the given system.

Continue this process until no nontrivial equations remain. The saved pivot equations (and bad equations, if any) from each iteration form the desired row echelon form.

The Gauss-Jordan Reduction of a Matrix (See Section 1.1)

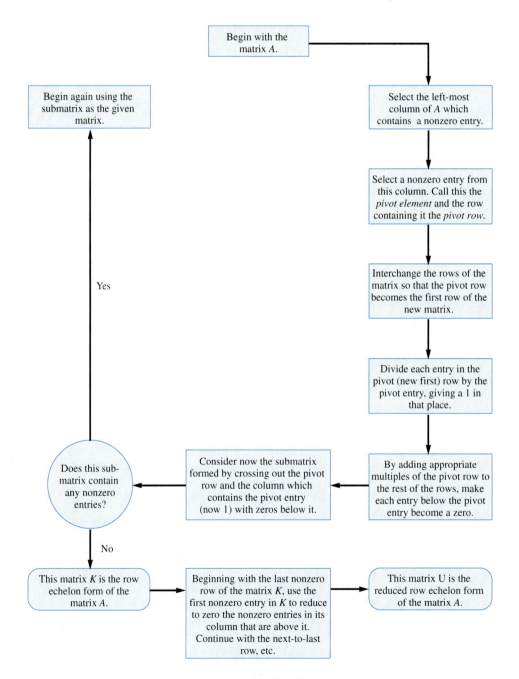

Begin with the matrix A.

Begin again using the submatrix as the given matrix.

Select the left-most column of A which contains a nonzero entry.

Select a nonzero entry from this column. Call this the *pivot element* and the row containing it the *pivot row*.

Interchange the rows of the matrix so that the pivot row becomes the first row of the new matrix.

Divide each entry in the pivot (new first) row by the pivot entry, giving a 1 in that place.

By adding appropriate multiples of the pivot row to the rest of the rows, make each entry below the pivot entry become a zero.

Consider now the submatrix formed by crossing out the pivot row and the column which contains the pivot entry (now 1) with zeros below it.

Yes

Does this sub-matrix contain any nonzero entries?

No

This matrix K is the row echelon form of the matrix A.

Beginning with the last nonzero row of the matrix K, use the first nonzero entry in K to reduce to zero the nonzero entries in its column that are above it. Continue with the next-to-last row, etc.

This matrix U is the reduced row echelon form of the matrix A.

CHART 1

The Matrix-Inversion Algorithm (See Section 1.5)

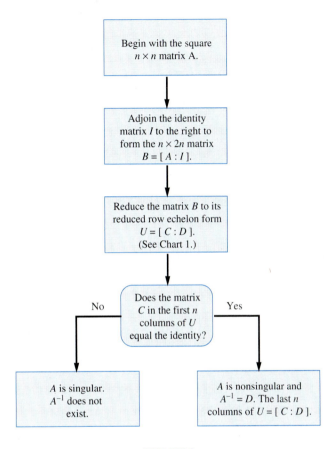

Begin with the square
$n \times n$ matrix A.

Adjoin the identity
matrix I to the right to
form the $n \times 2n$ matrix
$B = [A : I]$.

Reduce the matrix B to its
reduced row echelon form
$U = [C : D]$.
(See Chart 1.)

Does the matrix
C in the first n
columns of U
equal the identity?

No   Yes

A is singular.
A^{-1} does not
exist.

A is nonsingular and
$A^{-1} = D$. The last n
columns of $U = [C : D]$.

CHART 2

The L-U Decomposition (See Section 1.6)

Let A be an $n \times n$ matrix and assume that A can be row reduced to an upper-triangular matrix U in the usual way, but *without any interchanges of its rows.* Keep a record of the sequence $E_k \cdots E_3 E_2 E_1$ of elementary matrices multiplied by A to result in the matrix U. We assume that each of the elementary matrices involved is of Type III, so each E_i is lower triangular. $E_k \cdots E_3 E_2 E_1 A = U$. Therefore, $A = E_1^{-1} E_2^{-1} E_3^{-1} \cdots E_k^{-1} U = LU$.

To obtain L, directly apply the *inverse* of the operation used to row reduce A in the same order to the identity matrix I. That is, suppose that row j of U is formed by the elementary operations

$$\text{row } j \text{ of } A - \ell_{j1}(\text{row } 1 \text{ of } U) - \cdots - \ell_{jk}(\text{row } k \text{ of } U), \, k < j.$$

Then the main diagonal elements of L are *ones*, and the off-diagonal elements in row j are these numbers ℓ_{jk}.

$$L = \begin{bmatrix} 1 & 0 & \cdots & \cdots & 0 \\ \ell_{21} & 1 & \cdots & \cdots & 0 \\ \ell_{31} & \ell_{32} & 1 & \cdots & 0 \\ \cdot & \cdot & \cdots & \cdots & 0 \\ \cdot & \cdot & \cdots & \cdots & 0 \\ \cdot & \cdot & \cdots & \cdots & 0 \\ \ell_{n1} & \ell_{n2} & \cdots & \cdots & 1 \end{bmatrix}$$

Properties of Determinants (See Sections 2.3 & 2.4)

Let A and B be $n \times n$ matrices and A^T be the transpose of A. The following are true.

1. $\det A^T = \det A$.   (Theorem 2.2)
2. $\det (AB) = (\det A)(\det B)$.   (Theorem 2.7)
3. If B is obtained from A by interchanging any two rows (columns) of A, then

$$\det B = -\det A.$$

4. If B is obtained from A by multiplying any row (column) of A by the nonzero constant k, then

$$\det B = k(\det A).$$

5. If B is obtained from A by adding to any row (column) of A a multiple of any other row (column) of A, then

$$\det B = \det A.$$

6. If two rows (columns) of A are identical, then

$$\det A = 0.$$

7. If A has a zero row (column), then

$$\det A = 0.$$

8. A is nonsingular if, and only if, $\det A \neq 0$, and then

$$\det A^{-1} = \frac{1}{\det A}.$$

9. For every $n \times n$ matrix A, $A(\text{adj } A) = (\det A)I_n$, and if A is nonsingular,

$$A^{-1} = \frac{1}{\det A}(\text{adj } A),$$

where $(\text{adj } A)$ is the adjugate (classical adjoint) of A; that is, the transpose of the cofactor matrix of A.

Test for Linear Independence or Dependence (See Sections 3.2, 3.4, and 5.3)

Given is a set $S = \{v_1, v_2, \ldots, v_n\}$ of vectors from the vector space $\mathbb{R}^m$. Represent the vectors as n-tuples in $\mathbb{R}^m$ and consider these representations to be the vectors in S. To test S for linear dependence or independence:

1. Set the linear combination (with unknown scalars) equal to the zero vector.

$$c_1 v_1 + c_2 v_2 + \ldots + c_n v_n = 0.$$

2. Consider the scalars c_i by examining the resulting homogeneous system of linear equations $AC = 0$, where A is the matrix with $[v_i]$ as its ith column, in either one of the following ways.
   (a) Reduce the coefficient matrix A to its row echelon form K.
      (i) If K is I_n, then the system can only have the trivial solution, $C = A^{-1}0$, so the set S is *linearly independent*.
      (ii) If K is not I_n, the system has nontrivial solutions and the set S is *linearly dependent*.
   (b) Use the determinant of the matrix A.
      (i) If $\det A \neq 0$, then A is nonsingular and the system has only the trivial solution, so the set S is *linearly independent*.
      (ii) If $\det A = 0$, then A is singular, the system has nontrivial solutions, and the set S is *linearly dependent*.

Example   Let S be the set of vectors $\{v_1 = x - 1, v_2 = 2x + 1, v_3 = 4x^2 - 3\}$ from P_2, with standard basis $E = \{1, x, x^2\}$. Represent them as triples in $\mathbb{R}^3$ as follows:

$$v_1 = \begin{bmatrix} -1 \\ 1 \\ 0 \end{bmatrix}, \qquad v_2 = \begin{bmatrix} 1 \\ 2 \\ 0 \end{bmatrix}, \qquad \text{and} \qquad v_3 = \begin{bmatrix} -3 \\ 0 \\ 4 \end{bmatrix}.$$

1. Form the linear combination

$$c_1 v_1 + c_2 v_2 + c_3 v_3 = 0 = \begin{bmatrix} -c_1 \\ c_1 \\ 0 \end{bmatrix} + \begin{bmatrix} c_2 \\ 2c_2 \\ 0 \end{bmatrix} + \begin{bmatrix} -3c_3 \\ 0 \\ 4c_3 \end{bmatrix}.$$

2. The coefficient matrix for the resulting homogeneous system of equations $AC = 0$ is

$$A = \begin{bmatrix} -1 & 1 & -3 \\ 1 & 2 & 0 \\ 0 & 0 & 4 \end{bmatrix}.$$

(Note that the columns of A are the representations of the vectors v_1, v_2, and v_3 in S.)

A is row equivalent to I_3, so the set S is linearly independent. Alternatively, $\det A = -12 \neq 0$, so the set S is linearly independent.   ∎

Properties of an Inner Product (See Sections 4.3 and 4.6)

Real Inner Products   For each pair of vectors **u** and **v** in the real vector space V, $<$**u**, **v**$>$ is a real number such that:

1. $<$**u**, **v**$> = <$**v**, **u**$>$.
2. $<\alpha$**u** $+ \beta$**v**, **w**$> = \alpha <$**u**, **w**$> + \beta<$**v**, **w**$>$, for all real numbers α and β.
3. $<$**u**, **u**$> \geq 0$.
4. $<$**u**, **u**$> = 0$ if, and only if, **u** $= $ **0**.
5. $<$**u**, α**v** $+ \beta$**w**$> = \alpha<$**u**, **v**$> + \beta <$**u**, **w**$>$ (Theorem 4.4).

Complex Inner Products   For each pair of vectors **u** and **v** in the complex vector space V, $<$**u**, **v**$>$ is a complex number such that:

1. $<$**u**, **v**$> = \overline{(<$**v**, **u**$>)}$.
2. $<\alpha$**u** $+ \beta$**v**, **w**$> = \alpha<$**u**, **w**$> + \beta<$**v**, **w**$>$, for all complex scalars α and β.
3. $<$**u**, **u**$> \geq 0$; with equality if, and only if, **u** $= $ **0**.
4. $<$**u**, α**v** $+ \beta$**w**$> = \overline{\alpha}<$**u**, **v**$> + \overline{\beta} <$**u**, **w**$>$ (Theorem 4.12).

Norm (Length) in an Inner Product Space

$$\|\mathbf{u}\| = <\mathbf{u}, \mathbf{u}>^{1/2},$$
$$\|\alpha\mathbf{u}\| = |\alpha| \cdot \|\mathbf{u}\|.$$

Angle Between Two Vectors   Let θ be the angle between the nonzero vectors **u** and **v**. Then:

$$\cos \theta = \frac{<\mathbf{u}, \mathbf{v}>}{\|\mathbf{u}\| \, \|\mathbf{v}\|}.$$

The Cauchy-Schwarz Inequality   (Theorems 4.7 and 4.13)
   $|<$**u**, **v**$>| \leq \|$**u**$\| \, \|$**v**$\|$; with equality if, and only if, **u** and **v** are linearly dependent.

Orthonormal Set of Vectors $S = \{\mathbf{v}_1, \mathbf{v}_2, \ldots, \mathbf{v}_n\}$ is an orthonormal set if, and only if,

$$<\mathbf{v}_i, \mathbf{v}_j> = \begin{cases} 1 & \text{when } i = j \\ 0 & \text{when } i \neq j \end{cases}.$$

The Gram-Schmidt Process in $\mathbb{R}^3$ (See Section 4.5)

Step 1   Select a nonzero vector **v** in $\mathbb{R}^3$ and divide **v** by its norm, $\|\mathbf{v}\| = \sqrt{<\mathbf{v}, \mathbf{v}>}$, to obtain a unit vector $\mathbf{v}_1$ along **v**.

Step 2   Take any other vector **u** in $\mathbb{R}^3$ which is linearly independent of **v**; that is, is not a scalar multiple of **v**. Then project **u** along (onto) $\mathbf{v}_1$ to obtain the vector $\mathbf{w}_1 = <\mathbf{u}, \mathbf{v}_1> \mathbf{v}_1$.

Step 3   Find the projection $\mathbf{w}_2 = \mathbf{u} - \mathbf{w}_1$ of **u** orthogonal to $\mathbf{v}_1$ (orthogonal to **v**).

Step 4   Divide the orthogonal projection $\mathbf{w}_2$ by its norm $\|\mathbf{w}_2\| = \sqrt{<\mathbf{w}_2, \mathbf{w}_2>}$ to obtain the unit vector $\mathbf{v}_2$ orthogonal to $\mathbf{v}_1$.

The subspace $W = \mathbf{L}(\{\mathbf{u}, \mathbf{v}\})$ of $\mathbb{R}^3$, spanned by **u** and **v**, has $\{\mathbf{v}_1, \mathbf{v}_2\}$ as an *orthonormal basis,* where

$$\mathbf{v}_1 = \frac{\mathbf{v}}{\|\mathbf{v}\|} \quad \text{and} \quad \mathbf{v}_2 = \frac{\mathbf{u} - <\mathbf{u}, \mathbf{v}_1> \mathbf{v}_1}{\|\mathbf{u} - <\mathbf{u}, \mathbf{v}_1> \mathbf{v}_1\|}.$$

Step 5   Select any vector $\mathbf{u}'$ which is not in the subspace $W = \mathbf{L}(\{\mathbf{v}_1, \mathbf{v}_2\})$ spanned by this basis. Then project $\mathbf{u}'$ *onto* the subspace W to obtain the vector

$$\mathbf{w}_1' = \sum_{i=1}^{2} <\mathbf{u}', \mathbf{v}_i> \mathbf{v}_i.$$

The projection of $\mathbf{u}'$ onto the subspace $W = \mathbf{L}(\{\mathbf{v}_1, \mathbf{v}_2\})$ is a vector $\mathbf{w}_1'$ lying in W.

Step 6   Find the projection $\mathbf{w}_2' = \mathbf{u}' - \mathbf{w}_1'$ of $\mathbf{u}'$ *orthogonal to* W (orthogonal to the linear span $\mathbf{L}(\{\mathbf{v}_1, \mathbf{v}_2\})$).

Step 7   Divide the orthogonal projection $\mathbf{w}_2'$ by its norm $\|\mathbf{w}_2'\| = \sqrt{<\mathbf{w}_2', \mathbf{w}_2'>}$ to obtain the unit vector $\mathbf{v}_3$ orthogonal to W. The third orthonormal basis vector is

$$\mathbf{v}_3 = \frac{\mathbf{u}' - <\mathbf{u}', \mathbf{v}_1> \mathbf{v}_1 - <\mathbf{u}', \mathbf{v}_2> \mathbf{v}_2}{\|\mathbf{u}' - <\mathbf{u}', \mathbf{v}_1> \mathbf{v}_1 - <\mathbf{u}', \mathbf{v}_2> \mathbf{v}_2\|}.$$

The Matrix Presentation $[T]_H^G$ of a Linear Transformation (See Section 5.3)

Let $T: \mathbb{R}^n \rightarrow \mathbb{R}^m$ be a given linear transformation. Suppose that $G = \{\mathbf{u}_1, \mathbf{u}_2, \ldots, \mathbf{u}_n\}$ is a basis for $\mathbb{R}^n$, and $H = \{\mathbf{v}_1, \mathbf{v}_2, \ldots, \mathbf{v}_m\}$ is a basis for $\mathbb{R}^m$. To find the matrix which represents T with respect to these bases proceed as follows. Note that $[\mathbf{v}_i]$ is the standard representation of the vector $\mathbf{v}_i$.

1. Use the defining condition for the transformation T to find the image vectors $T(\mathbf{u}_1)$, $T(\mathbf{u}_2)$, ..., $T(\mathbf{u}_n)$ in $\mathbb{R}^m$ for each basis vector in G.
2. Form the $m \times n$ matrix K whose i^{th} column vector is the column vector $[T(\mathbf{u}_i)]$, $i = 1, 2, \ldots, n$.
3. Form the $m \times m$ matrix L whose j^{th} column vector is the column vector $[\mathbf{v}_i]$, $j = 1, 2, \ldots, m$.
4. Form the block matrix $M = [L : K]$ as follows.

$$\begin{array}{ccc} [\quad\quad L & : & K \quad\quad] \\ [\mathbf{v}_1]\,[\mathbf{v}_2]\ldots[\mathbf{v}_m] & & [T(\mathbf{u}_1)]\ldots[T(\mathbf{u}_n)] \end{array}$$

5. Use a Gauss-Jordan reduction to reduce $[L : K]$ to $[I_m : A]$. Then $A = [T]_H^G$.

Change of Basis and Similarity (See Section 5.5)

Let $G = \{\mathbf{u}_1, \mathbf{u}_2, \ldots, \mathbf{u}_n\}$ and $H = \{\mathbf{w}_1, \mathbf{w}_2, \ldots, \mathbf{w}_n\}$ be bases for $\mathbb{R}^n$. Then the matrix

$$S = [[\mathbf{u}_1]_H \, [\mathbf{u}_2]_H \ldots [\mathbf{u}_n]_H],$$

whose i^{th} column vector is the column vector $[\mathbf{u}_i]_H$, is the *transition matrix* or *change of basis matrix, from the basis G to the basis H*. Thus, for each vector $\mathbf{v}$ in $\mathbb{R}^n$,

$$[\mathbf{v}]_H = S[\mathbf{v}]_G.$$

The matrix S represents the identity transformation on $\mathbb{R}^n$, $S = [I]_H^G$ and is nonsingular. Its inverse $S^{-1} = [I]_G^H$ is the *change of basis (transition) matrix from the basis H to the basis G*. Thus, for each vector $\mathbf{v}$ in $\mathbb{R}^n$,

$$[\mathbf{v}]_G = S^{-1}[\mathbf{v}]_H.$$

Algorithm for Calculating a Change of Basis Matrix

1. Represent each of the basis vectors in G and in H with respect to the standard basis E for $\mathbb{R}^n$. As above, denote these by $[\mathbf{u}_i]$ and $[\mathbf{w}_i]$, respectively; $i = 1, 2, \ldots, n$.
2. Form the block matrix M with column vectors as follows.

$$M = [[\mathbf{w}_1]\ldots[\mathbf{w}_n] : [\mathbf{u}_1]\ldots[\mathbf{u}_n]]$$

3. Perform a Gauss-Jordan reduction on M to arrive at the matrix $K = [I_n : S]$.
4. The matrix S in the right-hand block of K is the desired change of basis matrix from G to H; that is, $S = [I]_H^G$ and is nonsingular.

5. To find the change of basis matrix $S^{-1} = [I]_G^H$ from H to G, either (a) invert the matrix S found in Step 4, or (b) use the matrix $M' = [[\mathbf{u}_1] \dots [\mathbf{u}_n] : [\mathbf{w}_1] \dots [\mathbf{w}_n]]$ in place of the matrix M in Step 2 and proceed.

Example   Let G and H be the two bases for $\mathbb{R}^3$ given below. Find the change of basis (transition) matrix from G to H.

$$G = \{\mathbf{u}_1 = (1, 0, 1), \mathbf{u}_2 = (1, -1, 0), \mathbf{u}_3 = (0, 0, -1)\},$$

and $$H = \{\mathbf{w}_1 = (1, 1, 1), \mathbf{w}_2 = (1, -1, -1), \mathbf{w}_3 = (1, 0, 1)\}$$

Using the algorithm, we set up the matrix M as follows.

$$M = \begin{bmatrix} 1 & 1 & 1 & : & 1 & 1 & 0 \\ 1 & -1 & 0 & : & 0 & -1 & 0 \\ 1 & -1 & 1 & : & 1 & 0 & -1 \end{bmatrix}$$
$$\quad\ \ \mathbf{w}_1 \quad \mathbf{w}_2 \quad \mathbf{w}_3 \ : \ \mathbf{u}_1 \quad \mathbf{u}_2 \quad \mathbf{u}_3$$

This matrix is row equivalent to the matrix

$$M = \begin{bmatrix} 1 & 0 & 0 & : & 0 & -\dfrac{1}{2} & \dfrac{1}{2} \\ 0 & 1 & 0 & : & 0 & \dfrac{1}{2} & \dfrac{1}{2} \\ 0 & 0 & 1 & : & 1 & 1 & -1 \end{bmatrix}$$

So the desired change of basis matrix (transition matrix) $S = [I]_H^G$ is the right-hand block:

$$S = [I]_H^G = \begin{bmatrix} 0 & -\dfrac{1}{2} & \dfrac{1}{2} \\ 0 & \dfrac{1}{2} & \dfrac{1}{2} \\ 1 & 1 & -1 \end{bmatrix}.$$

The change of basis matrix

$$S^{-1} = [I]_G^H = \begin{bmatrix} 2 & 0 & 1 \\ -1 & 1 & 0 \\ 1 & 1 & 0 \end{bmatrix}. \quad \blacksquare$$

Diagram for Similarity   Let $T: V \to V$ be a linear transformation (operator) on the n-dimensional vector space V with bases $G = \{\mathbf{u}_1, \mathbf{u}_2, \dots, \mathbf{u}_n\}$ and $H = \{\mathbf{w}_1, \mathbf{w}_2, \dots, \mathbf{w}_n\}$. Let $A = [T]_H$ and $B = [T]_G$ be the matrix representations of T relative to the bases G and H, respectively. Also let $S = [I]_H^G$ be the change of basis (transition) matrix from G to H. Then A and B are similar. (Theorem 5.9) $A = SBS^{-1}$. Chart 3 illustrates what happens to a given vector $\mathbf{x}$ from V.

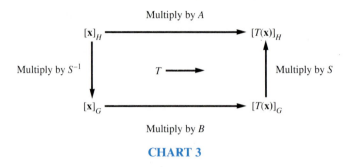

CHART 3

Eigenvalues, Eigenvectors, and Diagonalization (See Sections 6.1–6.3)

1. To compute the *eigenvalues* of a matrix A or of the transformation which it represents, find the roots λ_i of the *characteristic equation*

$$\det (\lambda I - A) = 0.$$

2. To compute the *eigenvectors* $\mathbf{p}_i$ of the matrix A (or transformation) associated with the eigenvalue λ_i, solve the homogeneous system of equations

$$(\lambda I - A)\mathbf{x} = \mathbf{0}.$$

Note that this system must have an infinite number of nontrivial solutions. Find a basis for the solution space.

Similar matrices have the same eigenvalues.

The Diagonalizing Matrix P   Suppose that A is an $n \times n$ real matrix with exactly n distinct eigenvectors

$$\mathbf{p}_1, \mathbf{p}_2, \ldots, \mathbf{p}_n.$$

Then the matrix P whose column vectors are the representations of these eigenvectors

$$P = [[\mathbf{p}_1]\,[\mathbf{p}_2] \ldots [\mathbf{p}_n]]$$

is such that

$$P^{-1}AP = D = \begin{bmatrix} \lambda_1 & 0 & 0 & \cdots & 0 \\ 0 & \lambda_2 & 0 & \cdots & 0 \\ \vdots & \vdots & \vdots & \cdots & \vdots \\ 0 & 0 & 0 & \cdots & \lambda_n \end{bmatrix},$$

where the diagonal entries λ_i in D are the not necessarily distinct eigenvalues of A corresponding to the respective eigenvectors $\mathbf{p}_i$; $i = 1, 2, \ldots, n$.

Note   Not every matrix is similar to a diagonal matrix. However, if A is real symmetric, $A^T = A$, then it is similar to a diagonal matrix and P can be an orthogonal matrix, $P^T = P^{-1}$. If A is a normal matrix, $AA^* = A^*A$, then there is a unitary matrix $U(U^* = U^{-1})$ so that U^*AU is diagonal (Theorem 6.16).

A.4 REFERENCES

Barnsley, Michael, *Fractals Everywhere,* Academic Press, Boston, 1988.

Coleman, J. S., *Introduction to Mathematical Sociology,* The Free Press, New York, 1964.

Cooper, L. and Steinberg, D., *Methods and Applications of Linear Programming,* Saunders, Philadelphia, 1974.

Councilman, Samuel, "Eigenvalues and Eigenvectors of '*N*-Matrices'." *American Mathematical Monthly,* **93**, No. 5, May, 1986, pp. 392–395.

Garner, Lynn E. "Notes for Users of the HP-28S™ in Mathematics Courses," Department of Mathematics, Brigham Young University, Provo, UT, 1989.

Horn, Roger A. and Johnson, Charles R., *Matrix Analysis,* Cambridge Univ. Press, Cambridge, 1985.

Lancaster P., *Mathematical Models of the Real World,* Prentice-Hall, Englewood Cliffs, N. J., 1976.

Maki, Daniel P. and Thompson, Maynard, *Mathematical Models and Applications,* Prentice-Hall, Englewood Cliffs, N. J., 1973.

Marcus, Marvin and Minc, Henryk, *A Survey of Matrix Theory and Matrix Inequalities,* Allyn and Bacon, Boston, 1964.

Matlab™, The Math Works, 21 Eliot St. South, Natick, MA. 01760.

Osborne, Anthony and Liebcck, Hans, "Orthogonal Bases for $\mathbb{R}^3$ with Integer Coordinates and Integer Lengths," *American Mathematical Monthly,* **96** (1989), pp. 49–53.

Skrien, Dale, *Matrix Works™*, Department of Mathematics, Colby College, Waterville, ME, 1990.

Solow, Daniel, *How to Read and Do Proofs,* John Wiley and Sons, New York, 1982.

Supprunenko, D. A. and Tyshkevich, R. I., *Commutative Matrices,* Academic Press, New York, 1968, pp. 26–30.

Watkins, William, "Similarity of Matrices," *American Mathematical Monthly,* **87**, No. 4, Apr. 1980, p. 300.

White, H. C., *An Anatomy of Kinship,* Prentice-Hall, Englewood Cliffs, N. J., 1963.

Williams, G. "Overdetermined Systems of Linear Equations," *American Mathematical Monthly,* **97**, No. 6, June-July 1990, pp. 511–513.

Wolfram, Stephen, *Mathematica, A System for Doing Mathematics by Computer,* Addison-Wesley, Redwood City, 1988. *Mathematica™*, Wolfram Research, Inc., Champaign, IL.

A.5 *THE GREEK ALPHABET AND FREQUENTLY USED SYMBOLS*

The Greek Alphabet

A	α	alpha	N	ν	nu
B	β	beta	Ξ	ξ	xi
Γ	γ	gamma	O	o	omicron
Δ	δ	delta	Π	π	pi
E	ε	epsilon	P	ρ	rho
Z	ζ	zeta	Σ	σ	sigma
H	η	eta	T	τ	tau
Θ	θ	theta	Υ	υ	upsilon
I	ι	iota	Φ	φ	phi
K	κ	kappa	X	χ	chi
Λ	λ	lambda	Ψ	ψ	psi
M	μ	mu	Ω	ω	omega

Frequently Used Symbols

Symbol	Meaning	Book Page
O	The zero matrix	32
$A = (a_{ij})$	An $m \times n$ matrix	30
$[A : \mathbf{b}]$	The augmented matrix	7, 69
$\mathbf{a}_j$	column j of the matrix $A = (a_{ij})$	48
$\mathbf{a}'_i$	row i of the matrix $A = (a_{ij})$	48
A^T	The transpose of A	46
$\overline{A}$	The (complex) conjugate matrix $\overline{A} = (\overline{a_{ij}})$	270
A^*	The conjugate transpose $(\overline{A})^T$	271
A^{-1}	The inverse matrix	52
adj A	The adjugate (classical adjoint) of A	135
A_{ij}	The cofactor of the element a_{ij} in $A = (a_{ij})$	135
cof A	The matrix of cofactors of the elements of A	
$A \oplus B$	direct sum of A and B, $A \oplus B = \begin{bmatrix} A & O \\ O & B \end{bmatrix}$ where O is an appropriate zero matrix	81

Symbol	Meaning	Book Page
$\Delta = \det A = |A|$	The determinant of A	108
tr C	The trace (sum of elements on main diagonal) of the matrix C	239
$\mathbb{A}$	The algebra of linear operators $T: V \to V$ on a vector space V	339
$\displaystyle\sum_{i=1}^{n} a_i$	Summation notation $= a_1 + a_2 + a_3 + \ldots + a_n$	40
$B = [\mathbf{b}_1 : \mathbf{b}_2 : \ldots : \mathbf{b}_n]$	A partitioned matrix, into columns	69
diag $[a_1, a_2, \ldots, a_n]$	A diagonal matrix	45
E	An elementary matrix	56
I_n	The $n \times n$ identity matrix $=$ diag $[1, 1, \ldots, 1]$	45
LU	LU decomposition of a matrix $A = LU$, L is upper-triangular, U is lower-triangular	75
$\beta \in S_n$	β is a permutation of n elements (numbers)	107
sgn (β)	The sign of the permutation $\beta = \begin{cases} 1 & \text{when } \beta \text{ is even} \\ -1 & \text{when } \beta \text{ is odd} \end{cases}$	108
$\mathbb{R}$	The set of all real numbers	153
$\mathbb{C}$	The set of all complex numbers	153
$\mathbb{C}^n$	The vector space of all complex n-tuples, **row** vectors	159
$\mathbb{R}^n$	The vector space of all real n-tuples, **row** vectors	156
$\overline{\mathbb{R}^n}$	The vector space of all n-dimensional **column** vectors of real numbers	243
$\overline{\mathbb{C}^n}$	The vector space of all n-dimensional **column** vectors of complex numbers	303
P_n	The vector space of all real polynomials of degree $\leq n$	157
$\mathbb{P}_n$	The vector space of all complex polynomials of degree $\leq n$	158
$M_n(\mathbb{R})$	The vector space of all $n \times n$ (square) matrices with entries from $\mathbb{R}$	158
$M_n(\mathbb{C})$	The vector space of all $n \times n$ (square) matrices with entries from $\mathbb{C}$	158
$M_{m \times n}(\mathbb{R})$	The vector space of all $m \times n$ matrices with entries from $\mathbb{R}$	158

Symbol	Meaning	Book Page
$M_{m \times n}(\mathbb{C})$	The vector space of all $m \times n$ matrices with entries from $\mathbb{C}$	158
solution space of A	The vector space of all solution vectors $\mathbf{x}$ to the homogeneous system $A\mathbf{x} = \mathbf{0}$	159
$\mathbf{x}, \mathbf{y}, \mathbf{u}, \mathbf{v}$, etc.	Vectors in a vector space	150
V, U, W, etc.	Vector spaces or subspaces	153
$\mathbf{e}_i$	The standard basis vector with "1" in position i and zeros everywhere else	189
$E = \{\mathbf{e}_1, \mathbf{e}_2, \ldots, \mathbf{e}_n\}$	The standard basis for the vector spaces $\mathbb{R}^n, \mathbb{C}^n, \overline{\mathbb{R}^n}$, or $\overline{\mathbb{C}^n}$	189
$\mathbf{L}(S)$	The linear span of a set S of vectors in a vector space V	180
$\text{Proj}_v\, \mathbf{u} = \dfrac{<\mathbf{u}, \mathbf{v}>}{<\mathbf{v}, \mathbf{v}>}\mathbf{v}$	The projection of the vector $\mathbf{u}$ *along* (or *onto* or *on*) the nonzero vector $\mathbf{v}$	261
$\mathbf{u} - \dfrac{<\mathbf{u}, \mathbf{v}>}{<\mathbf{v}, \mathbf{v}>}\mathbf{v}$	The projection of the vector $\mathbf{u}$ *orthogonal* to the nonzero vector $\mathbf{v}$	261
$\overline{z}$	The conjugate $x - iy$ of the complex number $z = x + iy$	270
$\overline{\mathbf{u}}$	The conjugate of the vector $\mathbf{u}$	270
$\mathbf{u}^*$	The conjugate transpose of $\mathbf{u} = (\overline{\mathbf{u}})^T$	271
$[\mathbf{u}]_B$	The representation (coordinatization) of the vector $\mathbf{u}$ as a column vector ($m \times 1$ matrix) with respect to a given basis B	322
$W^{\perp}$	The orthogonal complement of the subspace W	277
$A = [T]_H^G$	The matrix representation of a linear transformation $T: U \to V$, where G is a basis for U and H is a basis for V	326
$A = SBS^{-1}$	A and B are similar matrices	351
$T_A(\mathbf{u}) = A\mathbf{u}$	Matrix transformation $T_A: \overline{\mathbb{R}^m} \to \overline{\mathbb{R}^n}$ or $T_A: \overline{\mathbb{C}^m} \to \overline{\mathbb{C}^n}$	322
$T: U \to V$	A linear transformation from U to V	301
T^{-1}	The inverse (function) of a transformation	341
λ	An eigenvalue of a matrix or a transformation $A\mathbf{x} = \lambda\mathbf{x}$ or $T(\mathbf{x}) = \lambda\mathbf{x}$	376
$\{\mathbf{0}\}$	The vector space (subspace) consisting of just the zero vector	160

Symbol	Meaning	Book Page
$C[a, b]$	The vector space of all real valued functions continuous on the closed interval $[a, b]$	160
$\mathbf{u} \times \mathbf{v}$	The cross product of two vectors in $\mathbb{R}^3$	227
$<\mathbf{u}, \mathbf{v}>$	The inner (scalar) product of two vectors	236
$\|\mathbf{u}\|$	The norm $<\mathbf{u}, \mathbf{u}>^{1/2}$ of a vector $\mathbf{u}$	246
ker T	The kernel = null-space of a given (function) linear transformation T	309
range T	The range = image of a given (function) linear transformation T	311
$p_A(\lambda)$ or $p_A(t)$	The characteristic polynomial of the matrix A	384

Chapter 1

Section 1.1 (pp. 14–17)

1. Solution is $(3, 0)$. **3.** No solutions.

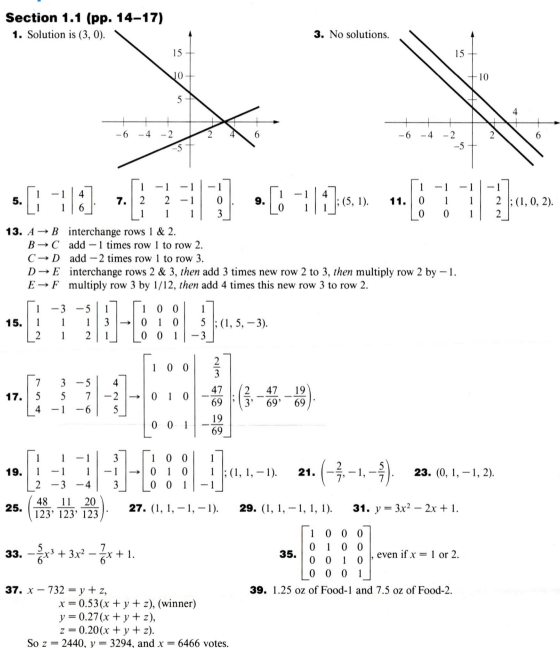

5. $\begin{bmatrix} 1 & -1 & | & 4 \\ 1 & 1 & | & 6 \end{bmatrix}$. **7.** $\begin{bmatrix} 1 & -1 & -1 & | & -1 \\ 2 & 2 & -1 & | & 0 \\ 1 & 1 & 1 & | & 3 \end{bmatrix}$. **9.** $\begin{bmatrix} 1 & -1 & | & 4 \\ 0 & 1 & | & 1 \end{bmatrix}$; $(5, 1)$. **11.** $\begin{bmatrix} 1 & -1 & -1 & | & -1 \\ 0 & 1 & 1 & | & 2 \\ 0 & 0 & 1 & | & 2 \end{bmatrix}$; $(1, 0, 2)$.

13. $A \to B$ interchange rows 1 & 2.
$B \to C$ add -1 times row 1 to row 2.
$C \to D$ add -2 times row 1 to row 3.
$D \to E$ interchange rows 2 & 3, *then* add 3 times new row 2 to 3, *then* multiply row 2 by -1.
$E \to F$ multiply row 3 by $1/12$, *then* add 4 times this new row 3 to row 2.

15. $\begin{bmatrix} 1 & -3 & -5 & | & 1 \\ 1 & 1 & 1 & | & 3 \\ 2 & 1 & 2 & | & 1 \end{bmatrix} \to \begin{bmatrix} 1 & 0 & 0 & | & 1 \\ 0 & 1 & 0 & | & 5 \\ 0 & 0 & 1 & | & -3 \end{bmatrix}$; $(1, 5, -3)$.

17. $\begin{bmatrix} 7 & 3 & -5 & | & 4 \\ 5 & 5 & 7 & | & -2 \\ 4 & -1 & -6 & | & 5 \end{bmatrix} \to \begin{bmatrix} 1 & 0 & 0 & | & \frac{2}{3} \\ 0 & 1 & 0 & | & -\frac{47}{69} \\ 0 & 0 & 1 & | & -\frac{19}{69} \end{bmatrix}$; $\left(\frac{2}{3}, -\frac{47}{69}, -\frac{19}{69}\right)$.

19. $\begin{bmatrix} 1 & 1 & -1 & | & 3 \\ 1 & -1 & 1 & | & -1 \\ 2 & -3 & -4 & | & 3 \end{bmatrix} \to \begin{bmatrix} 1 & 0 & 0 & | & 1 \\ 0 & 1 & 0 & | & 1 \\ 0 & 0 & 1 & | & -1 \end{bmatrix}$; $(1, 1, -1)$. **21.** $\left(-\frac{2}{7}, -1, -\frac{5}{7}\right)$. **23.** $(0, 1, -1, 2)$.

25. $\left(\frac{48}{123}, \frac{11}{123}, \frac{20}{123}\right)$. **27.** $(1, 1, -1, -1)$. **29.** $(1, 1, -1, 1, 1)$. **31.** $y = 3x^2 - 2x + 1$.

33. $-\frac{5}{6}x^3 + 3x^2 - \frac{7}{6}x + 1$. **35.** $\begin{bmatrix} 1 & 0 & 0 & 0 \\ 0 & 1 & 0 & 0 \\ 0 & 0 & 1 & 0 \\ 0 & 0 & 0 & 1 \end{bmatrix}$, even if $x = 1$ or 2.

37. $x - 732 = y + z$,
$x = 0.53(x + y + z)$, (winner)
$y = 0.27(x + y + z)$,
$z = 0.20(x + y + z)$.
So $z = 2440$, $y = 3294$, and $x = 6466$ votes.

39. 1.25 oz of Food-1 and 7.5 oz of Food-2.

41. $I_1 = -\dfrac{36}{17}$, $I_2 = \dfrac{12}{17}$, $I_3 = \dfrac{24}{17}$ amps.   **43.** $3a = 2c$,   (H)   So $a = 4$,
$\qquad\qquad\qquad\qquad\qquad\qquad\qquad\qquad\quad 2b = 2 + c$,   (O) $b = 5$,
$\qquad\qquad\qquad\qquad\qquad\qquad\qquad\qquad\qquad\; a = 2$.   (N)   and $c = 6$.

45. For each 10 grams use 6 gms X, 2 gms Y, and 2 gms Z.   **47.** Let $u = \sqrt{x}$, $v = \sqrt{y}$, $w = \sqrt{z}$; (4, 9, 16).

49. $(2, \sqrt{3})(-2, \sqrt{3})$.   **51.** $\left(\dfrac{2059}{4160}, \dfrac{2407}{4160}\right)$.

53–55. Substitute the given n-tuple into the given system of equations for each of these problems. We already know from system (†) that, for each $i = 1, 2, \ldots, n$,

$$a_{i1}s_1 + a_{i2}s_2 + \cdots + a_{in}s_n = b_i.$$

Use this fact and elementary algebra (factoring, etc.) to obtain the desired results in each exercise.

Section 1.2 (pp. 28–29)

1. This system is inconsistent.   **3.** Consistent; for each real number t, $(9 + 2t, t, 4)$ is a solution.
5. Inconsistent.   **7.** $(2, 3, -3)$.   **9.** $(3 + 2t, t, -1)$, each $t \in \mathbb{R}$.
11. $(3 - s + 2t, s, t, -2)$, each $s, t \in \mathbb{R}$.   **13.** $(-t, 0, t)$, each $t \in \mathbb{R}$.   **15.** $(2t, 7t, 5t)$, each $t \in \mathbb{R}$.

17. $(0, t, -t, t)$, each $t \in \mathbb{R}$.   **19.** $(1, 0, 1)$.   **21.** $\left(\dfrac{14 + 5t}{5}, t - \dfrac{1}{5}\right)$, each $t \in \mathbb{R}$.

23. Inconsistent.   **25.** $(0, t, t, 0)$, each $t \in \mathbb{R}$.   **27.** $\left(1 + \dfrac{2}{7}t, \dfrac{3}{7}t, t\right)$, each $t \in \mathbb{R}$.

29. For each $i = 1, 2$; $a_{i1}(d_1 + c_1) + a_{i2}(d_2 + c_2) = (a_{i1}d_1 + a_{i2}d_2) + (a_{i1}c_1 + a_{i2}c_2) = 0 + 0 = 0$.
31. Since the given system has two distinct solutions $S = (s_1, s_2, \ldots, s_n)$ and $T = (t_1, t_2, \ldots, t_n)$, Exercise 30 shows that for each real number k, $S + kT$ is a solution. There are infinitely many distinct possible values for k since if $S + k_1 T = S + k_2 T$, then for $j = 1, 2, \ldots, n$, $s_j + k_1 t_j = s_j + k_2 t_j$, so $k_1 = k_2$.

33. The coefficient matrix is $\begin{bmatrix} a - 1 & -3 & 5 \\ 0 & a & -3 \\ 0 & 0 & a + 1 \end{bmatrix}$. If $a = -1$, the system has infinitely many solutions of the form $(7t, -3t, t)$. If $a = 0$, the system has infinitely many solutions of the form $(-3t, t, 0)$. If $a = 1$, there are infinitely many solutions of the form $(t, 0, 0)$. For all other values of the real number a, the reduced row echelon form for this coefficient matrix is I_3, so the system has only the trivial solution $(0, 0, 0)$.

35. If the $n + 1$st row of the reduced row echelon form of the augmented matrix had a nonzero entry it would have occurred in column n (since there are but n columns) and would have been eliminated by the Gauss-Jordan process. However, the first n rows could each contain a one, and the reduced row echelon form would be I_n. In this case, only the trivial solution is possible.

Section 1.3 (pp. 39–41)

1. $x = 1$, $y = 2x = 2$, $w = 1$, $z = 3y = 6$.   **3.** $x = \dfrac{1}{2}(A - 3B) = \begin{bmatrix} -\dfrac{5}{2} & 2 \\ -\dfrac{13}{2} & -\dfrac{1}{2} \\ -9 & \dfrac{3}{2} \end{bmatrix}$.   **5.** $\begin{bmatrix} 5 & -1 \\ 7 & 4 \\ 12 & 3 \end{bmatrix}$.

7. $\begin{bmatrix} \dfrac{23}{15} & \dfrac{6}{15} & \dfrac{-17}{15} \\ \dfrac{14}{15} & \dfrac{-1}{15} & \dfrac{57}{15} \\ \dfrac{5}{15} & \dfrac{11}{15} & \dfrac{12}{15} \end{bmatrix}$.   **9.** Not defined; A is 3×2, but B is 3×2.   **11.** Not defined; A is 3×2, but F is 3×3.

13. $\begin{bmatrix} 1 & 1 & -6 \\ 11 & 5 & 0 \\ 12 & 6 & -6 \end{bmatrix}$.   **15.** $\begin{bmatrix} -2\pi & 4\pi \\ -4\pi & -4\pi \\ -6\pi & 0 \end{bmatrix}$.   **17.** $\begin{bmatrix} -8 & 0 \\ 0 & -8 \end{bmatrix}$.   **19.** $\begin{bmatrix} 1 & -2 & -3 \\ -4 & 9 & 13 \\ -1 & 3 & 4 \end{bmatrix}$.

21. $D^4 = \begin{bmatrix} 64 & 0 \\ 0 & 64 \end{bmatrix}$, $8D^2 = \begin{bmatrix} -64 & 0 \\ 0 & -64 \end{bmatrix}$. So $D^4 + 8D^2 = 0$.

23. $CE = \begin{bmatrix} -1 & 0 & 1 \\ 5 & 5 & 0 \end{bmatrix}$, AC is in Exercise 10. $A(CE) = \begin{bmatrix} 4 & 5 & 1 \\ 11 & 10 & -1 \\ 15 & 15 & 0 \end{bmatrix} = (AC)E$.

25. If $A + Z = A$, then $-A + (A + Z) = -A + A = 0$, or $(-A + A) + Z = 0$. So we have $Z = 0 + Z = 0$. Similarly, if $Z + A = 0$.

27. If $A = (a_{ij})$ and $B = (b_{ij})$ for $i = 1, 2, 3$, and $j = 1, 2, 3$; then $rB = (rb_{ij})$. Let $ArB = (g_{ij})$ and $AB = (c_{ij})$. Then $g_{ij} = \sum_{k=1}^{3} a_{ik}rb_{kj} = \sum_{k=1}^{3} ra_{ik}b_{kj} = r\sum_{k=1}^{3} a_{ik}b_{kj} = rc_{ij}$. But $r(AB) = (rc_{ij})$. Hence, $ArB = rAB$.

29. Write $A = (a_{ij})$, $B = (b_{ij})$, and $C = (c_{ij})$. So $A + (B + C) = (a_{ij}) + (b_{ij} + c_{ij}) = (a_{ij} + (b_{ij} + c_{ij})) = ((a_{ij} + b_{ij}) + c_{ij}) = (a_{ij} + b_{ij}) + (c_{ij}) = (A + B) + C$.

31. If $A = (a_{ij})$, then $(rs)A = ((rs)a_{ij}) = r(sa_{ij}) = r(sA)$.

33. Let $S = (s_{ir})$, $i = 1, 2, \ldots, n$, $r = 1, 2, \ldots, m$. Let $T = (t_{rj})$, $j = 1, 2, \ldots, k$. Then $ST = (g_{ij})$, where $g_{ij} = \sum_{r=1}^{m} s_{ir}t_{rj}$.

Therefore, row i of ST is a $1 \times k$ matrix G whose entries are g_{ij}, $j = 1, 2, \ldots, k$. Consider the ith row of the matrix S.

It is a $1 \times m$ matrix $S^{(i)} = (s_{ir})$ (i fixed here). The matrix $S^{(i)}T$ is a $1 \times k$ matrix whose entries are $h_{ij} = \sum_{r=1}^{m} s_{ir}t_{rj}$. But then $h_{ij} = g_{ij}$. So $G = S^{(i)}T$.

35. $\sum_{i=1}^{n} ra_i = ra_1 + ra_2 + \cdots + ra_n = r(a_1 + a_2 + \cdots + a_n) = r\sum_{i=1}^{n} a_i$.

37. $\sum_{i=1}^{n} (a_i + b_i) = (a_1 + b_1) + (a_2 + b_2) + \cdots + (a_n + b_n) = (a_1 + a_2 + \cdots + a_n) + (b_1 + b_2 + \cdots + b_n) = \sum_{i=1}^{n} a_i + \sum_{i=1}^{n} b_i$.

39. We omit the details of this proof. Write out the various sums involved and note their equality.

41. $E - iF = \begin{bmatrix} 1 - 3i & -i & -1 + 2i \\ -2 - 4i & 1 + i & 3 - 7i \\ 1 & 1 - i & -2i \end{bmatrix}$. **43.** $iCF = \begin{bmatrix} 10i & 0 & i \\ 13i & 6i & 9i \end{bmatrix}$. **45.** $FiE = \begin{bmatrix} -i & -i & 0 \\ 13i & 6i & -7i \\ 0 & 3i & 3i \end{bmatrix}$.

Section 1.4 (pp. 50–52)

1. $\begin{bmatrix} a_{11} & a_{12} & a_{13} \\ 0 & 0 & 0 \\ 0 & 0 & 0 \end{bmatrix}$. **3.** $\begin{bmatrix} a_{31} & a_{32} & a_{33} \\ 0 & 0 & 0 \\ 0 & 0 & 0 \end{bmatrix}$. **5.** $\begin{bmatrix} a_{11} & a_{12} & a_{13} \\ a_{11} & a_{12} & a_{13} \\ 0 & 0 & 0 \end{bmatrix}$. **7.** $\begin{bmatrix} a_{11} & a_{12} & a_{13} \\ -a_{21} & -a_{22} & -a_{23} \\ a_{31} & a_{32} & a_{33} \end{bmatrix}$.

9. $\begin{bmatrix} a_{11} - a_{21} + 2a_{31} & a_{12} - a_{22} + 2a_{32} & a_{13} - a_{23} + 2a_{33} \\ a_{21} - a_{31} & a_{22} - a_{32} & a_{23} - a_{33} \\ a_{31} & a_{32} & a_{33} \end{bmatrix}$.

11. Let $A = (a_{ij})$, $B = (b_{ij})$, and $C = (c_{ij})$; $i, j = 1, 2$. Then $AB = (\alpha_{ij})$, and $BC = (\beta_{ij})$; where $\alpha_{ij} = a_{i1}b_{1j} + a_{i2}b_{2j}$ and $\beta_{ij} = b_{i1}c_{1j} + b_{i2}c_{2j}$. Then $(AB)C = (\gamma_{ij})$, where $\gamma_{ij} = \sum_{r=1}^{2} \alpha_{ir}c_{rj} = \sum_{r=1}^{2} \left(\sum_{k=1}^{2} a_{ik}b_{kr} \right)c_{rj} = \left(\sum_{k=1}^{2} a_{ik}b_{k1} \right)c_{1j} + \left(\sum_{k=1}^{2} a_{ik}b_{k2} \right)c_{2j} = (a_{i1}b_{11} + a_{i2}b_{21})c_{1j} + (a_{i1}b_{12} + a_{i2}b_{22})c_{2j} = a_{i1}b_{11}c_{1j} + a_{i2}b_{21}c_{1j} + a_{i1}b_{12}c_{2j} + a_{i2}b_{22}c_{2j} = a_{i1}b_{11}c_{1j} + a_{i1}b_{12}c_{2j} + a_{i2}b_{21}c_{1j} + a_{i2}b_{22}c_{2j} = a_{i1}(b_{11}c_{1j} + b_{12}c_{2j}) + a_{i2}(b_{21}c_{1j} + b_{22}c_{2j}) = \sum_{k=1}^{2} a_{ik} \left(\sum_{r=1}^{2} b_{kr}c_{rj} \right) = \rho_{ij}$. Notice that $A(BC) = (\rho_{ij})$, where $\rho_{ij} = \sum_{k=1}^{2} a_{ik} \left(\sum_{r=1}^{2} b_{kr}c_{rj} \right)$. So $(AB)C = A(BC)$. Generalize this proof for any $n \geq 2$.

13. $AB = \begin{bmatrix} 1 & 0 \\ 0 & 1 \end{bmatrix}$. $BA = \begin{bmatrix} -5 & 2 & -4 \\ 9 & -2 & 6 \\ 12 & -4 & 9 \end{bmatrix}$.   **15.** $A^T = \begin{bmatrix} 1 & 2 \\ 2 & 0 \\ -1 & 1 \end{bmatrix}$. $B^T = \begin{bmatrix} 1 & -1 & -2 \\ -3 & 5 & 7 \end{bmatrix}$.

17. $A + B^T = \begin{bmatrix} 2 & 1 & -3 \\ -1 & 5 & 8 \end{bmatrix}$, $A^T + B = \begin{bmatrix} 2 & -1 \\ 1 & 5 \\ -3 & 8 \end{bmatrix}$.

19. There are many; for example, the matrix $\begin{bmatrix} a & b & d \\ b & c & e \\ d & e & f \end{bmatrix}$, with at least one of b, d, or e not zero.

21. $(AA^T)^T = (A^T)^T A^T = AA^T$.

23. Let $A = \begin{bmatrix} 1 & -1 & 1 \\ 0 & 2 & 3 \\ 1 & 0 & -1 \end{bmatrix}$ and $D = \begin{bmatrix} 2 & 0 & 0 \\ 0 & 3 & 0 \\ 0 & 0 & -1 \end{bmatrix}$, then $AD = \begin{bmatrix} 2 & -3 & -1 \\ 0 & 6 & -3 \\ 2 & 0 & 1 \end{bmatrix}$ while $DA = \begin{bmatrix} 2 & -2 & 2 \\ 0 & 6 & 9 \\ -1 & 0 & 1 \end{bmatrix}$.

25. If $A = (a_{ij})$, then $-A = (-a_{ij})$, so $-(-A) = -(-a_{ij}) = (a_{ij}) = A$.

27. Let $A = (a_{ij})$, $B = (b_{jk})$, and $C = (c_{jk})$, for $i = 1, 2, \ldots, m$; $j = 1, 2, \ldots, n$; and $k = 1, 2, \ldots, s$. Then

$B - C = (b_{jk} - c_{jk})$, so $A(B - C) = (\alpha_{ik})$, where $\alpha_{ik} = \sum_{r=1}^{n} a_{ir}(b_{rk} - c_{rk}) = \sum_{r=1}^{n} (a_{ir}b_{rk} - a_{ir}c_{rk}) = \sum_{r=1}^{n} a_{ir}b_{rk} - \sum_{r=1}^{n} a_{ir}c_{rk}$.

So $A(B - C) = AB - AC$, as desired.

29. $(r - s)A = (r + (-s))A = rA + (-s)A = rA - sA$.

31. Let $A = (a_{ij})$, $i = 1, 2, \ldots, m$; $j = 1, 2, \ldots, n$. The entry in row j column i of the transpose A^T is a_{ij}; the same entry as in row i and column j of A. The entry row i and column j of $(A^T)^T$ is the same as the entry in row j and column i of A^T; namely, a_{ij}. Hence, $A = (A^T)^T$.

33. See Exercise 31. The entry in row j column i of the matrix rA^T is ra_{ij}, which is precisely the entry in row j column i of $r(A^T)$.

35. See the notation in Exercise 31. Let $B = (b_{jt})$ and $C = (c_{jt})$, $t = 1, 2, \ldots, k$. So the matrix $A(B + C) = (\xi_{it})$ where

$\xi_{it} = \sum_{j=1}^{n} a_{ij}(b_{jt} + c_{jt})$. So $\xi_{it} = \sum_{j=1}^{n} (a_{ij}b_{jt} + a_{ij}c_{jt}) = \sum_{j=1}^{n} a_{ij}b_{jt} + \sum_{j=1}^{n} a_{ij}c_{jt} = \alpha_{it} + \beta_{it}$, where $AB = (\alpha_{it})$ and $AC = (\beta_{it})$.

37. Let $A = (a_{ij})$ and $B = (b_{jt})$, $i = 1, 2, \ldots, m$; $j = 1, 2, \ldots, n$; $t = 1, 2, \ldots, k$. Then the matrix $AB = (\alpha_{it})$, where

$\alpha_{it} = \sum_{j=1}^{n} a_{ij}b_{jt}$. So we have $rAB = (r\alpha_{it})$, and $r\alpha_{it} = r\sum_{j=1}^{n} a_{ij}b_{jt} = \sum_{j=1}^{n} r(a_{ij}b_{jt}) = \sum_{j=1}^{n} (ra_{ij})b_{jt} = \sum_{j=1}^{n} a_{ij}(rb_{jt})$. Hence,

$r(AB) = (rA)B = A(rB)$.

39. $\begin{bmatrix} 2 & 1 & 3 & 5 \\ 0 & 2 & 1 & 0 \\ 0 & 0 & 2 & 1 \\ 0 & 0 & 0 & 2 \end{bmatrix}$ is one example.   **41.** Diag $[1, 2, \ldots, -1]$ is an example.

43. Let $A = (a_{ij})$ and $B = (b_{ij})$, $i = 1, 2, \ldots, m$; $j = 1, 2, \ldots, n$; and $a_{ij} = b_{ij} = 0$ when $i > j$. So $A + B = (a_{ij} + b_{ij})$, with $a_{ij} + b_{ij} = 0$, when $i > j$. Therefore, $A + B$ is upper triangular.

45. Let $A = (a_{ij})$ and $B = (b_{jt})$, $i = 1, 2, \ldots, m$; $j = 1, 2, \ldots, n$; $t = 1, 2, \ldots, k$. Then the matrix $AB = (\alpha_{it})$, where

$\alpha_{it} = \sum_{j=1}^{n} a_{ij}b_{jt}$. Then for any $i > t$, it is either true that $i \leq j$ or $i > j$. In the case $i > j$, then $a_{ij} = 0$, so $\alpha_{it} = 0$. In the case $i \leq j$, then $t < i \leq j$, so $j > t$ and $b_{jt} = 0$, again forcing $\alpha_{it} = 0$. So AB is upper triangular.

47. $\begin{bmatrix} 0 & 1 & 2 \\ 0 & 0 & -1 \\ 0 & 0 & 0 \end{bmatrix}^2 = \begin{bmatrix} 0 & 0 & -1 \\ 0 & 0 & 0 \\ 0 & 0 & 0 \end{bmatrix}$. $\begin{bmatrix} 0 & 1 & 2 \\ 0 & 0 & -1 \\ 0 & 0 & 0 \end{bmatrix}^3 = \begin{bmatrix} 0 & 0 & 0 \\ 0 & 0 & 0 \\ 0 & 0 & 0 \end{bmatrix}$.

49. Let $A = (a_{ij}) \neq O$ be an $n \times n$ upper triangular matrix all of whose main diagonal elements $a_{ii} = 0$. Then $A^2 = (\alpha_{ij})$,

where $\alpha_{ij} = \sum_{t=1}^{n} (a_{it}a_{tj})$. Now, notice that $a_{i,i+1} = \sum_{t=1}^{n} a_{it}a_{t,i+1}$. When $t \leq i$, $a_{it} = 0$. But when $t > i$, then $a_{t,i+1} = 0$,

because A is upper triangular. Thus $\alpha_{i,i+1} = 0$, for all $i = 1, 2, \ldots, n$. That is, all of the elements on the diagonal and on the "super diagonal" are zero. Now suppose that $A^k = (\beta_{ij})$ and $\beta_{ij} = 0$, for $j = 1, 2, \ldots, i + k - 1 < n$. Then we

have $A^{k+1} = A(A^K) = (\xi_{ij})$, where $\xi_{ij} = \sum_{t=1}^{n} (a_{it}\beta_{tj}) = 0$, when $j \leq i + k$. Thus, $A^n = 0$. For lower triangular matrices A,

use the above proof for A^T and apply $(A^n)^T = (A^T)^n$.

51. By Theorem 1.5(3) and induction, as mentioned above, $(A^T)^k = (A^k)^T$, so A is nilpotent if, and only if, A^T is nilpotent.

53. Let $A = \begin{bmatrix} 0 & 1 \\ 0 & 0 \end{bmatrix}$ and $B = \begin{bmatrix} 0 & 0 \\ 1 & 0 \end{bmatrix}$. Both A and B are nilpotent, but the matrix sum $A + B = \begin{bmatrix} 0 & 1 \\ 1 & 0 \end{bmatrix}$ is not

nilpotent.

Section 1.5 (pp. 66–69)

1. $\begin{bmatrix} 0 & 1 \\ 1 & 0 \end{bmatrix}$. **3.** $\begin{bmatrix} 1 & 0 \\ 0 & -\frac{1}{5} \end{bmatrix}$. **5.** $\begin{bmatrix} 1 & 0 & 0 \\ 0 & \frac{1}{7} & 0 \\ 0 & 0 & 1 \end{bmatrix}$. **7.** $\begin{bmatrix} 0 & 0 & 1 & 0 \\ 0 & 1 & 0 & 0 \\ 1 & 0 & 0 & 0 \\ 0 & 0 & 0 & 1 \end{bmatrix}$.

9. $\begin{bmatrix} \frac{1}{7} & 0 & 0 \\ 0 & -\frac{1}{3} & 0 \\ 0 & 0 & \frac{1}{4} \end{bmatrix}$. **11.** $\begin{bmatrix} \frac{1}{4} & 0 & 0 \\ \frac{1}{6} & \frac{1}{3} & 0 \\ -\frac{4}{15} & \frac{1}{15} & \frac{3}{15} \end{bmatrix}$. **13.** $\begin{bmatrix} 0 & -1 & 0 & 1 \\ -1 & 0 & 0 & 1 \\ 0 & 1 & 1 & -1 \\ 1 & 1 & 0 & -1 \end{bmatrix}$.

15. $\begin{bmatrix} 2 & -1 & 3 \\ 0 & 1 & -4 \\ 2 & -1 & -2 \end{bmatrix} \cdot \begin{bmatrix} x_1 \\ x_2 \\ x_3 \end{bmatrix} = \begin{bmatrix} 2 \\ 5 \\ 7 \end{bmatrix}$, so $X = \begin{bmatrix} \frac{3}{5} & \frac{1}{2} & -\frac{1}{10} \\ \frac{4}{5} & 1 & -\frac{4}{5} \\ \frac{1}{5} & 0 & -\frac{1}{5} \end{bmatrix} \cdot \begin{bmatrix} 2 \\ 5 \\ 7 \end{bmatrix} = \begin{bmatrix} 3 \\ 1 \\ -1 \end{bmatrix}$.

17. $\begin{bmatrix} 1 & 2 & -1 \\ 2 & -3 & 1 \\ 5 & 7 & 2 \end{bmatrix} \cdot \begin{bmatrix} x_1 \\ x_2 \\ x_3 \end{bmatrix} = \begin{bmatrix} 4 \\ -1 \\ -1 \end{bmatrix}$, so $X = \begin{bmatrix} \frac{13}{40} & \frac{11}{40} & \frac{1}{40} \\ -\frac{1}{40} & -\frac{7}{40} & \frac{3}{40} \\ -\frac{29}{40} & -\frac{3}{40} & \frac{7}{40} \end{bmatrix} \cdot \begin{bmatrix} 4 \\ -1 \\ -1 \end{bmatrix} = \begin{bmatrix} 1 \\ 0 \\ -3 \end{bmatrix}$.

19. $\begin{bmatrix} -48 & 18 & 9 & -7 \\ 82 & -31 & -15 & 12 \\ 49 & -18 & -9 & 7 \\ 27 & -10 & -5 & 4 \end{bmatrix}$. **21.** $\begin{bmatrix} \frac{7}{3} & -\frac{1}{3} & -\frac{1}{3} & -\frac{2}{3} \\ \frac{4}{9} & -\frac{1}{9} & -\frac{4}{9} & \frac{1}{9} \\ -\frac{1}{9} & -\frac{2}{9} & \frac{1}{9} & \frac{2}{9} \\ -\frac{5}{3} & \frac{2}{3} & \frac{2}{3} & \frac{1}{3} \end{bmatrix}$.

23. Given $AB = I$ and $CA = I$, then $B = IB = (CA)B = C(AB) = CI = C$. So $AB = I = BA$ and $CA = I = AC$; therefore, $B = C = A^{-1}$.

25. If A^{-1} exists and $AX = O$, then $X = A^{-1}(AX) = A^{-1}(O) = O$. So $X = 0$ is the only possible solution. Conversely, if $X = 0$ is the only solution, the coefficient matrix A has I for its reduced row echelon form, so A is invertible. (Use the inversion algorithm to find A^{-1}.)

27. Suppose N is nilpotent and $N^k = 0$, while $N^{k-1} \neq 0$. If N were also nonsingular, then we would have $0 = N^k N^{-1} = N^{k-1} N N^{-1} = N^{k-1}$. This is a contradiction, so N is not invertible.

29. Since $AA^{-1} = I_3$, use Exercise 34 of Section 1.3 to verify that the jth column of I_3 is the product indicated for $j = 1, 2, 3$.

31. Let $A = I_2$ and $B = -I_2$. Then $A + B = 0$ is clearly not invertible even though both A and B are.

33. See the first page of Section 2.1 where this is done.

35. Since Taxes = Tax Matrix $\times$ Profits, or $T = MP$, we have $M^{-1}T = P$. Thus,

$$\text{Profits} = \begin{bmatrix} 29.02 & -5.05 & -21.14 \\ -5.50 & -3.47 & 16.72 \\ -26.18 & 13.25 & 17.98 \end{bmatrix} \begin{bmatrix} 9 \\ 11 \\ 8 \end{bmatrix} = \begin{bmatrix} 36593 \\ 50158 \\ 53943 \end{bmatrix}.$$

That is, Macht Corp. made approximately \$36.6 thousand in U.S. profits, \$50.2 thousand in Japanese profits, and \$53.9 thousand in European profits.

37. To undo multiplying row i by the nonzero constant c, multiply the new row i by the nonzero constant $\dfrac{1}{c}$.

39. Suppose that A is singular and that BA is nonsingular. Then $E_k E_{k-1} \cdots E_2 E_1 BA = I$. But then $(E_k E_{k-1} \cdots E_2 E_1 B) = A^{-1}$, contrary to the fact that A is singular.

41. The identity matrix is an elementary matrix (of either Type I or Type II, why?). Therefore, $A = IA$.

43. If A is row equivalent to B and B is row equivalent to C, there exist elementary matrices so that $A = E_1 E_2 \cdots E_k B$, and $B = F_1 F_2 \cdots F_j C$. Therefore, $A = E_1 E_2 \cdots E_k F_1 F_2 \cdots F_j C$. And A is row equivalent to C, as desired.

45. It is possible for A to be row equivalent to several different matrices each of which is in row echelon form. For example, let $A = \begin{bmatrix} 1 & -1 \\ 1 & 3 \end{bmatrix}$. Both of the matrices $M = \begin{bmatrix} 1 & -1 \\ 0 & 1 \end{bmatrix}$ and $K = \begin{bmatrix} 1 & 3 \\ 0 & 1 \end{bmatrix}$ are row echelon forms for A.

47. If A is strictly lower triangular, then A^T is strictly upper triangular. A is singular if, and only if, A^T is singular; which it always is, by Exercise 46.

49. Yes, $A^{-1} = \begin{bmatrix} 1 & -2 & 11 \\ 0 & 1 & -4 \\ 0 & 0 & 1 \end{bmatrix}$.   **51.** By matrix multiplication, $E^2 = E$.

53. If E is an $n \times n$ nilpotent and idempotent matrix, then $0 = E^k = E$. ($E^k = E$, by induction.) So the answer must be the zero matrix.

55. Some of the results are the following: $\varepsilon = 0.1$: $A^{-1} = \begin{bmatrix} 18.11 & -1.78 & -9.48 & 6.18 \\ -8.08 & 0.93 & 4.34 & -2.95 \\ -3.68 & -0.20 & 2.87 & -1.18 \\ -4.08 & 0.91 & 1.76 & -1.18 \end{bmatrix}.$

$$\varepsilon = -0.1: A^{-1} = \begin{bmatrix} 18.11 & -1.50 & -5.65 & 3.52 \\ -4.99 & 0.76 & 2.49 & -1.67 \\ -2.99 & -0.07 & 2.16 & -0.75 \\ -3.89 & 0.84 & 1.25 & -0.75 \end{bmatrix}. \quad \varepsilon = 0.3: A^{-1} = \begin{bmatrix} -33.96 & 2.11 & 20.84 & -14.06 \\ 16.91 & -0.09 & -10.27 & 6.78 \\ 5.99 & -1.13 & -2.40 & 2.45 \\ 4.56 & 0.30 & -3.83 & 2.45 \end{bmatrix}.$$

$$\varepsilon = -0.3: A^{-1} = \begin{bmatrix} 13.75 & -1.82 & -5.94 & 3.30 \\ -5.74 & 0.88 & 2.59 & -1.54 \\ -3.98 & 0.17 & 2.34 & -0.80 \\ -4.75 & 0.94 & 1.57 & -0.80 \end{bmatrix}. \quad \varepsilon = 1: A^{-1} = \begin{bmatrix} -5.57e18 & 5.57e18 & -5.57e18 & -1.41 \\ 5.31e18 & -5.31e18 & 5.31e18 & 0.86 \\ -1.48e19 & 1.48e19 & -1.48e19 & -0.85 \\ 9.22e18 & -9.22e18 & 9.22e18 & 0.75 \end{bmatrix};$$

where $-1.48e19$ means -1.48×10^{19}, etc. For $\varepsilon = -1$: $A^{-1} = \begin{bmatrix} -9.90 & -1.59 & -3.65 & 1.18 \\ -3.23 & 0.62 & 1.30 & -0.49 \\ -6.12 & 0.78 & 2.67 & -0.51 \\ -6.62 & 1.28 & 2.17 & -0.51 \end{bmatrix}.$

Section 1.6 (pp. 79–81)

1. Let $A = \begin{bmatrix} 3I_2 & K \\ 0 & I_3 \end{bmatrix}$. So $A^2 = \begin{bmatrix} 9I_2 & 4K \\ 0 & I_3 \end{bmatrix}$, where $K = \begin{bmatrix} 1 & 2 & -1 \\ 0 & 1 & 2 \end{bmatrix}$.

3. $AB = \begin{bmatrix} 6 & 7 & 12 & 5 & 15 & 3 \\ 5 & -3 & 12 & 1 & -4 & 0 \\ 4 & 1 & -1 & 0 & 1 & 0 \\ 0 & 0 & 1 & 1 & 0 & 0 \\ 1 & 0 & 1 & 0 & 1 & 0 \end{bmatrix}$.

5. $C^2 = \begin{bmatrix} 14 & 3 & 0 & 0 & 1 \\ 3 & 10 & 3 & 0 & 0 \\ 0 & 2 & 6 & 0 & -1 \\ -3 & -3 & -1 & 1 & 0 \\ -2 & 0 & -1 & 0 & 2 \end{bmatrix}$.

7. $M^4 = \begin{bmatrix} 1 & 0 & 0 & 0 & 0 & 0 \\ 0 & 1 & 0 & 0 & 0 & 0 \\ 0 & 0 & 1 & 0 & 0 & 0 \\ 0 & 0 & 0 & \frac{1}{16} & 0 & \frac{259}{16} \\ 0 & 0 & 0 & 0 & 81 & 0 \\ 0 & 0 & 0 & 0 & 0 & 81 \end{bmatrix}$.

9. $K = \begin{bmatrix} 2 & 1 & 0 & 0 & 0 & 0 & 0 & 0 & 0 & 0 & 0 & 0 & 0 \\ 0 & 2 & 1 & 0 & 0 & 0 & 0 & 0 & 0 & 0 & 0 & 0 & 0 \\ 0 & 0 & 2 & 0 & 0 & 0 & 0 & 0 & 0 & 0 & 0 & 0 & 0 \\ 0 & 0 & 0 & \frac{1}{3} & 1 & 0 & 0 & 0 & 0 & 0 & 0 & 0 & 0 \\ 0 & 0 & 0 & 0 & \frac{1}{3} & 0 & 0 & 0 & 0 & 0 & 0 & 0 & 0 \\ 0 & 0 & 0 & 0 & 0 & \frac{1}{2} & 1 & 0 & 0 & 0 & 0 & 0 & 0 \\ 0 & 0 & 0 & 0 & 0 & 0 & \frac{1}{2} & 1 & 0 & 0 & 0 & 0 & 0 \\ 0 & 0 & 0 & 0 & 0 & 0 & 0 & \frac{1}{2} & 0 & 0 & 0 & 0 & 0 \\ 0 & 0 & 0 & 0 & 0 & 0 & 0 & 0 & 1 & 0 & 0 & 0 & 0 \\ 0 & 0 & 0 & 0 & 0 & 0 & 0 & 0 & 0 & 1 & 0 & 0 & 0 \\ 0 & 0 & 0 & 0 & 0 & 0 & 0 & 0 & 0 & 0 & -1 & 0 & 0 \\ 0 & 0 & 0 & 0 & 0 & 0 & 0 & 0 & 0 & 0 & 0 & -1 & 0 \\ 0 & 0 & 0 & 0 & 0 & 0 & 0 & 0 & 0 & 0 & 0 & 0 & -1 \end{bmatrix}$.

11. Write it out; it is 13×13 also. See the matrix in Exercise 9 as an example.

13. $x_1 = -\frac{31}{15}, x_2 = \frac{44}{15}, x_3 = \frac{5}{3}$.

15. $y_1 = 0, y_2 = -3, y_3 = \frac{43}{5}$; so $x_1 = \frac{11}{56}, x_2 = \frac{61}{56}, x_3 = \frac{43}{56}$.

17. $y_1 = 1, y_2 = -5, y_3 = 1$; so $x_1 = \frac{8}{15}, x_2 = -\frac{22}{15}, x_3 = -\frac{1}{3}$.

19. $A = LU = \begin{bmatrix} 1 & 0 & 0 & 0 \\ 0 & 1 & 0 & 0 \\ 0 & \frac{1}{3} & 1 & 0 \\ \frac{3}{2} & -\frac{7}{6} & \frac{1}{4} & 1 \end{bmatrix} \begin{bmatrix} 2 & 3 & 1 & 0 \\ 0 & 3 & -1 & 0 \\ 0 & 0 & \frac{4}{3} & -1 \\ 0 & 0 & 0 & -\frac{7}{4} \end{bmatrix}$.

21. The LU display is $\begin{bmatrix} 2 & 3 & 1 & 0 \\ 0 & 3 & -1 & 0 \\ 0 & \frac{1}{3} & \frac{4}{3} & -1 \\ \frac{3}{2} & -\frac{7}{6} & \frac{1}{4} & -\frac{7}{4} \end{bmatrix}$.

23. The matrix A itself is a Type I elementary matrix, and $AA = I$. The identity matrix has an LU-decomposition $I = II$, where I is both L and U.

25. $\begin{bmatrix} 1 & 0 & 0 & 0 & 0 & 0 & 0 \\ 0 & 1 & 0 & 0 & 0 & 0 & 0 \\ 0 & 0 & 1 & 0 & 0 & 0 & 0 \\ 0 & 0 & 0 & 1 & -2 & 4 & 0 \\ 0 & 0 & 0 & 3 & 5 & 1 & 2 \\ 0 & 0 & 0 & 3 & -1 & 0 & 1 \end{bmatrix}$.

27. Yes, by block multiplication $\begin{bmatrix} C & 0 \\ 0 & D \end{bmatrix} \cdot \begin{bmatrix} A & 0 \\ 0 & B \end{bmatrix} = \begin{bmatrix} CA & 0 \\ 0 & DB \end{bmatrix}$.

Section 1.7 (pp. 94–96)

1. No; this is not the matrix representation of a diagraph since $a_{22} = 1 \neq 0$ means the presence of a loop.

3. Yes. In this situation 3 is the consensus leader.   **5.** There is no dominance.

7. This was not a matrix representation of a diagraph.

9. $M^2 = \begin{bmatrix} 0 & 0 & 1 \\ 1 & 1 & 0 \\ 0 & 1 & 1 \end{bmatrix}$; $M + M^2 = \begin{bmatrix} 0 & 1 & 1 \\ 1 & 1 & 1 \\ 1 & 2 & 1 \end{bmatrix} \begin{matrix} = 2 \\ = 3 \\ = 4 \end{matrix}$. Yes, Individual number 3 is the leader.

11. $M^2 = \begin{bmatrix} 0 & 1 & 0 \\ 0 & 0 & 1 \\ 1 & 0 & 0 \end{bmatrix}$; $M + M^2 = \begin{bmatrix} 0 & 1 & 1 \\ 1 & 0 & 1 \\ 1 & 1 & 0 \end{bmatrix} \begin{matrix} = 2 \\ = 2 \\ = 2 \end{matrix}$. No leader.

13. $\begin{matrix} & A & B & C & D \\ A \\ B \\ C \\ D \end{matrix} \begin{bmatrix} 0 & 1 & 0 & 1 \\ 0 & 0 & 1 & 0 \\ 1 & 0 & 0 & 0 \\ 0 & 1 & 0 & 0 \end{bmatrix} \begin{matrix} = 2 \\ = 1 \\ = 1 \\ = 1 \end{matrix}$.

15. $M^2 = \begin{bmatrix} 0 & 1 & 1 & 0 \\ 1 & 0 & 0 & 0 \\ 0 & 1 & 0 & 1 \\ 0 & 0 & 1 & 0 \end{bmatrix}$; $M + M^2 = \begin{bmatrix} 0 & 2 & 1 & 1 \\ 1 & 0 & 1 & 0 \\ 1 & 1 & 0 & 1 \\ 0 & 1 & 1 & 0 \end{bmatrix} \begin{matrix} = 4 \\ = 2 \\ = 3 \\ = 2 \end{matrix}$.

17. $\begin{matrix} & K & T & O'B & C \\ K \\ T \\ O'B \\ C \end{matrix} \begin{bmatrix} 0 & 1 & 1 & 0 \\ 0 & 0 & 0 & 0 \\ 0 & 1 & 0 & 0 \\ 1 & 1 & 1 & 0 \end{bmatrix}$. $M^2 = \begin{bmatrix} 0 & 1 & 0 & 0 \\ 0 & 0 & 0 & 0 \\ 0 & 0 & 0 & 1 \\ 0 & 2 & 1 & 0 \end{bmatrix}$; so $M + M^2 = \begin{bmatrix} 0 & 2 & 1 & 0 \\ 0 & 0 & 0 & 0 \\ 0 & 1 & 0 & 0 \\ 1 & 3 & 2 & 0 \end{bmatrix} \begin{matrix} = 3 \\ = 0 \\ = 1 \\ = 6 \end{matrix} \begin{matrix} \text{Kelly} \\ \text{Thompson} \\ \text{O'Brian} \\ \text{Carter} \end{matrix}$

19.

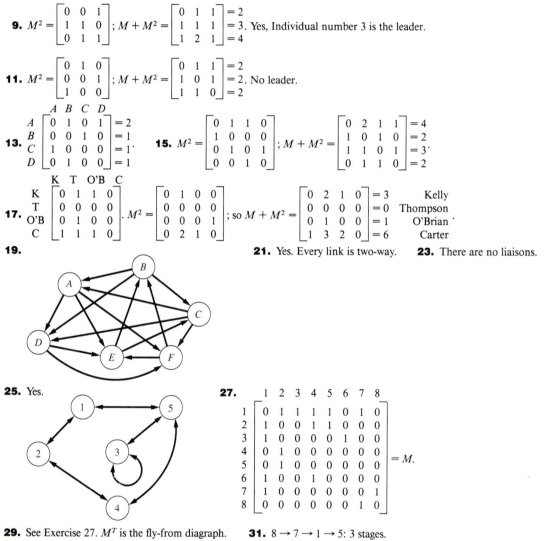

21. Yes. Every link is two-way.   **23.** There are no liaisons.

25. Yes.

27.

$$\begin{matrix} & 1 & 2 & 3 & 4 & 5 & 6 & 7 & 8 \\ 1 \\ 2 \\ 3 \\ 4 \\ 5 \\ 6 \\ 7 \\ 8 \end{matrix} \begin{bmatrix} 0 & 1 & 1 & 1 & 1 & 0 & 1 & 0 \\ 1 & 0 & 0 & 1 & 1 & 0 & 0 & 0 \\ 1 & 0 & 0 & 0 & 0 & 1 & 0 & 0 \\ 0 & 1 & 0 & 0 & 0 & 0 & 0 & 0 \\ 0 & 1 & 0 & 0 & 0 & 0 & 0 & 0 \\ 1 & 0 & 0 & 1 & 0 & 0 & 0 & 0 \\ 1 & 0 & 0 & 0 & 0 & 0 & 0 & 1 \\ 0 & 0 & 0 & 0 & 0 & 0 & 1 & 0 \end{bmatrix} = M.$$

29. See Exercise 27. M^T is the fly-from diagraph.   **31.** $8 \rightarrow 7 \rightarrow 1 \rightarrow 5$: 3 stages.

33. Cities 1, 2, 3, and 7 are liaison cities.

35. There are but $n - 1$ other individuals in the network, so there can be at most $n - 1$ intermediaries.

37. This diagraph will reflect your family authority structure.

Chapter 1 Review Exercises (p. 97)

1. $\begin{bmatrix} 1 & -1 & 1 & 1 \\ 1 & 2 & -1 & 2 \\ 2 & -3 & 3 & 1 \end{bmatrix}$ is the augmented matrix, $\begin{bmatrix} 1 & 0 & 0 & 2 \\ 0 & 1 & 0 & -1 \\ 0 & 0 & 1 & -2 \end{bmatrix}$ is its reduced row echelon form, and $x_1 = 2$, $x_2 = -1$, $x_3 = -2$ is the unique solution.

2. $\begin{bmatrix} 1 & 0 & 0 & -1 \\ 0 & 1 & 0 & 0 \\ 0 & 0 & 1 & 0 \\ 0 & 0 & 0 & 0 \end{bmatrix}$ is the reduced row echelon form, and $x_1 = -1$, $x_2 = 0$, $x_3 = 0$ is the unique solution.

3. $a = 0, 1, -1$.   **4.** For $a = 0$: $(t, -t, 0)$; t any real number. For $a = -1$: $(t, 4t, 2t)$. For $a = 1$: $(t, 0, 0)$.
5. For all a different from 0, 1, and -1, the system has the unique (trivial) solution $(0, 0, 0)$.

6. $A + B = \begin{bmatrix} 2 & 2 \\ 8 & 3 \end{bmatrix}$. $A - B = \begin{bmatrix} -12 & 2 \\ -2 & 5 \end{bmatrix}$. $AB = \begin{bmatrix} -25 & -2 \\ 41 & -4 \end{bmatrix}$, and $BA = \begin{bmatrix} -35 & 14 \\ -28 & 6 \end{bmatrix}$; $3A = \begin{bmatrix} -15 & 6 \\ 9 & 12 \end{bmatrix}$. AB and

BA are *not* equal.

7. A^{-1} exists. $A^{-1} = \begin{bmatrix} 1 & 0 & 0 \\ -2 & 1 & 0 \\ 7 & -3 & 1 \end{bmatrix}$.

8. $A^2 = \begin{bmatrix} 0 & 0 & 0 \\ 0 & 0 & 0 \\ 6 & 0 & 0 \end{bmatrix}$ and $A^3 = \begin{bmatrix} 0 & 0 & 0 \\ 0 & 0 & 0 \\ 0 & 0 & 0 \end{bmatrix}$. A is singular. Some fairly obvious reasons include: (1) A has a row of zeros
so cannot be reduced to I; (2) A is nilpotent.

9. $[A : I]$ reduces to $\begin{bmatrix} 1 & 0 & -1 & 0 & -2.67 & 1.67 \\ 0 & 1 & 2 & 0 & 2.33 & -1.33 \\ 0 & 0 & 0 & 1 & -2 & 1 \end{bmatrix}$, so A is singular.   **10.** Surely $A^{-1} = A$.

11. If $AA^{-1} = I_n$, then $(AA^{-1})^T = I_n{}^T = I_n$. But $(AA^{-1})^T = (A^{-1})^T A^T$; therefore, $(A^{-1})^T$ is the inverse of A^T:
$(A^{-1})^T = (A^T)^{-1}$.
12. Note that for each positive integer n, $(A^T)^n = (A^n)^T$. So $(A^T)^k = (A^k)^T = 0^T = 0$.

Chapter 2

Section 2.1 (pp. 113–115)
1. 10.   **3.** 6.   **5.** $a^2 + b^2$.   **7.** 72.
9. -6.   **11.** Even, sgn(β) $= 1$.   **13.** Odd, sgn(β) $= -1$.   **15.** Odd, sgn(β) $= -1$.
17. det $A = a_{11}a_{22}a_{33}a_{44} - a_{11}a_{22}a_{34}a_{43} - a_{11}a_{24}a_{33}a_{42} - a_{11}a_{23}a_{32}a_{44} + a_{11}a_{23}a_{34}a_{42} + a_{11}a_{24}a_{32}a_{43} - a_{12}a_{23}a_{34}a_{41}$
$\quad - a_{12}a_{21}a_{33}a_{44} + a_{12}a_{21}a_{34}a_{43} + a_{12}a_{23}a_{31}a_{44} - a_{12}a_{24}a_{31}a_{43} + a_{12}a_{24}a_{33}a_{41} + a_{13}a_{21}a_{32}a_{44} - a_{13}a_{21}a_{34}a_{42}$
$\quad - a_{13}a_{22}a_{31}a_{44} + a_{13}a_{22}a_{34}a_{41} - a_{13}a_{24}a_{32}a_{41} + a_{13}a_{24}a_{31}a_{42} - a_{14}a_{22}a_{33}a_{41} + a_{14}a_{22}a_{31}a_{43} - a_{14}a_{21}a_{32}a_{43}$
$\quad + a_{14}a_{21}a_{33}a_{42} - a_{14}a_{23}a_{31}a_{42} + a_{14}a_{23}a_{32}a_{41}$.

19.

Elementary Product	Rewritten Form	Row Permutation	Odd or Even	
$a_{11}a_{22}a_{33}$	$a_{11}a_{22}a_{33}$	{1, 2, 3}	even	
$a_{12}a_{23}a_{31}$	$a_{31}a_{12}a_{23}$	{3, 1, 2}	even	
$a_{13}a_{21}a_{32}$	$a_{21}a_{32}a_{13}$	{2, 3, 1}	even	Yes.
$a_{13}a_{22}a_{31}$	$a_{31}a_{22}a_{13}$	{3, 2, 1}	odd	
$a_{12}a_{21}a_{33}$	$a_{21}a_{12}a_{33}$	{2, 1, 3}	odd	
$a_{11}a_{23}a_{32}$	$a_{11}a_{32}a_{23}$	{1, 3, 2}	odd	

21. Each determinant is equal to $a_{11}a_{22} - a_{12}a_{21}$.
23. Each determinant is equal to $(a_1 + a_1')b_2 - (a_2 + a_2')b_1 = a_1b_2 - a_2b_1 + a_1'b_2 - a_2'b_1$.
25. (1, 1).   **27.** (0, 1, 0).   **29.** (3 cos θ $-$ 5 sin θ, 5 cos θ $+$ 3 sin θ).   **31.** $15 - i$.   **33.** $17 - 16i$.
35. Each elementary product has n factors. The first factor comes from the first row and any one of the columns. There
are n choices for this column. The next factor is from the second row and any of the remaining $n - 1$ choices of
columns. There are $n - 2$ remaining choices for the column for the third factor, and so on. Finally, there is but one
choice of a column remaining for the choice of the factor from the nth row. So the total number of choices is
$n(n - 1)(n - 2) \cdots (2)(1) = n!$.

37. $\dfrac{1}{A} + \dfrac{1}{B} = \dfrac{11}{60}$,

$\dfrac{1}{B} + \dfrac{1}{C} = \dfrac{3}{20}$,

$\dfrac{1}{A} + \dfrac{1}{C} = \dfrac{1}{6}$.

So $A = 10$ days, $B = 12$ days, $C = 15$ days.

39. If $\det A \neq 0$, there is just the trivial solution, so suppose that $\det A = 0 = a(a + 1)(a - 1)$. So $a = 0$, or 1, or -1.

41. Since $\det \begin{bmatrix} a+1 & -1 \\ 2 & a-1 \end{bmatrix} = (a + 1)(a - 1) + 2 = a^2 + 1 > 0$ for all real numbers a, by Cramer's rule $x_1 = 0$, and

$x_2 = 0$ is the only solution to the given system that is possible.

43.

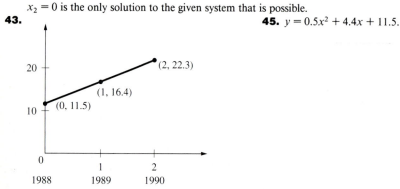

45. $y = 0.5x^2 + 4.4x + 11.5$.

47. Use the areas of the indicated trapezoids:

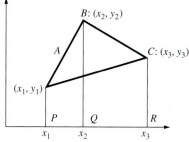

Trapezoid $APQB = \dfrac{1}{2}(y_1 + y_2)(x_2 - x_1) = M$.

Trapezoid $QBCR = \dfrac{1}{2}(y_2 + y_3)(x_3 - x_2) = N$.

Trapezoid $PACR = \dfrac{1}{2}(y_1 + y_3)(x_3 - x_1) = K$.

Area of triangle $ABC = M + N - K$

$$= \dfrac{1}{2}[(y_1 + y_2)(x_2 - x_1) + (y_2 + y_3)(x_3 - x_2)$$
$$\quad - (y_1 + y_3)(x_3 - x_1)]$$
$$= \left| \dfrac{1}{2} \det \begin{bmatrix} x_1 & y_1 & 1 \\ x_2 & y_2 & 1 \\ x_3 & y_3 & 1 \end{bmatrix} \right|.$$ (Verify the computation.)

Section 2.2 (pp. 121–123)

1. 1.   **3.** -4.   **5.** $\sin^2 x + \cos^2 x = 1$.   **7.** 0.   **9.** 4.

11. $-31 = \det \begin{bmatrix} 1 & 0 & 2 \\ -3 & 5 & 1 \\ -1 & 3 & -4 \end{bmatrix} = \det \begin{bmatrix} 1 & -3 & -1 \\ 0 & 5 & 3 \\ 2 & 1 & -4 \end{bmatrix}$.   **13.** $3^5 = 243$.   **15.** $2^5 = 32$.   **17.** -1.   **19.** -1.

21. $\dfrac{1}{2}$.   **23.** 1.   **25.** 1.   **27.** If $U = \begin{bmatrix} 1 & 3 & 3 \\ 0 & 7 & 2 \\ 0 & 0 & \frac{41}{7} \end{bmatrix}$, then $\det U = 41$.   **29.** $J = I_3$, so $\det J = 1$.

31. If $U = \begin{bmatrix} 1 & -2 & 1 \\ 0 & 7 & -2 \\ 0 & 0 & \frac{26}{7} \end{bmatrix}$, then det $U = 26$.   **33.** $J = I_3$, so det $J = 1$.

35. $\begin{bmatrix} 1 & 3 & 3 \\ 2 & -1 & 4 \\ 3 & -2 & 0 \end{bmatrix} + \begin{bmatrix} 2 & 3 & 0 \\ 1 & -2 & 1 \\ 4 & 0 & -2 \end{bmatrix} = \begin{bmatrix} 3 & 6 & 3 \\ 3 & -3 & 5 \\ 7 & -2 & -2 \end{bmatrix}$ $(\det A) + (\det B) \neq \det (A + B)$.

$\quad\quad\quad 41 \quad\quad + \quad\quad 26 \quad\quad \neq \quad\quad 339$

37. The result is obvious if $n = 2$. Suppose then that $n > 2$ and that A is an $n \times n$ matrix, with identical ith and jth rows. Assume that the theorem holds for any $(n-1) \times (n-1)$ matrix with this property. Expand det A across row s where $s \neq i$ and $s \neq j$. (Since $n > 2$, such an s exists.) Use Theorem 2.1 so that we get det $A = a_{s1}A_{s1} + a_{s2}A_{s2} + \cdots + a_{sn}A_{sn}$. But each A_{sk} ($k = 1, 2, \ldots, n$) is the determinant of an $(n-1) \times (n-1)$ matrix with two identical rows. Hence, by the induction assumption, $A_{sk} = 0$. Then det $A = 0$.

39. det $A' = 0$ by Exercise 37. Similarly det $A' = 2 \det \begin{bmatrix} -2 & 1 \\ 3 & 0 \end{bmatrix} - 3 \det \begin{bmatrix} 1 & 1 \\ 2 & 0 \end{bmatrix} + 0 \det \begin{bmatrix} 1 & -2 \\ 2 & 3 \end{bmatrix} = 0$.

41. $8 + 3i$.

43. The equation of line ℓ is $(y - b) = \dfrac{d - b}{c - a} (x - a)$ if $a \neq c$. So $x(b - d) - y(a - c) + (ab - bc + ad - ab) = 0$, or

$x \det \begin{bmatrix} b & 1 \\ d & 1 \end{bmatrix} - y \det \begin{bmatrix} a & 1 \\ c & 1 \end{bmatrix} + 1 \det \begin{bmatrix} a & b \\ c & d \end{bmatrix} = 0$. From this we have $\det \begin{bmatrix} x & y & 1 \\ a & b & 1 \\ c & d & 1 \end{bmatrix} = 0$, as desired.

45. Let $A = (a_{ij})$ be an $n \times n$ matrix and let A' be derived from the matrix A except that while all the other rows are the same as the rows of the matrix A, row j of A' is the same as row i, $j \neq i$. Both rows are identical with row i of A. Expand det A' across row $j = $ row i, det $A' = a_{i1}A_{j1} + a_{i2}A_{j2} + \cdots + a_{in}A_{jn}$. But this is precisely Δ'; while from Exercise 37, det $A' = 0$, so $\Delta' = 0$.

Section 2.3 (pp. 132–134)

1. By Property 1, interchanging column 2 and 3 changes the sign of the determinant.

3. By Property 3, the determinant does not change when a matrix is changed by adding (-2) times row 1 to row 2 and also by adding (-3) times row 1 to row 3.

5. Property 5, since the second column is entirely zero the determinant is zero.

7. Use Theorem 2.3. The determinant of an upper triangular matrix is equal to the product of its main diagonal elements.

9* $\det \begin{bmatrix} 1 & x & x^2 & x^3 \\ 1 & y & y^2 & y^3 \\ 1 & z & z^2 & z^3 \\ 1 & w & w^2 & w^3 \end{bmatrix} = \det \begin{bmatrix} 1 & 0 & 0 & 0 \\ 1 & y - x & y^2 - x^2 & y^3 - x^3 \\ 1 & z - x & z^2 - x^2 & z^3 - x^3 \\ 1 & w - x & w^2 - x^2 & w^3 - x^3 \end{bmatrix}$

$\quad\quad\quad = (y - x)(z - x)(w - x)(z - y)(w - y)[(w + y + x) - (z + y + x)]$
$\quad\quad\quad = (y - x)(z - x)(w - x)(z - y)(w - y)(w - z) = (x - y)(x - z)(x - w)(y - z)(y - w)(z - w)$.

11. 48.   **13.** $\dfrac{1}{60}$.   **15.** $3 - 3 \tan^2 x$.   **17.** 10.   **19.** $-4i$.

21. Interchange the two identical rows of A; this changes $(\det A)$ to $-(\det A)$ by Lemma 2.4. Yet the matrix remains the same. Hence, det $A = -\det A$, and thus, det $A = 0$.

23. When at least one of the entries on the main diagonal is zero (Theorem 2.3).

---

* See Rushanan, Joseph J. "On the Vandermonde matrix," *American Mathematical Monthly,* **96,** #10, Dec. 1989, pp. 921–924.

25. Let $A = \begin{bmatrix} a & b \\ c & d \end{bmatrix}$ and $B = \begin{bmatrix} e & f \\ g & h \end{bmatrix}$. Then $AB = \begin{bmatrix} ae + bg & af + bh \\ ce + dg & cf + dh \end{bmatrix}$, and det $AB = (ae + bg)(cf + dh)$

$- (af + bh)(ce + dg) = aecf + aedh + bgcf + bgdh - afce - afdg - bhce - bhdg = (ad - bc)(eh - fg) =$
(det A)(det B).

27. Since $AA^{-1} = I$, by Theorem 2.7 we have det $(AA^{-1}) = (\det A)(\det A^{-1}) = \det I = 1$. Thus, det $A \neq 0$.

29. Use Theorems 2.7 and 2.2. We see that $(\det AA^T) = (\det A)(\det A^T) = (\det A)(\det A) = (\det A)^2$. But if $AA^T = I$, then $(\det AA^T) = (\det A)^2 = 1$, so det $A = 1$ or -1.

31. Let A^* be the matrix in (2.24) preceding Lemma 2.6. Then det $A^* = 0$, because A^* has two identical rows. On the other hand, expanding det A^* across the jth row, we get the following since rows i and j are identical:
det $A^* = a_{i1}A_{j1} + a_{i2}A_{j2} + \cdots + a_{in}A_{jn} = 0$.

33. Adding columns $2, 3, \ldots, n$ of the matrix T_n to its first column does not change the value of the determinant

(Property 3). So we have det $T_n = \det \begin{bmatrix} a + (n-1)b & b & b & b & \cdots & b \\ a + (n-1)b & a & b & b & \cdots & b \\ a + (n-1)b & b & a & b & \cdots & b \\ \cdot & \cdot & \cdot & \cdot & \cdot & \cdot \\ \cdot & \cdot & \cdot & \cdot & \cdot & \cdot \\ \cdot & \cdot & \cdot & \cdot & \cdot & \cdot \\ a + (n-1)b & b & b & b & \cdots & a \end{bmatrix} =$

$(a + (n-1)b) \det \begin{bmatrix} 1 & b & b & b & \cdots & b \\ 0 & a - b & 0 & 0 & \cdots & 0 \\ 0 & 0 & a - b & 0 & \cdots & 0 \\ \cdot & \cdot & \cdot & \cdot & \cdot & \cdot \\ \cdot & \cdot & \cdot & \cdot & \cdot & \cdot \\ 0 & 0 & 0 & 0 & \cdots & a - b \end{bmatrix} = (a + (n-1)b)(a - b)^{n-1}$.

35. Write the system in matrix form $AX = 0$, and note that det $A = \det \begin{bmatrix} 1 & a & b \\ 0 & a + 1 & 2 \\ 0 & b^2 + 1 & 1 - a \end{bmatrix} = -1 - a^2 - 2b^2$ is never

zero, because a and b are real. Hence by Cramer's rule the only solution is the trivial solution $X = A^{-1}0$.

37. det $\begin{bmatrix} x + 1 & 1 & 1 \\ 1 & x + 1 & 1 \\ 1 & 1 & x + 1 \end{bmatrix} = (x + 1)^3 + 2 - 3(x + 1) = x^2(x + 3) = 0$, so $x = 0$ or -3.

39. $w(\sin x, \sin 2x, \pi/4) = \det \begin{bmatrix} \dfrac{\sqrt{2}}{2} & 1 \\ \dfrac{\sqrt{2}}{2} & 0 \end{bmatrix} = \dfrac{-\sqrt{2}}{2}$.   **41.** $w = \det \begin{bmatrix} 1 & 1 & 1 \\ 1 & 0 & 0 \\ 1 & 2 & 0 \end{bmatrix} = 2$.

43. $f(x) = \sin x; f'(x) = \cos x; f''(x) = -\sin x; f'''(x) = -\cos x$. $g(x) = \sin^2 x; g'(x) = 2 \sin x \cos x$;
$g''(x) = -2 \sin^2 x + 2 \cos^2 x = 2 \cos 2x; g'''(x) = -4 \sin 2x$. $h(x) = \sin^3 x, h'(x) = 3 \sin^2 x \cos x$;
$h''(x) = -3 \sin^3 x + 6 \sin x \cos^2 x; h'''(x) = 6 \cos^3 x - 21 \sin^2 x \cos x. k(x) = \sin^4 x$;
$k'(x) = 4 \sin^3 x \cos x; k''(x) = 12 \sin^2 x \cos^2 x - 4 \sin^4 x. k'''(x) = 24 \cos^3 x \sin x - 40 \sin^3 x \cos x$.

So, $w = \det \begin{bmatrix} \dfrac{\sqrt{2}}{2} & \dfrac{1}{2} & \dfrac{\sqrt{2}}{4} & \dfrac{1}{4} \\ \dfrac{\sqrt{2}}{2} & 1 & \dfrac{3\sqrt{2}}{4} & 1 \\ \dfrac{-\sqrt{2}}{2} & 0 & \dfrac{3\sqrt{2}}{4} & 2 \\ \dfrac{-\sqrt{2}}{2} & -4 & \dfrac{-15\sqrt{2}}{4} & -4 \end{bmatrix} = \dfrac{3}{8}$.

45. Let $D = \det F(x) = \det \begin{bmatrix} f_{11}(x) & f_{12}(x) & \cdots & f_{1n}(x) \\ f_{21}(x) & f_{22}(x) & \cdots & f_{2n}(x) \\ \cdot & \cdot & \cdot & \cdot \\ \cdot & \cdot & \cdot & \cdot \\ \cdot & \cdot & \cdot & \cdot \\ f_{n1}(x) & f_{n2}(x) & \cdots & f_{nn}(x) \end{bmatrix}$, $\dfrac{dD}{dx} = \sum_{i=1}^{n} \det \begin{bmatrix} f_{11}(x) & f_{12}(x) & \cdots & f'_{1i}(x) & \cdots & f_{1n}(x) \\ f_{21}(x) & f_{22}(x) & \cdots & f'_{2i}(x) & \cdots & f_{2n}(x) \\ \cdot & \cdot & \cdot & \cdot & \cdot & \cdot \\ \cdot & \cdot & \cdot & \cdot & \cdot & \cdot \\ \cdot & \cdot & \cdot & \cdot & \cdot & \cdot \\ f_{n1}(x) & f_{n2}(x) & \cdots & f'_{ni}(x) & \cdots & f_{nn}(x) \end{bmatrix}$.

Use induction on the size of the matrix. See Exercise 44 for the case $n = 2$. First expand D along the first row of the matrix $F(x)$ and write $D = f_{11}(x)D_{11} + f_{12}(x)D_{12} + \cdots + f_{1n}(x)D_{1n}$. Then $\dfrac{dD}{dx} = f_{11}(x)\dfrac{dD_{11}}{dx} + f'_{11}(x)D_{11} + \cdots$

$+ f_{1n}(x)\dfrac{dD_{1n}}{dx} + f'_{1n}(x)D_{1n}$. Regroup the terms and use the induction assumption about the cofactors $\dfrac{dD_{1i}}{dx}$;

$\dfrac{dD}{dx} = (f'_{11}(x)D_{11} + \cdots + f'_{1n}(x)D_{1n}) + \left(f_{11}(x)\dfrac{dD_{11}}{dx} + \cdots + f_{1n}(x)\dfrac{dD_{1n}}{dx}\right)$.

Section 2.4 (pp. 139–140)

1. $\det K = 14$, and hence K is nonsingular. $K(\text{adj } K) = \begin{bmatrix} 14 & 0 \\ 0 & 14 \end{bmatrix}$. $K^{-1} = \dfrac{1}{14}\begin{bmatrix} 2 & -3 \\ 4 & 1 \end{bmatrix}$.

3. $A(\text{adj } A) = \begin{bmatrix} -11 & 0 \\ 0 & -11 \end{bmatrix}$. $A^{-1} = \dfrac{1}{-11}\begin{bmatrix} 3 & -1 \\ -5 & -2 \end{bmatrix}$.   **5.** A^{-1} does not exist since $\det A = 0$. $A(\text{adj } A) = O$.

7. $A^{-1} = \dfrac{1}{15 - i}\begin{bmatrix} 1 - 2i & 3i \\ 4i & 1 + i \end{bmatrix}$.

9. $\det A = -35$; $\text{adj } A = \begin{bmatrix} -1 & -10 & 3 \\ -10 & 5 & -5 \\ 4 & 5 & -12 \end{bmatrix}$; $A(\text{adj } A) = \begin{bmatrix} -35 & 0 & 0 \\ 0 & -35 & 0 \\ 0 & 0 & -35 \end{bmatrix}$; $A^{-1} = \begin{bmatrix} \frac{1}{35} & \frac{2}{7} & \frac{-3}{35} \\ \frac{2}{7} & \frac{-1}{7} & \frac{1}{7} \\ \frac{-4}{35} & \frac{-1}{7} & \frac{12}{35} \end{bmatrix}$.

11. $\det A = 1 \neq 0$, so A^{-1} exists. Furthermore, $\text{adj } A = \begin{bmatrix} 1 & 0 & 0 & 0 \\ 0 & 0 & 0 & 1 \\ 0 & 1 & 0 & 0 \\ 0 & 0 & 1 & 0 \end{bmatrix} = A^{-1}$. (Since $\det A = 1$.)

13. By Theorem 2.3, $\det A = a^3 \neq 0$, since $a \neq 0$. $A^{-1} = \dfrac{1}{a^3}\begin{bmatrix} a^2 & -ab & bd - ac \\ 0 & a^2 & -ad \\ 0 & 0 & a^2 \end{bmatrix}$.

15. By Theorem 2.7, since $A^k = O$, $\det(A^k) = 0$. Therefore, $(\det A)^k = 0$, and $(\det A) = 0$. Therefore, by Theorem 2.9, A^{-1} does *not* exist.

17. By hypothesis, A^{-1} exists and, of course, $AA^{-1} = A^{-1}A = I_n$. Hence, $(AA^{-1})^T = (A^{-1}A)^T = (I_n)^T = I_n$. Thus, $(A^{-1})^T A^T = A^T (A^{-1})^T = I_n$, so $(A^T)^{-1}$ exists and is equal to $(A^{-1})^T$.

19. No. For example, take $A = I_n$ and $B = E_{11} = \begin{bmatrix} 1 & 0 & 0 & \cdots & 0 \\ 0 & 0 & 0 & \cdots & 0 \\ \cdot & \cdot & \cdot & & \cdot \\ \cdot & \cdot & \cdot & & \cdot \\ \cdot & \cdot & \cdot & & \cdot \\ 0 & 0 & 0 & \cdots & 0 \end{bmatrix}$. Then $AB = B$ is singular, but A is nonsingular ($\det B = 0$).

21. No, for suppose that A and B are both singular. Then $\det A = \det B = 0$. Hence, from Theorem 2.7, $\det(AB) = (\det A)(\det B) = 0$, so AB is singular.

23. Recall that $A(\text{adj } A) = (\det A)I_n$, and since A is singular, $\det A = 0$. Therefore, $A(\text{adj } A) = O$. Suppose first that $A \neq O$. If $\text{adj } A$ were nonsingular, then $A(\text{adj } A)(\text{adj } A)^{-1} = A = O(\text{adj } A)^{-1} = O$ is a contradiction, so $\text{adj } A$ must be singular and $\det(\text{adj } A) = 0 = (\det A)^k$ for any k, so clearly for $k = n - 1$. On the other hand, if $A = O$, then $\text{adj } A = O$ also and again we have $0 = \det(\text{adj } A) = (\det A)^{n-1}$.

25. By Exercises 22 and 23, det (adj A) = (det $A)^{n-1}$ for every $n \times n$ matrix A. If A is nonsingular, then det (adj A) $\neq 0$, because det $A \neq 0$, so adj A is nonsingular also. Conversely, if adj A is nonsingular, then det (adj A) $\neq 0$, so the same equation implies det $A \neq 0$, so A is nonsingular.

27. In Exercise 26 we proved that if A is nonsingular, then $1 = $ (det $A)^{n-2}$. In that case, when n is an odd positive integer the only possibility is that det $A = 1$; while if n is even, then det A could be -1. In the event that A is singular, then, since $A^2 = $ (det $A)I_n$, this implies that $A^2 = O$; that is, A is nilpotent of index 2, or $A = O$.

29. T_5 is nonsingular, if and only if, $a \neq -4b$ and $a \neq b$; a could be zero, but in that event $b \neq 0$.

31. det $M = 4256$.   **33.** $x_1 = (0.68, 0.64, 0.43, -0.45, -0.14, -0.06)^T$;
$x_2 = (0.88, 0.46, 0.21, -0.24, 0.17, 0.06)^T$;
$x_3 = (-2.75, -1.11, -1.06, 1.76, 1.21, -0.03)^T$;
$x_4 = (-1.07, -1.79, -0.18, -0.45, 1.08, 0.40)^T$.

Chapter 2 Review Exercises (p. 141)

1. 15 and 96.   **2.** $\begin{bmatrix} \frac{1}{5} & \frac{-3}{5} & 0 \\ \frac{1}{5} & \frac{2}{5} & 0 \\ \frac{-11}{15} & \frac{-2}{15} & \frac{1}{3} \end{bmatrix}$ and $\begin{bmatrix} \frac{1}{3} & \frac{-7}{3} & \frac{3}{4} & \frac{17}{96} \\ 0 & 1 & \frac{-1}{2} & \frac{1}{16} \\ 0 & 0 & \frac{1}{4} & \frac{-7}{32} \\ 0 & 0 & 0 & \frac{1}{8} \end{bmatrix}$.

3. $x = -18$, $y = 25$, $z = 28$.   **4.** $x = \frac{22}{29}$, $y = \frac{-8}{29}$, $z = \frac{-9}{29}$.   **5.** det $A = \pm 1$.

6. (det $A)^2 = (-1)^n$. If n is even, then det $A = \pm 1$. If n is odd, then det $A = \pm i$.

7. Yes. A^{-1} exists; in fact, $A^{-1} = A$.   **8.** Apply Exercise 9 of Section 2.3: 240.

9. $A(-A) = -A^2 = $ (det $A)I_n$. So $A^2 = (-\det A)I_n$. Thus, (det $A)^2 = (-\det A)^n = (-1)^n(\det A)^n$. All the possible values of det A are therefore 0, 1, or -1.

10. det $(A^T) = $ det $(-A)$, so det $A = (-1)^n(\det A)$. Since n is *odd*, this implies that det $A = -\det A$. Thus, det $A = 0$.

Chapter 3

Section 3.1 (pp. 162–163)

1.

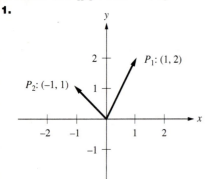

3. Let $\overrightarrow{OP_1} = \mathbf{u}$ and $\overrightarrow{OP_2} = \mathbf{v}$. We want $\mathbf{u} - \mathbf{v} = \mathbf{u} + (-\mathbf{v})$.

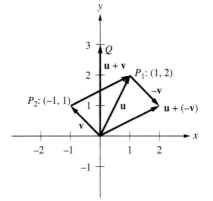

	$\mathbf{u} + \mathbf{v}$	$3\mathbf{u}$	$\frac{2}{3}\mathbf{v}$	$\mathbf{u} - \mathbf{v}$	$\frac{1}{2}\mathbf{u} - \frac{3}{2}\mathbf{v}$
5.	$(1, -1)$	$(6, 3)$	$\left(\frac{-2}{3}, \frac{-4}{3}\right)$	$(3, 3)$	$\left(\frac{5}{2}, \frac{7}{2}\right).$
7.	$(6, 1, -3)$	$(3, -3, -3)$	$\left(\frac{10}{3}, \frac{4}{3}, \frac{-4}{3}\right)$	$(-4, -3, 1)$	$\left(-7, \frac{-7}{2}, \frac{5}{2}\right).$
9.	$(-2, 4, 2, -3, 1)$	$(3, -3, 0, -3, 3)$	$\left(-2, \frac{10}{3}, \frac{4}{3}, \frac{-4}{3}, 0\right)$	$(4, -6, -2, 1, 1)$	$\left(5, -8, -3, \frac{5}{2}, \frac{1}{2}\right).$
11.	$(1 + 2i, 2 - 2i, -i)$	$(3, -3i, -3 + 3i)$	$\left(\frac{4}{3}i, \frac{4}{3} - \frac{2}{3}i, \frac{2}{3} - \frac{4}{3}i\right)$	$(1 - 2i, -2, -2 + 3i)$	$\left(\frac{1}{2} - 3i, -3 + i, -2 + \frac{7}{2}i\right).$
13.	$5x^3 - 4x^2 - 10x + 7$	$12x^3 - 6x^2 - 9x + 12$	$\frac{2}{3}x^3 - \frac{4}{3}x^2 - \frac{14}{3}x + 2$	$3x^3 + 4x + 1$	$\frac{1}{2}x^3 + 2x^2 + 9x - \frac{5}{2}.$
15.	$\begin{bmatrix} 3 & -1 \\ 1 & 2 \end{bmatrix}$	$\begin{bmatrix} 3 & -3 \\ 6 & 3 \end{bmatrix}$	$\begin{bmatrix} \frac{4}{3} & 0 \\ \frac{-2}{3} & \frac{2}{3} \end{bmatrix}$	$\begin{bmatrix} -1 & -1 \\ 3 & 0 \end{bmatrix}$	$\begin{bmatrix} \frac{-5}{2} & \frac{-1}{2} \\ \frac{5}{2} & -1 \end{bmatrix}.$
17.	$e^x + e^{-x}$	$3e^x$	$\frac{2}{3}e^{-x}$	$e^x - e^{-x}$	$\frac{1}{2}e^x - \frac{3}{2}e^{-x}.$

19.

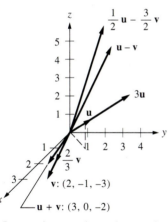

21. Let $\mathbf{u} = (u_1, u_2, u_3)$, $\mathbf{v} = (v_1, v_2, v_3)$, and $\mathbf{w} = (w_1, w_2, w_3)$. Then $\mathbf{u} + \mathbf{v} = (u_1 + v_1, u_2 + v_2, u_3 + v_3) = (v_1 + u_1, v_2 + u_2, v_3 + u_3) = \mathbf{v} + \mathbf{u}$, because complex number addition is commutative. Now

$$\begin{aligned}
(\mathbf{u} + \mathbf{v}) + \mathbf{w} &= ((u_1 + v_1) + w_1, (u_2 + v_2) + w_2, (u_3 + v_3) + w_3) \\
&= (u_1 + (v_1 + w_1), u_2 + (v_2 + w_2), u_3 + (v_3 + w_3)) \\
&= \mathbf{u} + (\mathbf{v} + \mathbf{w}),
\end{aligned}$$

because complex number addition is associative. The zero vector is $\mathbf{0} = (0, 0, 0)$, and $-\mathbf{u} = (-u_1, -u_2, -u_3)$ is the negative of $\mathbf{u}$. For all α in $\mathbb{C}$ (scalars), $\alpha\mathbf{u} = (\alpha u_1, \alpha u_2, \alpha u_3) \in \mathbb{C}^3$, $\alpha(\mathbf{u} + \mathbf{v}) = (\alpha(u_1 + v_1), \alpha(u_2 + v_2), \alpha(u_3 + v_3)) = (\alpha u_1 + \alpha v_1, \alpha u_2 + \alpha v_2, \alpha u_3 + \alpha v_3) = \alpha\mathbf{u} + \alpha\mathbf{v}$. Note that $(\alpha + \beta)\mathbf{u} = ((\alpha + \beta)u_1, (\alpha + \beta)u_2, (\alpha + \beta)u_3) = \alpha u_1 + \beta u_1, \alpha u_2 + \beta u_2, \alpha u_3 + \beta u_3) = \alpha\mathbf{u} + \beta\mathbf{u}$, also $(\alpha\beta)\mathbf{u} = ((\alpha\beta)u_1, (\alpha\beta)u_2, (\alpha\beta)u_3) = (\alpha(\beta u_1), (\alpha(\beta u_2), (\alpha(\beta u_3)) = \alpha(\beta\mathbf{u})$, for all scalars α and β in $\mathbb{C}$. Finally, $1\mathbf{u} = (1u_1, 1u_2, 1u_3) = (u_1, u_2, u_3) = \mathbf{u}$, 1 in $\mathbb{C}$. Thus, all of the axioms are satisfied.

23. See Theorems 1.1 and 1.2 in Section 1.3 of Chapter 1.

25. If $u(x)$, $v(x)$, and $w(x)$ are real valued functions which are continuous on the closed interval $[a, b]$ of the real line, then so are the functions $(u + v)(x) = u(x) + v(x)$ and $(\alpha u)(x) = \alpha(u(x))$. The remaining properties are also easily checked since $u(x)$, $v(x)$, $w(x)$, $\alpha(u(x))$, etc., are real numbers for each x in $[a, b]$.

27. This W' is *not* a vector space. For one thing, suppose that $\mathbf{u} = \left(\frac{1}{2}, 0\right)$. Then while $\mathbf{u}$ is in W', $2\mathbf{u} = (1, 0)$ is not in W'. Other axioms are also violated.

29. If $c = 0$, then U' is a vector space. However, for any nonzero c, the zero vector would not be in U'. So U' is not a vector space if $c \neq 0$.

31. This H is *not* a vector space with this definition of multiplication by a scalar. For one thing, the last axiom fails: $1(x, y) = (1x, 0) \neq (x, y)$ for any nonzero y.

33. $\mathbb{Q}$ satisfies the vector space axioms.

35. $\mathbb{Z}$ is not considered to be a vector space. The scalars are not a *field*. Such sets as $\mathbb{Z}$ are called *modules* ($\mathbb{Z}$-modules), rather than *vector* spaces.

37. V is a vector space with respect to the given operations.   **39.** K is a vector space with respect to the given operations.

41. Axioms 2, 3, 7, 8, and 9 hold because they are true for all continuous functions, and every differentiable function is continuous. The rest of the axioms also follow from theorems in calculus; e.g., the derivative of a sum is the sum of the derivatives, etc.

43. Axioms 2, 3, 7, 8, and 9 hold because they are true for all functions, and every integrable function is a function. The rest of the axioms also follow from theorems in calculus; e.g., the integral of a sum is the sum of the integrals, and the fact $\int_a^b \alpha f(x)\,dx = \alpha \int_a^b f(x)\,dx$, etc.

45. Suppose that both $\mathbf{z}$ and $\mathbf{0}$ are zero vectors in V. Then note that by Axioms 2 and 4 we have $\mathbf{0} = \mathbf{z} + \mathbf{0} = \mathbf{0} + \mathbf{z} = \mathbf{z}$.

47. Do part (iv) first as follows: $(0_S)\mathbf{u} = (0_S + 0_S)\mathbf{u} = (0_S)\mathbf{u} + (0_S)\mathbf{u}$. Then from Axiom 5, add $-(0_S)\mathbf{u}$ to the left and right members to obtain $\mathbf{0} = (0_S)\mathbf{u}$. Then $\mathbf{u} + (-1)\mathbf{u} = 1\mathbf{u} + (-1)\mathbf{u} = (1 + (-1))\mathbf{u} = 0\mathbf{u} = \mathbf{0}$. Hence $(-1)\mathbf{u} = -\mathbf{u}$.

49. Apply the uniqueness of the negative proved in Exercise 48 and note that it is true that $\mathbf{u} - \mathbf{v} + (\mathbf{v} - \mathbf{u}) = \mathbf{0}$.

51. Note that for any $\mathbf{u}$, $\alpha(\mathbf{0}_v) = \alpha(0_S\mathbf{u})$ from part (iv.). But $\alpha(0_S\mathbf{u}) = (\alpha 0_S)\mathbf{u}$ by Axiom 9, and $\alpha 0_S = 0_S$. Therefore, $\alpha(\mathbf{0}_v) = 0_S\mathbf{u} = \mathbf{0}_v$, by part (iv) again.

53. $(\alpha - \beta)\mathbf{u} = \alpha\mathbf{u} + (-\beta)\mathbf{u} = \alpha\mathbf{u} + (-1\beta)\mathbf{u} = \alpha\mathbf{u} + (-\beta\mathbf{u}) = \alpha\mathbf{u} - \beta\mathbf{u}$.

Section 3.2 (pp. 172–173)

1. If $a(1, 0) + b(0, 1) = (0, 0)$, then $a = b = 0$. Hence $\{(1, 0), (0, 1)\}$ is linearly independent.

3. If $a(1, 0, 0) + b(0, 1, 0) + c(0, 0, 1) = (a, b, c) = (0, 0, 0)$, then $a = b = c = 0$. The set $\{(1, 0, 0), (0, 1, 0), (0, 0, 1)\}$ is linearly independent.

5. The set $\{(1, i), (i, 1)\}$ is linearly independent.   **7.** The three given vectors form a linearly independent set.

9. The set $\{1, x, x^2, x^3\}$ is linearly independent.

11. If $a(\cos x) + b(\sin x)$ is the zero function, then for *every* x, $a\cos x + b\sin x = 0$. When $x = 0$ this is true, so $a(1) + b(0) = 0$, or $a = 0$. When $x = \pi/2$, $a(0) + b(1) = 0$, so $b = 0$. Thus, the two functions $\cos x$ and $\sin x$ are linearly independent in $C[-\pi, \pi]$.

13. If $a\mathbf{r}_1 + b\mathbf{r}_2 + c\mathbf{r}_3 = \mathbf{0}$, then we have $a(1, 2, -1) + b(3, 1, 4) + c(5, 6, -7) = (0, 0, 0)$. This results in the homogeneous system of equations

$$\begin{aligned} a + 3b + 5c &= 0, \\ 2a + b + 6c &= 0, \\ -a + 4b - 7c &= 0, \end{aligned}$$

which has only the trivial solution $a = b = c = 0$. Thus the three rows of A are linearly independent.

15. The four matrices form a linearly independent set.   **17.** Linearly dependent; e.g., set $a = -2$, $b = 1$.

19. Linearly dependent; e.g., note that $1(1, 1, 1) + 1(-1, 1, 0) - 1(0, 2, 1) = (0, 0, 0)$.

21. Linearly dependent; e.g., note that $i(1, i) - 1(i, -1) = (0, 0)$.

23. Linearly dependent; note that $1(1, 0, 1, 0, 0) + 1(0, 1, 0, 1, 1) - 1(1, 1, 1, 1, 1) = \mathbf{0}$.

25. Linearly independent.

27. Linearly dependent; note that $-7(1 - x) + 4(1 + 2x) + 3(1 - 5x) = 0 + 0x$.

29. Linearly dependent; note that $1(1, 2, 4) - 1(0, 3, 5) + 1(-1, 1, 1) = (0, 0, 0)$.

31. Linearly independent.

33. Linearly dependent since, for example, $0\begin{bmatrix} 1 & 2 \\ 2 & -1 \end{bmatrix} + 0\begin{bmatrix} 1 & 1 \\ 1 & 1 \end{bmatrix} + 1\begin{bmatrix} 0 & 0 \\ 0 & 0 \end{bmatrix} + 0\begin{bmatrix} 2 & -2 \\ 1 & 1 \end{bmatrix} = \begin{bmatrix} 0 & 0 \\ 0 & 0 \end{bmatrix}$. (Any set containing the zero matrix will be dependent. See Exercise 39.)

35. If x is a real number in $[0, 1]$, then $|x| = x$, and therefore $f_1(x) = 2x$, $f_2(x) = x$, in which cases $f_1(x) - 2f_2(x) = 0(x)$, the zero function, so f_1 and f_2 are linearly dependent in $C[0, 1]$. On the other hand, if x is a real number in $[-1, 1]$, then $f_1(1) + 2f_2(1) = 0$, or $2 + 2(1) = 0$, which is absurd. So we cannot find a scalar k so that for *all* x in $[-1, 1]$, $f_1(x) + kf_2(x)$ is the zero function. Hence, $af_1(x) + bf_2(x) = 0(x)$ can only be satisfied when $a = b = 0$. Thus, f_1 and f_2 are linearly independent in $C[-1, 1]$. (See also Exercise 37.)

37. Let $S = \{u, v\}$, $u \neq 0$, $v \neq 0$. Suppose that S is linearly dependent, so there are scalars a and b, not both zero such that $au + bv = 0$. If $a \neq 0$, then solving for u gives us $u = \dfrac{-b}{a}v$, or if $b \neq 0$, $v = \dfrac{-a}{b}u$. Conversely, suppose that one vector, say u, is a scalar multiple of the other, $u = kv$ (rename the two vectors if it is the other way around). Then $1u - kv = 0$. Therefore, $S = \{u, v\}$ is linearly dependent.

39. Suppose that $S = \{0, v_1, v_2, \ldots, v_n\}$ is any finite set of vectors from the vector space V. Note that the linear combination $10 + 0v_1 + 0v_2 + \cdots + 0v_n = 0$ is such that not all of the coefficients are zero. Hence, S is linearly dependent.

41. Suppose that $S = \{v_1, v_2, \ldots, v_n\}$ is linearly independent and that say $T = \{v_1, v_2, \ldots, v_k\}$, $1 \leq k < n$ is linearly dependent. (Relabel the vectors of S if necessary so that the first k are the purported dependent vectors.) Then there exists scalars, not all zero, such that $c_1v_1 + c_2v_2 + \cdots + c_kv_k = 0$. It follows that $c_1v_1 + c_2v_2 + \cdots + c_kv_k + 0v_{k+1} + \cdots + 0v_n = 0$ is a linear combination of the vectors of S in which not all of the coefficients are zero. This contradicts the independence of S. So it is not possible that T was dependent.

43. This statement is false. For example, take $u = (1, 0)$, $v = (0, 1)$, and $w = (1, 1)$ in $\mathbb{R}^2$. The set $\{u, v, w\}$ is linearly dependent, while the two sets $\{u, v\}$ and $\{v, w\}$ are linearly independent.

45. Let $f_1(x) = e^x$, $f_2(x) = e^{-x}$, and $f_3(x) = e^{2x}$. Note that for $x_0 = 0$, $\det \begin{bmatrix} e^0 & e^0 & e^0 \\ e^0 & -e^0 & 2e^0 \\ e^0 & e^0 & 4e^0 \end{bmatrix} = \det \begin{bmatrix} 1 & 1 & 1 \\ 1 & -1 & 2 \\ 1 & 1 & 4 \end{bmatrix} =$

$-6 \neq 0$, so the given set of functions is linearly independent.

47. Let $f_1(x) = e^x$, $f_2(x) = \sin x$, and $f_3(x) = \cos x$. Now $f'_1(x) = f''_1(x) = e^x$; $f'_2(x) = \cos x$, and $f''_2(x) = -\sin x$,

while $f'_3(x) = -\sin x$, and $f''_3(x) = -\cos x$. Note that for $x_0 = 0$, $\det \begin{bmatrix} 1 & 0 & 1 \\ 1 & 1 & 0 \\ 1 & 0 & -1 \end{bmatrix} = -2 \neq 0$, so the given set of functions is linearly independent.

Section 3.3 (pp. 183–185)

1. This subset is a subspace of $\mathbb{R}^2$.   **3.** This is a subspace.

5. This is not a subspace. Note that $(0, 1, 0) + (0, 1, 0) = (0, 2, 0)$ fails to satisfy the condition even though $(0, 1, 0)$ clearly does.

7. Is a subspace.   **9.** This is a subspace.   **11.** This is a subspace.

13. This subset is not a subspace. Note for example, that $\begin{bmatrix} 1 & 1 \\ 1 & 1 \end{bmatrix} + \begin{bmatrix} 1 & -1 \\ -1 & 1 \end{bmatrix} = 2I_2$ does not have determinant zero.

15. This subset is not a subspace. Note for example, that $\begin{bmatrix} 1 & 0 \\ 0 & 1 \end{bmatrix} + \begin{bmatrix} -1 & 0 \\ 0 & -1 \end{bmatrix} = O$ does not have a nonzero determinant.

17. This is not a subspace of $\mathbb{C}^2$ because complex scalars are allowed and $i(1, 1) = (i, i)$ does not belong to the subset.

19. This subset is not a subspace of $\mathbb{C}^2$ because complex scalars are allowed and $i(i, i) = (-1, -1)$ does not belong to the subset.

21. This subset of $C[0, 1]$ is not a subspace because the sum of the two functions $f(x) = x$ and $g(x) = 1 - x$ is the constant function 1 and is not in the subset while $g(1) = 0$ and $f(0) = 0$. So $f(x)$ and $g(x)$ are in the subset.

23. This subset is a subspace since the sum of two functions $f(x)$ and $g(x) = (f + g)(x)$ has the property that $(f + g)(0) = f(0) + g(0) = 0$ as does any scalar multiple $(\alpha f)(x) = \alpha f(x)$, because $\alpha f(0) = \alpha 0 = 0$.

25. This subset is a subspace. The sum of two differentiable functions $f(x)$ and $g(x)$ is differentiable as is the function $\alpha f(x)$.

27. This subset is a subspace.   **29.** It is easy to verify that this is a subspace.   **31.** No.

33. No; $(1, -2i)$ is not a linear combination of $(1, 1)$ and (i, i). In fact, since $(i, i) = i(1, 1)$, they both have the same linear span, and $(1, -2i)$ is not in that linear span.

35. No. If there are scalars α and β such that $(2, -1, -1) = \alpha(1, 0, 1) + \beta(0, 1, 1) = (\alpha, \beta, \alpha + \beta)$, then $\alpha = 2, \beta = -1$, and $1 = \alpha + \beta = -1$; but this is inconsistent. So no such scalars exist.
37. Yes. $(1, 5, -3) = -1(1, -1, 3) + 1(2, 4, 0)$.   **39.** No.   **41.** No.

43. Yes. $2 - x^2 = 2(1) + (-1)x^2$.   **45.** No.   **47.** Yes. $\begin{bmatrix} 1 & 2 \\ 0 & -1 \end{bmatrix} = (1)\begin{bmatrix} 2 & 4 \\ 0 & 1 \end{bmatrix} + (-1)\begin{bmatrix} 1 & 2 \\ 0 & 2 \end{bmatrix}$.

49. Yes. Note that any vector (a, b, c) in $\mathbb{R}^3$ is $(a + b - c)(1, 0, 0) + b(0, 1, 1) + (c - b)(1, 0, 1)$.
51. No. These vectors are linearly dependent.
53. Yes. Note that any vector (a, b, c) in $\mathbb{R}^3$ is in fact a linear combination of the first three.

55. No; the columns span $\overline{\mathbb{R}^3}$. Note that for any vector $\begin{bmatrix} a \\ b \\ c \end{bmatrix}$ in $\mathbb{R}^3$, $\begin{bmatrix} a \\ b \\ c \end{bmatrix} = \left(\dfrac{a}{3} + \dfrac{-2b}{3} + \dfrac{2c}{3}\right)\begin{bmatrix} 1 \\ 0 \\ 1 \end{bmatrix}$

$+ \left(\dfrac{-5a}{3} + \dfrac{b}{3} + \dfrac{5c}{3}\right)\begin{bmatrix} 2 \\ 3 \\ 2 \end{bmatrix} + (a - c)\begin{bmatrix} 4 \\ 5 \\ 3 \end{bmatrix}$.

57. If X_1 and X_2 are both solutions to $KX = B$, then $KX_1 = B$ and $KX_2 = B$. So $K(X_1 + X_2) = KX_1 + KX_2 = B + B = 2B = B$ if, and only if, $B = 0$. Since Axiom 1 fails for this set, it is not a vector space. Note also that the zero vector is not in this set of solutions either.
59. $x = z - y$, or $x + y - z = 0$.
61. Suppose that W has the given property. Then when $\alpha = \beta = 1$ we have the statement that when $\mathbf{u}$ and $\mathbf{v}$ are in W so is $\mathbf{u} + \mathbf{v}$ in W. Combine this with the result α arbitrary and $\beta = 0$, and we have that $\alpha\mathbf{u} \in W$ when $\mathbf{u} \in W$. Thus, the given property proves that W is a subspace by Theorem 3.3. Conversely, suppose that W is a subspace, then the given condition follows because all linear combinations of vectors in a subspace are in the subspace by Axioms 1 and 6.

Section 3.4 (pp. 193–195)

1. For each $\mathbf{x} = (a, b, c)$ in $\mathbb{R}^3$ we seek scalars $c_1, c_2,$ and c_3 so that $\mathbf{x} = c_1\mathbf{x}_1 + c_2\mathbf{x}_2 + c_3\mathbf{x}_3$. That is, $(a, b, c) = c_1(1, -2, 1) + c_2(1, 0, 1) + c_3(-1, 0, 1)$, which means that

$$\begin{aligned} c_1 + c_2 - c_3 &= a, \\ -2c_1 \quad\quad &= b, \\ c_1 + c_2 + c_3 &= c. \end{aligned}$$

and

Unique solutions are $c_1 = \dfrac{-b}{2}, c_2 = \dfrac{a + b + c}{2},$ and $c_3 = \dfrac{c - a}{2}$. Therefore, the three given vectors *span* $\mathbb{R}^3$. That these vectors are also *linearly independent* follows immediately from these calculations as well. For the vector $\mathbf{0} = (0, 0, 0)$, we have $a = b = c = 0$. Thus, $c_1 = c_2 = c_3 = 0$ is the only possibility.

3. No. Note that for any nonzero t, $-2t(1, 0, 0) - 2t(0, 0, 1) + t(1, 1, 1) + t(1, -1, 1) = (0, 0, 0)$, so the given set is linearly *dependent*.

5. Computation similar to Exercises 1–4 gives, in matrix form, $\begin{bmatrix} 1 & -10 & a \\ 1 & 11 & b \end{bmatrix} \rightarrow \begin{bmatrix} 1 & 0 & \dfrac{11a + 10b}{21} \\ 0 & 1 & \dfrac{b - a}{21} \end{bmatrix}$, so any vector

$\mathbf{x} = (a, b)$ in $\mathbb{R}^2$ can be expressed as a linear combination of the two given vectors in a unique way. Thus the set *spans* $\mathbb{R}^2$. When $(a, b) = (0, 0) = \mathbf{0}$, we see that the coefficients must be both zero, so the set is also linearly *independent*. Hence it is a basis.

7. Computation, in matrix form, gives $\begin{bmatrix} 1 & 1 & 2 & a \\ 1 & -1 & 1 & b \\ -1 & 1 & 2 & c \end{bmatrix} \rightarrow \begin{bmatrix} 1 & 0 & 0 & \dfrac{a - c}{2} \\ 0 & 1 & 0 & \dfrac{3a - 4b - c}{6} \\ 0 & 0 & 1 & \dfrac{b + c}{3} \end{bmatrix}$, so any vector $\mathbf{x} = (a, b, c)$ in

$\mathbb{R}^3$ can be expressed as a linear combination of the three given vectors in a unique way. Thus the set *spans* $\mathbb{R}^3$. When $(a, b, c) = (0, 0, 0) = \mathbf{0}$, we see that all of the coefficients must be zero, so the set is also linearly *independent*. Hence it is a basis.

9. Suppose that $p(x) = a + bx + cx^2$ is an arbitrary polynomial (vector) in P_2. Are there scalars c_1, c_2, and c_3 so that $p(x)$ can be written as $c_1(1 - x) + c_2(1 + x) + c_3(1 - x^2)$? That is, do the equations we get by equating coefficients of like powers of x have a solution? They are, as you can see,

$$a = c_1 + c_2 + c_3,$$
$$b = -c_1 + c_2,$$
and
$$c = -c_3.$$

Computation, in matrix form, gives $\begin{bmatrix} 1 & 1 & 1 & a \\ -1 & 1 & 0 & b \\ 0 & 0 & -1 & c \end{bmatrix} \rightarrow \begin{bmatrix} 1 & 0 & 0 & \frac{a-b+c}{2} \\ 0 & 1 & 0 & \frac{a+b+c}{2} \\ 0 & 0 & 1 & -c \end{bmatrix}$. So any polynomial (vector) in

P_2 can be expressed as a linear combination of the three given polynomials. Thus, the set *spans* P_2. Now note that when $a + bx + cx^2 = 0 + 0x + 0x^2 = \mathbf{0}$, then $c_1 = \frac{a-b+c}{2}$, $c_2 = \frac{a+b+c}{2}$, and $c_3 = -c$ must all be zero; so the given set of polynomials is also linearly *independent*. Hence it is a basis.

11. Note that any 2×2 real matrix $\begin{bmatrix} a & b \\ c & d \end{bmatrix} = a\begin{bmatrix} 1 & 0 \\ 0 & 0 \end{bmatrix} + b\begin{bmatrix} 0 & 1 \\ 0 & 0 \end{bmatrix} + c\begin{bmatrix} 0 & 0 \\ 1 & 0 \end{bmatrix} + d\begin{bmatrix} 0 & 0 \\ 0 & 1 \end{bmatrix}$. So the given set *spans*

$M_2(\mathbb{R})$. Furthermore, the zero matrix can only occur when $a = b = c = d = 0$, so the given set of matrices is also linearly *independent*. Hence it is a basis.

13. If we have $a(1, 1, -1) + b(0, 2, 1) = (0, 0, 0) = (a, a + 2b, -a + b)$, then it must be that $a = b = 0$. Thus, the two vectors are linearly independent. They do not form a basis for $\mathbb{R}^3$ because there are vectors such as $\mathbf{v} = (1, 0, 0)$ which are not in their linear span. Note that $S \cup \{\mathbf{v}\}$ is a basis for $\mathbb{R}^3$.

15. Let $\mathbf{u} = (a, b, c)$ and $\mathbf{v} = (a', b', c')$ be two vectors in W. Then $a + 2b - 3c = 0$ and also $a' + 2b' - 3c' = 0$. Then, add these equations, $(a + a') + 2(b + b') - 3(c + c') = 0$. So the vector $\mathbf{u} + \mathbf{v} = (a + a', b + b', c + c')$ is in W, as is the vector $\alpha\mathbf{u} = (\alpha a, \alpha b, \alpha c)$, because $\alpha(0) = \alpha(a + 2b - 3c) = \alpha a + 2\alpha b - 3\alpha c$. Since W is not empty, Theorem 3.3 assures us that W is a subspace of $\mathbb{R}^3$. It is fairly easy to verify that $B = \{(1, 1, 1,), (-2, 1, 0)\}$ is a basis for W.

17. Let $\mathbf{u} = (a, b, c)$ and $\mathbf{v} = (a', b', c')$ be two vectors in Ω. Then it follows that we have

$$a - b - c = 0, \qquad\qquad a + b + c = 0,$$
$$a' - b' - c' = 0, \quad \text{and} \quad a' + b' + c' = 0.$$

Then, $(a + a') - (b + b') - (c + c') = 0$ and $(a + a') + (b + b') + (c + c') = 0$. So the vector $\mathbf{u} + \mathbf{v} = (a + a', b + b', c + c')$ is in Ω, as is the vector $\alpha\mathbf{u} = (\alpha a, \alpha b, \alpha c)$, because $\alpha(0) = \alpha(a - b - c) = \alpha a - \alpha b - \alpha c = \alpha a + \alpha b + \alpha c$. Since Ω is not empty, Theorem 3.3 assures us that Ω is a subspace of $\mathbb{R}^3$. It is fairly easy to verify that every vector in Ω is of the form $(0, t, -t)$. So $B = \{(0, 1, -1)\}$ is a basis for Ω. This subspace is a straight line in three-space passing through the origin and the point $(0, 1, -1)$.

19. Any polynomial in P_n is of the form $a_0(1) + a_1(x) + a_2(x^2) + \cdots + a_n(x^n)$, so the set $\{1, x, x^2, \ldots, x^n\}$ spans P_n. That it is also linearly independent follows at once from noting that a combination $a_0 + a_1x + a_2x^2 + \cdots + a_nx^n$ is the zero polynomial if, and only if, each coefficient is zero. Hence the given set is a basis. The dimension of P_n is, therefore, $n + 1$.

21. $W = \{a(x + x^2) + b(2x) \,|\, a, b \in \mathbb{R}\}$ is the set of all those polynomials which are of the form $f(x) = (a + 2b)x + ax^2$. This describes any polynomial of degree two or less with zero as a constant term.

23. $B = \{x + x^2, 2x, 1\}$ is such a basis.

25. Use the techniques of Chapter 1 to show that the solution space $S(A)$ consists of all column vectors of the form

$\begin{bmatrix} -2t \\ -5t \\ 3t \end{bmatrix}$, $t \in \mathbb{R}$. A basis is $\left\{ \begin{bmatrix} -2 \\ -5 \\ 3 \end{bmatrix} \right\}$, $S(A)$ is one dimensional.

27. Suppose that $\mathbf{v}$ is any nonzero vector in V. Then the subspace $\mathbf{L}(\mathbf{v})$ of V, spanned by $\mathbf{v}$, is not the zero subspace, so it must be all of V. This is because V has only two subspaces. Since $\{\mathbf{v}\}$ spans V and is linearly independent, it must be a basis for V. So V is one dimensional.

29. Let $S = \{v_1, v_2, \ldots, v_n\}$ be linearly independent in the n-dimensional vector space V. Let $x \in V$, be different from any of the vectors in S. Then the set $S \cup \{x\} = \{v_1, v_2, \ldots, v_n, x\}$ contains more than n vectors. By Lemma 3.4 it is linearly dependent. So there exist scalars, $c_1, c_2, \ldots, c_n, c_{n+1}$, not all zero, so that

$$c_1 v_1 + c_2 v_2 + \cdots + c_n v_n + c_{n+1} x = 0. \tag{†}$$

If $c_{n+1} = 0$, then expression (†) implies $c_1 v_1 + c_2 v_2 + \cdots + c_n v_n = 0$, without all of the coefficients being zero. This cannot happen since S is linearly independent. Therefore, $c_{n+1} \neq 0$ and (†) can be solved for x as follows:

$$x = \frac{-c_1}{c_{n+1}} v_1 + \frac{-c_2}{c_{n+1}} v_2 + \cdots + \frac{-c_n}{c_{n+1}} v_n.$$ So each x different from the vectors in S is $L(S)$. Each v_i in S is already in $L(S)$. Thus, every vector in V is a linear combination of the vectors in S. S spans V and is a basis for V.

31. $S = \{v_1, v_2, \ldots, v_k\}$ and its linear span $L(S)$ is a proper subspace of V whose dimension is $k < n$. S is a basis for $L(S)$. Let v_{k+1} be a vector in V not in $L(S)$, $S \cup \{v_{k+1}\}$ is linearly independent. Suppose otherwise. Then there are scalars, $c_1, c_2, \ldots, c_k, c_{k+1}$, not all zero, such that $c_1 v_1 + c_2 v_2 + \cdots + c_n v_k + c_{k+1} v_{k+1} = 0$. But, $c_{k+1} \neq 0$, or we would have a contradiction to the linear independence of S. Then v_{k+1} can be expressed as a linear combination of the vectors of S, contrary to the fact that v_{k+1} is not in $L(S)$. So $S_1 = S \cup \{v_{k+1}\}$ is linearly independent. If $L(S_1)$ is not all of V, repeat the process of adding vectors until reaching the desired basis for V.

33. Given any matrix $A = (a_{ij})$ in $M_n(\mathbb{R})$, it can be seen that $A = \sum_{i,j=1}^{n} a_{ij} E_{i,j}$. So the set E spans $M_n(\mathbb{R})$. To prove that E is linearly independent, suppose that $\sum_{i,j=1}^{n} c_{ij} E_{i,j} = O$, the zero matrix. Equate entries on both sides of this equation and note that it is necessary that each $c_{ij} = 0$. So E is linearly independent, thus a basis.

35. Suppose that, for some scalars $\alpha, \beta,$ and γ, we have $\alpha(0, 1, 1) + \beta(1, 0, 1) + \gamma(1, 1, 0) = (0, 0, 0)$. Then

$$\beta + \gamma = 0,$$
$$\alpha + \gamma = 0,$$
$$\alpha + \beta = 0.$$

The only solution is the trivial one $\alpha = \beta = \gamma = 0$. So the set S is linearly independent.

37. All of the equational axioms hold by the usual rules of high school algebra. Also the sum of two polynomials is again a polynomial and so is the product of a polynomial and a scalar. Clearly the zero polynomial belongs to P and if $f \in P$, the $-f$ is also a polynomial in P. Thus, P is a vector space.

39. The vector space P of all polynomials is clearly a subspace of $C[0, 1]$ when considered as polynomial functions on the closed interval $[0, 1]$. Since P is not finite dimensional and $\dim P \leq \dim C[0, 1]$, the space $C[0, 1]$ cannot be finite dimensional either. See Exercise 38.

Section 3.5 (pp. 204–205)

1. Row echelon form: $\begin{bmatrix} 1 & 2 \\ 0 & 0 \end{bmatrix}$, row rank $= 1$. A^T row echelon form: $\begin{bmatrix} 1 & 4 \\ 0 & 0 \end{bmatrix}$, row rank $A^T = 1 =$ the column rank of A.

3. Row echelon form: $\begin{bmatrix} 1 & 2 & 0 \\ 0 & 0 & 0 \\ 0 & 0 & 0 \end{bmatrix}$, row rank $= 1$. A^T row echelon form: $\begin{bmatrix} 1 & 2 & 3 \\ 0 & 0 & 0 \\ 0 & 0 & 0 \end{bmatrix}$, so the row rank of $A^T = 1 =$ the column rank of A.

5. Row echelon form for A is $\begin{bmatrix} 1 & 2 & 4 & 0 \\ 0 & 1 & 11 & -2 \\ 0 & 0 & 1 & \frac{-12}{55} \end{bmatrix}$, so its row rank $= 3$. The row echelon form for A^T is $\begin{bmatrix} 1 & 3 & 3 \\ 0 & 1 & 6 \\ 0 & 0 & 1 \\ 0 & 0 & 0 \end{bmatrix}$, so the row rank of $A^T = 3 =$ the column rank of A.

7. Row echelon form for A is $\begin{bmatrix} 1 & -2 & -1 \\ 0 & 1 & 1 \\ 0 & 0 & 1 \\ 0 & 0 & 0 \end{bmatrix}$, so its row rank $= 3$. The row echelon form for A^T is $\begin{bmatrix} 1 & 0 & 1 & 0 \\ 0 & 1 & 4 & 1 \\ 0 & 0 & 0 & 1 \end{bmatrix}$, so the row rank of $A^T = 3 =$ the column rank of A.

9. Row echelon form for A is $\begin{bmatrix} 1 & -1 & 0 & -5 \\ 0 & 1 & -2 & -1 \\ 0 & 0 & 1 & -6 \\ 0 & 0 & 0 & 1 \end{bmatrix}$, so its row rank $= 4$. The row echelon form for A^T is

$\begin{bmatrix} 1 & 0 & -1 & 1 \\ 0 & 1 & 1 & -1 \\ 0 & 0 & 1 & 4 \\ 0 & 0 & 0 & 1 \end{bmatrix}$, so the row rank of $A^T = 4 =$ the column rank of A.

11. See Exercise 2. $\{(1, 2, 4), (2, 4, 7)\}$ or $\{(1, 2, 4), (0, 0, 1)\}$ is a basis.

13. See Exercise 4, $\{(1, 2, -1), (-1, 4, -3)\}$.   **15.** $\{(1, 0, -1, 0), (3, -1, 1, 2), (3, 0, 1, 0)\}$.

17. $B = \{(1, -1, 0), (-1, 0, 5), (0, 1, -2)\}$, or the standard basis $E = \{e_1, e_2, e_3\}$, where $e_1 = (1, 0, 0)$, $e_2 = (0, 1, 0)$, and $e_3 = (0, 0, 1)$.

19. See Exercise 1. One may use either of the two columns of A as a basis, say $\left\{ \begin{bmatrix} 1 \\ 4 \end{bmatrix} \right\}$. This is also the transpose of the one nonzero row of the row echelon form of A^T.

21. See Exercise 3. $\left\{ \begin{bmatrix} 1 \\ 2 \\ 3 \end{bmatrix} \right\}$.

23. $\left\{ \begin{bmatrix} 2 \\ 5 \\ 0 \end{bmatrix}, \begin{bmatrix} 4 \\ 1 \\ 1 \end{bmatrix}, \begin{bmatrix} 0 \\ 2 \\ 0 \end{bmatrix} \right\}$, or $\left\{ \begin{bmatrix} 1 \\ 3 \\ 3 \end{bmatrix}, \begin{bmatrix} 0 \\ 1 \\ 6 \end{bmatrix}, \begin{bmatrix} 0 \\ 0 \\ 1 \end{bmatrix} \right\}$, or the standard basis $E = \{e_1, e_2, e_3\}$, where

$e_1 = (1, 0, 0)^T$, $e_2 = (0, 1, 0)^T$, and $e_3 = (0, 0, 1)^T$. See Exercise 5.

25. $\left\{ \begin{bmatrix} 1 \\ 0 \\ 1 \\ 0 \end{bmatrix}, \begin{bmatrix} -2 \\ 1 \\ 2 \\ 1 \end{bmatrix}, \begin{bmatrix} -1 \\ 1 \\ 3 \\ -1 \end{bmatrix} \right\}$, or $\left\{ \begin{bmatrix} 1 \\ 0 \\ 1 \\ 0 \end{bmatrix}, \begin{bmatrix} 0 \\ 1 \\ 4 \\ 1 \end{bmatrix}, \begin{bmatrix} 0 \\ 0 \\ 0 \\ 1 \end{bmatrix} \right\}$, etc. See Exercise 7.

27. $\left\{ \begin{bmatrix} -1 \\ 0 \\ 1 \\ -1 \end{bmatrix}, \begin{bmatrix} 1 \\ 1 \\ 0 \\ 0 \end{bmatrix}, \begin{bmatrix} 0 \\ -2 \\ -3 \\ -2 \end{bmatrix}, \begin{bmatrix} 5 \\ -1 \\ 0 \\ 0 \end{bmatrix} \right\}$, or the standard basis $E = \{e_1, e_2, e_3, e_4\}$, where $e_1 = (1, 0, 0, 0)^T$, $e_2 = (0, 1, 0, 0)^T$,

$e_3 = (0, 0, 1, 0)^T$, and $e_4 = (0, 0, 0, 1)^T$, etc.

29. If A is $m \times n$ it has at most m linearly independent rows and at most n linearly independent columns. Let the rank of $A = r =$ row rank $=$ column rank. Then $r \leq m$ and $r \leq n$, so $r \leq \min \{m, n\}$.

31. Let $U = EA$, where E is the product of elementary matrices and U is a row echelon form for A. Since the nonzero row vectors of U are linearly independent, they form a basis for the row space of U. But U and A have the same row spaces, so the nonzero rows of U form a basis for the row space of A.

33. *Note:* Column rank $AC =$ row rank of $(AC)^T$
$= $ row rank of $C^T A^T$
$= $ row rank of A^T (by Exercise 32 applied to A^T)
$= $ column rank of A.

35. Suppose that A is nonsingular. Then $\det A \neq 0$, then $A\mathbf{x} = \mathbf{0}$ has only the trivial solution, and the columns of A are linearly independent; hence, rank $A = n$. Conversely, if A has rank n, then $A\mathbf{x} = \mathbf{0}$ has only the trivial solution. Thus, $\det A \neq 0$, and A is nonsingular.

37. Suppose rank $N = n$. Then $\det N \neq 0$. Since for each positive integer k $\det N^k = (\det N)^k$ it follows that $0 = \det O = \det N^k = (\det N)^k \neq 0$. This contradiction proves that rank $N < n$. Suppose $N^k = O$, but $N^{k-1} \neq O$. It does not necessarily follow that rank $N = k - 1$. You can verify that the matrix $N = \begin{bmatrix} 0 & 1 & 0 & 0 \\ 0 & 0 & 0 & 0 \\ 0 & 0 & 0 & 1 \\ 0 & 0 & 0 & 0 \end{bmatrix}$ satisfies $N^2 = O$,

while rank $N = 2 \neq 2 - 1$.

39. Note that rank $A = n$ if, and only if, the rows of A are linearly independent and form a basis for $\mathbb{R}^n$. Then A is row equivalent to the identity matrix I_n, whose rows are also a basis for $\mathbb{R}^n$.

41. Let $A = [\mathbf{a}_1 : \mathbf{a}_2 : \ldots : \mathbf{a}_n]$. Then the homogeneous system of equations $A\mathbf{x} = \mathbf{0}$ can be written in the form $x_1\mathbf{a}_1 + x_2\mathbf{a}_2 + \cdots + x_n\mathbf{a}_n = \mathbf{0}$. Then $A\mathbf{x} = \mathbf{0}$ has a nontrivial solution if, and only if, these column vectors are linearly dependent; hence, if and only if, rank $A < n$.

43. By elementary row and column operations we can transform the matrix A into the block matrix $M = \begin{bmatrix} I_r & O \\ O & O \end{bmatrix}$,

where r is the rank of A. $M = PAQ$ where both P and Q are products of elementary matrices and are therefore invertible. In Exercise 40 we proved that A and PA have the same rank. Set $\mathbf{y} = Q^{-1}\mathbf{x}$. Now $A\mathbf{x} = \mathbf{0}$ if, and only if, $PA\mathbf{x} = \mathbf{0}$ if, and only if, $PAQ\mathbf{y} = \mathbf{0}$; that is, $M\mathbf{y} = \mathbf{0}$. This last statement is true if, and only if, $y_1 = y_2 = \ldots = y_r = 0$. So the solution space for M has dimension $n - r$. But then the solution space of $A\mathbf{x} = \mathbf{0}$ must have the same dimension.

45. Let $\mathbf{a}_i$ be the rows of A and $\mathbf{e}_j$, the rows of the identity matrix I_n; $i = 1, 2, \ldots, m$ and $j = 1, 2, \ldots, n$. Since the rows of A span $\mathbb{R}^n$, there must exist scalars m_{ji} so that for each j the vector $\mathbf{e}_j$ is a linear combination of the rows of A. Thus, $\mathbf{e}_j = m_{j1}\mathbf{a}_1 + m_{j2}\mathbf{a}_2 + \cdots + m_{jm}\mathbf{a}_m$. Set $M = (m_{ji})$. Then row-by-row we see that $MA = I_n$, so A has a left inverse.

47. Rank $A = 4$, rank adj $A = 4$.   **49.** Rank $A = 1$, adj $A = O$, so rank adj $A = 0$.

Chapter 3 Review Exercises (pp. 206–207)

1. Yes. If $a(1, 2, -1) + b(3, 0, 2) + c(-1, 1, 0) = \mathbf{0} = (a + 3b - c, 2a + c, -a + 2b) = (0, 0, 0)$. Then $a = b = c = 0$ is the only solution.

2. No. The dimension of $\mathbb{R}^3$ is 3. Hence any set with more than three vectors from $\mathbb{R}^3$ is a linearly dependent set.

3. No. These vectors are linearly dependent since the dimension of P_2 is 3 and here we have four vectors.

4. Yes. Suppose that there are scalars a and b such that $a\begin{bmatrix} 1 & 0 \\ -1 & 0 \end{bmatrix} + b\begin{bmatrix} 2 & 0 \\ 3 & 0 \end{bmatrix} = \begin{bmatrix} 0 & 0 \\ 0 & 0 \end{bmatrix}$. Then $a + 2b = 0$ and $-a + 3b = 0$. This homogeneous system has only the trivial solution $a = b = 0$.

5. Let the set be $S = \{\mathbf{u}_0, \mathbf{u}_1, \mathbf{u}_3 \ldots\}$ with $\mathbf{u}_0 = \mathbf{0}$. Then $\mathbf{0} = 1\mathbf{u}_0 + 0\mathbf{u}_1 + 0\mathbf{u}_3 + \cdots + 0\mathbf{u}_k$ is a finite linear combination of the vectors from S, which gives the zero vector but not all of the coefficients are zero.

6. Let $\mathbf{u} = (x, y, z)$ and $\mathbf{v} = (r, s, t)$ be any two vectors in the given set W. Clearly if $2x - y + 3z = 0$ and $2r - s + 3t = 0$, then both $2\alpha x - \alpha y + 3\alpha z = \alpha 0 = 0$, for any α and $2(x + r) - (y + s) + 3(z + t) = 0$. So $\alpha\mathbf{u}$ and $\mathbf{u} + \mathbf{v}$ are in W. So W is a subspace of dimension 2 and $\{(3, 0, -2), (1, 2, 0)\}$ is a possible basis.

7. Select any vector which is not in the linear span of $H = \{(1, 2, -1), (-3, 0, 5)\}$. For example, take $\mathbf{e}_1 = (1, 0, 0)$. H is linearly independent and $\mathbf{e}_1$ is not in $L(H)$, so $H \cup \{\mathbf{e}_1\}$ is a linearly independent set of three vectors from $\mathbb{R}^3$; so it is a basis: $H \cup \{\mathbf{e}_1\} = \{(1, 2, -1), (-3, 0, 5), (1, 0, 0)\}$.

8. No. Since the dimension of $\mathbb{R}^4$ is 4, any set of more than four vectors cannot be linearly independent.

9. $a = \dfrac{-3}{2}$.   **10.** Row rank of $A =$ column rank of $A = 3$.

11. 3.   **12.** 1.   **13.** n.   **14.** n.

Chapter 4

Section 4.1 (pp. 224–226)

1. $\mathbf{x} = t(1, 2, 1) + (1, -t)(-1, -3, 0)$, so $x_1 = 2t - 1$, $x_2 = 5t - 3$, $x_3 = t$.

3. $x_1 = -7t + 5$, $x_2 = 4t - 4$, $x_3 = -6t + 7$.

5. $(x_1, x_2) = t(1, -2) + (1 - t)(-3, 1)$. Therefore, $x_1 = 4t - 3$, $x_2 = -3t + 1$.   **7.** 2.   **9.** 4.   **11.** -1.

13. $\|\mathbf{u}\| = \langle \mathbf{u}, \mathbf{u} \rangle^{1/2} = (1 + 1 + 1)^{1/2} = \sqrt{3}$, $\|\mathbf{v}\| = \langle \mathbf{v}, \mathbf{v} \rangle^{1/2} = (1 + 4 + 1)^{1/2} = \sqrt{6}$.

15. $\|\mathbf{u}\| = \sqrt{2}$ and $\|\mathbf{v}\| = 1$.   **17.** Yes; $\langle \mathbf{u}, \mathbf{v} \rangle = 1 - 2 + 1 = 0$.

19. $\langle \mathbf{u}, \mathbf{v} \rangle = u_1v_1 + u_2v_2 + u_3v_3 = \mathbf{u}\mathbf{v}^T$.   **21.** $\cos \theta = \dfrac{\langle \mathbf{u}, \mathbf{v} \rangle}{\|\mathbf{u}\| \, \|\mathbf{v}\|} = \dfrac{2}{(5)(\sqrt{5})} = \dfrac{2}{5\sqrt{5}}$.

23. $\cos \theta = \dfrac{4}{3\sqrt{2}}$.   **25.** $\cos \theta = \dfrac{10}{21}$.   **27.** $\sqrt{2^2 + (-2)^2 + (-2)^2} = 2\sqrt{3}$.

29. See Figures 4.5 and 4.6 in the text. Let the angle at A be θ, and note that $u = b \cos \theta$ while $v = b \sin \theta$. Thus from the distance formula, $a^2 = |CB|^2 = (u - c)^2 + (v - 0)^2 = (b \cos \theta - c)^2 + (b \sin \theta)^2 = b^2 \cos^2 \theta - 2bc \cos \theta + c^2 + b^2 \sin^2 \theta = b^2(\sin^2 \theta + \cos^2 \theta) - 2bc \cos \theta + c^2 = b^2 - 2bc \cos \theta + c^2$. Hence, $a^2 = b^2 - 2bc \cos \theta + c^2$, as desired. The other three forms are similar.

31. Given that $u = (u_1, u_2, u_3)$, then from Exercise 30, $\cos \theta_i = \frac{u_i}{||u||}$, $i = 1, 2, 3$, so that $u = ||u||(\cos \theta_1, \cos \theta_2, \cos \theta_3)$. Then it follows that $\cos^2 \theta_1 + \cos^2 \theta_2 \cos^2 \theta_3 = \frac{u_1^2}{||u||^2} + \frac{u_2^2}{||u||^2} + \frac{u_3^2}{||u||^2} = \frac{u_1^2 + u_2^2 + u_3^2}{||u||^2} = 1$.

33. No. Note that $\langle v, -x \rangle = -\langle v, x \rangle$ and one of these $\langle v, x \rangle$ or $-\langle v, x \rangle$ must, therefore, be positive if the other is negative.

35. In $\mathbb{R}^3$, $\langle 0, x \rangle = \langle (0, 0, 0), (x_1, x_2, x_3) \rangle = 0x_1 + 0x_2 + 0x_3 = 0$.

37. Take $x = (1, 1)$ and $y = (2, 2)$, then it follows that $\frac{x_1 x_2 + y_1 y_2}{\sqrt{x_1^2 + y_1^2}\sqrt{x_2^2 + y_2^2}} = 1$. Take $x = (1, -1)$ and $y = (-2, 2)$ for -1.

39. Given $u = (u_1, u_2, u_3)$ and $v = (v_1, v_2, v_3)$, then $\langle u, v \rangle = u_1 v_1 + u_2 v_2 + u_3 v_3 = v_1 u_1 + v_2 u_2 + v_3 u_3 = \langle v, u \rangle$, because real number multiplication is commutative.

41. Let $u = (u_1, u_2, u_3)$, so $ku = (ku_1, ku_2, ku_3)$. Then $\langle ku, v \rangle = ku_1 v_1 + ku_2 v_2 + ku_3 v_3 = k(u_1 v_1 + u_2 v_2 + u_3 v_3) = k\langle u, v \rangle$.

43. $\langle u, u \rangle = u_1^2 + u_2^2 + u_3^2 = 0$ if, and only if, each one of the real numbers $u_i^2 = 0$, hence if, and only if, $u_1 = u_2 = u_3 = 0$. So $u = 0$.

45. Let $k_2 = 0$ in part (iv) to get part (iii). Then let $k_1 = k_2 = 1$ in part (iv) to get part (ii).

47. Note that $||\vec{AB}|| = 13$, $||\vec{AC}|| = 13$, and $||\vec{BC}|| = 13\sqrt{2}$. These numbers satisfy the Pythagorean theorem. Furthermore, note that $\langle \vec{AB}, \vec{AC} \rangle = 0$, so these two vectors $\vec{AB} = (3, 4, 12)$ and $\vec{AC} = (-4, 12, -3)$, are orthogonal (perpendicular) and the right angle is at A.

49. In the figure, suppose that $||\vec{OB}|| = ||\vec{OA}|| = ||\vec{AC}|| = ||\vec{BC}||$. Then $\langle \vec{OC}, \vec{BA} \rangle = (a_1 + b_1)(a_1 - b_1) + (a_2 + b_2)(a_2 - b_2) = a_1^2 - b_1^2 + a_2^2 - b_2^2$. But $||\vec{OA}||^2 = a_1^2 + a_2^2 = b_1^2 + b_2^2 = ||\vec{OB}||^2$, so $\langle \vec{OC}, \vec{BA} \rangle = a_1^2 - b_1^2 + a_2^2 - b_2^2 = 0$, and the two diagonals are perpendicular.

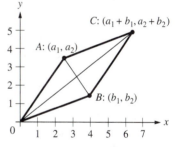

Section 4.2 (pp. 234–235)

1. $u \times v = \det \begin{bmatrix} e_1 & e_2 & e_3 \\ 1 & 0 & -1 \\ 0 & -1 & 1 \end{bmatrix} = (-1, -1, -1)$.   **3.** $(1, -4, -6)$.   **5.** $(-2, 3, 6)$.

7. $(-3, -1, -1)$.   **9.** $(-2, 6, 6)$.   **11.** $(-5, -1, -1)$.   **13.** $(-4, 3, 6)$.   **15.** $(-2, 0, -4)$.

17. $-k$.   **19.** $u \times v = \det \begin{bmatrix} i & j & k \\ 2 & 0 & -3 \\ 0 & -2 & 1 \end{bmatrix} = -6i - 2j - 4k$.

21. $u \times v = -4i - 7j - 6k$.   **23.** $u \times v = 5i - 35j - 20k$.

25. $\frac{a \times b}{||a \times b||}$ and $\frac{-(a \times b)}{||a \times b||} = \frac{b \times a}{||b \times a||}$. Thus, $\frac{\sqrt{2}}{6}(-1, -4, -1)$ and $\frac{\sqrt{2}}{6}(1, 4, 1)$ are the desired vectors.

27. $a \times b = (-4, -16, -4)$, so $-4x_1 - 16x_2 - 4x_3 = 0$, or, simplified, $x_1 + 4x_2 + x_3 = 0$.

29. $\langle a \times b, x \rangle = \langle a \times b, c \rangle$ so the scalar equation is $19x_1 + 6x_2 - 4x_3 = 45$.

31. $12x_1 - 7x_2 - 4x_3 = 18$.   **33.** See equation (4.12). $||a \times b|| = \sqrt{173}$.

35. $u \times v = \det \begin{bmatrix} e_1 & e_2 & e_3 \\ u_1 & u_2 & u_3 \\ v_1 & v_2 & v_3 \end{bmatrix} = -\det \begin{bmatrix} e_1 & e_2 & e_3 \\ v_1 & v_2 & v_3 \\ u_1 & u_2 & u_3 \end{bmatrix} = -(v \times u)$.

37. In the following expand the first determinant along its third row and compare.

$$\mathbf{u} \times (\mathbf{v} + \mathbf{w}) = \det \begin{bmatrix} \mathbf{e}_1 & \mathbf{e}_2 & \mathbf{e}_3 \\ u_1 & u_2 & u_3 \\ v_1 + w_1 & v_2 + w_2 & v_3 + w_3 \end{bmatrix}$$

$$= \det \begin{bmatrix} \mathbf{e}_1 & \mathbf{e}_2 & \mathbf{e}_3 \\ u_1 & u_2 & u_3 \\ v_1 & v_2 & v_3 \end{bmatrix} + \det \begin{bmatrix} \mathbf{e}_1 & \mathbf{e}_2 & \mathbf{e}_3 \\ u_1 & u_2 & u_3 \\ w_1 & w_2 & w_3 \end{bmatrix}$$

$$= (\mathbf{u} \times \mathbf{v}) + (\mathbf{u} \times \mathbf{w}).$$

39. Let $\mathbf{u} = \mathbf{i}$, $\mathbf{v} = \mathbf{i} + \mathbf{j}$, and $\mathbf{w} = \mathbf{i} + \mathbf{j} + \mathbf{k}$. Then $\mathbf{u} \times \mathbf{v} = \mathbf{k}$, so that $(\mathbf{u} \times \mathbf{v}) \times \mathbf{w} = -\mathbf{i} + \mathbf{j}$. But $\mathbf{u} \times (\mathbf{v} \times \mathbf{w}) = -\mathbf{k}$. Therefore, $(\mathbf{u} \times \mathbf{v}) \times \mathbf{w} \neq \mathbf{u} \times (\mathbf{v} \times \mathbf{w})$.

41. $\mathbf{u} \times \alpha\mathbf{v} = \det \begin{bmatrix} \mathbf{e}_1 & \mathbf{e}_2 & \mathbf{e}_3 \\ u_1 & u_2 & u_3 \\ \alpha v_1 & \alpha v_2 & \alpha v_3 \end{bmatrix} = \alpha \det \begin{bmatrix} \mathbf{e}_1 & \mathbf{e}_2 & \mathbf{e}_3 \\ u_1 & u_2 & u_3 \\ v_1 & v_2 & v_3 \end{bmatrix} = \alpha(\mathbf{u} \times \mathbf{v}).$

43. Let h be the height of the parallelepiped (box) in $\mathbb{R}^3$. Then the volume of this solid is $V = h \|\mathbf{v} \times \mathbf{w}\| = \|\mathbf{u}\| \cos\theta \|\mathbf{v} \times \mathbf{w}\|$ where θ is the angle between $\mathbf{u}$ and the vertical axis. Thus, the volume $V = <\mathbf{u}, \mathbf{v} \times \mathbf{w}> = \det M$. See Exercise 42.

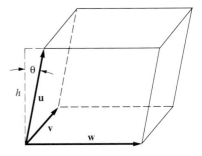

45. If $\mathbf{m}$ and $\mathbf{n}$ are parallel, then $\mathbf{m} = k\mathbf{n}$ for some nonzero scalar k. Thus, $<\mathbf{m}, \mathbf{x}> = <k\mathbf{n}, \mathbf{x}> = k<\mathbf{n}, \mathbf{x}>$ and $<\mathbf{m}, \mathbf{c}> = k<\mathbf{n}, \mathbf{c}>$ and the plane $<\mathbf{m}, \mathbf{x}> = <\mathbf{m}, \mathbf{c}>$ becomes $k<\mathbf{n}, \mathbf{x}> = k<\mathbf{n}, \mathbf{c}>$, so the two planes coincide. If the planes are distinct, they cannot coincide so $\mathbf{m} \neq k\mathbf{n}$ and $\mathbf{m}$ and $\mathbf{n}$ are not parallel. Therefore, the system of equations

$$n_1 x_1 + n_2 x_2 + n_3 x_3 = b_1,$$
$$m_1 x_1 + m_2 x_2 + m_3 x_3 = b_1$$

is consistent. Solve this system and note that the direction numbers for the line of intersection are the components of $\mathbf{n} \times \mathbf{m}$; i.e., $\det \begin{bmatrix} n_2 & n_3 \\ m_2 & m_3 \end{bmatrix}$, $-\det \begin{bmatrix} n_1 & n_3 \\ m_1 & m_3 \end{bmatrix}$, and $\det \begin{bmatrix} n_1 & n_2 \\ m_1 & m_2 \end{bmatrix}$.

47. Row reduce the matrix $\begin{bmatrix} 2 & -3 & 1 & 5 \\ 1 & 4 & -2 & -2 \end{bmatrix}$ to obtain $\begin{bmatrix} 1 & 0 & \frac{-2}{11} & \frac{14}{11} \\ 0 & 1 & \frac{-5}{11} & \frac{-9}{11} \end{bmatrix}$. The general solution to the system consisting of the two equations of the planes is therefore $x_1 = \frac{2}{11}t + \frac{14}{11}$, $x_2 = \frac{5}{11}t - \frac{9}{11}$, $x_3 = t$. Solve these three equations for the parameter t and obtain the result of Exercise 46: $\dfrac{x_1 - \frac{14}{11}}{2} = \dfrac{x_2 + \frac{9}{11}}{5} = \dfrac{x_3 - 0}{11}$.

49. $\mathbf{x} \times \mathbf{F} = \det \begin{bmatrix} \mathbf{i} & \mathbf{j} & \mathbf{k} \\ 9 & 0 & 1 \\ 1 & 1 & -1 \end{bmatrix} = -\mathbf{i} + 10\mathbf{j} + 9\mathbf{k} = (-1, 10, 9)$, $\|\mathbf{x} \times \mathbf{F}\| = \sqrt{182}$.

Section 4.3 (pp. 244–246)

1. $<u, v> = (1)(0) + (-1)(2) + (2)(0) = -2$. **3.** $<u, v> = 1 + 2 + 0 + 0 + (-3) = 0$.

5. $<u, v> = 0 + 0 + 0 + 0 + 0 + 0 = 0 = <u, 0>$. **7.** $<u, v> = -1 + 1 + 1 + 2 + 4 + 1 = 8$.

9. $<u, v> = u\,Dv^T = [1, -1, 2]\begin{bmatrix} 2 & 0 & 0 \\ 0 & 2 & 0 \\ 0 & 0 & 2 \end{bmatrix}\begin{bmatrix} 0 \\ 2 \\ 0 \end{bmatrix} = u(2I_2)v^T = 2uI_2v^T = 2uv^T = 2[1, -1, 2]\begin{bmatrix} 0 \\ 2 \\ 0 \end{bmatrix} = 2(-2) = -4$.

11. $<u, v> = u\,Dv^T = [1, -1, 2]\begin{bmatrix} 2 & 0 & 0 \\ 0 & 3 & 0 \\ 0 & 0 & 4 \end{bmatrix}\begin{bmatrix} 1 \\ 2 \\ -1 \end{bmatrix} = -12$.

13. $<A, B> = \mathrm{tr}(A^T B) = \mathrm{tr}\left(\begin{bmatrix} 0 & -1 \\ 1 & -1 \end{bmatrix}\begin{bmatrix} 2 & 3 \\ 1 & -1 \end{bmatrix}\right) = \mathrm{tr}\begin{bmatrix} -1 & 1 \\ 1 & 4 \end{bmatrix} = -1 + 4 = 3$.

15. $<A, B> = \mathrm{tr}(A^T B) = \mathrm{tr}\left(\begin{bmatrix} 1 & 0 & 1 & 0 \\ 0 & 1 & 1 & 1 \\ 0 & 1 & 0 & 1 \\ -1 & 0 & 1 & -1 \end{bmatrix}\begin{bmatrix} 1 & -1 & 1 & -1 \\ 0 & 1 & -1 & 1 \\ 0 & 0 & 1 & -1 \\ 0 & 0 & 0 & 1 \end{bmatrix}\right) = 1 + 1 - 1 - 1 = 0$. (Note: one can

avoid the total multiplication by merely multiplying row i of A^T by column i of B; that is, note that trace $A^T B =$

$\displaystyle\sum_{i=1}^{n} <a_i, b_i> = \sum_{i=1}^{n} a_i{}^T b_i$, where, as before, we have denoted the column vectors of A and B by a_i and b_i, respectively.)

17. $<p(x), q(x)> = <x^2, x - 3> = (1)(0) + (0)(1) + (0)(-3) = 0$. **19.** $<p(x), q(x)> = 12 + 10 = 22$.

21. $<X, Y> = X^T Y = [1, -1, 1, 2]\begin{bmatrix} 2 \\ -2 \\ 1 \\ 0 \end{bmatrix} = 5$. **23.** $<X, Y> = X^T Y = [2, 7, 1, -4, 1, 2]\begin{bmatrix} -10 \\ 0 \\ 2 \\ -2 \\ 1 \\ 0 \end{bmatrix} = -9$.

25. No. Notice that for this function $Q(2u, v) \neq 2Q(u, v)$. For example, let $u = (1, 1) = v$. Then $Q(2u, v) = (2)(2)(1)(1) = 4 \neq 2(1)(1)(1)(1) = 2Q(u, v)$.

27. No. If $u = (1, 0)$, then $Q(u, u) = uAu^T = -1 < 0$, for this function. This contradicts Property (3) of the definition of an inner-product function.

29. No. If $u = (1, 1)$, then $Q(u, u) = uAu^T = -1 < 0$, for this function. This contradicts Property (3) of the definition of an inner-product function.

31. Note that trace $A^T B = \displaystyle\sum_{i=1}^{n} <a_i, b_i>$, where, as before (see Exercise 15), we have denoted the column vectors of A and B by a_i and b_i, respectively. Therefore, $<A, B> = \displaystyle\sum_{i=1}^{n} <a_i, b_i> = \sum_{i=1}^{n} <b_i, a_i> = \mathrm{trace}\,(B^T A) = <B, A>$. This is Property (1). For (2) note that $<\alpha A + \beta B, C> = \mathrm{tr}((\alpha A + \beta B)^T C) = \mathrm{tr}(\alpha A^T C + \beta B^T C) = \mathrm{tr}(\alpha A^T C) + \mathrm{tr}(\beta B^T C) = \alpha\mathrm{tr}(A^T C) + \beta\mathrm{tr}(B^T C) = \alpha<A, C> + \beta<B, C>$. For (3) and (4) note that $<A, A> = \displaystyle\sum_{i=1}^{n} <a_i, a_i>$ is the sum of nonnegative real numbers and is zero if, and only if, each $<a_i, a_i> = 0$, which is true if, and only if, each column vector of A is zero. This is true if, and only if, A is the zero matrix O.

33. Those in Exercises 16 and 17. See the results given in the answers above.

35. Use equation (4.12). The area of the parallelogram is $\|a \times b\| = \|a\|\,\|b\| \sin\theta$. But note that $a \times b =$

$\det\begin{bmatrix} e_1 & e_2 & e_3 \\ a_1 & a_2 & 0 \\ b_1 & b_2 & 0 \end{bmatrix} = \det\begin{bmatrix} a_1 & a_2 \\ b_1 & b_2 \end{bmatrix} e_3 = \det\begin{bmatrix} a_1 & b_1 \\ a_2 & b_2 \end{bmatrix} e_3$. Hence, $\|a \times b\| = |\det [a : b]|$.

37. *Note:* $<u, \alpha v> = <\alpha v, u> = \alpha<v, u> = \alpha<u, v>$, using Properties (1) and (2) of the definition of an inner product.

39. See Exercise 37 and 38 in connection with Properties 1 and 2. We have $<a_1u_1 + a_2u_2, a_3u_3 + a_4u_4> =$
$a_1<u_1, a_3u_3 + a_4u_4> + a_2<u_2, a_3u_3 + a_4u_4> = a_1<a_3u_3 + a_4u_4, u_1> + a_2<a_3u_3 + a_4u_4, u_2> = a_1a_3<u_3, u_1>$
$+ a_1a_4<u_4, u_1> + a_2a_3<u_3, u_2> + a_2a_4<u_4, u_2> = a_1a_3<u_1, u_3> + a_1a_4<u_1, u_4> + a_2a_3<u_2, u_3>$
$+ a_2a_4<u_2, u_4>$, as desired.

41. $<f, g> = \int_{-1}^{1} x(1 + x) \, dx = \left[\frac{x^2}{2} + \frac{x^3}{3}\right]_{-1}^{1} = \frac{2}{3}.$ **43.** $<f(x), g(x)> = \int_{-1}^{1} 3x(1 + x^2) \, dx = 0.$

45. $<f(x), g(x)> = \int_{-1}^{1} (1 - x - 2x^2)(1 + x^3 + x^5) \, dx = \frac{-2}{105}.$ **47.** $<f(x), g(x)> = \int_{-1}^{1} (x - e^x)e^{-x} \, dx = -2e^{-1} - 2.$

49. $<f(x), g(x)> = \int_{-\pi}^{\pi} x \cos x \, dx = [\cos x + x \sin x]_{-\pi}^{\pi} = 0.$

51. For (1), note that $<f, g> = \int_{c}^{d} f(x) g(x) w(x) \, dx = \int_{c}^{d} g(x) f(x) w(x) \, dx = <g, f>.$ For (2),

$<af + bg, h> = \int_{c}^{d}(af(x) + bg(x))h(x) w(x) \, dx = \int_{c}^{d} af(x) h(x) w(x) \, dx + \int_{c}^{d} bg(x) h(x) w(x) \, dx =$

$a\int_{c}^{d} f(x) h(x) w(x) \, dx + b\int_{c}^{d} g(x) h(x) w(x) \, dx = a<f, h> + b<g, h>.$ For (3) and (4), $<f, f> =$

$\int_{c}^{d} f^2(x) w(x) \, dx \geq 0,$ since both f^2 and w are nonnegative functions. We have a zero integral if, and only if, the
integrand is identically zero on $[c, d]$. Since $w(x) > 0$ there, this can be true if, and only if, $f(x) = 0$ for all x in
$[c, d]$; that is, f is the zero function.

53. $<f(x), g(x)> = \int_{-1}^{1} x^2(1 + x + x^2)x^2 \, dx = \frac{24}{35}.$ **55.** $<f(x), g(x)> = \int_{-1}^{1} (x - e^x)(e^{-x})x^2 \, dx = -16e^{-1} + 2e - \frac{2}{3}.$

Section 4.4 (pp. 256–257)

1. $\|u\| = <u, u>^{1/2} = \sqrt{1 + 9} = \sqrt{10}.$ **3.** $\|u\| = <u, u>^{1/2} = \sqrt{1 + 1 + 1} = \sqrt{3}.$

5. $\|u\| = <u, u>^{1/2} = \sqrt{\frac{1}{3} + \frac{1}{3} + \frac{1}{3}} = 1.$ **7.** $\|u\| = <u, u>^{1/2} = \sqrt{1 + 1 + 1 + 1 + 0} = 2.$

9. $\|u\| = \sqrt{\frac{1}{5} + \frac{1}{5} + 0 + \frac{1}{5} + \frac{1}{5} + \frac{1}{5}} = 1.$

11. See the previous answers. Unit vectors are those in Exercises 5 and 9.

13. Take $\frac{1}{\|u\|}u$; e.g., $\frac{1}{\sqrt{3}}u = \left(\frac{1}{\sqrt{3}}, \frac{1}{\sqrt{3}}, \frac{1}{\sqrt{3}}\right).$ **15.** Take $\frac{1}{\|u\|}u$; e.g., $\frac{1}{2}u = \left(\frac{1}{2}, \frac{-1}{2}, \frac{1}{2}, \frac{-1}{2}, 0\right).$

17. $<u, v> = 2 - 2 = 0.$ **19.** $<u, v> = -2 + 3 - 1 = 0.$ **21.** $<u, v> = 1 - 1 + 2 - 2 = 0.$

23. $<u, v> = 6 - 3 + 0 = 3 \neq 0,$ so they are not orthogonal in the standard inner product. However, when we use the

inner product $<u, v> = u \, Dv^T = (3, -3, 1)\begin{bmatrix} 1 & 0 & 0 \\ 0 & 2 & 0 \\ 0 & 0 & 3 \end{bmatrix}\begin{bmatrix} 2 \\ 1 \\ 0 \end{bmatrix} = 0,$ they are orthogonal.

25. $<u, v> = 0 + 3 - 2 = 1 \neq 0,$ so they are not orthogonal in the standard inner product. However, when we use the

inner product $<u, v> = u \, Dv^T = (0, 3, 1)\begin{bmatrix} 1 & 0 & 0 \\ 0 & 2 & 0 \\ 0 & 0 & 3 \end{bmatrix}\begin{bmatrix} 19 \\ 1 \\ -2 \end{bmatrix} = 0,$ they are orthogonal.

27. $<x, y> = x \, Dy^T = (x, y, z)\begin{bmatrix} 1 & 0 & 0 \\ 0 & 2 & 0 \\ 0 & 0 & 3 \end{bmatrix}\begin{bmatrix} 1 \\ 1 \\ 1 \end{bmatrix} = x + 2y + 3z = 0$ if, and only if, $x = -2y - 3z.$ So all vectors

$y = (-2r - 3s, r, s)$ for any $r, s \in \mathbb{R}.$

29. A and B will be orthogonal provided $<A, B> = \text{tr}(A^T B) = 0$. Thus, $A = \begin{bmatrix} a & b \\ c & d \end{bmatrix}$ is orthogonal to $B = \begin{bmatrix} -1 & -3 \\ 5 & 8 \end{bmatrix}$

provided $\text{tr}\left(\begin{bmatrix} a & c \\ b & d \end{bmatrix} \begin{bmatrix} -1 & -3 \\ 5 & 8 \end{bmatrix} \right) = \text{tr} \begin{bmatrix} -a + 5c & -3a + 8c \\ -b + 5d & -3b + 8d \end{bmatrix} = -a + 5c - 3b + 8d = 0$. The matrix

$\begin{bmatrix} 8 & 0 \\ 0 & 1 \end{bmatrix}$ is one such example. In general, the matrix has the form $\begin{bmatrix} -3r + 5s + 8t & r \\ s & t \end{bmatrix}$.

31. $\|p(x)\|^2 = <p(x), p(x)> = \int_0^1 x^2 \, dx = \left[\frac{x^3}{3}\right]_0^1 = \frac{1}{3}$; so $\|p(x)\| = \frac{1}{\sqrt{3}}$. **33.** $\|p(x)\| = \left(\int_0^1 (1 - x^2)^2 \, dx\right)^{\frac{1}{2}} = \frac{2\sqrt{2}}{\sqrt{15}}$.

35. $\|p(x)\| = \left(\int_0^1 (4x^3 - 3x)^2 \, dx\right)^{\frac{1}{2}} = \frac{\sqrt{17}}{\sqrt{35}}$. **37.** $\|p(x)\| = \left(\int_0^1 \frac{1}{64}(35x^4 - 30x^2 + 3)^2 \, dx\right)^{\frac{1}{2}} = \frac{1}{3}$.

39. $q(x) = 3x - 2$. To obtain this, first solve $<q(x), p(x)> = \int_0^1 (ax + b)x \, dx = \left[\frac{a}{3}x^3 + \frac{b}{2}x^2\right]_0^1 = 0$ to obtain $b = \frac{-2a}{3}$.

Then solve $<q(x), q(x)> = 1$, when $q(x) = ax - \frac{2a}{3}$.

41. For these vectors note that $<\mathbf{u}, \mathbf{v}> = 0$, $\|\mathbf{u}\| = \sqrt{5} = \|\mathbf{v}\|$. So $|<\mathbf{u}, \mathbf{v}>| \le \|\mathbf{u}\| \, \|\mathbf{v}\| = \sqrt{5}\,\sqrt{5} = 5$.

43. For these vectors note that $<\mathbf{u}, \mathbf{v}> = \frac{-2}{5}$, $\|\mathbf{u}\| = \frac{2\sqrt{2}}{\sqrt{15}}$, and $\|\mathbf{v}\| = \frac{\sqrt{7}}{\sqrt{15}}$; and $\frac{2}{5} = |<\mathbf{u}, \mathbf{v}>| \le \|\mathbf{u}\| \, \|\mathbf{v}\| = 2\frac{\sqrt{14}}{15}$.

45. We illustrate the method with a typical computation. $<P_3, P_4> = \int_{-1}^1 \frac{1}{2}(5x^3 - 3x)\frac{1}{8}(35x^4 - 30x^2 + 3) \, dx =$

$\frac{1}{16}\int_{-1}^1 (175x^7 - 255x^5 + 105x^3 - 9x) \, dx = 0$, because the integrand is an odd function; that is, $h(x) = -h(-x)$.

47. Note that $<\mathbf{e}_i, \mathbf{e}_j> = \delta_{ij} = \begin{cases} 0 & i \ne j \\ 1 & i = j \end{cases}$, for $i, j = 1, 2, 3, 4$.

49. Note that $<\mathbf{e}_{i_s}, \mathbf{e}_{i_t}> = \delta_{st} = \begin{cases} 0 & s \ne t \\ 1 & s = t \end{cases}$, for $s, t = 1, 2, \ldots, n$.

51. Let $\mathbf{x}$ and $\mathbf{y}$ be vectors such that $<\mathbf{v}, \mathbf{x}> = 0$ and $<\mathbf{v}, \mathbf{y}> = 0$. Let α and β be arbitrary scalars, then
$<\mathbf{v}, \alpha\mathbf{x} + \beta\mathbf{y}> = \alpha<\mathbf{v}, \mathbf{x}> + \beta<\mathbf{v}, \mathbf{y}> = \alpha 0 + \beta 0 = \mathbf{0}$, also $<\mathbf{v}, \mathbf{0}> = 0$. Therefore, the collection of all vectors orthogonal to $\mathbf{v}$ is a subspace of V. In $\mathbb{R}^2$ the orthogonal complement of $\mathbf{v}$ is a line through the origin perpendicular (orthogonal) to $\mathbf{v}$. In $\mathbb{R}^3$ the orthogonal complement of $\mathbf{v}$ is a plane through the origin perpendicular (orthogonal) to $\mathbf{v}$.

53. $\mathbf{u} + \mathbf{v} = (19, 4, -1)$; $\|\mathbf{u} + \mathbf{v}\| = \sqrt{378}$. Also $\|\mathbf{u}\| = \sqrt{10}$, and $\|\mathbf{v}\| = \sqrt{366}$. Now take note that $\sqrt{378} \le \sqrt{10} + \sqrt{366}$, so $\|\mathbf{u} + \mathbf{v}\| \le \|\mathbf{u}\| + \|\mathbf{v}\|$. In $\mathbb{R}^3$ the triangle formed by $\mathbf{u}$, $\mathbf{v}$ and $\mathbf{u} + \mathbf{v}$, the length of the side $\mathbf{u} + \mathbf{v}$ is shorter than the sum of the lengths of the sides formed by $\mathbf{u}$ and $\mathbf{v}$.

55. $\mathbf{u} + \mathbf{v} = x^2$; $\|\mathbf{u} + \mathbf{v}\| = \left(\int_0^1 (x^2)^2 \, dx\right)^{\frac{1}{2}} = \frac{1}{\sqrt{5}}$. Also $\|\mathbf{u}\| = \sqrt{\frac{8}{15}}$, and $\|\mathbf{v}\| = \sqrt{\frac{7}{15}}$. Note that $\frac{1}{\sqrt{5}} \le \sqrt{\frac{8}{15}} + \sqrt{\frac{7}{15}}$, so $\|\mathbf{u} + \mathbf{v}\| \le \|\mathbf{u}\| + \|\mathbf{v}\|$.

Section 4.5 (pp. 268–270)

1. The orthonormal basis is $\left\{ \left(\frac{1}{\sqrt{2}}, \frac{1}{\sqrt{2}}\right), \left(\frac{1}{\sqrt{2}}, \frac{-1}{\sqrt{2}}\right) \right\}$. **3.** The orthonormal basis is $\{(0, 1), (1, 0)\}$.

5. Take $\mathbf{v} = (0, 1, 1)$, so $\mathbf{v}_1 = \left(0, \frac{1}{\sqrt{2}}, \frac{1}{\sqrt{2}}\right)$. With $\mathbf{u}_1 = (1, -1, 0)$, we have $\mathbf{w}_1 = \text{Proj}_{\mathbf{v}_1}\mathbf{u}_1 = <\mathbf{u}_1, \mathbf{v}_1>\mathbf{v}_1 = \frac{-1}{\sqrt{2}}\mathbf{v}_1 =$

$\left(0, \frac{-1}{2}, \frac{-1}{2}\right)$. Now $\mathbf{w}_2 = \mathbf{u}_1 - \mathbf{w}_1 = \left(1, \frac{-1}{2}, \frac{1}{2}\right)$, so $\mathbf{v}_2 = \frac{\mathbf{w}_2}{\|\mathbf{w}_2\|} = \frac{2}{\sqrt{6}}\left(1, \frac{-1}{2}, \frac{1}{2}\right) = \left(\frac{2}{\sqrt{6}}, \frac{-1}{\sqrt{6}}, \frac{1}{\sqrt{6}}\right)$. With $\mathbf{u}_2 =$

$(1, 0, -1)$. Take $\text{Proj}_{\mathbf{v}_1}\mathbf{u}_2 = <\mathbf{u}_2, \mathbf{v}_1>\mathbf{v}_1 = \frac{-1}{\sqrt{2}}\mathbf{v}_1 = \left(0, \frac{-1}{2}, \frac{-1}{2}\right)$ and $\text{Proj}_{\mathbf{v}_2}\mathbf{u}_2 = <\mathbf{u}_2, \mathbf{v}_2>\mathbf{v}_2 = \frac{1}{\sqrt{6}}\mathbf{v}_2 =$

$\left(\frac{2}{6}, \frac{-1}{6}, \frac{1}{6}\right)$; therefore, we have $\mathbf{w}_1' = <\mathbf{u}_1, \mathbf{v}_1>\mathbf{v}_1 + <\mathbf{u}_2, \mathbf{v}_2>\mathbf{v}_2 = \left(\frac{1}{3}, \frac{-2}{3}, \frac{-1}{3}\right)$, and so $\mathbf{w}_2' = \mathbf{u}_2 - \mathbf{w}_1' =$

$\left(\frac{2}{3}, \frac{2}{3}, \frac{-2}{3}\right)$. The third unit basis vector is, therefore, $\mathbf{v}_3 = \frac{\mathbf{w}_2'}{\|\mathbf{w}_2'\|} = \left(\frac{1}{\sqrt{3}}, \frac{1}{\sqrt{3}}, \frac{-1}{\sqrt{3}}\right)$. And the orthonormal basis is

$\left\{ \left(0, \frac{1}{\sqrt{2}}, \frac{1}{\sqrt{2}}\right), \left(\frac{2}{\sqrt{6}}, \frac{-1}{\sqrt{6}}, \frac{1}{\sqrt{6}}\right), \left(\frac{1}{\sqrt{3}}, \frac{1}{\sqrt{3}}, \frac{-1}{\sqrt{3}}\right) \right\}$.

7. The orthonormal basis is $\left\{\left(\frac{1}{\sqrt{3}}, \frac{1}{\sqrt{3}}, \frac{1}{\sqrt{3}}\right), \left(\frac{2}{\sqrt{6}}, \frac{-1}{\sqrt{6}}, \frac{-1}{\sqrt{6}}\right), \left(0, \frac{1}{\sqrt{2}}, \frac{-1}{\sqrt{2}}\right)\right\}$. The calculations are similar to Exercise 5.

9. $\mathbf{v}_1 = \left(\sqrt{\frac{6}{5}}, \sqrt{\frac{6}{5}}\right)$. $\mathbf{v}_2 = \left(\frac{2}{\sqrt{5}}, \frac{-3}{\sqrt{5}}\right)$. So the orthonormal basis is $\{\mathbf{v}_1, \mathbf{v}_2\}$.

11. The orthonormal basis is $\{\mathbf{v}_1, \mathbf{v}_2\}$, where $\mathbf{v}_1 = \left(\sqrt{\frac{6}{5}}, \sqrt{\frac{6}{5}}\right)$ and $\mathbf{v}_2 = \left(\frac{2}{\sqrt{5}}, \frac{-3}{\sqrt{5}}\right) = \frac{1}{\sqrt{5}}(2, -3)$.

13. The orthonormal basis is $\{\mathbf{v}_1, \mathbf{v}_2, \mathbf{v}_3\}$, where $\mathbf{v}_1 = \left(\frac{\sqrt{3}}{2}, 0, \frac{\sqrt{3}}{2}\right)$, $\mathbf{v}_2 = \left(\frac{1}{6}, \frac{2}{3}, \frac{-1}{2}\right)$, and $\mathbf{v}_3 = \left(\frac{-2}{3\sqrt{2}}, \frac{1}{3\sqrt{2}}, \frac{6}{3\sqrt{2}}\right)$.

15. The standard basis consists of the four matrices $E_{11} = \begin{bmatrix} 1 & 0 \\ 0 & 0 \end{bmatrix}$, $E_{12} = \begin{bmatrix} 0 & 1 \\ 0 & 0 \end{bmatrix}$, $E_{21} = \begin{bmatrix} 0 & 0 \\ 1 & 0 \end{bmatrix}$, and $E_{22} = \begin{bmatrix} 0 & 0 \\ 0 & 1 \end{bmatrix}$.

It can be easily calculated that $\text{tr}(E_{11}^T E_{12}) = \text{tr}\begin{bmatrix} 0 & 1 \\ 0 & 0 \end{bmatrix} = 0$, that $\text{tr}(E_{11}^T E_{21}) = \text{tr}\begin{bmatrix} 0 & 0 \\ 0 & 0 \end{bmatrix} = 0$, etc. Also note that $\text{tr}(E_{ij}^T E_{ij}) = \|E_{ij}\| = 1$, for each $i, j = 1, 2$.

17. With this inner product, since $1 = 1 + 0x + 0x^2 + 0x^3$ and $x = 0(1) + 1x + 0x^2 + 0x^3$, it is clear that $<1, x> = (1)(0) + (0)(1) + (0)(0) + (0)(0) = 0$. Similarly, $<1, x^2> = <1, x^3> = 0$ and $<1, 1> = 1$. In the same manner, $<x, x^2> = <x, x^3> = 0$ and $<x, x> = 1$; also $<x^2, x^3> = 0$ and $<x^2, x^2> = 1 = <x^3, x^3>$.

19. No. Note that $<x^2, 1> = \int_{-1}^{1} x^2 \, dx = \frac{2}{3} \neq 0$. They are not orthogonal under the integral inner product.

21. Use the integral inner product $<f, g> = \int_0^1 f(x) \, g(x) \, dx$. The orthonormal basis is
$$\left\{\sqrt{\frac{3}{7}}(1 + x), \frac{1}{\sqrt{7}}(-9x + 5), 6\sqrt{5}\left(x^2 - x + \frac{1}{6}\right)\right\}.$$

23. Suppose that there are scalars α and β such that $\alpha\mathbf{u} + \beta\mathbf{v} = \mathbf{0}$. We are given that $\mathbf{u}$ and $\mathbf{v}$ are nonzero orthogonal vectors. Then $0 = <\mathbf{0}, \mathbf{u}> = <\alpha\mathbf{u} + \beta\mathbf{v}, \mathbf{u}> = \alpha<\mathbf{u}, \mathbf{u}> + \beta<\mathbf{v}, \mathbf{u}> = \alpha(\|\mathbf{u}\|^2) + \beta(0)$, so $\alpha = 0$. In a similar way, $0 = <\alpha\mathbf{u} + \beta\mathbf{v}, \mathbf{v}> = \alpha<\mathbf{u}, \mathbf{v}> + \beta<\mathbf{v}, \mathbf{v}> = \alpha(0) + \beta(\|\mathbf{v}\|^2)$ forces $\beta = 0$. Hence, $\{\mathbf{u}, \mathbf{v}\}$ is linearly independent.

25. Let $\mathbf{u} = \alpha_1\mathbf{w}_1 + \alpha_2\mathbf{w}_2 + \cdots + \alpha_n\mathbf{w}_n$. Then for each $i = 1, 2, \ldots, n$ we have the equation $<\mathbf{u}, \mathbf{w}_i> = <\alpha_1\mathbf{w}_1 + \alpha_2\mathbf{w}_2 + \cdots + \alpha_n\mathbf{w}_n, \mathbf{w}_i> = \alpha_1<\mathbf{w}_1, \mathbf{w}_i> + \alpha_2<\mathbf{w}_2, \mathbf{w}_i> + \cdots + \alpha_n<\mathbf{w}_n, \mathbf{w}_i> = \alpha_i<\mathbf{w}_i, \mathbf{w}_i> = \alpha_i$, because $\{\mathbf{w}_1, \mathbf{w}_2, \ldots, \mathbf{w}_n\}$ is an *orthonormal* basis. Thus $\alpha_i = <\mathbf{u}, \mathbf{w}_i>$ and $\mathbf{u} = \sum_{i=1}^{n} <\mathbf{u}, \mathbf{w}_i>\mathbf{w}_i$.

27. The desired orthonormal basis is $B = \left\{\frac{1}{\sqrt{\pi}} \sin x, \frac{1}{\sqrt{\pi}} \cos x\right\}$. The two functions $\sin x$ and $\cos x$ were previously shown to be orthogonal.

29. Remember that $\mathbf{u} = \mathbf{w}_1 + \mathbf{w}_2$ is the vector sum of its projection onto and orthogonal to W. Also $\mathbf{w}_1$ and $\mathbf{w}_2$ are orthogonal, so $<\mathbf{w}_1, \mathbf{w}_2> = <\mathbf{w}_2, \mathbf{w}_1> = 0$. So we have $<\mathbf{u}, \mathbf{u}> = <\mathbf{w}_1 + \mathbf{w}_2, \mathbf{w}_1 + \mathbf{w}_2> = <\mathbf{w}_1, \mathbf{w}_1> + <\mathbf{w}_1, \mathbf{w}_2> + <\mathbf{w}_2, \mathbf{w}_1> + <\mathbf{w}_2, \mathbf{w}_2> = <\mathbf{w}_1, \mathbf{w}_1> + <\mathbf{w}_2, \mathbf{w}_2>$. Therefore, $\|\mathbf{u}\|^2 = \|\mathbf{w}_1\|^2 + \|\mathbf{w}_2\|^2$.

31. The modified formulas for the alternative method are as follows:

$$(4.42') \quad \mathbf{v}_1 = \mathbf{v}, \quad \mathbf{v}_2 = \mathbf{u} - \frac{<\mathbf{u}, \mathbf{v}_1>}{<\mathbf{v}_1, \mathbf{v}_1>}\mathbf{v}_1; \quad \mathbf{w}_1 = \frac{\mathbf{v}_1}{\|\mathbf{v}_1\|}, \quad \mathbf{w}_2 = \frac{\mathbf{v}_2}{\|\mathbf{v}_2\|}.$$

$$(4.43') \quad \mathbf{v}_3 = \mathbf{u}' - \frac{<\mathbf{u}', \mathbf{v}_1>}{<\mathbf{v}_1, \mathbf{v}_1>}\mathbf{v}_1 - \frac{<\mathbf{u}', \mathbf{v}_2>}{<\mathbf{v}_2, \mathbf{v}_2>}\mathbf{v}_2; \quad \mathbf{w}_3 = \frac{\mathbf{v}_3}{\|\mathbf{v}_3\|}.$$

Section 4.6 (pp. 278–279)

1. $M^* = \begin{bmatrix} 1 & 2i \\ -2i & 0 \end{bmatrix}$.   **3.** $M^* = \begin{bmatrix} 2 + i & i \\ 2 & 1 - i \end{bmatrix}$.   **5.** $M^* = \begin{bmatrix} 1 & 0 & 1 \\ 2 & 3 & 2 \\ 4 & 5 & 3 \end{bmatrix}$.

7. $<\mathbf{u}, \mathbf{v}> = \mathbf{u}\,\mathbf{v}^* = 7$.   **9.** $<\mathbf{u}, \mathbf{v}> = \mathbf{u}\,\mathbf{v}^* = 7$.   **11.** $<\mathbf{u}, \mathbf{v}> = \mathbf{u}\,\mathbf{v}^* = 9i$.   **13.** $<\mathbf{u}, \mathbf{v}> = \mathbf{u}\,\mathbf{v}^* = 9 + 8i$.

15. $<\mathbf{u}, \mathbf{u}>_2$ is an inner product if, and only if, each of λ_1, λ_2, and λ_3 is a positive real number. See Exercise 14 for the case $\lambda_1 = \lambda_2 = 1, \lambda_3 = i$.

17. No. $<\alpha\mathbf{u}, \mathbf{v}>_4 \neq \alpha<\mathbf{u}, \mathbf{v}>_4$ in general. In the case $\alpha = -1$, for example, $<-\mathbf{u}, \mathbf{v}>_4 = <\mathbf{u}, \mathbf{v}>_4$ rather than $-<\mathbf{u}, \mathbf{v}>_4$.

19. Yes. Let $\mathbf{u} = (u_1, u_2, u_3)$ and $\mathbf{v} = (v_1, v_2, v_3)$, each component a complex number. Then $<\mathbf{u}, \mathbf{v}>_6 = u_1\overline{v_1} + 3u_2\overline{v_2} + 5u_3\overline{v_3} = \overline{\overline{u_1}v_1 + 3\overline{u_2}v_2 + 5\overline{u_3}v_3} = \overline{<\mathbf{v}, \mathbf{u}>_6}$. It follows that $<\alpha\mathbf{u} + \beta\mathbf{v}, \mathbf{w}>_6 = (\alpha\mathbf{u} + \beta\mathbf{v})A\mathbf{w}^* = \alpha\mathbf{u}A\mathbf{w}^* + \beta\mathbf{v}A\mathbf{w}^* = \alpha<\mathbf{u}, \mathbf{w}>_6 + \beta<\mathbf{v}, \mathbf{w}>_6$. And, clearly, $<\mathbf{u}, \mathbf{u}>_6 = \mathbf{u}A\mathbf{u}^* \geq 0$, with equality if, and only if, $\mathbf{u} = \mathbf{0}$.

21. $<\mathbf{u}, \mathbf{v}>_6 = (1 + 3i, 1 - 2i, 2 - i)\begin{bmatrix} 1 & 0 & 0 \\ 0 & 3 & 0 \\ 0 & 0 & 5 \end{bmatrix}\begin{bmatrix} 2 - i \\ 1 - 3i \\ 2 + 5i \end{bmatrix} = 35 + 30i.$

23. $<\mathbf{u}, \mathbf{v}>_2 = 2(1 + 3i)\overline{(2 + i)} + (-2)(1 - 2i)\overline{(1 + 3i)} + 2i(2 - i)\overline{(2 - 5i)} = 4 + 38i.$ Note that $<\mathbf{u}, \mathbf{v}>_2$ is not an inner product, as seen in Exercises 15 and 22, since not each of λ_1, λ_2, and λ_3 is a positive real number.

25. See Exercise 8. $\|\mathbf{u}\| = \sqrt{5}$. $\|\mathbf{v}\| = \sqrt{17}$.   **27.** See Exercise 10. $\|\mathbf{u}\| = \sqrt{6}$. $\|\mathbf{v}\| = \sqrt{19}$.

29. See Exercise 12. $\|\mathbf{u}\| = 2\sqrt{2}$. $\|\mathbf{v}\| = 5$.

31. See Exercise 12. $\|\mathbf{u}\|^2 = <\mathbf{u}, \mathbf{u}>_6 = \mathbf{u}A\mathbf{u}^* = 22$, so $\|\mathbf{u}\| = \sqrt{22}$.   **33.** $<\mathbf{u}, \mathbf{v}> = (1, i, -i)\begin{bmatrix} 0 \\ 1 \\ 1 \end{bmatrix} = 0.$

35. This is false unless each of the scalars $\lambda_1, \lambda_2, \ldots, \lambda_n$ is a positive real number. See Exercises 15, 22, and 23. Given that each λ_i is a positive real number, we see that

$$<\mathbf{u}, \mathbf{v}>_\lambda = \lambda_1 u_1 \overline{v_1} + \lambda_2 u_2 \overline{v_2} + \cdots + \lambda_n u_n \overline{v_n}$$

$= \overline{(\lambda_1 u_1 v_1 + \lambda_2 u_2 v_2 + \cdots + \lambda_n u_n v_n)} = (\lambda_1 \overline{u_1} v_1 + \lambda_2 \overline{u_2} v_2 + \cdots + \lambda_n \overline{u_n} v_n) = \overline{<\mathbf{v}, \mathbf{u}>_\lambda}$ (Property 1). Now (Property 2) $<\alpha\mathbf{u} + \beta\mathbf{v}, \mathbf{w}>_\lambda = \lambda_1(\alpha u_1 + \beta v_1)\overline{w_1} + \lambda_2(\alpha u_2 + \beta v_2)\overline{w_2} + \cdots + \lambda_n(\alpha u_n + \beta v_n)\overline{w_n} = \alpha\lambda_1 u_1\overline{w_1} + \beta\lambda_1 v_1\overline{w_1} + \alpha\lambda_2 u_2\overline{w_2} + \beta\lambda_2 v_2\overline{w_2} + \cdots + \alpha\lambda_n u_n\overline{w_n} + \beta\lambda_n v_n \overline{w_n} = \alpha<\mathbf{u}, \mathbf{w}>_\lambda + \beta<\mathbf{v}, \mathbf{w}>_\lambda$. And for Properties (3) and (4) note that $<\mathbf{u}, \mathbf{u}>_\lambda = \lambda_1 u_1\overline{u_1} + \lambda_2 u_2\overline{u_2} + \cdots + \lambda_n u_n\overline{u_n} = \lambda_1|u_1|^2 + \lambda_2|u_2|^2 + \cdots + \lambda_n|u_n|^2 \geq 0$ with equality if, and only if, each complex number $u_i = 0$, hence if, and only if, $\mathbf{u} = \mathbf{0}$. See Exercise 47.

37. Let $A = (a_{ij})$. The (i, j)th entry (entry in row i and column j) of A^T, you will recall, is a_{ji}. Therefore, the (i, j)th entry in $\overline{A^T}$ is $\overline{a_{ji}}$. The (i, j)th entry in $\overline{A}$ is $\overline{a_{ij}}$, so the (i, j)th entry in $(\overline{A})^T$ is $\overline{a_{ji}}$. Thus, the two matrices $\overline{A^T}$ and $(\overline{A})^T$ are equal.

39. We already have the fact that $(AB)^T = B^TA^T$. Therefore, $(AB)^* = \overline{(AB)^T} = \overline{B^TA^T} = \overline{B^T}\,\overline{A^T} = B^*A^*.$

41. $<\mathbf{u}, \alpha\mathbf{v} + \beta\mathbf{w}> = \overline{<\alpha\mathbf{v} + \beta\mathbf{w}, \mathbf{u}>} = \overline{\alpha <\mathbf{v}, \mathbf{u}> + \beta<\mathbf{w}, \mathbf{u}>} = \overline{\alpha} \, \overline{<\mathbf{v}, \mathbf{u}>} + \overline{\beta} \, \overline{<\mathbf{w}, \mathbf{u}>} = \overline{\alpha}<\mathbf{u}, \mathbf{v}> + \overline{\beta}<\mathbf{u}, \mathbf{w}>.$

43. $<\mathbf{u}, <\mathbf{u}, \mathbf{v}>\mathbf{v}> = <\mathbf{u}, \mathbf{v}> \, \overline{<\mathbf{u}, \mathbf{v}>} = |<\mathbf{u}, \mathbf{v}>|^2.$

45. Suppose that $\mathbf{v} = \mathbf{u} + \mathbf{w}$ and $\mathbf{v} = \mathbf{u}' + \mathbf{w}'$ where $\mathbf{u}$ and $\mathbf{u}'$ are in U and $\mathbf{w}$, $\mathbf{w}' \in W$. Then since $\mathbf{u} + \mathbf{w} = \mathbf{u}' + \mathbf{w}'$, $\mathbf{u} - \mathbf{u}' = \mathbf{w}' - \mathbf{w}$. But since $\mathbf{u} - \mathbf{u}' \in U$ and $\mathbf{w}' - \mathbf{w} \in W$, and $U \cap W = \{\mathbf{0}\}$ so $\mathbf{u} - \mathbf{u}' = \mathbf{w}' - \mathbf{w} = \mathbf{0}$. This implies that $\mathbf{u} = \mathbf{u}'$ and $\mathbf{w}' = \mathbf{w}$, so the representation of $\mathbf{v}$ is indeed unique.

47. Note that all of the entries of D are *real* numbers, so $D = \overline{D} = D^*$. Therefore, because it is 1×1, $<\mathbf{u}, \mathbf{v}> = \mathbf{u} D\mathbf{v}^* = \overline{\mathbf{v} D\mathbf{u}^*} = \overline{<\mathbf{v}, \mathbf{u}>}$. Note also that $<\alpha\mathbf{u} + \beta\mathbf{v}, \mathbf{w}> = (\alpha\mathbf{u} + \beta\mathbf{v}) D\mathbf{w}^* = \alpha\mathbf{u} D\mathbf{w}^* + \beta\mathbf{v} D\mathbf{w}^* = \alpha<\mathbf{u}, \mathbf{w}> + \beta<\mathbf{v}, \mathbf{w}>$. Since each d_{ii} is a positive real number $<\mathbf{u}, \mathbf{u}> = \mathbf{u} D\mathbf{u}^* = d_{11}u_1 \overline{u_1} + d_{22}u_2 \overline{u_2} + \cdots + d_{nn}u_n \overline{u_n} = d_{11}|u_1|^2 + d_{22}|u_2|^2 + \cdots + d_{nn}|u_n|^2 \geq 0$ with equality if, and only if, each complex number $u_i = 0$, hence if, and only if, $\mathbf{u} = \mathbf{0}$.

49. $<\mathbf{u}, \mathbf{v}> = (1, 1 + i, i)$ diag $[2, 4, 6](0, -i, 2 - i)^T = 10 + 8i.$

51. $<\mathbf{u}, \mathbf{v}> = (1, 1 + i, i)$ diag $[3, 1, 5](0, -i, 2 - i)^T = 6 + 9i.$

53. If $\mathbf{u} = (0, 1, 0)$, then $<\mathbf{u}, \mathbf{u}> = ((0, 1, 0)$ diag $[1, 0, 6](0, 1, 0)^T) = 0$, but $\mathbf{u} \neq \mathbf{0}.$

Section 4.7 (p. 288)

1. $Y = MS = \begin{bmatrix} 1 \\ 2 \\ 4 \end{bmatrix} = \begin{bmatrix} 1 & 1 \\ 2 & 1 \\ 3 & 1 \end{bmatrix}\begin{bmatrix} a \\ b \end{bmatrix}$, so $M^TM = \begin{bmatrix} 14 & 6 \\ 6 & 3 \end{bmatrix}$ and $(M^TM)^{-1} = \begin{bmatrix} \frac{1}{2} & -1 \\ -1 & \frac{7}{3} \end{bmatrix}$. Thus, $S' = (M^TM)^{-1}M^TY =$

$\begin{bmatrix} \frac{1}{2} & -1 \\ -1 & \frac{7}{3} \end{bmatrix}\begin{bmatrix} 17 \\ 7 \end{bmatrix} = \begin{bmatrix} \frac{3}{2} \\ \frac{-2}{3} \end{bmatrix}$. The equation of the line is $y = 1.50x - 0.667.$

3. $y = 1.3x + 1.1.$ **5.** $y = 0.86x - 1.48.$ **7.** $y = 2.6x - 0.9.$ **9.** $y = -1.3571x^2 - 0.5x + 3.3143.$

11. $y = 0.10x^2 + 0.46x - 1.28$.   **13.** $y = 0.0176x^3 - 0.2151x^2 + 1.6673x - 3.8333$.

15. Let $K = [\mathbf{k}_1: \mathbf{k}_2: \ldots : \mathbf{k}_n]$ where the $\mathbf{k}_i$ denote the columns of the matrix K. Then $c_1\mathbf{k}_1 + c_2\mathbf{k}_2 + \cdots + c_n\mathbf{k}_n = \mathbf{0}$ if, and

only if, $(c_1, c_2, \ldots, c_n)K = O$. That is, if and only if, $\sum\limits_{j=1}^{n} c_j < \mathbf{x}_i, \mathbf{x}_j > = 0$, for $i = 1, 2, \ldots, n$.

17. If the $m \times n$ matrix A has linearly independent columns, then there exists an $n \times m$ matrix M such that $MA = I_n$. So $MAA^TM^T = I_n$ and AA^T has rank n. Since AA^T and its transpose A^TA have the same rank, it follows that the $n \times n$ matrix $K = A^TA$ is nonsingular. This also follows from Exercises 15 and 16.

19. This requires a proof.

21. $y = 0.37x + 3.55$, or $y = -0.1071x^2 + 1.0129x + 2.800$, or $y = -0.075x^3 + 0.5679x^2 - 0.7571x + 4.0600$.

23.

t	$e^{-\mu_1 t}$	$e^{-\mu_2 t}$	ρ
0	1	1	12.2
1	0.9920	0.9940	11.5
2	0.9841	0.9881	10.7
3	0.9763	0.9822	9.8
4	0.9685	0.9763	8.6
5	0.9608	0.9704	6.8
6	0.9531	0.9646	4.9

The least squares fit is given by $y = 582.5050x_1 - 569.6673$.

Chapter 4 Review Exercises (p. 289)

1. 1.   **2.** −4.

3. $\|\mathbf{u}\| = \sqrt{13}$ and $\|\mathbf{v}\| = \sqrt{26}$, $<\mathbf{u}, \mathbf{v}> = -17$. Note that $17 = |-17| \le \sqrt{13}\sqrt{26}$, verifying the Cauchy-Schwarz inequality.

4. $\mathbf{u} + \mathbf{v} = (1, 2)$, $\|\mathbf{u} + \mathbf{v}\| = \sqrt{5}$. Note that $\sqrt{5} \le \sqrt{13} + \sqrt{26}$, which verifies the triangle inequality in this example.

5. $\mathbf{u}_1 = (1, 1, 0)$, $\mathbf{u}_2 = (0, -1, 1)$, $\mathbf{u}_3 = (1, 0, -1)$, $\mathbf{v}_1 = \dfrac{\mathbf{u}_1}{\|\mathbf{u}_1\|} = \dfrac{1}{\sqrt{2}}(1, 1, 0) = \left(\dfrac{1}{\sqrt{2}}, \dfrac{1}{\sqrt{2}}, 0\right)$,

$\mathbf{v}_2 = \dfrac{\mathbf{u}_2 - <\mathbf{u}_2, \mathbf{v}_1>\mathbf{v}_1}{\|\mathbf{u}_2 - <\mathbf{u}_2, \mathbf{v}_1>\mathbf{v}_1\|} = \dfrac{\sqrt{2}}{\sqrt{3}}\left(\dfrac{1}{2}, \dfrac{-1}{2}, 1\right) = \left(\dfrac{1}{\sqrt{6}}, \dfrac{-1}{\sqrt{6}}, \dfrac{2}{\sqrt{6}}\right)$,

$\mathbf{v}_3 = \dfrac{\mathbf{u}_3 - <\mathbf{u}_3, \mathbf{v}_1>\mathbf{v}_1 - <\mathbf{u}_3, \mathbf{v}_2>\mathbf{v}_2}{\|\mathbf{u}_3 - <\mathbf{u}_3, \mathbf{v}_1>\mathbf{v}_1 - <\mathbf{u}_3, \mathbf{v}_2>\mathbf{v}_2\|} = \dfrac{\sqrt{3}}{3}(1, -1, -1) = \left(\dfrac{1}{\sqrt{3}}, \dfrac{-1}{\sqrt{3}}, \dfrac{-1}{\sqrt{3}}\right)$.

6. $A^* = \begin{bmatrix} 1 - i & 5 \\ 3 + i & 7 + 2i \end{bmatrix}$.   **7.** 0.

8. $\|\mathbf{u}\| = \sqrt{2}$ and $\|\mathbf{v}\| = \sqrt{2}$, $\mathbf{u} + \mathbf{v} = (1 + i, 1 + i)$. Note that $\|\mathbf{u} + \mathbf{v}\| = \sqrt{4} = 2$, and thus we have $\|\mathbf{u} + \mathbf{v}\| \le \sqrt{2} + \sqrt{2} = \|\mathbf{u}\| + \|\mathbf{v}\|$.

9. No. If $\mathbf{u} = \mathbf{v} = (0, 1)$, then $<\mathbf{u}, \mathbf{u}> = 0$, but $\mathbf{u} \ne \mathbf{0}$.

10. $<p, q> = \displaystyle\int_0^1 p(x)q(x)\, dx = \int_0^1 (1 + x^2)(-9 + 16x)\, dx = [-9x + 8x^2 - 3x^3 + 4x^4]_0^1 = -9 + 8 - 3 + 4 = 0$.

Hence, $p(x)$ and $q(x)$ are orthogonal in P_2 with respect to this inner product.

Chapter 5

Section 5.1 (pp. 306–308)

1. T is a linear transformation.

3. Let $\mathbf{x} = (x_1, x_2)$ and $\mathbf{y} = (y_1, y_2)$. Note that $T(\mathbf{x} + \mathbf{y}) = (x_1 + y_1, 1 - x_2 - y_2)$ is not the same as $T(\mathbf{x}) + T(\mathbf{y}) = (x_1, 1 - x_2) + (y_1, 1 - y_2) = (x_1 + y_1, 2 - x_2 - y_2)$. So T is *not* a linear transformation.

5. T is a linear transformation.   **7.** T is *not* a linear transformation, since $T(0, 0) = (1, -1) \ne \mathbf{0}$.

9. $T(\alpha\mathbf{x} + \beta\mathbf{y}) = (2\alpha x_1 + 2\beta y_1, 0) = \alpha T(\mathbf{x}) + \beta T(\mathbf{y})$. So T is a linear transformation.

11. T is a linear transformation.

13. T is *not* linear. For example, $T(\mathbf{x} + \mathbf{y}) = (x_1 + y_1, 1 - x_2 - y_2, x_3 + y_3)$, whereas $T(\mathbf{x}) + T(\mathbf{y}) = (x_1 + y_1, 2 - x_2 - y_2, x_3 + y_3)$.

15. T is a linear transformation.

17. $T(\alpha\mathbf{x} + \beta\mathbf{y}) = (2(\alpha a_1 + \beta a_2), 2(\alpha b_1 + \beta b_2)) = \alpha T(\mathbf{x}) + \beta T(\mathbf{y})$. So T is a linear transformation.

19. Clearly $I(\alpha\mathbf{x} + \beta\mathbf{y}) = \alpha\mathbf{x} + \beta\mathbf{y} = \alpha I(\mathbf{x}) + \beta I(\mathbf{y})$. So I is a linear transformation.   **21.** T is a linear transformation.

23. If T is linear, then $T(x\mathbf{u}) = xT(\mathbf{u})$ for any $\mathbf{u}$ in the domain space. Since in this case that is $\mathbb{R}^1$ this is true for $\mathbf{u} = 1$. So $T(x) = T(x1) = xT(1) = xm = mx$, where $m = T(1)$.

25. $A\mathbf{x} = \begin{bmatrix} -7 \\ 21 \\ 14 \end{bmatrix}$.   **27.** $\mathbf{x}A = (-4, 1, 17, 2)$.   **29.** $(1, -1, 5)\begin{bmatrix} 1 & -1 & 2 \\ 0 & -1 & -2 \\ 2 & -1 & 0 \end{bmatrix} = (11, -5, 4)$.

31. $K_A(\mathbf{x} + \mathbf{y}) = A(\mathbf{x} + \mathbf{y}) - (\mathbf{x} + \mathbf{y})A = A\mathbf{x} + A\mathbf{y} - \mathbf{x}A - \mathbf{y}A = A\mathbf{x} - \mathbf{x}A + A\mathbf{y} - \mathbf{y}A = K_A(\mathbf{x}) + K_A(\mathbf{y})$. Also $K_A(\alpha\mathbf{x}) = A(\alpha\mathbf{x}) - (\alpha\mathbf{x})A = A\alpha(\mathbf{x}) - \alpha(\mathbf{x}A) = \alpha(A\mathbf{x}) - \alpha(\mathbf{x}A) = \alpha(A\mathbf{x} - \mathbf{x}A) = \alpha K_A(\mathbf{x})$. So K is a linear transformation.

33. $A\mathbf{x} - \mathbf{x}A = \begin{bmatrix} -8 & -2 & 6 \\ 4 & -1 & -3 \\ -5 & -3 & 9 \end{bmatrix}$.   **35.** $\begin{bmatrix} 1 \\ -1 \\ -i \end{bmatrix}$.

37. Note $m(\alpha\mathbf{x}) + \mathbf{b} = (max_1 + b_1, max_2 + b_2)$, whereas $\alpha(m(\mathbf{x}) + \mathbf{b}) = (max_1 + \alpha b_1, max_2 + \alpha b_2)$. These are not the same unless $\mathbf{b} = \mathbf{0}$, in which case the transformation is linear.

39. $\begin{bmatrix} -2i & -1 - i & -1 \\ -2 + i & 0 & -1 - i \\ -1 - i & -1 & 2i \end{bmatrix}$.

41. Let $B = \{\mathbf{u}_1, \mathbf{u}_2, \ldots, \mathbf{u}_n\}$ be a basis for U. Then for any $\mathbf{x}$ in U, $\mathbf{x} = c_1\mathbf{u}_1 + c_2\mathbf{u}_2 + \cdots + c_n\mathbf{u}_n$, so we have the following equation.

$$(*) \qquad T(\mathbf{x}) = c_1 T(\mathbf{u}_1) + c_2 T(\mathbf{u}_2) + \cdots + c_n T(\mathbf{u}_n).$$

Now since $T(\mathbf{u}_i) = \mathbf{u}_i$ for each $i = 1, 2, \ldots, n$ we have $T(\mathbf{x}) = c_1\mathbf{u}_1 + c_2\mathbf{u}_2 + \cdots + c_n\mathbf{u}_n = \mathbf{x}$, for every $\mathbf{x} \in U$. So T is the identity transformation I.

43. No, it only spans the range of T. There may be some $\mathbf{v}$ in V such that $\mathbf{v} \neq T(\mathbf{x})$ for any $\mathbf{x} \in U$.

45. It may not be linearly independent, so it might not be a basis. For example, the transformation $T(x_1, x_2) = (2x_1, 0)$ is such that when operating on the standard basis gives $T(\mathbf{e}_1) = (2, 0) = 2\mathbf{e}_1$, and $T(\mathbf{e}_2) = \mathbf{0}$. The set $T(E) = \{T(\mathbf{e}_1), T(\mathbf{e}_2)\} = \{2\mathbf{e}_1, \mathbf{0}\}$ spans the range of T, but is linearly dependent. It is not a basis.

47. $(D^3 + 2D^2 - 3D)(f + g) = (D^3 + 2D^2 - 3D)(f) + (D^3 + 2D^2 - 3D)(g)$ and $(D^3 + 2D^2 - 3D)(\alpha f) = \alpha(D^3 + 2D^2 - 3D)(f)$ by well-known calculus theorems.

49. $D(\sin x) = \cos x$, $D^2(\sin x) = -\sin x$, and $D^3(\sin x) = -\cos x$. Thus, $T(\sin x) = (D^3 + 2D^2 - 3D)(\sin x) = -\cos x - 2\sin x - 3\cos x = -4\cos x - 2\sin x$.

51. From calculus, $L(\alpha f + \beta g) = \int_a^b (\alpha f + \beta g)(x)\, dx = \alpha \int_a^b f(x)\, dx + \beta \int_a^b g(x)\, dx = \alpha L(f) + \beta L(g)$.

53. $L_1(\sin x) = \int_0^1 \sin x\, dx = [-\cos x]_0^1 = -\cos(1) + \cos(0) = 1 - \cos(1)$.

55. $L_1(e^x) = \int_0^1 e^x\, dx = [e^x]_0^1 = e - e^0 = e - 1$.

Section 5.2 (pp. 320–321)

1. $T_A\begin{bmatrix} 12 \\ 4 \end{bmatrix} = A\begin{bmatrix} 12 \\ 4 \end{bmatrix} = \begin{bmatrix} 0 \\ 0 \end{bmatrix}$. So $\begin{bmatrix} 12 \\ 4 \end{bmatrix} \in \ker T_A$.

3. The augmented matrix $A = \begin{bmatrix} -1 & 3 & | & 2 \\ 2 & -6 & | & 5 \end{bmatrix}$ row reduces to $U = \begin{bmatrix} 1 & -3 & | & 0 \\ 0 & 0 & | & 1 \end{bmatrix}$. Hence $T_A(\mathbf{x}) = \begin{bmatrix} 2 \\ 5 \end{bmatrix}$ has no solutions.

Thus $\begin{bmatrix} 2 \\ 5 \end{bmatrix}$ is *not* in the range of T_A.

5. The range of T_A is the column space of A, by Theorem 3.14. That is, $\mathbf{v} \in$ range T_A if, and only if, there exists $\mathbf{x} \in \mathbb{R}^2$ so that $T_A(\mathbf{x}) = \mathbf{v}$. That is, $A\mathbf{x} = \mathbf{v}$. So a basis for the range of T_A is a basis for the column space of the matrix A; namely $\left\{ \begin{bmatrix} -1 \\ 2 \end{bmatrix} \right\}$.

7. (a, b) is in the range of T if, and only if, for some (x, y) in the domain of T, $T(x, y) = (a, b) = (2x, -y)$. That is, if, and only if, $b = -y$ and $a = 2x$. Hence for $y = -b$ and $x = \frac{a}{2}$, $T\left(\frac{a}{2}, -b\right) = (a, b)$. Thus, the range of T is $\mathbb{R}^2$, and $E = \{e_1, e_2\}$ is a basis for the range of T.

9. The range of T is $\mathbb{R}^2$ with the standard basis E as basis.

11. All vectors in range T are of the form $xe_1 + ye_2 = x(1, 0) + y(0, 1)$. Therefore, $E = \{e_1, e_2\}$ is a basis.

13. $T: \mathbb{R}^2 \to \mathbb{R}^3$. Let $v = (a, b, c) \in$ range T. Then $(a, b, c) = T(x, y)$ some $x, y, \in \mathbb{R}$, if, and only if, $a = x$, $b = -y$, and $c = x - y$, and hence $c = a + b$. So the vectors in range T all are of the form $(a, b, a + b) = a(1, 0, 1) + b(0, 1, 1)$. A basis for the range of T is $B = \{(1, 0, 1), (0, 1, 1)\}$.

15. For $T: \mathbb{R}^3 \to \mathbb{R}^3$ the identity map $T(x) = x$, we have ker $T = \{0\}$ and nullity of T is 0. Any v in $\mathbb{R}^3$ is in the range of T so range $T = \mathbb{R}^3$ and rank of T is 3. Thus, $3 + 0 = 3 = \dim \mathbb{R}^3$. A basis for the range of T is $\{e_1, e_2, e_3\}$, and a basis for the kernel of T is the empty set.

17. Ker $T = \{0\}$, nullity 0. Range $T = \mathbb{R}^2 =$ codomain; rank $T = 2$. $2 + 0 = 2 = \dim \mathbb{R}^2$. A basis for the range of T is $\{e_1, e_2\}$, and a basis for the kernel of T is the empty set.

19. Range $T = \{(a, b, 0)\}$ with basis $\{e_1, e_2\}$. Ker $T = \{(0, 0, t)$ any $t\}$ with basis $\{(0, 0, 1)\}$. Here $\dim \mathbb{R}^3 = 3 = 2 + 1$.

21. Write $T(x_1, x_2, x_3) = (a, b, c)$. Then, given arbitrary $a, b,$ and c, we seek a vector x in $\mathbb{R}^3$, $x = (x_1, x_2, x_3)$ such that

$$
\begin{aligned}
x_1 - x_2 &= a \\
x_2 + x_3 &= b \\
-x_1 \quad\ + x_3 &= c.
\end{aligned}
$$

and

Since the coefficient matrix for this system, $\begin{bmatrix} 1 & -1 & 0 \\ 0 & 1 & 1 \\ -1 & 0 & 1 \end{bmatrix}$ is nonsingular, the system always has a solution for any a, b, c. So the range of $T = \mathbb{R}^3$. Similarly, $T(x_1, x_2, x_3) = (0, 0, 0)$ results in a system with the same coefficient matrix and hence has only the trivial solution. So ker $T = \{0\}$. $3 = 3 + 0$ confirms Sylvester's law of nullity. A basis for the range of T is $\{e_1, e_2, e_3\}$, and a basis for the kernel of T is the empty set.

23. We wish to know for what u does $Ax = u$, for some x. The answer is when u is in the column space of the matrix A.

A basis for the column space of A is the set of vectors $\left\{ \begin{bmatrix} 1 \\ 0 \\ 0 \end{bmatrix}, \begin{bmatrix} 0 \\ 1 \\ 0 \end{bmatrix}, \begin{bmatrix} 0 \\ 0 \\ 1 \end{bmatrix} \right\} = \{e_1, e_2, e_3\}$. The kernel of T is the set of

vectors for which $T(x) = 0$; that is, the solutions to $Ax = 0$. These are all vectors of the form $\begin{bmatrix} 5 \\ -19 \\ 1 \\ 11 \end{bmatrix}$,

so $\{(5, -19, 1, 11)^T\}$ is a basis for ker T. Note we have $3 + 1 = 4 = \dim \overline{\mathbb{R}^4} = \dim$ domain T.

25. Note that $\dim M_3(\mathbb{R}) = 9$. Note also that $K_A(X) = O$ if, and only if, $AX - XA = O$; or $AX = XA$. Thus, the kernel of K_A is the set of all those matrices in $M_3(\mathbb{R})$ which commute with A. Using techniques of the next section one can find

that $\dim$ ker $K_A = 3$, and has a basis consisting of the matrices $I_3 = \begin{bmatrix} 1 & 0 & 0 \\ 0 & 1 & 0 \\ 0 & 0 & 1 \end{bmatrix}$, $K_1 = \begin{bmatrix} 0 & 3 & -6 \\ 4 & 4 & 0 \\ -4 & 2 & 0 \end{bmatrix}$, and

$K_2 = \begin{bmatrix} -2 & -1 & 2 \\ -4 & -2 & 4 \\ 0 & 0 & 0 \end{bmatrix}$. A basis for the six-dimensional range of K_A consists of the matrices $R_1 = \begin{bmatrix} 2 & 0 & 0 \\ 0 & 0 & 0 \\ 0 & 1 & -2 \end{bmatrix}$,

$R_2 = \begin{bmatrix} 0 & 3 & 0 \\ 0 & 0 & 0 \\ 2 & 2 & 0 \end{bmatrix}$, $R_3 = \begin{bmatrix} 0 & 0 & 3 \\ 0 & 0 & 0 \\ -2 & 1 & 0 \end{bmatrix}$, $R_4 = \begin{bmatrix} 0 & 0 & 0 \\ 2 & 0 & 0 \\ 1 & 0 & 0 \end{bmatrix}$, $R_5 = \begin{bmatrix} 0 & 0 & 0 \\ 0 & 6 & 0 \\ 4 & 1 & -6 \end{bmatrix}$, and $R_6 = \begin{bmatrix} 0 & 0 & 0 \\ 0 & 0 & 6 \\ 2 & -1 & 0 \end{bmatrix}$.

27. *Proof:* Let u and v be two vectors in the range of $T: U \to V$, and let α and β be any scalars. Then there exist vectors x and y in U (the domain of T) such that $u = T(x)$ and $v = T(y)$. Therefore, since T is linear we see that $\alpha u + \beta v = \alpha T(x) + \beta T(y) = T(\alpha x + \beta y)$. Thus, the vector $\alpha u + \beta v$ is in the range of T. Therefore, since it is not empty, the range of T is a subspace of V.

29. Since $D(p(x)) = 0$, the zero polynomial if, and only if, $p(x) = c$, a constant, the polynomial 1 makes up a basis $\{1\}$ for kernel D. Note then that $n + 1 = \dim P_n = (n) + 1 = \dim \text{range } T + \dim \ker T$.

31. The set $T(B) = \{T(\mathbf{u}_1), T(\mathbf{u}_2), \ldots, T(\mathbf{u}_n)\}$ spans the range of T. It is a basis for the range if, and only if, it is linearly independent. So if $\mathbf{x} \in \ker T$, then $\mathbf{x} = c_1\mathbf{u}_1 + c_2\mathbf{u}_2 + \cdots + c_n\mathbf{u}_n$ for some scalars c_i and $\mathbf{0} = T(c_1\mathbf{u}_1 + c_2\mathbf{u}_2 + \cdots + c_n\mathbf{u}_n) = c_1T(\mathbf{u}_1) + c_2T(\mathbf{u}_2) + \cdots + c_nT(\mathbf{u}_n)$. Since $T(B)$ is a basis, linear independence forces the coefficients to all be 0. So $\mathbf{x} = \mathbf{0}$ is the only vector in kernel T. Conversely, if kernel $T = \{\mathbf{0}\}$, then if $c_1T(\mathbf{u}_1) + c_2T(\mathbf{u}_2) + \cdots + c_nT(\mathbf{u}_n) = \mathbf{0}$, the vector $\mathbf{x} = c_1\mathbf{u}_1 + c_2\mathbf{u}_2 + \cdots + c_n\mathbf{u}_n$ lies in the kernel of T, so $\mathbf{x} = \mathbf{0}$. Therefore, each $c_i = 0$ because $B = \{\mathbf{u}_1, \mathbf{u}_2, \ldots, \mathbf{u}_n\}$ is a basis, hence, is linearly independent. So $T(B)$ is independent and therefore a basis for the range of T.

33. $T: U \to U$. First suppose that T is onto U. Then by Theorem 5.5 dim domain $= \dim U = \dim \text{range } T + \dim \ker T$. Since $U = $ range, $\dim \ker T$ is zero. Since $\ker T = \{\mathbf{0}\}$, T is one-to-one. Conversely, if T is one-to-one, $\dim \ker T = 0$, so $\dim \text{range } T = \dim U$ (domain). Thus, the range of T must also be all of U, and T is onto. This statement is not true if $U \neq V$. To see this suppose that $T: \mathbb{R}^2 \to \mathbb{R}^3$ is given by $T(x_1, x_2) = (x_1, x_2, 0)$. Then T is clearly one-to-one, but not onto $\mathbb{R}^3$.

Section 5.3 (pp. 333–335)

1. $(1, -2) = \mathbf{e}_1 - 2\mathbf{e}_2$, so $[(1, -2)]_E = \begin{bmatrix} 1 \\ -2 \end{bmatrix}$.   **3.** $[1 + x + x^3] = \begin{bmatrix} 1 \\ 1 \\ 0 \\ 1 \end{bmatrix}$.

5. $(1, 1 - i, 1 + i) = \mathbf{e}_1 + (1 - i)\mathbf{e}_2 + (1 + i)\mathbf{e}_3$, so $[(1, 1 - i, 1 + i)]_E = \begin{bmatrix} 1 \\ 1 - i \\ 1 + i \end{bmatrix}$.   **7.** $\begin{bmatrix} -i \\ 1 \\ 1 - 2i \\ 3 + 5i \end{bmatrix}$.

9. Note that $K = \begin{bmatrix} -1 & -3 \\ 5 & 8 \end{bmatrix} = -1\begin{bmatrix} 1 & 0 \\ 0 & 0 \end{bmatrix} - 3\begin{bmatrix} 0 & 1 \\ 0 & 0 \end{bmatrix} + 5\begin{bmatrix} 0 & 0 \\ 1 & 0 \end{bmatrix} + 8\begin{bmatrix} 0 & 0 \\ 0 & 1 \end{bmatrix}$, so the *matrix representation* $[K]_E$

is $\begin{bmatrix} -1 \\ -3 \\ 5 \\ 8 \end{bmatrix}$.

11. $T(1, 0, 0) = (1, -1, 0)$, $T(0, 1, 0) = (1, 0, 1)$, $T(0, 0, 1) = (0, 1, -1)$. So $[T] = \begin{bmatrix} 1 & 1 & 0 \\ -1 & 0 & 1 \\ 0 & 1 & -1 \end{bmatrix}$.

13. $[T] = \begin{bmatrix} 1 & 1 & 1 \\ -1 & 0 & 0 \\ 0 & 1 & -1 \end{bmatrix}$.   **15.** $[T] = \begin{bmatrix} 1 & 0 & 0 \\ 0 & 1 & 0 \\ 0 & 0 & 0 \end{bmatrix}$.

17. The basis is $S = \{\mathbf{v}_1 = (1, 1, 1), \mathbf{v}_2 = (0, 1, 1), \mathbf{v}_3 = (1, 0, -1)\}$. Since $\mathbf{u} = 1\mathbf{v}_1 + 0\mathbf{v}_2 + 0\mathbf{v}_3$ we have $[\mathbf{u}]_S = \begin{bmatrix} 1 \\ 0 \\ 0 \end{bmatrix}$.

19. The basis is the same as Exercise 17. Using methods of Chapter 1, etc., solve $(1, 2, 3) = c_1\mathbf{v}_1 + c_2\mathbf{v}_2 + c_3\mathbf{v}_3$, so

$$\begin{aligned} c_1 \quad\quad + c_3 &= 1 \\ c_1 + c_2 \quad\quad &= 2 \\ c_1 + c_2 - c_3 &= 3. \end{aligned}$$

We have $c_1 = 2$, $c_2 = 0$, and $c_3 = -1$. Therefore $[(1, 2, 3)]_S = \begin{bmatrix} 2 \\ 0 \\ -1 \end{bmatrix}$.

21. $[\mathbf{u}]_S = \begin{bmatrix} 1 \\ -1 \\ -1 \end{bmatrix}$.   **23.** $[T]_S = I_3 = \begin{bmatrix} 1 & 0 & 0 \\ 0 & 1 & 0 \\ 0 & 0 & 1 \end{bmatrix}$.

25. The basis is the same as Exercise 17. Note that $T(\mathbf{v}_1) = T(\mathbf{v}_2) = T(\mathbf{v}_3) = (0, 0, 0)$. So, in any basis $[T]_S$ is the zero matrix.

27. The basis is the same as Exercise 17. $T(\mathbf{v}_1) = (3, -1, 0) = 4\mathbf{v}_1 - 5\mathbf{v}_2 - 1\mathbf{v}_3$, $T(\mathbf{v}_2) = (2, 0, 0) = 2\mathbf{v}_1 - 2\mathbf{v}_2 + 0\mathbf{v}_3$,

$T(\mathbf{v}_3) = (0, -1, 1) = 2\mathbf{v}_1 - 3\mathbf{v}_2 - 2\mathbf{v}_3$. So $[T]_S = \begin{bmatrix} 4 & 2 & 2 \\ -5 & -2 & -3 \\ -1 & 0 & -2 \end{bmatrix}$.

29. $H(1) = x = 0 + 1x + 0x^2$, $H(x) = x^2 = 0 + 0x + 1x^2$. Therefore, $[H] = \begin{bmatrix} 0 & 0 \\ 1 & 0 \\ 0 & 1 \end{bmatrix}$.

31. $[H(1 + 4x)] = [H][1 + 4x] = \begin{bmatrix} 0 & 0 \\ 1 & 0 \\ 0 & 1 \end{bmatrix}\begin{bmatrix} 1 \\ 4 \end{bmatrix} = \begin{bmatrix} 0 \\ 1 \\ 4 \end{bmatrix}$. So $H(1 + 4x) = x + 4x^2$.

33. $D(1) = 0 = 0 + 0x$, $D(x) = 1 = 1 + 0x$, $D(x^2) = 2x = 0 + 2x$. Therefore, $[D] = \begin{bmatrix} 0 & 1 & 0 \\ 0 & 0 & 2 \end{bmatrix}$.

35. $H(1 - x) = x - x^2 = -1(1 - x) + 0(1 + x) + 1(1 - x^2)$, $H(1 + x) = x + x^2 = 0(1 - x) + 1(1 + x) - 1(1 - x^2)$.

So $[H]_K^G = \begin{bmatrix} -1 & 0 \\ 0 & 1 \\ 1 & -1 \end{bmatrix}$.

37. $[H(1 + 4x)]_K = [H]_K^G[1 + 4x]_G = \begin{bmatrix} -1 & 0 \\ 0 & 1 \\ 1 & -1 \end{bmatrix} \cdot \begin{bmatrix} -\frac{3}{2} \\ \frac{5}{2} \end{bmatrix} = \begin{bmatrix} \frac{3}{2} \\ \frac{5}{2} \\ -4 \end{bmatrix}$. Therefore, $H(1 + 4x) = \frac{3}{2}(1 - x) + \frac{5}{2}(1 + x)$

$- 4(1 - x^2) = 0 + x + 4x^2$, as before.

39. *Step 1* is justified by our discussion in this section. *Step 2:* The column vectors of a matrix in $M_n(\mathbb{R})$ can be considered as vectors in $\mathbb{R}^n$. See Section 3.5. *Step 3:* The rank of a matrix M is equal to the dimension of its column space which is the number of linearly independent columns in M. Therefore, if rank $M = n$, its column vectors $\mathbf{m}_i$ form a linearly independent set. But we have each $m_i = [\mathbf{v}_i]_E$. So the set $\{\mathbf{v}_1, \mathbf{v}_2, \ldots, \mathbf{v}_n\}$ is linearly independent. Since V is n-dimensional, S is a basis by Theorem 3.7.

41. Suppose $\mathbf{u} = c_1\mathbf{u}_1 + c_2\mathbf{u}_2 + \cdots + c_n\mathbf{u}_n$ and also $\mathbf{u} = b_1\mathbf{u}_1 + b_2\mathbf{u}_2 + \cdots + b_n\mathbf{u}_n$. Subtract these two expressions to obtain $\mathbf{0} = \mathbf{u} - \mathbf{u} = (c_1 - b_1)\mathbf{u}_1 + (c_2 - b_2)\mathbf{u}_2 + \cdots + (c_n - b_n)\mathbf{u}_n$. Since the vectors $\mathbf{u}_1, \mathbf{u}_2$, and $\mathbf{u}_n$ are linearly independent we must have zero coefficients in this expression. Therefore, $c_1 - b_1 = c_2 - b_2 = \cdots = c_n - b_n = 0$. So $c_1 = b_1, c_2 = b_2, \ldots, c_n = b_n$.

43. $[T]_E = \begin{bmatrix} 1 & -2 & 0 \\ -1 & 0 & 2 \\ 0 & 2 & -3 \end{bmatrix} = M$. Since M is nonsingular, its columns are linearly independent. They form a basis for

the column space of M. So $\{(1, -1, 0), (-2, 0, 2), (0, 2, -3)\}$ is a basis for the range of T in $\mathbb{R}^3$. So, of course, is the standard basis E also a basis for the range of T. By Theorem 5.5, then, ker $T = \{\mathbf{0}\}$, and it has the null basis.

45. $[T]_E = \begin{bmatrix} 1 & -2 & -2 \\ -1 & 2 & -3 \\ 0 & 0 & 0 \end{bmatrix} = M$ is singular, a basis for its column space will represent a basis for the range of T in $\mathbb{R}^3$.

So, row reduce M^T to obtain the basis $\{\mathbf{e}_1, \mathbf{e}_2\}$ or use the first and third columns of M (which are linearly independent) to obtain the basis $\{(1, -1, 0), (-2, -3, 0)\}$ for the range of T in $\mathbb{R}^3$. A basis for ker T is obtained by solving the homogeneous system $M\mathbf{x} = \mathbf{0}$. Thus, $\{(2, 1, 0)\}$ is such a basis for ker T.

47. T is the identity map, so $E = \{e_1, e_2, e_3\}$ is a basis for the range of T and ker $T = \{0\}$, with the null basis.

49. The column vectors of $[L]$ are linearly independent. Hence they represent the basis vectors for the range of L, namely $\{x, x^2\}$. Since range L has dimension $2 =$ dimension of the domain of L, ker L has dimension 0. So ker $L = \{0\}$, and has the null basis.

51. Given any basis $B = \{v_1, v_2, \ldots, v_n\}$ of the vector space U, $0 = 0v_1 + 0v_2 + \cdots + 0v_n$ is the unique representation of the zero vector. This is because B is a linearly independent set. So $[0]$ is always the zero (column) matrix. Thus,
$$\Phi_B(0) = 0.$$

53. If $x \in$ ker Φ_B, then $\Phi_B(x) = 0$. So $[x]_B = 0$. Hence, $x = 0v_1 + 0v_2 + \cdots + 0v_n = 0$.

55. Use Exercises 53 and 54. Since ker $\Phi_B = \{0\}$, Φ_B is one-to-one. (See Exercise 32 of Section 5.2.) And since range $\Phi_B = \overline{\mathbb{R}^n}$ (or $\overline{\mathbb{C}^n}$), Φ_B is an onto map.

Section 5.4 (pp. 344–346)

1. $T_1 T_2(x_1, x_2) = T_1(-x_1, 4x_1 + x_2) = (7x_1 + 2x_2, -4x_1 - x_2)$, and $T_2 T_1(x_1, x_2) = T_2(x_1 + 2x_2, -x_2)$
$$= (-x_1 - 2x_2, 4x_1 + 7x_2). \text{ Thus, } L(x_1, x_2) = \frac{1}{2}T_1 T_2(x_1, x_2) - 2T_2 T_1(x_1, x_2)$$
$$= \frac{1}{2}(7x_1 + 2x_2, -4x_1 - x_2) - 2(-x_1 - 2x_2, 4x_1 + 7x_2) = \left(\frac{11}{2}x_1 + 5x_2, -10x_1 - \frac{29}{2}x_2\right).$$

3. $T_2(1, 0) = (-1, 4)$; $T_2(0, 1) = (0, 1)$. So $B = [T_2]_E = \begin{bmatrix} -1 & 0 \\ 4 & 1 \end{bmatrix}$.

5. $G = [T_1 T_2] = \begin{bmatrix} 7 & 2 \\ -4 & -1 \end{bmatrix} = AB.$ **7.** $[T_1^{-1}]_E = \begin{bmatrix} 1 & 2 \\ 0 & -1 \end{bmatrix} = A^{-1}.$

9. $(a, b) \in$ range T_2 if, and only if, $T_2(x_1, x_2) = (a, b)$. Thus if, and only if, $a = -x_1$ and $b = 4x_1 + x_2$. So $x_1 = -a$ and $x_2 = b + 4a$. Therefore, $T_2^{-1}(a, b) = (-a, 4a + b)$.

11. Note that they are the same.

13. $[L] = \begin{bmatrix} \dfrac{11}{2} & 5 \\ -10 & -\dfrac{29}{2} \end{bmatrix} = \frac{1}{2}AB - 2BA = \frac{1}{2}\begin{bmatrix} 7 & 2 \\ -4 & -1 \end{bmatrix} - 2\begin{bmatrix} -1 & -2 \\ 4 & 7 \end{bmatrix}.$

15. $[L] = \begin{bmatrix} \dfrac{11}{2} & 5 \\ -10 & -\dfrac{29}{2} \end{bmatrix}$ is nonsingular, so range $L = \mathbb{R}^2$ with basis $E = \{e_1, e_2\}$.

17. $[T_2]_E = \begin{bmatrix} 1 & 0 & -1 \\ 0 & 1 & -1 \\ 0 & -2 & 1 \end{bmatrix}.$ **19.** $[3T_1 - 2T_2]_E = \begin{bmatrix} -2 & 0 & 5 \\ 3 & -2 & 2 \\ 0 & 4 & 1 \end{bmatrix}.$

21. $[T_2 T_1]_E = \begin{bmatrix} 0 & 0 & 0 \\ 1 & 0 & -1 \\ -2 & 0 & 1 \end{bmatrix}.$ **23.** $[T_2]^{-1} = [T_2^{-1}] = \begin{bmatrix} 1 & -2 & -1 \\ 0 & -1 & -1 \\ 0 & -2 & -1 \end{bmatrix}.$

25. The range of T_1 corresponds to the column space of $[T_1]$, so a basis is $\{(0, 1, 0), (1, 0, 1)\}$.

27. $[T_3]^{-1} = [T_3^{-1}] = \begin{bmatrix} 1 & -\dfrac{1}{2} & -\dfrac{1}{2} \\ 0 & \dfrac{1}{2} & -\dfrac{1}{2} \\ 0 & \dfrac{1}{2} & \dfrac{1}{2} \end{bmatrix}$. Therefore, $T_3^{-1}(x_1, x_2, x_3) = \left(x_1 - \dfrac{x_2}{2} - \dfrac{x_3}{2}, \dfrac{x_2}{2} - \dfrac{x_3}{2}, \dfrac{x_2}{2} + \dfrac{x_3}{2}\right).$

29. $[T] = \begin{bmatrix} 1 & -1 & 1 \\ -2 & 0 & 1 \end{bmatrix}$. T^{-1} does not exist because T is not one-to-one; ker $T = \{t(1, 3, 2)\}$.

31. On range F, $[TF] = \begin{bmatrix} 1 & 0 \\ 0 & 1 \end{bmatrix}$, and $[FT] = \begin{bmatrix} 3 & -1 & 0 \\ 6 & -2 & 0 \\ 4 & -2 & 1 \end{bmatrix}$.

33. $[FT]$ is not I_3. In fact, it is singular. The range of F is a 2-dimensional subspace of $\mathbb{R}^3$, so not every vector in $\mathbb{R}^3$ can be in this range. Range FT has basis $\{(1, 2, 0), (0, 0, 1)\}$. Also note that ker FT is 1-dimensional with basis $\{(1, 3, 2)\}$. Note that $FT(1, 3, 2) = (0, 0, 0)$; so FT is not the identity map on $\mathbb{R}^3$.

35. Given that T and $S \in \mathbb{A}$ are linear operators on V, and B is a basis for V, then $\Omega_B[T] = [T]_B$ and $\Omega_B[S] = [S]_B$ and furthermore, $\Omega_B[S + T] = [S + T]_B$. Now for each basis vector v_i in B we know that $(S + T)(v_i) = S(v_i) + T(v_i)$. So the ith column vector of $[S + T]_B$ is $[S(v_i)]_B + [T(v_i)]_B$; that is, the sum of the ith column vectors of $[S]_B$ and $[T]_B$. This being true for each of the columns we have the desired result $[S + T]_B = [S]_B + [T]_B$.

37. Read Exercise 35 for notation. Let $T \in \mathbb{A}$, so $T: U \to U$. Let k be any scalar, $\Omega_B(kT)$ is the matrix such that for any x in U, $\Omega_B(kT)[x]_B = [kT(x)]_B = [kT]_B[x]_B = k[T]_B[x]_B$. So $\Omega_B(kT) = k[T]_B = k\Omega_B(T)$, as desired.

39. See the proof of Exercise 32 and Exercise 38 in Section 5.2. ker $\Omega_B = \{O\}$ if, and only if, Ω_B is one-to-one.

41. $(T_1 + T_2)(\alpha u + \beta v) = T_1(\alpha u + \beta v) + T_2(\alpha u + \beta v) = T_1(\alpha u) + T_1(\beta v) + T_2(\alpha u) + T_2(\beta v) = \alpha T_1(u) + \beta T_1(v)$
$+ \alpha T_2(u) + \beta T_2(v) = \alpha T_1(u) + \alpha T_2(u) + \beta T_1(v) + \beta T_2(v) = \alpha(T_1(u) + T_2(u)) + \beta(T_1(v) + T_2(v))$
$= \alpha((T_1 + T_2)u) + \beta((T_1 + T_2)v)$.

43. $T_1 T_2(\alpha u + \beta v) = T_1(T_2(\alpha u + \beta v)) = T_1(\alpha T_2(u) + \beta T_2(v)) = \alpha T_1(T_2(u)) + \beta T_1(T_2(v)) = \alpha(T_1 T_2)(u) + \beta(T_1 T_2)(v)$.

45. Note $(S + (T + F))(u) = S(u) + (T + F)(u) = S(u) + (T(u) + F(u)) = (S(u) + T(u)) + F(u)$
$= (S + T)(u) + F(u) = ((S + T) + F)(u)$, for all vectors u in U.

47. $(S + O)(u) = S(u) + O(u) = S(u) + 0 = S(u)$ for all vectors u in U. So $S + O = S$. Also $(IS)(u) = I(S(u)) = S(u)$ for all vectors u in U. Also $(SI)(u) = S(I(u)) = S(u)$. Thus $IS = SI = S$.

49. For all vectors u in U, $(k(T + S))(u) = k(T(u) + S(u)) = kT(u) + kS(u) = (kT + kS)(u)$.

51. For all vectors u in U, $((km)T)(u) = (km)(T(u)) = k(mT(u)) = (k(mT))(u)$.

53. Let $T(x) = y$ in U. If T is one-to-one then one can define the function $L: U \to U$ by $L(y) = x$. Then $L(y) = L(T(x)) = x$ for all x in U. On the other hand, $TL(y) = T(x) = y$ for all y in range L. But range $L = U$ because of the result of Exercise 33 in Section 5.2. So $TL = LT = I$ and $L = T^{-1}$. Conversely, if T^{-1} exists and $T(x) = T(v)$, then $x = T^{-1}(T(x)) = T^{-1}(T(v)) = v$. So T is one-to-one.

55. Combine the results of Exercise 53 and Exercise 33 in Section 5.2. T^{-1} exists if, and only if, T is one-to-one, if, and only if, T is onto U.

57. Suppose $LT = TL = I$ and $MT = TM = I$. Then $L = L(I) = L(TM) = (LT)M = IM = M$.

59. Let B be any basis for U and let $A = [T]_B$. If A^{-1} exists then $A^{-1}[T]_B = I$. So, for any x in U, $[x]_B = I[x]_B = (A^{-1}A)[x]_B = A^{-1}([T]_B[x]_B) = A^{-1}[T(x)]_B$. From this we conclude that $A^{-1} = [T^{-1}]_B$, so T is invertible and T^{-1} is the matrix transformation determined by A^{-1}. Conversely, suppose that T is an invertible transformation. Then $[T^{-1}]_B[T]_B = [T^{-1}]_B A = I$. Thus $A^{-1} = [T^{-1}]_B$ and A is an invertible (nonsingular) matrix.

Section 5.5 (pp. 355–357)

1. Yes, $\underset{A}{\begin{bmatrix} 0 & 1 \\ 0 & 0 \end{bmatrix}} = \underset{S}{\begin{bmatrix} 1 & 1 \\ 1 & 0 \end{bmatrix}} \cdot \underset{B}{\begin{bmatrix} 0 & 0 \\ 1 & 0 \end{bmatrix}} \underset{S^{-1}}{\begin{bmatrix} 0 & 1 \\ 1 & -1 \end{bmatrix}}$.   **3.** Yes, $A = SBS^{-1}$, for $S = \begin{bmatrix} 0 & 1 \\ 1 & 0 \end{bmatrix}$.

5. Yes, $\underset{A}{\begin{bmatrix} 1 & 0 & 0 \\ 0 & -1 & 0 \\ 0 & 0 & -1 \end{bmatrix}} = \underset{S}{\begin{bmatrix} 0 & 0 & 1 \\ 0 & 1 & 0 \\ 1 & 0 & 0 \end{bmatrix}} \cdot \underset{B}{\begin{bmatrix} -1 & 0 & 0 \\ 0 & -1 & 0 \\ 0 & 0 & 1 \end{bmatrix}} \cdot \underset{S^{-1}}{\begin{bmatrix} 0 & 0 & 1 \\ 0 & 1 & 0 \\ 1 & 0 & 0 \end{bmatrix}}$.

7. Let $K = kI$. So if A is similar to K we have $A = SKS^{-1} = SkIS^{-1} = kSIS^{-1} = kSS^{-1} = kI = K$.

9. $A = SBS^{-1}$. So det $A =$ det $(SBS^{-1}) = ($det $S)($det $B)($det $S^{-1}) = ($det $B)($det $S)($det $S^{-1}) = ($det $B)($det $(SS^{-1})) =$ det B.

11. No. See Exercise 9. If A and B are similar, then det $A =$ det B. So if A is singular, then det $A = 0 =$ det B. Thus, B is also singular.

13. One approach is to think of the two similar matrices A and B as two representations of the same linear operator $T: U \to U$ with respect to two different bases for U. Then we have that rank $A =$ dim (range of T) $=$ rank B.

15. The matrices A and B of Example 4 are both row equivalent to I_2, but they are not similar.

17. $[T]_G = \begin{bmatrix} 1 & 1 & 0 \\ -1 & 0 & 1 \\ 0 & 0 & 2 \end{bmatrix}$ is also nonsingular. Hence it has rank = 3 and nullity = 0.

19. $\begin{bmatrix} 0 & -1 & -\frac{1}{2} & -1 \\ 0 & 1 & \frac{3}{2} & 0 \\ 0 & 0 & 0 & 0 \\ 0 & 0 & -1 & 1 \end{bmatrix} \rightarrow \begin{bmatrix} 0 & 1 & 0 & \frac{3}{2} \\ 0 & 0 & 1 & -1 \\ 0 & 0 & 0 & 0 \\ 0 & 0 & 0 & 0 \end{bmatrix}$. Therefore, rank = 2 and nullity = 2.

21. $[T]$ has rank 2 = row rank and nullity = 1. Note that ker $T = \{s(1, 1, 1),$ any $s\}$ and $\{(1, 0, 0), (0, 0, 1)\}$ is a basis for range T.

23. $S^{-1} = \begin{bmatrix} \frac{1}{2} & \frac{-1}{2} & \frac{1}{2} \\ \frac{1}{2} & \frac{-1}{2} & \frac{-1}{2} \\ \frac{1}{2} & \frac{1}{2} & \frac{-1}{2} \end{bmatrix}$. **25.** $[T]_B = \begin{bmatrix} \frac{-1}{2} & -1 & \frac{-1}{2} \\ \frac{-1}{2} & 0 & \frac{1}{2} \\ \frac{-1}{2} & 0 & \frac{1}{2} \end{bmatrix} = \begin{bmatrix} \frac{1}{2} & \frac{-1}{2} & \frac{1}{2} \\ \frac{1}{2} & \frac{-1}{2} & \frac{-1}{2} \\ \frac{1}{2} & \frac{1}{2} & \frac{-1}{2} \end{bmatrix} \begin{bmatrix} -1 & 1 & 0 \\ 0 & 0 & 0 \\ -1 & 0 & 1 \end{bmatrix} \begin{bmatrix} 1 & 0 & 1 \\ 0 & -1 & 1 \\ 1 & -1 & 0 \end{bmatrix}$.

$\qquad\qquad\qquad\qquad\qquad\qquad\qquad\qquad\qquad\qquad\qquad\quad S^{-1} \qquad\qquad\quad [T] \qquad\qquad\quad S$

27. $P^{-1} = \begin{bmatrix} \frac{1}{2} & \frac{1}{4} & \frac{-1}{4} \\ \frac{-1}{2} & \frac{1}{4} & \frac{-1}{4} \\ 0 & \frac{1}{2} & \frac{1}{2} \end{bmatrix}$. **29.** $P[T]_H P^{-1} = \begin{bmatrix} -1 & 1 & 0 \\ 0 & 0 & 0 \\ -1 & 0 & 1 \end{bmatrix} = [T]$.

31. $E = \{1, x, x^2\}$ is the standard basis for P_2. Note $T(1) = x - x^2$, $T(x) = -x^2$, $T(x^2) = -x^2$; so $[T] = \begin{bmatrix} 0 & 0 & 0 \\ 1 & 0 & 0 \\ -1 & -1 & -1 \end{bmatrix}$.

33. $S = \begin{bmatrix} 0 & -1 & 0 \\ 2 & 0 & 1 \\ 0 & 0 & 1 \end{bmatrix}$; det $S = 2$, so S is nonsingular.

35. $[T]_B = S^{-1}[T]S = \begin{bmatrix} 1 & -1 & 1 \\ 0 & 0 & 0 \\ -2 & 1 & -2 \end{bmatrix}$. S and S^{-1} are in Exercises 33 and 34.

37. $R = \begin{bmatrix} 0 & -1 & 0 \\ 0 & 0 & 1 \\ 2 & 1 & -1 \end{bmatrix}$; det $R = -2$, so R is nonsingular. See Exercise 38 for R^{-1}.

39. See the algorithm in Appendix A.3. $\begin{bmatrix} 0 & -1 & 0 & 0 & 0 & 0 \\ 0 & 0 & 1 & 0 & -1 & 0 \\ 2 & 1 & -1 & -2 & 0 & 0 \end{bmatrix} \rightarrow \begin{bmatrix} 1 & 0 & 0 & -1 & \frac{-1}{2} & 0 \\ 0 & 1 & 0 & 0 & 0 & 0 \\ 0 & 0 & 1 & 0 & -1 & 0 \end{bmatrix}$.

So $[T]_H = \begin{bmatrix} -1 & \frac{-1}{2} & 0 \\ 0 & 0 & 0 \\ 0 & -1 & 0 \end{bmatrix}$.

41. $[T]_B = S^{-1}R[T]_H R^{-1}S = K^{-1}[T]_H K = S^{-1}[T]_S$. Here $K = R^{-1}S = \begin{bmatrix} 1 & \frac{-1}{2} & 1 \\ 0 & 1 & 0 \\ 2 & 0 & 1 \end{bmatrix}$.

43. Clearly for any a, b, c, d in $\mathbb{R}$ (or $\mathbb{C}$) the matrix $\begin{bmatrix} a & b \\ c & d \end{bmatrix} = aE_{11} + bE_{12} + cE_{21} + dE_{22}$. We get the zero matrix if, and

only if, $a = b = c = d = 0$. So the set $E = \{E_{11}, E_{12}, E_{21}, E_{22}\}$ is linearly independent and spans $M_2(\mathbb{R})$.

45. $\det [T] = -1$. So range $T = M_2(\mathbb{R})$; rank $T = 4$, and nullity of $T = 0$.

47. $S^{-1} = \begin{bmatrix} \frac{1}{2} & \frac{1}{2} & \frac{1}{2} & \frac{1}{2} \\ \frac{-1}{2} & \frac{1}{2} & \frac{-1}{2} & \frac{1}{2} \\ \frac{-1}{2} & \frac{-1}{2} & \frac{-1}{2} & \frac{1}{2} \\ \frac{-1}{2} & \frac{1}{2} & \frac{1}{2} & \frac{1}{2} \end{bmatrix}$.   **49.** $[T] = S[T]_B S^{-1}$ where $S = \begin{bmatrix} 1 & 0 & 0 & -1 \\ 0 & 1 & -1 & 0 \\ 0 & -1 & 0 & 1 \\ 1 & 0 & 1 & 0 \end{bmatrix}$.

51. If $A = \begin{bmatrix} a & b \\ c & d \end{bmatrix}$, then $A^T = \begin{bmatrix} a & c \\ b & d \end{bmatrix}$, so we have three cases.

Case 1: If $b \neq 0$ and $c \neq 0$, then $A = \begin{bmatrix} 1 & 0 \\ 0 & \frac{c}{b} \end{bmatrix} A^T \begin{bmatrix} 1 & 0 \\ 0 & \frac{b}{c} \end{bmatrix} = SA^T S^{-1}$.

Case 2: If $b = 0$ and $c \neq 0$, then $A = \begin{bmatrix} \frac{a-d}{c} & 1 \\ 1 & 0 \end{bmatrix} A^T \begin{bmatrix} 0 & 1 \\ 1 & \frac{d-a}{c} \end{bmatrix} = SA^T S^{-1}$. Note that the case $c = 0$, $b \neq 0$ is

analogous to this one.

Case 3: If $b = c = 0$, then $A = A^T = IA^T I^{-1}$. Hence A is similar to $A^T (= A)$.

53. If A is diagonal then so is A^n for each positive integer n. In fact $A^n = \text{diag} [(a_{ii}^n)]$. And for each scalar k, $kA^n = \text{diag} [(ka_{ii}^n)]$. Since sums of diagonal matrices are diagonal, the result follows.

55. Clearly if $A = \text{diag} [(a_{ii})]$, then $c_j A^j = \text{diag} [(c_j a_{ii}^j)]$. So $p(A) = \text{diag} [(p(a_{ii})]$. If A is upper triangular it is still true that the diagonal elements of $c_j A^j$ are $c_j a_{ii}^j$, for each j, and the result follows.

Section 5.6 (pp. 363–365)

1. Is not orthogonal. Check the inner product of the columns. Note for example that the inner product of column vectors one and two is $-1/3$, not zero.

3. Is orthogonal.   **5.** Is not orthogonal.   **7.** Is orthogonal   **9.** Is orthogonal.

11. Let $\mathbf{u} = (u_1, u_2, u_3)$, $\mathbf{v} = (v_1, v_2, v_3)$, so that $<\mathbf{u}, \mathbf{v}> = u_1 v_1 + u_2 v_2 + u_3 v_3$. Then compute $<T(\mathbf{u}), T(\mathbf{v})> = u_3 v_3$
$+ \frac{3}{4} u_1 v_1 + \frac{\sqrt{3}}{4} u_1 v_2 + \frac{\sqrt{3}}{4} u_2 v_1 + \frac{1}{4} u_2 v_2 + \frac{1}{4} u_1 v_1 - \frac{\sqrt{3}}{4} u_1 v_2 - \frac{\sqrt{3}}{4} u_2 v_1 + \frac{3}{4} u_2 v_2 = u_3 v_3 + u_1 v_1 + u_2 v_2 = <\mathbf{u}, \mathbf{v}>$. Thus T
is orthogonal.

13. T is not orthogonal. For example, if $\mathbf{u} = (1, 1, 1)$ and $\mathbf{v} = (0, 1, -1)$, then $<\mathbf{u}, \mathbf{v}> = 0$; whereas $T(\mathbf{u}) = \left(\frac{17}{13}, \frac{7}{13}, 1 \right)$
and $T(\mathbf{v}) = \left(\frac{-12}{13}, \frac{17}{13}, 0 \right)$. So $<T(\mathbf{u}), T(\mathbf{v})> = \frac{-85}{169} \neq <\mathbf{u}, \mathbf{v}>$.

15. $\begin{bmatrix} 0 & 0 & 1 \\ \frac{\sqrt{3}}{2} & \frac{1}{2} & 0 \\ \frac{1}{2} & \frac{-\sqrt{3}}{2} & 0 \end{bmatrix}$.   **17.** $\begin{bmatrix} \frac{5}{13} & 0 & \frac{12}{13} \\ 0 & \frac{12}{13} & \frac{-5}{13} \\ 1 & 0 & 0 \end{bmatrix}$.

19. $<1, 1> = \int_0^1 1\, dx = x]_0^1 = 1$. $<q(x), q(x)> = \int_0^1 3(4x^2 - 4x + 1)\, dx = (4x^3 - 6x^2 + 3x)]_0^1 = 1$; and

$<p(x), q(x)> = \int_0^1 \sqrt{3}(2x - 1)\, dx = \sqrt{3}(x^2 - x)]_0^1 = \sqrt{3}(0) = 0$. So, yes it is orthonormal.

21. Yes; $<q_i, q_j> = d_{ij}$.

23. Let $A = \begin{bmatrix} \cos \phi & -\sin \phi \\ \sin \phi & \cos \phi \end{bmatrix}$. Then define the transformation

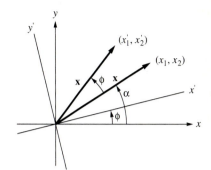

$T_A: \mathbb{R}^2 \to \mathbb{R}^2$ by $T_A(\mathbf{x}) = A[\mathbf{x}] = [\mathbf{x}'] = A\begin{bmatrix} x_1 \\ x_2 \end{bmatrix} =$

$\begin{bmatrix} x_1 \cos \phi - x_2 \sin \phi \\ x_1 \sin \phi + x_2 \cos \phi \end{bmatrix} = \begin{bmatrix} x'_1 \\ x'_2 \end{bmatrix}$ representing $\mathbf{x}' = (x', y')$. That is to say, for each $\mathbf{x}$ in $\mathbb{R}^2$, $T_A(\mathbf{x}) = (x', y')$, where $x' = x_1 \cos \phi - x_2 \sin \phi$ and $y' = x_1 \sin \phi + x_2 \cos \phi$. These are precisely the coordinates of the vector $\mathbf{x}$ rotated through and angle ϕ. See the figure at the right and Example 1 in the text.

25. Refer to the preceding exercises. We see that det $A = 1$ and det $B = -1$. Since these are the only two possible forms for a 2×2 orthogonal matrix P (Exercise 22), we must have P as a rotation if, and only if, it is of the form A; thus, if and only if, det $P = 1$.

27. $PP^T = I$, so det $(PP^T) = (\det P)(\det p^T) = \det I = 1$. But, det $P^T = \det P$, for any matrix P. So here we have $(\det P)^2 = 1$. Hence det $P = 1$ or -1.

29. $<T(\mathbf{u}), T(\mathbf{v})> = <k\mathbf{u}, k\mathbf{v}> = k^2 <\mathbf{u}, \mathbf{v}>$. $\| T(\mathbf{x}) \| = \| k\mathbf{x} \| = <k\mathbf{x}, k\mathbf{x}>^{1/2} = k \| \mathbf{x} \|$, for any $\mathbf{x}$. So $\dfrac{<T(\mathbf{u}), T(\mathbf{v})>}{\| T(\mathbf{u}) \| \, \| T(\mathbf{u}) \|} = \dfrac{k^2 <\mathbf{u}, \mathbf{v}>}{k \| \mathbf{u} \| \, k \| \mathbf{v} \|} = \dfrac{<\mathbf{u}, \mathbf{v}>}{\| \mathbf{u} \| \, \| \mathbf{v} \|}$.

31. By Exercises 29 and 30 we see that T is not orthogonal, since it does not preserve norms.

33. Yes, its matrix representation relative to the standard basis is an orthogonal matrix; or $<L(\mathbf{x}), L(\mathbf{y})> = x_2 y_2 + (-x_1)(-y_1) = x_1 y_1 + x_2 y_2 = <\mathbf{x}, \mathbf{y}>$.

35. $\| T(\mathbf{u} + \mathbf{v}) \|^2 = <T(\mathbf{u} + \mathbf{v}), T(\mathbf{u} + \mathbf{v})> = <T(\mathbf{u}) + T(\mathbf{v}), T(\mathbf{u}) + T(\mathbf{v})> = <T(\mathbf{u}), T(\mathbf{u})> + 2<T(\mathbf{u}), T(\mathbf{v})> + <T(\mathbf{v}), T(\mathbf{v})> = \| T(\mathbf{u}) \|^2 + \| T(\mathbf{v}) \|^2 + 2<T(\mathbf{u}), T(\mathbf{v})>$. Whereas, $\| \mathbf{u} + \mathbf{v} \|^2 = <(\mathbf{u} + \mathbf{v}), (\mathbf{u} + \mathbf{v})> = \| \mathbf{u} \|^2 + \| \mathbf{v} \|^2 + 2<\mathbf{u}, \mathbf{v}>$. Since $\| T(\mathbf{x}) \| = \| \mathbf{x} \|$ for all $\mathbf{x}$, $\| T(\mathbf{u} + \mathbf{v}) \|^2 = \| \mathbf{u} + \mathbf{v} \|^2$ and $\| T(\mathbf{u}) \|^2 + \| T(\mathbf{v}) \|^2 = \| \mathbf{u} \|^2 + \| \mathbf{v} \|^2$. Therefore, $2<T(\mathbf{u}), T(\mathbf{v})> = 2<\mathbf{u}, \mathbf{v}>$. Thus, $<T(\mathbf{u}), T(\mathbf{v})> = <\mathbf{u}, \mathbf{v}>$, and T is orthogonal. $<\mathbf{v}_i, \mathbf{v}_j> = \alpha_{i1} \alpha_{j1} + \cdots + \alpha_{in} \alpha_{jn}$. Therefore, $A = [T]_B$ is an orthogonal matrix if, and only if, $<\mathbf{v}_i, \mathbf{v}_j> = \delta_{ij} = \begin{cases} 0 \text{ if } i \neq j \\ 1 \text{ if } i = j \end{cases}$. Now, $<T(\mathbf{u}_i), T(\mathbf{u}_j)> = <[T]_B[\mathbf{u}_i]_B, [T]_B[\mathbf{u}_j]_B> = <\mathbf{v}_i, \mathbf{v}_j> = <\mathbf{u}_i, \mathbf{u}_j>$ if, and only if, $A = [T]_B$ is an orthogonal matrix.

37. Let T be any linear transformation and let $\mathbf{a}_i$, $i = 1, 2, \ldots, n$, denote the column vectors of the matrix representation $A = [T]_B$ of T with respect to the basis $B = \{\mathbf{u}_1, \mathbf{u}_2, \ldots, \mathbf{u}_n\}$. Then $\mathbf{a}_i = [\mathbf{v}_i]$ for some vector $\mathbf{v}_i = \alpha_{i1}\mathbf{u}_1 + \cdots + \alpha_{in}\mathbf{u}_n$ in U since B is orthonormal.

39. Let $\mathbf{p}_i$ denote the ith column vector of the matrix P. Now $\mathbf{p}_i \in \mathbb{R}^n$, set $\delta_{ij} = \begin{cases} 0 \text{ if } i \neq j \\ 1 \text{ if } i = j \end{cases}$. Then $<\mathbf{p}_i, \mathbf{p}_j> = \delta_{ij}$ if, and only if, the vectors $\mathbf{p}_i$ form an orthonormal set of vectors. But $<\mathbf{p}_i, \mathbf{p}_j> = \delta_{ij}$ if, and only if, $PP^T = P^TP = (\delta_{ij}) = I_n$.

41. No; The transformation $T(\mathbf{x}) = k\mathbf{x}$, with $k \neq 1$, has this property but is not an isometry.

43. $(PQ)(PQ)^T = PQQ^TP^T = PIP^T = PP^T = I$.

45. $AA^* = I_n = (\delta_{ij})$ if, and only if, $<\mathbf{a}_i, \mathbf{a}_j> = \mathbf{a}_i^* \mathbf{a}_j = \delta_{ij}$ (see Exercise 39) for each column vector $\mathbf{a}_i$ of the matrix A. The standard inner product for column vectors in $\mathbb{C}^n$ is $<\mathbf{u}, \mathbf{v}> = \mathbf{u}^* \mathbf{v}$.

47. Note that $\mathbf{y}$ is a column vector, so the inner product is $\mathbf{y}^* \mathbf{y}$. Thus $\| \mathbf{y} \|^2 = \| A\mathbf{x}^T \|^2 = <A\mathbf{x}^T, A\mathbf{x}^T> = (A\mathbf{x}^T)^*(A\mathbf{x}^T) = (\mathbf{x}^T)^* A^* A\mathbf{x}^T = (\mathbf{x}^T)^* A^* A\mathbf{x}^T = (\mathbf{x}^T)^* I_n \mathbf{x}^T = \mathbf{x}^{T*} \mathbf{x}^T = <\mathbf{x}^T, \mathbf{x}^T> = \| \mathbf{x}^T \|^2$. But $\mathbf{x}^{T*} \mathbf{x}^T = \| \mathbf{x}^T \| = \| \mathbf{x} \| = \mathbf{x}\mathbf{x}^*$.

49. $AA^T = I_n = A^2$, so $AA = AA^T$, A is nonsingular so $A = A^T$.

51. If $A = \text{diag} [a_1, a_2, \ldots, a_n]$, then $A^{-1} = A^T = A$ so $A^2 = \text{diag} [a_1^2, a_2^2, \ldots, a_n^2] = I_n$. Hence, $a_i^2 = 1$, so that if A is real its diagonal elements are $a_i = 1$ or -1. If A is complex, then $a_i = 1$ or -1 or i or $-i$.

Chapter 5 Review Exercises (p. 366)

1. Compute $T(\alpha\mathbf{u} + \beta\mathbf{v}) = (-\alpha u_1 - \beta v_1, \alpha u_2 + \beta v_2, \alpha u_2 + \beta v_2 - \alpha u_1 - \beta v_1)$
$= \alpha(-u_1, u_2, u_2 - u_1) + \beta(-v_1, v_2, v_2 - v_1) = \alpha T(\mathbf{u}) + \beta T(\mathbf{v})$.

2. $[T] = \begin{bmatrix} -1 & 0 \\ 0 & 1 \\ -1 & 1 \end{bmatrix}$ row reduces to $U = \begin{bmatrix} 1 & 0 \\ 0 & 1 \\ 0 & 0 \end{bmatrix}$, so the kernel of T is $\{\mathbf{0}\}$ with the null basis.

3. See Review Exercise 1. The matrix $[T]$ has two linearly independent columns so its rank is 2. The two linearly independent columns represent a basis for the range of T in $\mathbb{R}^3$. There are other possible bases for the range including $\{(1, 0, 1), (0, 1, 1)\}$. Why isn't $\{e_1, e_2\}$ a basis for the range of T? These are represented by the column vectors of U in Review Exercise 2.

4. The rank is 2, nullity is 0, dim $\mathbb{R}^2 = 2 = 2 + 0$.   **5.** This matrix is given in the answer to Review Exercise 1 above.

6. Note that $B = [T] = \begin{bmatrix} 1 & 1 & 0 \\ 0 & 1 & 1 \\ 1 & 0 & 1 \end{bmatrix}$. And B represents a transformation $T_B = Bx$.

7. Yes, $B^{-1} = [T]^{-1} = [T^{-1}] = \begin{bmatrix} \frac{1}{2} & \frac{-1}{2} & \frac{1}{2} \\ \frac{1}{2} & \frac{1}{2} & \frac{-1}{2} \\ \frac{-1}{2} & \frac{1}{2} & \frac{1}{2} \end{bmatrix}$.

8. $T^{-1}(\mathbf{x}) = T^{-1}(x_1, x_2, x_3) \rightarrow [T^{-1}][\mathbf{x}] = \begin{bmatrix} \frac{1}{2} & \frac{-1}{2} & \frac{1}{2} \\ \frac{1}{2} & \frac{1}{2} & \frac{-1}{2} \\ \frac{-1}{2} & \frac{1}{2} & \frac{1}{2} \end{bmatrix} \begin{bmatrix} x_1 \\ x_2 \\ x_3 \end{bmatrix}$

$= \begin{bmatrix} \frac{x_1 - x_2 + x_3}{2} \\ \frac{x_1 + x_2 - x_3}{2} \\ \frac{-x_1 + x_2 + x_3}{2} \end{bmatrix} \rightarrow \left(\frac{x_1 - x_2 + x_3}{2}, \frac{x_1 + x_2 - x_3}{2}, \frac{-x_1 + x_2 + x_3}{2} \right)$.

9. Yes, $A = SBS^{-1}$, where $S = \begin{bmatrix} 1 & 0 & -2 \\ 1 & 1 & 0 \\ 1 & 0 & 0 \end{bmatrix}$, or for that matter any nonsingular matrix of the form $S = \begin{bmatrix} a & 0 & -2a \\ b & c & 0 \\ d & e & 0 \end{bmatrix}$.

10. Both are. $K^{-1} = K^T = K$ and $M^{-1} = M^T = \begin{bmatrix} 0 & 0 & -1 \\ -1 & 0 & 0 \\ 0 & -1 & 0 \end{bmatrix}$.

Chapter 6

Section 6.1 (pp. 385–388)

1. (a) $\lambda^2 + 2\lambda - 3 = (\lambda + 3)(\lambda - 1)$. (b) -3, 1. (c) For -3: $(-t, t)$; for 1: (t, t); t any real number.

3. (a) $\lambda^2 - 2\lambda = \lambda(\lambda - 2)$. (b) 0, 2. (c) For 0: $(t, -t)$; for 2: (t, t); t any real number.

5. (a) $\lambda^2 - \frac{1}{2}$. (b) $\frac{1}{\sqrt{2}}$, $\frac{-1}{\sqrt{2}}$. (c) For $\frac{1}{\sqrt{2}}$: $\left(\frac{2\sqrt{2}t}{2 - \sqrt{2}}, 2t \right)$; for $\frac{-1}{\sqrt{2}}$: $\left(\frac{-2\sqrt{2}t}{2 + \sqrt{2}}, 2t \right)$; t any real number.

7. (a) $\lambda^3 - 3\lambda^2 + 3\lambda - 1 = (\lambda - 1)^3$. (b) 1 is the only eigenvalue. (c) For 1: $(t, 0, 0)$; t any real number. See Exercise 13.

9. (a) $\lambda^4 - 6\lambda^3 - 28\lambda^2 + 216\lambda - 288 = (\lambda - 6)(\lambda + 6)(\lambda - 4)(\lambda - 2)$. (b) 6, -6, 4, 2. (c) For 6: $(0, 0, t, 0)$; for -6: $(0, -t, 0, t)$; for 4: $(0, t, -4t, t)$; for 2: $(t, 0, 0, 0)$; t any real number.

11. A basis for the eigenspace associated with $\lambda = 0$ is $\{(2, 3)\}$; and with $\lambda = -1$ is $\{(1, 1)\}$.

13. A basis for the eigenspace associated with $\lambda = 1$ is $\{(1, 0, 0)\}$. It is one-dimensional.

15. A basis for the eigenspace associated with $\lambda = 6$ is $\{(0, 0, 1, 0)\}$; with $\lambda = -6$ is $\{(0, -1, 0, 1)\}$; with $\lambda = 4$ is $\{(0, 1, -4, 1)\}$; and with $\lambda = 2$ is $\{(1, 0, 0, 0)\}$.

17. On the other hand, if **x** lies along **w**, then $T(\mathbf{x}) = 0$, so the eigenvalue is 0. For no other vector **x** does $T(\mathbf{x}) = \lambda\mathbf{x}$.

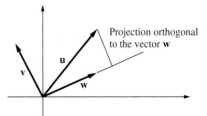

Projection orthogonal to the vector **w**

Note that if **v** is already orthogonal to **w** then $T(\mathbf{v}) = \mathbf{v}$, so we have an eigenvalue of 1.

19. If the vector $\mathbf{u} = \alpha\mathbf{w}$ already lies along **w**, then the projection $T(\mathbf{u}) = \mathbf{u}$, so its eigenvalue is 1 and {**w**} is a basis of eigenvectors for its eigenspace. If **u** is orthogonal to **w**, then its projection along **w** is **0** so the eigenvalue is 0. The eigenspace associated with 0 is a plane orthogonal to **w**, that is, it is the orthogonal compliment of **w**. If $\mathbf{w} = (a, b, c)$, then the equation of this plane is $ax + by + cz = 0$. A basis of eigenvectors for it is $\{(c, 0, -a), (0, c, -b)\}$.

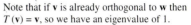

$T(\mathbf{u})$

Projection onto (along) **w**

21. If **u** lies at 45° (or 225°) to the positive x-axis, then it rotates to 135° (or 315°) and then reflects back on itself. The eigenvalue is 1, and $\{(1, 1)\}$ is a basis for the eigenspace. If **u** lies at 135° (or 315°) to the positive x-axis, then it rotates to 225° (or 405°) and then reflects to $-\mathbf{u}$: its eigenvalue is -1 and $\{(0, 1)\}$ is a basis for this eigenspace.

23. Eigenvalues: 2 only. Eigenspace\Basis: for 2: $\{(1, 1, -3)\}$; it is one-dimensional.

25. Eigenvalues: 4, 2, 0. Eigenspace\Bases: for 4: $\{(1, 0, -1)\}$; for 2: $\{(0, 1, 0)\}$; for 0: $\{(1, 1, 1)\}$.

27. Eigenvalues: $1, 3, \dfrac{1}{\sqrt{2}}, -\dfrac{1}{\sqrt{2}}$. Bases: for 1: $\{(-1, 2, 0, 0)\}$; for 3: $\{(1, 0, 0, 0)\}$; for $\dfrac{1}{\sqrt{2}}$: $\{(0, 0, 1, \sqrt{2} - 1)\}$; for $-\dfrac{1}{\sqrt{2}}$: $\{(0, 0, -1, \sqrt{2} + 1)\}$.

29. $\lambda(\lambda - 3)(\lambda - 2) = 0.$ **31.** For 3: $\{(0, 1, 0)\}$; for 2: $\{(1, 0, -1)\}$; and for 0: $\{(1, 0, 1)\}$.

33. $p_A(\lambda) = \det(\lambda I - A) = \lambda^3 - 6\lambda^2 + 11\lambda - 6 = (\lambda - 1)(\lambda - 2)(\lambda - 3).$

35. The matrix representation of eigenvectors corresponding to 1 is $\begin{bmatrix} 0 \\ t \\ 0 \end{bmatrix}$, so a corresponding eigenvector is the polynomial x. For 2, they are $\begin{bmatrix} -t \\ 2t \\ 2t \end{bmatrix}$, so an eigenvector is $-1 + 2x + 2x^2$. For 3, they are $\begin{bmatrix} -t \\ t \\ t \end{bmatrix}$, so an eigenvector is $-1 + x + x^2$.

37. $p_A(\lambda) = \lambda^4 - 8\lambda^3 + 7\lambda^2 + 72\lambda - 144 = (\lambda^2 - 9)(\lambda - 4)^2 = (\lambda - 3)(\lambda + 3)(\lambda - 4)^2.$

39. For 4: $\left\{\begin{bmatrix} 3 & 2 \\ 0 & 0 \end{bmatrix}\right\}$; for 3: $\left\{\begin{bmatrix} 0 & 0 \\ 1 & 1 \end{bmatrix}\right\}$; and for -3: $\left\{\begin{bmatrix} 0 & 0 \\ 5 & -1 \end{bmatrix}\right\}$.

41. The characteristic polynomial is $p_A(\lambda) = \lambda^2 + 3$. This has no real roots (zeros). But $i\sqrt{3}$ and $-i\sqrt{3}$ are complex roots and, therefore, (complex) eigenvalues for A. The corresponding eigenvectors are as follows: for $i\sqrt{3}$: $\begin{bmatrix} 2 \\ 1 + i\sqrt{3} \\ 1 \end{bmatrix}$, and for $-i\sqrt{3}$, $\begin{bmatrix} 2 \\ 1 - i\sqrt{3} \\ 1 \end{bmatrix}$.

43. Note that $A\mathbf{x} = 0\mathbf{x} = \mathbf{0}$ for a nonzero column vector **x** if, and only if, the matrix A is singular; that is, a homogeneous system of equations $A\mathbf{x} = \mathbf{0}$ has a nontrivial solution if, and only if, A is singular.

45. See Exercise 44. The eigenvalues of A^3 are $\lambda_1^3 = 2^3 = 8$ and $\lambda_2^3 = -1^3 = -1$.

47. See Exercise 44. If λ is an eigenvalue of A, then λ^k is an eigenvalue of A^k. If $A^k = 0$, then it must follow that $\lambda^k = 0$; consequently, $\lambda = 0$.

49. Write out the messy algebra and verify that $\det(\lambda I - A) = \det \begin{bmatrix} \lambda - a_{11} & -a_{12} & -a_{13} \\ -a_{21} & \lambda - a_{22} & -a_{23} \\ -a_{31} & -a_{32} & \lambda - a_{33} \end{bmatrix} = \lambda^3 - (a_{11} + a_{22} + a_{22})\lambda^2$

$$+ \left\{ \det \begin{bmatrix} a_{22} & a_{23} \\ a_{32} & a_{33} \end{bmatrix} + \det \begin{bmatrix} a_{11} & a_{13} \\ a_{31} & a_{33} \end{bmatrix} + \det \begin{bmatrix} a_{11} & a_{12} \\ a_{21} & a_{22} \end{bmatrix} \right\} \lambda - \det \begin{bmatrix} a_{11} & a_{12} & a_{13} \\ a_{21} & a_{22} & a_{23} \\ a_{31} & a_{32} & a_{33} \end{bmatrix}.$$

51. $p_A(x) = \det(xI - J_2) = x^2 - 1$; eigenvalues 1 and -1.
53. $p_A(x) = \det(xI - J_4) = (x^2 - 1)^2$; eigenvalues 1 and -1.   **55.** Eigenvalues 1 and -1.
57. See the answers to Exercises 1–9. Note that computer answers to Exercise 5 are 0.70711 and -0.70711.
59. For 1: basis $\{(0, 0, 1, 0, 0, 0), (0, 0, 0, 0, 1, 0)\}$; for -1: $\{(0, 0, 0, 1, 0, 0)\}$; for 2: $\{(0, 1, 10, 0, 0, 0)\}$; for -2: $\{(12, -3, -28, -84, -4, 12)\}$; and for 4: $\{(-30, -45, 10, -6, 30, 30)\}$.

Section 6.2 (pp. 398–400)

1. $p_A(\lambda) = \lambda^2 - 2\lambda - 3$. Eigenvalues 3 and -1. Therefore, $P = \begin{bmatrix} 3 & 1 \\ 1 & -1 \end{bmatrix}$, $P^{-1} = \begin{bmatrix} \frac{1}{4} & \frac{1}{4} \\ \frac{1}{4} & \frac{-3}{4} \end{bmatrix}$, and $P^{-1}AP = \begin{bmatrix} 3 & 0 \\ 0 & -1 \end{bmatrix}$.

3. $\det(\lambda I - A) = \lambda^2 - \lambda$; eigenvalues 0 and 1. The corresponding basic eigenvectors are, for $\lambda = 0$, $(1, -1)$ and for

$\lambda = 1$, $(1, 1)$. Therefore, $P = \begin{bmatrix} 1 & 1 \\ -1 & 1 \end{bmatrix}$, $P^{-1} = \begin{bmatrix} \frac{1}{2} & \frac{-1}{2} \\ \frac{1}{2} & \frac{1}{2} \end{bmatrix}$, and $P^{-1}AP = \begin{bmatrix} 0 & 0 \\ 0 & 1 \end{bmatrix}$.

5. The zero matrix is a diagonal matrix. $P = I$.

7. $P = \begin{bmatrix} 1 & 0 & 1 \\ 1 & 0 & -1 \\ 0 & 1 & 0 \end{bmatrix}$, $P^{-1} = \begin{bmatrix} \frac{1}{2} & \frac{1}{2} & 0 \\ 0 & 0 & 1 \\ \frac{1}{2} & \frac{-1}{2} & 0 \end{bmatrix}$, $P^{-1}AP = \begin{bmatrix} 3 & 0 & 0 \\ 0 & -2 & 0 \\ 0 & 0 & -1 \end{bmatrix}$.

9. Eigenvalues are 2 and -1. For $\lambda = 2$, eigenspace basis $\{(3, 1, -3)\}$. For $\lambda = -1$, eigenvectors have the form

$(s + t, -s, -t)$; eigenspace has a basis $\{(1, 0, -1), (1, -1, 0)\}$. $P = \begin{bmatrix} 3 & 1 & 1 \\ 1 & 0 & -1 \\ -3 & -1 & 0 \end{bmatrix}$, and $D = P^{-1}AP =$

$\begin{bmatrix} 2 & 0 & 0 \\ 0 & -1 & 0 \\ 0 & 0 & -1 \end{bmatrix}$.

11. The characteristic polynomial is $\lambda^2 + \lambda + 1$, so C has no real eigenvalues. Similar matrices have the same eigenvalues, so C is not similar to any diagonal matrix with *real* eigenvalues.

13. The characteristic polynomial is $\lambda^2 - \lambda + 1$, so this matrix has no real eigenvalues. Hence this matrix is not similar to any diagonal matrix with *real* eigenvalues. See Exercise 11.

15. The eigenvalues are -3, 2, and 1, so let $P = \begin{bmatrix} 1 & 0 & 1 \\ 0 & 1 & -2 \\ -1 & 0 & 1 \end{bmatrix}$. Then $P^{-1} = \begin{bmatrix} \frac{1}{2} & 0 & \frac{-1}{2} \\ 1 & 1 & 1 \\ \frac{1}{2} & 0 & \frac{1}{2} \end{bmatrix}$ and $P^{-1}AP =$

$\begin{bmatrix} -3 & 0 & 0 \\ 0 & 2 & 0 \\ 0 & 0 & 1 \end{bmatrix}$.

17. Let $P = \begin{bmatrix} 1 & -11 & 1 \\ 3 & 7 & 0 \\ -10 & 0 & 0 \end{bmatrix}$. So $P^{-1}AP = \begin{bmatrix} 6 & 0 & 0 \\ 0 & -4 & 0 \\ 0 & 0 & 3 \end{bmatrix}$.

19. $p(\lambda) = (\lambda - 1)^3(\lambda + 3)$. Eigenspace for $\lambda = 1$ is $\{(s, -t, u, t)\}$; s, t, and u arbitrary real numbers. A basis is $\{(1, 0, 0, 0), (0, -1, 0, 1), (0, 0, 1, 0)\}$. The eigenspace for $\lambda = -3$ has a basis $\{(0, 2, -3, 2)\}$. So $P =$

$$\begin{bmatrix} 1 & 0 & 0 & 0 \\ 0 & -1 & 0 & 2 \\ 0 & 0 & 1 & -3 \\ 0 & 1 & 0 & 2 \end{bmatrix} \text{gives } P^{-1}AP = \begin{bmatrix} 1 & 0 & 0 & 0 \\ 0 & 1 & 0 & 0 \\ 0 & 0 & 1 & 0 \\ 0 & 0 & 0 & -3 \end{bmatrix}.$$

21. $p(\lambda) = (\lambda + 1)^2(\lambda - 4)(\lambda - 2)$. Eigenspace for $\lambda = -1$ has basis $\{(1, 0, 0, 0), (0, 1, 0, 0)\}$. For $\lambda = 4$, a basis is $\{(0, 0, 1, 1)\}$, and for $\lambda = 2$ a basis is $\{(0, 0, 1, -1)\}$. So $P = \begin{bmatrix} 1 & 0 & 0 & 0 \\ 0 & 1 & 0 & 0 \\ 0 & 0 & 1 & 1 \\ 0 & 0 & 1 & -1 \end{bmatrix}$ gives $P^{-1}AP = \begin{bmatrix} -1 & 0 & 0 & 0 \\ 0 & -1 & 0 & 0 \\ 0 & 0 & 4 & 0 \\ 0 & 0 & 0 & 2 \end{bmatrix}$.

23. $[T] = \begin{bmatrix} 1 & 0 \\ 0 & 0 \end{bmatrix}$ is already diagonal, so the standard basis does it.

25. Note $[T] = \begin{bmatrix} 3 & 1 \\ 1 & 3 \end{bmatrix}$. The characteristic polynomial is $p(\lambda) = \lambda^2 - 6\lambda + 8 = (\lambda - 4)(\lambda - 2)$. A basis made up of the bases for the two eigenspaces, namely, $B = \{(1, 1), (1, -1)\}$ for $\mathbb{R}^2$ has the property that

$$D = [T]_B = \begin{bmatrix} 4 & 0 \\ 0 & 2 \end{bmatrix} = P^{-1}[T]P.$$

27. $[T] = \begin{bmatrix} -1 & 0 & 2 \\ 1 & 2 & 1 \\ 2 & 0 & -1 \end{bmatrix}$. Basis of eigenvectors is $B = \{(1, 0, -1), (0, 1, 0), (1, -2, 1)\}$. Diagonal matrix:

$$[T]_B = \begin{bmatrix} -3 & 0 & 0 \\ 0 & 2 & 0 \\ 0 & 0 & 1 \end{bmatrix} = P^{-1}[T]P.$$

29. $p_A(\lambda) = (\lambda - 3)^2(\lambda - 7)$. Eigenvalues, $\lambda_1 = 3$, $\lambda_2 = 7$. Eigenspace for $\lambda_1 = 3$ has basis $B_1 = \{1 - x^2, x\}$. Eigenspace for $\lambda_2 = 7$ has basis $B_2 = \{2 - x + 2x^2\}$. Let $B = B_1 \cup B_2$. $[T]_B = \begin{bmatrix} 3 & 0 & 0 \\ 0 & 3 & 0 \\ 0 & 0 & 7 \end{bmatrix}$.

31. The eigenvalues of A are 2 and 1, so $A = P^{-1}\begin{bmatrix} 2 & 0 \\ 0 & 1 \end{bmatrix}P$. $A^9 = P^{-1}\begin{bmatrix} 2 & 0 \\ 0 & 1 \end{bmatrix}^9 P$. $P = \begin{bmatrix} 1 & 0 \\ 1 & 1 \end{bmatrix}$, $P^{-1} = \begin{bmatrix} 1 & 0 \\ -1 & 1 \end{bmatrix}$, so

$$A^9 = \begin{bmatrix} 1 & 0 \\ -1 & 1 \end{bmatrix}\begin{bmatrix} 2^9 & 0 \\ 0 & 1 \end{bmatrix}\begin{bmatrix} 1 & 0 \\ 1 & 1 \end{bmatrix} = \begin{bmatrix} 512 & 0 \\ -511 & 1 \end{bmatrix}.$$

33. $A^9 = \begin{bmatrix} 1 & 0 & 1 \\ 0 & 1 & 4 \\ -1 & 0 & 1 \end{bmatrix}\begin{bmatrix} 2^9 & 0 & 0 \\ 0 & (-1)^9 & 0 \\ 0 & 0 & 0 \end{bmatrix}\begin{bmatrix} \frac{1}{2} & 0 & \frac{-1}{2} \\ -2 & 1 & -2 \\ \frac{1}{2} & 0 & \frac{1}{2} \end{bmatrix} = \begin{bmatrix} 256 & 0 & -256 \\ 2 & -1 & 2 \\ -256 & 0 & 256 \end{bmatrix}.$

35. If $D = aI$, then only D is similar to D. This is because $A = P^{-1}aIP = aP^{-1}IP = aI$. So A must have been $D = aI$ already.

37. If $D = P^{-1}AP$, then $A = PDP^{-1}$, so $A^{-1} = PD^{-1}P^{-1}$. For $D = \begin{bmatrix} 3 & 0 \\ 0 & -1 \end{bmatrix}$, $D^{-1} = \begin{bmatrix} \frac{1}{3} & 0 \\ 0 & -1 \end{bmatrix}$, so

$$A^{-1} = \begin{bmatrix} 3 & 1 \\ 1 & -1 \end{bmatrix} \begin{bmatrix} \frac{1}{3} & 0 \\ 0 & -1 \end{bmatrix} \begin{bmatrix} \frac{1}{4} & \frac{1}{4} \\ \frac{1}{4} & -\frac{3}{4} \end{bmatrix} = \begin{bmatrix} 0 & 1 \\ \frac{1}{3} & -\frac{2}{3} \end{bmatrix}.$$

39. $A^{-1} = \begin{bmatrix} \sqrt{2}-1 & -(\sqrt{2}+1) \\ 1 & 1 \end{bmatrix} \begin{bmatrix} 1 & 0 \\ 0 & -1 \end{bmatrix} \begin{bmatrix} \frac{\sqrt{2}}{4} & \frac{2+\sqrt{2}}{4} \\ -\frac{\sqrt{2}}{4} & \frac{2-\sqrt{2}}{4} \end{bmatrix} = \begin{bmatrix} \frac{-1}{\sqrt{2}} & \frac{1}{\sqrt{2}} \\ \frac{1}{\sqrt{2}} & \frac{1}{\sqrt{2}} \end{bmatrix} = A^T = A$, because A is orthogonal.

41. $A^{-1} = \begin{bmatrix} 3 & 1 & 1 \\ 1 & 0 & -1 \\ -3 & -1 & 0 \end{bmatrix} \begin{bmatrix} \frac{1}{2} & 0 & 0 \\ 0 & -1 & 0 \\ 0 & 0 & -1 \end{bmatrix} \begin{bmatrix} 1 & 1 & 1 \\ -3 & -3 & -4 \\ 1 & 0 & 1 \end{bmatrix} = \begin{bmatrix} \frac{7}{2} & \frac{9}{2} & \frac{9}{2} \\ \frac{3}{2} & \frac{1}{2} & \frac{3}{2} \\ \frac{-9}{2} & \frac{-9}{2} & \frac{-11}{2} \end{bmatrix}.$

43. If A is upper triangular and $a_{ii} \neq a_{jj}$ when $i \neq j$, then $p_A(\lambda) = \det(\lambda I - A) = \prod_{i=1}^{n}(I - a_{ii})$. So the distinct diagonal entries a_{ii} are the eigenvalues of A. The result then follows immediately from Theorem 6.5.

45. $A^2 + 2A - 3I = \begin{bmatrix} 1 & 2 & 0 \\ 2 & 1 & 0 \\ 0 & 0 & -2 \end{bmatrix}^2 + 2\begin{bmatrix} 1 & 2 & 0 \\ 2 & 1 & 0 \\ 0 & 0 & -2 \end{bmatrix} - \begin{bmatrix} 3 & 0 & 0 \\ 0 & 3 & 0 \\ 0 & 0 & 3 \end{bmatrix} = \begin{bmatrix} 4 & 8 & 0 \\ 8 & 4 & 0 \\ 0 & 0 & -3 \end{bmatrix}.$ **47.** $\begin{bmatrix} -4 & 0 & 0 & 0 \\ 0 & -4 & 0 & 0 \\ 0 & 0 & 13 & 8 \\ 0 & 0 & 8 & 13 \end{bmatrix}.$

49. Recall Exercise 44 of Section 6.1 $A^k\mathbf{x} = \lambda^k\mathbf{x}$. $p(A)\mathbf{x} = a_m A^m\mathbf{x} + a_{m-1}A^{m-1}\mathbf{x} + \cdots + a_1 A\mathbf{x}$
$+ a_0 I\mathbf{x} = a_m\lambda^m\mathbf{x} + a_{m-1}\lambda^{m-1}\mathbf{x} + \cdots + a_1\lambda\mathbf{x} + a_0 I\mathbf{x} = p(\lambda)\mathbf{x}.$

51. $p_A(A) = A^2 - 2A - 3I = \begin{bmatrix} -1 & 2 \\ 0 & 3 \end{bmatrix}^2 - 2\begin{bmatrix} -1 & 2 \\ 0 & 3 \end{bmatrix} - \begin{bmatrix} 3 & 0 \\ 0 & 3 \end{bmatrix} = \begin{bmatrix} 1 & 4 \\ 0 & 9 \end{bmatrix} - \begin{bmatrix} -2 & 4 \\ 0 & 6 \end{bmatrix} - \begin{bmatrix} 3 & 0 \\ 0 & 3 \end{bmatrix} = \begin{bmatrix} 0 & 0 \\ 0 & 0 \end{bmatrix}.$

53. See Exercise 50. Here is an alternate way. $p_A(t) = \det(tI = A) = (t - d_1)(t - d_2)(t - d_3)$, so $p_A(A) =$

$$(A - d_1 I)(A - d_2 I)(A - d_3 I) = \begin{bmatrix} 0 & 0 & 0 \\ 0 & d_2 - d_1 & 0 \\ 0 & 0 & d_3 - d_1 \end{bmatrix} \begin{bmatrix} d_1 - d_2 & 0 & 0 \\ 0 & 0 & 0 \\ 0 & 0 & d_3 - d_2 \end{bmatrix} \begin{bmatrix} d_1 - d_3 & 0 & 0 \\ 0 & d_2 - d_3 & 0 \\ 0 & 0 & 0 \end{bmatrix} = \begin{bmatrix} 0 & 0 & 0 \\ 0 & 0 & 0 \\ 0 & 0 & 0 \end{bmatrix}.$$

Section 6.3 (pp. 407–408)

1. Eigenvalue   Basis for Eigenspace
-4 $\{(1, -1)^T\}$
2 $\{(1, 1)^T\}$

3. Eigenvalue   Basis for Eigenspace
3 $\{(1, -1)^T\}$
1 $\{(1, 1)^T\}$

5. Eigenvalue   Basis for Eigenspace
$\dfrac{1+3\sqrt{5}}{2}$ $\{(1 + \sqrt{5}, 2)^T\}$
$\dfrac{1-3\sqrt{5}}{2}$ $\{(1 - \sqrt{5}, 2)^T\}$

7. Eigenvalue   Basis for Eigenspace
3 $\{(1, 1, 1)^T\}$
0 $\{(1, 0, -1)^T, (1, -1, 0)^T\}$

9. Eigenvalue   Basis for Eigenspace
4 $\{(0, 0, 1)^T\}$
-3 $\{(1, -1, 0)^T\}$
1 $\{(1, 1, 0)^T\}$

11. Eigenvalue   Basis for Eigenspace
3 $\{(0, 0, 1, 1)^T\}$
-3 $\{(1, -1, 0, 0)^T\}$
-5 $\{(0, 0, 1, -1)^T\}$
1 $\{(1, 1, 0, 0)^T\}$

13. Eigenvalue   Basis for Eigenspace
$a + b$ $\{(1, 1)^T\}$ or $\left\{\left(\frac{1}{\sqrt{2}}, \frac{1}{\sqrt{2}}\right)^T\right\}$
$a - b$ $\{(1, -1)^T\}$ or $\left\{\left(\frac{1}{\sqrt{2}}, \frac{-1}{\sqrt{2}}\right)^T\right\}$

15. Use the eigenvectors in Exercise 1 and divide by their norms. Thus, the orthogonal matrix $P = \begin{bmatrix} \frac{1}{\sqrt{2}} & \frac{1}{\sqrt{2}} \\ \frac{-1}{\sqrt{2}} & \frac{1}{\sqrt{2}} \end{bmatrix}$ and

$$P^T A P = \begin{bmatrix} -4 & 0 \\ 0 & 2 \end{bmatrix} = D.$$

17. The orthogonal matrix $P = \begin{bmatrix} \frac{1}{\sqrt{2}} & \frac{1}{\sqrt{2}} \\ \frac{-1}{\sqrt{2}} & \frac{1}{\sqrt{2}} \end{bmatrix}$ and $P^T A P = \begin{bmatrix} 3 & 0 \\ 0 & 1 \end{bmatrix} = D.$

19. Let $c = \sqrt{10 + 2\sqrt{5}}$ and $d = \sqrt{10 - 2\sqrt{5}}$. Then the orthogonal matrix $P = \begin{bmatrix} \frac{1+\sqrt{5}}{c} & \frac{1-\sqrt{5}}{d} \\ \frac{2}{c} & \frac{2}{d} \end{bmatrix}$,

and $D = \begin{bmatrix} \frac{1+3\sqrt{5}}{2} & 0 \\ 0 & \frac{1-3\sqrt{5}}{2} \end{bmatrix} = P^T A P.$

21. Orthogonal matrix $P = \begin{bmatrix} \frac{1}{\sqrt{3}} & \frac{1}{\sqrt{2}} & \frac{1}{\sqrt{6}} \\ \frac{1}{\sqrt{3}} & 0 & \frac{-2}{\sqrt{6}} \\ \frac{1}{\sqrt{3}} & \frac{-1}{\sqrt{2}} & \frac{1}{\sqrt{6}} \end{bmatrix}$, $P^T A P = \begin{bmatrix} 3 & 0 & 0 \\ 0 & 0 & 0 \\ 0 & 0 & 0 \end{bmatrix} = D.$

23. $P = \begin{bmatrix} 0 & \frac{1}{\sqrt{2}} & \frac{1}{\sqrt{2}} \\ 0 & \frac{-1}{\sqrt{2}} & \frac{1}{\sqrt{2}} \\ 1 & 0 & 0 \end{bmatrix}$, $P^T A P = \begin{bmatrix} 4 & 0 & 0 \\ 0 & -3 & 0 \\ 0 & 0 & 1 \end{bmatrix} = D.$

25. Orthogonal matrix $P = \begin{bmatrix} 0 & \frac{1}{\sqrt{2}} & 0 & \frac{1}{\sqrt{2}} \\ 0 & \frac{-1}{\sqrt{2}} & 0 & \frac{1}{\sqrt{2}} \\ \frac{1}{\sqrt{2}} & 0 & \frac{1}{\sqrt{2}} & 0 \\ \frac{1}{\sqrt{2}} & 0 & \frac{-1}{\sqrt{2}} & 0 \end{bmatrix}$, $P^T A P = \begin{bmatrix} 3 & 0 & 0 & 0 \\ 0 & -3 & 0 & 0 \\ 0 & 0 & -5 & 0 \\ 0 & 0 & 0 & 1 \end{bmatrix} = D.$

27. For any real $n \times n$ matrix A (symmetric or not) notice that $(A^T A)^T = A^T (A^T)^T = A^T A$ and $(A A^T)^T = (A^T)^T A^T = A A^T$. Also $(A + A^T)^T = A^T + (A^T)^T = A^T + A = A + A^T$.

29. Let $A = (a_{ij})$ be skew symmetric. The diagonal entries in A and A^T are a_{ii}. If $A^T = -A$, then $a_{ii} = -a_{ii}$ for each $i = 1, 2, \ldots, n$. Thus each $a_{ij} = 0$.

31. $A = \frac{1}{2}(A + A^T) + \frac{1}{2}(A - A^T).$

33. $D^k = P^{-1}A^kP$, so $A^k = PD^kP^{-1}$. If $D = \begin{bmatrix} 3 & 0 \\ 0 & 1 \end{bmatrix}$, and $P = \begin{bmatrix} \frac{1}{\sqrt{2}} & \frac{1}{\sqrt{2}} \\ \frac{-1}{\sqrt{2}} & \frac{1}{\sqrt{2}} \end{bmatrix}$. Then $D^{10} = \begin{bmatrix} 59049 & 0 \\ 0 & 1 \end{bmatrix}$, so

$$A^{10} = \begin{bmatrix} 29525 & -29524 \\ -29524 & 29525 \end{bmatrix}.$$

35. Let $A = \begin{bmatrix} a & b \\ b & d \end{bmatrix}$. Note that $p_A(\lambda)$ is a quadratic whose discriminant is $(a - d)^2 + 4b^2 \geq 0$. So A has real eigenvalues $\lambda = \frac{1}{2}[(a + d) \pm \sqrt{(a - d)^2 + 4b^2}]$. These are distinct unless $a = d$ and $b = 0$. But when $b = 0$, A is already diagonal.

37. $B = P^TAP$ is given. So $B^T = (P^TAP)^T = P^TA^T P$. So if $A^T = A$, then $B^T = P^TAP = B$ and B is symmetric when A is. Conversely, if $B = P^TAP$, then $A = PBP^T$ because P being orthogonal is invertible. So if B is symmetric, then $A^T = (PBP^T)^T = PB^TP^T = PBP^T = A$. Thus A is symmetric.

39. Yes, to diag $[7.54, 4.42, -4.09, 3, 59, -3.46]$.

41. $\left\{ \begin{bmatrix} 1.00 \\ 0.97 \\ 0.88 \\ 0.36 \\ 0.36 \end{bmatrix}, \begin{bmatrix} 1.00 \\ -2.42 \\ 1.87 \\ -1.09 \\ 0.26 \end{bmatrix}, \begin{bmatrix} 1.00 \\ -0.04 \\ -0.64 \\ -0.28 \\ -0.82 \end{bmatrix}, \begin{bmatrix} 1.00 \\ 0.11 \\ -1.90 \\ -1.79 \\ 3.33 \end{bmatrix}, \begin{bmatrix} 1.00 \\ -1.71 \\ -0.98 \\ 3.27 \\ 0.96 \end{bmatrix} \right\}$

Section 6.4 (pp. 421–423)

1. xy, $A = \begin{bmatrix} 0 & \frac{1}{2} \\ \frac{1}{2} & 0 \end{bmatrix}$.

3. $7xy$, $A = \begin{bmatrix} 0 & \frac{7}{2} \\ \frac{7}{2} & 0 \end{bmatrix}$.

5. $53x^2 - 72xy + 32y^2$, $A = \begin{bmatrix} 53 & -36 \\ -36 & 32 \end{bmatrix}$.

7. $16x^2 - 24xy + 9y^2$, $A = \begin{bmatrix} 16 & -12 \\ -12 & 9 \end{bmatrix}$.

9. $2x^2 - 3xy + 4y^2$, $A = \begin{bmatrix} 2 & \frac{-3}{2} \\ \frac{-3}{2} & 4 \end{bmatrix}$.

11. $x^2 + y^2 + z^2$, $A = \begin{bmatrix} 1 & 0 & 0 \\ 0 & 1 & 0 \\ 0 & 0 & 1 \end{bmatrix} = I_3$.

13. $x^2 - y^2 - z^2$, $A = \begin{bmatrix} 1 & 0 & 0 \\ 0 & -1 & 0 \\ 0 & 0 & -1 \end{bmatrix}$.

15. $4x^2 + 4y^2 + 4z^2 + 4xy + 4xz + 4yz$, $A = \begin{bmatrix} 4 & 2 & 2 \\ 2 & 4 & 2 \\ 2 & 2 & 4 \end{bmatrix}$.

17. $2x^2 - 3y^2 + 6z^2 - 2xy + 2xz + 4yz$, $A = \begin{bmatrix} 2 & -1 & 1 \\ -1 & -3 & 2 \\ 1 & 2 & 6 \end{bmatrix}$.

19. $X^TAX = 3$, or $[x, y]\begin{bmatrix} 0 & \frac{1}{2} \\ \frac{1}{2} & 0 \end{bmatrix}\begin{bmatrix} x \\ y \end{bmatrix} = [3]$.

21. $X^TAX = -2$, or $[x, y]\begin{bmatrix} 0 & \frac{7}{2} \\ \frac{7}{2} & 0 \end{bmatrix}\begin{bmatrix} x \\ y \end{bmatrix} = [-2]$.

23. $X^TAX = 80$, or $[x, y]\begin{bmatrix} 53 & -36 \\ -36 & 32 \end{bmatrix}\begin{bmatrix} x \\ y \end{bmatrix} = [80]$.

25. $X^T A X + [-60, -80]X + 100 = 0$, or $[x, y] \begin{bmatrix} 16 & -12 \\ -12 & 9 \end{bmatrix} \begin{bmatrix} x \\ y \end{bmatrix} + [-60, -80] \begin{bmatrix} x \\ y \end{bmatrix} = [-100]$.

27. $X^T A X + [-7, 2]X + 7 = 0$, or $[x, y] \begin{bmatrix} 2 & \dfrac{-3}{2} \\ \dfrac{-3}{2} & 4 \end{bmatrix} \begin{bmatrix} x \\ y \end{bmatrix} + [-7, 2] \begin{bmatrix} x \\ y \end{bmatrix} = [-7]$.

29. $X^T X = 16$, or $[x, y, x] \begin{bmatrix} 1 & 0 & 0 \\ 0 & 1 & 0 \\ 0 & 0 & 1 \end{bmatrix} \begin{bmatrix} x \\ y \\ z \end{bmatrix} = [16]$. **31.** $X^T A X = 16$, or $[x, y, z] \begin{bmatrix} 1 & 0 & 0 \\ 0 & -1 & 0 \\ 0 & 0 & -1 \end{bmatrix} \begin{bmatrix} x \\ y \\ z \end{bmatrix} = [16]$.

33. $X^T A X = 7$, or $[x, y, z] \begin{bmatrix} 4 & 2 & 2 \\ 2 & 4 & 2 \\ 2 & 2 & 4 \end{bmatrix} \begin{bmatrix} x \\ y \\ z \end{bmatrix} = [7]$. **35.** $X^T A X = 18$, or $[x, y, z] \begin{bmatrix} 2 & -1 & 1 \\ -1 & -3 & 2 \\ 1 & 2 & 6 \end{bmatrix} \begin{bmatrix} x \\ y \\ z \end{bmatrix} = [18]$.

37. It's a hyperbola with transformed equation $\dfrac{x'^2}{6} - \dfrac{y'^2}{6} = 1$. The orthogonal transformation matrix to a basis of eigenvectors is a $45°$ (counter-clockwise) rotation.

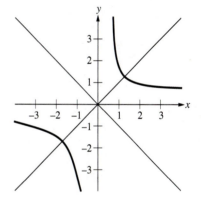

39. It is a hyperbola, transformed equation: $\dfrac{y'^2}{4/7} - \dfrac{x'^2}{4/7} = 1$.

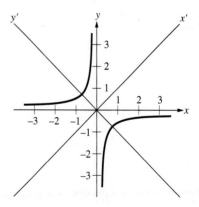

41. It is an ellipse. The orthogonal matrix is $P = \begin{bmatrix} \frac{3}{5} & \frac{-4}{5} \\ \frac{4}{5} & \frac{3}{5} \end{bmatrix}$. The trans-

formed equation is $\dfrac{x'^2}{16} + \dfrac{y'^2}{1} = 1$.

43. Parabola. $P = \begin{bmatrix} \frac{4}{5} & \frac{3}{5} \\ \frac{-3}{5} & \frac{4}{5} \end{bmatrix}$. The transformed equation is $x'^2 = 4(y' - 1)$.

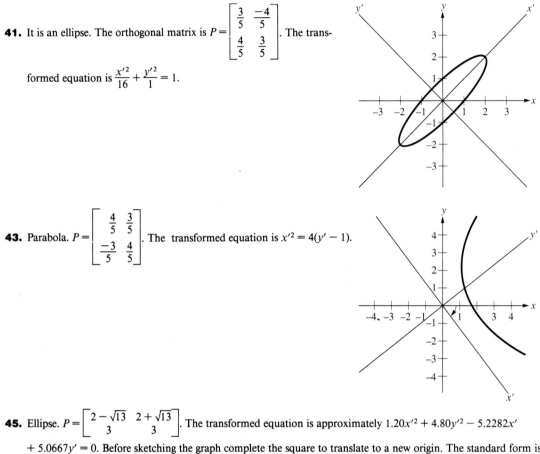

45. Ellipse. $P = \begin{bmatrix} \frac{2 - \sqrt{13}}{3} & \frac{2 + \sqrt{13}}{3} \end{bmatrix}$. The transformed equation is approximately $1.20x'^2 + 4.80y'^2 - 5.2282x'$
$+\ 5.0667y' = 0$. Before sketching the graph complete the square to translate to a new origin. The standard form is
(approximately) $\dfrac{(x' - 2.18)^2}{5.86} + \dfrac{(y' + 0.53)^2}{1.47} = 1$.

47. A sphere center $(0, 0, 0)$ radius 4. Equation $x^2 + y^2 + z^2 = 4^2$.

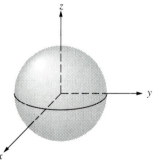

49. A hyperboloid of two sheets with circular cross section. See Figure 6.8. Transformed equation is $\dfrac{x'^2}{16} - \dfrac{y'^2}{16} - \dfrac{z'^2}{16} = 1$.

51. An ellipsoid. See Figure 6.5. The transformed equation is $\dfrac{x'^2}{7/8} + \dfrac{y'^2}{7/2} + \dfrac{z'^2}{7/2} = 1$.

53. The eigenvalues of A are irrational, the characteristic polynomial is $\lambda^3 - 5\lambda^2 - 18\lambda + 51$. You will need to estimate
the roots; $\lambda_1 \approx 2.12$, $\lambda_2 \approx 6.56$, and $\lambda_3 \approx -3.67$. The transformed equation is $\lambda_1 x'^2 + \lambda_2 y'^2 + \lambda_3 z'^2 = 18$. This is an
elliptic hyperboloid of one sheet. See Figure 6.6 in the text.

55. Let $X = (x_1, x_2, x_3)^T$ and let $A = (a_{ij})$ be a 3×3 real symmetric matrix. Then there exists an orthogonal 3×3 matrix P such that $P^TAP = \text{diag } [\lambda_1, \lambda_2, \lambda_3]$. Now let $X = PY$ where $Y = (y_1, y_2, y_3)^T$. Then it follows that

$$X^TAX = (PY)^TA(PY) = Y^TP^TAPY = [y_1, y_2, y_3] \begin{bmatrix} \lambda_1 & 0 & 0 \\ 0 & \lambda_2 & 0 \\ 0 & 0 & \lambda_3 \end{bmatrix} \begin{bmatrix} y_1 \\ y_2 \\ y_3 \end{bmatrix} = \lambda_1 y_1^2 + \lambda_2 y_2^2 + \lambda_3 y_3^2.$$

57. Given $ax^2 + 2bxy + cy^2 + dx + ey + f = 0$ or $[x, y] \begin{bmatrix} a & b \\ b & c \end{bmatrix} \begin{bmatrix} x \\ y \end{bmatrix} + [d, e] \begin{bmatrix} x \\ y \end{bmatrix} + [f] = [0]$. If $R\phi$ is a

rotation of the plane, then the matrix representation of $R\phi$ is $P = \begin{bmatrix} \cos \phi & -\sin \phi \\ \sin \phi & \cos \phi \end{bmatrix}$. Therefore, for any

vector $\mathbf{x}$ in the plane, $R\phi(\mathbf{x}) = P[\mathbf{x}] = PX = Y$. Thus $X = P^TY$ and so $x^TAX = Y^TPAP^TY$. Now

$PAP^T = \begin{bmatrix} \cos \phi & -\sin \phi \\ \sin \phi & \cos \phi \end{bmatrix} \begin{bmatrix} a & b \\ b & c \end{bmatrix} \begin{bmatrix} \cos \phi & \sin \phi \\ -\sin \phi & \cos \phi \end{bmatrix} = \begin{bmatrix} a' & b' \\ b' & c' \end{bmatrix}$. One may directly compute that

$b'^2 - a'c' = b^2 - ac$ using trigonometric identities. An easier way is to use determinants. Note that $b'^2 - a'c' = -\det (PAP^T) = -(\det P)(\det A)(\det P^T)$. But $\det P = \det P^T = 1$, so $-\det (PAP^T) = -\det A = b^2 - ac$, as desired.

59. In Exercises 57 and 58 we saw that the discriminant R is invariant under a rotation, so we may without loss of generality suppose that the conic is in standard form. For an ellipse we have $\lambda_1 x^2 + \lambda_2 y^2 + dx + ey = f$, with $\lambda_1 \lambda_2 > 0$, so $-ac = -(\lambda_1 \lambda_2) < 0$. For a parabola we have one of λ_1 or $\lambda_2 = 0$, so $-ac = 0$. For a hyperbola the form is $\lambda_1 x^2 - \lambda_2 y^2 + \ldots$, ($\lambda_1$ and λ_2 have opposite signs), so $-ac = -(\lambda_1 \lambda_2) > 0$.

61. Two lines $y = \pm x$.   **63.** The origin $(0, 0)$ only.

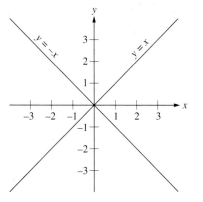

65. Rotated equation is $13x'^2 = 52$, so the two lines $x' = \pm 2$.

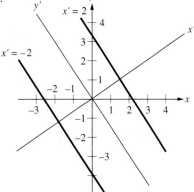

67. The one line $y = x$.   **69.** The origin $(0, 0, 0)$ only.

Section 6.5 (pp. 435–438)

1. An absorbing chain. $\begin{bmatrix} 1 & \frac{2}{3} \\ 0 & \frac{1}{3} \end{bmatrix}$. State 2 is the absorbing state.

3. An absorbing chain. State 1 is the absorbing state and the matrix is already in canonical form.

5. Neither state is absorbing. There is constant movement out of a state. This is not a regular Markov chain. $A^2 = I$.

7. The original state 3 is the absorbing state. Canonical form is $\begin{bmatrix} 1 & 0 & \frac{2}{3} & 0 \\ 0 & \frac{1}{2} & 0 & \frac{3}{4} \\ 0 & \frac{1}{2} & 0 & \frac{1}{4} \\ 0 & 0 & \frac{1}{3} & 0 \end{bmatrix}$.

9. Not a transition matrix; column 2 has sum 2, not 1.

11. All entries are already positive. This is a regular chain. There are no absorbing states.

13. State 3 is absorbing. The canonical form is $\begin{bmatrix} 1 & 0.3 & 0 & 0.4 & 0 \\ 0 & 0 & 1 & 0.1 & 0.2 \\ 0 & 0.1 & 0 & 0.2 & 0.3 \\ 0 & 0.2 & 0 & 0.3 & 0.4 \\ 0 & 0.4 & 0 & 0 & 0.1 \end{bmatrix}$.

15. States 1, 2, and 3 are all absorbing. The matrix is already in canonical form.

17. It doesn't have one. None of its eigenvalues is 1. They are approximately 1.29, $0.022 + 0.043i$, $0.022 - 0.043i$, and 0.

19. $\begin{bmatrix} 0.2 \\ 0.4 \\ 0.4 \end{bmatrix}$.

21. $\begin{bmatrix} \dfrac{bd}{ac + 2ad + bd} \\ \dfrac{ad}{ac + 2ad + bd} \\ \dfrac{ad}{ac + 2ad + bd} \\ \dfrac{ac}{ac + 2ad + bd} \end{bmatrix}$.

23. $\begin{bmatrix} \frac{1}{2} & \frac{1}{2} \\ \frac{1}{2} & \frac{1}{2} \end{bmatrix}$.

25. $\begin{bmatrix} 0.5 & 0.5 \\ 0.5 & 0.5 \end{bmatrix}$.

27. $\begin{bmatrix} \frac{8}{11} & \frac{8}{11} \\ \frac{3}{11} & \frac{3}{11} \end{bmatrix}$.

29. $\begin{bmatrix} 0.38 & 0.38 & 0.38 \\ 0.48 & 0.48 & 0.48 \\ 0.14 & 0.14 & 0.14 \end{bmatrix}$.

31. $S = \begin{bmatrix} \frac{1}{3} \end{bmatrix}$, $I - S = \begin{bmatrix} \frac{2}{3} \end{bmatrix}$, $F = (I - S)^{-1} = \begin{bmatrix} \frac{3}{2} \end{bmatrix}$.

33. $S = \begin{bmatrix} \frac{1}{2} & 0 & \frac{3}{4} \\ \frac{1}{2} & 0 & \frac{1}{4} \\ 0 & \frac{1}{3} & 0 \end{bmatrix}$, $I - S = \begin{bmatrix} \frac{1}{2} & 0 & \frac{-3}{4} \\ \frac{-1}{2} & 1 & \frac{-1}{4} \\ 0 & \frac{-1}{3} & 1 \end{bmatrix}$, $F(I - S)^{-1} = \begin{bmatrix} \frac{11}{4} & \frac{3}{4} & \frac{9}{4} \\ \frac{3}{2} & \frac{3}{2} & \frac{3}{2} \\ \frac{1}{2} & \frac{1}{2} & \frac{3}{2} \end{bmatrix}$.

35. $S = \begin{bmatrix} 0 & 1 & 0.1 & 0.2 \\ 0.1 & 0 & 0.2 & 0.3 \\ 0.2 & 0 & 0.7 & 0.4 \\ 0.4 & 0 & 0 & 0.1 \end{bmatrix}$, $I - S = \begin{bmatrix} 1 & -1 & -0.1 & -0.2 \\ -0.1 & 1 & -0.2 & -0.3 \\ -0.2 & 0 & 0.8 & -0.4 \\ -0.4 & 0 & 0 & 0.9 \end{bmatrix}$, $F = (I - S)^{-1} \approx \begin{bmatrix} 1.94 & 1.94 & 0.83 & 1.45 \\ 0.66 & 1.66 & 0.57 & 0.95 \\ 1.05 & 1.05 & 1.88 & 1.42 \\ 0.86 & 0.86 & 0.37 & 1.75 \end{bmatrix}$.

37. $\mathbf{n} = (1)\begin{bmatrix} \frac{3}{2} \end{bmatrix} = \left(\frac{3}{2} \right).$ **39.** $\mathbf{n} = \left(\frac{19}{4}, \frac{11}{4}, \frac{21}{4} \right).$ **41.** $\mathbf{n} = (4.51, 5.51, 3.65, 5.57).$

43. $1I - P = \begin{bmatrix} a+b & -c & -e \\ -a & c+d & -f \\ -b & -d & e+f \end{bmatrix}$ is singular; its determinant is zero.

45. $T = \begin{bmatrix} 1 & \frac{1}{2} & 0 & 0 & 0 \\ 0 & 0 & \frac{1}{2} & 0 & 0 \\ 0 & \frac{1}{2} & 0 & \frac{1}{2} & 0 \\ 0 & 0 & \frac{1}{2} & 0 & 0 \\ 0 & 0 & 0 & \frac{1}{2} & 1 \end{bmatrix}. \ F = 2 \begin{bmatrix} \frac{3}{4} & \frac{1}{2} & \frac{1}{4} \\ \frac{1}{2} & 1 & \frac{1}{2} \\ \frac{1}{4} & \frac{1}{2} & \frac{3}{4} \end{bmatrix} = \begin{bmatrix} \frac{3}{2} & 1 & \frac{1}{2} \\ 1 & 2 & 1 \\ \frac{1}{2} & 1 & \frac{3}{2} \end{bmatrix}.$ The average number of times A can play is 4.

47. Springfield: 0.3234; Portland: 0.3536; Middletown: 0.323.

49. The matrices are

$$\text{Group 1: } A = \begin{array}{c} \\ \text{boom} \\ \text{recession} \end{array} \begin{array}{cc} \text{boom recession} \\ \begin{bmatrix} 0.4 & 0.7 \\ 0.6 & 0.3 \end{bmatrix} \end{array}. \qquad \text{Group 2: } B = \begin{array}{c} \\ \text{boom} \\ \text{recession} \end{array} \begin{array}{cc} \text{boom recession} \\ \begin{bmatrix} 0.7 & 0.5 \\ 0.3 & 0.5 \end{bmatrix} \end{array}.$$

Since $A^3 = \begin{bmatrix} 0.53 & 0.55 \\ 0.47 & 0.45 \end{bmatrix}$ and $B^3 = \begin{bmatrix} 0.63 & 0.62 \\ 0.37 & 0.38 \end{bmatrix}$, choose Group 2.

51. $P = \begin{bmatrix} 0.8 & 0.2 \\ 0.2 & 0.8 \end{bmatrix}, \ P^3 = \begin{bmatrix} 0.61 & 0.39 \\ 0.32 & 0.61 \end{bmatrix},$ and $P^n \rightarrow \begin{bmatrix} 0.5 & 0.5 \\ 0.5 & 0.5 \end{bmatrix}.$ After $n = 10$ the matrix P^n is within 0.001 of this, so for $n > 10$, the probability is 1/2.

53. $S = \begin{bmatrix} 0 & \frac{1}{2} & 0 & 0 \\ \frac{1}{2} & 0 & \frac{1}{2} & 0 \\ 0 & \frac{1}{2} & 0 & \frac{1}{2} \\ 0 & 0 & \frac{1}{2} & 0 \end{bmatrix}, I - S = \begin{bmatrix} 1 & \frac{-1}{2} & 0 & 0 \\ \frac{-1}{2} & 1 & \frac{-1}{2} & 0 \\ 0 & \frac{-1}{2} & 1 & \frac{-1}{2} \\ 0 & 0 & \frac{-1}{2} & 1 \end{bmatrix}, R = \begin{bmatrix} 0 & 0 & 0 & \frac{1}{2} \\ \frac{1}{2} & 0 & 0 & 0 \end{bmatrix}, F = (I - S)^{-1} = \begin{bmatrix} \frac{8}{5} & \frac{6}{5} & \frac{4}{5} & \frac{2}{5} \\ \frac{6}{5} & \frac{12}{5} & \frac{8}{5} & \frac{4}{5} \\ \frac{4}{5} & \frac{8}{5} & \frac{12}{5} & \frac{6}{5} \\ \frac{2}{5} & \frac{4}{5} & \frac{6}{5} & \frac{8}{5} \end{bmatrix},$

$\mathbf{n} = [4, 6, 6, 4] = \mathbf{c}F. \ B = RF = \begin{bmatrix} \frac{1}{5} & \frac{2}{5} & \frac{3}{5} & \frac{4}{5} \\ \frac{4}{5} & \frac{3}{5} & \frac{2}{5} & \frac{1}{5} \end{bmatrix}.$

55. The probability that A, who began with \$3, will end in state 1 and thus win is $3/5 = b_{13}$. The probability that player A will end in state 2 with no money and lose is $b_{23} = 2/5$. This is the probability that player B wins.

57. Canonical form $A = \begin{array}{c} N \\ M \\ K \\ L \end{array}\begin{bmatrix} 1 & 0 & \frac{1}{2} & \frac{1}{3} \\ 0 & \frac{1}{3} & 0 & \frac{1}{3} \\ 0 & \frac{2}{3} & 0 & \frac{1}{3} \\ 0 & 0 & \frac{1}{2} & 0 \end{bmatrix}$. $F = (I - S)^{-1} = \begin{bmatrix} \frac{15}{8} & \frac{3}{8} & \frac{3}{4} \\ \frac{3}{2} & \frac{3}{2} & 1 \\ \frac{3}{4} & \frac{3}{4} & \frac{3}{2} \end{bmatrix}$. $\mathbf{n} = \left\{\frac{33}{8}, \frac{21}{8}, \frac{13}{4}\right\} = \mathbf{c}F$ gives the average

number of years before Mr. Brown settles into staying in Hotel N provided he starts by staying in each of other hotels. For example, a little over 4 years if he stays first in Hotel M. The maximum numbers are, therefore, 5, 3, and 4 years respectively beginning in M, K, or L. Since $B = (1, 1, 1)$ the probability of staying eventually in Hotel N is 1. This is an absorbing Markov chain.

59. The canonical form of the transition matrix is Child $\begin{array}{c} \\ C \\ D \\ S \end{array}\begin{array}{c} \text{Parent} \\ \begin{array}{ccc} C & D & S \end{array} \\ \begin{bmatrix} 0.8 & 0.2 & 0.2 \\ 0.2 & 0.7 & 0.1 \\ 0 & 0.1 & 0.7 \end{bmatrix} \end{array}$.

Section 6.6 (pp. 447–449)

1. $A_1^* = \begin{bmatrix} 1 & -i \\ -i & 1 \end{bmatrix}$, $A_1 A_1^* = \begin{bmatrix} 2 & 0 \\ 0 & 2 \end{bmatrix} = A_1^* A_1$. But $A_1 \neq A_1^*$.

3. $A^* = \begin{bmatrix} 1 & 1 & -i \\ 0 & -i & 0 \\ -i & 1 & 1 \end{bmatrix}$, so A^* is neither Hermitian nor skew-Hermitian, or normal: $A^*A = \begin{bmatrix} 3 & i & 1 \\ -i & 1 & -i \\ 1 & i & 3 \end{bmatrix} \neq AA^*$.

5. $A^* = \begin{bmatrix} \frac{-i}{\sqrt{2}} & \frac{-i}{\sqrt{2}} \\ \frac{-i}{\sqrt{2}} & \frac{i}{\sqrt{2}} \end{bmatrix} = -A$, so A is skew-Hermitian and normal. **7.** Let $U = \begin{bmatrix} \frac{1}{\sqrt{2}} & \frac{1}{\sqrt{2}} \\ \frac{1}{\sqrt{2}} & \frac{-1}{\sqrt{2}} \end{bmatrix} = U^*$, so $U^*MU = \begin{bmatrix} 0 & 0 \\ 0 & 2 \end{bmatrix}$.

9. Eigenvectors are $\begin{bmatrix} 1 + \sqrt{2} \\ 1 \end{bmatrix}$ and $\begin{bmatrix} 1 - \sqrt{2} \\ 1 \end{bmatrix}$. Normalize to obtain the columns of U.

$U = \begin{bmatrix} \frac{1 + \sqrt{2}}{\sqrt{4 + 2\sqrt{2}}} & \frac{1 - \sqrt{2}}{\sqrt{4 - 2\sqrt{2}}} \\ \frac{1}{\sqrt{4 + 2\sqrt{2}}} & \frac{1}{\sqrt{4 - 2\sqrt{2}}} \end{bmatrix}$, $U^*MU = \begin{bmatrix} 1 & 0 \\ 0 & -1 \end{bmatrix}$.

11. $U = \begin{bmatrix} 1 & 0 & 0 \\ 0 & \frac{1}{\sqrt{2}} & \frac{1}{\sqrt{2}} \\ 0 & \frac{-i}{\sqrt{2}} & \frac{i}{\sqrt{2}} \end{bmatrix}$; $D = \begin{bmatrix} -2 & 0 & 0 \\ 0 & \frac{1+i}{\sqrt{2}} & 0 \\ 0 & 0 & \frac{-1+i}{\sqrt{2}} \end{bmatrix}$.

13. $U = \begin{bmatrix} 0 & \frac{1}{\sqrt{2}} & \frac{-1}{\sqrt{2}} \\ 1 & 0 & 0 \\ 0 & \frac{1}{\sqrt{2}} & \frac{1}{\sqrt{2}} \end{bmatrix}$; upper-triangular matrix is $\begin{bmatrix} 1 & 0 & 0 \\ 0 & 1 & 0 \\ 0 & 0 & -1 \end{bmatrix}$.

15. $U = \begin{bmatrix} \frac{1}{\sqrt{3}} & \frac{1}{\sqrt{2}} & \frac{1}{\sqrt{2}} \\ \frac{1}{\sqrt{3}} & 0 & \frac{-1}{\sqrt{2}} \\ \frac{1}{\sqrt{3}} & \frac{-1}{\sqrt{2}} & 0 \end{bmatrix}$; upper-triangular matrix is $\begin{bmatrix} 3 & 0 & 0 \\ 0 & 0 & 0 \\ 0 & 0 & 0 \end{bmatrix}$.

17. $U = \begin{bmatrix} 1 & 0 & 0 \\ 0 & \frac{i}{\sqrt{2}} & \frac{-i}{\sqrt{2}} \\ 0 & \frac{1}{\sqrt{2}} & \frac{1}{\sqrt{2}} \end{bmatrix}$; upper-triangular matrix is $\begin{bmatrix} -2 & 0 & 0 \\ 0 & \sqrt{2} & 0 \\ 0 & 0 & 0 \end{bmatrix}$.   **19.** $A^*A = (-A)(A) = -A^2 = A(-A) = AA^*$.

21. See Theorem 6.7. Let $\lambda_1 \neq \lambda_2$ be two eigenvalues of the Hermitian matrix A. Let $\mathbf{v}_1$ and $\mathbf{v}_2$ be the associated eigenvectors. Thus we have $A\mathbf{v}_1 = \lambda_1\mathbf{v}_1$ and $A\mathbf{v}_2 = \lambda_2\mathbf{v}_2$. Furthermore, since A is Hermitian its eigenvalues are real, so $\overline{\lambda_2} = \lambda_2$. Let $k = \mathbf{v}_1^*A\mathbf{v}_2$ which is a 1×1 matrix, so $k^* = \overline{k}$. Note also that $\mathbf{v}_1^*\mathbf{v}_2$ is also 1×1, so we have $\mathbf{v}_2^*\mathbf{v}_1 = (\mathbf{v}_1^*\mathbf{v}_2)^* = \overline{\mathbf{v}_1^*\mathbf{v}_2} = \mathbf{v}_1^* \, \mathbf{v}_2$. From this we compute $(\mathbf{v}_1^*A\mathbf{v}_2)^* = (\mathbf{v}_1^*\lambda_2\mathbf{v}_2)^* = \overline{\mathbf{v}_1^*}\lambda_2\mathbf{v}_2 = \lambda_2\mathbf{v}_1^* \, \mathbf{v}_2 = \lambda_2\mathbf{v}_2^*\mathbf{v}_1$. But it is also true that $(\mathbf{v}_1^*A\mathbf{v}_2)^* = \mathbf{v}_2^*A^*\mathbf{v}_1 = \mathbf{v}_2^*A\mathbf{v}_1 = \mathbf{v}_2^*\lambda_1\mathbf{v}_1 = \lambda_1\mathbf{v}_2^*\mathbf{v}_1$. Therefore, $\lambda_2\mathbf{v}_2^*\mathbf{v}_1 = \lambda_1\mathbf{v}_2^*\mathbf{v}_1$, or $\lambda_2\mathbf{v}_2^*\mathbf{v}_1 - \lambda_1\mathbf{v}_2^*\mathbf{v}_1 = (\lambda_2 - \lambda_1)\mathbf{v}_2^*\mathbf{v}_1 = 0$. But $\lambda_1 \neq \lambda_2$, so $(\lambda_2 - \lambda_1) \neq 0$. Therefore $\mathbf{v}_2^*\mathbf{v}_1 = 0$, they are orthogonal.

23. $\|U\mathbf{x}\| = <U\mathbf{x}, U\mathbf{x}>^{1/2} = ((U\mathbf{x})^*U\mathbf{x})^{1/2} = (\mathbf{x}^*U^*U\mathbf{x})^{1/2} = (\mathbf{x}^*I\mathbf{x})^{1/2} = (\mathbf{x}^*\mathbf{x})^{1/2} = \|\mathbf{x}\|$.

25. From Exercise 23, $\|U\mathbf{x}\| = \|\mathbf{x}\|$. But if $U\mathbf{x} = \lambda\mathbf{x}$, then $\|U\mathbf{x}\| = \|\lambda\mathbf{x}\| = |\lambda| \, \|\mathbf{x}\| = \|\mathbf{x}\|$. So $|\lambda| = 1$.

27. Let $H = (h_{ij})$ be an Hermitian matrix, $H = H^*$, so $h_{ii} = \overline{h_{ii}}$ for each $i = 1, 2, \ldots, n$. Thus each h_{ii} is real.

29. $\|N\mathbf{x}\|^2 = <N\mathbf{x}, N\mathbf{x}> = (N\mathbf{x})^*N\mathbf{x} = \mathbf{x}^*N^*N\mathbf{x} = \mathbf{x}^*NN^*\mathbf{x} = (N^*\mathbf{x})^*(N^*\mathbf{x}) = \|N^*\mathbf{x}\|^2$.

31. Let A be a normal matrix, then by Theorem 6.15 there is an upper triangular matrix M and a unitary matrix U so that $M = U^*AU$. But by Theorem 6.16, U can be chosen so that M is the diagonal matrix diag $[\lambda_1, \lambda_2, \ldots, \lambda_n]$, with the eigenvalues of A on the main diagonal. In Exercise 22 we showed that the columns of U are mutually orthogonal unit vectors. Since $UM = AU$ we have that the columns of U are the eigenvectors of A. Conversely, if $M = U^*AU =$ diag $[\lambda_1, \lambda_2, \ldots, \lambda_n]$, then M is normal because $MM^* = M^*M =$ diag $[|\lambda_1|^2, |\lambda_2|^2, \ldots, |\lambda_n|^2]$, so by Exercise 26, A is normal also.

33. $<\mathbf{u}, \mathbf{v}> = \mathbf{v}^*A\mathbf{u} = [-i, 0, -1]\begin{bmatrix} 1 & 0 & 0 \\ 0 & 5 & 3-2i \\ 0 & 3+2i & 1 \end{bmatrix}\begin{bmatrix} 1 \\ i \\ 1 \end{bmatrix} = 1 - 4i$.

$<\mathbf{u}, \mathbf{u}> = \mathbf{u}^*A\mathbf{u} = [1, -i, 1]\begin{bmatrix} 1 & 0 & 0 \\ 0 & 5 & 3-2i \\ 0 & 3+2i & 1 \end{bmatrix}\begin{bmatrix} 1 \\ i \\ 1 \end{bmatrix} = 3$, so $\|\mathbf{u}\| = \sqrt{3}$.

$<\mathbf{v}, \mathbf{v}> = \mathbf{v}^*A\mathbf{v} = [-i, 0, -1]\begin{bmatrix} 1 & 0 & 0 \\ 0 & 5 & 3-2i \\ 0 & 3+2i & 1 \end{bmatrix}\begin{bmatrix} i \\ 0 \\ -1 \end{bmatrix} = 2$, so $\|\mathbf{v}\| = \sqrt{2}$.

35. $<\mathbf{u}, \mathbf{v}> = [1, i, -i]\begin{bmatrix} 1 & 0 & 0 \\ 0 & 5 & 3-2i \\ 0 & 3+2i & 1 \end{bmatrix}\begin{bmatrix} 1 \\ 1-i \\ 1+i \end{bmatrix} = 5 + 4i$; $<\mathbf{u}, \mathbf{u}> = 1$, so $\|\mathbf{u}\| = 1$; $<\mathbf{v}, \mathbf{v}> = 21$, so $\|\mathbf{v}\| = \sqrt{21}$.

Section 6.7 (pp. 453–454)

1. $Y' = AY = \begin{bmatrix} 3 & -1 \\ 2 & 0 \end{bmatrix}Y$, so the transformed equation is $Z' = \begin{bmatrix} 2 & 0 \\ 0 & 1 \end{bmatrix}Z$. The solutions then are $z_1 = c_1e^{2x}$, $z_2 = c_2e^x$.

Since $P = \begin{bmatrix} 1 & 1 \\ 1 & 2 \end{bmatrix}$, the solution is $Y = PZ = \begin{bmatrix} 1 & 1 \\ 1 & 2 \end{bmatrix}\begin{bmatrix} c_1e^{2x} \\ c_2e^x \end{bmatrix} = \begin{bmatrix} c_1e^{2x} + c_2e^x \\ c_1e^{2x} + 2c_2e^x \end{bmatrix}$. Thus $y_1 = c_1e^{2x} + c_2e^x$ and $y_2 = c_1e^{2x} + 2c_2e^x$.

3. $y_1 = c_1e^{5x} - 2c_2e^{-x}$ and $y_2 = c_1e^{5x} + c_2e^{-x}$.

5. $Y' = \begin{bmatrix} -1 & 2 & 2 \\ 2 & -1 & 2 \\ 2 & 2 & -1 \end{bmatrix} Y$, so the transformed equation is $Z' = \begin{bmatrix} 3 & 0 & 0 \\ 0 & -3 & 0 \\ 0 & 0 & -3 \end{bmatrix} Z$, whose solutions then are $z_1 = c_1 e^{3x}$,

$z_2 = c_2 e^{-3x}$, and $z_3 = c_3 e^{-3x}$. The solution is $Y = PZ$, with $P = \begin{bmatrix} 1 & -1 & -1 \\ 1 & 0 & 1 \\ 1 & 1 & 0 \end{bmatrix}$. Thus $y_1 = c_1 e^{3x} - (c_2 + c_3)e^{-3x}$,

$y_2 = c_1 e^{3x} + c_3 e^{-3x}$, and $y_3 = c_1 e^{3x} + c_2 e^{-3x}$.

7. $y_1(0) = c_1 e^0 + c_2 e^0 = c_1 + c_2 = 1$ and $y_2(0) = c_1 e^0 + 2c_2 e^0 = c_1 + 2c_2 = -1$. So $\begin{bmatrix} 1 & 1 \\ 1 & 2 \end{bmatrix}\begin{bmatrix} c_1 \\ c_2 \end{bmatrix} = \begin{bmatrix} 1 \\ -1 \end{bmatrix}$

so $\begin{bmatrix} c_1 \\ c_2 \end{bmatrix} = \begin{bmatrix} 3 \\ -2 \end{bmatrix}$. Thus, $y_1 = 3e^{2x} - 2e^x$ and $y_2 = 3e^{2x} - 4e^x$.

9. $y_1(0) = c_1 - 2c_2 = -1$ and $y_2(0) = c_1 + c_2 = 5$. So $\begin{bmatrix} 1 & 1 & -1 \\ 1 & \frac{-1}{2} & 5 \end{bmatrix} \rightarrow \begin{bmatrix} 1 & 0 & 3 \\ 0 & 1 & -4 \end{bmatrix}$. Thus, $y_1 = 3e^{5x} - 4e^{-x}$ and

$y_2 = 3e^{5x} + 2e^{-x}$.

11. $y_1(0) = c_1 - c_2 - c_3 = 1$, $y_2(0) = c_1 + c_3 = 0$, $y_3(0) = c_1 + c_2 = 1$. Thus, the particular solution

$y_1 = y_3 = \frac{2}{3}e^{3x} + \frac{1}{3}e^{-3x}$, $y_2 = \frac{2}{3}e^{3x} - \frac{2}{3}e^{-3x}$.

13. $Y' = \begin{bmatrix} 18 & 9 & 9 \\ 3 & 2 & 3 \\ -9 & -9 & -10 \end{bmatrix} Y$, so $Z' = \begin{bmatrix} 2 & 0 & 0 \\ 0 & -1 & 0 \\ 0 & 0 & -1 \end{bmatrix} Z$. $z_1 = c_1 e^{2x}$, $z_2 = c_2 e^{-x}$, and $z_3 = c_3 e^{-x}$. $P = \begin{bmatrix} 3 & 1 & 1 \\ 1 & 0 & -1 \\ -3 & -1 & 0 \end{bmatrix}$;

$y_1 = 3c_1 e^{2x} + (c_2 + c_3)e^{-x}$, $y_2 = c_1 e^{2x} - c_3 e^{-x}$, and $y_3 = -3c_1 e^{2x} - c_2 e^{-x}$.

15. $Y' = \begin{bmatrix} 2 & -1 & 0 & 0 \\ -1 & 2 & 0 & 0 \\ 0 & 0 & 1 & -1 \\ 0 & 0 & -1 & 1 \end{bmatrix} Y$, so $Z' = \begin{bmatrix} 3 & 0 & 0 & 0 \\ 0 & 1 & 0 & 0 \\ 0 & 0 & 2 & 0 \\ 0 & 0 & 0 & 0 \end{bmatrix} Z$. $z_1 = c_1 e^{3x}$, $z_2 = c_2 e^x$, $z_3 = c_3 e^{2x}$, and $z_4 = c_4$.

$P = \begin{bmatrix} 1 & 1 & 0 & 0 \\ -1 & 1 & 0 & 0 \\ 0 & 0 & 1 & 1 \\ 0 & 0 & -1 & 1 \end{bmatrix}$. Hence, $y_1 = c_1 e^{3x} + c_2 e^x$, $y_2 = -c_1 e^{3x} + c_2 e^x$, $y_3 = c_3 e^{2x} + c_4$, and $y_4 = -c_3 e^{2x} + c_4$.

17. Set $y_1 = y$, $y_2 = y'$, then $y'_1 = y_2$, and $y'' = y'_2 = 8y_1 - 2y_2$. So $Y' = \begin{bmatrix} 0 & 1 \\ 8 & -2 \end{bmatrix} Y$. Hence, $Z' = \begin{bmatrix} -4 & 0 \\ 0 & 2 \end{bmatrix} Z$;

$z_1 = c_1 e^{-4x}$, $z_2 = c_2 e^{2x}$. Thus $y_1 = y = c_1 e^{-4x} + c_2 e^{2x}$.

19. From calculus, $D(f + g) = D(f) + D(g)$ and $D(\alpha f) = \alpha D(f)$. See any calculus text. So D is linear.

Chapter 6 Review Exercises (p. 455)

1. $T(1, -1) = (-18(1) - 20(-1), 15(1) + 17(-1)) = (2, -2) = 2(1, -1)$. So $T(\mathbf{v}) = 2\mathbf{v}$.
2. Since $T(1, -1) = 2(1, -1)$, $\lambda = 2$ and $(1, -1)$ is its associated eigenvector. See Exercise 1.

3. $[T(1, 0)] = \begin{bmatrix} -18 \\ 15 \end{bmatrix}$, $[T(0, 1)] = \begin{bmatrix} -20 \\ 17 \end{bmatrix}$, so $[T] = \begin{bmatrix} -18 & -20 \\ 15 & 17 \end{bmatrix}$. Eigenvalues are 2 and -3. Bases for each

eigenspace are as follows. For 2: $\left\{ \begin{bmatrix} 1 \\ -1 \end{bmatrix} \right\}$; for -3: $\left\{ \begin{bmatrix} 4 \\ -3 \end{bmatrix} \right\}$.

4. $\lambda I - A = \begin{bmatrix} \lambda - 2 & 0 & 0 \\ 1 & \lambda + 1 & -6 \\ 0 & -8 & \lambda - 1 \end{bmatrix}$. $\det(\lambda I - A) = (\lambda - 2)(\lambda - 7)(\lambda + 7)$, so $\lambda_1 = 2, \lambda_2 = 7, \lambda_3 = -7$.

5. Yes. $D = \begin{bmatrix} 2 & 0 & 0 \\ 0 & 7 & 0 \\ 0 & 0 & -7 \end{bmatrix}$.

6. $p_A(\lambda) = \lambda^3 - 2\lambda^2 - 49\lambda + 98$, so $p_A(A) = 0 = A^3 - 2A^2 - 49A + 98I$. Thus, $I = \frac{-1}{98}(A^2 - 2A - 49I)A$,

so $A^{-1} = \frac{-1}{98}(A^2 - 2A - 49I) = \begin{bmatrix} \frac{1}{2} & 0 & 0 \\ \frac{-1}{98} & \frac{-2}{98} & \frac{12}{98} \\ \frac{8}{98} & \frac{16}{98} & \frac{2}{98} \end{bmatrix} = \frac{1}{98}\begin{bmatrix} 49 & 0 & 0 \\ -1 & -2 & -12 \\ 8 & 16 & 2 \end{bmatrix}$; $\det A = -98$.

7. Eigenvalues -3 and 0. For 0, eigenspace spanned by eigenvectors $\mathbf{v}_1 = \begin{bmatrix} 1 \\ 0 \\ -1 \end{bmatrix}$, and $\mathbf{v}_2 = \begin{bmatrix} 1 \\ -1 \\ 0 \end{bmatrix}$. For 3, eigenspace has

the eigenvector $\mathbf{u} = \begin{bmatrix} 1 \\ 1 \\ 1 \end{bmatrix}$ as a basis.

8. Yes. See Exercise 7. We need an orthonormal basis for the eigenspace of $\lambda_1 = 0$. Set $\mathbf{u}_1 = \begin{bmatrix} \frac{1}{\sqrt{2}} \\ 0 \\ \frac{-1}{\sqrt{2}} \end{bmatrix} = \frac{\mathbf{v}_1}{\|\mathbf{v}_1\|}$ and use the

Gram-Schmidt process to arrive at $\mathbf{u}_1 = \begin{bmatrix} \frac{1}{\sqrt{6}} \\ \frac{-2}{\sqrt{6}} \\ \frac{1}{\sqrt{6}} \end{bmatrix}$. Thus $P = \begin{bmatrix} \frac{1}{\sqrt{2}} & \frac{1}{\sqrt{6}} & \frac{1}{\sqrt{3}} \\ 0 & \frac{-2}{\sqrt{6}} & \frac{1}{\sqrt{3}} \\ \frac{1}{\sqrt{2}} & \frac{1}{\sqrt{6}} & \frac{1}{\sqrt{3}} \end{bmatrix}$ is such that $P^T K P = D = \text{diag} [0, 0, 3]$.

9. Matrix form is $X^T \begin{bmatrix} 0 & 2 \\ 2 & 0 \end{bmatrix} X = 9$. Eigenvalues are 2 and -2,

so $X'^T \begin{bmatrix} 2 & 0 \\ 0 & -2 \end{bmatrix} X' = 9$. Scalar equation is $2x'^2 - 2y'^2 = 9$,

or $\frac{x'^2}{9/2} - \frac{y'^2}{9/2} = 1$, an hyperbola. $P = \begin{bmatrix} \frac{1}{\sqrt{2}} & \frac{-1}{\sqrt{2}} \\ \frac{1}{\sqrt{2}} & \frac{1}{\sqrt{2}} \end{bmatrix}$.

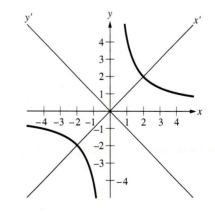

10. $X^T \begin{bmatrix} 1 & -1 \\ -1 & 3 \end{bmatrix} X = 1$. Eigenvalues are $2 + \sqrt{2}$ and $2 - \sqrt{2}$,

so $X'^T \begin{bmatrix} 2 + \sqrt{2} & 0 \\ 0 & 2 - \sqrt{2} \end{bmatrix} X' = 1$. This is the ellipse

$(2 + \sqrt{2})x'^2 + (2 - \sqrt{2})y'^2 = 1$.

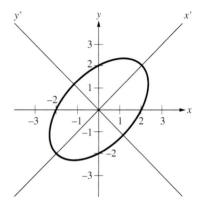

11. Eigenvalues and corresponding basic eigenvectors are as follows. $-i:\begin{bmatrix} 1 \\ 0 \\ 0 \end{bmatrix}, -5:\begin{bmatrix} 0 \\ 1 \\ -1 \end{bmatrix}, -1:\begin{bmatrix} 0 \\ 1 \\ 1 \end{bmatrix}$.

Let $U = \begin{bmatrix} 1 & 0 & 0 \\ 0 & \frac{1}{\sqrt{2}} & \frac{1}{\sqrt{2}} \\ 0 & \frac{-1}{\sqrt{2}} & \frac{1}{\sqrt{2}} \end{bmatrix}$. Then $U^*MU = \begin{bmatrix} -i & 0 & 0 \\ 0 & -5 & 0 \\ 0 & 0 & -1 \end{bmatrix}$ is diagonal, hence upper-triangular.

12. $Z' = \begin{bmatrix} 2 & 0 \\ 0 & -1 \end{bmatrix} Z$. First $z_1 = c_1 e^{2x}$, $z_2 = c_2 e^{-x}$, so $P = \begin{bmatrix} 1 & 0 \\ 1 & 1 \end{bmatrix}$. Therefore, $y_1 = c_1 e^{2x}$ and $y_2 = c_1 e^{2x} + c_2 e^{-x}$.

INDEX

3.4 **Dimension, p. 189.** The number of vectors in any ordered basis.

3.5 **Row and Column Space of a Matrix, pp. 197–203.** For an $m \times n$ matrix A, the row space of A is the subspace of $\mathbb{R}^m$ (or $\mathbb{C}^m$) spanned by the row vectors of A. The column space is the subspace of $\mathbb{R}^n$ (or $\mathbb{C}^n$) spanned by the column vectors of A = the row space of A^T.

3.5 **Rank of a Matrix, p. 203.** See Theorem 3.12. For an $m \times n$ matrix A, row rank = column rank = rank of the matrix.

4.1 **Dot Product, p. 220.** $\mathbf{u} \cdot \mathbf{v}$.

4.2 **Cross Product in $\mathbb{R}^3$, p. 227.** $\mathbf{u} \times \mathbf{v} = \det \begin{bmatrix} \mathbf{i} & \mathbf{j} & \mathbf{k} \\ u_1 & u_2 & u_3 \\ v_1 & v_2 & v_3 \end{bmatrix}$.

4.3 **Real Inner Product, p. 236.** $<\mathbf{u}, \mathbf{v}> \in \mathbb{R}$.

4.3 **Standard Real Inner Product, p. 237.** $<\mathbf{u}, \mathbf{v}> = u_1 v_1 + u_2 v_2 + \cdots + u_n v_n = \sum_{i=1}^{n} u_i v_i$.

4.3 **Trace Inner Product, p. 239.** On $M_n(\mathbb{R})$: $<A, B> = \operatorname{tr}(A^T B)$.

4.3 **Integral Inner Product, p. 240.** On $C[a, b]$ or P_n: $<f, g> = \int_c^d f(x)\, g(x)\, dx$.

4.3 **Inner-Product Spaces.** See Chapter 4 and Appendix A.3. Let V be any real or complex vector space. A function which assigns to each pair of vectors $\mathbf{u}$ and $\mathbf{v}$ in V a unique complex or real number, $<\mathbf{u}, \mathbf{v}>$, is called an **inner product** on V provided it has the following properties. For all vectors $\mathbf{u}, \mathbf{v}, \mathbf{w}$ and scalars a, b, c:

1. $<\mathbf{u}, \mathbf{v}> = \overline{<\mathbf{u}, \mathbf{v}>}$.
2. $<a\mathbf{u} + b\mathbf{v}, \mathbf{w}> = a<\mathbf{u}, \mathbf{w}> + b<\mathbf{v}, \mathbf{w}>$.
3. $<\mathbf{u}, \mathbf{u}> \geq 0$.
4. $<\mathbf{u}, \mathbf{u}> = 0$ if, and only if, $\mathbf{u} = \mathbf{0}$.

4.4 **Vector Norm, p. 246.** $\|\mathbf{v}\| = <\mathbf{v}, \mathbf{v}>^{1/2}$.

4.4 **Orthogonal Vectors, p. 253.** $\mathbf{u}$ and $\mathbf{v}$ are orthogonal if, and only if, $<\mathbf{u}, \mathbf{v}> = 0$.

4.4 **Unit Vector, p. 255.** $\|\mathbf{v}\| = <\mathbf{v}, \mathbf{v}>^{1/2} = 1$ or $<\mathbf{v}, \mathbf{v}> = 1$.

4.4 **Orthonormal Basis, p. 255.** $S = \{\mathbf{v}_1, \mathbf{v}_2 \ldots, \mathbf{v}_n\}$ is a **basis**, and $<\mathbf{v}_i, \mathbf{v}_i> = 1$ while $<\mathbf{v}_i, \mathbf{v}_j> = 0$, $i \neq j$.

4.4, 4.6 **Cauchy-Schwarz Inequality, pp. 249–275.** See Theorems 4.7 and 4.13.
$$|<\mathbf{u}, \mathbf{v}>| \leq \|\mathbf{u}\|\, \|\mathbf{v}\|.$$

4.4, 4.6 **Triangle Inequality, pp. 250, 276.** See Corollaries.
$$\|\mathbf{u} + \mathbf{v}\| \leq \|\mathbf{u}\| + \|\mathbf{v}\|.$$